INNOVATION AND TECHNOLOGICAL ADVANCES FOR SUSTAINABILITY

The proceedings publication of the International Conference on Innovation and Technological Advances for Sustainability (ITAS 2023) captures the essence of a dynamic international forum dedicated to advancing the United Nations Sustainable Development Goals (UN-SDGs).

It serves as a comprehensive repository of cutting-edge research, innovative strategies, and transformative tools discussed by a diverse community of participants, including researchers, academics, students, policymakers, industry leaders, and government officials. Encompassing local, regional, and international perspectives, the proceedings delve into critical global issues such as food security, environmental preservation, energy sustainability, economic resilience, and the role of digital technologies in fostering sustainable development.

This book emphasizes showcasing national and international accomplishments, fostering global collaborations, exploring future challenges and opportunities, introducing state-of-the-art technologies, and providing policy recommendations for building a sustainable society. It acts as a bridge between research and practice, promoting the dissemination of knowledge that will contribute to the achievement of UN-SDGs.

PROCEEDINGS OF THE INTERNATIONAL CONFERENCE ON INNOVATION AND TECHNOLOGICAL ADVANCES FOR SUSTAINABILITY, (ITAS 2023), 1–3 MARCH 2023, DOHA, QATAR

Innovation and Technological Advances for Sustainability

Edited by

Salem Al-Naemi
University of Doha for Science and Technology, Qatar

Rachid Benlamri
University of Doha for Science and Technology, Qatar

Rehan Sadiq
University of British Columbia, Canada

Aitazaz Farooque
University of Prince Edward Island, Canada

Michael Phillips
University of Doha for Science and Technology, Qatar

CRC Press
Taylor & Francis Group
Boca Raton London New York Leiden

CRC Press is an imprint of the
Taylor & Francis Group, an **informa** business
A BALKEMA BOOK

First published 2025
by CRC Press/Balkema
4 Park Square, Milton Park, Abingdon, Oxon, OX14 4RN

and by CRC Press/Balkema
2385 NW Executive Center Drive, Suite 320, Boca Raton FL 33431

CRC Press/Balkema is an imprint of the Taylor & Francis Group, an informa business

British Library Cataloguing-in-Publication Data
A catalogue record for this book is available from the British Library

Library of Congress Cataloging-in-Publication Data
A catalog record has been requested for this book

ISBN: 978-1-032-80372-2 (hbk)
ISBN: 978-1-032-80402-6 (pbk)
ISBN: 978-1-003-49672-4 (ebk)

DOI: 10.1201/9781003496724

Typeset in Times New Roman
by MPS Limited, Chennai, India

Innovation and Technological Advances for Sustainability – Al-Naemi et al. (Eds)
© 2025 The Editor(s), ISBN 978-1-032-80372-2
Open Access: www.taylorfrancis.com, CC BY-NC-ND 4.0 license

Table of Contents

Innovation and Technological Advances for Sustainability – Al-Naemi et al. (Eds)

Preface

The International Conference on Innovation and Technological Advances for Sustainability (ITAS 2023) proceedings book is an extensive archive of state-of-the-art studies, creative approaches, and game-changing instruments shared by a varied range of participants committed to achieving the Sustainable Development Goals (SDGs) of the United Nations.

The manuscripts in this conference proceedings book were each peer-reviewed and approved by at least two members of the ITAS 2023 scientific committee. This book explores and discusses important global concerns, such as food security, environmental preservation, energy sustainability, economic resilience, and the role of digital technologies in fostering sustainable development.

This publication is meant for researchers, academics, students, policymakers, business executives, and government officials who are interested in promoting sustainable development and helping to fulfill the UN Sustainable Development Goals.

Prof. Rachid Benlamri, Ph.D.
Conference Chair of ITAS 2023
Vice-President – Academic
University of Doha for Science and Technology, Qatar

Innovation and Technological Advances for Sustainability – Al-Naemi et al. (Eds)
© 2025 The Editor(s), ISBN 978-1-032-80372-2
Open Access: www.taylorfrancis.com, CC BY-NC-ND 4.0 license

Acknowledgements

We extend our heartfelt gratitude to all those who have contributed to the realization of this publication, the proceedings of the International Conference on Innovation and Technological Advances for Sustainability (ITAS 2023).

First and foremost, we would like to convey our sincere gratitude to the distinguished authors whose perceptive analysis and creative solutions serve as the cornerstone of this extensive archive. It is quite admirable how committed they are to using their work to further the United Nations Sustainable Development Goals (UN-SDGs).

We want to express our gratitude to the wide range of attendees at ITAS 2023, including researchers, academics, students, legislators, business executives, and public servants, all of whom actively participated in the stimulating conversations. Your combined knowledge, viewpoints, and excitement have enhanced the discourse during the conference.

Further, we thank the organizing committee, manuscript reviewers from the scientific committee, editors, and technical personnel for their support and contributions, which helped to assure the proceedings' integrity and quality. This publication would not exist without your dedication to quality.

Special thanks to the sponsors and partners whose kindness and cooperation made ITAS 2023 possible. Your assistance has made it easier to share knowledge, investigate cutting-edge technology, and establish international partnerships with the goal of resolving pressing sustainability-related global concerns.

Furthermore, we would like to thank all the stakeholders and readers who will interact with this publication. We sincerely hope that the ideas, suggestions, and innovations shared here will spur action and support efforts to realize the UN-SDGs and create a sustainable society.

Acknowledgements

Innovation and Technological Advances for Sustainability – Al-Naemi et al. (Eds)
© 2025 The Editor(s), ISBN 978-1-032-80372-2
Open Access: www.taylorfrancis.com, CC BY-NC-ND 4.0 license

Editor Biographies

Salem Al-Naemi is the First President of the University of Doha for Science and Technology (UDST), and has more than 30 years of work experience in both government and private sectors. He is a successful executive with considerable leadership experience in education, government, and business organizations. Dr. Al-Naemi holds an MBA and PhD in Computer Science. He also has a master's in mathematics and a Bachelor's degree with a double major in Mathematics and Computer Science. He began his career as a university academic. His expertise lies in organizational strategies, managerial leadership, communication, and operational efficiency to accomplish goals and achieve success for an organization. He has worked with a diverse workforce and in many multi-national organizations in key positions.

Rachid Benlamri is the Vice President, Academic at the University of Doha for Science and Technology (UDST) and holds a rich academic experience of over 30 years. Prior to joining UDST, Prof. Benlamri worked as the Vice President Academic at the Canadian University of Dubai. He also worked as Professor of Software Engineering and Head of the Department of Software Engineering in the Faculty of Engineering at Lakehead University, Canada. He earned his master's and PhD in Computer Science from the University of Manchester, UK. Dr. Benlamri was the Head of the Data Science and Artificial Intelligence Lab at Lakehead University. His research interests are in the areas of artificial intelligence, data science, semantic web, knowledge engineering, health informatics, and mobile & ubiquitous computing. His research is supported by many funding institutions. He has supervised over 70 students and postdoctoral fellows. He served as keynote speaker and as the general chair for many international conferences. Prof. Benlamri serves on the editorial board of many Q1 journals. Moreover, he published over 120 papers, mainly in Q1 journals and peer-reviewed international conference proceedings.

Rehan Sadiq is Provost and Vice-President, Academic and a Professor of Civil Engineering at UBC's Okanagan campus. He is a world-leading researcher in the areas of asset management of water supply systems, environmental risk analysis, and lifecycle assessment of built environment. He is the author of more than 600 peer-reviewed journal and conference articles, book chapters, and technical reports. With nearly 16,000 citations, he is among the top cited environmental engineering researchers globally; consistently ranked in the top 2% of scientists in the category of environmental science. He has served on and chaired numerous Canadian and international scientific committees and conferences. He sits on the editorial board of many international journals in his field. He has been registered as a Professional Engineer in BC (since 2010). He is also a Fellow of the Canadian Academy of Engineering (CAE), the Canadian Society of Civil Engineering (FCSE), Engineers Canada (FEC) and the Engineering Institute of Canada (FEIC).

Aitazaz A. Farooque is working as a Professor & Associate Dean at the School of Climate Change and Adaptation, University of Prince Edward Island (UPEI). He is also leading the Centre of Excellence in Food Security and Sustainability at UPEI. His research focuses on the fundamental understanding and development of state-of-the-art climate-smart agriculture technologies to improve soil health, increase farm profitability, and mitigate environmental risks for Canadian agriculture industry. He has supervised/supervising 13 PhD students, 16 MSc students, 3 post-doctoral fellows, 8 research associates, and over 30 undergraduate research students. He has authored/co-authored over 100 peer-reviewed journal articles, 8 book chapters, 80 conference proceedings, 70 industry presentations, and

30 keynote/invited talks at various conferences and professional events. He has received various national awards in recognition of his excellent contributions to research, teaching, and services. He has garnered over $8M in research funding from various provincial, federal, and industry funding agencies in Canada.

Michael Phillips is the Director of Applied Research, Innovation and Economic Development at the University of Doha for Science and Technology and has a degree in Civil Engineering, a Masters in Environmental Management and PhD in Coastal Morphodynamics. Other roles include UNESCO Chair on TVET and Sustainable Development and QAPCO Professional Chair in Vocational Studies (TVET). He has served as an Adjunct Professor in Canada, Iceland and Malaysia. His TEDx talk entitled: 'Living on the coast, climate change and acceptable risk' in response to economic and business loss from climate change impacts, focused on improving resilience. Dr. Phillips has published >100 academic articles and supervised >30 successful research degree students. He has also been involved in interdisciplinary research and, through a design education lens has helped develop strategies to deliver effective entrepreneurial education. He has long been involved in STEM and TVET education and has always emphasized its importance for the economy and employability. He encourages faculty to engage as many students as possible in applied research projects, increasing the number of proposals submitted in response to research calls and capacity-building programs, as well as engaging them in entrepreneurship.

Innovation and Technological Advances for Sustainability – Al-Naemi et al. (Eds)
© 2025 The Editor(s), ISBN 978-1-032-80372-2
Open Access: www.taylorfrancis.com, CC BY-NC-ND 4.0 license

Committee

General Executive Chair

Dr. Salem Al-Naemi (University of Doha for Science and Technology, Qatar)
Conference Chair
Dr. Rachid Benlamri (University of Doha for Science and Technology, Qatar)
Industrial Chair
Dr. Mohammed Yousef Al-Mulla (Qatar Petrochemical Company (QAPCO), Qatar)
Innovation Chair
Dr. Belgacem Haba (Adeia, Silicon Valley, USA)
Technical Program Committee Co-Chairs
Dr. Aitazaz Farooque (University of Prince Edward Island, Canada)
Dr. Mohamed Razi Nalim (Indiana University – Purdue University, USA)
Publication Co-Chairs
Dr. Ridha Hamila (IEEE Qatar Chapter Chair, Qatar University, Qatar)
Dr. Abdulhalim Dandoush (University of Doha for Science and Technology, Qatar)
Dr. Ala Gouissem (University of Doha for Science and Technology, Qatar)
Finance Chair
James Poirier (University of Doha for Science and Technology, Qatar)
Marketing Chair
Fatema Al-Hadad (University of Doha for Science and Technology, Qatar)
Communication Chair
Gisele Achkar (University of Doha for Science and Technology, Qatar)
Organization Committee Chair
Abitha John (University of Doha for Science and Technology, Qatar)

Program Committee

Dr. Belgacem Haba (Adeia, Silicon Valley, USA)
Dr. Kamal Youcef-Toumi (Massachusetts Institute of Technology, USA)
Dr. Rehan Sadiq (University of British Columbia, Canada)
Dr. Mohamed Razi Nalim (Indiana University – Purdue University, USA)
Dr. Carson Kai-Sang Leung (University of Manitoba, Canada)
Dr. Rachid Benlamri (University of Doha for Science and Technology, Qatar)
Ms. Alisa Pinchuk (Belarusian National Technical University, Belarus)
Mr. Romain Herault (Linnaeus University Sweden)
Dr. Benaoumeur Senouci (University of Southern Denmark, Denmark)
Dr. Mohammed Boulmalf (International University of Rabat, Morocco)
Dr. Amin Gasmi (Claude Bernard University, France)
Dr. Venkatesh Meda (University of Saskatchewan, Canada)
Dr. Youcef Brik (University of Msila, Algeria)
Dr. Bishnu Acharya (University of Saskatchewan, Canada)
Dr. Michael Phillips (University of Doha for Science and Technology, Qatar)
Dr. Mumtaz Khan (Sultan Qaboos University, Oman)
Dr. David J. Russomanno (Indiana University – Purdue University, USA)

Dr. Ziad Said (UNESCO Chair – Sustainable Development, Qatar)
Dr. Prof. Jilali Antari (Ibn Zohr University, Morocco)
Dr. Ismael Hinti (Al Hussein Technical University, Jordan)
Dr. Awni Al-Otoom (University of Doha for Science and Technology, Qatar)
Dr. Ridha Hamila (Qatar University, Qatar)
Dr. David J. Mulla (University of Minnesota, USA)
Dr. Helena Calado (University of the Azores, Portugal)
Dr. Sanjeev Bokhoree (University Technology Mauritius, Mauritius)
Ms. Helena Belfrage (Linnaeus University Sweden)
Dr. Abdallah Abdallah (Al Hussein Technical University, Jordan)
Dr. Ying Chen (University of Manitoba, Canada)
Dr. Rosie Bateson (University of Doha for Science and Technology, Qatar)
Dr. Marcelo Milrad (Linnaeus University, Sweden)
Dr. JoAnne MacDonald (University of Doha for Science and Technology, Qatar)
Mr. Moussa Habib (Al Hussein Technical University, Jordan)
Ms. Haneen Al Hmimat (Al Hussein Technical University, Jordan)
Dr. Amin Esmaeili Khalil Saraei (University of Doha for Science and Technology, Qatar)
Dr. Gautam Srivastava (Brandon University, Canada)
Dr. Kishwar Ali (University of Doha for Science and Technology, Qatar)
Dr. Anna Kobusinska (Poznan University of Technology, Poland)
Dr. Hassan Ali (University of Doha for Science and Technology, Qatar)
Dr. Oscar Lin (Athabasca University, Canada)
Dr. Wesam Mohamed Rohouma (University of Doha for Science and Technology, Qatar)
Dr. Yacine Atif (University of Skovde, Sweden)
Dr. Mohannad AlJarrah (University of Doha for Science and Technology, Qatar)
Dr. Mohamed Elhoseny (University of Sharjah, UAE)
Dr. Hassan Mahasneh (University of Doha for Science and Technology, Qatar)
Dr. M. Razi Nalim (Indiana University – Purdue University, USA)
Dr. Abdullatif Shikfa (University of Doha for Science and Technology, Qatar)
Dr. Priyadarsi De (Indian Institute of Science Education and Research, India)
Dr. Salama Ikki (Lakehead University, Canada)
Dr. Farhat Abbas (University of Doha for Science and Technology, Qatar)
Dr. Arafat Al-Deweik (Khalifa University, UAE)
Dr. Jawad Berri (King Saud University, KSA)
Dr. Paulo Pires (Federal University of Rio de Janeiro, Brazil)
Dr. Aitazaz Farooque (University of Prince Edward Island, Canada)
Dr. Mabrouk Ouederni (Qatar Petrochemical Company (QAPCO), Qatar)
Dr. Feng Li (Indiana University – Purdue University, USA)
Dr. Hanan Farhat (Qatar Environment and Energy Research Institute, Qatar)
Dr. Mutasim Ibrahim Nour (Herriot Watt University, Dubai)
Mr. Joseph Yang (Sweden)
Mr. Beni Suryadi (ASEAN Centre for Energy, Jakarta, Indonesia)
Dr. Tharinya Supasa (Agora Energiewende, Denmark)
Dr. Robert S. Weissback (Indiana University – Purdue University, USA)
Dr. Anser Shakoor (Energy Portfolio Management (EMEA), ABB)

Environment & climate change

Innovation and Technological Advances for Sustainability – Al-Naemi et al. (Eds)

Scalable machine learning infrastructure for wildfire prediction and mitigation[1]

Sean Choi
Santa Clara University

Son Le, Maxime Dion, Alexandra Puchko & Matt Lindeboom
Kettle Inc.

ABSTRACT: Ever-accelerating climate change is causing dramatic increases in scale and frequency of wildfires in many parts of the world, more notably in the western United States over the last few decades. Due to such increase in severity, scientists are working to better predict and model wildfire behaviors using both traditional and newer methods. However, modeling wildfire involves a large amount of data consisting of multiple types of data over multiple decades. For example, types of data often used for wildfire prediction include measured features such as temperature, precipitation and vegetation, along with unstructured data such as satellite imagery and weather mapping. Given the size of the data and complexity of the task, building models for weather or wildfire patterns traditionally required use of super computers that are not generally accessible to the public, making the research out of reach for individuals or smaller organizations.

This work introduces a scalable infrastructure for data processing, storage and machine learning that is specialized for wildfire prediction. The infrastructure is built on readily-available tools on a public cloud and is capable of ingesting multi-terabytes of data to be used to train various types of modern machine learning models, such as regularizing gradient boosting and neural network. Given such infrastructure, this paper presents the training pipeline by building models that can estimate the risk of wildfire in California while considering effects of climate change. The experiments show that our infrastructure is able to train models using large (TBs) of data, automatically improve the model performance from 0.75 to 0.80 AUC, all the while showing storage saving of at most 59% without using expensive supercomputers, suggesting that it is suitable as a wildfire predictor that can help mitigate the damage of wildfires in high risk areas and ultimately save human lives at a cost, resource and time optimal way.

1 INTRODUCTION

Over the past few decades, the world is seeing an increasing number of wildfires. Not only there are more of them, but the severity of each wildfire is significantly increasing year-over-year. As a perspective, in the year 2020 in California alone, there have been 9,917 wildfires that have ravaged through 4,397,809 acres of land, which is about 4% of the state's total land area, destroying over 10,000 structures and causing more than $12.079B in damages [1,2].

Climate affects the chances and severity of wildfires primarily through its effects on moisture availability. Wet conditions during growing seasons accumulate fuel via the growth of vegetation and dry conditions during and prior to the fire season increase flammability of vegetation that fuels wildfires. Moisture availability has a direct relation to amount of

[1]Work done in collaboration with Kettle Inc. Special thanks and acknowledgement to Nat Manning and Andrew Engler for full support of this work.

DOI: 10.1201/9781003496724-1

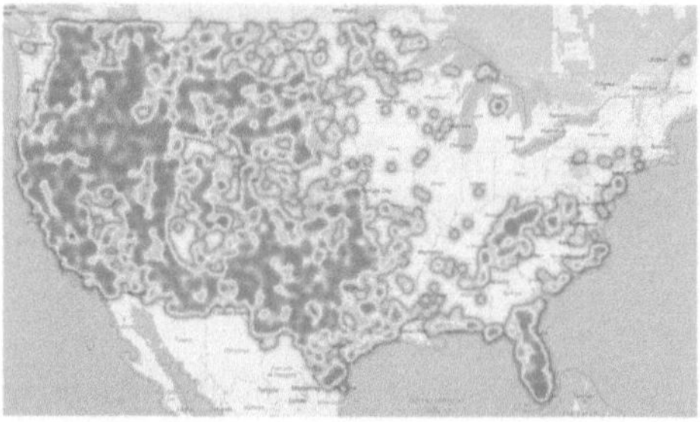

Figure 1. A Heat map of wildfires that burned more than 1,000 acres in continental United States, which shows that western United States has a high concentration of large wildfires.

precipitation and yearly temperature, as higher temperature reduces moisture availability via increased evaporation, reduced snow packs, and earlier snow melt [3]. It is widely accepted by the body of environmental research that climate change is increasing both frequency and severity of wildfires [3–9]. Due to climate change, many areas that used to experience a large amount of precipitation is receiving less and less precipitation over the years. For example, it is observed that cumulative drying effects of atmospheric aridity and precipitation deficits are particularly more pronounced in the western United States in the recent years [10,11]. Thus, the combination of amassed fuel and the reduced moisture availability have caused multiple wildfires wildfires in California at a historic proportion, as seen by Figure 1, which shows that major wildfires are highly concentrated in the areas of western United States.

Given such increasing frequency and severity of wildfires, it is more critical than ever to predict the time and location of the next wildfire. However, modeling and predicting wildfire behavior is a very difficult endeavor with multiple challenges due to the nature of ingesting, cleansing and analyzing large amounts and types of data over multiple decades from multiple non-uniform sources. Elaborating on these challenges further, first of such challenges is overcoming data quality issues. Although there exists a long history of data on wildfires, majority of the older data sets have not been modernized and digitized. Thus, researchers often struggle to work with such data sets that are hard to consume and analyze. To add to the difficulties, even the modern data sets are independently generated by multiple agencies and research groups without a generalized schema or approaches in formatting and storing of the data. Therefore, it is infeasible for most individuals to manually browse through the data and cleanse it, given that there is a large amount and types of data that must be cleansed and formatted in a uniform manner. Second of such challenges is to obtain enough computing resources to analyze the data. For example, given such requirements for analyzing a large amount of data, scientists and government agencies rely on computing powers that can be found at the level of supercomputers to perform such analytical and modeling tasks [12]. The need for computing power is only increasing with the advent of new machine learning algorithms such as deep neural networks that require even more calculations. Unfortunately, a single supercomputer generally costs in the order of millions of dollars, which means that such computing resources are either out of reach for most individuals and small team of researchers, or available in a way that a very limited set of resources are shared across multiple agencies. Such constraint causes a huge problem in developing accurate models that can accommodate quickly to the ever-changing environment, as new models may need to be experimented and trained continuously with the latest set of data to achieve the highest predictive performance. This problem becomes more pronounced when the problems researchers are trying to solve involve complex modeling and/or machine learning algorithms that require a large number of simulations or large amount of computing resources, such as deep neural networks.

The recent arrival of cloud computing, which provides the on-demand access to variable, and seemingly infinite, amount of computing services, such as servers, storage and databases, alleviates the above problem by enabling researchers in all fields to easily replace traditional supercomputers with computing resources on cloud [13]. The most important benefit of cloud computing comes with proven track record for scalability. Many of the largest firms run their entire operations on the cloud, processing petabytes of data generated from hundreds of millions of people around the world [14]. This shows that using cloud is not only feasible but also provides increased model update frequency, model accuracy and ease of model deployments in wildfire prediction. Therefore, a cloud-based approach is an inevitable next step in building better models and increasing the rate of innovation in wildfire prediction.

As a first-step towards a cloud-base climate modeling, this paper presents a generalizable and highly scalable architecture for climate scientists to build models that can predict wildfires as well as other weather patterns. In doing so, this paper discuss specific characteristics of various existing climate data sets and present solutions to process the data in a more efficient manner in the cloud. Using this architecture, this paper present a unique wildfire predictive model and show how this can be achieved without using a supercomputer. In summary, the main contributions of this paper are as follows:

- Data optimization methods for wildfire modeling.
- Architecture for cloud-based wildfire prediction that easily accommodates multiple data sets and modeling algorithms.
- Automatic tuning of wildfire prediction models.

The paper is structured as follows. First, the paper presents the background on wildfire prediction and cloud compute (§2), followed by the architecture of the wildfire prediction infrastructure(§3), then an evaluation of a generated model using the proposed infrastructure (§4) and concludes by discussing possible future directions (§5).

2 BACKGROUND

This section first discuss additional details on wildfire and climate data characteristics, current methods of predictions using such public data sets and the limitations of the existing methods. Then, it presents the background information on a cloud computing provider, its services and how each service can be used for the given purpose.

2.1 *Wildfire and climate data characteristics*

The data that the paper uses for wildfire and climate prediction are all public, retrieved from various government and academic agencies such as National Oceanic and Atmospheric Administration (NOAA) [15], National Interagency Fire Center (NIFC) [16], Wisconsin Silvis Lab [17] and CalFire [1]. These agencies provide a variety of data ranging from climate, such as North American Regional Reanalysis (NARR) [18], census and vegetation, such as (WUI) [19] and even to burn area maps for each wildfire that has occurred in the last 70 years [20]. One example data set is from the national Fire Program Analysis (FPA) system called fire-occurrence database (FPA FOD), which has been acquiring records from federal, state, and local systems since 1992, that contains 1.88 million wildfires, representing a total of 140 million acres burned during the 24-year period in the U.S. from 1992 to 2015.

However, major issues in such data sets are that, due to the nature of the data set, it is either in outdated formats that are difficult for modern data consumption and/or the formats are not uniform across multiple data sets of similar characteristics. Thus researchers generally favor using modern data, such as the FPA FOD, due to better data uniformity and minimal noise as they are recorded in accordance with modern data recording standards, often times it is not enough as it is common in practice to remove data dated far back to eliminate excessive noise.

The most prevalent characteristics of data sets often used for wildfire prediction is that data set often consists of data value associated with a geographical location and time. For example, Figure 4, which was obtained from FPA FOD, shows that wildfires are highly seasonal in nature and is highly dependent on location of lightning strikes, human activities or man-made power

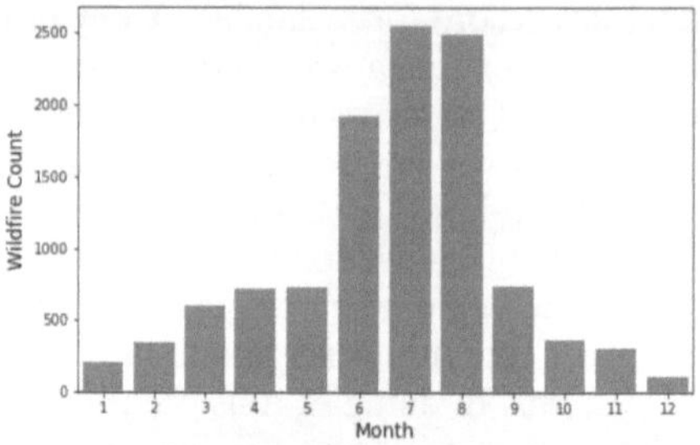

(a) Seasonality of wildfires in United States showing that wildfires are more prevalent in summer months.

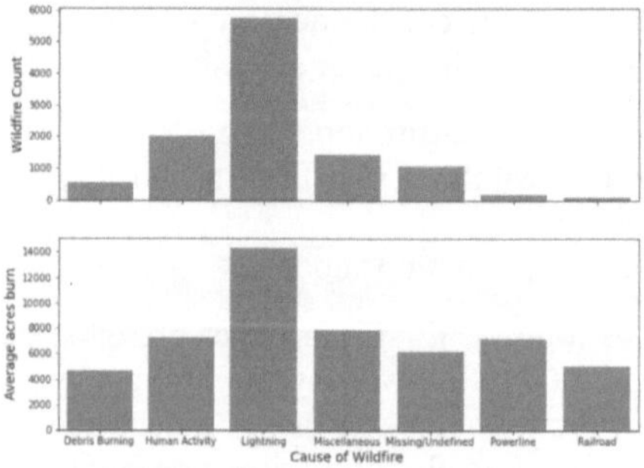

(b) Causes of wildfire attributing to the count and severity. As shown, natural lightning is a major cause of large wildfires.

Figure 2. Month and causes of wildfires as provided by FPA.

lines. This is mainly because climate conditions, which is one of the most important feature in predicting wildfires, are temporal and spatial in nature, as it varies greatly between different time of the year and the location. Such characteristics make it difficult to standardize the data set, as there are multiple formats for storing geography and time. To illustrate how complex this can be, there is an entire organization called International Association of Oil & Gas Producers (IOGP) who maintains the European Petroleum Survey Group (EPSG) Geodetic Parameter Dataset, which is a public registry of geodetic datums, spatial reference systems, Earth ellipsoids, coordinate transformations and related units of measurement [21]. EPSG Geodetic Parameter data set consists of coordinate system with code ranging from 1024 to 32767, and there are more than 5,000 different coordinate systems that exist. In addition to coordinate system registered to EPSG, countries and organizations can choose to create and use their own coordinate system. Time has a similar issue where data sets can be stored in various time zones in various formats, such as *DD.MM.YYYY* versus *MM-DD-YYYY* and more. This is a common issue when dealing with data sets from multiple sources from various agencies, thus the approaches to unify data formats shared in this paper can be generalized to other scenarios as well.

2.2 *Existing methods for climate and wildfire modeling*

There are two popular methods for wildfire modeling: Rothermel model and machine learning based model. The Rothermel surface fire spread model [22], created in 1972, is the most commonly used in U.S. fire management systems, despite the fact that there exists many other models for surface fire spread [23]. The model attempts to describe mathematically the physical and chemical processes of fire by reducing forest fires to a set of mathematcial equations operating in a hypothetical universe in which fires burn only small, uniform, dead fuels on the forest floor. Initially used as a quantitative way to calculate U.S. fire-danger rating indices, the Rothermel model gradually became a tool used mostly to predict the behavior of an ongoing wildfire [22,24] and now its extension can be found in many arenas such as fire and fuel management, fire ecology, climate change, suppression, smoke, fuel hazard assessment, and budget planning. Implementation of Rothermel fire model could be observed through the evolution of technology, as it is now embedded in complex wildfire spread estimation systems such as BehavePlus or Farsite [25]. Given that it tries to capture the forest fire using static mathematical equations, it is simpler to implement, but does not make any data-based decisions, reducing its accuracy and usefulness.

Since the early 2000s, which marked the beginning of an era of rapid growth in computational capacity and satellite imagery technology, there have been exciting research projects that investigate use of neural networks trained with satellite imagery for wildfire prediction. For example, scientists trained neural networks to learn the signatures of wildfires using data from a combination of Geostationary Operational Environmental Satellite (GOES), Advanced Very High Resolution Radiometer (AVHRR) and Moderate Resolution Imaging Spectroradiometer (MODIS) satellite data [26], providing promising results. The proposed infrastructure is complementary to such efforts, allowing such work to advance faster by providing flexible infrastructures and uniform data.

2.3 *Public cloud providers*

In the context of this paper, a public cloud is a platform that enables cloud computing resources, such as servers machines, applications or storage, available to users remotely with high availability. Public clouds provide services categorized into three types as follows: Infrastructure as a service (IaaS) is where a provider hosts infrastructure components, such as servers and storage, Platform as a service (PaaS) is where a provider delivers hardware and software tools and Software as a service (SaaS) is where a provider hosts applications and makes them available to customers over the internet. Public cloud providers can either provide these services for free or through a variety of subscription or on-demand pricing schemes, including a pay-per-usage model.

There are many public cloud providers that are available in the world. Out of them all, the three largest and most prominent providers are Amazon Web Services, Microsoft Azure and Google Cloud, ordered by the decreasing market share [27]. All three providers provide similar types offerings with different names that serve exactly the same purpose. This section will introduce a set of services that are relevant to the proposed infrastructure, such as cloud object storage, distributed query engine, machine learning and serverless frameworks.

2.3.1 *Cloud object storage*

Object based storage is a data storage architecture that manages data as objects, as opposed to files [28]. Each object includes the data, a metadata associated with the data and a globally unique identifier. Object based storage is optimized for storing massive amounts of unstructured data in which data is written once and read at least once, which is a popular use case in data analytic and machine learning. Given such benefits, cloud providers leverage object storage to implement cloud storage as an IaaS, where users can store seemingly unlimited amount of data in a remote data center. Some examples of cloud object storage services includes Amazon S3 [29], Microsoft Azure Blob Storage [30] and Google Cloud Storage [31].

2.3.2 *Distributed query engine*

Retrieving a large amount of data, with a set of user-defined requirements, from a distributed set of data sources is a difficult task to perform. Thus, various open-source software have been developed in order to facilitate such complex operations. For example, query engines allow users to simply enter Structured Query Language (SQL) commands that specify data sources to retrieve the data from, processes for filtering the data, along with data transformation and analytic to apply to the retrieved data. At a high-level, distributed query engine allows easy query and manipulation on a large amount of data stored at multiple remote locations. Some examples of distributed query engines are AWS Athena [32], Hive

[33] and Google BigQuery [34]. Discussing the entire detailed architectures of various distributed query engines is out of scope of this paper, thus this paper will focus mostly on Athena as it is used throughout the proposed architecture.

AWS Athena is based on a open-source software called Presto [35], which originally was developed to run interactive SQL queries over a distributed file system called Hadoop File System and is known to be successful in querying data in the order of petabytes in size with high performance [36]. Athena extended Presto's features by enabling data retrieval from S3, stored in various formats such as comma-separated values (CSV), Apache Parquet [37] and JSON. In order to query a a set of data in Athena, users must create a data table specifying the location of the data in S3 and the schema of the data stored in the given location. Once the table is created, users can query the data from the table, as if the data is being retrieved from a relational database.

2.3.4 *Machine learning and serverless framework*

Building a machine learning algorithm from scratch is a daunting task that a only few set of researchers can accomplish. If one is interested only in building a machine learning model rather than improving the modeling algorithm, the easiest, and the most widely accepted, approach is to use machine learning algorithms that are available in pre-built software packages like XGBoost [38], PyTorch [39] and TensorFlow [40]. However, often times these packages are configured to run in a set of local machines, making it quite difficult to simply use these packages in a cloud environment. In order to alleviate such issue, cloud providers have developed and provides a set of SaaS framework that allows users to build machine learning algorithms at scale in the given providers' native cloud environments. Such frameworks include Amazon Sagemaker [41], Azure Machine Learning [42] and Google Vertex AI [43].

Once the model is built using such frameworks, there are multiple options to use the model for inference. One of the popular choice is to deploy the model in a serverless framework, where the model would be deployed to the cloud and is accessed in a dynamic fashion whenever the model is needed for inference. Serverless frameworks have been developed by cloud providers to allow users to save cost in deploying their custom programs that, while enabling cloud providers to recoup idle resources for other users. Serverless frameworks are often used in scenarios where some program receives requests in a random manner, reducing the need to have a server running all the time. Machine learning inference is a good use case of such framework, as many organizations do not have live users continually making requests for model inference. Some examples of serverless frameworks include AWS Lambda [44], Azure Functions [45] and Google Functions [46].

The proposed infrastructure is based on AWS, utilizing S3 for storage, consuming Athena query results from S3 for the machine learning frameworks, AWS Sagemaker for distributed machine learning and Lambda for the online inference engine. The details of the the infrastructure is discussed in §3.

3 ARCHITECTURE OVERVIEW

We now present the data and machine learning architecture that enables modeling wildfire and climate behaviors at scale. Our proposed architecture comprises of four main components that are working in series: data pipeline, framework for model training, model inference and continuous integration/deployment. Each component is scaled on its own, meaning that we can allocate different type and number of hardware that will be used for each purpose. Given the overview, we now present each component in detail and how they are realized on cloud.

3.1 *Data pipeline*

Data pipeline consists of logical units that are responsible for retrieving data from the data sources, formatting, cleansing and optimizing the data for storage, replicated storage systems and query engine that allows easy access to the data that is stored in the storage. The entire pipeline is illustrated on the left of Figure 3 and each unit is described in details below.

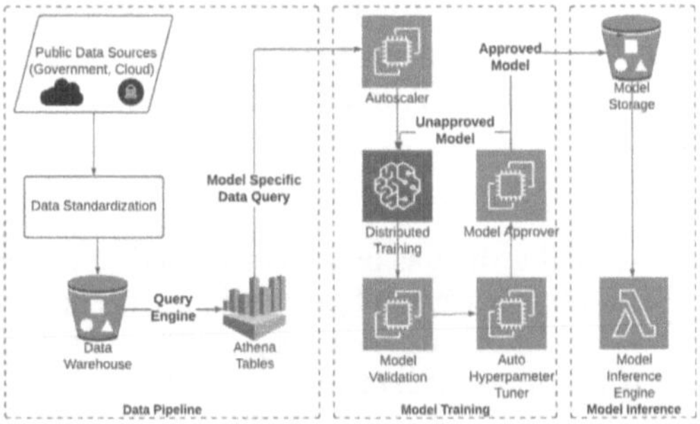

Figure 3. Overall architecture of the data pipeline and machine learning framework that automatically scales based on the size of the data.

3.1.1 *Data formatting and cleansing*

The main difference between a climate modeling infrastructure versus a standard machine learning infrastructure is the types of data that are available. As mentioned in §2, public climate data is stored in multiple formats that often consists of data value associated with a geographical location and time. Therefore, it is crucial that the data is stored in a consistent manner, where the geographical coordinate systems, data types and time representations are uniform across different data sets. In order to unify these values, here are a set of rules that we implement, along with libraries that allow easy transformation as follows:

3.1.1.1 *Geographical coordinate system* We convert all spatial information to EPSG:4326, which is a coordinate system on the surface of a sphere or ellipsoid of reference. Our library utilizes PROJ [47] to perform the coordinate system conversion and to compute distance based metrics, such as distance between two coordinates and area.

3.1.1.2 *Geographical data type and storage* We store all geometric data as Well-Known Type (WKT) strings as defined by the Open Geospatial Consortium (OGC) and described in their Simple Feature Access [48]. The main reason behind this is to enable easy transition between various file formats and binaries and also allow easier indexing and search into the data. While WKT format may take up more space, we believe benefits in flexibility and search outweighs platform specific formats. We also store latitude and longitude values in floats to enable quick sorting and boundary box calculations.

3.1.1.3 *Date and time* We store all data and time in a *YYYY-MM-DD HH:MM:SS* format. The main reason why we did not use Unix epoch is because timestamps based on Unix epoch starts in January 1st, 1970, where as we can have data points that precedes that.

Besides these generic rules, we make sure we do not store any data in binary format, as it gets very difficult to decode binary data in cases where the data loses its context.

3.1.2 *Data storage and optimizations*

Once the data format has been unified as mentioned above, we store the data in parquet [37] format, compressed using GZIP [49]. The main benefit of using parquet is the storage format, faster query speed and reduced storage size. Parquet is a column-format storage and allows storing chunks of data, which makes it possible to query and modify a subset of the data and also add more data if needed. This also means that most queries are much faster, as we do not need to read the entire data set first. In addition, parquet with GZIP compression greatly reduces the amount of space that the data takes up. For example, we have seen roughly from 2 up to 50 compression ratio, which is the ratio between the size of unprocessed and the processed files, going from raw data to compressed parquet format. This means that not only it is faster to retrieve the data, but costs less to store the file in the cloud, as cloud storage prices depend on the storage consumed by the user. Given the format, column names and types in the parquet files are standardized so that they are uniform across different data sets. This is performed using a data pipeline library, where it contains a custom logic for column conversions.

3.1.3 *Query engine*

Once the data is stored in parquet format as objects in a cloud storage, preferably in a single folder per data set, we define multiple schema for each data set. The schema structure closely follows the SQL format and depending on the query engine used, the data types can vary. For our infrastructure, we conform to the Athena [32] specification for defining the schema. Once the data schema is defined, we can easily query over the data set using the SQL-like language provided by the underlying query engine. The query engine can be used to run analytical queries or specify a set of train, test and validation data sets to be fed into the model training framework discussed in §3.2.

3.2 *Model training framework*

The model training framework consists of five components: autoscaler, distributed training engine, model validator, hyperparameter tuner and model approver. First, the autoscaler is responsible for analyzing the data set size, training parameters and type of algorithm, and deciding the number and the type of machines to use for the training steps. This step is crucial in automating the training framework, as incorrect estimation on the number of machines will cause the training to fail. A simple algorithm for finding the number of machine is to first get the size of the training data per year or per logical division and provision 3x the total RAM of all machines combined. For example, if a block of data is about 10GB, then the autoscaler will provision enough machines to store 30GB of data. Second, the distributed training engine is where each machine dedicated for training receives part of the data and trains the model. For models that do not support parallel training, the training engine will launch only one machine. Third, the model validator will perform model performance checks on the model generated by the distributed training engine to see if it meets the minimum requirements that the user can specify. For example, if we want to make sure that our model always exceeds 70% accuracy, the validator will make sure that the model generated is only accepted only if the model accuracy is above 70%. Fourth, the hyperparameter tuner can take the output of the validator and run multiple training jobs to provide a model that is better than the user provided model. This step is optional and can be run if the user is unsure of which hyperparameters to use. In addition, the hyperparameter tuner can tune the model based on different model performance metrics, such as accuracy, root mean squared error (RMSE), Area under the ROC curve (AUC), precision and recall and more. Also, you can specify the hyperparameter tuner to perform tuning in a randomized fashion or using a Bayesian based method [50]. Lastly, there is a model approver where the user will manually inspect the final output of the training framework and approve the model to be served by the inference framework.

3.3 *Model inference framework*

The last piece of the infrastructure involves the model inference framework that allows users to utilize the models that has been built using the training framework. The inference framework is given an approved model from the model approver and can serve the model inference in two different ways: offline and online. For offline model inference, the user provides a query for a set of data that you want to use for inference, then the inference framework generates a batch of inference results to the user. For online inference, the framework starts a AWS Lambda [44] endpoint which receives the features as HTTP data and provides the inference results.

4 EVALUATIONS

The following metrics are provided to evaluate the infrastructure: data optimization and query performance, model training performance and hyperparameter tuning performance.

4.1 *Data optimization and query performance*

The major piece of the contribution comes from unifying the data schema and optimizing the storage size. The following two metrics are provided in order to provide a perspective in on the performance of the given approaches: data compression ratio and query performance. First, the compression ratio for the following sets of data are provided: raw NARR daily monolevel data [18] and Palmer Drought Severity Index (PDSI) [51] converted from NetCDF file and CALFIRE Fire Perimeters [52] converted from Shapefile format [53] in Table 1. The compression ratio is quite impressive from data without geographical geometries, as the method both removing unnecessary columns and compressing the data. As for data with geometries, the compression ratio is less impressive, since the data is inefficiently stored as string in the data warehouse.

Given the optimized data stored in the data warehouse, the section now provides a view of the query performance. First, querying all the features in NARR for a given month took on average 3 seconds. In comparison, to perform a similar query using NetCDF file, one must load the entire data and search for a specific month, which can take minutes based on the size of the data and how it is stored. We expect the query performances to be greater in cases where one must join multiple data sources and/or apply complex filters.

Table 1. Table of data compression ratios for NARR, PDSI and CALFIRE Fire Perimeter data.

	Unoptimized (MB)	Optimized (MB)	Compression Ratio (%)
NARR	28100	572.4	49.09
PDSI	6300	344.3	18.30
Fire Perimeter	100.9	45.9	2.20

4.2 *Model training performance*

This section presents the performance of the training framework by providing the model training time and the model performance. For the purpose of measuring the performance of the framework, the following features in the data set are utilized: climate data, such as air temperatures, precipitation wind direction, wind speed, relative humidity and more from NARR, PDSI, vegetation cover and Ecoregions data. Then these features are group in geographical regions by rounding the latitude and longitude value to the tenth decimal place, effectively creating a grid where the side of each grid equals 0.1 latitude or longitude values, and select the grids where the grids belong in California to utilize the CALFIRE data set. Then each grid gets a label, where it is 1 if the grid has ever had a wildfire based on the fire perimeter data and 0 if the grid never had a wildfire. Then the data is split into train, test and validation, where the training data is the data from 1985 to 2018, validation data is the data from 2018 to 2020 and the test data is data in 2021. Using such features, labels and data sets, the infrastructure generates a XGBoost [38] model with a set of hyperparameters that are manually selected, where the model optimizes the Poisson count regression objective function, using 500 rounds at 0.018 learning rate. The training job runs on up to 20 *ml.m5.xlarge* type machine on AWS EC2, which consists of 4 vCPUs and 16 GB RAM, as the number of machines required for training jobs varies based on the amount of input data. The average training time over 10 runs given such data and hyperparameters is around 36 minutes. The model performed with 56.1% accuracy of predicting if a given grid will burn with the weighted average precision at 97.7% and recall at 56.1% with 0.75 AUC. Given that only a small percentage of the state has ever burned, the model performs well in finding the features that causes wildfires.

4.3 *Hyperparameter tuner*

An added benefit of the proposed infrastructure is that the infrastructure can find the best hyperparameters automatically by using the provided hyperparameter tuner using the features provided in Sagemaker [41]. To evaluate the performance of the hyperparameter tuner, the model performance based on different tuning parameters are provided, which can be seen in Figure 4. The experiment trained two models in addition to the manually optimized model, where each model optimizes hyperparameters based on different objective functions: AUC and RMSE. The hyperparameter tuning was done using a Bayesian-based method [50], which is a method that performs hyperparameter tuning by solving a regression problem. The experiment ran 30 different training jobs in parallel, allowing for faster tuning performance. As shown in Figure 4, it is clearly shown that the AUC optimized model produced the best AUC value at 0.80, whereas the RMSE optimized model produced a lower AUC of 0.74. In both models, hyperparameter tuning took about 14 minutes using *ml. m5.xlarge*. The tuning framework was able to reduce the training time and optimize the training process by finding the optimal number of rounds and learning rate.

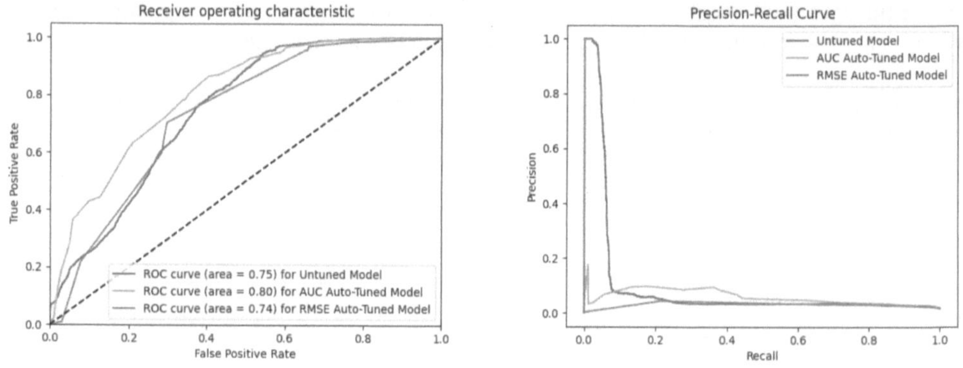

(a) ROC curves and AOC values three models

(b) Precision recall curves of three models

Figure 4. Performance comparison of three models that are manually optimized, optimized on RMSE and optimized on AUC. (a) ROC curves and AOC values three models. (b) Precision recall curves of three models.

5 DISCUSSIONS & FUTURE WORKS

This section discusses potential use cases and future directions of the proposed infrastructure and the generated model.

5.1 *Potential use cases*

The infrastructure is easily adaptable to be used by various agencies. Given that, first obvious use case is for wildfire prediction, prevention and damage mitigation. Given that the infrastructure is able to provide the most up-to-date model for wildfire prediction, government agencies can build a system that notifies the fire department or residents in high risk areas when the fire risk becomes high. In addition, given that the model can predict the possibility of wildfires given upcoming weather forecasts, it is possible to prevent wildfires by performing preventive actions, such as bush removal, limiting campfires and inspecting power lines. Another use case is for predicting other weather events, such as drought or flood. Since the infrastructure is designed for climate modeling, it is easy to build new models to predict and mitigate other extreme weather events.

5.2 *Additional machine learning algorithms*

Currently, this infrastructure is capable of creating a short list of machine learning algorithms including: XGBoost, K-nearest-neighbor, image classification and linear learner. To extend this, the plan is to add more algorithms that support more complex use cases, such as deep learning or time series forecasting algorithms. The main reason behind this is that, since wildfire events are seasonal in nature, there is a good chance that performing time-series based forecasting using deep learning on satellite images can lead to a better result than a regression or classification based methods using just climate data. Finally, in addition to the hyperparameter tuning, machine learning community is moving towards completely automated modeling, where the machine selects the correct model as well as the hyperparameter for the given problem. One example of such approach is the wide adoption of AutoML [54], where the machine learning algorithm chooses not only the hyperparameters, but also the model as well. Future work will add this feature further automate and increase the performance of the model building process for wildfire prediction.

6 CONCLUSION

This paper aims to accomplish the goal of providing a generalizable architecture for climate researchers to produce large scale machine learning models without needing complex and expensive hardware. The proposed infrastructure is able to train models using a large amount of data and automatically improve the model performance using the hyperparameter tuning capabilities and is able to save storage size and cost. These results suggest that this infrastructure is suitable as a wildfire predictor that can help mitigate the damage of wildfires in high risk areas. This paper opens ways for individuals and government agencies to adopt and benefit from the proposed approach and ultimately save human lives at a cost and time efficient way

REFERENCES

[1] Fire C., *2020 Incident Archive*, https://www.fire.ca.gov/incidents/2020/, 2020.

[2] N.I.F. Center, (2020). *National Large Incident Year-to-Date Report*, https://gacc.nifc.gov/sacc/predictive/intelligence/NationalLargeIncidentYTDReport.pdf, National Interagency Fire Center, Tech. Rep.

[3] Westerling A. *et al.*, (03 2008). Climate change and wildfire in California, *Climatic Change*, vol. 87, pp. 231–249.

[4] Abatzoglou J. T. *et al.*, (2016). Impact of anthropogenic climate change on wildfire across western us forests, *Proceedings of the National Academy of Sciences*, vol. 113, no. 42, pp. 11 770–11 775. https://www.pnas.org/doi/abs/10.1073/pnas.1607171113

[5] Marlon J. R. *et al.*, (2009). Wildfire responses to abrupt climate change in north america, *Proceedings of the National Academy of Sciences*, vol. 106, no. 8, pp. 2519–2524. https://www.pnas.org/doi/abs/10.1073/pnas.0808212106

[6] Westerling A. *et al.*, (2011). Climate change and growth scenarios for california wildfire, *Climatic Change*, vol. 109, no. 1, pp. 445–463.

[7] Fried J. *et al.*, (01 2004). The impact of climate change on wildfire severity: A regional forecast for northern california, *Climatic Change*, vol. 64, pp. 169–191.

[8] Spracklen D. *et al.*, (10 2009). Impacts of climate change from 2000 to 2050 on wildfire activity and carbonaceous aerosol concentrations in the western united states, *Journal of Geophysical Research*, vol. 114.

[9] Liu Y. *et al.*, (2010). Trends in global wildfire potential in a changing climate, *Forest Ecology and Management*, vol. 259, no. 4, pp. 685–697, adaptation of Forests and Forest Management to Changing Climate. https://www.sciencedirect.com/science/article/pii/S0378112709006148
[10] Keeley J. *et al.*, (08 2016). Climate change and future fire regimes: Examples from california, *Geosciences*, vol. 6, p. 37.
[11] Nauslar N. *et al.*, (06 2018). The 2017 north bay and southern california fires: A case study, *Fire*, vol. 1, p. 18.
[12] Altintas I. *et al.*, (2015). Towards an integrated cyberinfrastructure for scalable data-driven monitoring, dynamic prediction and resilience of wildfires, *Procedia Computer Science*, vol. 51, pp. 1633–1642.
[13] Poort J., (Nov 2021). How cloud-based supercomputing is changing r&d, https://hbr.org/2021/11/how-cloud-based-supercomputing-is-changing-rd.
[14] AWS, (Mar 2022). Aws customer success stories, https://aws.amazon.com/solutions/case-studies/.
[15] U.D. of Commerce, National oceanic and atmospheric administration, https://www.noaa.gov/, 2022.
[16] N.I.F. Center, (2022). National interagency fire center, https://www.nifc.gov/.
[17] U. of Wisconsin Madison, (2022). *Silvis Lab – Spatial Analysis for Conservation and Sustainability*, http://silvis.forest.wisc.edu/.
[18] Mesinger F. *et al.*, (2006). North american regional reanalysis, *Bulletin of the American Meteorological Society*, vol. 87, no. 3, pp. 343 – 360. https://journals.ametsoc.org/view/journals/bams/87/3/bams-87-3-343.xml
[19] Radeloff V. *et al.*, (03 2018). Rapid growth of the us wildland-urban interface raises wildfire risk, *Proceedings of the National Academy of Sciences*, vol. 115, p. 201718850.
[20] CALFIRE, (2022). Fire perimeters dataset, https://frap.fire.ca.gov/frap-projects/fire-perimeters/.
[21] I.A. of Oil & Gas Producers (IOGP), Epsg geodetic parameter dataset, https://epsg.org/home.html, 2022.
[22] Rothermel R., (1972). *A Mathematical Model for Predicting Fire Spread in Wildland Fuels*, ser. USDA Forest Service research paper INT. 1em plus 0.5em minus 0.4em Intermountain Forest & Range Experiment Station, Forest Service, U.S. Department of Agriculture. https://books.google.com/books?id=27n_RugVVK0C
[23] Sullivan A., (07 2009). Wildland surface fire spread modelling, 1990–2007. 1: Physical and quasi-physical models, *International Journal of Wildland Fire*, vol. 18, pp. 349–368.
[24] Andrews P.L., (2018). The rothermel surface fire spread model and associated developments: A comprehensive explanation, *Gen. Tech. Rep. RMRS-GTR-371. Fort Collins, CO: US Department of Agriculture, Forest Service, Rocky Mountain Research Station. 121 p.*, vol. 371.
[25] Andrews P., (01 2014). Current status and future needs of the behaveplus fire modeling system, *International Journal of Wildland Fire*, vol. 23, p. 21.
[26] Miller J. *et al.*, (2013). *Automated Wildfire Detection Through Artificial Neural Networks*. Berlin, Heidelberg: Springer Berlin Heidelberg, pp. 293–304. https://doi.org/10.1007/978-3-642-32530-4_20
[27] Stalcup K., (May 2021). *Aws vs Azure vs Google Cloud Market Share 2021: What the Latest Data Shows*, https://www.parkmycloud.com/blog/aws-vs-azure-vs-google-cloud-market-share/.
[28] Mesnier M. *et al.*, (2003). Object-based storage, *IEEE Communications Magazine*, vol. 41, no. 8, pp. 84–90.
[29] AWS, (Mar 2006). *Amazon Web Services Launches Amazon s3*, https://press.aboutamazon.com/news-releases/news-release-details/amazon-web-services-launches-amazon-s3-simple-storage-service.
[30] M. Azure, (Mar 2022). Introduction to azure blob storage, https://docs.microsoft.com/en-us/azure/storage/blobs/storage-blobs-introduction.
[31] Google, (2010). Google cloud storage – getting started, https://cloud.google.com/storage/docs.
[32] AWS, (2022). Aws athena documentation, https://docs.aws.amazon.com/athena/index.html.
[33] Camacho-Rodrguez J. *et al.*, Apache hive: From mapreduce to enterprise-grade big data warehousing, in *Proceedings of the 2019 SIGMOD*, ser. SIGMOD '19. ACM, 2019, p. 1773–1786.
[34] Google, (2010). Google bigquery documentation, https://cloud.google.com/bigquery/docs/reference.
[35] Sethi R. *et al.*, Presto: Sql on everything, in *2019 IEEE 35th International Conference on Data Engineering (ICDE)*, 2019, pp. 1802–1813.
[36] Tse E. *et al.*, (Oct 2014). Using presto in our big data platform on aws, https://netflixtechblog.com/using-presto-in-our-big-data-platform-on-aws-938035909fd4.
[37] Vohra D., (09 2016). *Apache Parquet*. Berkeley, CA, USA: Apress, pp. 325–335.
[38] Chen T. *et al.*, (2016). Xgboost: A scalable tree boosting system, in *Proceedings of the 22nd acm sigkdd international conference on knowledge discovery and data mining*, pp. 785–794.
[39] Paszke A. *et al.*, (2019). Pytorch: An imperative style, high-performance deep learning library, in *Advances in Neural Information Processing Systems 32*, H. Wallach *et al.*, Eds. Curran Associates, Inc., pp. 8024–8035.

[40] Abadi M. *et al.*, TensorFlow: Large-scale machine learning on heterogeneous systems, 2015, software available from tensorflow.org. https://www.tensorflow.org/
[41] AWS, (2022). Aws sagemaker documentation, https://docs.aws.amazon.com/sagemaker/index.html.
[42] M. Azure, (2022). *Azure Machine Learning*, https://azure.microsoft.com/en-us/services/machine-learning/.
[43] Google, (2010). Google vertex ai, https://cloud.google.com/vertex-ai.
[44] AWS, (2022). *AWS Lambda Developer Guide*, https://docs.aws.amazon.com/lambda/latest/dg/welcome.html.
[45] M. Azure, (2022). *Introduction to Azure Blob Storage*, https://docs.microsoft.com/en-us/azure/azure-functions/.
[46] Google, (2022). *Google Cloud Functions*, https://cloud.google.com/functions.
[47] PROJ contributors, *PROJ coordinate transformation software library*, Open Source Geospatial Foundation, 2022. https://proj.org/
[48] Herring, John R., (2022). *Simple Feature Access – Part 1: Common Architecture*, Open Geospatial Consortium. https://www.ogc.org/standards/sfa
[49] Deutsch P., (May 1996). GZIP file format specification version 4.3, RFC 1952 (Informational), *Internet Engineering Task Force*. http://www.ietf.org/rfc/rfc1952.txt
[50] Snoek J. *et al.*, (2012). Practical bayesian optimization of machine learning algorithms, in *Proceedings of the 25th International Conference on Neural Information Processing Systems – Volume 2*, ser. NIPS'12. Red Hook, NY, USA: Curran Associates Inc., p. 2951–2959.
[51] N.C. for Atmospheric Research, Palmer drought severity index (pdsi), https://climatedataguide.ucar.edu/climate-data/palmer-drought-severity-index-pdsi, 2022.
[52] Fire C. *et al.*, (2022). *Fire Perimeters*, https://frap.fire.ca.gov/frap-projects/fire-perimeters/.
[53] A. R. Center, Shapefile file extensions, https://help.arcgis.com/en/arcgisdesktop/10.0/help/index.html#/Shapefile_file_extensions/005600000003000000/, 2 2011.
[54] Guyon I. *et al.*, (2015). AutoML challenge 2015: Design and first results, in *Proc. of AutoML 2015@ICML*. https://drive.google.com/file/d/0BzRGLkqgrI-qWkpzcGw4bFpBMUk/view

Innovation and Technological Advances for Sustainability – Al-Naemi et al. (Eds)

Comparative evaluation of the impact of roofing systems on the energy efficiency of low-rise buildings in the various climatic conditions in India

Farheen Bano
Faculty of Architecture and Planning, Dr APJ Abdul Kalam Technical University, Lucknow, India

Mohd Tahseen
Department of Architecture, Government Polytechnic, Lucknow, India

Vandana Sehgal
Faculty of Architecture and Planning, Dr APJ Abdul Kalam Technical University, Lucknow, India

ABSTRACT: The building envelope is the interface between the external atmospheric conditions and the indoor environment. It contributes to approximately 60–70% of the heating and cooling load. The building envelope constitutes walls, fenestrations, and roofs from which thermal transfer from the roof is significantly very high compared to walls in low-rise buildings. This research paper aims to provide recommendations for energy-efficient roof design considering roof form, material, and external coating suitable for various climate conditions in India. The paper's methodology consists of a literature review of related research papers, learnings from the roofing system of energy-efficient building case studies, and a simulation of all possible energy-efficient roofing systems for energy efficiency and thermal comfort in five climate types in India. The expected outcome of this paper would be in the form of recommendations for roofing systems in different climate type and their potential for energy saving from the highest to the lowest. This research has been performed in three stages viz, Step-1 Literature Review and case study of best practices: A review of the related literature and best practices examples help formulate the maximum possible cases for the simulation. Step-2: The roofing systems, including building form, material (Thermal and insulation), and finishes (reflectance), would be simulated one by one for various climate types of India (composite, hot & dry, warm and humid, moderate and cold) to optimize energy efficiency. Step 3: Special roof types like Green roofs, cool roofs, Photovoltaic roofs, roof ponds, and the like would also be simulated and compared for their effectiveness in selected climate types. Based on the literature review, simulation results, and analysis, the recommendation would be framed for energy-efficient roofing systems in the specified climate type concerning heat transfer through the roof and thermal comfort.

Keywords: energy efficiency, roofing system, thermal comfort, building energy simulation

1 INTRODUCTION

Buildings are significant energy consumers in their construction, running, and maintenance phases. In India, 31% of energy is used in residential and commercial buildings from the total energy consumption [1]. An air- conditioned building consumes 55% of its energy in its use phase to make the building comfortable with the heating, ventilation, and air conditioning (HVAC) system. For instance, the load on HVAC comes from the heat gain or losses through the building envelope comprising the roof, wall, and windows. Heat transfer from the roof is

 DOI: 10.1201/9781003496724-2

very significant in low-rise buildings compared to high-rise. It contributes approximately 26% to the HVAC load (refer Figure 1). Hence, energy-efficient roof design can reduce heat transfer, thus, the load on HVAC. It has been estimated that a 10 to 40% reduction in air conditioning expenditure can be achieved through the climate-responsive design of a roof [2].

India has diversified climate types distributed all over. Hence, one roof type could not be effective in all climates. This research paper aims to find a suitable roofing system for various climate types of India, viz. hot and dry, warm and humid, temperate, composite and cold (ECBC, 2017). The roofing system comprises roof form, material, and finishes. The simulation was carried out initially for optimum roof form, and then materials and finishes were identified through a literature review. The cases made for the simulation were checked for all the climate types and cities selected, as given in Table 1.

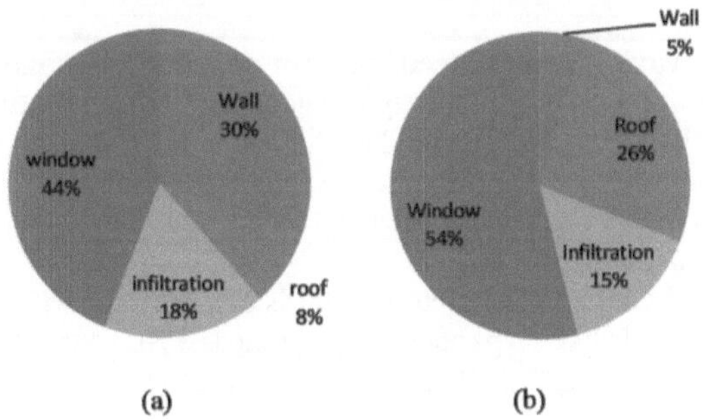

Figure 1. Heat gain through various building envelope components (a) in multistoried buildings (b) in a single storied building.

Table 1. Climate type and selected cities.

Climate Type	City
Hot and Dry	Ahmedabad
Warm and Humid	Mumbai
Temperate	Pune
Composite	New Delhi
Cold	Srinagar

Source: [3]

2 METHODOLOGY

The research has been carried out in the following steps:

- Initially, a literature review of research papers from reputed journals, codes, and standards were performed to identify the prevalent energy–efficient roofing systems and their significance. Moreover, it also explains the efficiency of roofing systems in different climate types.
- Then, selected roof form, materials, and finishing techniques were simulated in different climate types one at a time.

3 LITERATURE REVIEW

No two buildings are equivalent in energy consumption considering residential and commercial buildings and their climate. As a result, multiple types of roofs are designed to suit various building designs and climatic conditions. Moreover, several engineers and architects work on the efficiency of a roof in a particular climate. To understand the thermal performance of different roof types, a study from 5–10 sources from each climate type has been carried out in this research paper, which includes published research papers in reputed journals.

Th.data has been broadly categorized into various parameters like location, climate, roof type, roof form, and energy saving. Further comparative analysis of all these studies on these parameters provides us with the roofing systems' design idea for selected climate types in India. The inferences drawn from the literature review have been listed below according to the climate type.

3.1 *Hot and dry*

- Roof Form- A flat, vaulted, and domed roof can be used in hot and dry climates, out of which vaulted and domed roof is energy-efficient roof forms (without any insulation). Heat gain decreased by 25% in the domed roof, despite the dome covering only 60 to 70% of the roof [4]. A vaulted roof is more efficient than a domed roof, with an energy- saving of 40%, and covers 100% of the roof area [5]. It led to a decrease in the average indoor temperature in August to 1.5°C.
- Material- Reinforced Cement Concrete (RCC) is the primary structural material of the roof. Insulating materials like Expanded polystyrene (EPS), Extruded polystyrene (XPS), polyurethane foam (PUF), Glass fiber, Rockwool, and water have been used above the roof to reduce heat gain [2,6,7]. The lowest U-value can be achieved through PUF, and an economical insulated roof solution can be achieved using Rockwool [8,9]. Phase change material is also used as an insulating material above or below the RCC roof for heat exchange and reduces energy consumption by 13 to 40% with variation in PCM depth [10,11].
- Finishes- Cool Roofs, Green roofs, Roof ponds, shaded roof ponds, Photovoltaic roofs, Biosolar roofs, and double-skin roofs are the options for energy-efficient finishes over the roof to reduce heat transfer in hot and dry climates [2]. Out of these, Roof ponds, Green roofs, and Cool roofs are the most efficient roofing type. Cool roof and Photovoltaic roof has the lowest payback time. The cool roof is cheaper and can reduce 50 to 65% of energy consumption in a building. A green roof reduces 30 to 50% energy consumption and minimizes ambient and indoor temperature in a building. On the contrary, it has a high cost and high payback time.

3.2 *Warm and humid*

- Roof Form- Pitched and flat roofs are the best options in humid climates due to high humidity and rainfall [12].
- Materials- We can use concrete as a base for the cool roof and insulating materials like PUF, Rockwool, and water [6]. Lesser the U-value higher the energy efficiency in the building, while Rockwool turns out to be the economical solution with optimum energy efficiency [13].
- Finishes- Cool Roofs, green roofs, ventilated roofs, Photovoltaic roofs, and biosolar roofs are the options in which cool Roofs, ventilated roofs, and Photovoltaic roofs are the most energy-efficient roofing system [2]. Green roofs minimize ambient and indoor temperature in a building. Ventilated roofs allow the hot air to escape, keeping the roof cooler. Energy saving is slightly lower in this climate than in the hot and dry climate because uncomfortable hours are low. Ventilated roofs can reduce up to 50% energy consumption. Cool roofs are more efficient than green roofs with a 40% reduction in energy consumption.

3.3 *Temperate*

- Roof Form- Pitched and flat roofs are the best options in moderate climates due to rainfall [14].
- Materials- Concrete slab (100–200MM) and metal deck roof with insulating material can be used [15].
- Finishes- Cool roofs and green roofs work best in this climate type. A cool roof saves 10 to 20% and Green roof reduces 10 to 15% energy consumptions in the building [14]. Biosolar roof is very useful and can also be used [16]. Composite
- Roof Form- A flat roof is preferred in composite climates to reduce heat loss in winter and heat gain in the summer [6,17].
- Materials- Insulated roofs can reduce heat transfer in this climate type [15,18].
- Roof Finishes – Cool and green roofs are extensively used in composite climates coupled with PV panels [19]. Cold Climate
- Roof Form- A flat and pitched roof is preferred in a cold climate for passive heating [20].
- Materials- Insulated and double-skin roofs can reduce heat transfer in this climate type. An airtight envelope and good insulation can reduce heat loss in cold climates. Insulation is essential in cold climate roofs, for instance, a concrete slab with insulation of extended polystyrene or PUF [21].
- Roof Finishes- Skylight roofs with louvers could increase heat gain in a building [22].

4 SIMULATION

After a literature review of the roofing system in different climate types in India, a simulation has been conducted to estimate the best possible roof form, material, and finishes. The simulation was carried out initially for optimum roof form and then materials and finishes (identified through a literature review) given in Table 2. The cases made for the simulation were checked for all the climate types listed above and cities selected as given in Table 2. A classroom for 40 students has been considered a base-case with simulation input parameters adopted from codes and standards as listed in Table 3.

Table 2. Simulation cases: Variations in base case to achieve energy-efficient roofs.

	Type	Variations
Roof Forms	Flat	Basecase
	Pitched	30° inclined (NS and EW orientation)
	Vault	NS and EW orientation
	Dome	In the center
Material	Insulating Material	50mm PUF and Rockwool above deck
	Double skin roof	100mm air
	water	50mm water

Table 3. Simulation inputs for a classroom for 40 students.

Room Type	Classroom for class 5 students	Details	Reference
Occupancy	40 students		
Occupancy hours	8am–5pm		
Area	72–80m^2	1.8–2m^2/pupil	(Neufert *et al.*, 2012)
Dimensions	7.2mx10m	If the windows are all on one side, the max. room depth is 7.20 m	(Neufert *et al.*, 2012)
Infiltration rate	13 cfm/person, 6.5 l/s person	For ages 9 years plus	(NBC, 2016)
LPD	15.1 W/m^2		(NBC, 2016)

5 SIMULATION CASES

The base-case model of the classroom has been simulated in different climate types of India discussed above to identify the optimum roof type in terms of roof type, material, and finishes. The analysis was based on overall heat transfer in kWh from the roof and the predicted percentage of dissatisfaction (PPD) index. It estimates how many occupants in space would feel dissatisfied with the thermal conditions. All occupied areas in a space should be kept below 20% PPD to ensure thermal comfort according to the known standards [23].

6 RESULT AND DISCUSSIONS

6.1 *Roof form simulations results and discussion*

Simulation of roof form was performed to calculate the variation in heat transfer through flat, pitched, vaulted, and dome roofs (Figure 3) in various climate types of India and associated PPD values.

Figure 2. Variations in roof form (a) pitch (b) vault (c) dome simulated in design builder.

The primary purpose of energy efficiency is to restrict heat transfer through the roof and to make indoor conditions comfortable. Table 4 shows the simulation results on altering selected roof forms in a hot and dry, warm and humid, moderate, composite, and cold climate.

Pitched roof in NS orientation has the most negligible heat gain in hot climates like hot and dry, warm and humid, and temperate climate types; however, if PPD is compared, NS vaulted and dome roofs seem the most appropriate option for hot climates to make indoor conditions comfortable. The composite climate has extreme summers and winter seasons; thus, heat gain in summer and heat losses in the winter season is calculated through simulation. On simulation, it is observed that apart from the flat roof, all roof forms are energy efficient in a composite climate when compared based on heat transfer. However, a dome roof could be optimum for heat transfer and thermal comfort in both hot and cold seasons. In a cold climate, NS oriented pitched roof is the most efficient roof based on heat transfer during cold seasons and thermal comfort.

Table 4. Simulation results for annual heat gain through roof forms in different climate types.

Simulation result Data

Hot and dry- Ahmedabad		
Roof form	Heat gain through the roof in kWh	PPD%
Flat	3531.52	56.85
Pitched roof NS	467.76	59.07
Pitched roof EW	899.18	58.86
Vaulted roof NS	4530.38	51.79
Vaulted roof EW	5001.29	51.91
Dome	3,227.30	52.96

Charts

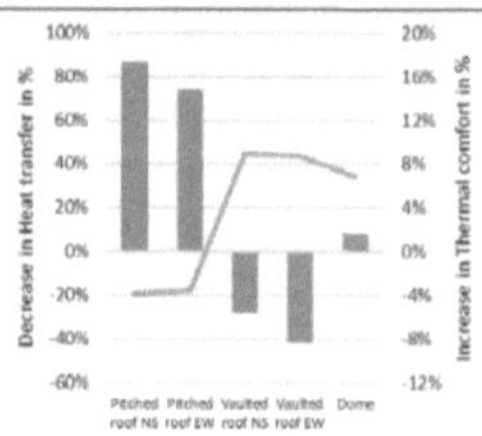

(continued)

Table 4. Continued

Warm and Humid- Mumbai		
Roof form	Heat gain through the roof in kWh	PPD%
Flat	5721.67	63.33
Pitched roof NS	2128.65	65.14
Pitched roof EW	2553.5	65.57
Vaulted roof NS	6391	56.43
Vaulted roof EW	6965.13	57.29
Dome	4,806.88	58.04

Temperate- Bengaluru		
Roof form	Heat gain through the roof in kWh	PPD%
Flat	5374.18	46.43
Pitched roof NS	2520.77	46.13
Pitched roof EW	2547.75	48.05
Vaulted roof NS	5853.89	38.2
Vaulted roof EW	6056.01	38.47
Dome	4671.18	39.1

Composite- New Delhi			
Roof form	Heat gain through the roof in kWh	PPD%	Heat losses through the roof in kWh
Flat	5374.81	50.03	2144.18
Pitched roof NS	1492.75	51.07	1868.3
Pitched roof EW	1584.79	51.76	1391.53
Vaulted roof NS	2612.46	44.28	1561.18
Vaulted roof EW	2645.95	45.47	1244.17
Dome	1817.24	45.28	2152.99

Cold Srinagar			
Roof form	Heat gain through the roof in kWh	PPD%	Heat losses through the roof in kWh
Flat	2876.37	48.4	4921.64
Pitched roof NS	2250.42	49.1	3028.46
Pitched roof EW	2340.75	46.3	3884.42
Vaulted roof NS	5062.28	42.01	5011.32
Vaulted roof EW	6710.24	44.26	4361.27
Dome	3562.84	42.88	4960.54

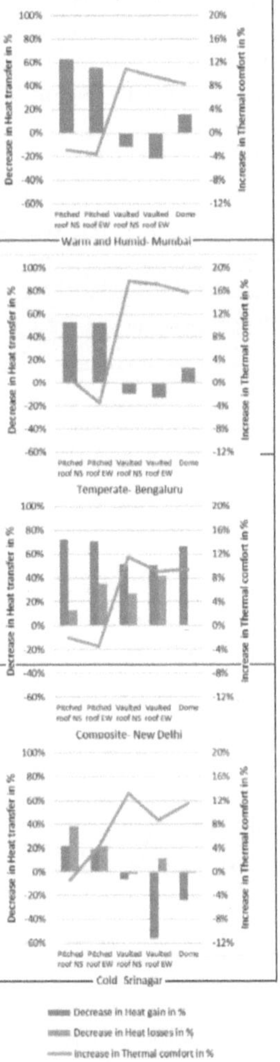

Table 5. Roof assembly in the simulation program.

Roof Assembly		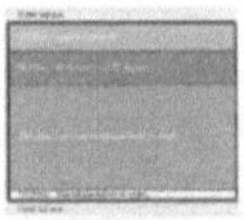		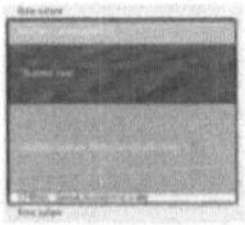
Insulating material	Polyurethane foam (PUF) (50mm thickness)	Rockwool (50mm thickness)	Air (100mm thickness)	Water (100mm thickness)
U-value of material (W/mK)	0.028	0.033	0.3	0.6
Specific heat (J/kgK)	1470	710	1000	4190
Density Kg/m^3	30	100	1000	1000
U value of roof assembly (W/m^{2}K)	0.492	0.568	1.728	2.428

6.2 *Roofing material simulation results and discussion*

As observed through the literature review, polyurethane foam (PUF), Rockwool, air, and water are readily used insulating materials and thus selected as roof insulation in simulation software (Table 5). The thickness of these materials has been kept as per the best practices, i.e., the thickness for PUF and Rockwool is kept at 50mm, whereas air and water are 100m.

Table 6. Simulation results for annual heat gain through. Roofing materials in different climate types.

Simulation result Data	Charts

Hot and dry - Ahmedabad		
Insulating material	Heat gain through the roof in kWh	PPD%
no insulation	3531.52	56.85
PUF	313.25	53.38
Rockwool	365.53	53.45
Air	1372.63	54.55
Water	2247.04	55.74

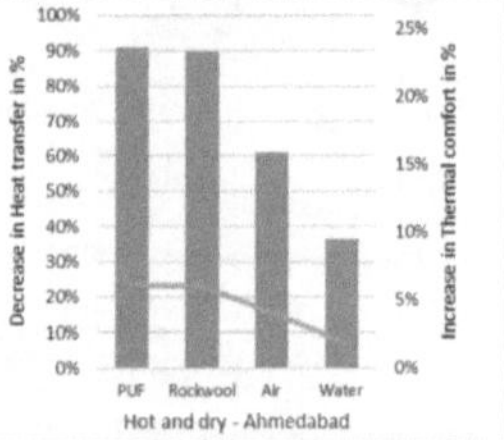

Warm and Humid- Mumbai		
Insulating material	Heat gain through roof in kWh	PPD%
no insulation	5721.67	63.33
PUF	689.64	57.21
Rockwool	760.01	57.43
Air	2491.46	60.78
Water	3797.3	62.55

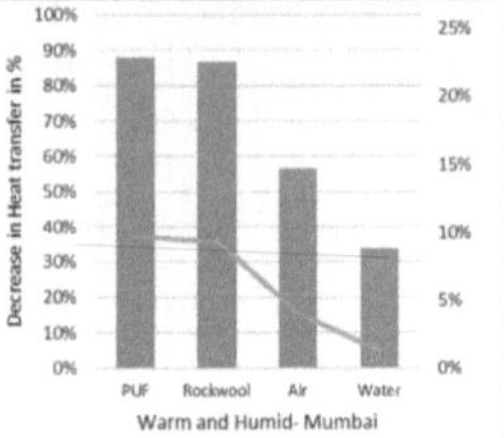

Temperate- Bengaluru		
Insulating material	Heat gain through roof in kWh	PPD%
no insulation	5468.79	46.43
PUF	663.06	36.71
Rockwool	763.74	37.02
Air	2425.86	42.14
Water	3595.32	44.43

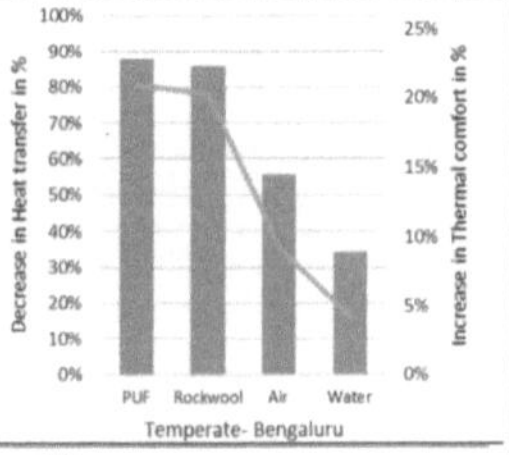

Composite - New Delhi			
Roof form	Heat gain through roof in kWh	PPD%	Heat losses through roof in kWh
no insulation	5374.81	50.03	2144.18
PUF	573.71	42.6	486.18
Rockwool	778.18	42.83	667.24
Air	1477.45	45.65	2411.61
Water	3559.08	47.22	1653.49

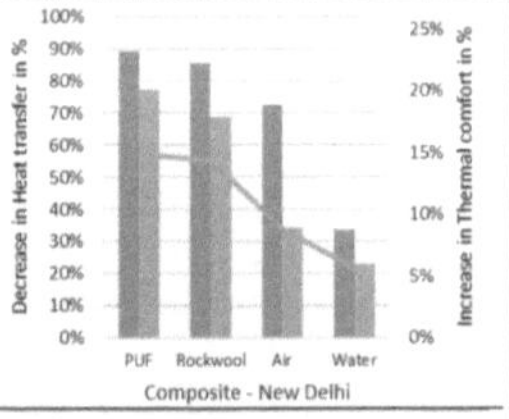

Cold - Srinagar			
Roof form	Heat gain through roof in kWh	PPD%	Heat losses through roof in kWh
no insulation	2876.37	48.46	4921.64
PUF	374.65	35.95	1,421.86
Rockwool	430.17	36.4	1,599.93
Air	1195.06	41.56	3233.76
Water	2110.79	43.53	4009.83

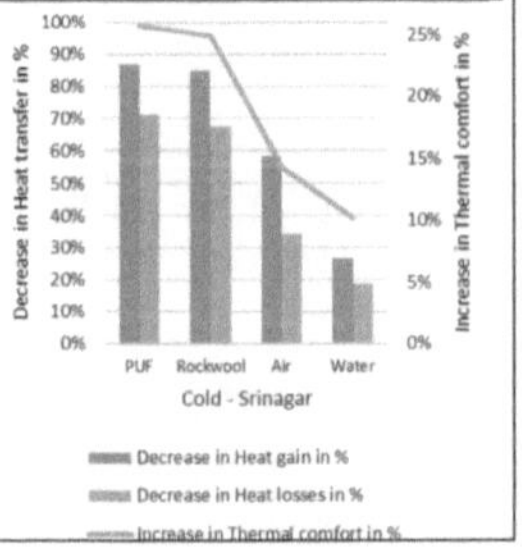

Simulation shows that PUF is the best insulation material in all climate types for reducing heat transfer and thermal comfort (Table 6). In contrast, water is the most inefficient insulating material. An air gap as insulation can be an economical solution to reduce heat gain and increase thermal comfort in all climate types. In a hot and dry climate, the variation in thermal comfort on adding insulation is low compared to temperate and cold climates.

6.3 *Simulation results of roofing finishes*

Over flat roofs, three types of roof finishes have been selected for comparative study: cool roof, green roof, and roof pond. Cool and green roofs are popularly used in hot climates to reduce heat gain in buildings through roofs. Roof pond is also a passive strategy to reduce heat gain through roofs. The properties of these types of roofs are given in Table 7. These strategies are compared through simulation by placing them above the flat roofs in selected climate types.

Table 7. Roofing finishes in the simulation program.

	Cool Roof	Green Roof			Roof Pond		
Roofing Finishes	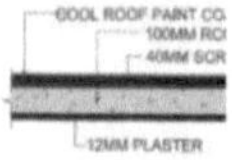				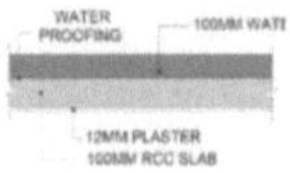		
U-value of Roof (W/m^2-K)	4.078			1.4			1.78
Properties of Roof Finishes	Solar Reflectance: .95 Thermal Emittance: 0.91 Solar Reflective Index: 122	Thermal Conductivity (W/m.K)	:	0.9	Thermal Conductivit y (W/m.K)	:	0.6
		Density (kg/m3)	:	1850			
		Specific Heat (J/kg.K)	:	850	Density (kg/m3)	:	997
		Thermal emissivity	:	0.65			
		Minimal stomatal resistance (s/m)	:	180	Specific Heat (J/kg.K)	:	1000
		Solar Absorptance	:	0.35			
		Leaf reflectivity	:	0.22			
		Leaf emissivity	:	0.95			
		Soil thickness (m)	:	0.3			
		Height of plant (m)	:	0.3			

Table 8 shows the simulation results and its analysis regarding the percentage of decrease in heat transfer through roofs and increase in thermal comfort. Graphs show that a decrease in heat transfer is directly proportional to an increase in thermal comfort. Cool roofs perform outstandingly in hot climates like hot and dry, warm and humid, and temperate, followed by roof ponds. Heat transfer reduction and thermal comfort achieved through a green roof in a composite and cold climate are the highest. Hence cool roof is suitable for hot and dry, warm and humid, and temperate climates, while green is suitable for composite and cold climate types.

Table 8. Simulation results for annual heat transfer and thermal comfort through roof finish in different climate types.

Simulation result Data | **Charts**

Hot and dry - Ahmeda ad		
Insulating material	Heat gain through the roof in kWh	PPD%
Flat roof	3531.52	56.85
Cool roof	-5123.15	45.1
Green roof	-1759.68	49.43
Roof Pond	-2716.79	47.4

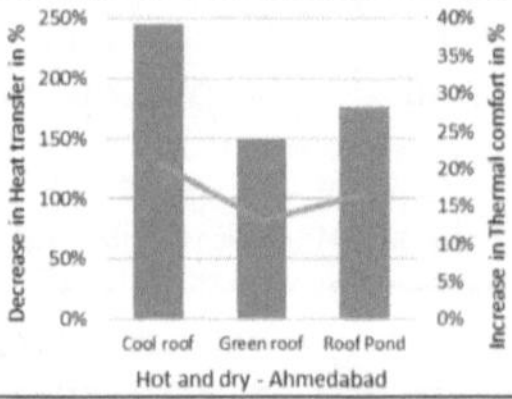

Warm and Humid- Mur bai		
Insulating material	Heat gain through the roof in kWh	PPD%
Flat roof	5721.67	63.33
Cool roof	-3472.86	48.46
Green roof	-1033.22	53.61
Roof Pond	-1813.76	51.04

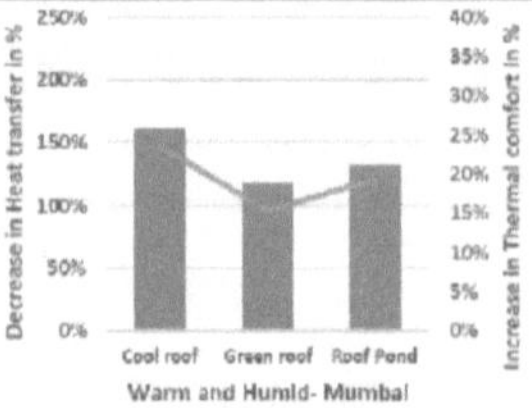

Te perate- Bengaluru		
Insulating material	Heat gain through the roof in kWh	PPD%
Flat roof	5468.79	46.43
Cool roof	-4438.91	27.83
Green roof	-1453.46	30.86
Roof Pond	-2371.04	29.06

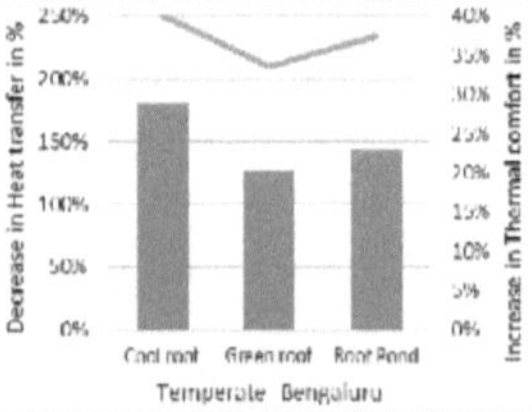

Composite - New Delh			
Roof form	Heat gain through the roof in kWh	PPD%	Heat losses through the roof in kWh
Flat roof	5374.81	50.03	2144.18
Cool roof	-791.98	42.23	4119.05
Green roof	-704.63	39.89	1348.37
Roof Pond	-401.12	40.58	2333.12

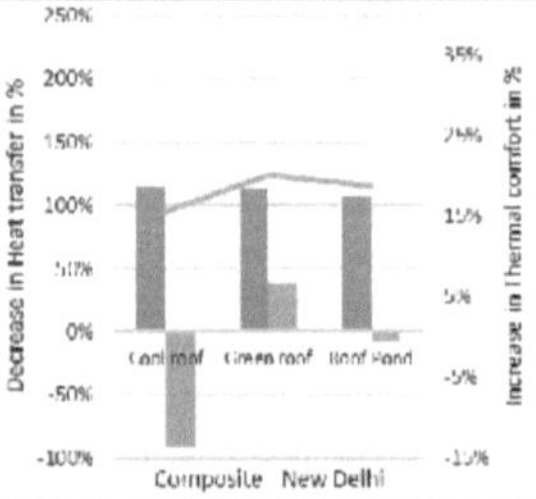

Roof form	Heat gain through the roof in kWh	PPD%	Heat losses through the roof in kWh
Flat roof	2876.37	48.46	4921.64
Green roof	-440.69	36.64	2,834.26
Roof Pond	-666.94	40.64	5820.13

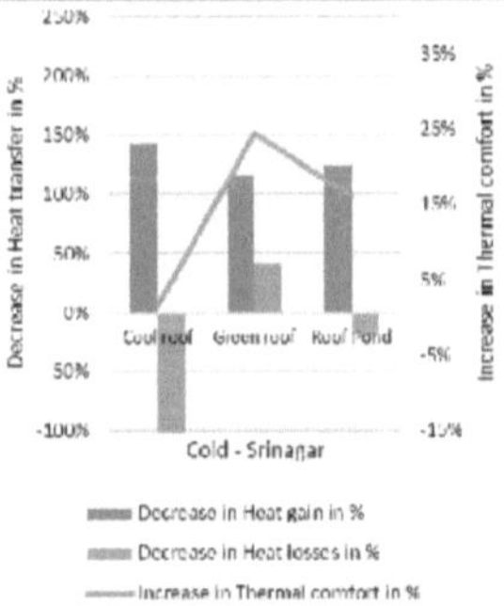

7 CONCLUSION

This research establishes that roof form, material, and finishes contribute significantly to the heat transfer through the roof inside the building and thermal comfort. The literature review, followed by the simulation, helps formulate the recommendation for roof design in different climate types of India in Table 9. The total U-value of the roof requirements for ECBC Compliant Building should be 0.33 W/m2 .K except for the cold climate. The U-value for roof in cold climate is recommended 0.28 W/m2 .K.

Table 9. Recommended roof forms, materials and finishes in different climate types of India.

Climate type	Hot and dry	Warm and Humid	Temperate	Composite	Cold
ROOF FORM					
Flat	N	N	N	N	N
Pitch	L	V.H	V.H	L	V.H
Vault	H	H	H	H	H
Dome	V.H	M	M	V.H	M
ROOF MATERIAL					
No insulation	N	N	N	N	N
PUF	H	H	H	H	H
Rockwool	M	M	M	M	M
Air	M	M	M	M	M
Water	L	L	L	L	L
ROOF FINISHES					
Cool Roof	V.H	V.H	V.H	H	L
Green Roof	M	M	M	V.H	V.H
Roof Pond	M	M	M	M	M

*V- Very, H – Highly efficient, M – Moderately efficient, L – Less efficiently, N – Not efficient

Vaulted and domed roofs are the most efficient roof forms in hot and dry climates (without any insulation). Insulators (Resistive insulators) like polyurethane, Rockwool, air and water can be used. The lowest U-value can be achieved by corrugated fibrous cement with 50mm PUF. Green roofs and Cool roofs are the most efficient roofing type. Cool roof with finishes like high albedo bright white paint, ceramic roof tile reddish or white coloured is recommended. The total U-value of the roof requirements for ECBC Compliant Building should be 0.33 W/m2 .K

The pitched roof is the best option in humid climates due to high humidity and rainfall with light material and ventilated roof. Cool roofs and PUF insulation reduce heat transfer inside the building. The total U-value of the roof requirements for ECBC Compliant Building should be 0.33 W/m2 .K. Energy saving is slightly lower in this climate than in the hot and dry climate.

The pitched roof is the best option in moderate climates due to rainfall with insulation, or a double roof finished with cool roof technique. The total U-value of the roof in requirements for ECBC Compliant Building should be 0.33 W/m2 .K.

In a composite climate, there are both extreme heat and cold, so there should be a high reflectance surface on the top of the roof along with high thermal mass and high insulation protected by hot and cold both. A highly insulted dome roof with a cool roof coating or green roof is recommended in this climate type.

An airtight envelope and sound insulation can reduce heat loss in cold climates. Green roof works best in this climate. Pitched roofs and flat roofs are the best options in the cold climate.

ACKNOWLEDGMENT

I would like to thank the Principal and Dean of our institute, Dr Vandana Sehgal and the Honourable Vice Chancellor of Dr APJ Abdul Kalam Technical University for giving financial assistance in conducting this research.

REFERENCES

[1] Bano F. and Kamal M. A., (2016). Examining the role of building envelope for energy efficiency in office buildings in India, *Archit. Res.*, vol. 6, no. 5, pp. 107–115, doi: 10.5923/j.arch.20160605.01.

[2] Abuseif M. and Gou Z., (2018). A review of roofing methods: Construction features, heat reduction, payback period and climatic responsiveness, *Energies*, vol. 11, no. 11, doi: 10.3390/en11113196.

[3] ECBC, *Energy Conservation Building Code*, chrome-extension://efaidnbmnnnibpcajpcglclefindmkaj/https://beeindia.gov.in/sites/default/files/ECBC%20book%20final%20one%20%202017.pdf, New Delhi, 2017. Accessed: May 09, 2022. [Online]. Available: chrome-extension://efaidnbmnnnibpcajpcglcle-findmkaj/https://beeindia.gov.in/sites/default/files/ECBC book final one 2017.pdf.

[4] Rawat M. and Singh R. N., (2022). A study on the comparative review of cool roof thermal performance in various regions, *Energy and Built Environment*, vol. 3, no. 3. Elsevier B.V., pp. 327–347, doi: 10.1016/j.enbenv.2021.03.001.

[5] Dabaieh M., Wanas O., Hegazy M. A., and Johansson E., (Feb. 2015). Reducing cooling demands in a hot dry climate: A simulation study for non-insulated passive cool roof thermal performance in residential buildings, *Energy Build.*, vol. 89, pp. 142–152, doi: 10.1016/j.enbuild.2014.12.034.

[6] Dhaka S., Mathur J., and Garg V., (Dec. 2012). Combined effect of energy efficiency measures and thermal adaptation on air conditioned building in warm climatic conditions of India, *Energy Build.*, vol. 55, pp. 351–360, doi: 10.1016/j.enbuild.2012.09.038.

[7] Goudarzi H. and Mostafaeipour A., (Feb. 01, 2017). Energy saving evaluation of passive systems for residential buildings in hot and dry regions, *Renewable and Sustainable Energy Reviews*, vol. 68. Elsevier Ltd, pp. 432–446, doi: 10.1016/j.rser.2016.10.002.

[8] Hasan A. S., Ali O. M., and Hussein A. A., (2021). Comparative study of the different materials combinations used for roof insulation in Iraq, in *Materials Today: Proceedings*, vol. 42, pp. 2285–2289, doi: 10.1016/j.matpr.2020.12.317.

[9] Najim K. B. and Fadhil O. T., (Feb. 2015). Assessing and improving the thermal performance of reinforced concrete-based roofing systems in Iraq, *Energy Build.*, vol. 89, pp. 213–221, doi: 10.1016/j.enbuild.2014.12.049.

[10] Elawady N., Bekheit M., Sultan A. A., and Radwan A., (Feb. 2022). Energy assessment of a roof-integrated phase change materials, long-term numerical analysis with experimental validation, *Appl. Therm. Eng.*, vol. 202, doi: 10.1016/j.applthermaleng.2021.117773.

[11] Al-Yasiri Q. and Szabó M., (Jun. 2021). Case study on the optimal thickness of phase change material incorporated composite roof under hot climate conditions, *Case Stud. Constr. Mater.*, vol. 14, doi: 10.1016/j.cscm.2021.e00522.

[12] Cassar J., Galdies C., and Muscat Azzopardi E., (2021). A new approach to studying traditional roof behaviour in a changing climate—A case study from the mediterranean Island of Malta, *Heritage*, vol. 4, no. 4, pp. 3543–3571, doi: 10.3390/heritage4040196.

[13] Muhieldeen M. W., Lye L. C., Sameer M., Kassim S., and Yen W., (2022). The optimum thickness of rockwool as roof thermal insulation : An experimental and numerical study, *J. Adv. Res. Fluid Mech. Therm. Sci.*, vol. 89, no. 1, pp. 77–91, doi: https://doi.org/10.37934/arfmts.89.1.7791.

[14] Pisello A. L., Cotana F., and Brinchi L., (2014). On a cool coating for roof clay tiles: Development of the prototype and thermal-energy assessment, *Energy Procedia*, vol. 45, pp. 453–462, doi: 10.1016/j.egypro.2014.01.049.

[15] Bhatia A., Mathur J., and Garg V., (2011). Calibrated simulation for estimating energy savings by the use of cool roof in five Indian climatic zones, *J. Renew. Sustain. Energy*, vol. 3, no. 2, doi: 10.1063/1. 3582768.
[16] Ascione F., Bianco N., de' Rossi F., Turni G., and Vanoli G. P., (2013). Green roofs in European climates. Are effective solutions for the energy savings in air-conditioning?, *Appl. Energy*, vol. 104, pp. 845–859, doi: 10.1016/j.apenergy.2012.11.068.
[17] Alexandri E. and Jones P., (2008). Temperature decreases in an urban canyon due to green walls and green roofs in diverse climates, *Build. Environ.*, vol. 43, no. 4, pp. 480–493, doi: 10.1016/j.buildenv.2006. 10.055.
[18] Arumugam R., Garg V., Mathur J., Reddy N., Gandhi J., and Fischer M. L., (2013). Experimental determination of comfort benefits from cool-roof application to an un-conditioned building in India 1 Rathish Arumugam <rathish.iiit@gmail, *Adv. Build. Energy Res.*, vol. 8, no. 1, pp. 14–27.
[19] Poptani H. and Bandyopadhyay A., (2014). Extensive green roofs: Potential for Thermal and Energy benefits in buildings in central India, *30th Int. PLEA Conf. Sustain. Habitat Dev. Soc. Choos. W. Forw. – Proc.*, vol. 2, no. September, pp. 154–161.
[20] Szagri D. and Nagy B., (2021). Experimental and numerical hygrothermal analysis of a refurbished double-skin flat roof, *Case Stud. Therm. Eng.*, vol. 25, no. March, doi: 10.1016/j.csite.2021.100941.
[21] Molleti S., Lefebvre D.a, and van Reenen D., (2018). Long-term in-situ assessment of vacuum insulation panels for integration into roofing systems: Five years of field-performance, *Energy Build.*, vol. 168, pp. 97–105, doi: 10.1016/j.enbuild.2018.03.010.
[22] Stockton G. R., (2013). The case for using a sacrificial layer of absorbent insulation in the design of flat and low-sloped roofing, *Thermosense Therm. Infrared Appl. XXXV*, vol. 8705, no. 800, p. 870504, doi: 10.1117/12.2018294.
[23] Ashrae 55, Thermal Environmental Conditions for Human Occupancy, Atlanta, GA, (2017).

Innovation and Technological Advances for Sustainability – Al-Naemi et al. (Eds)

Increase in indoor CO_2 in occupational settings within Qatar: Implications for student health and climate change mitigation strategies

Ravi Rangarajan, Abdullah Ansari & Nashwa Khadeej
Department of Environmental Health and Safety, College of Health Sciences, University of Doha for Science and Technology, Doha, Qatar

ABSTRACT: Impact of high indoor CO_2 exposure has not been studied in detail in an academic occupational setting, especially in Qatar, albeit the existence of a clear cause and effect between high CO_2 concentration and declining global health. In comparison to the outdoor CO_2 concentration, the average indoor CO_2 concentration in occupational environment (offices and schools), typically ranges from 600 to 1,000 ppm. This indoor concentration can exceed 2,000 ppm with increase in room occupancies and reduction in building ventilation rates, and can result in a multitude of public health issues. In here we studied the variability in CO_2 concentration and try to model its change as a function of population density mimicking the global changes in climate and population. Results indicated that in places where population density crosses a threshold value, the indoor CO_2 concentration can increase as much as ~9 ppm/min and can have enormous health impact even in short durations of acute exposure.

Keywords: CO2, indoor pollution, population density, green buildings, climate change

1 INTRODUCTION

Globally, most of the funded research on climate change focuses on atmospheric CO_2 increase and its mitigation strategies, with less emphasis on increasing indoor CO_2 concentrations and its effect on human health [1–4]. Qatar is in the forefront of studying CO_2 driven climate change and its mitigation research strategies. Impact of high indoor CO2 exposure has not been studied in detail in an academic occupational setting, especially in Qatar, albeit the existence of a clear cause and effect between high CO_2 concentration and declining global health [5–7]. According to the World Meteorological Organization (WMO), globally averaged CO_2 concentration in the atmosphere reached a significant milestone of 400 ppm for the first time in April 2015 [8], and has reached 413.6 ppm as of October 2021. An average increase of 2.1 ppm year on year was observed during this period and the overall 6-year change of 13.6 ppm would translate to ~1.2°C increase in average global temperature estimates [8]. In comparison to the outdoor CO_2 concentration, the average indoor CO2 concentration in occupational environment (offices and schools), typically ranges from 600 to 1,000 ppm. This indoor concentration can exceed 2,000 ppm with increase in room occupancies and reduction in building ventilation rates [9,10].

A recent review on studies investigating CO_2 concentrations in classrooms showed that most studies reported median peak CO_2 values above 1,000 ppm, with many above 2,000 ppm, suggesting a widespread inadequacy of ventilation in schools [9]. Furthermore, conditions like sedentary desk work, less movement of occupants and air conditioning units that ventilate air closer to the ceiling can cause relatively static air, resulting in higher CO_2

DOI: 10.1201/9781003496724-3

concentrations and making the occupants rebreathe their own exhaled CO_2 [9–11]. The source of CO_2 in non-industrial indoor/occupational environment is majorly human metabolism. However, other sources of CO2 like fossil fuel emissions, local combustion, industrial production and plant respired CO2 can also find its way into indoor spaces as CO_2 is a rapidly mixing trace gas [12].

In countries with high development index, people spend >75% of their time indoors, and in an educational facility the time students spend indoors is always considerably high. As a result, indoor air quality is of critical importance to public health and overall well-being in such occupational settings [4,13]. Continued atmospheric CO_2 increase will need more demanding mitigation strategies as it is essential to prevent concomitant rise in indoor CO_2 concentrations, especially in the occupational environment settings.

Very few studies have focused on the direct effects of increasing indoor CO_2 concentrations on human health. Cognitive impact of high indoor CO_2 exposure has been studied at different concentrations in the literature and the findings vary considerably between investigations with no clear consensus within the broader scientific community. However, a clear cause and effect has been stablished for high CO_2 concentrations on human health [5,14] and a summary of the adverse health effects associated with these high CO_2 concentrations in the indoor environment are shown in Table 1. The adverse health effects can range from observed physiological effects like calcification in kidneys [15,16] to neurological effect like attention deficit, behavioral changes and cognitive decline [17–21], a primary health concern in student population. Furthermore, these health effects vary as a function of exposure duration with chronic exposure resulting in physiological problems like respiratory acidosis [22,23], bone demineralization [15,16] and oxidative stress [22–25], and acute exposure causing inflammatory effects [26–28] and cognitive dysfunctions. Furthermore, more recent studies have explored the possible association between obesity and increasing outdoor/indoor CO_2 concentrations [29–31].

In order to understand the enormity of health effect that arises from increasing CO_2 concentrations, especially within indoor settings, there is a need to look at it from the perspective of global changes in climate and population. In this context, the current study focuses on continuous CO_2 concentration monitoring at different indoor locations within an occupational setting to see the variability in concentration and model the observed variability as a function of population density, mimicking the global changes in climate and population in line with the current trends in global Green House Gas (GHG) Emissions.

Table 1. Summary of health outcomes in response to increased CO_2 concentration within indoor settings as a function of exposure levels and exposure duration [Adapted, modified and summarized from ref 5,15–28].

Health Outcomes	Levels (in ppm)	Exposure Type	Exposure Time
Increased Blood pressure	>500	-	-
CO2 retention in body	1000–5000	Acute	<4 hours
Inflammation	2000-4000	Acute	2 hours
Cognitive Effects	1000–2700	Acute	1–6 hours
Attention Deficit	3000	Acute	2–3 hours
Low grade Systemic Inflammation	3000	Sub-chronic	13 days
Decision Making	2500	Sub-chronic	14 days
Behavioral Changes and Stress	700–3000	Sub-chronic	13–15 days
Bone Demineralization and Kidney Calcification	2000–3000	Chronic	2 to 3 months
Oxidative Stress and Endothelial Dysfunction	3000–5000	Chronic	0.5 to 6 months
Respiratory Acidosis	>5000	Chronic	>10 years

2 METHODS AND MATERIALS

2.1 *CO_2 concentration measurements*

A CO_2 Gas Sensor measures gaseous CO_2 by way of monitoring the amount of infrared radiation absorbed by carbon dioxide molecules and are an important component of climate change studies [32]. CO_2 gas concentrations were measured in this study using calibrated TSI IAQ-CALC 7545 instrument. The device uses a dual wavelength (3900 nm and 4200 nm) Non-Dispersive Infra-Red sensor with an accuracy of 50 ppm and a resolution of 1 ppm for CO_2 concentration in the range of 0 to 5000 ppm [33]. Data validation was done using concurrent measurements with two sensors and measurement of outdoor ambient CO_2, which was also used to carry out back ground correction. This approach of simple validation was used as indoor CO_2 cannot be lower than outdoor CO_2 levels. Furthermore, alongside the actual measured values, the objective was to understand the relative changes on indoor CO_2 in order to model these changes. Routine measurement of indoor CO_2 was carried out in pre-identified at 12 pre-identified study sites within the University of Doha for Science and Technology (UDST) campus at classrooms of different densities at different times of the day.

Analysis of the collected data was done using Sigma Plot statistical package v.14.

2.2 *Modelling of density-concentration factor as function of time*

CO_2 gas concentration were measured in pockets of 2-hour clusters at pre-identified study sites within the UDST campus. The timing and sites were selected based on the population density index in a given area. Background CO_2 concentration was also measured at all these sites for offset corrections and for developing concentration anomaly curves. The measured background concentrations on all measurement days ranged between ~430 ppm to 475 ppm with an average value of 455 ± 28 ppm the Using the obtained CO_2 concentration at each study site and population density data at each of these sites, the concomitant increase in CO_2 was estimated and then extrapolated to predict the rates of CO_2 increase as a function of time. This approach of modelling is commonly followed with the research community and was done with an aim to develop more complex real time air transport models of CO_2 within occupational settings.

3 RESULTS AND DISCUSSIONS

3.1 *Variability in CO_2 concentration at study sites*

Based on the population density data at study sites, these sites were first classified into low density sites and high-density sites. In parallel with the identification and grouping of study sites, CO_2 concentration were measured in these low-density factor distributions and high-density factor distribution sites as shown in Figure 1A and 1B.

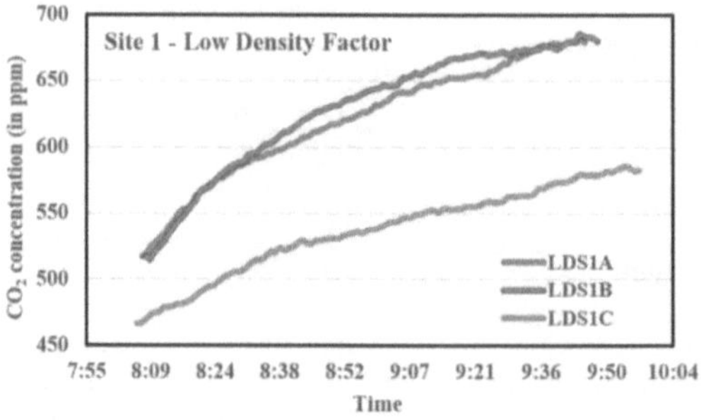

Figure 1A. The increase in CO_2 concentration at study sites with low density factor distribution. The individual lines represent the concurrent measurements taken at same time of the day on different days.

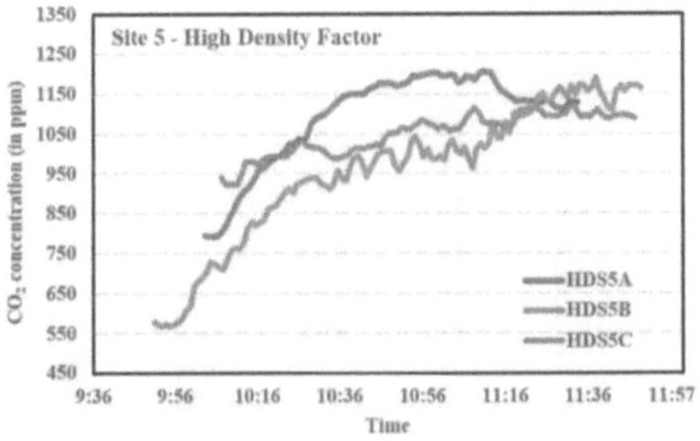

Figure 1B. The increase in CO_2 concentration at study sites with high-density factor distribution. The individual lines represent the concurrent measurements taken at same time of the day on different days.

The individual lines in Figures 1A and 1B represent the concurrent measurements taken at same time of the day on different days. This was done to understand the inherent variability in CO_2 mixing in a given area. The preliminary data indicated a significant increase in indoor CO_2 in the sites within the occupational setting studied. In sites with low density factor, an average increase of 100–150 ppm was observed in the measured time span. In contrast, the measured concentration in the high-density study site increased from 550 ppm to ~1150 ppm within a short span of 100–120 minutes indicating high unprecedented levels of exposure within the occupational setting.

Corresponding to the concurrent measurements of CO_2 concentration done at all the study sites, anomaly plots were generated to understand the actual increase in CO_2 as a function of time. These anomaly levels refer to the rate at which CO_2 fluxes happen and are denoted as change in CO_2 concentration (ΔCO_2). The anomaly plot obtained for one of the high-density factor study sites is shown in Figure 2. At site 6, the highest anomaly was observed and the value was as high as ~700 ppm at the end of a 120-minute measurement period. In general, at all high-density sites, these anomaly plots indicated a rapid increase in CO_2 over a very small duration of time. These anomaly concentrations which indicate the actual increase in the CO_2 concentration were used as input factors in the modelling estimates. The obtained anomaly concentration especially in high density regions were relatively high and unexpected.

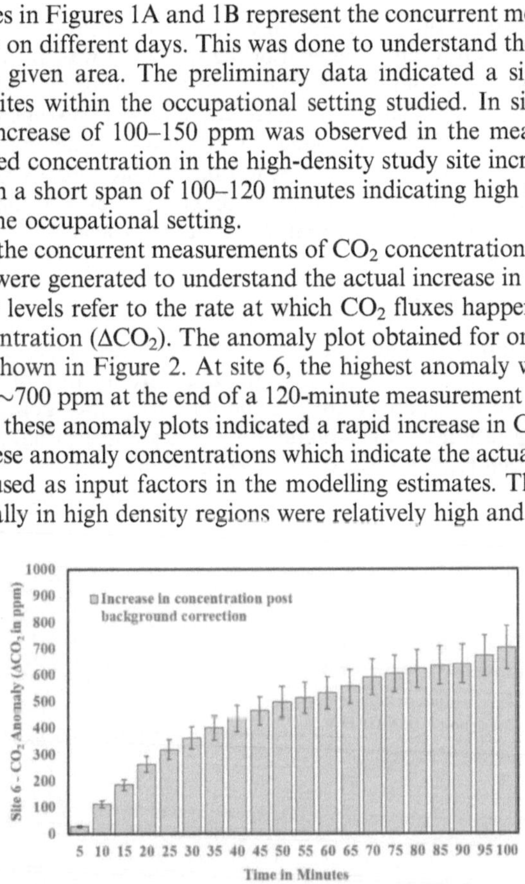

Figure 2. Anomalous increase in indoor CO_2 concentration in a classroom within the UDST campus post background correction of 455 ppm.

3.2 *Modelling of density-concentration factor of CO_2*

The inputs for the multi-factor modelling that was carried out include the derived CO_2 anomaly rates, population density factor and the time interval used for obtaining these measurements. The CO_2 anomaly data were extracted from the anomaly plots that were generated. Population density data was obtained considering the number of people in the

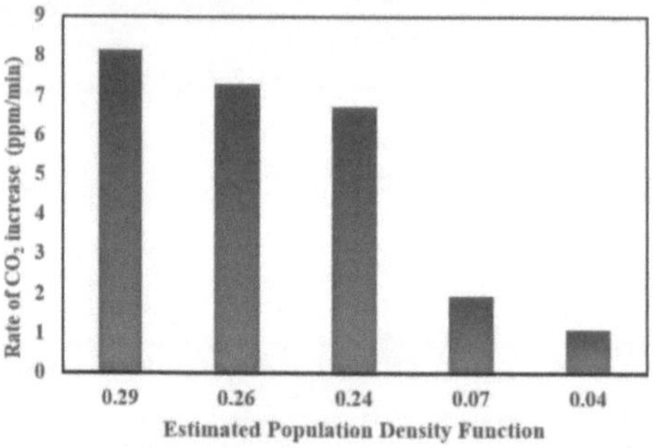

Figure 3. Estimated changes in rate of indoor CO_2 as a function of population density.

study site at the beginning of the 120-minute CO_2 concentration measurement and the approximate area of the study site. The area of the study site was in turn obtained by measuring the dimensions of the study sites prior to start of the study. The model used was a simple multi-variate regression analysis to arrive at the rate of CO2 increase as a function of population density in each of the study area. The obtained modelling results are shown in Figure 3. The results indicated that in places where population density crosses a threshold value, the indoor CO2 concentration can increase as much as ~9 ppm/min and can have enormous health impact even in short durations of acute exposure.

3.3 *Green buildings and rate of CO_2 increase*

Green buildings are commonly defined as energy- efficient, healthy, and sustainability driven buildings from the context of urban life and global environment [34]. One of the key sustainability drivers of such buildings are the indoor air quality comprising of CO_2 concentration and concentration of other pollutants (VOCs and Formaldehyde). Roughly, in 65% of the existing green building certification programmes worldwide, CO_2 concentration is one of the most important parameter in defining the standards for certifications [35]. However, these certifications do not consider the changes in the rate of CO_2 increase and mostly have set thresholds for concentrations. The current study indicated us that the rate of increase in CO_2 could be very high in buildings with higher density and that such increases will be an important factor and can contribute substantially within the framework of global public health, owing to the multitude health effects shown in Table 1. Given the fact that in any newly designed building the population density would tend to vary as function of occupancy rate and the concomitant increase in the rate of CO_2, the use of set thresholds alone will not be enough in the global future scenarios with increasing population, increasing global CO_2 and the push to achieve sustainability.

4 CONCLUSION

In this study, the variability in the increase in CO_2 concentration was studied using experimental approach in view of the increasing public health issues associated with rapid increase in indoor CO_2 concentrations. The obtained experimental data was used for developing a density function model and the empirical model was used for estimating the anomalous CO_2 increase in high density as well as low density areas. The developed model could be fine-tuned for global future scenarios where drastic changes in population density is expected and the need for closely monitoring the possible health effects of this increase in density and its relation to increasing CO_2 need to be monitored closely.

ACKNOWLEDGMENT

This study was funded by the Early SEED grant awarded to Dr. Ravi Rangarajan. The authors sincerely thank the Office of Applied Research, Innovation and Economic Development at UDST and the Vice President of Academics, Dr. Rachid Benlamri for their continuous support.

REFERENCES

[1] McMichael A. J., Haines J., Slooff R. *et al.*, (1996). Climate change and human health: an assessment. *World Health Organization* 1996 report.

[2] Haines A., Kovats R. S., Campbell-Lendrum D. *et al.*, (2006). Climate change and human health: impacts, vulnerability and public health. *Public Health.* 120(7):585–96.

[3] Rocque R. J., Beaudoin C., Ndjaboue R. *et al.*, (2020). Health effects of climate change: an overview of systematic reviews. *BMJ Open*, 11(6).

[4] Spengler J. D., (2012). Climate change, indoor environment, and health. *Indoor Air*, 22, 89–95.

[5] Jacobson T. A., Kler J. S., Hernke M. T. *et al.*, (2019). Direct human health risks of increased atmospheric carbon dioxide, *Nature Sustainability*, 2, 691–701

[6] Duarte C. M., Jaremko L., Jaremko M., (2020). Hypothesis: Potentially systematic impacts of elevated CO_2 on the human proteome and health. *Front. Public Helath*, https://doi.org/10.3389/fpubh.2020.543322.

[7] Robertson D. S., (2006). Health effects of increase in concentration of carbon dioxide in the atmosphere, *Current Science*, 90, 12, 1607–1609

[8] World Meteorological Organization (WMO, 2016). The state of greenhouse gases in the atmosphere based on global observations through 2015. *WMO Bull.*, 12 1–8.

[9] Fisk WJ., 2017. The ventilation problem in schools: literature review. *Indoor Air* 27, 1039–1051.

[10] Seppänen O. A., Fisk W. J., Mendell M. J., (2004). Association of ventilation rates and CO_2 concentrations with health and other responses in commercial and institutional buildings, *Indoor Air*, https://doi.org/10.1111/j.1600-0668.1999.00003.x

[11] Ghahramani A., Pantelic J., Vannuci M. *et al.*, (2019). Personal CO_2 bubble: context-dependent variations and wearable sensors usability. *J. Build. Eng.* 22, 295–304.

[12] Bergeron, O. and Strachan, I. B., (2011). CO_2 sources and sinks in urban and suburban areas of a northern mid-latitude city. *Atmos. Environ.* 45, 1564–1573.

[13] Schweizer C., Edwards RD., Bayer-Oglesby L. *et al.*, (2007). Indoor time–microenvironment–activity patterns in seven regions of Europe, *J. of Exposure Science & Environmental Epidemiology*, 17, 170–181.

[14] Snow S., Boyson A. S., Paas K. H. W. *et al.*, (2019). Exploring the physiological, neurophysiological and cognitive performance effects of elevated carbon dioxide concentrations indoors. *Build. Environ.* 156, 243–252.

[15] Schaefer, K. E., Douglas, W. H. J., Messier, A. A. *et al.*, (1979). Effect of prolonged exposure to 0.5% CO_2 on kidney calcification and ultrastructure of lungs. *Undersea Biomed. Res.* 6, S155–S161.

[16] Wade, C. E., Wang, T. J., Lang, K. C. *et al.*, (2000). Rat growth, body composition, and renal function during 30 days increased ambient CO_2 exposure. *Aviat. Space Environ. Med.* 71, 599–609.

[17] Vehviläinen T., Lindholm H., Rintamäki H. *et al.*, (2016). High indoor CO_2 concentrations in an office environment increases the transcutaneous CO_2 level and sleepiness during cognitive work. *J. Occ. And Env. Hygiene*, https://doi.org/10.1080/15459624.2015.1076160.

[18] Du B., Tandoc M. C., Mack M. L. *et al.*, (2020). Indoor CO_2 concentrations and cognitive function: critical review. *Indoor Air*, https://doi.org/10.1111/ina.12706.

[19] Kajtar, L. and Herczeg, L., (2012). Influence of carbon-dioxide concentration on human well-being and intensity of mental work. *Idojaras* 116, 145–169.

[20] Hutter, H. P., Haluza D., Piegler K., *et al.*, (2013). Semivolatile compounds in schools and their influence on cognitive performance of children. *Int. J. Occup. Med. Environ. Health* 26, 628–635.

[21] Satish, U., Mendell N. J. And Shekhat K., (2012). Is CO_2 an indoor pollutant? Direct effects of low-to-moderate CO_2 concentrations on human decision-making performance. *Environ. Health Perspect.* 120, 1671–1677.

[22] Veselá, A. and Wilhelm, J. (2002). The role of carbon dioxide in free radical reactions of the organism. *Physiol. Res.* 51, 335–339.

[23] Ezraty, B., Chabalier, M., Ducret, A., *et al.*, (2011). CO_2 exacerbates oxygen toxicity. *EMBO Rep.* 12, 321–326.

[24] Martrette, J. M., Egloff C., Clement C. *et al.*, (2017). Effects of prolonged exposure to CO_2 on behaviour, hormone secretion and respiratory muscles in young female rats. *Physiol. Behav.* 177, 257–262.

[25] Hacquemand, R., Buron G., Pourie G., *et al.*, (2010). Effects of CO_2 inhalation exposure on mice vomeronasal epithelium. *Cell Biol. Toxicol.* 26, 309–317.

[26] Zappulla, D. (2008). Environmental stress, erythrocyte dysfunctions, inflammation, and the metabolic syndrome: adaptations to CO_2 increases? *J. Cardiometab. Syndr.* 3, 30–34.

[27] Miller, J., Edwards L. D., Agusti A. *et al.*, (2013). Comorbidity, systemic inflammation and outcomes in the ECLIPSE cohort. *Resp. Med.* 107, 1376–1384.

[28] Thom, S. R., Bhopale, V. M., Hu, J. and Yang, M., (2017). Inflammatory responses to acute elevations of carbon dioxide in mice. *J. Appl. Physiol.* 123, 297–302.

[29] Hersoug, L. G., Sjödin, A. and Astrup, A., (2012). A proposed potential role for increasing atmospheric CO_2 as a promoter of weight gain and obesity. *Nutr. Diabetes 2*, e31.

[30] Zheutlin, A. R., Adar, S. D. and Park, S. K., (2014). Carbon dioxide emissions and change in prevalence of obesity and diabetes in the United States: an ecological study. *Environ. Int.* 73, 111–116.

[31] Kikuchi, R., Tsuji T., Watanabe O., *et al.*, (2017). Hypercapnia accelerates adipogenesis: a novel role of high CO_2 in exacerbating obesity. *Am. J. Resp. Cell Mol. Biol.* 57, 570–580.

[32] Scholz L., Perez A. O., Bierer B. *et al.*, (2018). Gas Sensors for Climate research, *J. Sens. Sens. Syst.*, 7, 535–541.

[33] Chen, X., Li, J., Zhang, X. *et al.*, (2021). The effects of warm air heater on the dispersion and deposition of particles in an enclosed environment. *Aerosol Air Qual. Res.* 21, 200620. https://doi.org/10.4209/aaqr.200620.

[34] Kibert C. J., (2012). *Sustainable construction: green building design and delivery*. 3rd ed. *John Wiley & Sons*, Inc; ISBN 978-1-118-33013-5.

[35] Wei W., Ramalho O. and Mandin C., (2015). Indoor air quality requirements in green building certifications. *Build. Environ.* 92, 10–19.

Innovation and Technological Advances for Sustainability – Al-Naemi et al. (Eds)

Traffic air pollution emissions in Qatar: Actual, projections and mitigation

Hanadi Al Thani
Division of Sustainable Development, College of Science and Engineering, Hamad Bin Khalifa University, Qatar Foundation Doha, Qatar
ORCID ID: 0000-0003-1735-5732

Muammer Koç
Division of Sustainable Development, College of Science and Engineering, Hamad Bin Khalifa University, Qatar Foundation Doha, Qatar
ORCID ID: 0000-0001-6543-8116

Rima J. Isaifan
Division of Sustainable Development, College of Science and Engineering, Hamad Bin Khalifa University, Qatar Foundation Doha, Qatar
ORCID ID: 0000-0002-9590-4939

ABSTRACT: The diesel vehicle count data from the traffic department was used along with the annual traveled distance to calculate the actual emissions for the diesel fleet in Qatar in 2017. The projection of diesel fleet emissions in 2030 was then estimated based on three scenarios: (1) if the fleet growth rate continued at the same rate as per the historical data; (2) if the rate of vehicle count increased by 10% reference to 2017 count; and (3) if there was a decline in the growth rate of vehicle count by 20% since major infrastructure projects shall be accomplished soon. The results show that the annual projected CO_2 emissions amounted to 4171, 1800, and 1750 kt for the three scenarios; respectively. Similarly, the annual projected NOx emissions were about 100, 43, and 42 tons for the three scenarios; respectively. The results showed that if no policy is put in place, there will be around 170,000 diesel vehicles contributing to huge amounts of particulate and gaseous emissions. However, under the current plans for the electrification of 5% of cars and 25% of public buses, a 10% and 5% reduction in CO_2 and NOx emissions will be attained, respectively. The impact of public transport bus electrification on this category is not as significant as it could be since the majority of emissions are due to heavy-duty vehicles, which are not impacted by this policy.

Keywords: Air pollution, emissions, diesel vehicle, annual travel, projection

1 INTRODUCTION

Urban air quality has attained global attention due to the significant impact on human health as gaseous and particulate pollutants concentration increases. The other reason for the increased attention in air quality studies is the increasing share of the world's population living mainly in urban cities and hence, the demand for a cleaner urban environment. The United Nations has estimated that 4.9 billion inhabitants out of 8.1 billion will be living in cities by 2030 [1]. This will directly cause elevated air pollution emissions due to the increase in demand for power generation from fossil fuels and industrial activities. Moreover, transportation (including intense motor and diesel-driven vehicle use), and construction projects have added to the emissions of particulate matter, especially in developing countries. Air pollutants such as NOx CO_2, PM, and SO_2 and their relation with transport has been well documented in a wide range of case studies [2–4]. Reports indicate that on-road vehicle emissions constitute the major source of atmospheric NO_2 and CO_2 in urban areas [5]. Carbon dioxide emissions from traffic contribute around 10% of the total global and 20% of the European atmospheric CO_2 emissions. Recently, we have reported on the first study to evaluate PM traffic emissions in Doha. The results showed that most PM was of fine size, which significantly impacts human health. Moreover, the sulfur elemental fingerprint in the collected samples was 5 times higher in traffic-heavy roads compared with less congested ones. When exhaust and non-exhaust PM were evaluated, it was also found that tyre wear generates the majority of road dust compared with brake wear and exhaust emissions [6]. In this research, the growth in vehicle fleet count in Qatar up to 2030 was estimated. Moreover, the actual emissions for all registered diesel vehicles as per the actual mileage driven in 2017 were evaluated. In addition, the expected emissions based on the current situation and in the case of implementing two mitigation policies will be assessed. The built scenarios are based on the government's plans to make 25% of Qatar's public transport buses run electric and have up to 5% of the cars in the country go green by 2022.

2 METHODS

The actual vehicle count was obtained from the counts from the Department of Traffic on all registered vehicles in 2017. The actual mileage of the registered vehicles in 2017 was taken from Fahes for buses, trucks, heavy-duty vehicles, and other types, which were mainly cars. Then, data cleaning was performed for both data sets, the count and the mileage, to eliminate any inconsistent entries to avoid their impact on the results. Data was then analyzed and put into categories based on vehicle and fuel types. Python programming was performed to align the mileage of each registered diesel vehicle with the chassis number (VIN) of each vehicle to get the distance traveled per year. The VIN, also known as the Chassis Number, is a 17-digit number that distinguishes each vehicle from other models and is an exclusive car identification number allocated to each vehicle by the manufacturer. Then the average mileage for each of the four types of vehicles was calculated and used as the basis of the actual air emissions for diesel vehicles. The emissions were evaluated based on the assumption that no significant change would impact the average annual distance driven by each of the 4 main types of vehicles (bus, truck, heavy-duty, and cars) for 2030. The emission factors from the Greenhouse gases, Regulated Emissions, and Energy use in Transportation (GREET) were then used to evaluate air pollution emissions of CO_2, PM_{10}, $PM_{2.5}$, SOx, BC, OC, and NOx. GREET is a tool sponsored by the U.S. Department of Energy (DOE) and was developed by Argon National Laboratory. Three scenarios were proposed: 1- Emissions in 2030 were projected based on the historical growth rate of vehicle count – First Scenario. 2- Emissions in 2030 were projected based on a 10% growth rate in vehicle registration – Second Scenario. 3- Emissions in 2030 were projected based on a 20% decline rate in vehicle registration – Third Scenario.

In order to estimate the impact of the newly proposed policies (5% and 25% of cars and buses go electric; respectively), the emissions were obtained according to each percentage.

3 RESULT AND DISCUSSION

3.1 *Actual fleet and emissions in 2017*

Figure 1 shows the vehicle count as registered in 2017 for the most dominant vehicle type categories (bus, truck, heavy-duty, and cars). It is shown that heavy-duty vehicles make up 33% of the total diesel fleet. Along with the cars, these two categories make up about 60% of the total registered diesel vehicles.

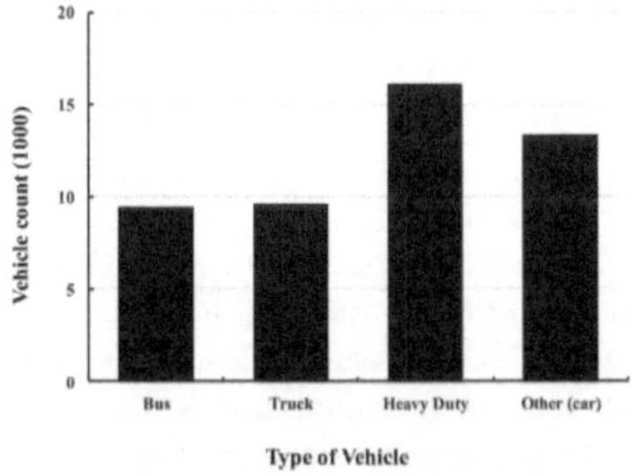

Figure 1. Diesel registered vehicle count by 2017 of the main four vehicle types.

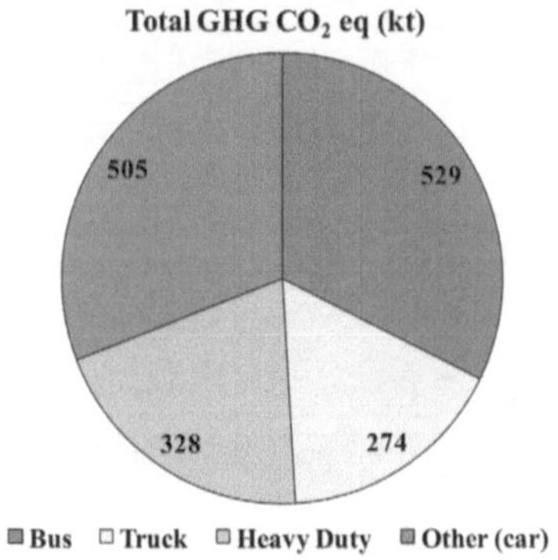

Figure 2. Actual annual emissions based on the average actual annual distance travelled by the four types of registered diesel vehicles (bus, truck, heavy-duty, and cars) in 2017.

Figure 2 shows that CO_2 traffic emissions from diesel vehicles are dominated by buses and cars. The major source of anthropogenic carbon dioxide in metropolitan cities is due to vehicle exhaust. Kakouei *et al.*, estimated CO_2 emissions from the urban fleet in 22 municipality regions of Tehran which was comprised of about 3.5 million vehicles (2.4 million private cars, 560,000 motorcycles, 78,000 buses, and 82,358 taxis) in 2012 [7]. Their results showed that the contribution of CO_2 emissions increased in the following order: private cars, motorcycles, buses, and taxis. The overall average contribution of CO_2 emissions in the private car, motorcycles, buses, and taxis was 26372, 1648, 1433, and 374 tons per day; respectively. This amounts to almost 10 million tons of CO_2 annually from those types of vehicles. The average contribution of CO_2 emissions of private cars in Tehran was higher (88%) than other vehicles.

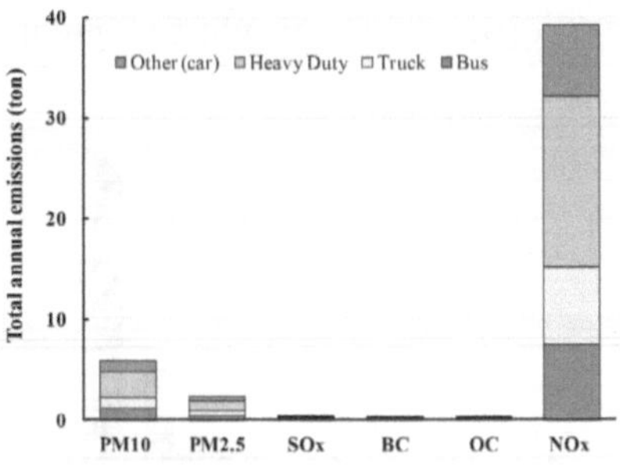

Figure 3. Actual annual emissions based on the average actual annual distance traveled by the four types of registered diesel vehicles (bus, truck, heavy-duty and other (car)).

Figure 3 shows that the highest emission of traffic is NOx which is in agreement with previous studies for different parts of the world [8,9]. In 2016, the EU and many European cities still regularly exceeded current EU limits for NO_2, which was responsible for 68,000 premature deaths [10].

Figures 4 and 5 show the average annual driven distance per person and identify the main sources of NO_2 pollution for each city examined in Europe, helping policymakers to design actions that target them. For the 30 European cities analyzed in the report, the average contribution of transport to overall nitrogen oxide (NOx) emissions was 47%, confirming that NOx emissions from roads have dominated other sources. In Athens and Milan, over 70% of emissions come from transport, while in Lisbon, where shipping emissions are high, road transport is only responsible for 20% of NOx pollution. Hence, targeted actions could include restricting the access of certain types of vehicles to inner city areas or applying environmental charges to the largest polluters.

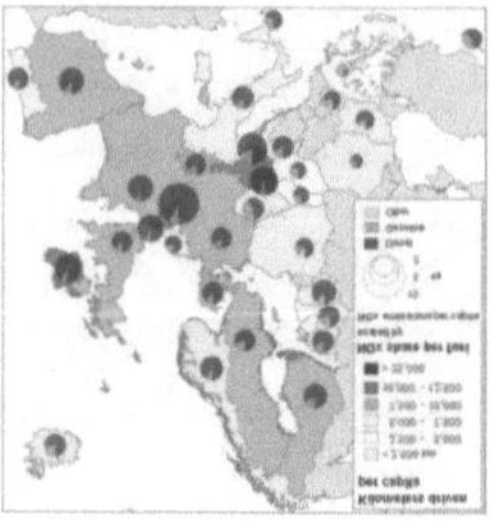

Figure 4. Country share of the NOx emissions per type of fuel (diesel, gasoline, and other), correlated with kilometers driven per capita [10].

Figure 5. Share of different sectors in NOx emissions for the 30 European cities analyzed [10].

3.2 *Projection fleet and emissions in 2030*

Figure 6 shows the projection of all vehicle counts up to 2030. It can be seen that if no policy was put in place, about 170,000 diesel vehicles would be registered.

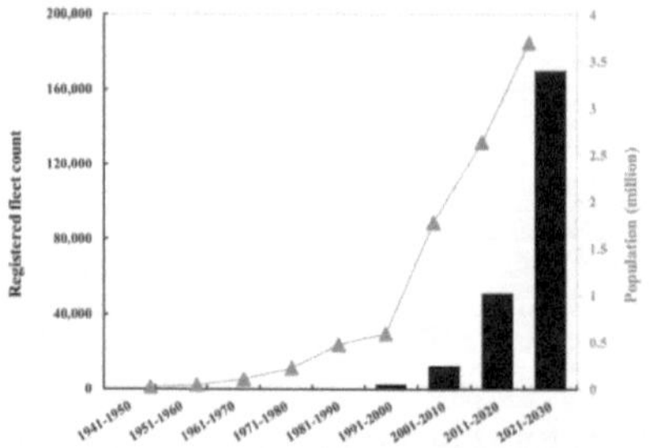

Figure 6. Count of vehicles and population projections in Qatar up to 2030 [194].

The growth in the four specific types of diesel vehicles related to this work is shown in Figure 7. It can be seen that heavy-duty vehicles and cars can reach up to 40,000 and 35,000 vehicles; respectively.

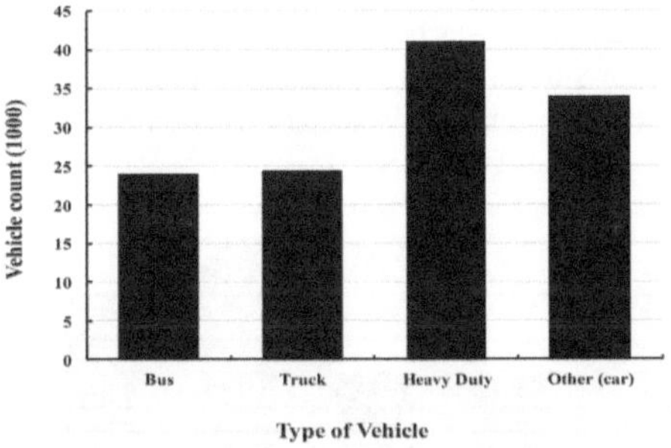

Figure 7. Projected count of the four types of diesel vehicles in 2030 based on the historical growth rate of vehicle count – Scenario 1.

The projected CO_2 emissions from the four main diesel vehicles are shown in Figure 8 which accounts in total to 4,162 kt annually by 2030.

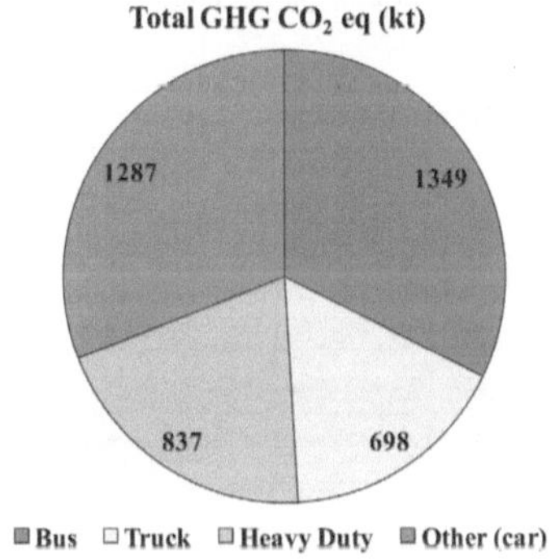

Figure 8. Projected total annual CO_2 from the four main types of diesel vehicles in 2030 based on the historical growth rate of vehicle count – Scenario 1.

The results in Figure 9 show that the traffic emissions are dominated by NO_x (100 tons) with the heavy-duty vehicles contributing the most. PM_{10} and $PM_{2.5}$ emissions come next as the current exhaust technologies have been developed to abate higher emissions of particulates and other gases [11,12].

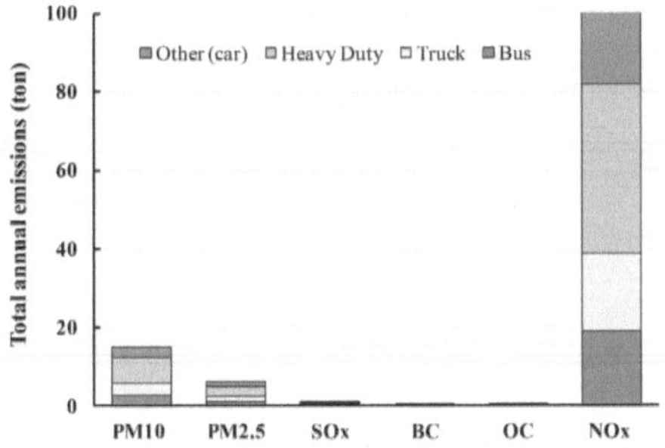

Figure 9. Actual annual emissions based on the average actual annual distance travelled by the four types of diesel registered vehicles (bus, truck, heavy duty and cars) in 2030 based on the historical growth rate of vehicle count – Scenario 1.

The results in Figures 10, 11, and 12 show the projected vehicle count, CO_2 emissions, and other emissions in 2030 based on a 10% growth rate of the vehicle registration scenario. This slight increase would be due to the possible halt in major infrastructure projects as being accomplished by 2022 or slightly after. Hence, fewer buses are needed to commute laborers from and to the project's sites. Moreover, fewer heavy-duty vehicles would be needed to transport raw and other materials for major infrastructure locations. The total annual CO_2 emissions were about 1,800 kt annually. The total annual projected NO_x emissions were about 43 tons.

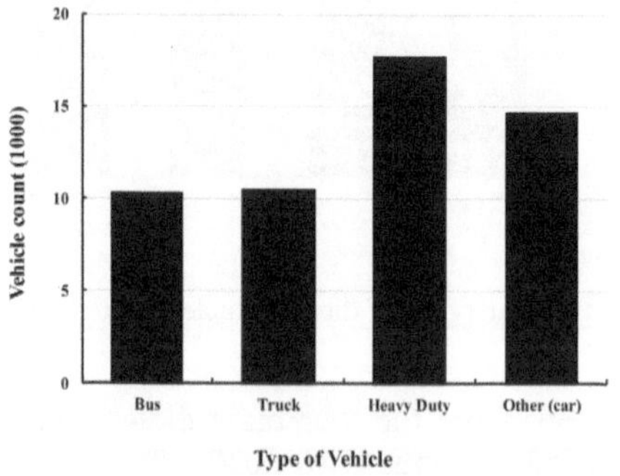

Figure 10. Projected count of the four types of diesel vehicles in 2030 based on 10% growth rate of vehicle registration – Scenario 2.

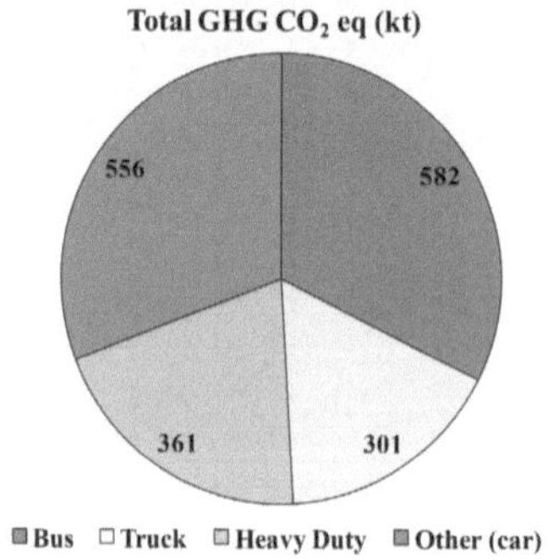

Figure 11. Projected total annual CO_2 from the main four types of diesel vehicles in 2030 in 2030 based on 10% growth rate of vehicle registration – Scenario 2.

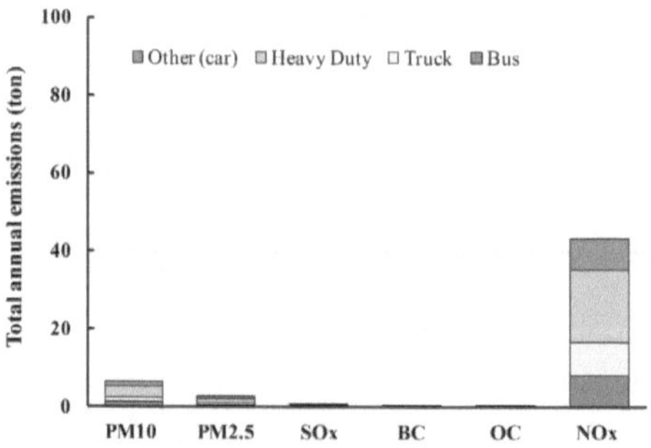

Figure 12. Actual annual emissions based on the average actual annual distance travelled by the four types of diesel registered vehicles (bus, truck, heavy duty and cars) in 2030 based on 10% growth rate of vehicle registration – Scenario 2.

The results in Figures 13, 14 and 15 show the projected vehicle count, CO_2 emissions as well as other emissions in 2030 based on 20% decline rate in vehicle registration scenario. This slight decline would be justified by further reduction in the need to purchase diesel vehicles by 2030 after the hosting of the World Cup. The total annual CO_2 emissions are about 1,750 kt annually. The total annual projected NO_x emissions were about 42 tons.

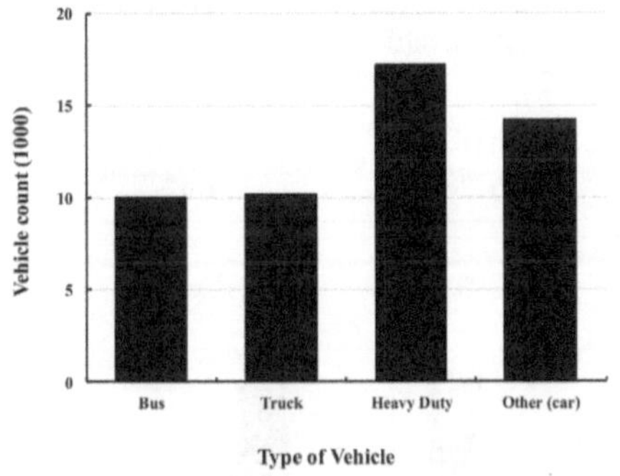

Figure 13. Projected count of the four types of diesel vehicles in 2030 based on a 20% decline rate in vehicle registration – Scenario 3.

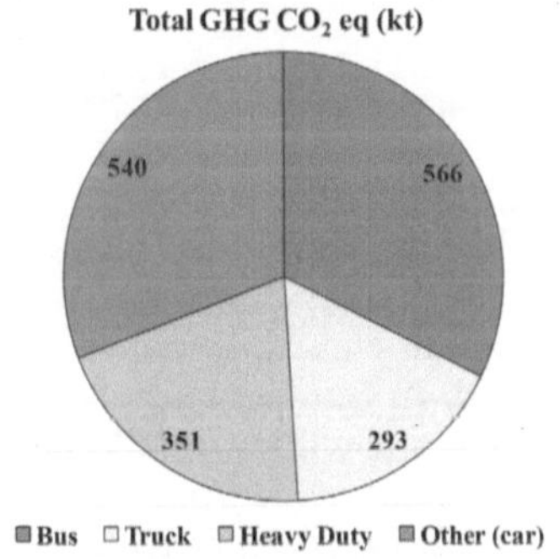

Figure 14. Projected total annual CO_2 from the main four types of diesel vehicles in 2030 in 2030 based on a 20% decline rate in vehicle registration – Scenario 3.

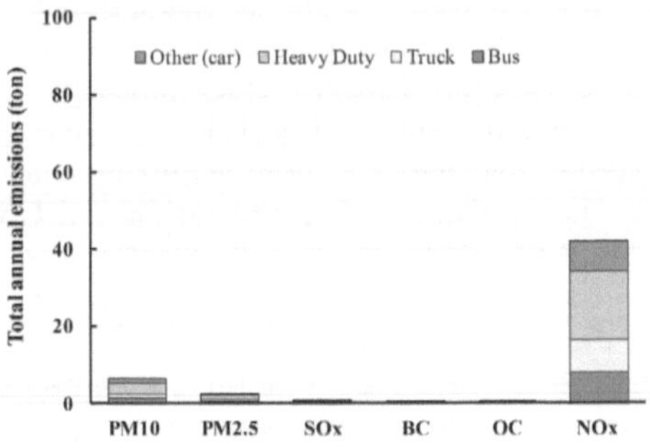

Figure 15. Actual annual emissions based on the average actual annual distance travelled by the four types of diesel-registered vehicles (bus, truck, heavy duty, and cars) in 2030 based on a 20% decline rate in vehicle registration – Scenario 3.

Table 1 and Figure 16 show the non-exhaust PM_{10} emissions based on the historical growth in vehicle count rate.

Table 1. Total non-exhust PM_{10} transport emissions (2017) and the projection (2030) for all registered vehicles.

	Total non-exhaust PM_{10} Transport Emissions		
Source	*EF* (g/km.veh)	*Total emissions* (2017) kg/km	*Total emissions* (2030) kg/km
Road dust resuspension	0.038	2.537	6.464
Tyre wear	0.04	2.670	6.804
Brake wear	0.0228	1.522	3.878

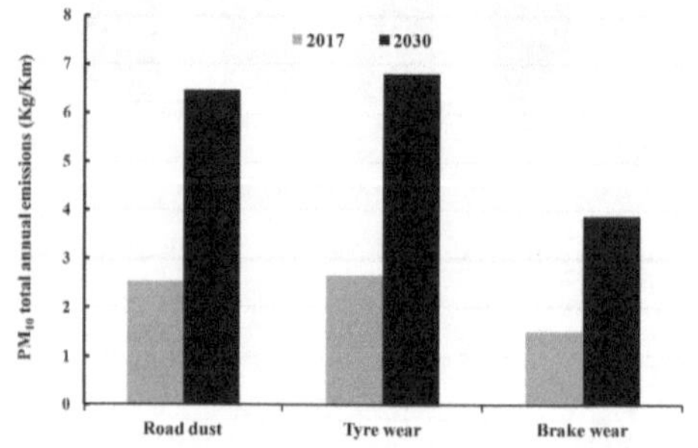

Figure 16. Total non-exhaust PM_{10} transport emissions 2017 and projection 2030 for all registered vehicles.

4 MITIGATION OF DIESEL VEHICLE POLLUTION

In order to reduce traffic air pollution emissions, policy change has been reported to be effective in major parts of the world. In Europe, scientists estimate that, through reducing the flow of NOx-emitting traffic, cities could lower NO_2 emissions by an average of 40%. Around 15%, 6% and 13% reduction could come from passenger diesel cars, trucks and from vans, respectively. NOx-emitting traffic flows can be reduced by limiting the access of highly pollutant vehicles – primarily older diesel cars – towards the center of cities. The same result can be also achieved by encouraging electric vehicles and the use of public transport, bikes and walking. These measures would not only improve air quality; but would also mean less noise, fewer accidents and improved quality of life (EU Science Hub 2019).

Recently, Qatar has announced that 25% of public buses, public schools buses, and Doha Metro feeder buses will be gradually transformed into electric buses in order to achieve the required percentage to reduce the rate of carbon emissions caused by traditional buses by 2030 and to promote environmental sustainability [13].

The plan is shown in Figure 17 and comes within the implementation of the strategy of the electric vehicle prepared by the Ministry of Transportation and Communications in cooperation with the concerned authorities.

The electric buses will be used in the main services during the 2022 FIFA World Cup, to be the first tournament to use electric mass transit buses, clearly demonstrating the commitment of the State of Qatar to clean energy solutions for transportation in the world and supporting the direction of hosting an exceptional world cup tournament that is environmentally friendly and "carbon neutral."

As part of the strategic plans, the Qatar General Electricity and Water Corporation (KAHRAMAA), in coordination with the Ministry of Transportation and

Communications, is working on establishing an integrated network of electric vehicle charging devices to support the Ministry's plan to gradually transform the electric transmission system. It is expected that about 30 electric chargers will be operating during the current year, and nearly 100 chargers by next year. To date, 11 chargers have been installed and are working uninterrupted. The largest electric charging station in the region will be established in Lusail City, running on solar energy and linked with all stations.

Moreover, a comprehensive plan to convert all taxis to electric-powered cars in the coming years is being prepared by The Ministry of Transportation and Communications in collaboration with Mowasalat Corporation (Karwa). The company plans to purchase about 140 electric taxis this year, to be put into operation to service Hamad International Airport as a first stage. The cost of this transformation rounds to about QR6bn [14].

Based on this scenario, the projection of CO_2 and other pollutant emissions are shown in Figures 18 and 19; respectively.

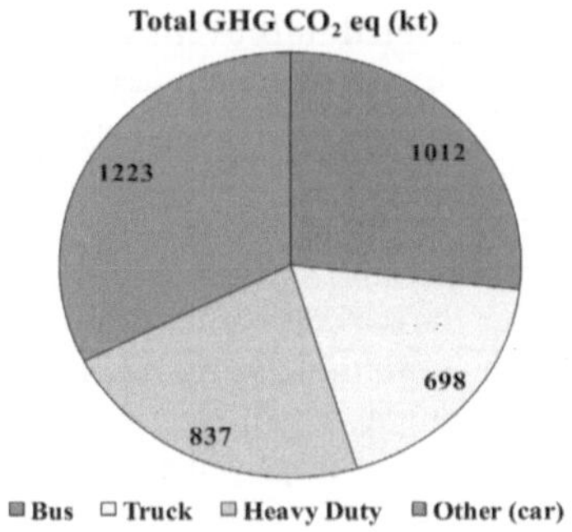

Figure 17. Projected total annual CO_2 from the main four types of diesel vehicles in 2030 when 25% of buses and 5% of cars are electric.

Figure 18 shows that, under the electrification of 5% of cars and 25% of buses, the total CO_2 emissions in 2030 amount to 3,761 kt annually. This makes a reduction of about 10% of CO_2 emissions compared with Scenario 1. However, the total reduction in NO_2 emissions under this scenario is about 5% since the major NO_x emissions are from heavy duty trucks, which are not affected by this policy.

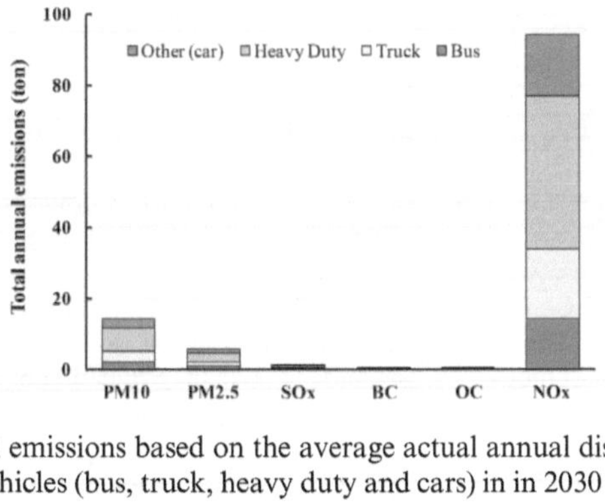

Figure 18. Actual annual emissions based on the average actual annual distance travelled by the four types of diesel registered vehicles (bus, truck, heavy duty and cars) in in 2030 when 25% of buses and 5% of cars are electric.

5 CONCLUSIONS AND RECOMMENDATIONS

Several mitigation or technical improvements can be suggested to reduce transportation emissions, such as the use of specialized materials to achieve specific performance goals in the transportation sector. Examples of technical advancements that have elevated the performance of vehicles include the use of platinum group metal nanoparticles to improve catalyst efficiency and the removal of air pollutants with a high efficiency at low temperatures. Another example is the use of lightweight materials in manufacturing of vehicles to minimize fuel consumption. Other more straightforward mitigation steps can include choosing smaller vehicles and driving fewer kilometers by using public transport more frequently. In addition, clean fuel, or fuel with an improved standard of composition can be used to reduce exhaust and evaporative emissions. For instance, in the United States, although there has been a tremendous increase in the number of vehicle registrations and vehicle kilometers traveled per year, air pollutant concentrations have been improving in most urban areas. This is due to several improvements in light drive vehicle emissions of conventional gasoline/diesel and associated improvements in fuel quality.

Several mitigation measures have been recommended to reduce CO_2 emissions from the transport sector in addition to adopting carbon sequestration measures. These include reducing vehicle kilometers traveled and carpooling. In addition, the fuel consumption of vehicles can be lowered by improvements in vehicle design. These changes could directly impact vehicle economics, performance, and consumer attractiveness. Another improvement to reduce CO_2 emissions is to lower the carbon content of fuel by moving from gasoline and petrol fuels to renewable fuels, such as ethanol. Other fuel options that may reduce emissions are compressed natural gas (CNG) or the use of electricity or hydrogen vehicles. In urban environments, carbon reduction from anthropogenic activities has also been achieved by green structures and planting trees in the streets.

ACKNOWLEDGMENT

We acknowledge the Traffic Department at the Ministry of Interior in Qatar for providing the vehicle count data. We would also like to acknowledge the customer service at Fahes for providing the mileage data for the traffic, with special thanks to Mr. Mario Fernandes.

REFERENCES

[1] UNCSD, (2001). *"Protection of the Atmosphere,"* New York, USA.

[2] Al-Thani H., Koç M., and Isaifan R. J., (2018). A review on the direct effect of particulate atmospheric pollution on materials and its mitigation for sustainable cities and societies A review on the direct effect of particulate atmospheric pollution on materials and its mitigation for sustainable citi, *Environ. Sci. Pollut. Res.*, vol. 25, no. August, pp. 27839–27857.

[3] Schipa I. and Tanzarella A., (2009). Differences Between Weekend and Weekday Ozone Levels Over Rural and Urban Sites in Southern Italy, pp. 509–523.

[4] Pérez-martínez A., Miranda P. J., Nogueira R. M., Guardani T., Fornaro M. L. et al., (2014). Emission factors of air pollutants from vehicles measured inside road tunnels in São Paulo: case study comparison., *Environ. Sci. Technol.*, vol. 11, pp. 2155–2168.

[5] Gokce H. B., Arıoğlu E., Copty N. K., Onay T. T., and Gun B., (2020). Exterior air quality monitoring for the Eurasia Tunnel in Istanbul, Turkey, *Sci. Total Environ.*, vol. 699.

[6] Al-Thani H., Koc M., Fountoukis C., Isaifan R. J., (2020). Evaluation of particulate matter emissions from non-passenger diesel vehicles in Qatar, *Journal of Air and Waste Management Association*, 70, 228–242.

[7] Kakouei A., Vatani A., and Bin Idris A. K., (2012). An estimation of traffic related CO2 emissions from motor vehicles in the capital city of, Iran, *Iran. J. Environ. Heal. Sci. Eng.*, vol. 9, no. 13, p. 1.

[8] Al-Jeelani H. A., (2013). *The Impact of Traffic Emission on Air Quality in an Urban Environment*, vol. 2013, no. February, pp. 205–217.

[9] Al Katheeri E., Jallad F., and Al Omar M., (2012). Aassesmnet of gaseous and particlulate pollutants in the ambient air of Al Mirfa City in the United Arab Emirates, *J. Environ. Prot. (Irvine,. Calif).*, vol. 3, pp. 640–647.

[10] EU Science Hub, (2019). Air quality: traffic measures could effectively reduce NO2 concentrations by 40% in Europe's cities. [Online]. Available: https://ec.europa.eu/jrc/en/news/air-quality-traffic-measures-could-effectively-reduce-no2-concentrations-40-europe-s-cities.

[11] Myung C. L. and Park S., (2012). Exhaust nanoparticle emissions from internal combustion engines: A review, vol. 13, no. 1, pp. 9–22.

[12] Prasad R. and Bella V. R., (2010). *A Review on Diesel Soot Emission, its Effect and Control A Review on Diesel Soot Emission, its Effect and Control*, no. October 2015.

[13] Ministry of Transportation, (2020). *25% of Public Transit Bus Fleet Electric by 2022*. [Online]. Available: https://thepeninsulaqatar.com/article/01/12/2020/Qatar-signs-nearly-QR6bn-contracts-for-transformation-to-eco-friendly-vehicles.

[14] Peninsula, (2020). *25% of Qatar's public transport buses will be electric by 2022*. [Online]. Available: https://thepeninsulaqatar.com/article/23/09/2020/25-of-Qatar-s-public-transport-buses-will-be-electric-by-2022.

Innovation and Technological Advances for Sustainability – Al-Naemi et al. (Eds)

Technical review of atmospheric water harvesting in a harsh environment (in Qatar)

Aiyad Gannan
Department of Mechanical Engineering, University of Doha for Science and Technology, Doha, Qatar

Mohannad Aljarrah
Department of Chemical Engineering, University of Doha for Science and Technology, Doha, Qatar

ABSTRACT: This study is being carried out to review and investigate several sustainable water harvesting techniques from atmospheric humidity, fogs, and dew. This investigation will show how well different fog/dew collectors' function in collecting water. Furthermore, a feasibility study will be carried out to obtain the technical details of the fog collector and efficiency enhancements. Of course, the prevalence of fog globally places a restriction on fog harvesting technologies. Dew water harvesters, on the other hand, are universally accessible but call for a cooled condensing surface. The dew water collection systems will also be investigated in this study. The main goal of all these strategies is the creation of an atmospheric water collector that can produce water regardless of the humidity level, location, and cost, or materials available. This project aims to adapt Atmospheric Water Harvesting (AWH) system design and material to ensure excellent performance and reliability of AWH in harsh conditions such as in Qatar. It will be particularly focused on the development of passive system. High efficiency solar cell technologies are proposed here as selected candidates for desert AWH development.

Keywords: Atmospheric Water Harvesting, Material, Passive System, Solar Energy, Sustainability

1 INTRODUCTION

This study was carried out to investigate and evaluate various sustainable water collection techniques from air fogs and dew [1]. This review study will show how well different fog/dew collectors throughout the world work at collecting water. Furthermore, a feasibility study should be investigated in more technical details of the fog collector and efficiency enhancements. Innovations in modern fog harvesting are frequently based on biotechnology.

Of course, the prevalence of fog globally places a restriction on fog harvesting technologies. Dew water harvesters, on the other hand, are universally accessible but call for a cooled condensing surface. The dew water collection systems will also be investigated in this study.

The main goal of all these strategies is the development of an atmospheric water collector that can create water regardless of the humidity level, location, cost, or materials available.

This project aims to adapt Atmospheric Water Harvesting (AWH) system design and material to ensure excellent performance and reliability of AWH in harsh conditions such as Qatar [2]. This research will be particularly focus on the development of passive system [3]. High efficiency solar cell technologies are proposed here as selected candidates for desert AWH development [4].

DOI: 10.1201/9781003496724-5

The obtained knowledge will be the basis for the development of new candidate system for desert AWH with enhanced reliability and later transferred to local and international standards and certification of desert systems [5].

For the outdoor AWH performance assessment and reliability, AWH will be installed at real operating conditions at the UDST.

2 OBJECTIVES

- Design a reliable AWH with high water production suited for high RH and temperature
- Benchmark high standard for testing desert AWH systems
- The proposed research project is expected to improve the fabrication of AWH and develop a low-cost, high production and concept that suits GCC conditions.
- The project opens new opportunities for further development of the technology and materials for desert-specific AWH in this constellation also after finishing the project (networking).
- Reduce the carbon footprint with every sip (Figure 1)
- Water is uncompromising on quality, safety, accessibility, and sustainability.
- Reinventing drinking water with technology that keeps both people and planet healthy and thriving.

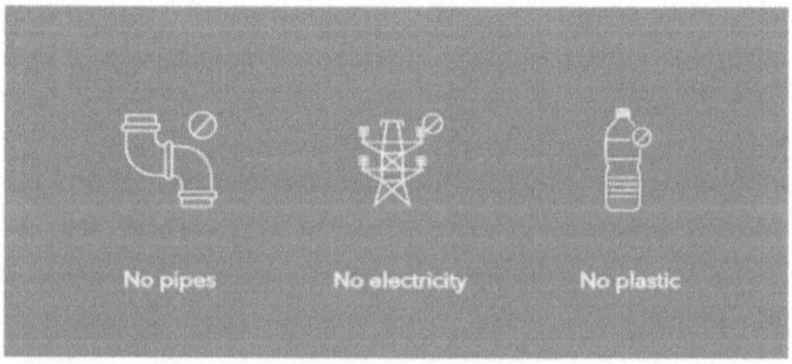

Figure 1. Environmental impact.

2.1 *Sustainable development goals, more water, less plastic*

2.1.1 *Mission*

Provide access to quality drinking water, at low cost, in a sustainable way and in situations without access to water or energy supply

2.1.2 *Vision*

Develop knowledge of heat transfer, control and water treatment technologies; to implement reliable solutions for the distributed generation of drinking water

2.1.3 *Values*

Technology, Innovation, Sustainability.

Figure 2. AWH sustainable development.

3 METHODOLOGY

This technology works in as little as 10% RH and requires ZERO electricity as the panels are completely self-powered. With 80% RH in Qatar the overall efficiency of this system and the output expected to be higher.

The output from the panels is a purified, mineralized drinking water that requires no additional filtration or processing. This can be achieved by using a passive condensation technology [6], [7].

The overall idea is that a combination of solar electric power harvested through a PV panel can be used as well as solar thermal power to heat up the air as is passes through the system and then enters a cooler area and condenses [8].

Once the water is condensed it is collected in a reservoir that circulates the water through an ozonation machine to ensure the sterility. Minerals are added to give the health and taste benefits [9].

Figure 3. AWH powered with solar.

This Hydropanel showing in Figure 3 is powered by an integral combination of solar photovoltaics and high- efficiency solar thermal. Moreover, in some of the world's driest deserts, the electrical and thermal energy is employed to produce water in an efficient modified psychrometric cycle.

3.1 *How does it work?*

- It draws in ambient air through fans and gathers water vapor from it onto a hygroscopic substance using solar PV.
- With heat from the sun, it converts water vapor collected into liquid water, made pure
- The water vapor is extracted and passively condenses into liquid
- The pure water is mineralized with magnesium and calcium to achieve an ideal taste profile
- Hydro panels are plumbed to a Home Reservoir or Access Station where the water can be accessed.
- Hydro panels are monitored in real-time for production and water quality
- It is fully self-sustained and off-grid.

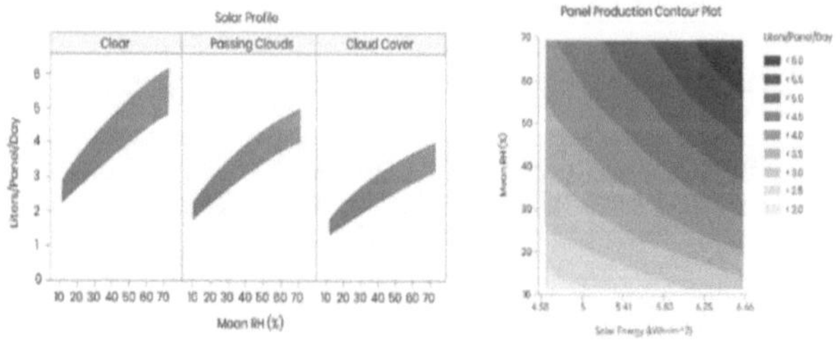

Figure 4. Humidity effect on AWH.

As can be seen in Figure 5, the process can be summarized as follows:

(1) The desiccant bed will absorb moisture from humid air during the first stage of the water absorption process, which occurs at night
(2) The second stage is water desorption during the day by heating the bed with solar radiation, which will regenerate the desiccant by driving out water vapor.
(3) The evaporated water will then be condensed into water droplets in the third and final stage, and collected in a tank.

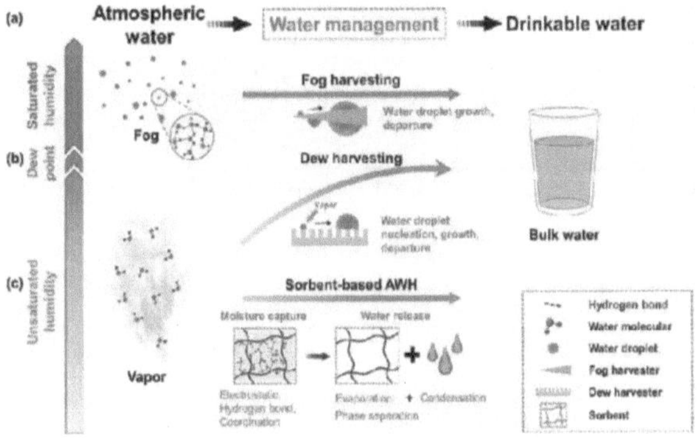

Figure 5. AWH process.

The Hydropanel, is a solar powered technology needing only sunlight and air to make drinking water. It can be installed on the roof or on the ground, extracts water vapor from the air to make, mineralize, and deliver drinking water to a tap or dispenser. Uniquely independent of infrastructure, this technology will be capable of making high-quality, delicious drinking water without plugging into an additional electric or water supply. its purpose is to make water from the air.

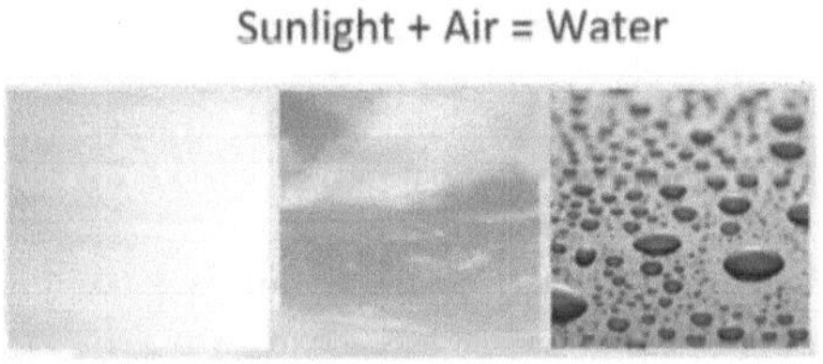

Figure 6. Atmospheric water generation.

4 THE TECHNOLOGY LANDSCAPE

The field of atmospheric water harvesting can be broken down into two primary methods: condensation and absorption. Condensation works by using a machine that acts like a dehumidifier, pulling in air, filtering it, and cooling it until the water condenses. Although this method can produce a large amount of water, it can be energy-intensive and may require high temperatures and humidity levels to function effectively. In comparison, the absorption process involves passing air over a material that captures the water, which is then heated to extract the moisture. This method is typically more energy-efficient, but currently has a limited daily output.

Recent advancements in the sector are improving the efficiency of both methods. A notable example is Drupps, a company specializing in atmospheric water generation that uses a patented heat exchanger made from food-grade polymer. This system has the advantage of producing water with no metal contamination, and can reportedly generate water even in low temperature and humidity conditions, although the yield may be reduced. Some companies now offer a range of machines that can produce up to 10,000 liters of water per day, which works with the condensation-based method.

4.1 *The headline opportunities*

AWHs can already produce water more cheaply than the bottled water industry and without the vast plastic waste. Exploiting mobile water and industrial opportunities can depend upon improved energy efficiency (see Figure 7).

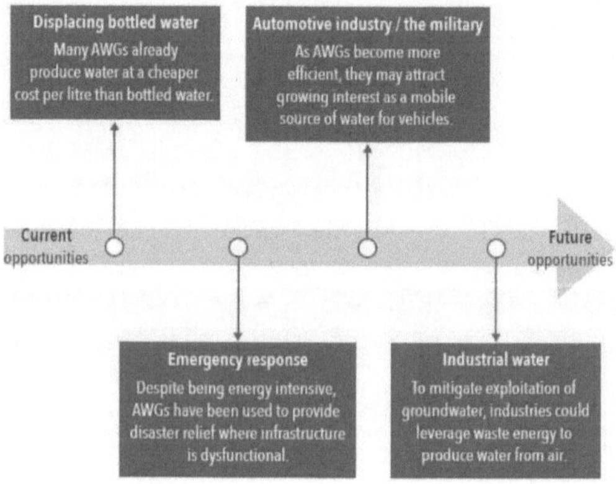

Figure 7. The important future of AWH.

4.2 *The heat and humidity distributions*

AWHs, especially those which use the condensation process, often need a relative humidity of at least 60% to function properly. As the map below shows, ideal conditions for AWHs do not always correlate with the areas which are most water stressed, such as central Africa, central Asia and the southwest United States (see Figure 8).

4.3 *Terminology*

4.3.1 *Absorption*

A method of atmospheric water harvesting where air is passed over a highly absorbent material that traps moisture, which is then heated to extract the water.

4.3.2 *Atmospheric water generator*

A machine that collects moisture from the air and converts it into potable water using either the condensation or absorption process.

4.3.3 *Condensation*

The most commonly used method of atmospheric water generation. It works by drawing air into the machine, filtering and cooling it until the water condenses.

4.3.4 *Desiccant*

A substance used to absorb moisture and act as a drying agent, and is the opposite of a humectant.

4.3.5 *Metal-organic frameworks*

Highly porous compounds created by linking inorganic and organic components with strong bonds. They are used in the absorption process of atmospheric water generation.

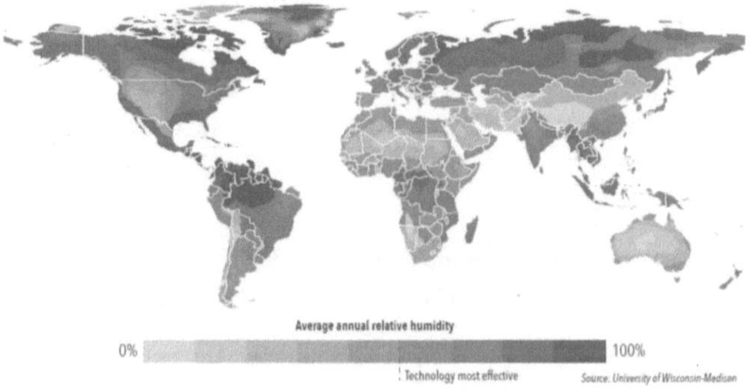

Figure 8. Map of humidity distributions in the world.

5 CONCLUSION

This project aims to adapt Atmospheric Water Harvesting (AWH) system design and material to ensure excellent performance and reliability of AWH in harsh conditions such as in Qatar. It will be particularly focused on the development of passive system. High efficiency solar cell technologies are proposed here as selected candidates for desert AWH development. The obtained knowledge will be the basis for the development of new candidate system for desert AWH with enhanced reliability and later transferred to local and international standards and certification of desert systems.

This research will contribute to the understanding of atmospheric water harvesting and its potential as a sustainable water source in desert environments. The study will provide valuable information on the feasibility of atmospheric water harvesting in desert environments. It will be also demonstrated to identify the factors that influence the ability to collect water from the air and determine the most promising locations for further study. Furthermore, intensive study will be conducted to evaluate the performance of different technologies and identify which ones have the greatest potential for providing a sustainable source of water in desert environments

REFERENCES

[1] Yaghi, O. M.; Kalmutzki, M. J. and Diercks, C. S. (2019). *Introduction to Reticular Chemistry: Metal-organic Frameworks and Covalent Organic Frameworks*; Wiley VCH: Weinheim.

[2] Wentao Xu and Omar M. Yaghi. (2020). Metal organic frameworks for water harvesting from air, anywhere, anytime. *ACS Cent. Sci.* 6, 1348–1354. https://dx.doi.org/10.1021/acscentsci.0c00678

[3] Kim, H.; Yang, S.; Rao, S. R.; Narayanan, S.; Kapustin, E. A.; Furukawa, H.; Umans, A. S.; Yaghi, O. M. and Wang, E. N. (2017). Water harvesting from air with metal-organic frameworks powered by natural sunlight. *Science*, 356, 430–434

[4] Yang Wang, Shouwei Gao, Hongmei Zhong, Baoping Zhang, Miaomiao Cui, Mengnan Jiang, Steven Wang, and Zuankai Wang. (May 18, 2022). Heterogeneous wettability and radiative cooling for efficient deliquescent sorbents-based atmospheric water harvesting, *Cell Reports Physical Science 3*, 100879

[5] Lord, J., Thomas, A., Treat, N., Forkin, M., Bain, R., Dulac, P., Behroozi, C. H., Mamutov, T., Fongheiser, J., Kobilansky, N., *et al.* (2021). Global potential for harvesting drinking water from air using solar energy. *Nature* 598, 611–617. https://doi.org/10.1038/s41586-021-03900-w.

[6] Khalil B., Adamowski J., Shabbir A., Jang C., Rojas M., Reilly K., Ozga-Zielinski B., (2016). A review: Dew water collection from radiative passive collectors to recent developments of active collectors. Sustain. *Water Resour. Manag.* 2, 71–86.

Innovation and Technological Advances for Sustainability – Al-Naemi et al. (Eds)

Testing constraints: Developing temporary sustainable housing models for American families

Shireen Kanakri, Delana Noil & Haitham Mohammad
Department of Interior Design and Construction Management, Ball State University, Muncie, Indiana

ABSTRACT: The purpose of this creative study is to understand the spatial needs and requirements of the thousands of individuals seeking housing and a more structured way of life for themselves and their families in America, primarily in the state of Florida. Reviewing studies of shelters and similar housing facilities in America that were designed as a refuge for homeless families will offer fresh direction for the design of a well-kept, functional housing model. Incorporating main ideas such as spatial limitation and functional area into this program will not only serve as a model for future designs in this architectural category but will support the ideas of government affordability and limiting environmental consumption as much as possible. This proposed model gives homeless, American families the necessary tools they need to reside comfortably until the establishment of permanent residency can become obtained

Keywords: homeless, shelters, families, transition, temporary, support services, facilities, community, residence, small-scale

1 INTRODUCTION

1.1 *Family homelessness: Sampling of the crisis*

The topic of family homelessness has been a growing and prevalent topic in recent years as it has been reported that more than 1.6 million children have joined the large masses of homeless people in the United States since the year 2010 (Bassuk, 2010), a 38% increase from the year 2007. Research data dictates that the homeless population often look just like the average person; while limited with hygiene, their bodily characteristics are more like yours and mine. Attached to the growth of the homeless population, specifically homeless families, are topics such as a lack of affordable housing, economic insecurity, domestic violence, behavioral health, and many other underlying causes (Aratani 2009, Bassuk 1988). Whether they are families or single adults, many of the homeless population do suffer from serious personal difficulties and defeats which are important underlying causes of their homelessness.

Poverty is one of the most accepted and intertwined structural causes of family homelessness. These structural factors also include any economic or societal problems that affect a particular person or group of people's physical environment (Gaetz 2014). As a family comes to terms with their poverty level, certain choices are made to protect their children and loved ones, and the house or current residence is the one physical entity that is subject to be removed from the equation. Other causes of family homelessness can still arise for large segments of America's population.

It can be safe to say that American citizens rely on their wages to provide financial support for themselves and their families. When these resources are limited, or unavailable, relying on government assistance is often the last resort. Studies have shown that unemployment

 DOI: 10.1201/9781003496724-6

rates have served as a predictor for poverty. While this issue is still being tested, recent welfare policies have not been adjusted even while unemployment numbers rise. A recent report done on 24 of the 50 states showed 58% of state departments did not make the proper amendment and program policy changes to their policies to account for the rising numbers of homeless people and families (Ehrenreich & Edelman 2009). While the caseload on the subject of health and welfare steadily increases, the unemployment benefits that are not accurately budgeted for will begin to run out.

There's also lies a large population of homeless families living openly and honestly on the streets, which is how most homeless people and families are characterized, (Snow *et al.* 1994). The Ottawa Panel Study suggests that visible homelessness is more prevalent among families than among single adults (Hulchanski *et al.* 2008). In a more common participant group, the homeless parent (typically women) with children will testify to current living conditions that fall under the hidden homeless category, rather than the single homeless individual (typically men) that will openly testify to being a part of a more visible classification of homelessness.

1.2 *Family homelessness: Individuals*

The overall population of those experiencing homelessness can be divided into three main groups: adults without children (mostly men), families (mostly young women with children under age 10), and adolescents on their own (Burt *et al.* 2001). If you were to start dissecting some of America's most largely distinct groups of homeless individuals, it would make up a never-ending trail of family separation and domestic unrest cases that this country must dissect before the programming of any shelter structure can start.

Disproportionately, homeless families are seeing a shift in single parenthood. Where there were more single women living on the shelters with their babies, single men are now being seen in homeless situations with their children alongside them without the other partner in the picture. What remains is the increasing number of men and women in a state of absolute homelessness with young (preschool) children. No matter the state of homelessness a couple is in, the child is in the center of it all. The child is no more than 6 years of age in most homeless family situations. Furthermore, the risk of a family being homeless increases for younger children, with the highest rate of risk among infants under the age of 1 (Culhane and Metraux 1999), suggesting that most families are at their most chronic point of homelessness when supporting a household with young children.

The prevalence of residential moves among American families has been a concern since the 1980s, and many studies have examined the consequences of residential mobility for children (Jelleyman and Spencer 2008). The negative effects of residential mobility are felt by families, local institutions such as schools and health clinics, and the communities that have high mobility rates. Residential instability among low-income households has been linked to several childhood risk factors, such as lack of school readiness, child maltreatment, poor health, and behavioral problems (Leventhal and Newman 2010).

1.3 *Family homelessness: The sampling of design*

Designing housing models for America's homeless family population is more of a scientific process now than it was during the most heavily trafficked times for shelter and asylum in American history. Lindsey (1998) reports that life in the shelters sometimes helped to bring families closer together emotionally, but that policies and routines in emergency shelters often disrupted their roles as disciplinarians and providers / caretakers. Davis (2004) notes that some shelters have day rooms where parents who are waiting for assistance can be with their children. Sometimes provided are both a quiet room and a separate room for conversation, with some providing a separate play area. Shelters that serve families usually have sleeping quarters for families separate from single adults.

2 RESEARCH MOTIVER AND METHODS

2.1 *Research question and hypothesis*

After carefully going over the sampling of the crisis, individuals, and design of homeless families this analysis led to the following research question: How can the study of current homeless shelters and support services designed for American citizens begins to develop into a conceptual framework to benefit the development of a temporary housing model for homeless families?

A hypothesis was created to propose if homeless families who acquired their support services from the same community they are temporarily residing in, then there is a greater chance of securing permanent placement in a residence at a later date. Through research this date could be anywhere from 6–12 months after the family has been provided temporary housing accommodations.

2.2 *Survey reports*

A total of three distinct survey reports were examined. Each survey report was collected from a different state in America: Livermore, CA, Las Vegas, NV, and Minneapolis/St. Paul area of MN. The findings in these reports give countable sources and numerical data collected face-to-face via convenience sampling.

The Livermore study had a total of 129 participants. The Las Vegas study had 406 participants. The Minneapolis study was able to gather a total of 4,181 participant responses for their research. All reports had requirements and had a select pool of qualifier questions that each participant had to answer to confirm they were: 18 or over, identified as homeless or housing insecure, and did not have any signs of intoxication or mental impairment that would prohibit them from providing accurate information. Participant responses that ended up as duplicate entries or did not meet the qualifying questions were removed and not included in the final statistical analysis.

2.3 *Questions and procedures*

The three survey reports collected for this study offered participants a physical survey to complete. The limitation of conducting an online survey with these types of participants is that you cannot guarantee that all participants would have access to a computer or mobile device to complete a survey. There was a range of 32 to 58 questions administered to participants during these three survey periods that covered several topics: demographics and background, history and current situation with homelessness, and support service preferences.

The questions on demographics and background focused more on family background and the issues children were exposed to homelessness. This type of questioning not only ensures that respondents have met all the requirements necessary to fully complete a survey, but also gives future studies ground to come up with a more universal solution the problem.

3 RESULTS

3.1 *Survey results*

The three surveys concluded that what participants needed the most was either free meals, access to a food pantry, or access to a meal program. 49% of participants agreed that these programs and pantries would have helped prevent their homelessness. The Livermore study concluded that most participants wanted access to housing assistance. The Las Vegas survey reported that many of their participants wanted and needed access to legal aid. The Minneapolis study had most participants (32%) identify that they would have preferred free meals and access to food programs over any other service. After meal support (49%) and housing support (31%), the next most beneficial service participants would have appreciated would have been some form of medical and/or dental care (29%). Among the services participants noted were important to them, travel and transportation assistance was among the top 5 (27%).

If homeless individuals must travel far to get to these other services and does not have a personal means of transportation, the likelihood of the individual successfully receiving these services is quite low. These findings demonstrate the importance of introducing programs near the participant's place of habitation.

3.2 *Case studies: Results*

There was a total of five cases utilized to benefit this study. The first being the Third Avenue Apartments, an apartment complex is in the Bronx in New York City. The second case is the New Genesis Apartments, located in Los Angeles, California. The third case the Austin Resource Center for the Homeless located in Austin, Texas. The Bakhita Gardens is the fourth case used in this study, which is a complex located in Seattle, Washington. The fifth and final case analyzed in this study is The Booth Centre; the only international case used, located in Southampton, UK. All of these facilities were constructed in urban areas within their city limits, construction ended for all of them between 2004 and 2015.

The goals of this study are to develop a temporary housing model and community that is set to provide restitution for American families. While many of the cases are permanent supportive housing facilities, they still offer opportunities for design innovation. The Third Avenue Apartments offer admission to their residential and support services primarily to a chronic homeless population; single individuals typically come into the facility with mental health issues and/or substance abuse problems. The New Genesis Apartments and the Booth Centre vary from all other facilities because admission is client funded; clients offer 30–60% of their annual median income. Costs for the Booth Centre are met through the housing benefit system in the United Kingdom, but there is also a small percentage fee charged for residents. Due to the fact that New Genesis is an example of a permanent supportive housing facility and the Booth Centre is an example of a temporary, transitional facility, it is vitally important that the programming involved in this design considers government, organization and/or client funding for the operations and upkeep of the design delivered.

Out of the five facilities, the Bakhita Gardens in Seattle, WA offered their residents and staff the most on-site services and amenities. With a total of fifteen reported services and amenities, the Bakhita Gardens was used as base line for the types of services and amenities this proposed design should offer residents.

The top 5 services and amenities were access to a place to for resident's to do their laundry (5 out of 5 facilities), access to case management offices and services (5 out of 5 facilities), administrative offices for re-occurring or full time staff to use (4 out of 5 facilities), communal dining and kitchen locations for residents to use (4 out of 5 facilities), and access to a porch or deck outside the facility for residents to use to recharge and unwind (4 out of 5 facilities).

4 ANALYSIS

This analysis was given to summarize the main concerns found within all five case studies. Analyzing these concerns will help provide direction as to the types of issues the design should be focused on. The first concerns deal with creating a sense of security, creating a sense of individual ownership, creating modular interior environments, creating a home-like environment, and finally creating a manageable community.

4.1 *Creating a sense of security*

Of the five reported concerns dealing with safety throughout these case studies (Figure 1), two dealt with implementing systems or equipment into the facilities to better protect staff and residents. The Austin Resource Center did not have a workable and flexible interior floor plan in place to prepare for new HIPPA privacy practices and guidelines. Since these privacy and safety requirements took precedence over the aesthetics, the operations and interior beauty of the facility suffered. On the other hand, New Genesis apartments concluded that the involving a coordinated entry system early into their design process would help ensure success and security for their facility and would help reduce evictions and minimize disruptions that might arise between residents. These cases demonstrate the importance of designing with flexibility in mind, so that when safety practices change the design does not have to suffer.

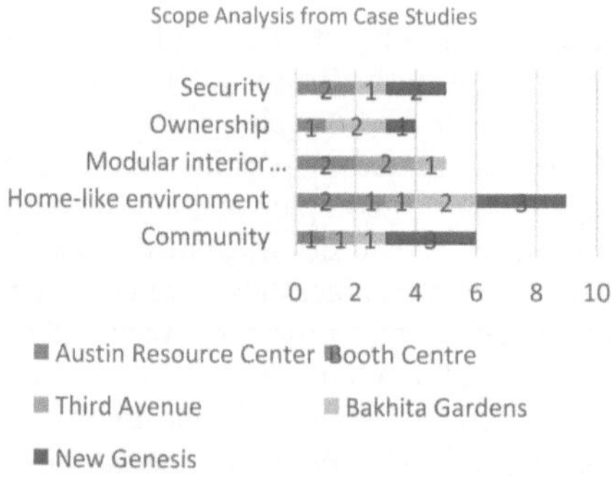

Figure 1. Chart describing main concerns found in five case studies.

The other three reported concerns dealing with safety came from case studies that were considering emotional safety as well as intellectual safety. The Austin Resource Center focuses again on safety, but in a way that deals with evoking moments of transparency and visibility towards its residents from staff. The Bakhita Gardens case depicts safety by utilizing volumetric design in their facility. The use of high ceilings and sensitive color palettes help achieve a calming mood and creates a sense of security for residents. The New Genesis case offers one other moment of security and that deals with creating permanent residence security measures, offering residents on-site case management support and on-demand clinical support. This ensures that residents are remaining safe and secure throughout their time in the facility.

4.2 *Creating a sense of individual ownership*

There were four reported concerns dealing with creating a sense of ownership for residents (Figure 1). One stemmed from the Austin Resource Center case and dealt with selecting certain building materials and finishes that were both durable and aesthetically pleasing, appealing to crowds that needed to feel certain emotion (ex: a sense of nurturement, insightfulness, or freedom). Two examples of individual ownership came from the Bakhita Gardens case. One example from this study showed that performance and longevity of furnishings not only resulted in a feeling of ownership for the resident, but also promoted better hygiene and environmental safety practices for residents.

4.3 *Creating a modular interior environment*

Having five reported concerns with creating a modular interior environment (Figure 1), four are unique and provide creative insight. One example stems from the Austin Resource Center; facility occupancy did not align with facility demand, which was far higher. It is important as the need increases, that the site or the interior of the facility has the room to expand. Two examples came from the Third Avenue case; material and finish flexibility were indeed some of the more important concerns. The interior's neutral color palette with the use of strategic art placement was found to complement the longevity of the facility and the durability of its higher traffic areas. This case also determined that items such as wall art and garden vegetation can and should change as the seasons do, creating a modular interior environment that appeals to different moments in the year. The Bakhita Gardens case found that creating private and public areas within individual spaces will help satisfy the needs of different types of residents to feel comfortable in, and satisfy a space's flexibility.

4.4 *Creating a home-like environment*

Out of all five concerns, creating a home-like environment was among the most important. Every case offered this as a concern and item they examined very closely when designing or analyzing their existing facilities (Figure 1). Of the nine examples that fall under creating a home-like environment, five examples are specific to only this concern. For the Austin Resource Center case, evoking moments of friendliness towards residents from staff was a main concern to making home-like environments. This involves making sure the design of the reception desk, or even signage displaying guidelines for the operations of the facility, are designed and placed in strategic locations that do not create any feeling that there is a hard barrier between the staff and the resident. An example from the Booth Centre deals with maintaining a calm environment that promotes an at-ease state of mind for residents, emphasizing a sense of welcome and promoting dignity, autonomy, and self-esteem. Deep care for residents' wellbeing through the design process helped the Third Avenue case usher more residents back to a place of equality in personhood. Lastly, the New Genesis case offered two strong.

4.5 *Creating a manageable community*

As seen in Figure 1, there are two are unique to the section on building and creating a manageable community for previously homeless residents. The example of designing a facility and its many programs to benefit the needs of the residents and the community comes from the Third Avenue case. In this case, Third Avenue's designs offered both a payoff for the designers, and for their community by giving neighboring areas the idea that this city was taking care of its society and people, in hopes those cities will follow suit. The second example came from New Genesis; integrating back into the community not only helped bring good staff and neighborhood people into contact with residents, but often helped community engagement efforts become more of an opportunity for residents rather than a chore.

5 DISCUSSION

5.1 *Summary of key findings*

The primary purpose of this study was to understand the spatial needs and requirements that needed to be considered in the development of a temporary housing model designed to benefit the needs of homeless families. The analysis of the literature (Walsh *et al.* 2009; p. 308) and the results both confirm that women are less likely than men to use homeless shelters since they do not conform to their idea of home, but single women and their young children are among the top category of family homelessness. Creating a home-like environment and a manageable community were the most important concerns found in the results of this project (Figure 1). *Interpretations*

The hypothesis proposed for this study dealt with the idea that there might be a positive effect to providing the same programming found in a transitional or permanent supportive housing developments and providing them in a temporary housing model designed for homeless families. When analyzing the results, the hypothesis stood firm as it related to the overarching categories of services these residents should provide (Figure 2), but when examining supply and demand, the differences were much easier to identify.

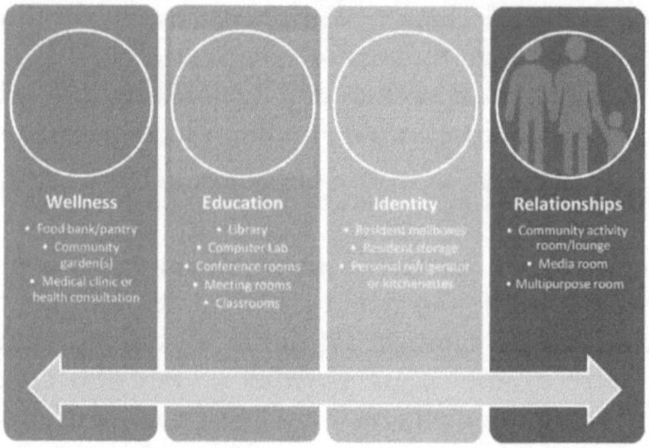

Figure 2. Four service goals the proposed design should promote and the services/amenities that fall under each category.

5.2 *Limitations and recommendations*

The main goal when determining the best methodological practices to use for this study was to comprehend the different spatial factors that lead to a more sustainable and well-composed temporary living facility. Although the study conducted thorough survey and case study research, there were certain limitations while exploring the goals of this study. The points mentioned below are likely to better future studies and to avoid researchers from facing similar shortcomings:

Disclaimers offered participants the ability to select more than one answer, going against the intent to get a clear and directional statistic. As a result, certain percentages were speculated and not fully proven, notes were given under these figures. Future studies involving surveys may consider interviewing the data collectors or organizations to understand how they organize their report data.

There was a total of five case studies utilized in this study, and all of these studies were used because they were located in different geographical locations, administered different types of living accommodations and services to their residents, and varied in the types of facility it chose to be run as. While multiple types of cases offer the ability for this study to be more diverse in its results and conclusions, the cases examined did not explicitly cater to a homeless family clientele. Future studies where case studies are used, it is best that the researcher further examine the facility being described in the study, whether it be through on-site observation or interviews with the facility directors or staff on how they accommodate housing homeless families and individuals.

6 DESIGN DELIVERABLES AND CONCLUSION

6.1 *The brand model*

Creating a welcoming, home-like atmosphere is the first step to develop a community that evokes a friendly environment for residents, and a feeling of safety to empower staff and residents to take care of the environment that surrounds them. The community must have a brand that echoes this statement.

6.2 *The site model*

There are three key areas utilized in a manageable community: residential units for families, residential opportunities for staff, and a community center (Figure 3).

The proposed site will offer (16) duplex-style residences, each unit is joined by one covered deck space, known as the residences' "core". These (8) duplexes are surrounded by a central green (#6 Figure 3). It is proposed that the central green as well as certain spaces within the community center (#2 Figure 19) will be open to the residents living in this community and to reoccurring. The central green is divided into five separate green spaces, utilizing sidewalks to provide families a direct path to the community center. A designated parking lot (#1 Figure 3), as well as an outdoor garden and playground (#3 Figure 3) are located near the community center to become an inviting entrance and exit experience for staff and visitors.

Offering twelve duplex-style residences (#4 Figure 3) to this community's staff will not only make the community center function and run more smoothly, but also help residents feel more secure and at ease in their new environment. After examining the case studies, giving staff their own connection to the community center is important to distinguish their role.

It is proposed that residential units are to only be permissible to families and staff with the greatest need. A proper approach to admission and costs for this type of community was gathered by examining.

Residences offered to staff will involve a monthly rent and utility charge, and have access to all services and amenities in the community. In return is they will be employed or working with the community's managing office or corporation.

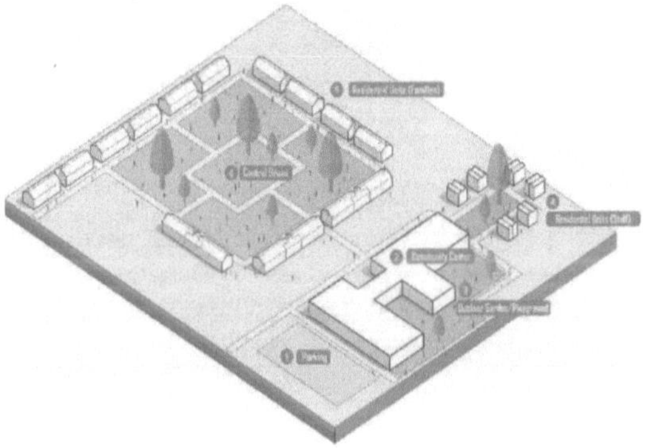

Figure 3. Connection between residences, community center, and other amenities/support services.

6.3 *The residence model*

In order to keep with the original plans to create a functional residence that meets the needs of an average homeless family and organizes all the necessities they require utilizing as little space as possible, shipping containers were used as a primary method to get the residence overall shape and size determined. A few trials and test fits were administered in this small-scale operation, and the final solution was to utilize three 10-foot long shipping containers and two 20-foot long shipping containers. The remaining space (63 sq. ft.) needed to be in place to ensure the central core connection and the extension of the outdoor deck could be maintained.

When examining Figure 5, the 378 sq. ft. living space includes a designated entry from the core, a modular study nook, a kitchenette on the other side of the study nook with a fitted dining nook and living room. The residences are all situated on a single level in order to create a more universally friendly and welcoming interior environment for different types of residents, especially those with more mobility impairments than others. Allowing the core and the outdoor deck to be divided by a bi-folding glass door will allow light to carry through, create a sense of privacy, but can also be fully opened to maximize outdoor access and sightlines.

When a family enters one of the units they are greeted by a double-height entrance. This height is provided to allow the small-scale floor space to not feel too constricting. The interior space looks quite large given its use of pale plywood applications and ivory walls. In the idea of individual ownership, families will be able to personalize the interior the way their family desires. A floor-to-ceiling, equally spaced partition separates the entryway with the study nook. Most of the furniture pieces in each unit are built-in to offer residences a stable and secure set of pieces to retain and re-use as families come in. The different plantings and hanging vegetation in these spaces, as well as others in the unit, is a subtle inclusion of the exterior's natural forms within the interior.

To the left of the study nook (Figure 5) is the kitchenette, dining nook and living room. For similar reasons in the entry and study nook areas, the living room and kitchen/dining spaces are divided by a similar floor-to-ceiling partition w/ a storage credenza underneath. The light from the corner window encasing the banquette dining feature is carried all the way to the end of the living room. The kitchenette offers families a built-in refrigerator/freezer combo. Also included in the kitchenette is a two-burner electric stovetop, microwave, sink, and countertop surface for light prep and food-handling. The dining nook features a suitable built in banquette, with an upholstered back rest. This nook is equipped with enough space to fit a family of four or even larger with enough space to dine together.

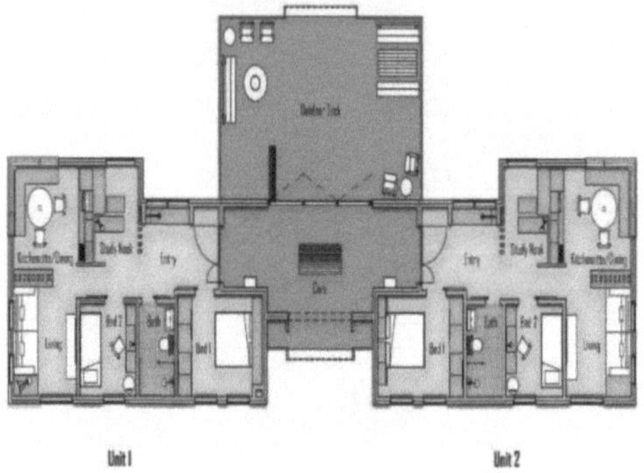

Figure 4. Floorplan of residential unit for families.

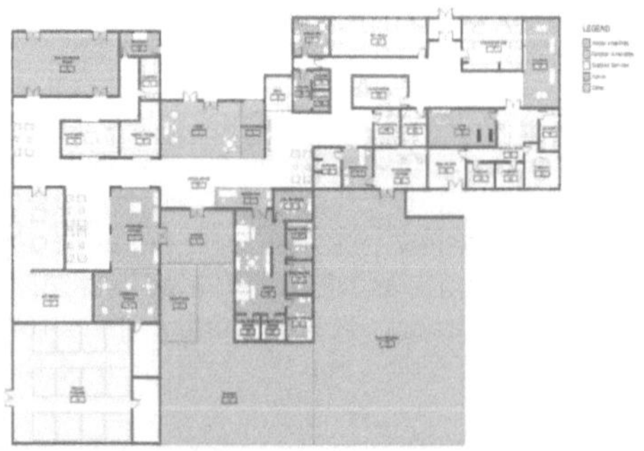

Figure 5. Color-coded area plan for community center.

The six duplexes that are situated along the right-hand side of the community center (Figure 3) are to be utilized by staff working inside the community. It is proposed that exceptions can be made on who can live in these residences, especially if there is an overflow of families in need of housing. These residences will utilize less ground space and more height;

a lofted, small-scale residence structure. These residences will eliminate the shared core space and simply create an exterior entrance into these residences. The number of bedrooms, bathrooms and types of spaces will remain to provide equal services to its residents.

6.4 *The community center model*

The entirety of the community center is located on a single level. After analyzing the services and amenities offered in the five case studies and cross-examining them with the needs and wishes of the homeless participants in the survey reports, there are five main categories to how these spaces involved in the community center were categorized: those classified as overall building requirements (Figure 4), indoor amenities, outdoor amenities, support services, and administrative spaces.

6.5 *Conclusion*

Applying the four service goals and top services and amenities inside this community center not only proposes a more universal design, but also ensures that residences and staff are provided some of the same luxuries and amenities. The survey reports and case studies analyzed have proven to offer residences the identity they need to thrive in this community and sustain a more permanent place in society later in time.

REFERENCES

Aratani, Y. (2009, September). Homeless children and youth: Causes and consequences. *National Center for Children in Poverty*. Retrieved September 8, 2019, from https://academiccommons.columbia.edu/doi/10.7916/D83F4ZCG

Bassuk, E. L., Weinreb, L., Buckner, J., Browne, A., Salomon, A., and Bassuk, S. (1996). The characteristics and needs of sheltered homeless and low-income housed mothers. *Journal of the American Medical Association*, 22, 3–20.

Burt, M. R., Aron, L. Y., Lee, E., and Valente, J. (2001). *Helping America's Homeless: Emergency Shelter or Affordable Housing?* Washington, DC: Urban Institute Press.

Cross, J. E. (2001). *What is Sense of Place? In Archives of the Twelfth Headwaters Conference*. Gunnison, CO: Western State Colorado University. Found at http://western.edu/sites/default/files/documents/cross_headwatersXII.pdf

Culhane, D. P., and Metraux, S. (1999). One-year prevalence rates of public shelter utilization by race, sex, age and poverty status for New York City (1990, 1995) and Philadelphia (1995). *Population Research and Policy Review*, 18 (3).

Davis, S. (2004). *Designing for the homeless: Architecture that works*. Berkeley, CA: University of California Press.

Design Resources for Homelessness. (2017). *Austin Resource Center for the Homeless: Austin, Texas*. Retrieved from http://designresourcesforhomelessness.org/foundation-information/

Design Resources for Homelessness. (2017). *Bakhita Gardens: Seattle, WA*. Retrieved from http://designresourcesforhomelessness.org/foundation-information/

Design Resources for Homelessness. (2017). *New Genesis Apartments: Los Angeles, CA*. Retrieved from http://designresourcesforhomelessness.org/foundation-information/

Design Resources for Homelessness. (2017). *The Booth Centre: Salvation Army: Southampton, United Kingdom*. Retrieved from http://designresourcesforhomelessness.org/foundationinformation/

Innovation and Technological Advances for Sustainability – Al-Naemi et al. (Eds)

Development of a VLF receiver for lightning detection, and monitoring natural & man-made electromagnetic emissions in the VLF band

Warren Hansen
Department of Electrical, Electronic and Computer Engineering, Cape Peninsula University of Technology, Cape Town, South Africa

Ali Almaktoof
Department of Electrical, Electronic and Computer Engineering, Cape Peninsula University of Technology, Cape Town, South Africa
ORCID ID: 0000-0002-1134-5587

ABSTRACT: The very low frequency (VLF) band of frequencies reveal useful details about the earth and is an important area of research. In this paper, VLF receiver is designed and fabricated to detect, monitor lightning strikes, and store this data as well as other emissions within the VLF band consisting of natural, and man-made VLF signals. A data acquisition system is designed and built which can receive signals in the VLF band. An antenna is built for the reception of VLF signals by winding 40 turns of 21 AWG copper wire on a wooden structure. A second order inductor-capacitor (LC) bandpass filter is designed with lower and upper cut off frequencies of ~700Hz and ~50kHz respectively. A pre-amplifier using the OP27 as the active amplifying component is used to amplify the filtered signal. The analog signal is digitized using the IQaudio Codec Zero soundcard HAT, which is attached to a Raspberry Pi 4B. The IQaudio Codec Zero soundcard acquires audio data at a sample rate of 96kHz and at a bit resolution of 24 bits. The Raspberry Pi runs the Raspberry Pi OS, which is an operating system based on the Linux Debian operating system. The Raspberry Pi runs a software tool package called vlfrx_tools which is made expressly for the processing, monitoring, and recording of VLF signals. Bash scripts are used to run the essential commands. The data stream is also passed via a local network and received by a program named Spectrum Lab, which runs on a Windows PC for more accurate processing and displaying of signals, if need be. The VLF receiver is intended to create awareness, educate society, and encourage further research about natural VLF radio, the geophysics and space science research fields.

Keywords: VLF, VLF receiver, VLF antenna, lightning, lightning detector, natural radio, Raspberry Pi, IQaudio Codec Zero, vlfrx_tools, spectrogram, FFT, fast Fourier transform, VLF station, SIDs, spherics

1 INTRODUCTION

Since the International Geophysical Year of 1957, many countries are actively carrying out research pertaining to electromagnetic (EM) signals in the VLF band (3–30kHz). Many different types of natural signals originate from the earth, the earth's magnetic field, the layers of the atmosphere (the ionosphere and the earth-ionosphere cavity). One of the main sources of these natural signals are lightning discharges [1]. These natural VLF signals comprise of what is known as 'radio atmospherics' or 'spherics'. A spheric originates from a

DOI: 10.1201/9781003496724-7

great pulse of EM energy. This large amount of EM energy originates from a lightning discharge and travels around the world. This EM energy bounces between the earth and the ionosphere [1]. In the case of VLF EM waves, the cavity between the earth and ionosphere acts as a waveguide as the wavelengths of the VLF waves were in the range of 100000km to 10000km [1]. The electromagnetic energy created by the lightning discharges are altered in different ways by the earth-ionosphere waveguide to create different types of spherics such as whistlers, tweaks, and chorus [1], [2]. These different types of spherics are mainly characterized by how their signal power varies with time, and what they sound like when the EM energy was converted to mechanical vibrations (sound) [1], [3].

A combination of man-made and natural VLF signals was used to detect and monitor Sudden Ionospheric Disturbances (SIDs). SIDs occur when solar radiation/wind contacts the ionosphere [4]. This creates detectable perturbations in the ionosphere [4]. SIDs are monitored using a weak VLF signal from a VLF station, and a VLF receiver that is set to monitor the received power level at the frequency of the VLF station [4]. Other events that occur in the atmosphere which generate natural signals that occur within the VLF band are called transient luminous events (TLEs) [5]. TLEs are atmospheric phenomena that usually occur above large thunderstorm clouds [5]. The main types of TLEs are called sprites, elves, and jets [5]. All these natural signals translate into valuable information about the how the earth and the surrounding region in space functions together as a whole. Knowing this helps scientists to better understand how the earth functions. Scientific organizations involved with research in the VLF fields include:

- AARDVARK [6]
- Worldwide Lightning Location Network (WWLLN) [7]
- The lightning geolocation web-community: blitzortung.org [8]
- Stanford VLF Group [9]
- NASA INSPIRE Project [10]
- Vaisala professional lightning detection network [11].

A VLF receiver can monitor and track local lightning storms which is helpful in regions that are prone to electrical storms. Such a device, in conjunction with local weather reports, will provide early warning of lightning in these regions, and can be used to advise residents and industry to install lightning surge/strike protection devices. This can mitigate damage to property including the power distribution, telecommunications, and maritime industries. Desktop computers are the main hardware used to monitor the VLF spectrum using their built-in soundcards, however desktop computers are heavy and bulky and are therefore not suitable in scenarios where a portable VLF receiving device is required. There is a need for a portable VLF receiver. This offers the ability to easily move the receiver's location when needed. Insurance companies don't usually compensate their customers for any type of damage caused by a lightning strike (electrical equipment failure, and/or physical damage due to a direct strike). If their customers have a VLF receiver/monitor installed on their premises, the customer would be able to provide evidence of the lighting strike to the insurance company, which could assist in the customer being more fairly compensated for the damage. It was found that not many good quality, cost effective, turn-key solutions for lightning detection and natural VLF radio monitoring are readily available on the market. Therefore, there is a need for an affordable and simple to use VLF receiver. An artefact that meets these criteria will ensure that it available to a larger audience. Additionally, it was found that the subject of natural VLF radio, TLEs and SIDs, EM seismic signals, and EM meteor signals are not widely known topics [12] [13]. This VLF receiver can serve as an educational tool as it is both cost effective and mobile, and therefore will be able to be easily transported and utilized at educational institutions. Along with the ability to detect and monitor lightning storms, a VLF receiver can be used for monitoring the previously described natural and man-made VLF signals. A low cost, mobile VLF receiver will increase the availability of such devices, which in turn will enable more people to be involved with and engage in VLF studies, thereby furthering education of the VLF field of research.

2 VLF RECEIVER SYSTEM DESIGN AND OPERATION

The VLF receiver system is represented and illustrated by the diagram in Figure 1. The components/parts, as well as reasons for choice of certain parts are outlined. The following sections 'System design', 'Practical setup', and 'Software configuration' deal with technical design details, the physical final artefact, and the software setup respectively.

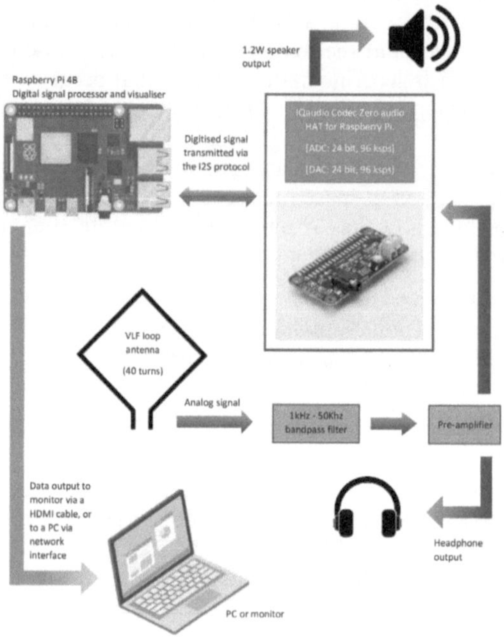

Figure 1. Complete VLF receiver system overview.

2.1 *System design*

The antenna receives the electromagnetic signal which is then converted to an electrical signal by the copper antenna wire. The frequency components not within the 1–50kHz range are then filtered out by the bandpass filter. This means that any powerline noise and the corresponding harmonics are eliminated, as well as any higher frequency FM radio station interference is eliminated. The filtered signal is then amplified by the pre-amplifier so that its amplitude is increased to a level large enough for it to be accurately sampled, as well as for the signal to be sampled with the highest dynamic range possible. This amplified signal is routed to a headphone output so that the raw signal can be listened to in real-time. The amplified signal is routed to the IQaudio Codec Zero audio data acquisition board where the signal is then digitized by an ADC at a 96kHz sample rate with a resolution of 24 bits. The digital data is communicated via the I2S protocol to the Raspberry Pi. The Raspberry Pi receives the data stream into a buffer, where the vlfrx_tools software package captures, processes, stores, and displays the received VLF signal in the form of a real-time spectrogram waterfall. The IQaudio board also supports a 1.2W mono 8Ω speaker output that is used if the user wishes to listen to previously recorded data. The Raspberry Pi desktop can be accessed via a local Wi-Fi network using the VNC software, or alternatively via a monitor, keyboard and mouse connected directly to the Raspberry Pi. Both options allow the user to view and interact with the software. If it is required, the received and digitized data stream is communicated via a local area network to an external computer running the Spectrum Lab software where the VLF signal data stream can be processed and displayed [14].

2.1.1 *Antenna design*

Wood pieces and wooden dowels were used to fabricate the structure of the antenna as seen in Figure 2. 40 notches were cut 6 mm apart on 30mm long dowels for the wire to run through to keep each parallel winding at least 6mm apart. The spacing between the wires keeps the capacitance between each winding at a minimum. If the antenna exhibits too much of a capacitance, it can change the frequency response of the filter to which it is connected. 40 turns of 21 AWG (0.72mm diameter) wire is used as the receiving antenna. Thick wire is used for the electromagnetic radio waves to couple to the wire more easily. Strips of PVC are inserted on all sides between the wire to stabilise the wire so that it cannot move due to wind. The loop antenna exhibits a nulling effect through the centre of the loop and exhibits maximum sensitivity perpendicular to the flat face of wire presented by the loops of wire. This is useful when there is an electromagnetic noise source that must be diminished. One simply rotates the antenna so that the signal nulling end is facing the noise source.

Figure 2. The VLF antenna.

2.1.2 *Filter design*

Figure 3 depicts the schematic of the filter design.

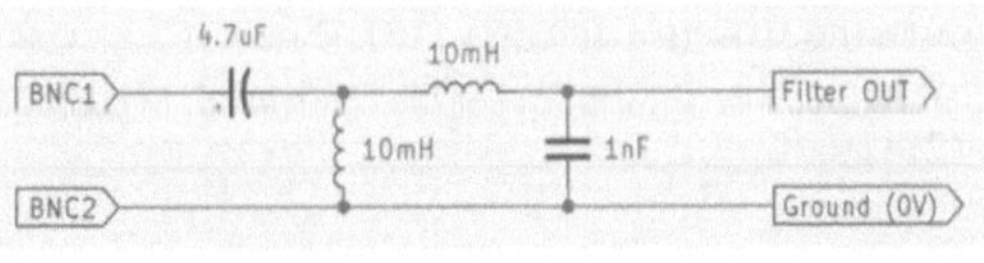

Figure 3. Filter schematic.

The chosen filter is a second order passive bandpass filter and is designed to pass signal frequencies between approximately 1kHz and 50kHz. The upper limit of approximately 50kHz was chosen to maximize the amount of spectrum that can be recorded and viewed. This is due to the frequency limit of the IQaudio Codec Zero soundcard which has a maximum sample rate of 96kHz. This corresponds to a maximum acquirable frequency of 48kHz due to the Nyquist theorem. Any other unwanted signals above 50kHz must be filtered out. An example of such signals are FM transmissions, and other high frequency noise. The lower limit of approximately 1kHz was chosen to reject the local 50Hz powerline interference and attenuate the corresponding harmonics to an acceptable level, as well as to save power in the amplifier, and to mitigate saturation of the ADC input. The bandpass filter is designed with capacitors and inductors for both the low-pass and high-pass filter sections. LC filters were chosen for their ability to pass signals with minimal attenuation, as opposed to an RC or RL filter configuration. The required values of capacitors for the low-pass and high-pass filter were calculated using (1) [15], [16].

$$f_{cut-off} = \frac{1}{2\pi\sqrt{LC}} \quad (Hz) \tag{1}$$

The inductor value used for both low-pass and high-pass filters is 10mH. Using a capacitor value of 4.7uF, a cut-off frequency of 734Hz is obtained for the high-pass filter section. Using a capacitor value of 1nF, a cut-off frequency of 50.33kHz is obtained for the low-pass filter section. Both capacitors were measured to obtain their real-world values. The value of the 1nF capacitor was measured to be correct, and the 4.7μF capacitor was measured to have a value of 4.1μF, which resulted in a cut-off frequency of 786Hz.

2.1.3 *Pre-amplifier design*

Figure 4 depicts the schematic of the pre-amplifier design. Reference [17] aided in the design of the pre-amplifier by providing an idea of what type of amplifier is required to amplify a VLF signal.

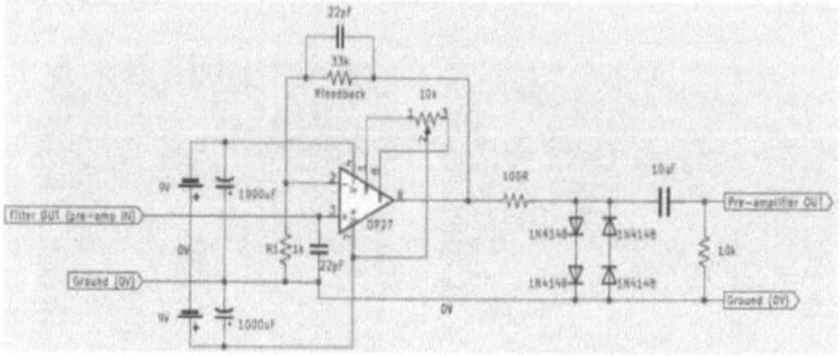

Figure 4. Pre-amplifier schematic.

Since a large portion of the concerned range of frequencies are within the range of audio frequencies, an operational amplifier would suffice. The OP27 operational amplifier was chosen as it is a high performance op-amp with an input sensitivity of 30μV (smallest voltage the op-amp is able to amplify), exhibits low noise characteristics of approximately 3nV/$\sqrt{}$Hz, has a low drift rate of 0.2μV/°C, and exhibits a high input impedance ensuring input signal is not loaded [18]. The OP27 is set up in non-inverting mode, the gain is set to be 34 by using a 33kΩ feedback resistor, and R1 = 1kΩ (see Figure 4) according to (2) [19].

$$G = 1 + \frac{R_{feedback}}{R_1} \tag{2}$$

Two 1000μF electrolytic capacitors are connected from the positive and negative rails to ground (0V) to eliminate any noise that might occur in the circuit or enter the batteries. A 22pF capacitor is connected between the input pin of the op-amp and ground to conduct any high frequency noise away from the op-amp input. Similarly, a 22pF capacitor A 10kΩ potentiometer is used to set the output offset DC voltage to 0V. An output protection network consisting of; a 100Ω resistor, four 1N4148 diodes and an output decoupling capacitor, is utilized to protect the soundcard input (ADC) by limiting the output current, and by limiting both negative and positive output voltage peaks to 1.4V peak (each 1N4148 diode conducts at 0.7V and conducts at a fast rate).

2.2 *Practical setup*

The final setup excluding peripherals is depicted in Figure 5. Note that the unit is powered by a 5VDC 2.1A power bank for portability. The total cost of the final artefact is approximately R3000.

Figure 5. The assembled VLF receiver.

The interior of the VLF receiver unit us shown in Figure 6 below. The enclosure houses two 9V batteries (power supply for the pre-amplifier), BNC connectors, the filter, pre-amplifier, Raspberry Pi, and the IQaudio Codec Zero soundcard. The enclosure is used to house and protect the parts. The coaxial cable runs from the antenna (situated outdoors) to the enclosure (situated indoors).

Figure 6. VLF receiver enclosure.

2.3 *Software configuration*

The open-source Linux software toolset named vlfrx_tools is used to acquire data from the soundcard [20]. The vlfrx_tools help webpage was used to select and configure each command. A 3rd party program that displays a live spectrogram view of the data acquired by the vlfrx_tools software was used. The commands were collated using Bash files to make them easier to run when required. The data acquisition starts up main bash file as shown below:

```
#!/bin/bash
  Begin data acquisition from sound card:
  sudo lxterminal -e vtcard -d hw:0,0 -b24 -r96000 -vv @data &
  Wait for the @data buffer to be ready:
    echo "waiting for the @data buffer..."
    vtwait -t @data || {
      echo "buffer @data invalid" >&2
      exit 1}
    Open real-time spectrum analyzer display:
  vtcat -a 7200 -v @data:1 | vtlivesgram
```

The remainder of the Bash code is omitted due to space constraints, and these commands were saved as an executable so that it is easy to run when needed.

2.4 *System operation*

To operate the VLF receiver, the steps are as follows:

(1) The antenna is positioned outdoors, on the ground, and preferably at least 1m above the ground.
(2) The coaxial cables are connected from the BNCs on the enclosure, to the antenna via its connector block.
(3) The power to the Raspberry Pi is applied by either plugging in the power supply or alternatively by connecting a 5V power bank.
(4) Once the Raspberry Pi has booted the desktop, the data acquisition, processing, storing, and spectrogram rendering programs are contained within their respective bash files.
(5) First run the bash file named 'VLFRX_tools_commands.sh' which calibrates the soundcard, creates a buffer, and starts data acquisition which fills the buffer with data.
(6) Then the bash files can be run as needed.
(7) To view the visual output from the Raspberry Pi, the VNC can be used to login to the desktop.
(8) If it is required, digitized data stream can be received via a local area network and be processed and displayed in the Spectrum Lab software running on a PC.

2.5 *Project costs*

The components used to build the VLF receiver, as well as their associated costs, are outlined in Figure 7. The total cost is R2986.50 (ZAR) for the receiver unit, antenna, and cables. The final realized VLF receiver is robust, functional, and cost effective.

Item	Cost (incl. 15% VAT)	Retailer	Quantity
Raspberry Pi 4B 8GB	R 1499	www.pishop.co.za	1 unit
Power supply for Raspberry Pi	R 160	www.pishop.co.za	1 unit
Micro SD card 32GB	R 139	www.pishop.co.za	1 card
IQaudio Codec Zero HAT	R 432	RS Components	1 board
Micro-HDMI to HDMI cable	R 100	RS Components	1 cable
0.7mm enamelled copper wire	R 290	Wilec	284 m
Wood (antenna frame)	R 120	Brights hardware store	1 unit
Project enclosure	R 160	Mantech	1 unit
Prototyping PCB (90x90mm)	R 25	Mantech	1 board
OP27G operational amplifier	R 16.50	Mantech	1 IC
Resistors	R 15	Mantech	10 units
Capacitors (1, 10, 100μF)	R 30	Mantech	6 units
TOTAL	R 2986.50		

Figure 7. Cost of components to build the VLF receiver.

3 RESULTS AND DISCUSSIONS

Below are screenshots of spectrograms produced by the VLF receiver system.

3.1 *VLF antenna test*

The first four screenshots are produced by Spectrum Lab as this was used to separately test the antenna using a desktop PC and its built-in soundcard (sample rate: 192kHz, highest acquirable frequency: 96kHz). A lightning storm occurred during the antenna testing. This enabled the reception and visualisation of lightning discharges, proving that the VLF antenna does indeed function as expected. The bright orange horizontal bars are created by local noise sources and are not lightning strikes, the brighter orange vertical bars are detected lightning strikes. In Figure 8, one can see a bright orange stripe near the middle of the spectrogram. This is what a nearby lightning strike looks like on a spectrogram. It is noticed that the received energy from the lightning discharge consists of a large amount of power, and is dissipated over time i.e., is not immediate. Many distant lightning discharges are depicted in the spectrogram in Figure 9. Compared to the discharge in Figure 8, the vertical lines are thinner and are less intense in color which means the discharges are occurring further away from the point of detection. Figure 10 shows a lightning discharge, with a damped reflection of the initial discharge. The reason for this is speculated to be that as the EM pulse decayed, it bounced inside the earth-ionosphere cavity.

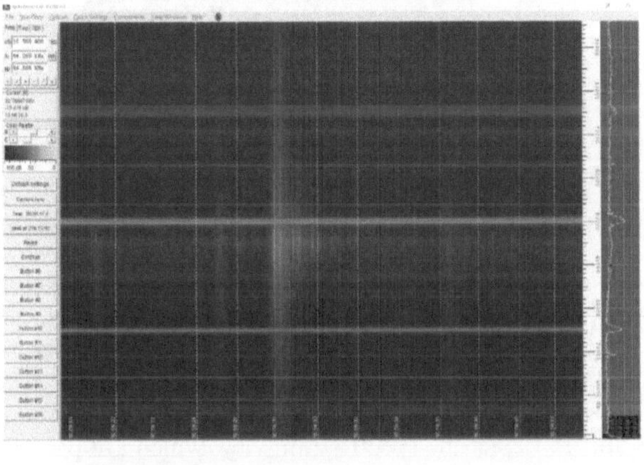

Figure 8. Spectrogram of nearby lightning discharges.

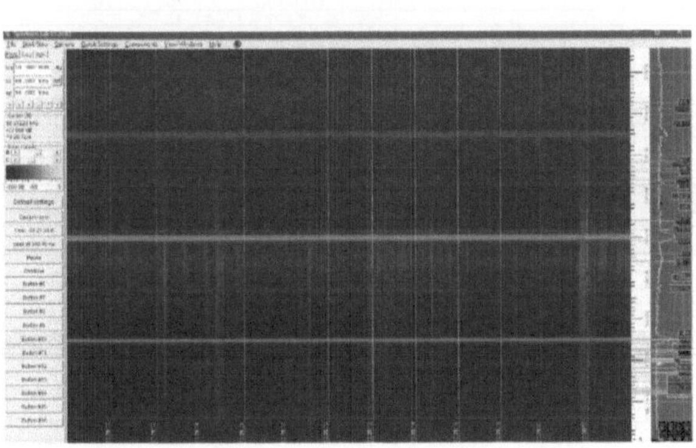

Figure 9. Spectrogram of lightning discharge from a distant location.

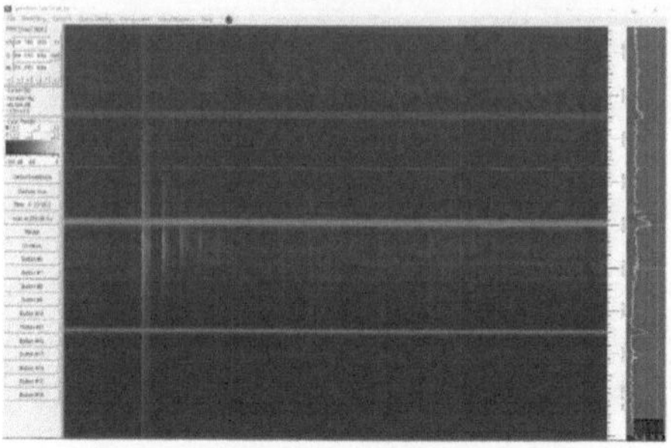

Figure 10. Spectrogram of a lightning discharge that exhibits what looks like a damped wave response.

In all the spectrograms, a constant line is situated at 19.2 kHz. This is a VLF station. Figure 11 below depicts a plot of the received power of only the 19.2kHz signal. The signal power drops abruptly from −62dB to −72dB.

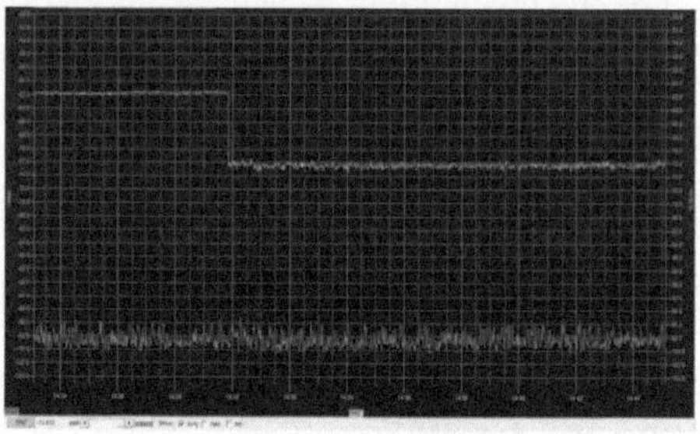

Figure 11. Plot of power of a 19.2kHz VLF signal over time.

3.2 *Final VLF receiver unit test*

The exact setup as seen in Figure 5 was used to obtain the screenshots that follow. Figures 12 to 14 are screenshots of the live (real-time) spectrogram waterfall displayed from the Raspberry Pi. The top part of the window shows a basic spectrum analyzer, and beneath it the spectrogram waterfall. The spectrogram moves from bottom to top, with the frequency situated on the horizontal axis (left to right = 0 to 48kHz), and time situated on the vertical axis. Figures 12, 13 and 14 depict the spectrogram view which are processed and displayed from the Raspberry Pi itself.

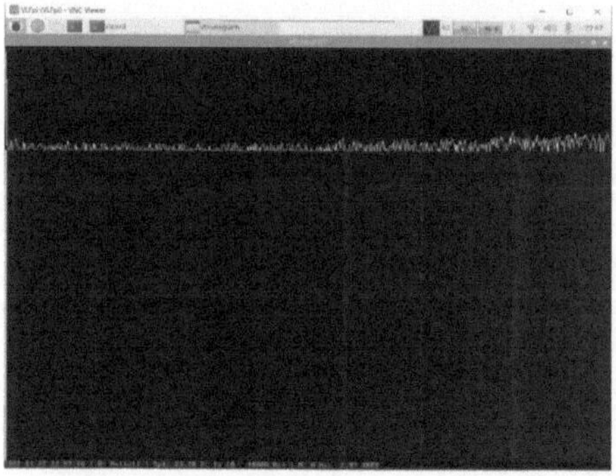

Figure 12. Spectrogram of noise floor when mains power is OFF.

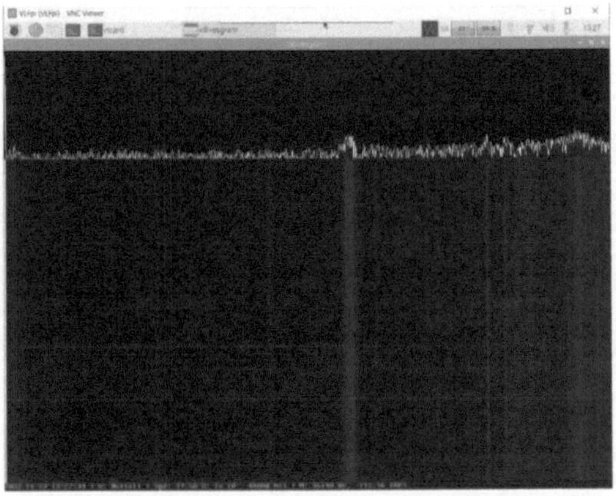

Figure 13. Spectrogram of noise floor when mains power is ON.

Loadshedding, also known as rolling blackouts, are prevalent in South Africa. Figure 12 below shows the background noise when the AC mains power is switched off. Most of the display is dark which corresponds to a very low noise floor due to the absence of the 50Hz power grid and the associated harmonics. The faint signals received are assumed to be artifacts from the clock circuity within the Raspberry Pi and IQaudio Codec Zero soundcard. Figure 13 shows the noise floor while the mains power is on, it can be seen clearly that there is increase in intensity of color which corresponds to an increase in received EM power.

An electric insect swatter was used to create arcs near the antenna. In Figure 14, these arcs were used to simulate lightning discharges. The bright horizontal lines are the detected and displayed arcs. This confirms the correct functioning of the VLF receiver unit.

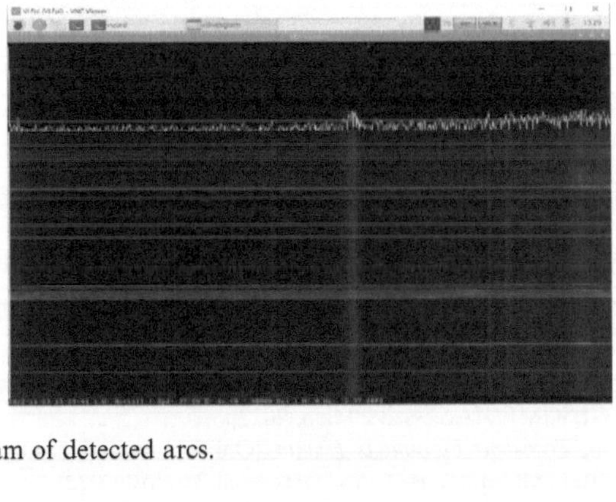

Figure 14. Spectrogram of detected arcs.

The above method was repeated as depicted in Figure 15, but this time a Bash script that captures 5 seconds of data and saves a PNG image of the spectrogram was used. The below image is what was captured. Note: This program sets the horizontal axis as time and the vertical axis as frequency (bottom to top = 0 to 48kHz). The arcs are detected are clearly seen. This test proves that the built VLF receiver functions as intended and will successfully detect lightning.

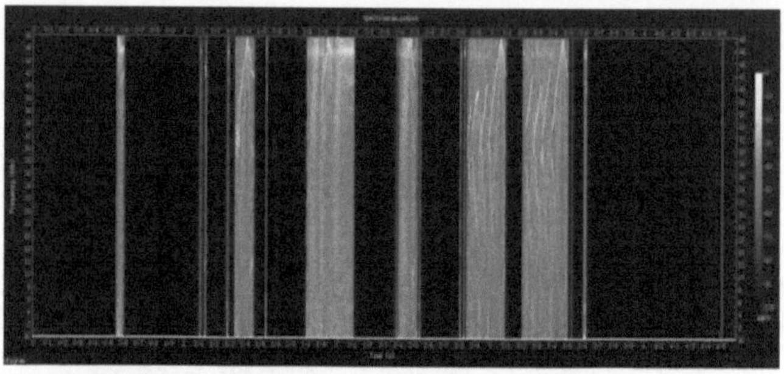

Figure 15. Generated 5 second spectrogram PNG image of detected arcs.

4 CONCLUSION

A lightning detector was successfully designed, fabricated, and tested to function correctly in real world conditions. The VLF receiver is capable of detecting, monitoring, storing and visualizing lightning discharges, and other types of natural and man-made VLF signals. The VLF receiver is cost effective and intuitive to use which makes it available to a wider audience. The device is portable and is easily moved from one location to another if desired. Using both local weather forecasts and the VLF receiver provides early warning of electrical storms in real-time. In the case of insurance claims for damage caused by a lightning strike, spectrograms of local lightning strikes are produced to be used as claim evidence. The VLF receiver enables anyone to easily participate in and contribute to VLF community observations and become involved with VLF research. In its entirety, the VLF receiver will provide meaningful data about the earth-atmosphere environment which can be used in many ways to study the earth, biosphere, atmosphere, and electrical storms. The VLF receiver is an integral tool that exposes the public and upcoming scientists to the fields of geoscience, space science, and remote sensing. With all matters considered, the final artifact is a significant step in assisting in the safety of fellow humans and providing valuable insights about this planet and the environment.

REFERENCES

[1] Stanford VLF Group, *Introduction to VLF*, [Online]. Available: https://vlfstanford.ku.edu.tr/research_topic_inlin/introduction-vlf/. [Accessed 28 Nov. 2022].

[2] Kramer L., *VLFradio.com*, (2022). [Online]. Available: http://www.home.pon.net/785/. [Accessed 13 Oct. 2022].

[3] www.vlf.it, *Radio waves below 22kHz*, (2022). [Online]. Available: http://www.vlf.it/. [Accessed 28 April 2022].

[4] Stanford SOLAR Center, The AWESOME Monitor, (2015). [Online]. Available: http://solar-center.stanford.edu/SID/AWESOME/. [Accessed 29 Sep. 2022].

[5] Stanford VLF Group, *Transient Luminious Events*, [Online]. Available: https://vlfstanford.ku.edu.tr/research_topic_inlin/transient-luminous-events/. [Accessed 28 Nov. 2022].

[6] AARDDVARK, *AARDDVARK homepage*, (2021). [Online]. Available: http://www.physics.otago.ac.nz/space/AARDDVARK_homepage.htm. [Accessed 22 Sep. 2022].

[7] Holzworth P. R., *WWLLN World Wide Lightning Location Network* (wwlln.net), (2022). [Online]. Available: http://wwlln.net/. [Accessed 21 Sep. 2022].

[8] Blitzortung.org, *Project Description*, (2022). [Online]. Available: https://www.blitzortung.org/en/cover_your_area.php. [Accessed 28 April 2022].

[9] Stanford VLF Group, *Stanford VLF Group*, Stanford VLF Group, [Online]. Available: https://vlfstanford.ku.edu.tr/. [Accessed 28 Nov. 2022].

[10] The INSPIRE Project, *INSPIRE VLF-3 RADIO RECEIVERS*, (2022). [Online]. Available: https://theinspireproject.org/default.asp?contentID=1. [Accessed 28 April 2022].

[11] Vaisala, *Superior Lightning Detection, Anywhere*, (2022). [Online]. Available: https://www.vaisala.com/en/digital-and-data-services/lightning. [Accessed 22 Sep. 2022].

[12] Gao Y., Chen X, Hu H., Wen J., Tang J. and Fang G., (2014). Induced electromagnetic field by seismic waves in Earth's magnetic field, *Journal of Geophysical Research: Solid Earth*, vol. 119, no. 7, pp. 5651–5685.

[13] Rault J.-L., Searching for meteor ELF /VLF signatures, WGN, *Journal of the International Meteor Organization*, vol. 38, pp. 67–75, 2010.

[14] DL4YHF, *Spectrum Analyzer with Waterfall Display and Real-time Audio Processing*, [Online]. Available: https://www.qsl.net/dl4yhf/spectra1.html. [Accessed 30 Nov. 2022].

[15] ElectronicBase, *Low Pass Filter Calculator*, [Online]. Available: https://electronicbase.net/low-pass-filter-calculator/. [Accessed 28 Nov. 2022].

[16] ElectronicBase, *High Pass Filter Calculator*, [Online]. Available: https://electronicbase.net/high-pass-filter-calculator/. [Accessed 28 Nov. 2022].

[17] Morgan D., *VLF Receivers for SIDs*, (11 February 2016). [Online]. Available: https://sites.google.com/site/radioastronomydm2/sids/vlf-receivers-for-sids. [Accessed 28 Nov. 2022].

[18] Analog Devices, Low Noise, Precision Operational Amplifier, (2015). [Online]. Available: https://www.analog.com/media/en/technical-documentation/data-sheets/op27.pdf. [Accessed 28 Nov. 2022].

[19] AspenCore, Non-inverting Operational Amplifier, *AspenCore*, (2022). [Online]. Available: https://www.electronics-tutorials.ws/opamp/opamp_3.html. [Accessed 28 November 2022].

[20] Nicholson P., *VLF Receiver Software Toolkit*, [Online]. Available: http://abelian.org/vlfrx-tools/. [Accessed 2022 Nov. 30].

Innovation and Technological Advances for Sustainability – Al-Naemi et al. (Eds)

Gas turbine degradation

Aiyad Gannan
Department of Mechanical Engeering, University of Doha for Science and Technology, Doha, Qatar

ABSTRACT: In this study the gas turbine compressor fouling mechanism and Online/Offline cleaning have been investigated. Gas turbine compressor fouling is a serious issue that can reduce the efficiency and performance of gas turbine engines. The problem occurs when foreign particles, such as dust, dirt, and debris, accumulate on the blades of the compressor, causing a reduction in airflow and an increase in compressor outlet temperature. This can lead to decreased output power, increased fuel consumption, and shortened component life. To mitigate the effects of compressor fouling, regular maintenance and cleaning of the compressor blades is required. Additionally, filtration systems can be installed to prevent foreign particles from entering the engine in the first place. Research is ongoing to develop new technologies and methods for detecting and preventing compressor fouling in gas turbine engines. This study examined the economic impact of compressor fouling on operating costs, as well as the effectiveness of various compressor washing techniques. Through the analysis of real engine data from a gas turbine over a four-month period, it was found that compressor fouling led to a 2.5% decrease in efficiency, a 7% increase in heat rate, and a 10% decrease in power output. The study also includes an analysis of the economic impact of compressor fouling on operating costs, as well as the effectiveness of various compressor washing techniques.

Keywords: gas turbine fouling, degradation, gas turbine compressor cleaning, online cleaning, offline cleaning

1 INTRODUCTION

This work discusses the fouling effect and fouling mechanisms on gas turbine performance and, in particular, on the axial compressor. An analysis of the impact of environmental conditions, and consequent fouling of a stationary gas turbine have been investigated to determine the optimum online compressor wash frequency. Furthermore, analysis of the impact of blade roughness on the compressor efficiency and the pressure ratio as a function of the location of the fouling has been completed.

Typical large gas turbines for power generation can have an inlet mass flow of over 600 kg/s of air and a power output of nearly 300MW. Under on-site conditions this could mean up to 800 m^3/s of air entering the gas turbine, carrying with it fine sand and dust particles, insects, oil, soot and unburned hydrocarbons, even sea water and salt. Some of the larger and stickier particles adhere to the compressor blade surfaces which changes the shape of the blade and its angle of attack. Such an adverse change in the aerodynamics of the system will result in decrease of the mass flow rate and the isentropic compressor efficiency. In turn this will lead to reduced power output and thermal efficiency and an increase in heat rate (turbine temperature) which will reduce the creep life of the turbine blades.

Usually in an attempt to recover these losses of power and reduce the threat to creep life, a compressor cleaning system is used.

DOI: 10.1201/9781003496724-8

2 BACKGROUND

The gas turbine operates using the Brayton thermodynamic cycle and converts the heat generated from combustion into mechanical energy. This process utilizes air as the fluid and is made up of three main stages:

2.1 *Compression*

Air is sucked into the compressor and as it is forced through the compression stage its pressure rises significantly, possibly by a factor of 30.

2.2 *Combustion*

The high-pressure air passes into the combustion stage where injectors inject a steady flow of fuel. After mixing with fuel the combustible mixture moves to the combustion chamber where combustion takes place. The energy from combustion increases both the enthalpy and temperature of the air.

2.3 *Expansion*

The final turbine section transforms the enthalpy of hot products of combustion is into mechanical work by moving the rotor blades and thus driving the shaft. The rotating shaft allows work to be extracted from the engine and is also used to drive the compressor and necessary external equipment.

The massive ingestion of air by a gas turbine means that even if the filters reduced the contamination rate to 3ppm, the turbine could take in – on a daily basis – as much as 4.5kg of foreign material, and this shows why airborne particles such as sand are so important in performance deterioration.

The axial compressors have many advantages but tend to be more complex than other designs with the result that they are very sensitive to the small changes in airfoil shape that can result from fouling.

Performance degradation due to fouling generally occurs because of build-up on blade surfaces changing their aerodynamic shape and roughening the surfaces, and erosion and abrasion of the blade giving an increased tip clearance. However, rain water will be ingested, and in coastal regions saltwater spray. Diakunchak (1991) has estimated compressor fouling is responsible for about 75% of all gas turbine performance loss [1].

Fouling is one the most important of a number of contributors to the degradation of gas turbine performance degradation [2]. This is particularly true of a desert environment where sand "gets everywhere" and mixes with oil vapour before entering the turbine. Figure 1 shows turbine compressor blade fouling that occurred in just such an environment. Fortunately, the deterioration of gas turbine performance deterioration due to fouling can be partially retrieved through washing and thus auxiliary systems have been developed specifically for this purpose.

Figure 1. Heavy fouled compressor blades in the desert environment.

With a fouled compressor a gas turbine is more difficult to start, the peak power is reduced as is the thermal efficiency. Importantly the compressor stall margin is also reduced, which may lead to serious engine damage. Unfortunately, even a badly fouled compressor may not show signs of fouling, so any parameter intended for use as a fouling indicator must be very sensitive to fouling.

It should also be a reliable method which detects fouling sufficiently early for turbine performance to be restored by the relatively simple and direct process of compressor cleaning [3]; and [4].

The full mechanisms of fouling in gas turbine compressors have not yet been adequately investigated.

The washing of the compressor is, of course, performed with the gas turbine off-line. The essence of the procedure is for the turbine to rotate slowly while a detergent solution is sprayed in. After that water is sprayed in to remove the detergent.

3 AIMS AND OBJECTIVES

The aims of this work is to investigate fouling mechanisms within the compressor section of a gas turbine engine,

To achieve these objectives, this research established the following programme:

The simulation and diagnostics of an industrial gas turbine engine using the GT performance code GasTurb 10.

(1) Specification of operating conditions of an industrial gas turbine.
(2) Analysis and simulation of the engine in both clean and degraded cases.
(3) Analysis of ambient temperature effect on GT engine performance.
(4) Investigation of the compressor fouling effect on GT engine performance.
(5) Estimation of turbine blade creep life.
(6) An economic analysis based on increased TET, fuel flow rate and power loss.
(7) less maintenance is performed for sewers unless there is a damage.

4 METHODOLOGY

4.1 *Compressor fouling*

Fouling is a significant issue, characterized by the decline in flow capacity and efficiency due to the buildup of contaminants on the surfaces of the gas turbine engine's airfoils and annuli. This results in the accumulation of unwanted materials which negatively alter the shape of the airfoils and inlet angles, increase surface roughness and decrease the open area of the airfoil throat, leading to a decrease in performance.

Fouling can occur in both the compressor and the turbine, with compressor fouling more common and with more serious effects on engine performance. Typically, between 70% and 85% of all gas turbine engine performance losses accumulated during operation are attributable to compressor fouling [5].

Fouling can be partially reversed by cleaning, but this will require turbine downtime. Unfortunately even when using high effectiveness air filters, fine particles less than 5μm will combine with oil vapour, sea salt, water, etc and adhere as deposits on the compressor blade surface. This changes their aerofoil shape, reduces airflow and increases temperature which could result in over-heating of the blades. Not only is power output reduced but turbine blade life shortened.

Fortunately, compressor washing may restore all or part of the loss in performance. Figure 2 shows the performance degradation as a result of compressor fouling – which begins as soon as the engine is in operation – and the effect of cleaning by washing.

All compressors are susceptible to fouling. The degree and rate of fouling depend on many factors including: type and condition of the air-borne contaminants, the local environmental conditions (e.g. presence of sand, humidity, etc.), compressor design, compressor airfoil loading, and airfoil surface smoothness/coating [6].

Thus fouling rates vary significantly with to site location due to the surrounding environment and climate. Indeed, weather has been shown to have a substantial effect on fouling rate. Fouling not only reduces compressor efficiency, but also reduces the compressor surge margin and may result in compressor surge. [7].

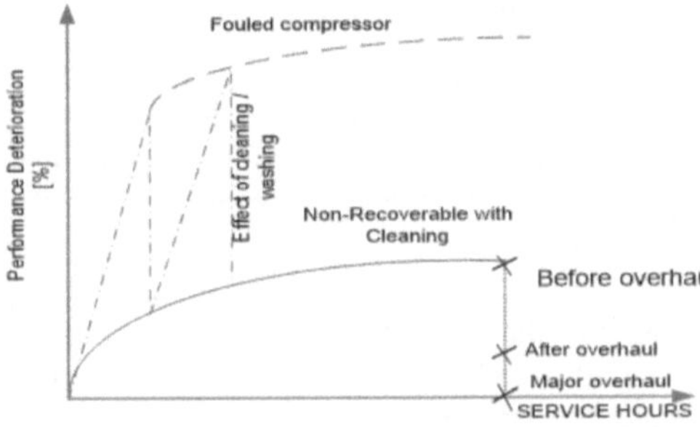

Figure 2. Typical compressor performance deterioration and recovery with cleaning.

4.2 *State of the art in compressor fouling, detection and localization*

Performance degradation of gas turbine engines increases with operating time because the condition of engine components that become fouled, eroded, corroded, etc., will get progressively worse with increasing operating time. Usually fouling is gradual and the consequent reduction in gas turbine performance is also gradual, but there comes a level of fouling when, if the compressor isn't washed, deterioration of performance accelerate [8], [9] and [10].

Three types of engine performance deterioration may be listed:

- Performance deterioration recoverable with cleaning/washing, such as deposits on the blades;
- Performance deterioration which is not-recoverable with cleaning/washing, due to baked-on contaminants that must be physically scraped off the airfoils (during an overhaul); and
- Permanent performance deterioration such as blade leading edge and tip erosion, which is not recoverable even after an overhaul and the refurbishment of all clearances, replacement of damaged parts, etc.

4.3 *Recoverable deterioration*

Performance deterioration due to fouling is mostly recoverable [8,11] and [12]. Recoverable performance loss is due mainly to compressor fouling which reduces flow capacity and compressor efficiency.

Recoverable performance loss can be restored through online or offline compressor washing [13]. It has been reported that after 700 operation hours for a 21.9MW gas turbine engine typically 1% of power loss was fully restored using a crank wash. There are several methods that can be used to clean the flow path components of gas turbine engines without the need for disassembling the engine, such as:

- On-line; Using dry cleaning materials like rice husks or pecan shells on the compressor while the engine is running.

- Online; Cleaning the compressor with water or a water-detergent mixture while the engine is running at full load or partial power.
- Off-line; Soaking the components in a cleaning solution while the engine is turned off.

This last method is particularly effective for removing deposits on hot-end components caused by running on crude oil fuel without opening the engine. Smooth airfoil surfaces or those with coatings that reduce surface roughness are less prone to fouling and respond better to cleaning or washing.

4.4 *The effect of fouling on the compressor performance*

When a single-shaft gas turbine experiences typical fouling in the front stages of the compressor, the percentage decrease in air inlet mass flow is greater than the percentage change in compressor pressure ratio, which is in turn greater than the percentage decrease in compressor adiabatic efficiency. Typically, for any gas turbine experiencing fouling in the front stages, the percentage decrease in airflow is 2–3 times greater than the percentage decrease in efficiency [14] and [15].

The increasing of fuel cost and the more request of electric power on the energy market pushed the owners of power plants to look for more power and reliability from their plant. The gas turbine plants nowadays have the need to control the performance of the engine, to define the performance trend during the normal operation and to try to define the health condition of each component.

In order to avoid failures of power plant component, the gas turbine monitoring became an important topic of research for many companies and universities. The monitoring of compressor condition is important research to increase the reliability of this component and to improve the knowledge of compressor maintenance systems.

The result of fouling is the building up of material on the compressor components, whose effects are:

- Abrasion of rotating surface and material corrosion (if particles are above 10 to 20),
- Wet corrosion, caused by atmospherics interactions between humidity and hydrous acid, salt and other corrosive substances,
- high temperature corrosion, caused by fuel contamination, sulfates, metals and metal oxides,
- Change in the shape and of the inlet angle of the airfoil,
- Increases surface roughness,
- Reduction of mass flow rate, caused by reduction of effective flow area, and then a reduction of mass flow coefficient Φ,
- Variation of air-stream deviation angle,
- Variation of speed profile and thickness of boundary layer,
- Air stream 3D separation.
- In some cases, blade natural frequencies can be affected by the increase in mass due to dirt build up. Excessive dirt on the blades can also cause imbalance, and a consequent increase in running speed vibration.
- Partial blockage of cooling passages of hot section stators and blades can be caused by fine particle size
- Foulants (typically less than 5 μm). Cement dust, coal dust and fly ash can be responsible for this problem

In other words, the effect of compressor fouling is a drop in airflow, compressor ratio and compressor isentropic efficiency, which results in a "re-matching" of the gas turbine and compressor causing a drop in power output and thermal efficiency. In extreme cases, fouling can also result in surge problems, as it tends to move the compressor surge line to the right; i.e. toward the operating line.

4.5 *Compressor cleaning*

Previously GT cleaning was achieved by crank soak washing and/or by the injection of solid substances such as rice husks or nutshells with the unit on-line and running at full speed.

This is known as cleaning by soft erosion, but now with the inclusion of coated compressor blades for further protection against pitting corrosion, wet cleaning is predominantly used. In addition, ashes and solid cleaning substances may block the complex turbine cooling system if they enter the GT air cooling stream.

It is also believed by many users that on-line washing could completely take over from off-line washing. It has been investigated that reduction of inlet airflow by 5% because of fouled compressor blades, reduced the output by 13% coupled with an increase in the heat rate by 5.5%.

4.6 *Compressor washing techniques*

The benefits of compressor cleaning have long been recognized and there have been many research studies published in this field. Crank soak cleaning and the use of solid compounds such as nutshells or rice husks have been replaced by wet on-line cleaning. An important factor in this process is the influence of the time interval between on-line washing events [16].

Compressor fouling has a significant impact on GT operation because the costs of power loss and increased fuel have increased dramatically. To avoid this penalty completely, compressor washing using both on-line and off-line techniques must be employed. This section deals with the different kinds of compressor cleaning and the effect of the various detergents used. In particular, the advantages and disadvantages of these techniques and the best solution of the problem are discussed.

The section is divided into sections as follows:

- Types of compressor cleaning,
- Washing materials,
- Liquid cleaning, and
- Compressor cleaning frequency,

4.7 *Types of compressor cleaning*

Several types of compressor cleaning have been developed over recent years. The initial method required that the engine is disassembled and cleaned manually. This is very time consuming and expensive. However, there are times when this method must be used, in particular in overhauls. However complementary methods are available:

- Abrasive compressor cleaning, and
- Wet compressor washing.

4.7.1 *Off-line crank washing*

Off-line washing can be more closely controlled than on-line washing and so is considered more effective, but is more time intensive and disruptive to operations. With this washing process the GT is turned over at very low speed and detergent injected into the inlet. There then follows a prescribed period called the "soaking period" after which there is a DI water rinse (Figure 3).

The cleaner is mainly surfactant (a wetting agent) which reduces the surface tension of the cleaning solution enabling it to wet, penetrate and disperse unwanted deposits. The water injected into the GT drains off the blades rapidly, reducing the contact time during the off-line soaking period.

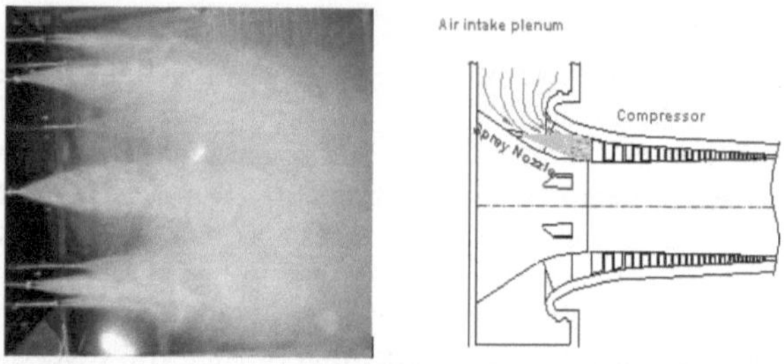

Figure 3. Cleaning fluid spraying through the air intake.

4.7.2 *On-line washing*

During part-load operation, DI water and/or cleaning solution is injected into the gas turbine at the inlet (as shown in Figure 4). This method is less disruptive and requires less time than offline crank washing, as it requires little to no cleanup. However, it is less effective than crank washing as the increased rotational speed causes the solution to be spread to the casing surface after just a few stages, which is more similar to a "steam cleaning" effect [17] and [18].

Figure 4. Spray nozzles online-offline washing.

Frequent online water washing can greatly reduce the rate of fouling, but more significant improvement in performance can be achieved through offline crank washing. However, it should be noted that no matter how effective the washing is, the rear stages of the compressor will not be cleaned, and manual cleaning or overhaul is necessary in those cases [17].

5 RESULT AND DISCUSSING

5.1 *Gas turbine performance simulation using GT performance code GasTurb 10*

5.1.1 *Introduction of the engine working environment*

In hostile environments the compressor component of a gas turbine is susceptible to fouling through deposits of oil and fine sand particles that pass through the filter into the compressor and possibly combined with salt. The combination over time decreases the compressor efficiency increasing compressor power demand and reducing output power demand proportionately.

The power loss can be as much as 10% over a 12 month period of the engine operating without cleaning. This fact is increased in hot environment typically in the Middle East where the ambient temperature riches 50 degree Celsius. The only ways to recover power loss are to clean the compressor regularly or increase the firing temperature.

The impact of increasing firing temperature is to increase fuel flow and reducing the turbine blade creep life. The reduction in creep life is more significant than fuel flow increase.

Indeed typically of 25 degree which may recover the full power will half the life of high pressure turbine rotor. This cost the primary reason for the compressor cleaning, whilst the additional benefits of fuel flow cost offset this to some extent (Figure 5). Through this work a more accurate prediction of axial compressor fouling has been produced and aligned to observations seen in the field (GT Sulzer type 7).

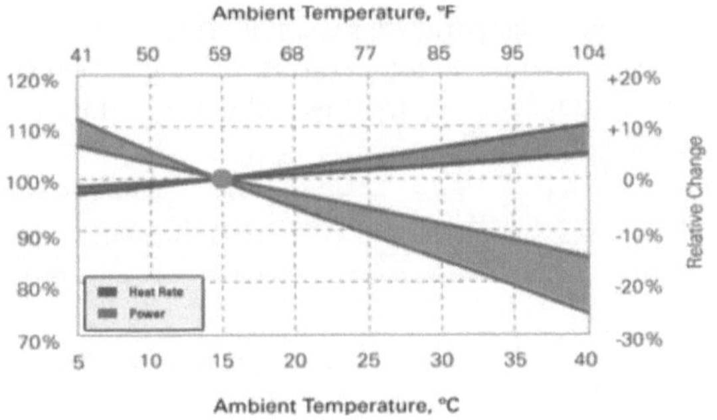

Figure 5. Ambient temperature effect on gas turbine performance.

5.2 *Influence of ambient temperature on the engine performance*

The variation of ambient temperature has a significant effect on the change of gas turbine performance as shown in Figure 6. A different simulation of 11MW engine (Sulzer Type 7) has been studied with the GT Performance Code.

This study shows as the ambient temperature increases to 45°C the power output decreases dramatically by approximately 25% and the heat rate increases by 10%, where the turbine inlet temperature remaining constant.

The variation of the ambient temperature in the engine operating location (desert environment) is shown in Figure 6 over the whole year.

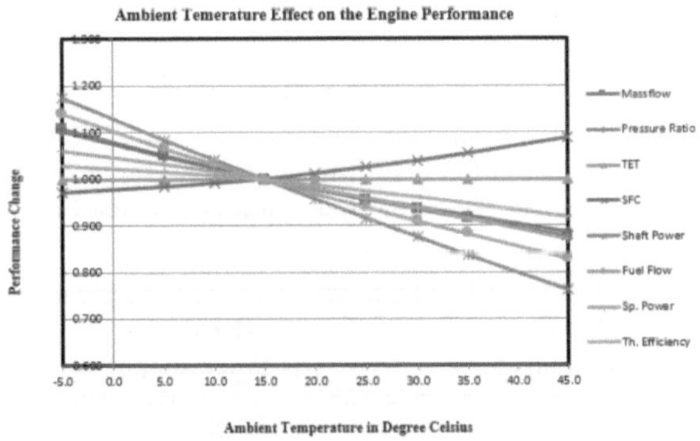

Figure 6. Effect of ambient temperature on GT9-Sulzer type 7 performance.

From Figure 6 the follows can be observed:

- The mass flow falls quickly by approximately 12%.
- The pressure drops by approximately 11 % as a result of the reduction in mass flow.
- The SFC increases slightly from to 10%
- The shaft power drops faster around 25%

There are two likely ways to react the changes in ambient temperature, by continuing the same power output, or by increasing the TET by increase the fuel flow in order to recompense the mass flow losses, that's occurs due to the reduction in the air density which decreases as the ambient temperature increase. The compressor needs more power from the turbine to compress the hot air, the fuel consumption increase and the thermal efficiency drops as a result of the reduction in power output.

All the oil fields are located in the desert environment, which is the weather is hot and dry during the summer with ambient temperature reaching around 45°C to 50°C sometimes and very cold during the winter with ambient temperature varying between 50°C to 10°C during the nights see (Figure 7). In addition, the environment is very sandy for several months in a year, especially from January to April.

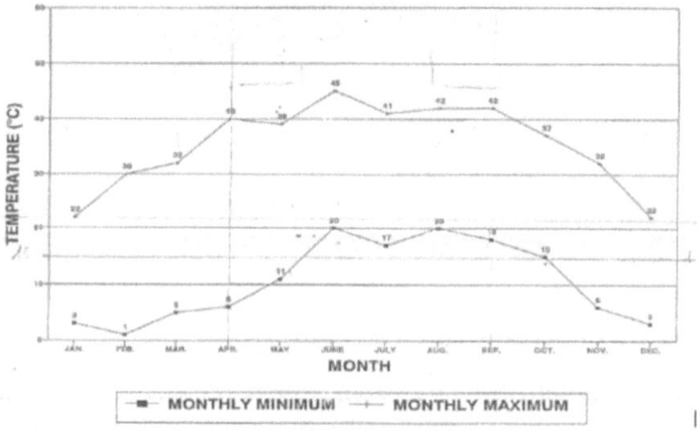

Figure 7. Ambient temperature variation in a desert environment.

These environmental conditions have a great influence on gas turbine engine performance in two ways.

Firstly, the change in ambient temperature and secondly the dust and sand that leads to severe compressor fouling, corrosion and erosion. In addition, the atmosphere is laden with hydrocarbon particles from neighbouring plants.

5.3 *Field data analysis*

The data of the gas turbine engine (11MW) has been collected from the power station in oil field. An actual data reading of power outlet, temperatures and pressures were taken and compared with the simulation data of GT Performance code. It has been demonstrated that the results of the code are closer to the actual field data (see Table 1).

Table 1. Power output compression.

Ambient Temp. [C]	Field Data Power output	GT Code Power output	Deviation in %
15	10.24	10.40	1.56
30	8.89	8.61	−3.31
45	7.75	7.46	−3.67
50	7.37	6.95	−5.68

5.5 *Observation of exhaust gas temperature*

As it has been discussed in the previous section the indication of compressor fouling observed with the increase of exhaust gas temperature (EGT), increase of turbine entry temperature (TET) and decrease in combustion chamber pressure (CCP) see (Figure 8).

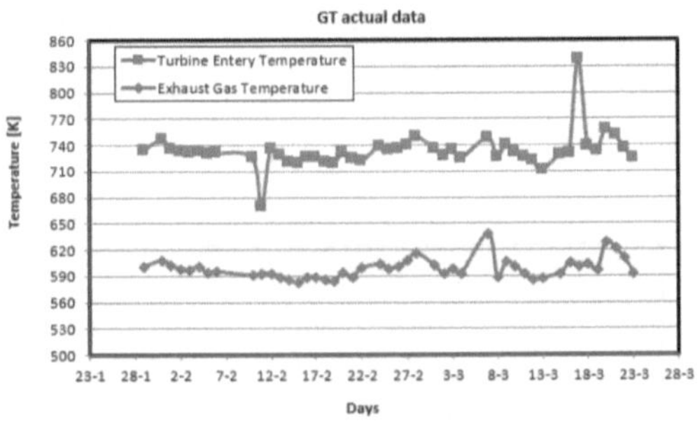

Figure 8. Actual reading of the GT temperature variation.

5.6 *Simulation of clean and fouled gas turbine*

Gas turbine engine performance is extremely sensitive to the degradation of the components such as the compressor, combustor and turbine. The performance starts to deteriorate as soon as the engine is started for the first time, the inlet mass flow and the isentropic efficiency will degraded and is simulated separately. The representation of component degradation cases in Table 2 implanted in the GT Perfromance code to predict the effect of the compressor fouling and erosion on engine performance.

Table 2. compressor implanted degradation.

Physical Fault	NDFM	EFF	Ratio	Range
Compressor fouling	Drop	Drop	(1.0-0.5) %	0.0-(-5.0) %
Compressor EFF	–	Drop	–	0.0-(-2.5) %
Compressor erosion	Drop	Drop	(1.0-0.5) %	0.0-(-5.0) %

5.6.1 *Compressor isentropic efficiency degradation*

The Figure 9 displays the effect of the deterioration in compressor isentropic efficiency and the mass flow in the Ambient condition, the inlet mass flow drops by approximately 5%, which results in increased TET about 5.9% and reduces the thermal efficiency by approximately 6%. In the other hand the specific fuel consumption (SFC) has been increased approximately 6.5%.

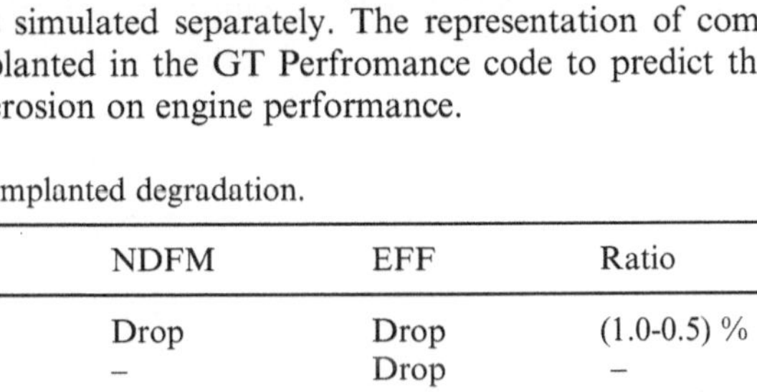

Figure 9. Compressor efficiency and mass flow deterioration.

5.6.2 *Degradation performance at fixed power output*

The assumed operation mode in this work that the power output of the Gas Turbine cycle is maintained constant.

This means that as the compressor inlet mass flow is reduced, the firing temperature must be increased to the given power outlet. As shown in Figure 10 the turbine entry temperature (TET) increases around 10% at ambient temperature 15°C than increases rapidly with increase of ambient temperature.

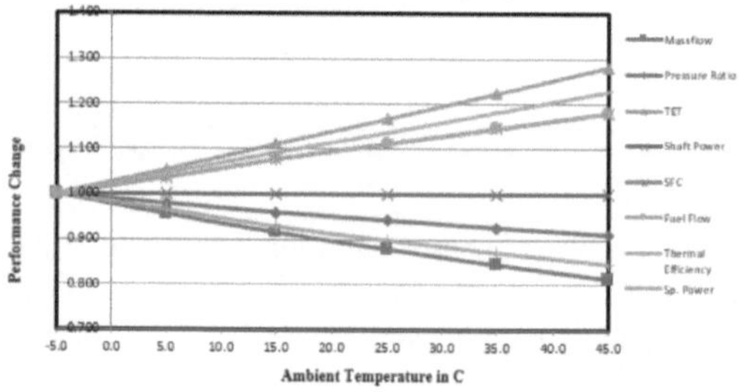

Figure 10. Compressor efficiency degradation at different ambient temperature.

The Non-Dimensional mass flow (NDMF) deterioration has a strong effect on the engine performance especially the thermal efficiency which drops by about 8% as can be seen in Figure 11.

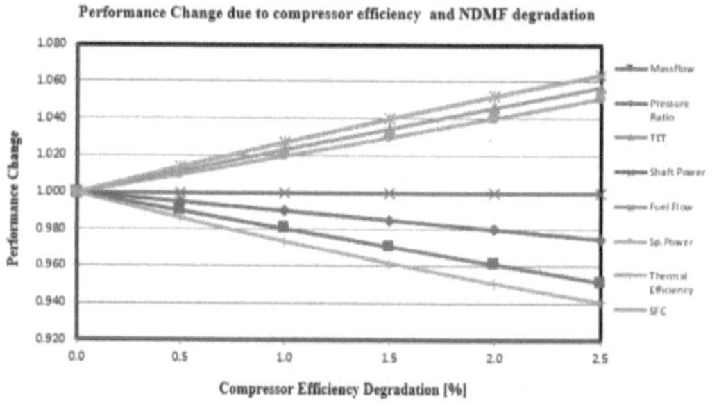

Figure 11. Non-dimensional mass flow deterioration.

5.6.3 *Degradation performance at fixed turbine entry temperature*

Figure 12 shows the variation of the gas turbine performance at constant turbine entry temperature (TET) with 2.5% compressor efficiency deterioration. It has been clearly illustrated that the power output decreases rapidly by increasing the ambient temperature. at 25°C power output drops 25% and the specific fuel consumption rises by 10%.

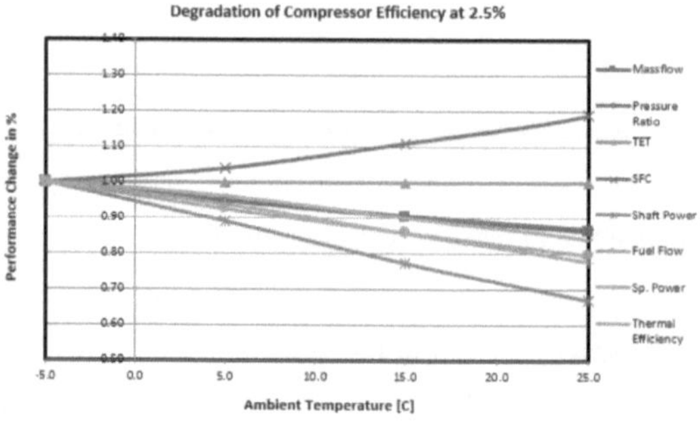

Figure 12. Performance deterioration by compressor fouling.

5.6.4 *Clean engine performance*

As it has been illustrated in Figure 13 the increase of ambient temperature will increase the specific fuel consumption (SFC). The specific power raises with increase of turbine entry temperature (TET) and this cause of drop in specific fuel consumption.

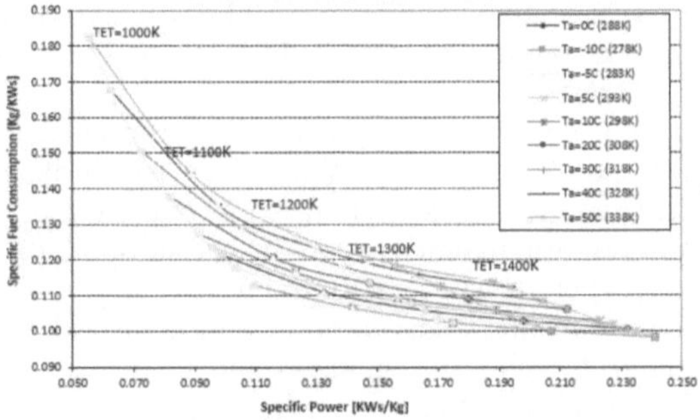

Figure 13. The effect of ambient temperature on the engine performance.

Figure 14 shows clearly the increase of the shaft power with increasing the pressure ratio and as the ambient temperature drops the power output will increase to reach the maximum power at the lowest ambient temperature.

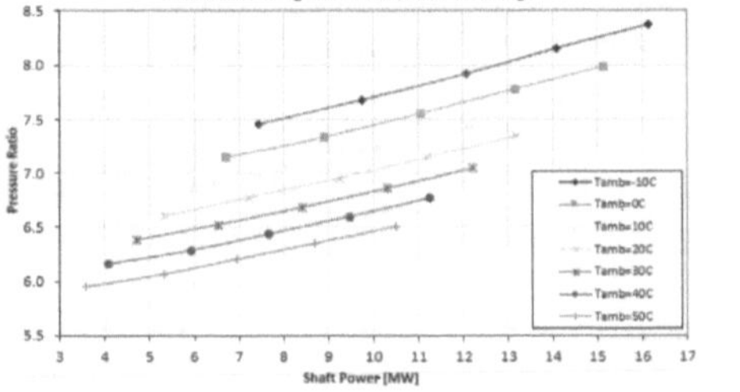

Figure 14. Increasing of power with the pressure ratio.

5.6.5 *Compression between the clean and fouled engine*

The performance changes of clean and fouled engine at compressor efficiency degradation of 3% are illustrated in Figure 15. As shown in Figure 15 the degraded engine consumes more fuel then clean engine therefore compressor cleaning is important to keep the engine running at lower cost.

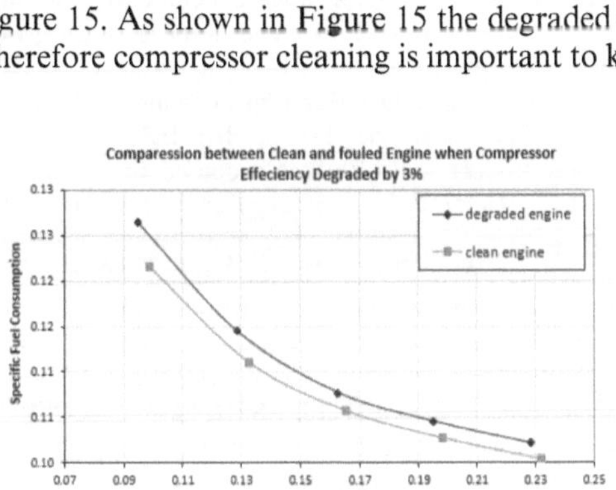

Figure 15. Compression between clean and fouled engine sulzer type 7.

6 CONCLUSIONS

The most important mechanism for degradation has been discussed in this research to show the variety of possible underlying physical reasons. The design point was first simulated successfully using a well-established performance code. The results are very close to those available to the user in the field. Off-design engine performance is carried out over a variety of ambient temperatures. It is demonstrated, as is well known, that at a given turbine entry temperature (TET) the power output decreases significantly and heat rate increases slightly as ambient temperature increases.

In addition, simulation of compressor degradation is examined. It is concluded that compressor cleaning leads to dramatic improvements in gas turbine performance since power loss can be recovered and fuel consumption can be reduced more significantly than is commonly assumed. Furthermore, maintenance costs are substantially reduced with compressor cleaning, especially when the engine is shut down.

An important contribution in this study is that actual industrial engine data is used to enhance the understanding of fouling mechanisms and to indicate the true variation of compressor fouling with the time. So often, in reality, the gas turbine user when deciding on whether to install cleaning equipment bases conclusions on unsubstantiated claims as to benefits by the vendor.

This work assists the user in improving his confidence in the out come of such purchases. This is now possible because the work reported here is based on the field observations. The field data used included daily readings of engine temperatures and pressures a period of time simultaneously with observations of exhaust gas temperature increases. Knowledge of the latter facilitates an estimation of firing temperature. This then has allowed the changing fouled state of the compressor to be quantitatively identified over a period of time. In consequence, the resulting increase in fuel consumption at a fixed power demand has been identified.

REFERENCES

[1] Diakunchak, I. S. (1991). Performance deterioration in industrial gas turbines. *Proceeding form the 1991 International Gas Turbine and Aeroengine Congress and Exposition.* June 1991, ASME Paper No. 91-GT-228.

[2] Boyce, M. P. (1994). Modeling and analysis of gas turbine performance deterioration. *Journal of Engineering for Gas Turbines and Power*, ASME.

[3] Boyce, M. P. (2007). A study of on-line and off-line turbine washing to optimize the operation of a gas turbine. *Journal of Engineering for Gas Turbines and Power*, ASME. Vol 129, pp. 114–122.

[4] Saravanamuttoo, H., and Seddigh, F. (1990). Proposed method for assessing the susceptibility of axial compressors to fouling. *International Gas Turbine and Aeroengine Congress and Exposition.* Brussels: ASME Paper #GT-348.

[5] Diakunchak, I. S. (1992). Performance Deterioration in Industrial Gas Turbines. ASME *Journal of Engineering for Gas Turbines and Power*, Vol. 114, pp. 161–168.

[6] Seddigh, F., and Saravanamuttoo, H. I. H. (1991). A proposed method for assessing the susceptibility of axial compressors to fouling. *ASME Journal of Engineering for Gas Turbines and Power*, Vol. 113, pp. 595–601.

[7] Mezheritsky, A. D., and Sudarev, A. V. (1990). The mechanism of fouling and the cleaning technique in application to flow parts of the power generation plant compressors. *Proceedings from the 1990 International Gas Turbine and Aeroengine Congress and Exposition.* ASME Paper No. 90-GT-103.

[8] Boyce et al. (1994). Modeling and analysis of gas turbine performance deterioration. *Journal of Engineering for Gas Turbines and Power*, vol. 116, ASME Paper 92-GT-395. 7.

[9] Meher-Homji, C. B. (1990). Gas Turbine Axial Compressor Fouling – A Unified Treatment of its Effects, Detection, and Control. ASME IGTI-Vol. 5.

[10] Stalder, J. P. (1998). Gas turbine compressor washing state of the art – field experiences. *Proceedings from the 1998 International Gas Turbine & Aeroengine Congress and Exhibition.* ASME Paper No. 98-GT-420.
[11] Boyce, M. e. (1993). *Condition Monitoring and Diagnostic Approaches for Advanced Gas Turbines.* IGTI Publication, vol. 8, Cogen-Turbo Conference 1993.
[12] Kurz, R. and Brun, K. (2000). *Degradation in Gas Turbine Systems.* ASME Paper No. 2000-GT-345.
[13] Peltier, R. V. (1995). LM2500 Recoverable and Non-Recoverable Power Loss. *Proceedings from the 1995 ASME Cogen-Turbo Power Conference.* ASME Paper No. 95-CTP-104.
[14] Boyce, M. e. (1993). *Condition Monitoring and Diagnostic Approaches for Advanced Gas Turbines.* IGTI Publication, vol. 8, Cogen-Turbo Conference 1993.
[15] Saravanamuttoo, H.I.H.; Muir, D. E. and Marshall, D. J. (1988). Health monitoring of variable geometry gas turbines for the canadian navy. *International Gas Turbine and Aeroengine Congress.* Amsterdam, Netherlands: ASME Paper # 88-GT-77.
[16] Stalder, J.-P. (2001). Gas turbine Compressor Washing State of the Art: Field experience. *Journal of Engineering for gas Turbines and Power, ASME J. Eng. Gas Turbines Power*, 123, pp. 363–370.
[17] Haub, B. L. (1990). Field evaluation of the on-line compressor cleaning in heavy duty industrial gas turbines. *Proceedings from the 1990 International Gas Turbine and Aeroengine Congress and Exposition.*
[18] Stalder, J. P. (1998). Gas turbine compressor washing state of the art – field experiences. *Proceedings from the 1998 International Gas Turbine & Aeroengine Congress and Exhibition.* ASME Paper No. 98-GT-420.

Innovation and Technological Advances for Sustainability – Al-Naemi et al. (Eds)

Machine learning techniques for forecasting the effect of climate change and anthropogenic pressures on coastal wetlands (Ichkeul Lake, Ramsar site)

Sabrine Sahbani
National Institute of Agronomy of Tunisia, Tunis, Tunisia

Ennio Ottaviani
OnAIR srl, Genoa, Italy

Hechmi Missaoui & Béchir Béjaoui
National Institute of Marine Sciences and Technology, La Goulette, Tunisia

Eva Riccomagno
University of Genova, Genova, Italy

Sihem Benabdallah
Center for Water Research and Technology, Borj Cedria, Tunisia

Enrico Prampolini
OnAIR srl, Genoa, Italy

ABSTRACT: The present study aims to forecast the effect of climate change and anthropogenic pressures on Tunisia's Ichkeul Lake. This lake is among the most productive ecosystems in the Mediterranean and is particularly important for the non-migratory European eel, which serves as an excellent bioindicator of the ecosystem's health.

Random Forest (RF) and Cubist (CB) approaches were developed for this purpose using a dataset from the long-term (2010–2020) environmental and biological monitoring of Lake Ichkeul.

The results revealed pronounced seasonal variability of the environmental parameters. While the models' outcomes highlighted their best predictive performance and revealed that the most important predictors of eel landings appear to be, salinity, turbidity, and water level. These results are consistent with the relationships found with Pearson correlation.

Our results proved that these variables should always be included in management measures and can be widely extended to other Tunisian coastal ecosystems.

This study has also proven that the sustainable approach, which combines the two models, can be used in decision making by civil authorities and other interested stakeholders.

Keywords: Anguilla Anguilla, Ichkeul Lake, Climate Change effects, random forest, Cubist model, cross validation

1 INTRODUCTION

Coastal ecosystems are renowned for their rich biodiversity and biogeochemical processes, making them one of the world's most valuable environments [1]. However, the rapid growth of human activities poses the most significant threat to the ecological quality of these ecosystems [1]. Despite being identified as some of the world's primary biodiversity hotspots, Mediterranean ecosystems are not immune to unfavorable biological diversity, and they are

DOI: 10.1201/9781003496724-9

currently experiencing a sharp decline [2]. In North Africa, specifically in Tunisia, ecosystems are vulnerable to water quality degradation caused by industrial development, tourism activities, and agriculture, further exacerbating the already challenging situation [3].

Predicting the impact of climate change on ecosystems is a complex process due to the interplay among various factors [4]. Therefore, numerous models have been developed to address this issue, including the Random Forest (RF) and Cubist (CB) techniques, which have recently gained popularity in marine ecology [5].

In this study, we propose that the diadromous species, the European eel, serves as a pertinent bioindicator to evaluate the ecological condition of the ecosystem. We utilized an extensive collection of data to construct predictive models that estimate the effects of climate change on the coastal area using both the Random Forest (RF) and Cubist (CB) models. The study has three primary objectives: (i) to characterize the seasonal changes in the Ichkeul Lake ecosystem; (ii) to develop two distinct machine learning models to assess the impact of climatic and anthropogenic pressures on non-migratory eel landings by ranking the importance of the variables that affect the species' presence, and (iii) to evaluate the performance of the RF and CB models by comparing their respective results.

2 MATERIALS AND METHODS

2.1 *Study area*

Ichkeul Lake is a vital ecosystem located in the northeastern part of Tunisia (as shown in Figure 1), surrounded by a temporary swamp and bounded to the south by the mountain Jebel Ichkeul. The hydrological regime of the lake is controlled by freshwater from precipitation and the watershed area of 2600 km^2, as well as by marine water from Bizerte Lagoon. The lake is fed by six rivers that have been gradually blocked by dams since the 1990s, affecting the lake's hydrology [6].

Due to its unique hydrology and ecological significance, Ichkeul Lake has been designated as a UNESCO Biosphere Reserve, UNESCO World Heritage Element, Wetland of International Importance under the RAMSAR Convention, National Park of Tunisia, and an Important Wetland for Bird Conservation (ZICO) [7]. Despite its protected status, the lake has been subjected to significant anthropogenic pressures such as damming of main rivers, agricultural intensification, and urbanization, which have adversely affected its ecological health [7]. As a result, it is imperative to implement sustainable management practices that promote the conservation of this valuable ecosystem for future generations.

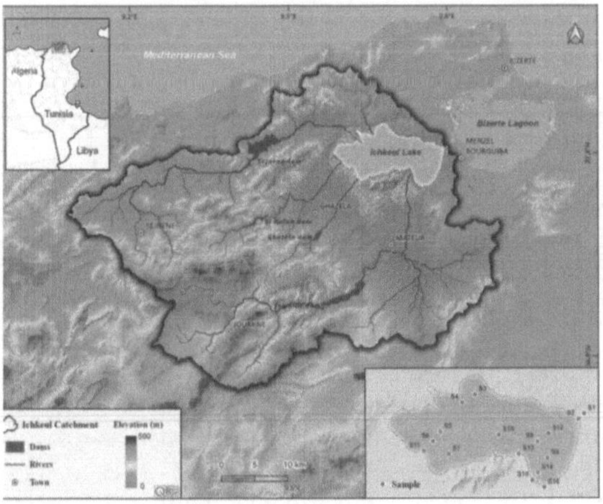

Figure 1. Geographical location of Lake Ichkeul and sampling stations.

2.2 *Data collection*

The dataset used in the present study consists of 13 environmental variables at 16 sites for the period 2010–2020 (Figure 1). It was collected from different sources; including the DGPA (General Direction of Fisheries and Aquaculture, Tunisia), BASSIANA database platform, which is performed within the framework of the IMAS-Ichkeul project, and a sampling campaign conducted in 2020. The sampling points were selected at the mouths of the tributary rivers, covering 16 sites in the lake.

In-situ measurements were performed using a WTW multiparameter (Multi 340i) to obtain physico-chemical parameters. Water samples from the surface were analyzed for chemical variables using an Autoanalyser [8]. The spectrophotometric method [9], as described by [10], was used to measure chlorophyll a (Chl.a) concentrations.

2.3 *Data analysis*

The dataset used in the present study is a combination of data from various sources, leading to insufficient information on some parameters and resulting in 5% missing values. To avoid the adverse negative effect of any missing information [11], we statistically imputed the missing values (NA) with a machine learning algorithm called "missForest" [12]. This method is based on prediction and estimation from the non-missing value available from the original dataset [12]. The performance of this algorithm is assessed using the normalized root mean squared error NRMSE [12]. This indicator calculates the difference between the actual data (measured values) and the imputed data (missing values). Considering that the mean and the variance are calculated on missing data only, the NRMSE is expressed as follows:

$$NRMSE = \sqrt{\frac{mean\left((X^{true} - X^{imp})^2\right)}{\mathrm{var}(X^{true})}} \tag{1}$$

Where X^{true} is the measured data and X^{imp} is the imputed data. When the estimation is accurate, NRMSE approaches its minimum value 0. While when the estimation is too poor or when the noise involved is too large, NRMSE approaches a value of 1.

The derived dataset consists of 12 predictors and 1 response variable. To conform to the requirements of normality when using regression models, the target variable was transformed using the Box-Cox transformation [11].

Additionally, to assess the underlying patterns of eel distribution and environmental variables, Pearson correlation between variables and box and violin plots were performed.

2.4 Machine *learning algorithms*

2.4.1 *Random Forest (RF) model*

The RF model is a statistical technique introduced by Breiman [13], and it is a popular algorithm that aims primarily to categorization by classification, then to obtain a consensus prediction by regression [13]. RF combine several decisions trees to achieve a reliable regression. In general, two-thirds of the data are used as a training data, while one-third is the out-of-bag OOB, used for testing set. An interest feature of RF is the use of OOB to evaluate the model and to estimate the OOB error [13].

2.4.2 *Cubist (CB) model*

CB model is developed by Quinlan and is the extension of the M5 tree model [14]. Cubist is a regression tree-based model, in which the trees are reformulated as rules [14], ending in a multivariate linear model relevant to each leaf. The set of "if-then" rules are pruned or possibly combined for simplification [15]. This model is easier to interpret than RF because it has access to the expressions that define the rules sets and to the regression equations fitted to each rule set [16].

2.4.3 *Optimizing the hyperparameters of RF and CB models*
To improve the accuracy of our models and avoid overfitting, we utilized hyperparameter tuning and k-fold cross-validation methods.

For the RF model, the tuning parameters were the *node size*, which protects against overfitting, and the m_{try}, which determines the number of randomly selected predictors at each split [13]. Meanwhile, for the CB model, the tuning parameters were the number of rule sets (committees) and the number of neighbors used to amend the rule-based prediction [14].

The optimum hyperparameters were obtained using a 3-fold cross-validation procedure on different training sets. All data processing and analysis were conducted using the R software [17].

To evaluate the performance of the RF and CB models, we calculated the coefficient of determination (R^2), which reflects the degree of fit between observed and predicted values [18,19], and it defines as:

$$R^2 = \frac{\sum_{i=1}^{n}\left(Y_i^{obs} - Y^{-obs}\right)^2 - \sum_{i=1}^{n}\left(Y_i^{obs} - Y_i^{pred}\right)^2}{\sum_{i=1}^{n}\left(Y_i^{obs} - Y^{-obs}\right)^2} \in [0, 1] \tag{2}$$

Where n is the total number of data, and are the predicted and the measured eels landing of i observation, respectively, and Y^{-obs} is the mean of all observed responses.

3 RESULTS

3.1 *Stattistical analysis*

The missForest algorithm was successfully used in the present study to effectively deal with missing values, and it provided an NRMSE of 0.22, indicating a sufficiently good performance. Correlation analysis using Pearson correlation coefficient (Figure 2) revealed that eel landing was positively correlated with water level (r = 0.64) and turbidity (r = 0.55), while negatively correlated with salinity (r = –0.51) and chlorophyll a (r = –0.42).

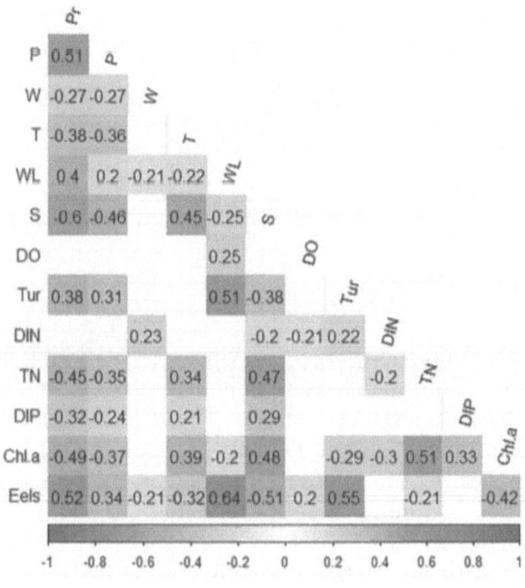

Figure 2. Lower triangular correlation matrix of the predictors and Eels landing in Ichkeul Lake.

3.2 *Parameter's properties*

Table 1 provides a comprehensive overview of the seasonal variability of the environmental parameters of Ichkeul Lake. The meteorological parameters were of particular interest, and it was found that the mean annual precipitation for the period 2010–2020 was 59.28 mm. The rainfall was unevenly distributed throughout the year, with only 9.42% occurring during the dry period and the remaining 90.57% occurring during the wet period. Moreover, the mean wind intensity during the dry and wet periods was approximately 6.01 m. s^{-1} and 5.22 m. s^{-1}, respectively. It was observed that the dominant wind direction was originating from the north-western side.

Regarding the physico-chemical parameters, the water level ranged from 44.29 cm during the dry season to 67.44 cm during the wet season. The average annual water temperature was 21.11°C, with a maximum of 28.60°C and a minimum of 12.76°C. The salinity level showed homogenous distribution with a mean of 44.68 psu and 21.42 psu during the dry and wet seasons, respectively. The mean annual dissolved oxygen levels were approximately 7 mg. l^{-1}, and mean turbidity varied between 18.90 NTU and 27.34 NTU.

The chemical parameters in the lake showed a clear seasonal trend, with two episodes during the year: high levels during the wet period and low concentrations during the dry period. Overall, the chemical parameters in the lake were mainly high in the estuaries of rivers.

The chlorophyll-a concentration, which approximates the phytoplankton biomass, ranged from 6.80 $\mu g.\ l^{-1}$ to 3.36 $\mu g.\ l^{-1}$. While the highest quantity of European eels in the lake was found during the period between November and March, with a peak in December (45 tons).

Table 1. Summary of environmental parameters in Ichkeul Lake for the period 2010–2020.

Parameter	Abbreviation and Unit	Mean	
Period	Pr	Dry	Wet
Precipitation	P (mm)	15.96	82.61
Wind Intensity	W ($m.s^{-1}$)	6.01	5.22
Temperature	T (°C)	21.11	15.39
Water Level	WL (cm)	44.29	67.44
Salinity	S (psu)	44.68	21.42
Dissolved Oxygen	DO ($mg.\ l^{-1}$)	7.29	7.13
Turbidity	Tur (NTU)	18.9	27.34
Total Nitrogen	DIN (μM)	22.05	29.46
Dissolved Inorganic Nitrogen	TN (μM)	22.66	20.97
Dissolved Inorganic Phosphorus	DIP (μM)	2.08	1.06
Chlorophyll a	Chl.a ($\mu g.\ l^{-1}$)	6.8	3.36
Eels landing	Eels (tons)	16.84	38.34

3.3 *Results of hyper-parameters tuning*

The input parameters (i.e., meteorological, physicochemical, and chemical parameters) are the variables that affect the prediction target (European eels landing) in both the RF and CB models. In supervised regression, the dataset is divided into training set and test set. In this study, a dataset of 100 observations was randomly split into training set (80 observations) and testing set (20 observations) using the R package "Caret" [20].

For the RF model, the tuning of hyperparameters was performed using the OOB procedure, which resulted in the best tuning values of m_{try} equal to 3 and *min node size* equal to 3, as depicted in Figure 3. To assess the model's performance, the RMSE value was used, and the simplified diagram of the cross-validated RMSE for the RF model is shown in Figure 3. Lower RMSE values are generally indicative of better-optimized models.

The CB model was implemented by fitting the two hyperparameters, committees, and neighbors. The relationship between committees and neighbors is illustrated in Figure 3. To cross-validate the model, the "repeatedcv" function of the trcontrol function was used. Finally, the optimal values for the cubist regression model were *committees* equal to 10 and *neighbors* (or *instances*) equal to 9 (Figure 3).

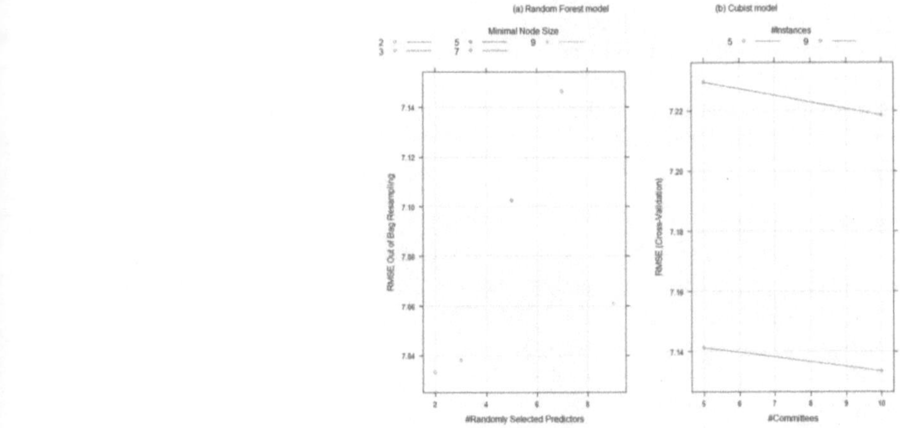

Figure 3. Five-fold cross-validated RMSE profiles for determining the optimal tuning parameters for RF (a) and CB (b) models.

The R^2 values obtained for the RF and CB models were 0.55 and 0.56, respectively, indicating that both models have moderate predictive power. Figure 4 provides a visual representation of the performance of the models through actual vs. predicted regression plots. The plots show that the predicted values of both models generally follow the trend of the actual values, with some deviations at higher or lower values. This indicates that the models may have some degree of error in predicting the eel landing values, but overall, they provide a reasonable estimation of the target variable. It is worth noting that the CB model seems to have slightly better performance compared to the RF model, as indicated by its slightly higher R^2 value and the slightly closer fit of the predicted values to the actual values in the regression plot.

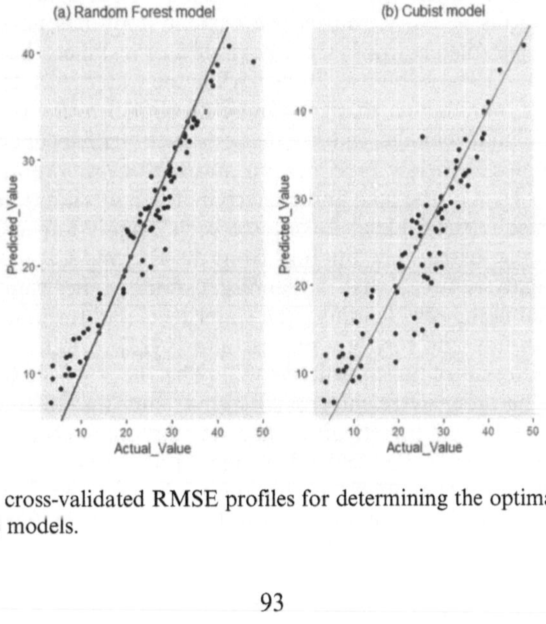

Figure 4. Three-fold cross-validated RMSE profiles for determining the optimal tuning parameters for RF (a) and CB (b) models.

3.4 *Variables importance and interpretation*

Using 12 indicators and a response variable for the period 2010–2020, the results of RF and CB models are presented in Figure 5, showing the ranking of descriptors.

The salinity, turbidity and water level were the most important predictors in the final RF model, followed by chlorophyll a, dissolved oxygen and nutrients. Similarly, in the CB model, salinity, water level, and turbidity were the most inferential parameters, followed by temperature, chlorophyll a, nutrients, and dissolved oxygen.

The strong relationship between eel's biomass and water level, salinity, turbidity, and chlorophyll a was also shown by Pearson correlation result (Figure 2). Thus, Pearson correlation quantitatively confirmed the outcomes of the RF and CB models. Interestingly, the precipitation did not contribute to the performance of the proposed models.

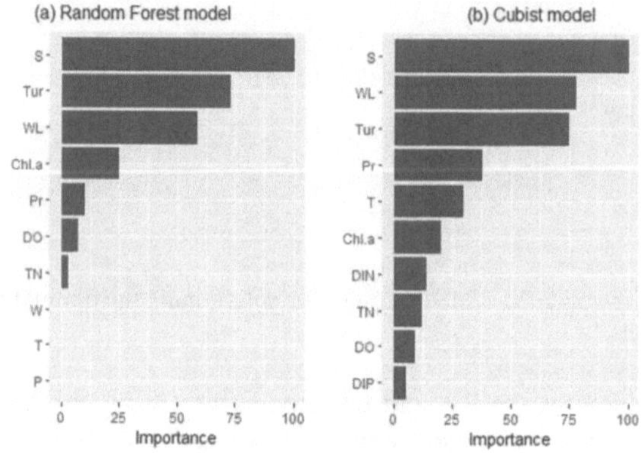

Figure 5. Variable importance scores for the predictors in the proposed models for Eels Landing: (a) RF model, and (b) Cubist model.

4 DISCUSSIONS

The analysis of data gathered from the Ichkeul ecosystem reveals that the region exhibits seasonal variations in its environmental factors, which are similar to those found in typical Mediterranean ecosystems [4,21]. However, the current study has uncovered some exceptional values for these parameters within the lake itself. For instance, the salinity level during the dry period was found to be as high as 70 psu, while the water level decreased by 30 cm in the same season. These anomalous changes can be attributed to a combination of anthropogenic stressors and climate change. Indeed, the construction of six dams on the main rivers feeding the lake, coupled with low precipitation and high evaporation rates, has led to a reduction in the water level of the lake and a sharp increase in salinity levels.

The marine resources in the Ichkeul Lake have been equally impacted by both climatic conditions and human activities. The findings from the RF and CB model, along with the Pearson correlation analysis, reveal that the salinity, turbidity, and water level, are the most crucial variables affecting the eels' landing. The water level and turbidity are crucial factors that facilitate the migration and foraging process of eels [22]. Additionally, salinity plays a pivotal role in the life cycle of European eels [22]. The order of importance of the other predictors, in decreasing order, were chlorophyll a, nutrients, and dissolved oxygen.

This study represents the first investigation into the relationship between abiotic variables and ichthyological resources in the Ichkeul wetland using machine learning algorithms. While previous studies have examined the relationship between environmental parameters and chlorophyll-a concentration in other Tunisian lagoons, such as Bizerte Lagoon [4] and

Ghar El Melh Lagoon [21]. This study fills an important knowledge gap in the understanding of Ichkeul Lake. The random forest (RF) model used in this study was found to be similar to that found in Ghar El Melh Lagoon ($R^2 = 0.54$), but superior to the model used in Bizerte Lagoon ($R^2 = 0.45$). This difference is likely attributable to variations in the datasets used, as well as to the hydrological processes and different impacts affecting each ecosystem.

The quality of the water body can be inferred from the presence and trophic strategies of eels, as they are sensitive to changes in their environment. In eutrophic conditions, for example, eels have been observed to expand their trophic niche and shift from pelagic (fish) to benthic (invertebrates) prey, as demonstrated in Ichkeul Lake [23]. This behavior confirms the high eutrophic conditions in this ecosystem. Overall, these findings highlight the importance of maintaining water quality to support the growth and survival of eels, as well as other aquatic organisms.

5 CONCLUSION

This paper investigates the potential of two machine learning algorithms, Random Forest and Cubist, in identifying the impact of climatic and anthropogenic stressors on the production of ichtyofauna in coastal ecosystems. The results of this study suggest that local management agencies can use these smart technologies in the monitoring system of the trophic and ecological status of the lake, as they offer a reliable and efficient means of maintaining its ecological conditions in the future. The study also concludes that the Random Forest and Cubist algorithms are appropriate models for the type of dataset used in this research. However, further experimentation using larger datasets is recommended to evaluate the performance of these models in predicting the effects of climate and anthropogenic pressures on biota in coastal ecosystems. Overall, this research highlights the importance of utilizing advanced analytical tools to better understand the complex relationships between environmental variables and biotic communities in coastal ecosystems, and to support evidence-based decision making for sustainable resource management.

REFERENCES

[1] Newton, A., Brito, A.C., Icely, J. D., Derolez, V., Clara, I., Angus, S., Schernewski, G., Inácio, M., Lillebø, A. I., Sousa, A. I., Béjaoui, B., Solidoro, C., Tosic, M., Cañedo-Argüelles, M., Yamamuro, M., Reizopoulou, S., Tseng, H. C., Canu, D., Roselli, L., Maanan, M., Cristina, S., Ruiz-Fernández, A. C., Lima, R., Kjerfve, B., RubioCisneros, N., Pérez-Ruzafa, A., Marcos, C., Pastres, R., Pranovi, F., Snoussi, M., Turpie, J., Tuchkovenko, Y., Dyack, B., Brookes, J., Povilanskas, R. and Khokhlov, V. (2018). Assessing, quantifying and valuing the ecosystem services of coastal lagoons. *J. Nat. Conserv.* https://doi.org/10.1016/j.jnc.2018.02.009.

[2] Taylor N. G., Grillas P., Al Hreisha H., Balkız O., Borie M. and Boutron O. (2021). The future for Mediterranean Wetlands: 50 key issues and 50 important conservation research questions, *Regional Environmental Change*, 21.2, 1–17 https://doi.org/10.1007/s10113-020-01743-1

[3] Abidi M., Ben Amor R. and Gueddari M. (2018). Assessment of the Trophic Status of the South Lagoon of Tunis (Tunisia, Mediterranean Sea): Geochemical and Statistical Approaches, *Journal of Chemistry* https://doi.org/10.1155/2018/9859546

[4] Béjaoui, B., Armi, Z., Ottaviani, E., Barelli, E., Gargouri-Ellouz, E., Chérif, R., Turki, S., Solidoro, C. and Aleya, L. (2016). Random Forest model and TRIX used in combination to assess and diagnose the trophic status of Bizerte Lagoon, southern Mediterranean. *Ecol. Indic.* 71, 293–301.

[5] Cutler, D. R., Edwards, T. C., Beard, K. H., Cutler, A., Hess, K. T., Gibson, J. and Lawler, J. J. (2007). Random forests for classification in ecology. *Ecology* 88, 2783–2792.

[6] Ben M'Barek, N. (2001). Etude de l'écosystème du Lac Ichkeul et de son bassin versant: Caractéristiques physiques et géophysiques des eaux et des sédiments. Thèse de Doctorat. *Faculté des Sciences de Tunis.* 235 p.

[7] Boukari, A., Benabdallah, S., Everbecq, E., Magermans, P., Grard, A., Habaieb, H. and Deliège, J. F. (2019). Assessment of agriculture pressures impact on the Joumine river water quality using the PEGASE Model. *Environ. Manage.* 64, 520–535. https://doi.org/10.1007/s00267-019-01207-1

[8] APHA, (1992). American Public Health Association. *Standard Methods for the Examination of Water and Wastewater*. American Public Health Association Washington.DC

[9] Lorenzen C. J. (1967). Determination of Chlorophyll and PheoPigments: Spectrophotometric Equations, *Limnology and Oceanography*, 12.2, 343–346 https://doi.org/10.4319/lo.1967.12.2.0343

[10] Parsons, T. R., Maita, Y. and Lalli, C. M. (1984). *A Manual of Chemical and Biological Methods for Seawater Analysis. Pergamon, Oxford Sized Algae and Natural Seston Size Fractions*. Pergamon Press, Oxford Oxfordshire; New York. P

[11] Kuhn M. and Johnson K. (2013)- Applied Predictive Modeling, Applied Predictive Modeling https://doi.org/10.1007/978-1-4614-6849-3

[12] Starkweather J. (2014). A New Recommended Way of Dealing with Multiple Missing Values: Using MissForest for All Your Imputation Needs, July 2014

[13] Breiman L. (2001). Random forests. *Mach. Learn.* 45, 5–32. https://doi.org/10.1201/9780429469275-8

[14] John, K., Isong, I. A., Kebonye, N. M., Ayito, E. O., Agyeman, P. C. and Afu, S. M. (2020). Using machine learning algorithms to estimate soil organic carbon variability with environmental variables and soil nutrient indicators in an alluvial soil. *Land*, 9, 487. doi: 10.3390/land9120487

[15] Das, S., Sun, X., and Sun, M. (2020). International journal of transportation rule-based safety prediction models for rural two-lane run-off-road crashes. *International Journal of Transportation Science and Technology*, xxxx, 1–10. https://doi.org/10.1016/j.ijtst.2020.08.001

[16] Kumar, C., Podesta, G., Kilpatrick, K. and Minnett, P. (2021). A machine learning approach to estimating the error in satellite sea surface temperature retrievals. *Remote Sensing of Environment* 255,112227. pp. 1–14. 10.1016/j.rse.2020.112227.

[17] R Core Team (2022). R: A Language and Environment for Statistical Computing. R Foundation for Statistical Computing, Vienna. URL https://www.R-project.org/

[18] Zhou, J., Li, E., Wei, H., Li, C., Qiao, Q. and Armaghani, DJ. (2019). Random Forests and Cubist Algorithms for Predicting Shear Strengths of Rockfill Materials. *Appl. Sci.*, 9, 1621. pp. 1–16. doi: 10.3390/app9081621

[19] Ben Hadid, N., Goyet, C., Chaar, H., Ben Maiz, N., Guglielmi, V., and Shili, A. (2021). Machine learning modeling techniques for forecasting the trophic level in a restored south mediterranean lagoon using chlorophyll-a. *Wetlands*, 41(8), 1–30. https://doi.org/10.1007/s13157-021-01479-6

[20] Kuhn, M., Wing, J., Weston, S., Williams, A., Keefer, C., Engelhardt, A., Cooper, T., Mayer, Z. and Kenkel,B., *R Core Team. (2022). Package 'Caret' Classification and Regression Training*, Version 6.0-93. pp 1–224.

[21] Béjaoui B., Ottaviani E., Enrico Barelli E., Ziadi B., Dhib A. and Michel Lavoie M. (2018). Machine Learning Predictions of Trophic Status Indicators and Plankton Dynamic in Coastal Lagoons, *Ecological Indicators*, 95. February, 765–74 https://doi.org/10.1016/j.ecolind.2018.08.041

[22] Lagarde R., Peyre J., Amilhat E., Bourrin F., Prellwitz F. and Simon G. (2021). *Movements of Non-Migrant European Eels in an Urbanised Channel Linking a Mediterranean Lagoon to the Sea, Water (Switzerland)*, 1–14 https://doi.org/10.3390/w13060839

[23] Shaiek M., Romdhane M.S. and Le Loc'h F. (2015). *Study of the Ichthyofauna Diet in the Ichkeul Lake (Tunisia), Cybium*, 39.3, 193–210

Innovation and Technological Advances for Sustainability – Al-Naemi et al. (Eds)

Computational analysis and optimization of photocatalytic degradation of organic dyes using ZnO nanoparticles

Morteza Golmohammadi
Department of Chemical Engineering, Birjand University of Technology, Birjand, Iran

Amin Esmaeili
Department of Chemical Engineering, School of Engineering Technology and Industrial Trades, College of the North Atlantic – Qatar, Doha, Qatar

Rasool Safaee
Department of Chemical Engineering, Amirkabir University of Technology, Tehran, Iran

ABSTRACT: Dye removal from industrial wastewater is critical. Photocatalytic degradation of dyes is an excellent approach to this separation problem. Large-scale dye removal in an open tank based on the photocatalytic reaction of Methyl Blue with ZnO nanoparticles as catalyst under UV/solar irradiation was investigated in this work using advanced computational methods. A working model was developed that can simulate the system. The transient and steady-state behaviour of the process at various operating capacities and feed concentrations was analyzed. The velocity, concentration, reaction rate, turbulent kinetic energy, turbulent dynamic viscosity, concentration, and mass flux profiles were obtained over time. The process was optimized for maximum dye degradation efficiency with a target product concentration of 10 ppm and a correlation for the estimation of the operating capacity for a given feed concentration was also developed. Furthermore, a dimensional analysis was also performed, and the dimensionless numbers of the process were determined. These numbers will be used for the scale-up of the process in future research. This model is a platform for future research and large-scale experiments for industrial applications.

Keywords: CFD, reacting flow, nanoparticles, photocatalyst, dye degradation

1 INTRODUCTION

Industrial wastewater has always been a crucial environmental challenge, especially with increasing population and industrial production. The release of industrial effluents, which often contain toxic chemicals, will cause irreparable damage to the environment and human life.[1] Organic dyes are one of the most widely used chemicals in various industries such as textile, food, cosmetics, leather, pharmaceutical and paper industries [2]. Therefore, these industrial effluents contain varied amounts of these dyes. Effluents containing synthetic dyes are hazardous to ecological systems and to public health. These substances may affect the optical activity of aquatic plants by reducing the infiltration of light, causing eutrophication, and increasing suspended solids and turbidity of water [3]. Moreover, most organic dyes are carcinogenic and mutagenic, causing allergic reactions [4]. As a result of the enactment of strict environmental laws, on the one hand, and the importance of public opinion to environmental challenges, on the other, the removal of dyes from wastewater with a suitable treatment process is necessary and inevitable.

Various methods have been reported for the treatment of industrial wastewater containing dye including adsorption [5], coagulation [6], electrocoagulation [7], various oxidation

DOI: 10.1201/9781003496724-10

techniques [8], and biological methods [9]. Most of these methods are based on the adsorption of dye and not their destruction. Therefore, each of these methods requires a secondary step to destroy the dye molecules. Advanced oxidation processes are one of the purification technologies that have been developed in recent years [10–14]. The photocatalytic process is categorized as an advanced oxidation process in which a metal oxide with photocatalytic properties generates a series of active free radicals when it is exposed to ultraviolet irradiation. The generated active free radicals play an important role in the destruction of pollutant molecules [15]. In many applications, solar irradiation can be sufficient for this process, which considerably improves the process economics [16]. Due to their very small size, nanoparticles (NPs) have a high specific surface area, which increases the adsorption capacity of the reactants and thereby promotes the probability of the reaction [17]. Therefore, the use of heterogeneous nano-sized catalysts improves the efficiency of catalytic reactions. One of the relatively new and environmentally friendly methods is the fabrication of NPs in the presence of plant extract as a reducing and stabilizing agent. In recent years, various research groups have synthesized NPs with photocatalytic properties using this green method and have investigated their performance in the degradation of various pollutants. Golmohammadi *et al.* [18,19] synthesized various types of ZnO and SnO_2 photocatalysts using the extract of *jujube* fruits as well as *Acroptilon repens* flower and examined their performance in the degradation of methylene blue (MB) and Eriochrome black T. Their results indicated the high capability of the nano-photocatalyst in degrading these organic dyes in an aqueous solution. According to their results, the degradation efficiency reached 98 % after 4 hours of reaction under sunlight. Karthik *et al.* [20] synthesized ZnO NPs in the presence of *Synadium grantii* leaf extract with varying Cu-dopant concentrations. They found that ZnO NPs with 3 and 5% Cu doping resulted in superior photocatalytic activities for MB, Indigo Carmine, and Rhodamine B.

Reactor modelling and simulation are a very useful tool for the design, development, analysis, optimization, and scale-up of industrial processes [21]. Since physical and chemical phenomena in most reactors are coupled, it is very important to implement a physics-based model as well as proper mathematical solvers to account for the multi-physics couplings. Besides proper models and solvers, reliable experimental data and parameter estimation are critical in a simulation. To be more specific, determining accurate and reliable kinetic parameters of a photocatalytic reaction system is an important step in developing a simulation and model verification. These parameters are often determined by fitting a dynamic model on an experimental data set obtained at the laboratory scale under transient conditions. Although various studies have been conducted to find the kinetic parameters related to the photodegradation reactions with various photocatalysts, little effort has been made to evaluate their performance on an industrial scale.

This work is focused on the computational design of a large-scale industrial dye removal process using the experimental results achieved in previous work by the authors [22]. The contamination of water with dye is a serious issue in local industries, and the development of an innovative solution would be of interest. This work is dedicated to using advanced computational methods to investigate the feasibility of dye removal using ZnO NPs in a photocatalytic process at a large scale. It is very wise to utilize the scientific and engineering tools at hand to find an optimal design and investigate the important design parameters prior to building expensive large-scale or pilot-scale equipment for testing. A turbulent reacting flow in an open tank is simulated in this work, and the transport phenomena are thoroughly investigated. The open tank is equipped with a custom-designed array of UV lamps, and the geometry of the vessel is optimized for optimal residence time to achieve the required dye degradation efficiency. Furthermore, the tank is designed to utilize solar irradiation in addition to UV irradiation using UV lamps. A parametric study was performed to investigate the impact of operating parameters and the geometry of the open tank on the process. Optimization was then employed to determine the operational capacity required to achieve a desired dye removal efficiency. A dimensional analysis was also performed, and the system's dimensionless numbers were determined using the Pi theorem. The results achieved in this work provided solid grounds for the scale-up of the process and large-scale experimentation in the next stages of research and development.

2 MATERIALS AND METHODS

This work is computational based on the experimental data obtained in a previous work by the authors [22]. Photocatalytic degradation of MB using ZnO NPs was investigated, and excellent results were achieved. Process scale-up for large-scale treatment of contaminated waters with MB is investigated in this work using advanced computational methods. This section illustrates the development and implementation of the model.

2.1 *Process description*

The process considered in this work is simple and easy to operate. The photocatalytic reaction takes place in an open tank with a design that is essentially like that of a clarifier. However, the difference is the turbulent reacting flow under solar/UV irradiation. In addition, no mechanical mixer was considered. Figure 1 shows the 3D model of the reaction tank. The tank is initially filled with the aqueous solution of colloidal ZnO NPs. The feed stream (contaminated water with the dye) enters the tank from the bottom with a known dye concentration at a constant flow rate. The flow is then diverted using a diverter, which is equipped with a circular array of UV lamps for enhanced UV irradiation at the deeper section of the tank. This effective design feature not only improves the turbulent mixing of the reacting fluid with the photocatalyst NPs but also enhances the reaction rate. This is because the inlet zone has the highest reactant concentration. Clean water, with a lower dye concentration, overflows through the small channel around the edge of the tank. A notched weir is installed before the outlet to stop any floating or large objects at the water surface. Since the tank is open, solar irradiation is also in effect, especially at zones closer to the surface. The process can operate continuously in a steady-state mode with facile control

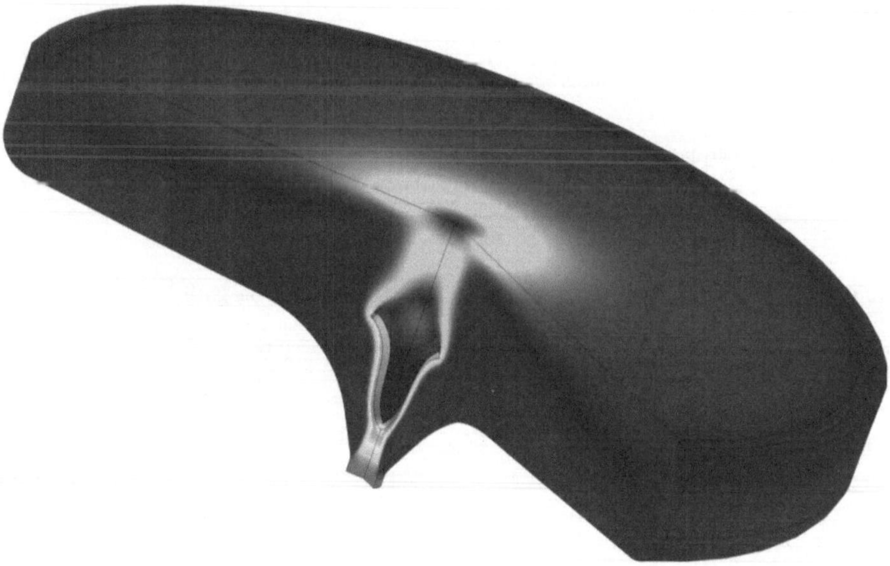

Figure 1. Overview of the designed open tank.

systems. It is important to coat the internal surface of the tank with a suitable material to prevent the adsorption of NPs on the surface.

2.2 *Model development*

The model equations for the flow, mass transfer, and reaction are discussed here. Every modeling has its assumptions and the assumptions made in this work are as follows:

1) 2D axisymmetric geometry
2) Isothermal condition
3) Turbulent flow in the tank
4) Isotropic rate of reaction
5) No slip condition for walls

The Navier-Stokes and the k – ε turbulent model equations describe the fluid mechanics of the system. Equation 1 shows the famous Navier-Stokes equation along with the continuity equation [23]:

$$\begin{gathered}\rho\frac{\partial u}{\partial t}+\rho(u.\nabla)u=\nabla.[-PI+\boldsymbol{K}]+\rho g\\ \boldsymbol{K}=(\mu+\mu_T)\left(\nabla u+(\nabla u)^T\right)\\ \rho\nabla.u=0\end{gathered} \tag{1}$$

where u is velocity, P is pressure, ρ is density, μ is dynamic viscosity, and μ_T is the turbulent dynamic viscosity. The $k-\varepsilon$ equation, which is the most used turbulent model in industrial applications, introduces a new transport equation for turbulent kinetic energy (k) and a new transport equation for turbulent dissipation rate (ε). The k equation reads [24]:

$$\rho\frac{\partial k}{\partial t}+\rho u.\nabla k=\nabla.((\mu+\mu_T)\nabla k)+P_k-\rho\varepsilon \tag{2}$$

The production term is:

$$P_k=\mu_T\left\{\nabla u:\left(\nabla u+(\nabla u)^T\right)-\frac{2}{3}(\nabla.u)^2\right\}-\frac{2}{3}\rho k\nabla.u \tag{3}$$

The dissipation (ε) equation takes the following form:

$$\rho\frac{\partial\varepsilon}{\partial t}+\rho u.\nabla\varepsilon=\nabla.((\mu+\mu_T)\nabla\varepsilon)+1.44\frac{\varepsilon}{k}P_k-1.92\rho\frac{\varepsilon^2}{k} \tag{4}$$

where $\mu_T=0.9\rho\frac{k^2}{e}$. Initial values for the model dependent variables are:

$$\mathrm{u}=0,\mathrm{p}=p_{atm},\ k=\left(\frac{10\mu}{\rho l_{\max}^{\lim}}\right)^2,\varepsilon=\frac{0.9k}{l_{\max}^{\lim}} \tag{5}$$

where the mixing length limit reads:

$$l_{\max}^{\lim}\leq\sqrt{\frac{2}{3}}\frac{\sqrt{k}}{\sqrt{S_{ij}S_{ij}}} \tag{6}$$

The strain rate tensor S_{ij} is calculated using the velocity field. The transport equation reads [25]:

$$\rho\frac{\partial\omega_i}{\partial t}+\nabla.j_i+\rho(u.\nabla)\omega_i=R_i$$
$$j_i=-\left(\rho D_i^m\nabla\omega_i+\rho D_i^m\omega_i\frac{\nabla M_n}{M_n}-j_{c,i}\right.$$
$$D_i^m=\frac{1-\omega_i}{\sum_{k\neq i}\frac{x_k}{D_{ik}}} \tag{7}$$
$$j_{c,i}=\rho\omega_i\sum_k\frac{M_i}{M_n}D_k^m\nabla x_k,\quad M_n=\left(\sum_i\frac{\omega_i}{M_i}\right)^{-1}$$

Where ω_i is the mass fraction of component i, R_i is the reaction term, D_i^m is molecular diffusivity, x_k is mole fraction of component k, M_i is the molecular weight of component i, and M_n is the mixture molecular weight.

According to the previous experimental data [22], the reaction rate in this work is expressed using a pseudo first-order kinetic model, which takes the following form:

$$\frac{dc}{dt}=-k_rc \tag{8}$$

Where c is the concentration of dye in the tank and k_r is the isotropic reaction rate constant. The boundary conditions are illustrated in the next section based on the model geometry. The diffusion coefficient of MB was obtained from reference [26].

2.3 *Model geometry and computational methods*

Figure 2 shows the model 2D asymmetric geometry and the mesh.

Symmetry boundary condition is considered at $r=0$:

$$u.n=0,\quad K-(K.n)n=0,\quad \nabla k.n=0,\quad \nabla\varepsilon.n=0,\quad n.j_i=0 \tag{9}$$

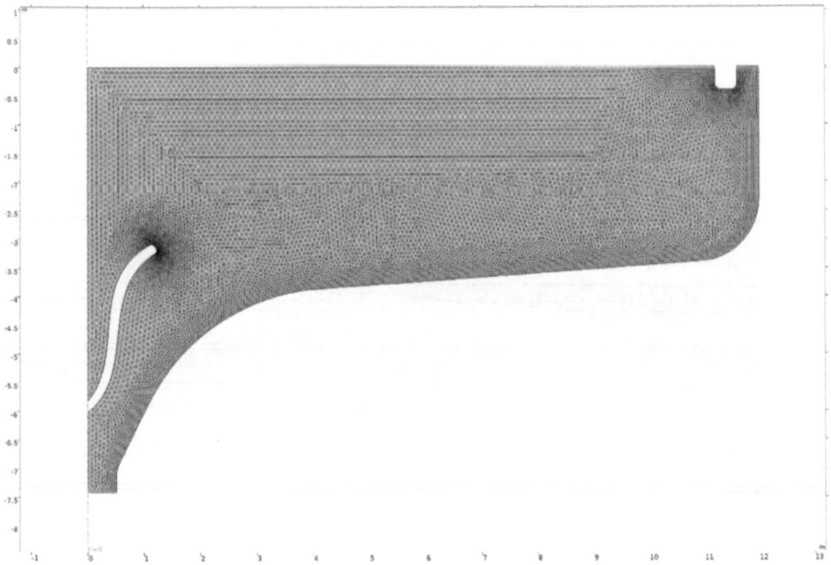

Figure 2. 2D asymmetric geometry of the open tank and the generated mesh for the simulation.

The feed conditions are known; hence, the inlet boundary conditions are: $u = u_{in}, c = c_{in}$. The tank is free of dye at t = 0 ($c = 0$). The model equations were solved using COMSOL Multiphysics solvers. 8 CPU cores (i9-9900K, 4.2 GHz) and 32 GB of processing memory were available for the computations.

3 RESULTS AND DISCUSSION

The computational results are systematically presented in this section starting with the fluid mechanics. The velocity profile is firstly discussed and analyzed. The turbulent flow parameters and their effect on the reacting flow are also discussed. Then, the mass transfer of species in the process along with the photocatalytic reaction is investigated. To determine the optimal operating capacity of the process, the computational results were used to optimize the process and to develop a correlation for a nominal operation.

3.1 *Analysis of the fluid mechanics*

Figure 3 demonstrates the velocity profile in the tank at t = 5 hours where the process reached a steady-state condition. In this figure, the velocity magnitude is shown using a color legend, while the streamlines are shown using normalized white arrows. It can be noticed that the highest velocities are in the zone around the diverter. From the tip of the diverter, the flow tends to move upward as a result of the vertical pressure gradient. At the free surface, the water flows towards the outlet and into the eddy current zone as the streamlines demonstrate. This flow pattern is ideal for the photocatalytic reaction. The inflow with the highest concentration of dye passes close to the UV lamps installed on the diverter; and then flows upward to the surface where solar irradiation is higher than the other zones in the tank. Moreover, some streamlines creep across the flat section at the bottom of the tank and move towards the inlet, where they are mixed with the high velocity inlet stream. This flow pattern is also suitable for the reaction as it increases the residence time of some dye particles in the tank and allows them to pass in the vicinity of the UV lamps for reaction more than once. Along this path from the inlet towards the outlet, the dye is degraded by the reaction catalyzed by the ZnO NPs under UV/solar irradiation resulting in a cleaner outlet stream.

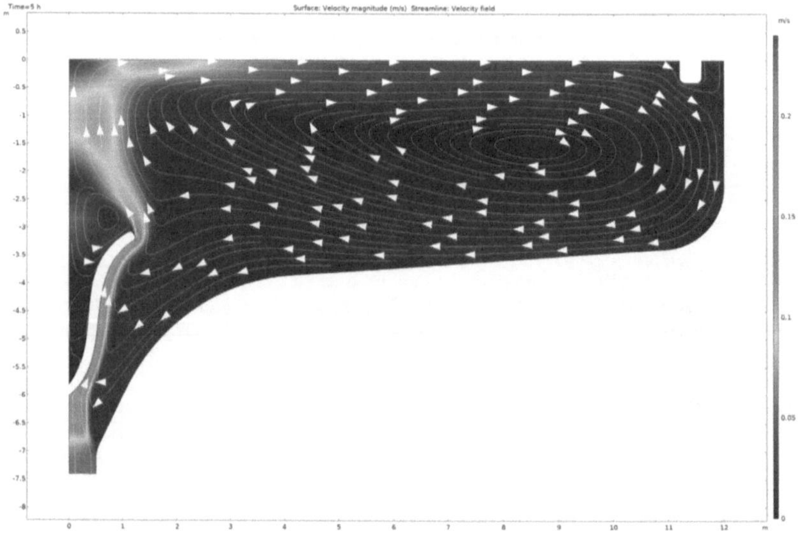

Figure 3. Velocity profiles and streamlines in the open tank at t = 5h; capacity: 130 tons/day.

Figures 4A and B show the velocity field and streamlines at t = 1 hour and t = 2.5 hours. The center of the eddy currents' region moves towards the outlet over time. For interested readers, a video showing the entire velocity field and streamlines over five hours is available in the supplementary information. Figure 1 is a 3D visualization of the velocity magnitude in the tank.

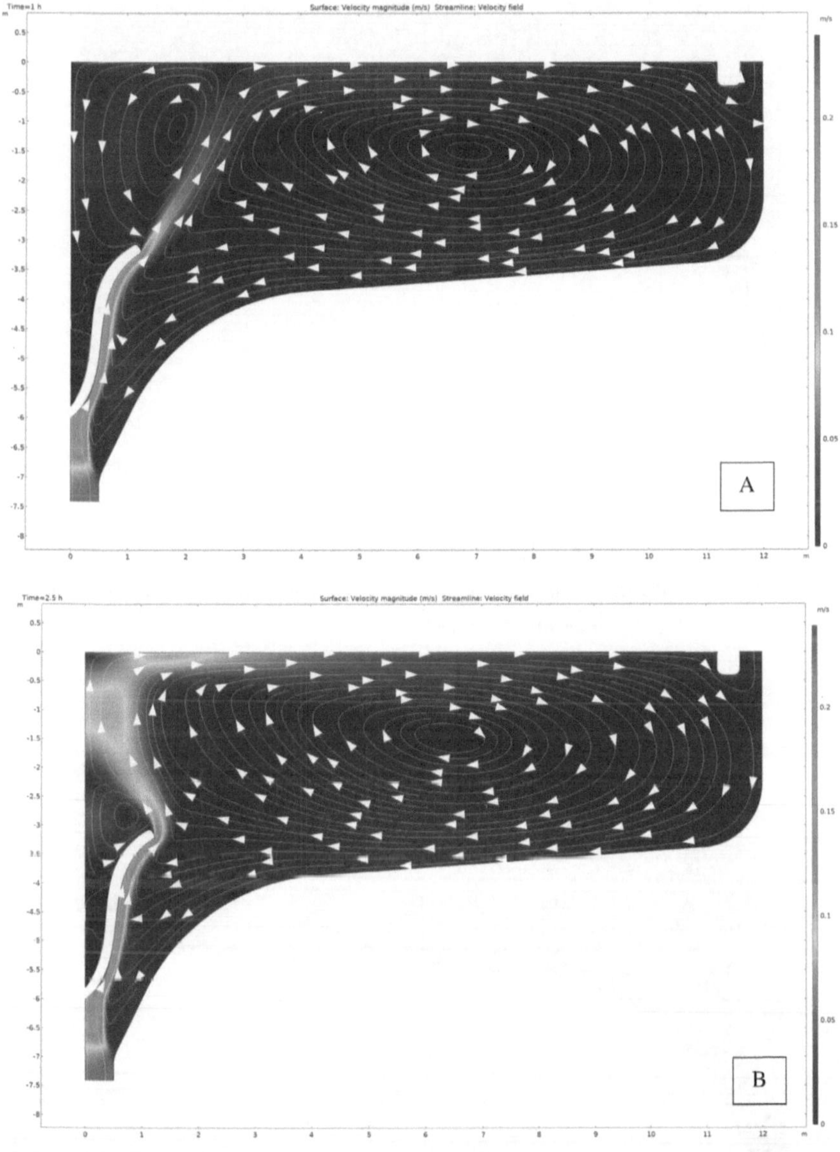

Figure 4. Transient behavior of the system at two different time steps in the simulation; A: 1 h; B: 2.5 h; capacity: 130 tons/day.

The turbulent kinetic energy in the tank is plotted in Figure 5. The zones around the diverter and at the center of the tank on the open surface have the highest mean kinetic energy per unit mass as a result of eddies in the turbulence and the boundary layer separation [23]. The eddies in the other regions (see streamlines in Figure 2) have lower velocities; and hence, lower turbulent kinetic energy.

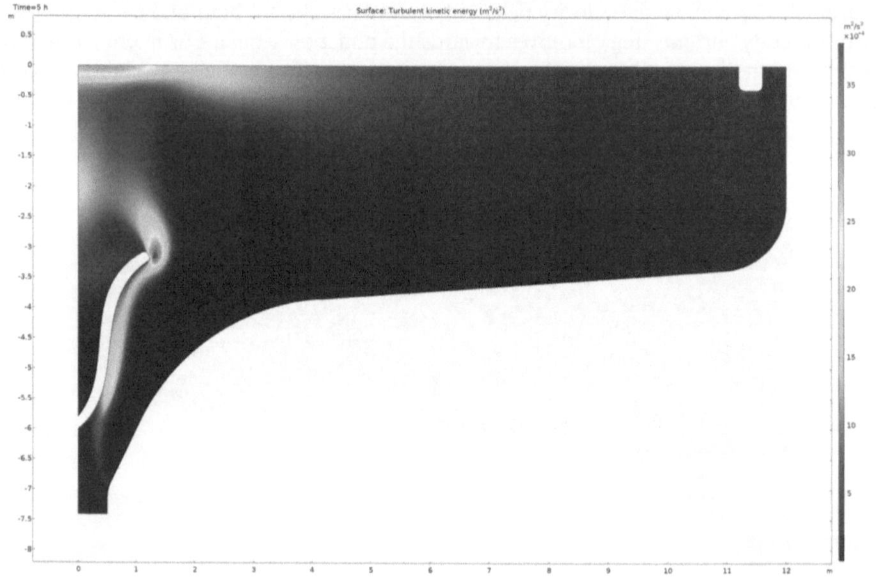

Figure 5. Turbulent kinetic energy profile in the open tank at t = 5 h; capacity: 130 tons/day.

The turbulent dynamic viscosity is plotted in Figure 6. It can be noticed that the values of turbulent dynamic viscosity in the tank are almost three orders of magnitude higher than the molecular viscosity of water in laminar flow (8.9 × 10-4 Pa.s at 25°C) because eddies give rise to fluid friction at a much larger scale [23].

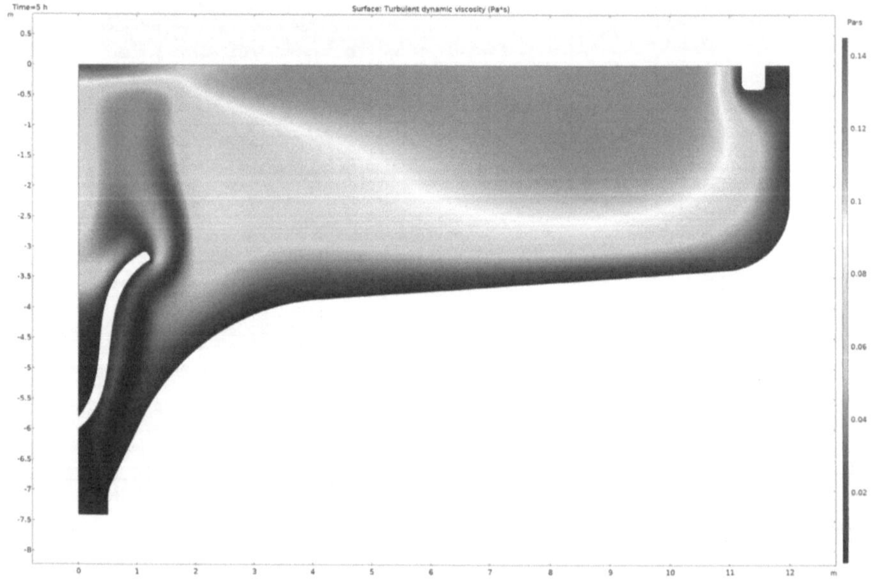

Figure 6. Turbulent dynamic viscosity profile in the open tank at t = 5 h; capacity: 130 tons/day.

Figure 7 also shows the zones with the highest shear stress. These values, along with the pressure values, are useful for the mechanical design of the diverter and the tank body, especially for engineering a suitable coating inside the tank, as coatings wear off at high shear rates.

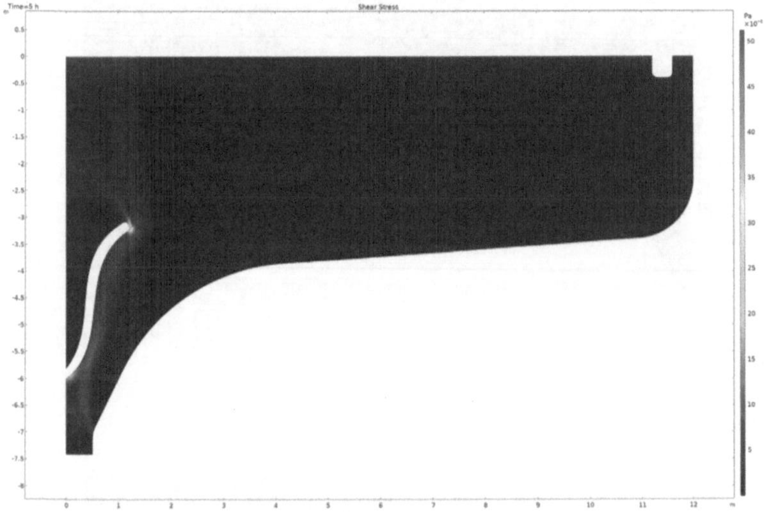

Figure 7. Shear Stress profile in the open tank at t = 5 h; capacity: 130 tons/day.

3.2 *Analysis of the mass transfer and reaction*

The concentration profile and the total flux streamlines at t = 1, 2.5, and 5 hours are shown in Figure 8. The capacity of the unit was 130 tons/day, and the inlet dye concentration was 100 ppm. Because of the photocatalytic reaction of dye molecules with the ZnO nanocatalysts in the zone near the diverter/UV lamps, the dye concentration decreases rapidly as the

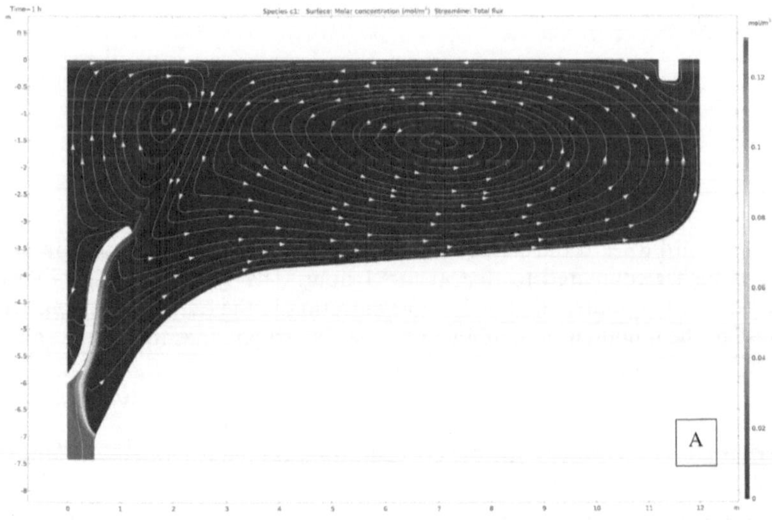

Figure 8. Concentration profile and mass streamlines in the open tank at t = 1 h (A); t = 2.5 h (B); t = 5 h (C); inlet dye concentration: 100 ppm; capacity: 130 tons/day.

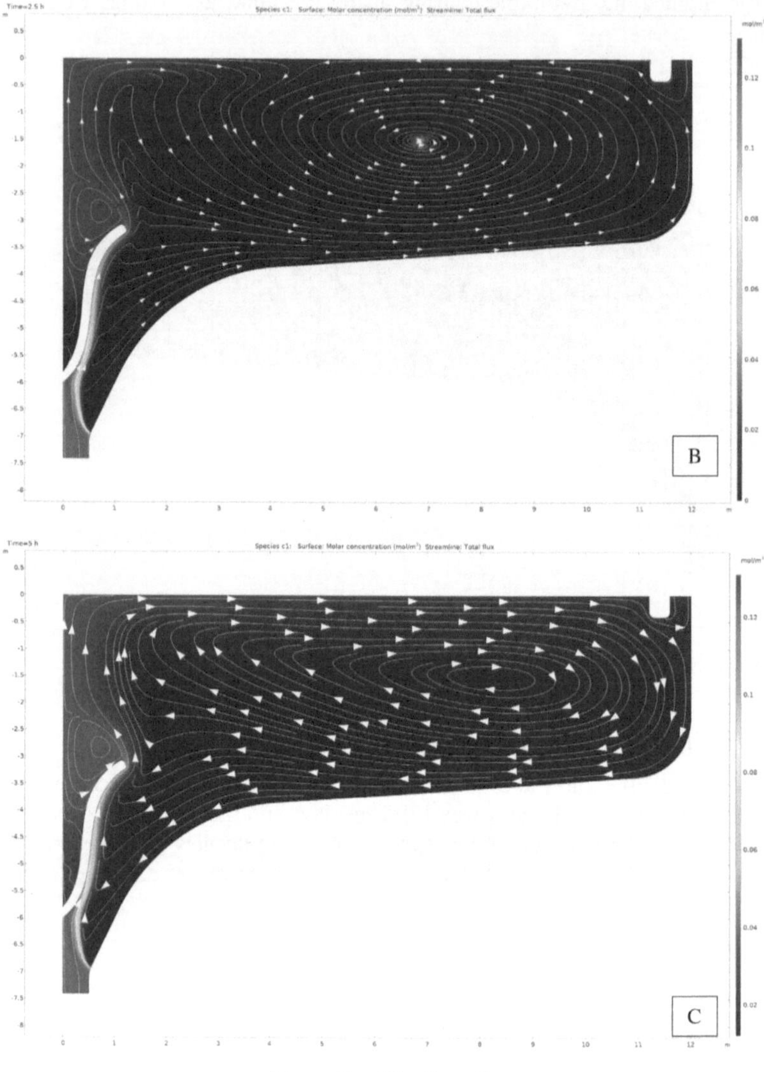

Figure 8. (Continued)

stream flows upward toward the surface. At t = 5 hours, the dye concentration around the free surface is higher compared to that at t = 1 hour. The system reached a steady-state condition after about 5 hours, and the dye concentration in the outlet stream was 10.5 ppm, which is close to the minimum required value. The flux streamlines follow the same pattern as the velocity streamlines seen in Figures 3 and 4.

The center of eddies in the middle zone of the tank similarly moves towards the outlet of the tank.

To better visualize the process, the 3D concentration profile is plotted in Figure 9.

The reaction rate at t = 5 hours is demonstrated in Figure 10. The reaction rate is in harmony with the concentration profile, where the highest reaction rates are around the inlet zone and the diverter/UV lamps. The maximum reaction rate was about 0.25 mg/m^3/s.

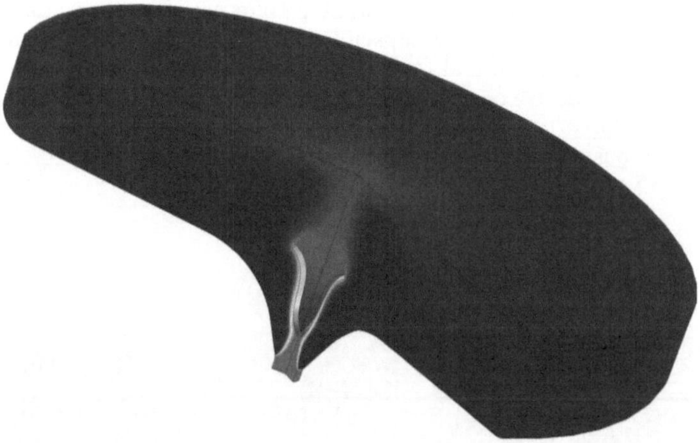

Figure 9. Dye concentration profile in the open tank under steady-state operating conditions.

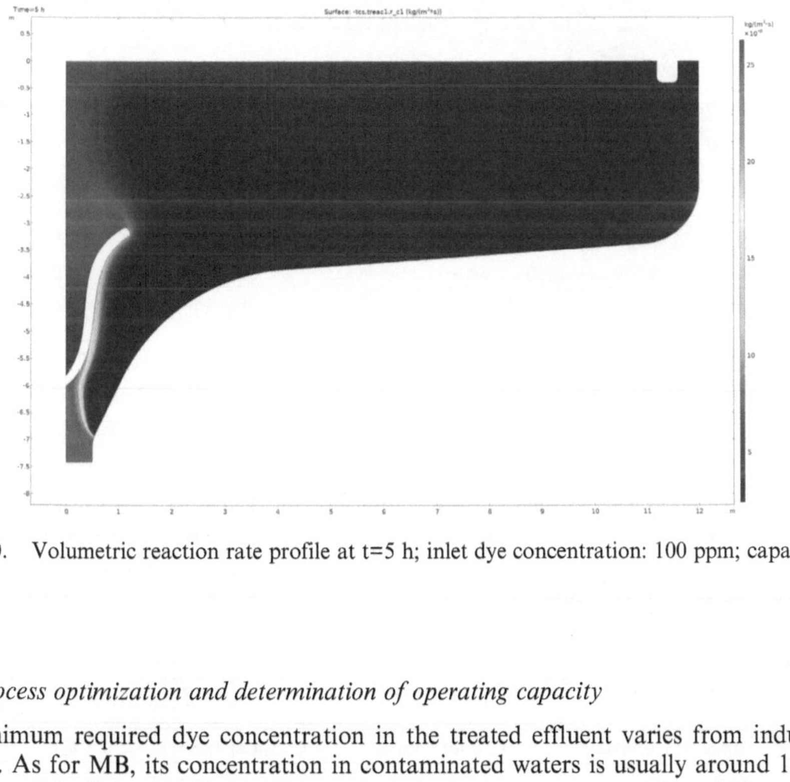

Figure 10. Volumetric reaction rate profile at t=5 h; inlet dye concentration: 100 ppm; capacity: 130 tons/day.

3.3 *Process optimization and determination of operating capacity*

The minimum required dye concentration in the treated effluent varies from industry to industry. As for MB, its concentration in contaminated waters is usually around 100 ppm and a removal efficiency of 90% or higher is desired. Hence, a target value of 10 ppm in the product stream was considered in this work, and the process was optimized to determine the suitable operating capacity. It is obvious that the dye removal efficiency decreases with increasing capacity, and the concentration of dye in the product stream increases. The developed model and computational methods were used in this work to determine the operating capacity of the process and to develop a correlation for rapid estimation of the operating capacity.

Figure 11 shows the concentration profile at t = 5 hours when the concentration of the dye in the feed stream was 50 ppm, and the capacity was 467 tons/day. The dye concentration in the product was about 30 ppm, and the removal efficiency was less than 50%. It can be noticed in Table 1 that the dye concentration in the product reached to about 11 ppm when the capacity was reduced to 190 tons/day. Hence, it is essential to have a correlation for industrial applications in order to estimate the operating capacity for a given feed concentration.

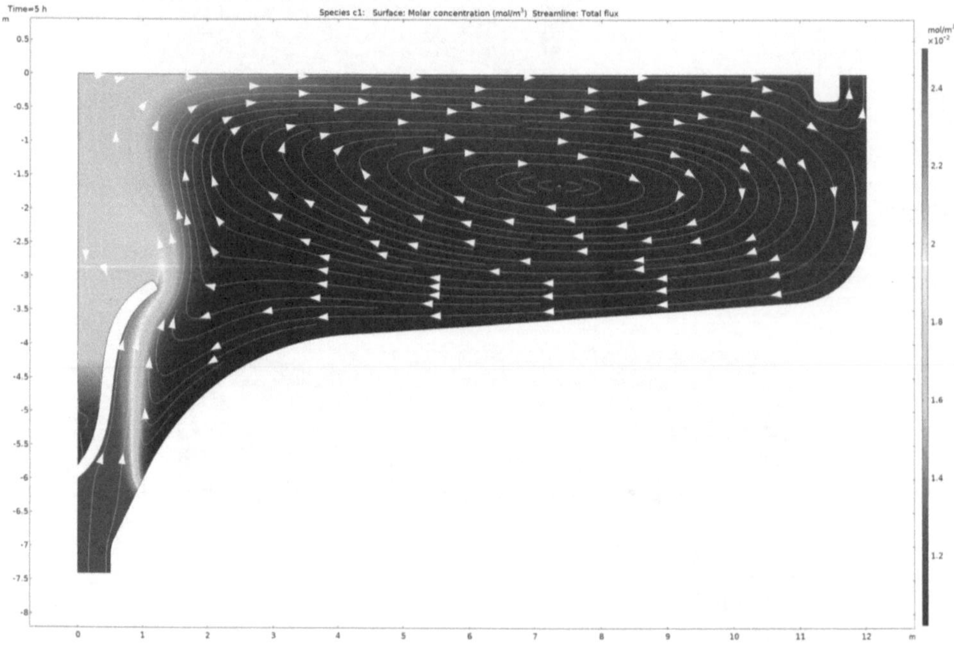

Figure 11. Dye concentration profile at t = 5 h; inlet dye concentration: 50 ppm; capacity: 467 tons/day.

A parametric study was performed, and the results are shown in Table 1. The operating capacity for cases where the dye concentration in the product was lower than 10 ppm is plotted as a function of feed concentration. As expected, the operating capacity decreases with increasing feed concentration.

Table 1. Parametric study results; COP is the coefficient of performance.

capacity (tons/day)	C inlet (ppm)	C outlet (ppm)	COP
160	500	103.51	79.3
160	100	20.14	79.9
160	50	0	100
160	20	0	100
130	100	10.5	89.5
190	50	11.1	77.8
467	20	8.6	57

A simple power trendline is fitted on the data, and the coefficients are shown in Figure 12. This simple equation can be used to estimate the operating capacity based on a given feed concentration. This equation was obtained using three data points. To verify this equation, two more simulations were run and the results from the simulation and equation are compared. Assuming a feed concentration of 80 ppm, the equation in Figure 12 results in a capacity of 146 tons/day to achieve a product concentration lower than 10 ppm. The simulation was run again based on these parameters, and a product value of 0 ppm was achieved. This equation was obtained based on the data points within the 0 – 100 ppm range. To test if the equation can be used outside of its range, the simulation was run for 500 ppm that is outside this range. A capacity of 146 tons/day is estimated using the equation; however, a product concentration of 25.67 ppm was achieved in the simulation, which is higher than the 10 ppm target point. The degradation efficiency is still quite high (95%). It can be concluded that the equation does not estimate correct operating capacities outside of its range (0 – 100 ppm). This equation can be further improved and later verified using experimental data; yet, at this point in the research and development, it is very useful to have some initial estimates and understanding of the process prior to running expensive large-scale experiments in industry.

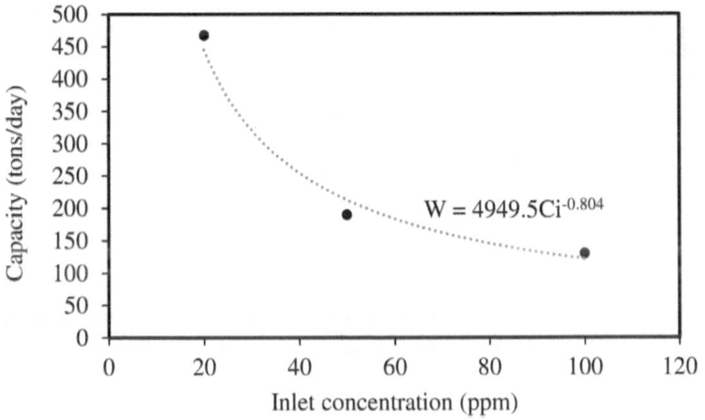

Figure 12. Correlation between the capacity and inlet dye concentration.

3.4 *Dimensional analysis of the process*

A dimensional analysis is performed here using the Pi theorem [27]. The flow condition in the tank is represented by the following parameters (u, μ, ρ, L_c). u is velocity, μ is viscosity, ρ is density, and L_c is the characteristic length of the tank. Characteristic length is volume divided by surface, which depends on each tank design and its irregular shape. Effective diffusivity (*D*) and residence time (τ) are the parameters for the mass transfer in the tank. As for the chemical reaction, the inlet and outlet concentrations, as well as the reaction rate constant k, are the parameters. Since the isothermal condition is considered, the parameters related to heat transfer are not considered here. Therefore, four dimensionless numbers can be formed from the eight-dimensional parameters (U, μ, ρ, Lc, c_{in}, c_{out}, k, D). These dimensionless numbers are ($Re = \rho u L_c/\mu$, $Sc = \mu/(\rho D)$, $k\tau$, C_{in}/C_{out}), including the anticipated trivial number c_{in}/c_{out}. These dimensionless numbers can be used for scaling up the process and design of systems with various capacities. In future steps of our research, these numbers, along with the model developed in this work, will be used in the design of equipment and large-scale experiments. Interested readers can refer to reference [27] for further information on the derivation of these numbers.

4 CONCLUSIONS

Large-scale dye removal in an open tank based on the photocatalytic reaction of Methyl Blue with ZnO nanoparticles as catalyst under UV/solar irradiation was investigated in this work using advanced computational methods. The objective of this work was to establish a working model based on experimental data to investigate the important parameters affecting the process and to optimize the process. Excellent results were achieved in previous works by the authors, and the next step of research and development is the scale-up of the process. Large-scale dye removal experiments are expensive and difficult to design. The model and computational results achieved in this work can facilitate these tasks. A suitable geometry for maximum reaction yield was designed for the open tank, and the equations of the k-ε turbulent flow model, along with the convection-diffusion mass transfer with the reaction model, were numerically solved. The pressure, velocity, concentration, reaction rate, turbulent kinetic energy, turbulent dynamic viscosity, and mass flux profiles were obtained over time. The transient and steady-state behaviour of the process at various operating capacities and feed concentrations was analyzed. The process was optimized for maximum dye degradation efficiency with a target product concentration of 10 ppm. A correlation for the estimation of the operating capacity for a given feed concentration was also developed. Furthermore, a dimensional analysis was also performed, and the dimensionless numbers of the process were determined. These numbers will be used for scaling up the process in future research. Overall, based on the concentration, mass flux, and velocity profiles, the process performance and the geometry design of the open tank were suitable. Acceptable degradation efficiencies were achieved. Startup of large-scale experiments is very challenging, and this time-dependent model can be used in the operation and control of the process. This model is a platform for future research and large-scale experiments for industrial applications.

Funding Declaration: None

Author Contribution: ***Project administration:*** Morteza Golmohammadi; ***Conceptualization:*** Rasool Safaee and Amin Esmaeili; ***Methodology:*** Morteza Golmohammadi; ***Formal analysis and investigation:*** Morteza Golmohammadi, Rasool Safaee, and Amin Esmaeili; ***Writing – original draft preparation:*** Morteza Golmohammadi, Rasool Safaee, and Amin Esmaeili; ***Writing – review and editing:*** Morteza Golmohammadi, Rasool Safaee, and Amin Esmaeili; ***Funding acquisition:*** Morteza Golmohammadi, Rasool Safaee; ***Resources:*** Morteza Golmohammadi, Rasool Safaee, and Amin Esmaeili; ***Supervision:*** Morteza Golmohammadi; ***Visualization:*** Morteza Golmohammadi and Rasool Safaee.

NOMENCLATURE

c	concentration of dye	kg/m^3
D^m	molecular diffusivity	m^2/s
k_r	isotropic reaction rate constant	1/s
k	turbulent kinetic energy	m^2/s^2
l_{max}^{lim}	mixing length limit	m
M	molecular weight	kg/kmole
p	pressure	N/m^2
p_{atm}	atmospheric pressure	N/m^2
R	reaction rate	$kg/m^3.s$
S_{ij}	strain rate tensor	N/m^2
t	time	s
u	velocity	m/s
x	mole fraction	

GREEK LETTERS

μ	dynamic viscosity	$N.s/m^2$
ρ	density	kg/m^3
ω	mass fraction	
ε	turbulent dissipation	m^2/s^3

SUBSCRIPTS

i	component
k	component
n	mixture
T	turbulent

REFERENCES

[1] Ajiboye, T. O., Oyewo O. A., and Onwudiwe D. C., (2021). Simultaneous removal of organics and heavy metals from industrial wastewater: A review. *Chemosphere.* **262**: p. 128379.

[2] Hanafi, M. F. and Sapawe N., (2020). A review on the water problem associate with organic pollutants derived from phenol, methyl orange, and remazol brilliant blue dyes. *Materials Today: Proceedings.* **31**: p. A141–A150.

[3] Sherin, L., *et al.*, (2020). Facile green synthesis of silver nanoparticles using Terminalia bellerica kernel extract for catalytic reduction of anthropogenic water pollutants. *Colloid and Interface Science Communications.* **37**: p. 100276.

[4] Chung, K.-T., (2016). Azo dyes and human health: a review. *Journal of Environmental Science and Health, Part C,* **34**(4): p. 233–261.

[5] Bensalah, H., *et al.*, (2020). Azo dye adsorption on an industrial waste-transformed hydroxyapatite adsorbent: kinetics, isotherms, mechanism and regeneration studies. *Journal of Environmental Chemical Engineering.* **8**(3): p. 103807.

[6] do Vale-Júnior, E., *et al.*, (2018). Treatment of an azo dye effluent by peroxi-coagulation and its comparison to traditional electrochemical advanced processes. *Chemosphere.* **204**: p. 548–555.

[7] Lach, C. E., *et al.*, (2022). Investigating the process of electrocoagulation in the removal of azo dye from synthetic textile effluents and the effects of acute toxicity on Daphnia magna test organisms. *Journal of Water Process Engineering.* **45**: p. 102485.

[8] Azimi, S. C., Shirini F., and Pendashteh A., (2021). Advanced oxidation process as a green technology for dyes removal from wastewater: A review. *Iranian Journal of Chemistry and Chemical Engineering (IJCCE).* **40**(5): p. 1467–1489.

[9] Wang, Y., *et al.*, (2020). Resuscitation, isolation and immobilization of bacterial species for efficient textile wastewater treatment: a critical review and update. *Science of the Total Environment,* **730**: p. 139034.

[10] Golmohammadi, M., Ahmadi S. J., and Towfighi J., (2018). Catalytic supercritical water destructive oxidation of tributyl phosphate: Study on the effect of operational parameters. *The Journal of Supercritical Fluids.* **140**: p. 32–40.

[11] Xiong, Z., *et al.*, (2021). Synthesis strategies and emerging mechanisms of metal-organic frameworks for sulfate radical-based advanced oxidation process: A review. *Chemical Engineering Journal.* **421**: p. 127863.

[12] Sun, P., *et al.*, (2019). Degradation of organic micropollutants in UV/NH2Cl advanced oxidation process. *Environmental Science & Technology.* **53**(15): p. 9024–9033.

[13] Monteoliva-García, A., *et al.*, (2020). Effects of carrier addition on water quality and pharmaceutical removal capacity of a membrane bioreactor–Advanced oxidation process combined treatment. *Science of the Total Environment.* **708**: p. 135104.

[14] Muniyasamy, A., *et al.*, Process development for the degradation of textile azo dyes (mono-, di-, poly-) by advanced oxidation process-Ozonation: Experimental & partial derivative modelling approach. *Journal of Environmental Management,* 2020. **265**: p. 110397.

[15] Koe, W. S., *et al.*, (2020). An overview of photocatalytic degradation: photocatalysts, mechanisms, and development of photocatalytic membrane. *Environmental Science and Pollution Research.* **27**(3): p. 2522–2565.
[16] Pandey, A., *et al.*, (2021). Utilization of solar energy for wastewater treatment: Challenges and progressive research trends. *Journal of Environmental Management.* **297**: p. 113300.
[17] Devendran, P., *et al.*, (2022). A novel visible light active rare earth doped CdS nanoparticles decorated reduced graphene oxide sheets for the degradation of cationic dye from wastewater. *Chemosphere.* **287**: p. 132091.
[18] Golmohammadi, M., Hassankiadeh M. N., and Zhang L., Facile biosynthesis of SnO2/ZnO nanocomposite using Acroptilon repens flower extract and evaluation of their photocatalytic activity. *Ceramics International*, 2021. **47**(20): p. 29303–29308.
[19] Golmohammadi, Honarmand M., M., and Esmaeili A., (2021). Biosynthesis of ZnO nanoparticles supported on bentonite and the evaluation of its photocatalytic activity. *Materials Research Bulletin*: p. 111714.
[20] Karthik, K., *et al.*, (2022). Green synthesis of Cu-doped ZnO nanoparticles and its application for the photocatalytic degradation of hazardous organic pollutants. *Chemosphere.* **287**: p. 132081.
[21] Ranade, V. V., *Computational Flow Modeling for Chemical Reactor Engineering.* Vol. 5. 2002: Academic press.
[22] Golmohammadi, M., Honarmand M., and Ghanbari S., (2020). A green approach to synthesis of ZnO nanoparticles using jujube fruit extract and their application in photocatalytic degradation of organic dyes. *Spectrochimica Acta Part A: Molecular and Biomolecular Spectroscopy.* **229**: p. 117961.
[23] Munson, B. R., *et al.*, *Fluid Mechanics.* 2013: Wiley Singapore.
[24] Multiphysics, C., *v. 5.4. Comsol AB, Stockholm, Sweden.* Google Scholar There is no corresponding record for this reference, 2019.
[25] Bird, R. B., Stewart W. E., and Lightfoot E. N., *Transport Phenomena.* Vol. 1. 2006: John Wiley & Sons.
[26] Selifonov, A., *et al.*, (2019). Determination of the diffusion coefficient of methylene blue solutions in dentin of a human tooth using reflectance spectroscopy and their antibacterial activity during laser exposure. *Optics and Spectroscopy.* **126**(6): p. 758–768.
[27] Zlokarnik, M., (2006). *Scale-up in Chemical Engineering.* John Wiley & Sons.

Innovation and Technological Advances for Sustainability – Al-Naemi et al. (Eds)
© 2025 The Author(s), ISBN 978-1-032-80372-2
Open Access: www.taylorfrancis.com, CC BY-NC-ND 4.0 license

Review of large-scale potential for carbon fiber from gas liquids for the carbon-hydrogen economy

M. Razi Nalim & Shahrzad Ghadiri
Department of Mechanical Engineering, Purdue University Indianapolis, Indianapolis, USA

ABSTRACT: This paper reviews the potential for production of low-cost carbon fiber (CF) from gas liquids in large scale comparable to other tensile structural materials such as steel and aluminum. Comparisons are made between the production of CFs from the most widely used polyacrylonitrile (PAN) precursor and low-cost precursors such as polyethylene (PE), that can be produced from gas liquids, particularly ethane and propane. The challenges, viability, and energy intensity of the use of PE precursor are discussed. The production of carbon materials such as structural fiber from gas hydrocarbons may open up large markets for higher-value use of carbon than energy generation from combustion. By thus avoiding the emissions of carbon dioxide the world may transition to a future 'carbon-hydrogen' economy.

Keywords: carbon fiber, olefins, polyethylene, hydrogen

1 INTRODUCTION

Among the elements, the carbon atom has unique bonding characteristics that allows it to form millions of organic compounds, and makes carbon the basis of life, along with water. Carbon exists in more allotropic forms than any other element; each form has vastly different properties, abundance, and value, as shown in Table 1. Among natural forms near the Earth's surface, coal is the most abundant and least expensive, while diamond is the rarest and costliest, with graphite in between. Among useful man-made forms, the price of carbon fiber is six orders of magnitude higher than that of coal, and carbon nanotubes (CNT) cost another four orders higher. Carbon black, graphite, and recently, carbon fiber (CF) are elemental forms consumed in relatively large amounts, after coal.

Carbon has not yet achieved its full potential as a versatile material. In the three decades since the C_{60} "buckyball" was discovered, research into CNT, graphene, and other carbon allotropes with elegant structures and symmetries has led to multiple Nobel prizes and niche applications. New allotropic forms of pure carbon were synthesized with nanoscale dimensions (e.g. fullerenes, nanotubes, graphene), exhibiting novel physio-chemical properties [1]. Researchers have made CNT-based materials that are extremely light with distinct mechanical, thermal, catalytic, adsorption, and electromagnetic properties. Graphene, CNT and diamond are highly valuable and may represent large markets by value but not by mass. More scalably, the forms of elemental carbon that can be used as structural materials are CNTs and CFs. Single-wall CNTs with a single perfect graphene layer may be considered the ultimate carbon fiber, due to light weight and remarkable engineering properties [2]. Multi-wall CNTs have the highest tensile strength of any material yet measured, up to 63 GPa and Young's modulus of 270–950 Gpa [3]. Thus, CNTs can be 1000 times as strong as low-carbon steel, but are currently too expensive to scale up, except as an additive to strengthen other polymers.

DOI: 10.1201/9781003496724-11

Table 1. Elemental carbon in various forms and comparative pricing and usage scale of products.

Elemental C forms and comparable products	Rough Price in US $/kg	Global consumption or emissions (Mg/yr)	Major consumption / emission drivers
Coal	~$0.00001	7,000,000,000	Electricity, steelmaking
Carbon black	~$1	13,000,000	Tires, paints, coatings
Natural graphite	~$1	2,800,000	Batteries, steel, brakes
Carbon Fiber	**$10**	**70,000**	**Airplanes, sports, wind turbines, automobiles**
Graphene	$10,000	~20,000	Batteries
Carbon nanotubes	$100,000	~10,000	Polymers
Diamond ½ carat	$10,000,000	~20	Jewelry, tools
Steel	*$0.6*	*2,000,000,000*	*Vehicles, machines, buildings, bridges,*
Natural gas	*$0.2*	*3,000,000,000*	*Electricity, heating*
Crude oil	*$0.4*	*4,000,000,000*	*Transportation, plastics*
Fuel CO_2 emissions	-	*33,000,000,000*	*Coal, oil, and gas combustion*

The current work reviews how CF could provide a practical leap forward as an affordable, strong, light-weight structural material that can replace steel and aluminum at scale. Although more expensive per unit mass than steel (Table 1), it may be made lower cost than steel per unit load supported. For many uses, even a moderate cost reduction might create significant new markets. Therefore, this work reviews CF from PE and other low-cost commodity polymers derived from gas liquids.

If CF and other forms of elemental carbon are needed in large volumes for high-value structural applications with long life and pathways for recycling and safe sequestering, there is potentially a significant opportunity to divert the carbon content of fossil fuels away from combustion and greenhouse gas emission. For comparison of scales, the global consumption of oil and gas and global CO_2 emissions from all fossil fuel combustion is shown in Table 1. Although coal remains the biggest source of both elemental carbon extraction and greenhouse gas (GHG) emissions, its use is rapidly falling while gas and oil use continues to grow. In some nations, gas from fracking shale has a much higher fraction of gas liquids than conventional natural gas. This is an additional reason to attempt diversion of the higher carbon content gas liquids from being burned and generating CO_2, provided this does not increase emissions of methane, a much more potent GHG.

1.1 *Carbon-hydrogen economy*

A livable Earth requires CO_2 emissions to be severely curtailed, but renewable energy such as solar and wind are not sufficient or reliable to meet energy demand, and other large energy sources such as nuclear energy are not practical or affordable. Much effort has gone to research and develop carbon (CO_2) capture and storage (CCS) methods, assuming fossil fuels will continue to be burned [4]. However, there has been limited success of CCS, mainly limited to pumping for oil extraction. CCS has challenges in the fundamental thermodynamics of separation, underground gas storage options, technical issues in scaling, and high cost [5,6]. An alternate approach may be to utilize the plentiful supply of natural gas and shale gas, but to sequester its carbon content as elemental carbon through diversion to high-value products, avoiding CO_2 emissions. In this case, energy value represented by oxidizing carbon is traded for higher value as elemental carbon products. To retain the energy value of the hydrogen component of gas, appropriate petrochemical engineering approaches must be developed to efficiently accomplish H_2 generation with low CO_2 emissions [7]. Whether carbon is sequestered as elemental C or as CO_2 for such 'blue' hydrogen production, the challenges of GHG reduction due to fugitive methane emissions must be considered [8]. Initially, the relevant technologies may be developed for diverting gas liquids, and eventually similar methods could be developed for methane as well, for example using electrochemical methods [9]. Bio-based olefins and other biopolymers can also be used, being derived from sugarcane, beet, various grains, and other lignocellulosic materials [10].

The envisioned carbon-hydrogen economy is depicted simplistically in Figure 1. Available low-cost fossil/biofuels would be processed to separate the hydrogen and carbon components. The carbon content of hydrocarbons would become high-value products such as CF to be used extensively as structural material, effectively sequestered long enough during the imminent critical period of climate change. The hydrogen content would contribute energy recovered in CF manufacture, or supplied as H_2 for engines, power plants, or (with purification) for electrochemical cells, or for other purposes.

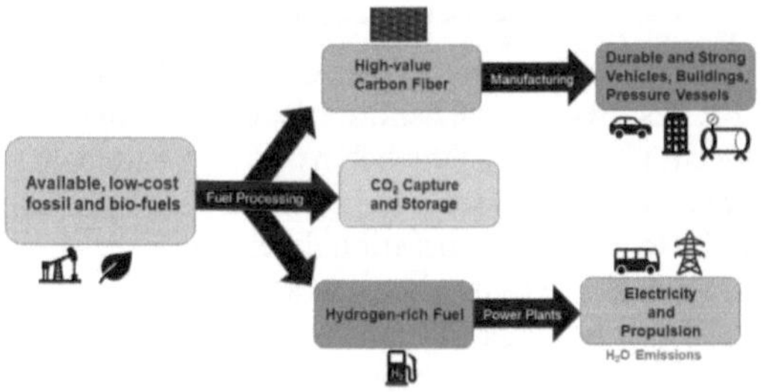

Figure 1. Concept of hydrogen-carbon economy.

Natural gas production has been increasing globally while coal use has been reduced. In the US and some nations, hydraulic fracturing is used to produce shale gas [11], which has high amounts of natural gas liquids (NGLs) such as ethane and propane [12]. Thermodynamically, the hydrogen and carbon constituents of ethane share its energy content of natural or shale gas about equally (Figure 2) [13]. Although this means only half of the energy of ethane can be converted to usable energy through hydrogen, the high-value carbon products such as CF would be worth more than the foregone energy value. It is noted that CO_2 and H_2O have zero useful energy value. Thermodynamic laws dictate that capturing and converting CO_2 to fuel or some other useful product would take more energy than that supplied when the CO_2 was produced. Rather than creating CO_2, the carbon content of hydrocarbons can instead form products such as CF.

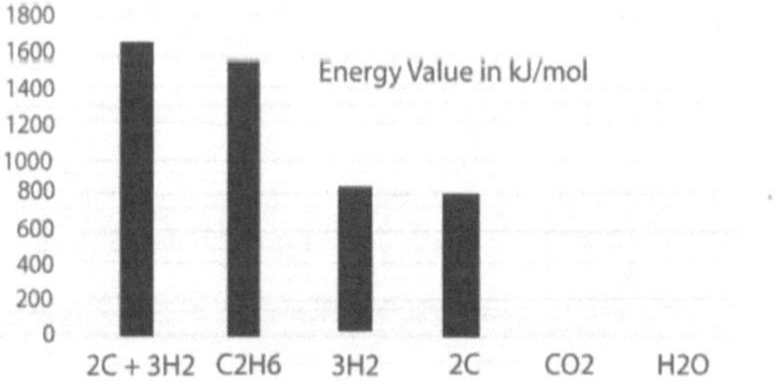

Figure 2. Energy value of ethane, its components and products.

Although current CF production involves significant CO_2 emissions, CF can have lower lifecycle greenhouse gas (GHG) impact than conventional materials [14]. It avoids energy-intensive extraction of metal from ores. Vehicles made with carbon fiber reinforced polymer (CFRP) composites can improve fuel consumption and associated emissions [15].

1.2 *Carbon fiber properties and applications*

Pyrolysis of a suitable precursor can yield CF with at least 90% carbon by weight [2], in filaments of diameter 5–10 micrometers [16]. CFs have exceptional mechanical properties such as high stiffness and high tensile strength along with other desirable material properties such as low density, high chemical resistance, high-temperature endurance and low thermal expansion [17]. Commercial grade CFs have high tensile modulus measured up to 900 GPa and tensile strength up to 7 GPa with very good creep resistance [18]. Due to its light weight, CF tensile specific strength is about 2.5–3.5 MN-m/kg, which is similar to or better than Kevlar, over 20 times that of aluminum alloys, and over 40 times that of stainless steel.

Carbon fiber reinforced polymer (CFRP) has applications in aerospace, automotive, sports equipment, infrastructure, pressure vessels, wind turbines and other industries. CFs for structural applications account for 51% of the total CF market in the U.S. [19]. Amongst these, wind energy and aerospace applications account for 42% of the overall CF production. These demands for carbon fibers have been growing rapidly, especially after the introduction into new commercial aircraft. To minimize weight, the Airbus A380 structures incorporate a range of new materials [20]. CFRP is used for the central box of the wings, the horizontal stabilizers, the fin, the rear fuselage section and for ceiling beams. An estimated 40 percent of the aircraft's structure and components are made from lightweight carbon composites and advanced metallic materials, which also improve operational reliability, maintainability, and ease of repair. Pressure vessels with CF composites as a structural material, usually with an impermeable metal (Type III) or polymer (Type IV) liner, are much lighter than those made of metal (Type I or II), and not as subject to embrittlement. For gas-fueled vehicles, lighter fuel tanks and structural components can result in substantial lifecycle reduction of fuel use and emissions [21]. By 2025, the automotive and wind turbine blade industries are likely to consume the most CF, and by 2030, the needs of H_2 storage tanks may double CF demand [22]. Commercial grade CF with high tensile strength and moderate modulus costs about $30/kg [23], which is only affordable in high-end and racing cars. CF would find a larger market application if the structural components of typical automobiles could be produced with carbon fiber composites instead of steel. This paper mainly aims to explore the utilization of carbon as a high-value product derived from major sources of carbon that today serve as fuels.

2 CARBON FIBER MANUFACTURING AND USE

High-quality CF has been made from polyacrylonitrile (PAN) precursor [22], and less commonly from other organic polymers such as petroleum pitch and rayon. The polymer is usually spun into fibers, thermally stabilized and then subject to carbonization and other final processing, as shown in Figure 3 [24].

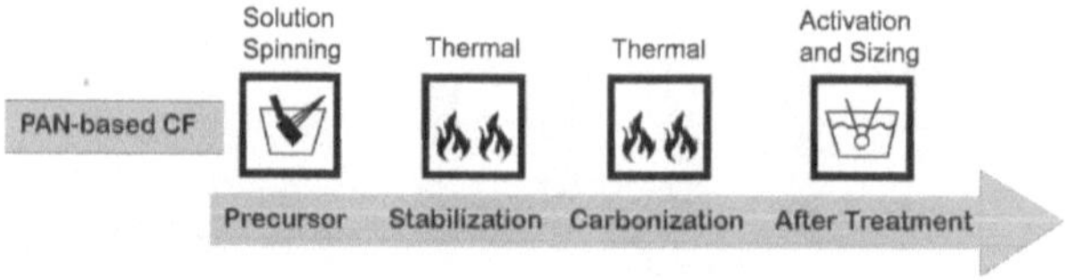

Figure 3. Steps in manufacturing of carbon fiber from PAN precursor.

The main monomer unit of PAN is acrylonitrile, made by the catalytic reaction of propylene, ammonia, and oxygen [25]. Other co-monomers are added [26] to provide the desired internal structure of fiber filaments, usually spun in a temperature-controlled bath to avoid degradation [17]. After solution spinning, washed fibers are stretched to align the molecules and control fiber thickness. Stretched PAN fiber bundles carefully heated below its 300°C melting point allows stable cyclic structure to form with dehydrogenation and oxidation, releasing thermal energy that can be recovered [18]. Further heating without oxygen increases carbon content to above 90% [2] and graphitic structures are formed above 2000°C. Fiber surfaces are oxidized with air and other chemicals to reduce surface defects and improve bonding in composites.

The PAN precursor accounts for 50–65% of CF production cost [24], including the expensive solution spinning process needed for PAN, but PAN yields only about half of its mass in CF. Low-cost precursors such as textile grade PAN, lignin, and polyolefins such as polyethylene (PE) are the focus of studies to lower the cost of CF for larger-scale application. Also promising are new processing technologies, such as plasma-assisted stabilization and carbonization, electron beam irradiation as pre-stabilization, and ultraviolet irradiation-assisted stabilization [27]. Oxidized PAN fiber is used to produce inherently flame-resistant fabrics, the demand for which makes PAN fibers more expensive. The properties of CFs produced from different precursors are shown in Table 2 [28]. It is seen that PAN results in stronger fiber compared to other precursor fibers.

Table 2. Carbon fiber precursors and achieved average strength.

Precursors	Source	Average tensile strength (GPa)	Spinning method, Yield
PAN	Propylene from crude oil	6.6	Wet, >50%
Sulfonated polyethylene	Natural gas, petroleum	2.5	Melt, 60–80%
Rayon	Cellulose (wood pulp)	0.4	Wet, 10%
Pitch, mesophase pitch	Petroleum, coal tar	0.6, 3.5	Melt, 60–80%
Lignin, lignin + PAN blend	Cellulose (wood pulp)	1.2, 2.5	Electro, wet, 10%,28%

2.1 *Automobile applications of carbon fiber*

Increased fuel economy requirements for automobiles require substantial weight reduction [15]. The high strength and light weight of CFRP composites replacing steel can lower vehicle mass by 50–70% [29], reducing fuel use and CO_2 emissions. By adding CF in appropriate structures to absorb impact energy, passenger safety can be improved [30]. Different applications requiring substantially different CF properties (Figure 4) [27,31].

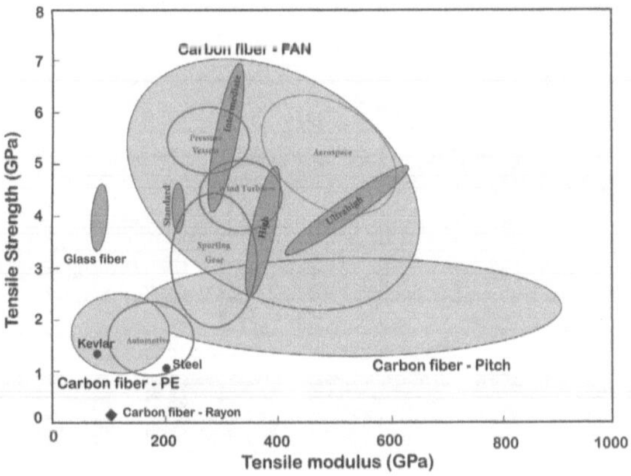

Figure 4. Properties of carbon fiber for various applications

PAN-based CFs meet current aerospace needs for CFs with very high tensile modulus and tensile strength [31]. Pressure vessels also require high tensile strength, but automotive structures are less demanding. CFs with moderate mechanical properties around 170 GPa modulus and 1.7 GPa tensile strength are adequate for automobiles [27], with low cost being the priority. Thus, other precursors are needed, such as biomass-derived PAN or other compounds, and commodity polymers including polyethylene and polypropylene. Even if not as strong or stiff as PAN-based CFs, such alternatives may be adequate for automotive, construction, and other structural applications. With new vehicle design using lightweight CFRP, one carmaker has achieved significant mass production [30].

3 GAS LIQUIDS TO CARBON FIBER

The continued availability and affordability of natural gas have supported its use as a feedstock for production of hydrogen, plastics, and other petrochemicals. Gas production through shale rock hydraulic fracturing has increased in recent years, especially in the U.S. With larger fractions of gas liquids ethane and propane than well gas, shale gas can help create high-value high-demand chemicals like ethylene and propylene. The trend to liquifying natural gas for trans-ocean shipment also creates opportunity to separate gas liquids. Steam cracking of ethane is now common in the U.S., with cracking of naphtha and gas oils more common elsewhere [32]. Other promising non-oxidative and alternative conversion methods have been proposed, some with potential for hydrogen generation [9,33]. Ethylene is the feedstock for making polyethylene (PE), with nearly 30% of all plastics being some form of PE [27].

PE is a widely available commodity polymer and has stable pricing, further assured by increased availability of gas liquids. As a potential CF precursor, PE fibers have low cost, high carbon content, ease of manufacture, and highly ordered and oriented fibers. While PAN precursor can cost ~\$10/kg, PE costs ~1/kg. With ~86% carbon content, PE can yield up to 70% CF in carbonization after stabilization with sulfonation [17,34].

The stages of CF production from gas liquids are shown in Figure 5. Natural gas liquids (NGLs) are separated cryogenically from methane, which is valued as pipeline fuel gas. The NGLs, mainly ethane and propane, are processed through steam cracking furnace and the ethylene and propylene generated are separated from hydrogen, methane, ethane, propane and larger molecule products. Ethylene (ethene) is then polymerized to form PE pellets. The process to convert PE to CF is described in more detail below. Another major commodity olefin, propylene (propene), can form polypropylene as another CF precursor but needs more research and is not discussed in this work.

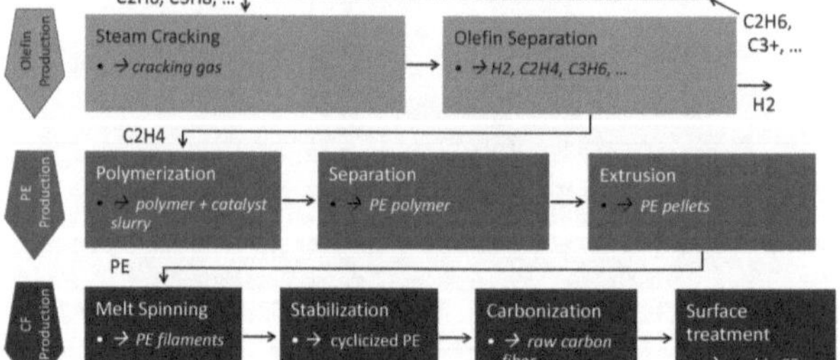

Figure 5. Gas Liquids to Carbon Fiber processes.

Lacking cross connections between its molecular chains, PE is a thermoplastic with low melting point of 140°C. Thus, unlike the solution spinning process used for PAN filaments, PE filaments can be made with a melt spinning process, which is simpler and cheaper [35]. Cross-linking of molecules is necessary to create a cyclic graphitic structure of carbon fiber that is stable when carbonized at high temperature, while retaining longitudinal structure. PE filaments can be functionalized and cyclized by heating the filaments in contact with sulfuric acid [36] or bromine [37]. Electron-beam irradiation can reduce the complexity of environmental controls of acid fumes [34]. Long stabilization time is a challenge in commercializing PE-based carbon fiber. Lower crystallinity and orientation of the PE fiber are beneficial for stabilization time, but at the expense of reducing CF yield [38]. With optimal crystal size, the overall stabilization time may be decreased.

A detailed analysis of costs of PAN-based and PE-based CF [27] shows that the overall cost of PE-based CFs is substantially lower than the PAN-based CFs. Typically, PE feedstock costs less than PAN, but processing costs may be higher [38]. PE-based CF can also be more sustainable due than the wet-spinning and conversion processes used in PAN-based CF [22].

3.1 *Energy intensity of carbon fibers manufacture from PE and PAN precursors*

The energy intensity of the PE-based CF is substantially lower than for PAN-based CF production [19] (Figure 6). Thus, the final CF made from PE may have very low energy intensity compared with CF from PAN, while also being a high-value form of carbon diverted from combustion. Energy intensity is also an indication of CO_2 emissions associated with energy use.

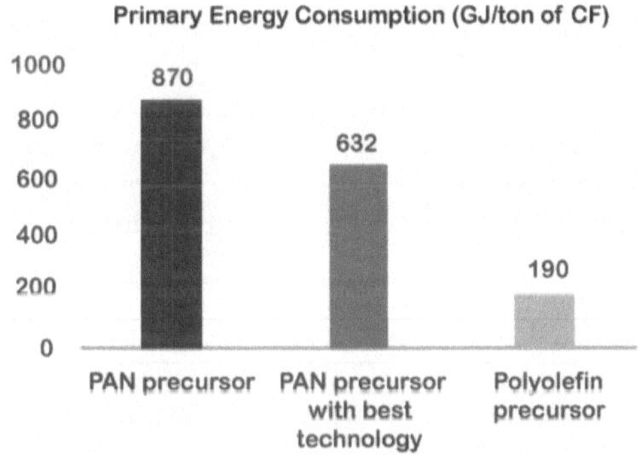

Figure 6. Comparison of energy intensity values.

4 RECENT DEVELOPMENTS AND CONCLUSIONS

PE-based CF is now being actively studied by researchers in Europe, Asia and the US [39,40]. A highly instrumented, highly flexible carbon fiber line has been established at Oak Ridge National Laboratory in the U.S. that can process 65 tons/year of melt-spun PE precursor fiber. Novel approaches can allow cyclization at lower temperature and without sulfur emissions [34,41], including bromination [37]. Boron-catalyzed graphitization process can yield tensile modulus >400 GPa [42]. A co-polymer precursor containing PE backbone and pitch side chains with some free pitch molecules was shown to yield high-strength, low-cost CFs suitable for pressure vessel applications [31]. Thus, there are now viable approaches to enable scaleup and commercialization of PE-based low-cost CF.

The continued abundance of natural gas concurrent with the urgent need to reduce CO2 emissions creates a dilemma that could be somewhat resolved if the higher carbon-content gas liquids could be diverted to high-value products rather than being burned. The symmetrical, and very strong crystalline structures possible with elemental carbon provide fiber with very high tensile modulus, and tensile specific strength. The high cost of PAN precursor and its processes can be overcome with the use of PE-based CF. The reduced cost can justify the economics of the use of carbon as a structural material in aerospace, automotive, building and infrastructure applications. Among emerging applications, high-pressure storage tanks for CNG or hydrogen fuel are needed by aircraft, automobiles, and gas-fueled vessels to save substantial operational energy by light-weighting over their lifetimes. If CF replaces aluminum and steel as a strong, low-cost, light-weight structural material, the potential for large volume applications would divert carbon away from combustion and GHG emissions.

Currently, CF has a high amount of embodied energy, about 200 MJ/kg, comparable to aluminum and higher than steel [43]. The energy costs of producing the feedstocks, creating the precursor polymer, and the CF baking process all contribute to this total. The embodied energy could be reduced by improving process efficiencies, and by recovering waste energy and 'blue' hydrogen, as much as possible. Research also continues to lower the cost of PAN-based CF by managing the energy-intensive areas, especially during the thermal stabilization step [44].

It is acknowledged that all aspects of environmental impact must be considered for any proposed technology. In addition to embodied energy use, various emissions, and environmental impact involved in the extraction, processing and manufacturing of olefins, a special concern is the potential for greater fugitive methane emissions, and CO_2 emissions associated with energy use [45,46], consistent with the justification based on climate change. Any significant increase in methane emissions from olefin production for CF could substantially negate the benefits of carbon diversion. While CF typically reduces operational energy use and environmental impact over the lifetime of the products or structures, there may be less examined impacts on energy and environment, such as the disposal or recycling of CF at end of life. The solid waste impact of CF disposal must be studied further, particularly the disposal or recycling of CFRP and other composites made with CF. Also, the potential for increased spillage or PE in processing or transport must be considered, as plastics pollution is already a major environmental problem. All these require further study.

ACKNOWLEDGMENT

The assistance of Ms. B. Sule in preparing portions of this manuscript is acknowledged.

REFERENCES

[1] Zhang J. *et al.*, (2016). Carbon science in 2016: Status, challenges and perspectives, *Carbon*, vol. 98, no. 70, pp. 708–732, doi: 10.1016/j.carbon.2015.11.060.

[2] Minus M. L. and Kumar S., (2000). Carbon Fibers, in *Kirk-Othmer Encyclopedia of Chemical Technology* vol. 26, ed.Hoboken, NJ, USA: John Wiley & Sons, pp. 729–749.

[3] Kim H.-I. *et al.*, (2017). Tensile properties of millimeter-long multi-walled carbon nanotubes, *Scientific Reports*, vol. 7, no. 1, pp. 1–7, Art no. 9512, doi: 10.1038/s41598-017-10279-0.

[4] Kazemifar F., (2022). A review of technologies for carbon capture, sequestration, and utilization: Cost, capacity, and technology readiness, *Greenhouse Gases: Science and Technology*, vol. 12, no. 1, pp. 200–230, doi:https://doi.org/10.1002/ghg.2131.

[5] Akerboom S., Waldmann S., Mukherjee A., Agaton C., Sanders M., and Kramer G. J., (2021). Different This Time? The Prospects of CCS in the Netherlands in the 2020s, *Frontiers in Energy Research*, vol. 9, p. 17, doi: https://doi.org/10.3389/fenrg.2021.644796.

[6] Biello D., (2016). The Carbon Capture fallacy, *Scientific American*, vol. 314, p. 8, [Online]. Available: https://www.jstor.org/stable/26046808.

[7] Bauer C. *et al.*, (2021). On the climate impacts of blue hydrogen production, *Sustainable Energy & Fuels*, no. 1, doi: https://doi.org/10.1039/D1SE01508G.

[8] Howarth R. W. and Jacobson M. Z., (2021). How green is blue hydrogen?, *Energy Science & Engineering*, vol. 9, no. 10, p. 12, doi: https://doi.org/10.1002/ese3.956.

[9] Zhu C., Hou S., Hu X., Lu J., Chen F., and Xie K., (2019). Electrochemical conversion of methane to ethylene in a solid oxide electrolyzer, *Nature Communications*, vol. 10, no. 1, p. 1173, 2019/03/12, doi: 10.1038/s41467-019-09083-3.

[10] Siracusa V. and Blanco I., (2020). Bio-Polyethylene (Bio-PE), Bio-Polypropylene (Bio-PP) and Bio-Poly (ethylene terephthalate) (Bio-PET): Recent Developments in Bio-Based Polymers Analogous to Petroleum-Derived Ones for Packaging and Engineering Applications, *Polymers*, vol. 12, no. 8, p. 1641. [Online]. Available: https://www.mdpi.com/2073-4360/12/8/1641.

[11] Yao Y., Graziano D. J., Riddle M., Cresko J., and Masanet E., (2015). Understanding variability to reduce the energy and GHG footprints of US ethylene production, *Environ. Sci. Technol.*, vol. 49, no. 24, pp. 14704–14716, doi: 10.1021/acs.est.5b03851.

[12] He C. and You F., (2016). Deciphering the true life cycle environmental impacts and costs of the mega-scale shale gas-to-olefins projects in the United States, *Energy Environ. Sci.*, vol. 9, no. 3, pp. 820–840, doi: 10.1039/c5ee02365c.

[13] Moran M. J., Shapiro H. N., Boettner D. D., and Bailey M. B., (2014). *Fundamentals of Engineering Thermodynamics*, 8th ed. John Wiley & Sons Inc.

[14] Das S., Life cycle assessment of carbon fiber-reinforced polymer composites, (2011). *The International Journal of Life Cycle Assessment*, vol. 16, no. 3, pp. 268–282, 2011/03/01, doi:10.1007/s11367-011-0264-z.

[15] Sunter D. A., Morrow W. R. III, Cresko J. W., and Liddell H. P. H., 2015. The manufacturing energy intensity of carbon fiber reinforced polymer composites and its effect on life cycle energy use for vehicle door lightweighting, in *20th International Conference on Composite Materials*, Copenhagen, 19th–24th July 2015.

[16] Liu Y., Zwingmann B., and Schlaich M., (2015). Carbon fiber reinforced polymer for cable structures—A review, *Polymers*, vol. 7, no. 10, pp. 2078–2099, doi: 10.3390/polym7101501.

[17] Liu Y. and Kumar S., (2012). Recent progress in fabrication, structure, and properties of carbon fibers, *Polymer Reviews*, vol. 52, no. 3, pp. 234–258, doi: 10.1080/15583724.2012.705410.

[18] Frank E., Hermanutz F., and Buchmeiser M. R., (2012). Carbon fibers: Precursors, manufacturing, and properties, *Macromol. Mater. Eng.*, vol. 297, no. 6, pp. 493–501, doi: 10.1002/mame.201100406.

[19] Energetics Incorporated, (2017). *Bandwidth study on energy use and potential energy saving opportunities in U.S. carbon fiber reinforced polymer manufacturing*, U.S. Department of Energy.

[20] Jerome P., (2001). Composite materials in the airbus A380-from history to future, in *Proceedings of the 13th International Conference on Composite Materials (ICCM-13), Beijing, China*, pp. 25–29.

[21] Das S., Warren J., West D., and Schexnayder S. M., (2016). *Global carbon fiber composites supply chain competitiveness analysis*, Oak Ridge National Laboratory; The University of Tennessee, Knoxville.

[22] Roding T., Langer J., Barbosa T. M., Bouhrara M., and Gries T., (2022). A review of polyethylene-based carbon fiber manufacturing, *Applied Research*, doi: 10.1002/appl.202100013.

[23] Toray. *Carbon Fiber*. https://www.toraycma.com/products/carbon-fiber/ (accessed).

[24] De Palmenaer A., Wortberg G., Merke M., Roeding T., Gries T. G., and Seide G. H. (2017) Carbon fiber development: From polyethylene-based precursors to carbon fibers. *Chemical Fibers International*

[25] Brazdil J. F., (2012). Acrylonitrile, in *Ullmann's Encyclopedia of Industrial Chemistry*, ed. Weinheim: Wiley-VCH Verlag GmbH & Co. KGaA.

[26] Johnson M. C. and Sullivan J. L., (2014). *Lightweight materials for automotive application: An assessment of material production data for magnesium and carbon fiber*, Argonne National Lab (ANL), Argonne, IL (United States).

[27] Choi D., Kil H.-S., and Lee S., (2019). Fabrication of low-cost carbon fibers using economical precursors and advanced processing technologies, *Carbon*, vol. 142, pp. 610–649, doi: 10.1016/j.carbon.2018.10.028.

[28] Kaur J., Millington K., and Smith S., (2016). Producing high quality precursor polymer and fibers to achieve theoretical strength in carbon fibers: A review, *J. Appl. Polym. Sci.*, vol. 133, no. 38, doi: 10.1002/APP.43963.

[29] Wheatley A., Warren D., and Das S., (2014). Development of low-cost carbon fibre for automotive applications, in *Advanced Composite Materials for Automotive Applications: Structural Integrity and Crashworthiness*, 1st ed., A. Elmarakbi, Ed.: John Wiley & Sons Ltd., ch. 3, pp. 51–73.

[30] Wakekar A. D., Patil S., Deshmukh S. P., and Thakar C. M., (2022). The review of Carbon Fiber Materials in Automative Industry, *ECS Transactions*, vol. 107, p. 9.

[31] Chung M., (2020). Developing a new polyolefin precursor for low-cost, high-strength carbon fiber, ed. 2020 DOE hydrogen program annual merit review.

[32] Worrell E., Price L., Neelis M., Galitsky C., and Nan Z., *World Best Practice Energy Intensity Values for Selected Industrial Sectors*, Ernesto Orlando Lawrence Berkeley National Laboratory, Berkeley, CA, USA, 2007.

[33] Gerzeliev I. M., Fairuzov D. K., Gerzelieva Z. I., and Maksimov A. L., (2019). Production of Ethylene from Ethane Fraction by a Method Alternative to Steam Cracking, *Russian Journal of Applied Chemistry*, vol. 92, no. 11, pp. 1549–1557, 2019/11/01, doi:10.1134/S1070427219110120.

[34] Choi D., Yoo S. H., and Lee S., (2019). Safer and more effective route for polyethylene-derived carbon fiber fabrication using electron beam irradiation, *Carbon*, vol. 146, pp. 9–16, doi: 10.1016/j.carbon.2019.01.061.

[35] De Palmenaer A., Wortberg G., Drissen F., Seide G., and Gries T., (2015). Production of polyethylene based carbon fibres, *Chemical Engineering Transactions*, vol. 43, pp. 1699–1704, doi: 10.3303/CET1543284.

[36] Younker J. M., Saito T., Hunt M. A., Naskar A. K., and Beste A., (2013). Pyrolysis pathways of sulfonated polyethylene, an alternative carbon fiber precursor, *J. Am. Chem. Soc.*, vol. 135, no. 16, pp. 6130–6141, doi: 10.1021/ja3121845.

[37] Laycock B. *et al.*, (2020). Pyrolysis of brominated polyethylene as an alternative carbon fibre precursor, *Polymer Degradation and Stability*, vol. 172, Art no. 109057, doi: 10.1016/j.polymdegradstab.2019.109057.

[38] Peijs T., Kirschbaum R., and Lemstra P. J., (2022). Chapter 5: A critical review of carbon fiber and related products from an indusrial perspective, *Advanced Industrial and Engineering Polymer Research*, vol. 5, p. 17, doi: https://doi.org/10.1016/j.aiepr.2022.03.008.

[39] Brown K. R., Harrell T. M., Skrzypczak L., Scherschel A., Wu H. F., and Li X., (2022). Carbon fibers derived from commodity polymers: A review, *Carbon*, vol. 196, pp. 422–439, 2022/08/30/, doi: https://doi.org/10.1016/j.carbon.2022.05.005.

[40] Aldosari S. M., Khan M., and Rahatekar S., (2020). Manufacturing carbon fibres from pitch and polyethylene blend precursors: a review, *J. Mater. Res. Technol.*, vol. 9, no. 4, pp. 7786–7806, doi: 10.1016/j.jmrt.2020.05.037.

[41] Frank E., Muks E., Ota A., Herrmann T., Hunger M., and Buchmeiser M. R., (2021). Structure evolution in polyethylene-derived carbon fiber using a combined electron beam-stabilization-sulphurization approach, *Macromolecular Materials and Engineering*, vol. 306, no. 10, p. 2100280, doi: https://doi.org/10.1002/mame.202100280.

[42] Barton B. E. *et al.*, (2017). High-modulus low-cost carbon fibers from polyethylene enabled by boron catalyzed graphitization, *Small*, vol. 13, no. 36, p. 1701926, doi: 10.1002/smll.201701926.

[43] Howarth J., Mareddy S. S. R., and Mativenga P. T., (2014). Energy intensity and environmental analysis of mechanical recycling of carbon fibre composite, *Journal of Cleaner Production*, vol. 81, pp. 46–50, 2014/10/15/, doi:https://doi.org/10.1016/j.jclepro.2014.06.023.

[44] Nunna S., Maghe M., Fakhrhoseini S. M., Polisetti B., and Naebe M., A pathway to Reduce Energy Consumption in the Thermal Stabilization Process of Carbon Fiber Production, *Energies*, vol. 11, 2018, doi: 10.3390/en11051145.

[45] Yang M. and You F., (2017). Comparative techno-economic and environmental analysis of ethylene and propylene manufacturing from wet shale gas and naphtha, *Ind. Eng. Chem. Res.*, vol. 56, no. 14, pp. 4038–4051, doi: 10.1021/acs.iecr.7b00354.

[46] Yao Y., Graziano D. J., Riddle M., Cresko J., and Masanet E., (2016). Prospective energy analysis of emerging technology options for the United States ethylene industry, *Ind. Eng. Chem. Res.*, vol. 55, no. 12, pp. 3493–3505, doi: 10.1021/acs.iecr.5b03413.

Innovation and Technological Advances for Sustainability – Al-Naemi et al. (Eds)

Corrosion deterioration of suspension bridge cable wires-case study

Abdulwahab Ibrahim
Mechanical Engineering, University of Doha for Science and Technology, Doha, Qatar

Scott MacIntyre
Forensic & Metallurgical Engineering, Atlantic Metallurgical Consulting, Dartmouth, Canada

ABSTRACT: Corrosion is a natural process that has a significant impact on our life. The negative impact of corrosion covers economy, security, infrastructure, energy, environment, health, and safety. Infrastructure corrosion deterioration is influenced not only by the materials used or the construction technique, but also by the effect of climate change and environmental variables over the structure's life cycle. Bridges failures are frequently costly in terms of lost business, fatalities, environment, and the replacement costs required to reconstruct the damaged bridges. Increased service life and efficiency are necessary to accomplish and improve the sustainability of steel-constructed bridges. Corrosion and corrosion failures have a significant impact on the durability of steel structures such as bridges. In this paper, a corrosion failure case study is presented. A suspension bridge shows signs of corrosion on the wrapping protection wires and the cable wires. The investigation involved visual examination, chemical analysis of the corrosion product, metallographic analysis, SEM, and EDS analysis. It is found that the deterioration of the protective galvanized is related to the chloride containing environment. The corrosion process started as pitting and eventually cracking and final failure.

Keywords: Corrosion, Galvanized Steel, Failure, Suspension wires

1 ANALYSIS INTRODUCTION AND BACKGROUND INFORMATION

Corrosion has a significant economic, environmental, and long-term impact on all our aspects of life. The consequence of corrosion affects our society, economy, and environment. Infra structure such as highways, bridges, buildings, as well as oil and gas and chemical processing industry suffers from corrosion [1,2]. Corrosion disrupts operations and necessitates significant and costly repair and replacement of damaged assets. In addition to creating serious environmental damage and dangers to public safety, the world losses a lot of its resources because of corrosion. The US Federal Highway Administration (FHWA) released a study on the direct cost of metallic corrosion in 2002, which was projected to be $276 billion, or nearly 3.1 percent of the nation's Gross Domestic Product (GDP) [4].

The deterioration of infrastructure is not only affected by the type of materials used or the construction procedure, but by the effect of climate change and the environmental variables during the service time over the structure life span. Accelerated high level of carbon dioxide (CO_2), high temperature, acidic rain, and halogen gases are known as one of the main factor altering our surrounding environment and consequently, making it more corrosive and hastening degradation. This deterioration has a great impact on infrastructure and defiantly would lead to negative impact on sustainability, durability, and shorten the life cycle [5–7]. To reduce the enormous economic and environmental impact of corrosion and to improve the sustainability of construction materials, existing corrosion control measures such as good material design and selection, use of corrosion inhibitors, coatings, cathodic protection, are applied. The catastrophic impact of highway bridges have brought the attention of scientific and safety community to address the problem and study the vulnerability of existing bridges and come up with different approaches to minimize and even stop the corrosion of bridges.

A suspension bridge is a continuous girder hung by suspension cables. Th cables pass through the main towers via a special structure known as saddle. The major load carrying members in the suspension bridges are the cables. The cables are high strength steel tension members and act as tension elements that extend from anchors at each of their ends. Bridge failures are traditionally associated with economic, environmental consequences and even life losses. There has been plenty of studies and scientific reports that investigated bridges related accidents. Researchers and engineers have paid close attention to the characteristics and causes of bridge failures. Bridges failures causes can be classified as into natural causes and human causes. In addition to design error and construction related mistakes, factors such as structural type, use, construction material, and service age were reported as a failure causes [8–10]. Bridges collapses have highlighted corrosion-related infrastructure failure on a global scale. Corroded bridge failures have occurred in the past and unfortunately associated with resulting in fatalities [8,11,12].

Operating and environmental conditions bring a real challenge of suspension bridges, specifically corrosion related challenges. The surrounding environment of most of the suspension bridges provides moisture, aggressive corrosive species, acid rain, and salty water. Pollutants such as SO_2, H_2S, and CO_2 are very corrosive when reacts with steel in an aqueous environment. Other factor is the high number of vehicles related pollution which may accelerate the corrosion rate of bridges. Researchers have revealed different corrosion mechanisms involved in corrosion of suspension bridges. Stress corrosion cracking (SCC), pitting corrosion, hydrogen cracking, crevice, and corrosion fatigue [13]. Because of its nature as cracking processes, SCC of suspension wires is one of the failures that investigated by many researchers. Morgado *et al* conducted a failure analysis study of steel cable used in a suspension bridge. They concluded that stress corrosion cracking (SCC) was the main cause of the failure. The failure happened into two stages, in the first stage about 50% the wires cracked due to SCC, in the second stage the remaining wires were not able to withstand the load and eventually broke by ductile fracture [14]. Suspension wires are mainly made from high strength galvanized steel. Galvanizing is the most common corrosion protection method due to its affordability, applicability, and effectiveness. The galvanized layer act as a barrier preventing corrosion species such as salt and moisture from reacting with the steel matrix [15]. Rou *et al* conducted an experimental study on the effect of pH and chloride concentration on stress corrosion behaviour of galvanized steel wires. The authors found that corrosion rate is related to pH and Cl^- concentration. However, the relation is not linear. The degradation rate was found to increase with the increase of the strength values. One approach for corrosion testing is to perform accelerated corrosion testing which has been developed to provide information regarding the degradation mechanism and provide comparative data in a short time. Betti *et al* studied the deterioration of high strength steel alloy using accelerated corrosion testing approach. The authors used cyclic corrosion tests to evaluate the effect of corrosion on galvanized and ungalvanized suspension bridge wires. The samples were tested under variety of conditions such as acidic salt spray, dry conditions, 100% relative humidity, and different levels of stresses. The results show that the samples show significant embrittlement. It is found that the ductility of the galvanized steel is related to the roughness of the external surface of the wire induced by the corrosion process [16]. Deeble *et al* have used nondestructive testing (NDT) methods to monitor the corrosion of high strength steel wires in a suspension bridge. The authors used a sensor network to monitor and acquire information about the environmental conditions and the corrosivity of the suspension bridge cables. The data recorded are relative humidity, temperature, and corrosion rate. The corrosion rate is affected by the relative humidity levels, significant increase in corrosion recorded when the RH level is greater than 50%. As expected, corrosion rate as measured using the liner polarization method, found to have strong correlation with the temperature [17].

This paper presents a case study summarizes the findings from a corrosion failure analysis investigation conducted on a suspension bridge. There were 61 high strength steel galvanized cables protected by galvanized wrapping wires. The corroded wrapping protection wires as well the failed cable wires were examined. The investigation includes visual examination, fractography, metallographic analysis, SEM, EDS, and metallographic analysis.

2 EXPERIMENTAL WORK AND ANALYSIS

2.1 *Visual examination*

Failed wrapping wires and cable wires were initially visually examined in as received conditions. In some cases, failed parts were cleaned with running water, acetone, ultrasound cleaner, and rust removal. The corrosion product was collected from the failed parts before any cleaning was carried out.

2.2 *Metallography and microstructural analysis*

Samples cut from the wrapping wires and the tension cable were prepared for metallographic analysis by hot mounting, grinding, polishing, and etching. In the first step, the samples were ground using SiC paper of 320, 400, 600, and 1000 grit. Polishing was performed using alumina suspension solution with a particle size of 9, 6, 3, and 1 μm. An optical microscope with a magnification power up to 1000 X attached to a digital camera and image analyzer software was used for optical microscopy analysis.

2.3 *Chemical analysis and Scanning Electron Microscopy (SEM)*

Chemical analysis of representative samples cut from the wrapping protection wires and the cable wires were analyzed using a Spectro test spectrometer equipped with ultraviolet spark probe under argon gas. SEM examination was performed using a Tescan field emission scanning electron microscope (FESEM), while EDS was used for micro analysis.

3 RESULTS AND DISCUSSION

3.1 *Inspection & visual examination*

To establish a base line of the structure, the first inspection was performed on the bridge. The wrapping protection wires (PW) and the cable wires (CW) were examined carefully in the lab using a stereoscopic microscope. Five sections close to the wire band were cut from the bridge wrapping wires and subjected to analysis and examination. Figure 1 shows a picture of wrapping wires and the cable band, respectively. The wrapping wires were made from galvanized steel and coated with an orange paint for more protection against environmental conditions. The wrapping wires were used to provide protection of the high strength steel suspension cables/wires.

Figure 1. Close up picture shows; (a) wrapping wires, (b) conditions of cable wires.

In general, the wrapping wires are affected by atmosphere that contains acid rain, moisture, and dissolved gases. The exposure of the wrapping wires to the de–icing salt and the splash are considered from the factors that affected the wires. In addition to the wrapping wires, the suspension cable wires were inspected for corrosion and damage. Similar to the wrapping wires, the suspension cable wires show sign of corrosion at 6:00 O'clock position as shown in Figure 2. Other locations had less severe corrosion because certain wires showed slightly damaged galvanized layer, possibly from what is known as a white rust layer formation due to the reaction between the zinc layer and the surrounding environment. Figure 3 shows the samples collected from the suspension cable wires. The failed wires from the suspension cable examined show that gradual corrosion and erosion due to the environmental conditions that created a very thin fracture surface. The main factor that contributed to corrosion of the cable wires is the water that enters the wrapping wires/ cables. Water may enter the wires either as a liquid or as a vapour which may condensate when the temperature drops. There was clear evidence that the wrapping wires did not provide enough protection to the cables and acted as a source for water penetration. The thinning which is attributed to corrosion has led to cross sectional reduction, and consequently reduction of load bearing ability and failure. The wire diameter at the fracture surface tip recorded to be 1.21 mm, while it recorded to be 4.72 mm and 3.56 mm at 18 mm away from the fracture surface.

Figure 2. Corrosion of suspension cable wires.

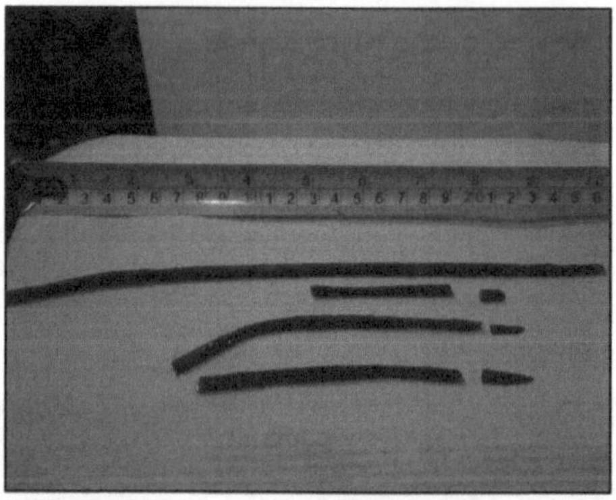

Figure 3. Failed wires after samples were cut for SEM examination.

Figure 4 shows the fracture surface of the suspension cable wire. The guidelines for inspection and strength evaluation of suspension bridge parallel wire cables classified cable wires corrosion severity into 4 stages [18]. The 4 failed cable wires examined can be classified into stage 4 which is the more sever stage. The onsite inspection of the cable wires still in use show different stages of corrosion in which stage 1, stage 2, and stage 3, and stage 4 where seen.

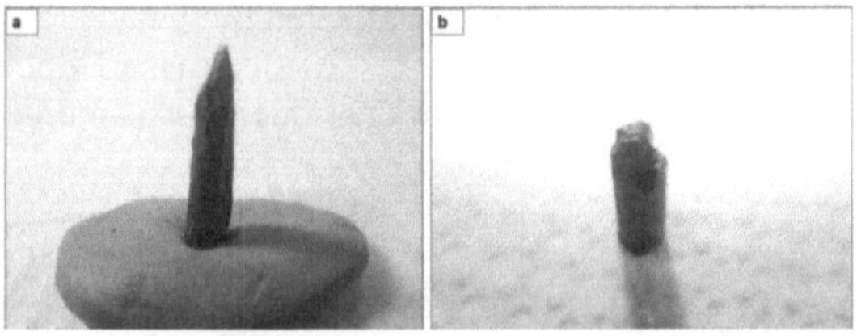

Figure 4. Fracture surface of the suspension wire shows (a) ductile fracture, (b) thinking.

3.2 *Metallography and zinc layer measurements*

By far, the most common, effective, and affordable corrosion prevention method for steel is galvanizing. The galvanizing layer serves as a barrier coating, preventing corrosive chemicals such as Cl-ions from entering the steel matrix. However, the galvanized protection layer deteriorates under harsh environment such as salty water, high chloride, acid rain, and SO2 containing environment [19]. Chemical analysis results of the corrosion product collected from the failed wires show high level of chlorides (37 mg/KG) that would negatively affect the protection ability of the zinc layer. Samples collected from different locations/positions were prepared for metallographic analysis in which each sample consist of 5 wires mounted together. The zinc layer thickness measurements were performed at the outer diameter (OD) and the inner diameter (ID). The results show that the zinc layer thickness is higher on the OD than ID, logically because of the extra paint applied on the outer surface. Figure 5 shows the zinc layer measurement of two wrapping wires sections. There was no clear trend on how the sections and the wire position could affect the zinc layer thickness. It appears that both sections were exposed to the same conditions and the zinc layer deteriorated at a similar rate. It is worth noting that some zinc thickness measurements showed complete depletion of the zinc layer. Most of the samples showed the complete deterioration were taken from 6:00 O'clock position.

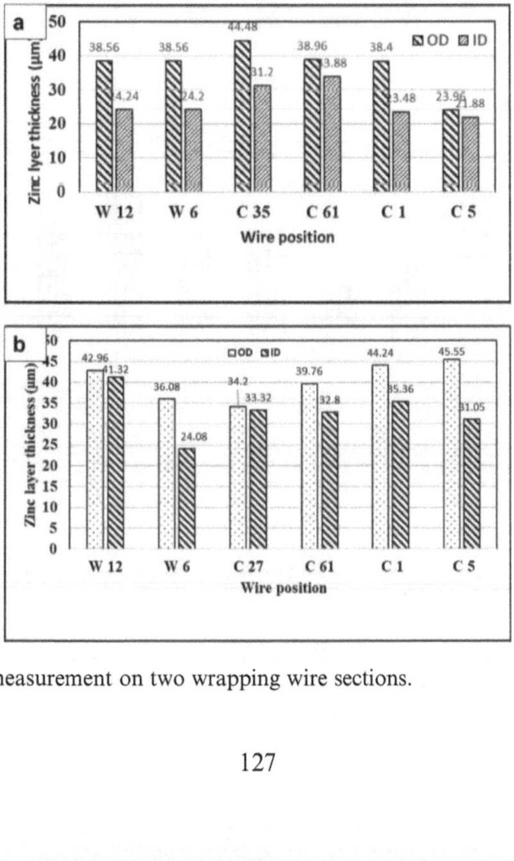

Figure 5. Zinc layer measurement on two wrapping wire sections.

The fact that the galvanized layer thickness is affected by the wire position was confirmed from the optical micrographs obtained. Figure 6 shows a cracked zinc layer, other samples show clear signs of pitting. In some cases, more than one pit was merged together to form a large pit that propagated in the wire material.

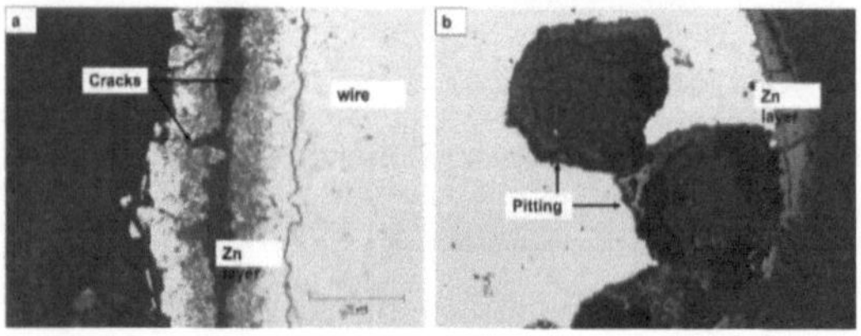

Figure 6. OM showing zinc layer deterioration. (a) cracking, (b)pitting.

In addition to wrapping wire zinc layer measurements, galvanized thickness measurements were conducted on the suspension cable wires. Result is shown in Figure 7. It appears that the smallest zinc layer thickness was associated with strand 5 and strand 1. It is worth noting that strand 1 and strand 5 are at the bottom of the cable while strand 35 and 27 are at the middle and strand 57 and strand 61 are at the top. This fundings is in a good agreement with the fact that the wrapping wires and the tension cable wires corrode the most at the bottom position (6:00 O'clock position).

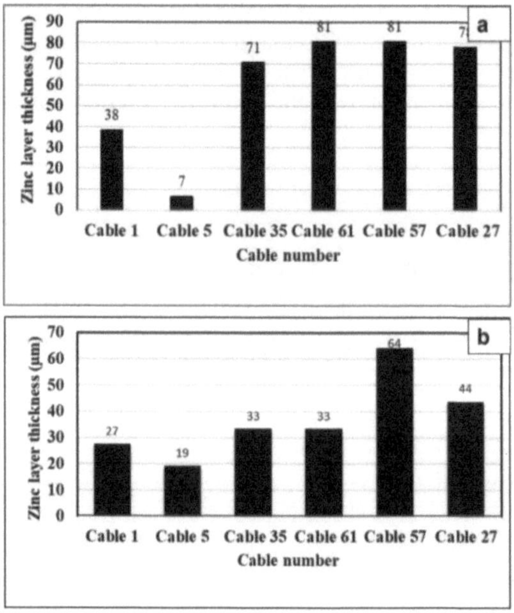

Figure 7. Zinc layer thickness measurements of two different sections.

3.3 *Scanning Electron Microscopy (SEM) & fractographic analysis of cable wires*

Fractographic analysis on the failed suspension cable wires was performed using SEM. Figure 8 shows an SEM micrograph of the fracture surface of sample 1. Higher magnification of the fracture surface shows presence of pitting which is most likely a result of corrosion as shown in Figure 8b and c. Additionally, it shows a secondary crack propagates from the surface along the transverse section of the wire. Moreover, evidence of deformation van be seen as shown in Figure 9d. The deformation is associated with the outer surface of the wire and appears to be a result of bending/rotation forces applied on the wire.

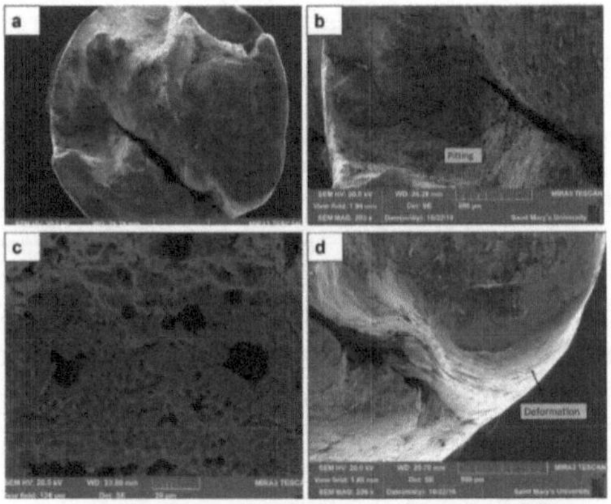

Figure 8. SEM observation of the suspension wire; (a) ductile type fracture; (b) and (c) pitting;(d) deformation.

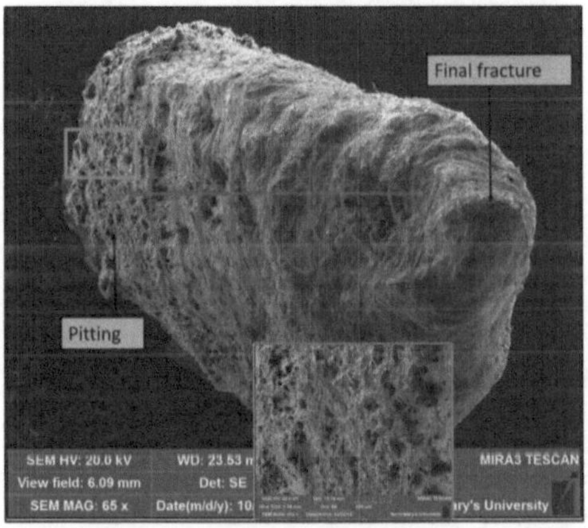

Figure 9. SEM micrograph showing ductile fracture initiated by pitting.

Corrosion product collected from the suspension wires was analyzed using EDS analysis. Table 1 shows the chemical analysis result obtained from three spectra acquired from the corrosion product. The analysis shows presence of noticeable amount of chloride, oxygen, and zinc. The presence of chloride is an indication that the wires were in contact with salt water. Sample 2 that was examined using EDS showed different fracture surface. It shows an evidence of circular crack progression from the outer surface towards the wire center. The

Table 1. EDS analysis of the corrosion product extracted from the fractured suspension wires.

Element-Wt% Spectrum number	O	Cl	Ca	Fe	**Zn**
1	45.57	3.13	——	44.18	7.12
2	29.11	2.37	——	53.1	15.42
3	51.44	0.99	——	35.9	11.67

wire outer surface shows pitting which was not seen at the wire center. This is an indication that corrosion/pitting that happens at the outer surface may initiated cracks that started initially at the pitting area and then propagated toward the center. Chemical analysis of the corrosion product extracted from the sample showed similar result to the other samples in which high chloride concentration exist.

More evidence of the ductile mode of fracture was obtained from the fractography analysis of wire #3. High magnification shows presence of elongated dimples as an indication of the ductile failure, Figure 10. Elongated dimples structure is an indication of the loading direction which is in this case a tensile force along the wire longitudinal axis. It is obvious that the pitting started at the outer surface of the wire which propagated in a form of a crack and eventually reached its critical length, and the wire failed. Figure 11 shows a fracture surface of one of the failed wires. It shows a similar pitting formation in addition to a transfer crack. It is worth noting that among the 4 wire examined, no evidence of fatigue fracture or brittle fracture have been observed.

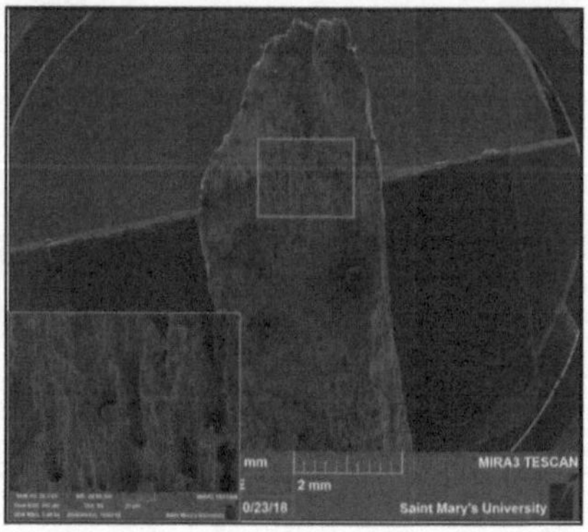

Figure 10. Fracture surface showing dimple-like structure.

Figure 11. SEM micrograph showing pitting and cacking.

4 CONCLUSION

(1) The main reason of the corrosion of the cable wire is the exposure to salt water/humidity. The corrosion attack resulted in thinning and reduction in cross diameter which eventually could not bear the load applied and fractured.

(2) The corrosion stage of the failed cable wires classified as stage 4.
(3) The wrapping wires will help to protect the cable wires as far as the wrapping wires able to prevent water and humidity from getting into contact with the cable wires.
(4) Due to the exposure to sale water, zinc layer deteriorated at different levels. Some wrapping wires show complete depletion of the zinc layer.
(5) One of the option to prevent/minimize corrosion of the wires is to use dehumidification. Using dehumidification should be combined with using wrapping wires and elastomeric membrane.

REFERENCES

[1] Hou, B., Li, X., Ma, X., Du, C., Zhang, D., Zheng, M., Xu, W., Lu, D. and Ma, F., (2017). The cost of corrosion in China. *Npj Materials Degradation*, 1(1), pp.1–10.

[2] Cherry, Brian Wilson, and Brian S. Skerry. (1983). *Corrosion in Australia: The Report of the Australian National Centre for Corrosion Prevention and Control Feasibility Study*, June 1983: BW Cherry and BS Skerry. Department of Materials Engineering, Monash University.

[3] Johnsen, Roy. (2015). Corrosion challenges for the oil and gas industry in the State of Qatar. In *Proceedings of the TMS Middle East—Mediterranean Materials Congress on Energy and Infrastructure Systems (MEMA 2015)*, pp. 3–12. Springer, Cham.

[4] Koch, G. H., Brongers, M. P., Thompson, N. G., Virmani, Y. P., and Payer, J. H. (2002). *Corrosion cost and preventive strategies in the United States (No. FHWA-RD-01-156,, R315-01)*. United States. Federal Highway Administration.

[5] Romero, N., de Rincón, O. T., Sanchez, M., Rincon, A., Paz, S., Montes, P., and Millano, V. A. L. E. N. T. I. N. A. (2006). Behavior of galvanized steel in tropical marine environments. *Materials performance*, 45(2), 24.

[6] Al-Sherrawi, M. H., Lyashenko, V., Edaan, E. M., and Sotnik, S. (2018). *Corrosion as a Source of Destruction in Construction.*

[7] McDonald, D. B. (2011). Corrosion Protection for Concrete Structures in Marine Environments. *In Coastal Engineering Practice* (2011) (pp. 78–88).

[8] Zhang, G., Liu, Y., Liu, J., Lan, S., and Yang, J. (2022). Causes and statistical characteristics of bridge failures: A review. *Journal of Traffic and Transportation Engineering* (English Edition).

[9] Rajendran, R. (2019). *Bridge Failures Case Studies in India* (2016–2019).

[10] Sharma, P., and Kumar, J. S. The Collapse of Jahu Bridge (Himachal Pradesh) A Case Study.

[11] Deng, L., Wang, W., and Yu, Y. (2016). State-of-the-art review on the causes and mechanisms of bridge collapse. *Journal of Performance of Constructed Facilities*, 30(2), 04015005.

[12] Milillo, P., Giardina, G., Perissin, D., Milillo, G., Coletta, A., and Terranova, C. (2019). Pre-collapse space geodetic observations of critical infrastructure: the Morandi Bridge, Genoa, Italy. *Remote Sensing*, 11(12), 1403.

[13] Furuya, K., Kitagawa, M., Nakamura, S. I., and Suzumura, K. (2000). Corrosion mechanism and protection methods for suspension bridge cables. *Structural Engineering International*, 10(3), 189–193.

[14] Morgado, T. L. M., and e Brito, A. S. (2015). A failure analysis study of a prestressed steel cable of a suspension bridge. *Case Studies in Construction Materials*, 3, 40–47.

[15] Li, R., Miao, C., and Wei, T. (2020). Experimental study on corrosion behaviour of galvanized steel wires under stress. *Corrosion Engineering, Science and Technology*, 55(8), 622–633.

[16] Betti, R., West, A. C., Vermaas, G., and Cao, Y. (2005). Corrosion and embrittlement in high-strength wires of suspension bridge cables. *Journal of Bridge Engineering*, 10(2), 151–162.

[17] Deeble Sloane, M. J., Betti, R., Marconi, G., Hong, A. L., and Khazem, D. (2013). Experimental analysis of a nondestructive corrosion monitoring system for main cables of suspension bridges. *Journal of Bridge Engineering*, 18(7), 653–662.

[18] Mayrbaurl, R. M., and Camo, S. (2004). Guidelines for inspection and strength evaluation of suspension bridge parallel-wire cables (Vol. 534). *Transportation Research Board.*

[19] Yu, Z., Hu, J., and Meng, H. (2020). A review of recent developments in coating systems for hot-dip galvanized steel. *Frontiers in Materials*, 7, 74.

Food security & sustainability

Innovation and Technological Advances for Sustainability – Al-Naemi et al. (Eds)

Ecofriendly method to control fusarium wilt on tomato

Halima Hussein & Shaker Al-Dulaimi
Department of Plant Protection, University of Baghdad, Baghdad, Iraq

Mustafa Adhab
Department of Plant Protection, University of Baghdad, Baghdad, Iraq
ORCID ID: 0000-0003-0579-9838

ABSTRACT: Fusarium wilt causes economic losses on tomatoes every year. Thus, a variety of chemicals have been used to combat the disease. Pesticides have been effective in managing the disease, but they keep damaging the environment. Recently, eco-friendly approaches have been used to control plant diseases. This study aimed to achieve an environmentally safe solution using biological agents to induce systemic resistance in tomato plants to control Fusarium wilt disease caused by *Fusarium oxysporum* f.sp. *lycopersici* (FOL) in the greenhouse. The pathogen (FOL) has been molecularly confirmed and the biological agents have been isolated from the Iraqi environment. The effectiveness of the biological agents has been tested and confirmed. Results showed that the treatment of combined *Pseudomonas fluorescence* and *Streptomyces* sp. decreased the infection percentage and disease severity on tomato plants significantly more than the single treatment with either Streptomyces or *Pseudomonas fluorescence*. These results indicate a potential eco-friendly solution to control fusarium wilt on tomato plants.

Keywords: Biological control, Fusarium oxysporum, Induced Systemic Resistance, tomato wilt

1 INTRODUCTION

Tomato is an important crop in Iraq and has been the target of different pathogens in the field [1]. Different pathogens belonging to different groups have been confirmed to cause serious damage to tomatoes in both open and protected farming; among these pathogens are several fungi, bacteria, and viruses that cause a significant decrease in tomato yield [2–5]. Fusarium oxysporum f.sp. lycopersici (FOL) causes significant losses to tomato plants annually; it can alone decrease the yield of tomatoes to about 60% in India and in some circumstances, the damage may reach 80% of the yield [6–8].

FOL is found to attack the root epidermis, from which it penetrates and moves into the vascular tissues. In the xylem vessels, FOL colonizes and produces mycelium and conidia. As a result of this activity, vessel clogging happens and water lack and wilt symptoms appear on the infected plants [9–11]. To date, there are three distinguished physiological races (1, 2 and 3) known for FOL on tomato cultivars [12,13]. Since FOL reproduces asexually, genetic exchange occurs via somatic fusion and heterokaryon format between vegetative compatible strains [14].

Traditional pesticides have been regularly used to combat FOL even though the pesticides are known to pollute the environment. To avoid the negative effects of chemicals, scientists work actively to find safe alternatives to control diseases. Biological agents are in many cases the best eco-friendly and viable replacement for pesticides on different crops and pathogens

DOI: 10.1201/9781003496724-13

[15–20]. The latter studies aimed to use Plant Growth Promotion Rhizobacteria (PGPR) and to propose it as a replacement for pesticides in the management of plant disease programs. PGPR is a group of eco-friendly microorganisms, found surrounding the roots and promote growth, and provide protection to plants in different ways; some of these ways are nitrogen fixing, producing iron-bearing materials, phosphorus thawing, and volatile organic compounds. Other activities include the production of enzymes that stimulates induced systemic resistance (ISR), such as Chitinase, Lipase, Glucanase, Protease and Amylases, which provide protection to the plants against pathogens [21–23]. This study aimed to assess the validity of some PGPR bacteria in controlling pathogenic fungi FOL through enhancing plant defenses.

2 MATERIALS AND METHODS

2.1 *Pathogenicity test on tomato seeds under greenhouse conditions*

Characterization of FOL has been done morphologically and molecularly at Mycotoxins Laboratory and the sequence obtained was dispensed into GenBank under accession number MH458918 [24]. The pathogenicity test was done according to the process in Figure 1.

$$\text{Germination } \% = \frac{\text{Total \# of seed germinated}}{\text{Total \# of seeds sown}} \times 100$$

Figure 1. Steps of testing pathogenicity of FOL using tomato seeds in the greenhouse.

2.2 *Testing PGPR bacteria antifungal effectiveness against FOL*

The Raspor *et al.*, [25] method was followed to assess the efficacy of PGPR bacteria as shown in Figure 2. Each treatment in this experiment was replicated three times.

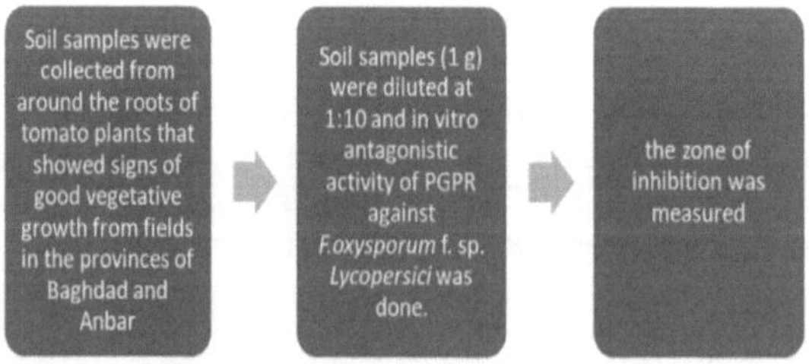

Figure 2. Steps of isolating PGPR from the Iraqi environment.

The following formula was used to measure the inhibition zone:

Biological control activity = $A/(A + B) \times 100$

A = area of inhibition (the distance between the biocontrol agent line and the end of the fungal growth).

B = the fungal expansion towards the biocontrol agent line [25].

2.3 *Characterization of PGPR and testing its activity in combating FOL*

Morphological characterization of Streptomyces spp. was done according to [26], while the identification of *Pseudomonas fluorescence* was done by Central Public Health Laboratory in Iraq using the Analytical Profile Index 20 Enterobacteriaceae.

To evaluate PGPR activity against FOL, steps in Figure 3 were followed.

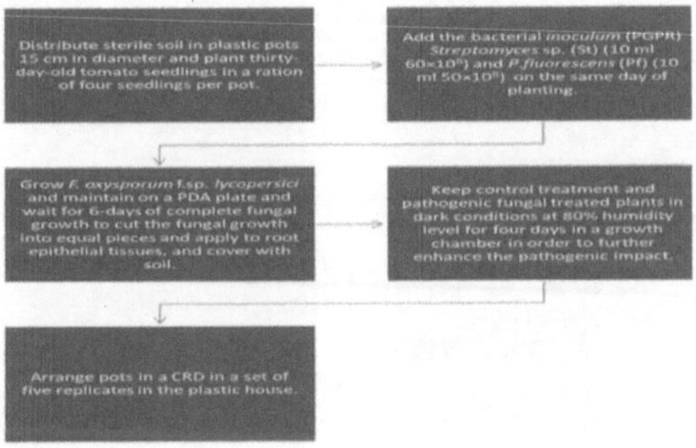

Figure 3. Steps to test the efficacy of PGPR bacteria against *Fusarium oxysporum* f.sp. lycopersici.

The following treatments were examined:

(1) Control: (No FOL; No biocontrol agent).
(2) FOL: (Pathogenic fungus only).
(3) St: (Streptomyces sp. treatment).
(4) Pf: (*P.fluorescens* treatment).
(5) FOL/St
(6) FOL/Pf
(7) St/Pf
(8) FOL/St/Pf

2.4 *Enzyme activities and Phenol content test*

Peroxidase enzyme activity was tested according to [27], enzyme PAL [28], Phenol content [29]. All reads were taken 7, 14 and 21-days post FOL inoculation in tomato plants.

2.5 *Measuring disease severity and infections*

Symptom severity on the shoot system was analyzed 6-weeks post fusarium inoculation following Souza *et al.*, [30] index:

0 = no symptoms.
1 = leaves show yellowing in 1–25% of the total leaves near the stem base.
2 = leaves show yellowing in 26–50% of the total leaves and wilt with a simple brown discoloration in xylem vessels.
3 = 51–75% of the leaves show yellowing and wilt with dark brown coloring in xylem vessels.
4 = 76–100% of leaves show wilt and death.

The formula below was used to assess disease severity index [31]:
Disease severity index (%) $= \frac{ni.si}{4n} \times 100$ [32].
The infection Percentage was calculated using the formula:

$$\textbf{Infection}\ \% = \frac{\boldsymbol{ni}}{\boldsymbol{n}} \times 100$$

Where ***ni***: is the number of plants affected by each degree of severity.

si: the degree of severity of the attack (0 – 4).
n: the total number of plants used for each energy level applied.

2.6 *Statistical analysis*

All data were analyzed by ANOVA using SAS statistical program [33]. Results were compared using the L.S.D. test at P 5%. Each experiment was repeated three times.

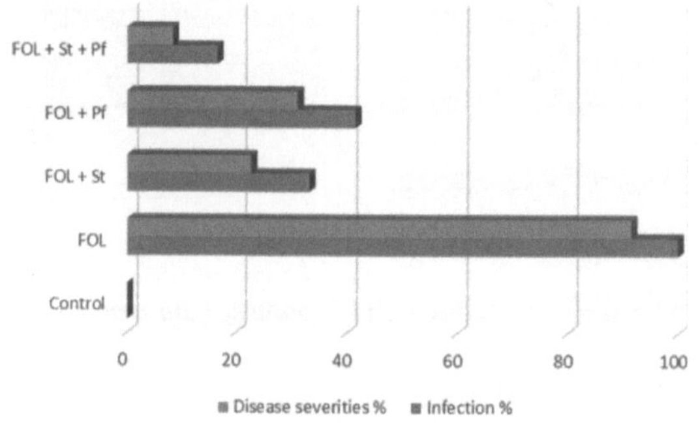

Figure 4. PGPR bacteria effect on Fusarium wilt disease on tomato.

3 RESULTS

3.1 *Pathogenicity of FOL and antifungal activity of PGPR*

According to the results obtained in this study, the isolate of pathogenic fungi reduced the germination of tomato seeds significantly. The germination rate in tomato seeds reached 16.55% in the fusarium treatment compared to the control treatment (no fusarium) where the germination rate was 100% with L.S.D. for this experiment 24.49.

It is believed that FOL reduces the ability of seeds to secrete the enzymes necessary for cell wall degradation such as Cellulase and Chitinase, so the pathogen severely suppresses the germination of seeds. FOL also manipulates the production of toxic metabolites such as Dehydro Fusaric acid, Lycomarsmin and Fusaric acid that reduce respiratory rates in the infected plant cell and affect the permeability of membranes and reduce respiratory rates, which eventually cause an imbalance in other plant biological processes [10,34,35].

For the antifungal activity of PGPR, forty-five isolates of PGPR were collected from the soil of tomato fields in Baghdad and Anbar regions in the middle of Iraq. Two isolates of soil PGPR inhibited 100% of FOL colony growth. These isolates were identified as Streptomyces spp (St) and *P. fluorescence* (Pf).

3.2 *Morphological and chemotaxonomy identification of the effective soil PGPR*

White colonies on PDA medium and red to pink ones on ISP2 and ISP4 media with a dusty surface indicated that the bacteria under study was Streptomyces spp., which was able to produce spores that are responsible for red color [36,37]. For F11 bacterial isolate, the Analytical Profile Index 20 Enterobacteriaceae (API 20E) results indicated that the isolate belongs to Pseudomonas fluorescence, which is an important species of the PGPR.

3.3 *Effectiveness of PGPR on FOL in tomato*

All treatments reduced the percentage of disease severity and infection percentage of Fusarium wilt (FW) disease in tomato plants significantly. As shown in Figure 4, treatment FOL/S/Pf has the best rate of preventing infection with 16.6% disease severity and 8.3% infection percentage followed by the treatment FOL/St which scored 33.3% and 22.8% then FOL/Pf that showed 41.6% and 31.2%, compared with the control (FOL alone) which scored 100 and 91.6%, respectively. Every treatment has been repeated three times and significance has been calculated using L.S.D.

There are two directions of PGPR resistance to pathogens, by which bacteria decrease disease severity and infection percentage of FW disease. First, is the direct effect of inhibiting pathogen mycelium radial growth and the spores' germination.

Secondly, the production of antibiotics and degradation enzymes by PGPR; these products include Lipase, Amylase, Proteases, Chitinase, and HCN, which all affect the pathogen lifecycle.

3.4 *Systemic resistance induction in tomato plants*

Phenol content, PAL and Peroxidase are linked to defense activity in plants. The activity of all three indicators was found to be increased post inoculation with PGPR on FOL-infected plants (Figures 5 and 6). At 14 days post inoculation with FOL, high Phenol content, PAL and Peroxidase were recorded. For the treatment (FOL/St/Pf), results showed 63.8, 32.6 and 3.8 for the three indicators respectively. Whereas for (FOL/St) treatment the scores were 48.6, 27.4 and 3.2, and for FOL/Pf treatment, readings were 43.6, 26.3 and 3.1 which is considered higher than the control level of the same indicators which were 22.3, 16.6 and 2.0 enzyme activity, respectively. There are many studies referred to different biological agents as a successful replacement for chemicals in the field [15–20].

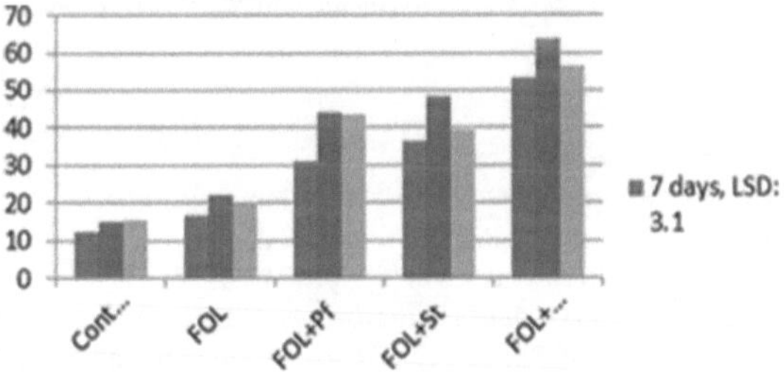

Figure 5. Peroxidase activity in tomato plants for different biological treatments.

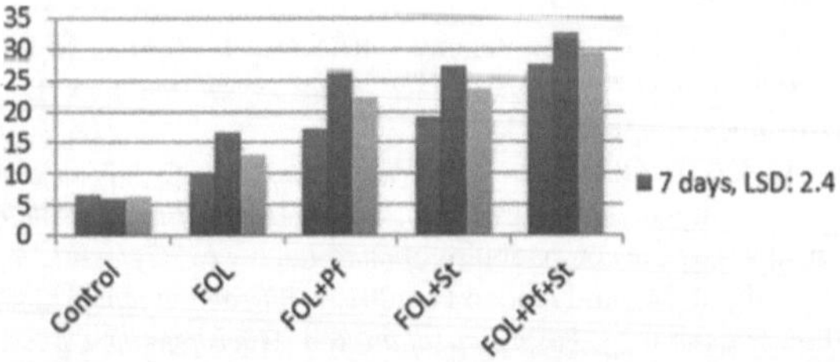

Figure 6. PAL activity in tomato plants for different biological treatments.

The level of host resistance is determined by the Peroxidase and PAL enzymes activity that are restored in plant tissues when exposed to stress. They also have roles in the deposition of the Suberin, the biosynthesis of lignin, and the intensity of the cell wall which increased structural defenses [38–40].

4 CONCLUSION

Two local isolates of PGPR were identified as successful biological agents to control FW disease on tomatoes in Iraq. The identified bacteria-induced systemic resistance in tomato plants, which can be applied to combat other pathogens.

ACKNOWLEDGMENT

The authors would like to acknowledge the University of Baghdad for support.

REFERENCES

[1] Adhab M., Al-Kuwaiti N., and Al-Ani R., (November 2021). Biodiversity and occurrence of plant viruses over four decades: Case study for Iraq, *In 2021 Third International Sustainability and Resilience Conference: Climate Change*, pp. 159–163.

[2] Al-ani R., Adhab M., *et al.*, (2011). Tomato yellow leaf curl virus (TYLCV), identification, virus vector relationship, strains characterization and a suggestion for its control with plant extracts in Iraq, *African Journal of Agricultural Research*, vol. 6, pp. 5149–5155.

[3] Adhab M., and Alkuwaiti N. (2022). Geminiviruses occurrence in the middle east and their impact on agriculture in Iraq. In *Geminivirus: Detection, Diagnosis and Management*, pp. 171–185. Academic Press.

[4] Obaid H. and Adhab M., (2024). Outbreak of tobamoviruses and potexviruses associated with disease epidemics in tomato production area of Iraq, *Iraqi Journal of Agricultural Sciences*, (in press).

[5] Dam B., Goffau M., Lidth de Jeude J., and Naika S., (2005). *Cultivation of Tomato: Production, Processing and Marketing*, Agromisa/CTA.

[6] Bawa I., (2016). Management strategies of Fusarium wilt disease of tomato incited by *Fusarium oxysporum* f.sp. lycopersici (Sacc.): A review, *International Journal of Advanced Academic Research*, vol. 2, pp. 32–42.

[7] Nirmaladevi D., Venkataramana M., Srivastava R., Uppalapati S., Gupta V., Yli-Mattila T., and Chandra N., (2016). Molecular phylogeny, pathogenicity and toxigenicity of *Fusarium oxysporum* f. sp. lycopersici, *Scientific Reports*, vol. 6, pp. 1–30.

[8] Sidharthan K., Rashmi K., *et al.*, (2018). Selection and characterization of the virulent *Fusariumoxysporum* f.sp. lycopersici isolate inciting vascular wilt of tomato, *International Journal of Current Microbiology and Applied Sciences*, vol. 7, pp. 1749–1756.

[9] Kennelly M., (2007). *Wilt, Nematode, and Virus Diseases*, Kansas State University, pp. 723.

[10] Ramanathan G., Banupriya S., and Abirami D., (2010). Production and optimization of cellulase from *Fusarium oxysporum* by submerged fermentation, *J. of Scientific and Industrial Research*, vol. 69, pp. 454–459.

[11] McGovern R., (2015). Management of tomato diseases caused by *Fusarium oxysporum*, *Crop Protection*, vol. 73, pp. 78–92.

[12] Biju V., Fokkens I., Houterman P., Rep M., and Cornelissen B., (2017). Multiple evolutionary trajectories have led to the emergence of races in *Fusarium oxysporum* f.sp. Lycopersici," *Applied and Environmental Microbiology*, vol. 83, pp. 1–20.

[13] Reis A., and Boiteux L., (2007). Outbreak of *Fusarium oxysporum* f.sp. lycopersici race 3 in commercial fresh-market tomato fields in Rio de Janeiro State, Brazil, *Horticultura Brasileira*, vol. 25, pp. 451–454.

[14] Leslie J., (1993). Fungal vegetative compatibility, *Annual Review of Phytopathology*, vol. 31, pp. 127–150.

[15] Al-Ani R., Adhab M., Mahdi M., and Abood H., (2012). *Rhizobium japonicum* as a biocontrol agent of soybean root rot disease caused by *Fusarium solani* and *Macrophomina phaseolina*, *Plant Protection Science*, vol. 48, pp. 149–155.

[16] Adhab M., Zhang Y., and Schoelz J. (2023). Transient expression of cauliflower mosaic virus (CaMV) P6-GFP complements a defective CaMV replicon to facilitate viral gene expression, replication and virion formation. *Virology* 587, 109854.

[17] Al-Ani R., Adhab M., El-Muadhidi M., and Al-Fahad M., (2011). Induced systemic resistance and promotion of Wheat and barley plants growth by biotic and non-biotic agents against Barley yellow dwarf virus, *African Journal of Biotechnology*, vol. 10, pp. 12079–12084.

[18] Al-Ani R., Athab M., and Matny O., (2013). Management of potato virus Y (PVY) in potato by some biocontrol agents under field conditions, *Advances in Environmental Biology*, vol. 7, pp. 441–444.

[19] Adhab M., (2021). Be smart to survive: virus-host relationships in nature. *Journal of Microbiology, Biotechnology and Food Sciences* 10, (6), e3422.

[20] Al-Ani R., and Adhab M., (2013). Bean yellow mosaic virus (BYMV) on broadbean: Characterization and resistance induced by *Rhizobium leguminosarum*, *Journal of Pure and Applied Microbiology*, vol. 7, pp. 135–142.

[21] Vejan P., Abdullah R., Khadiran T., Ismail S., and Boyce A., (2016). Role of plant growth promoting rhizobacteria in agricultural sustainability, *Molecules*, vol. 21, pp. 1–17.

[22] Zouari M., Elloumi N., Ahmed C., Delmail D., Rouina B., Abdallah F., and Labrousse P., (2016). Exogenous proline enhances growth, mineral uptake, antioxidant defense, and reduces cadmium-induced oxidative damage in young date palm (*Phoenix dactylifera* L.), *Ecological Engineering*, vol. 86, pp. 202–209.

[23] Mhlongo M., Piater L., Madala N., Labuschagne N., and Dubery I., (2018). The chemistry of plant-microbe interactions in the rhizosphere and the potential for metabolomics to reveal signaling related to defense priming and induced systemic resistance, *Frontiers in Plant Science*, vol. 9, pp. 1–17.

[24] Al-Dulaimi S., and Hussein H., (2019). Molecular diagnosis of *Fusarium oxysporum* f.sp. lycopersici and proving its pathogenic capacity on the incidence of Fusarium wilt disease, *Biochemical and Cellular Archives*, vol. 19, pp. 333–335.

[25] Raspor P., Milek D., Avbelj M., and Cadez N., (2010). Biocontrol of grey mold disease on grape caused by *Botrytis cinerea* with autochthonous wine yeasts, *Food Technology and Biotechnology*, vol. 48, pp. 336–343.

[26] Breed R., Murray E., and Smith N., (1957). *Bergey's Manual of Determinative Bacteriology*, The Williams & Wilkins Company, pp. 1094.

[27] Hammerschmidt R., Nuckles E., and Kuc J., (1989). Association of enhanced peroxidase activity with induced systemic resistance of cucumber to *Colletotrichum lagemarium*, *Physiology and Plant Pathology*, vol. 20, pp. 73–82.

[28] Dickerson D., Pascholati S., Ann E., Butler L., and Nicholson R., (1984). Phenylalanine ammonia-lyase and hydroxycinnamate: CoA ligase in maize mesocotyls inoculated with Helminthosporium maydis or *Helminthosporium carbonum*, *Physiological Plant Pathology*, vol. 25, pp. 111–123.

[29] Rishi K., Vidya P., and Arora D., (2008). Study on phenolic and their oxidative enzyme in *Capsicum annuum* L. infected with geminivirus, *Asian Journal of Exp. Sci.*, vol. 22, pp. 307–310.

[30] Souza L., Michereff S., *et al.*, (2010). Reaction of tomato genotypes to races 2 and 3 of *Fusarium oxysporum* f.sp. lycopersici, *Horticultura Brasileira*, vol. 28, pp. 102–106.

[31] Mohammed D., Adhab M., and Al-Kuwaiti N. (2021). Molecular characterization of viruses associated to leaf curl disease complex on zucchini squash in Iraq reveals Deng primer set could distinguish between New and Old World Begomoviruses. *Anais da Academia Brasileira de Ciências* 93, e20210050.

[32] Mckinney H., (1923). Influence of soil temperature and moisture on infection of wheat seedling by *Helminthosporium sativum*, *Journal of Agricultural Research*, vol. 26, pp. 195–217.

[33] SAS, Statistical Analysis System, User's Guide. Statistical, Version 9.1th ed., SAS. Inst. Inc. Cary. N.C. USA, 2012.

[34] Langcalces P., and Drysdale R., (2005). Response of tomato to infection by *Fusarium oxysporum* f. sp. lycopersici, *Phytopathology*, vol. 68, pp. 422–437.

[35] Calero-Nieto F., Pietro A., Roncero M., and Hera C., (2007). Role of the transcriptional activator XlnR of *Fusarium oxysporum* in regulation of xylanase genes and virulence, *MPMI*, vol. 20, pp. 977–985.

[36] Kieser M., Bibb-Buttner K., and Hopwood D., (2000). *Practical Streptomyces Genetics*, The John I nnes Foundation Bugay, Suffolk.UK.

[37] Kampfer P., (2006). The family Streptomycetaceae, part I: Taxonomy The Prokaryotes, pp. 538–604.

[38] Thakker J., Patel S., and Dhandhukia P., (2013). Induction of defense-related enzymes in banana plants: Effect of live and dead pathogenic strain of Fusarium oxysporum, *ISRN Biotech.*, vol. 13, pp. 1–6.

[39] Cass C., Antoine P., *et al.*, (2015). Effects of phenylalanine ammonia-lyase (PAL) knockdown on cell wall composition, biomass digestibility, and biotic and abiotic stress responses in Brachypodium, *Journal of Experimental Botany*, vol. 66, pp. 4317–4335.

[40] Chen Y., Fengjiao L., *et al.*, (2017). The phenylalanine ammonia-lyase gene LjPAL1 is involved 1 in plant defense responses to pathogens and plays diverse roles in Lotus japonicus-rhizobium symbioses, *Molecular Plant Microbe Interact.*, vol. 30, pp. 739–753.

Innovation and Technological Advances for Sustainability – Al-Naemi et al. (Eds)

Merging farms to enhance agricultural sustainability: An energy efficiency perspective

Amar Oukil
Department of Operations Management & Business Statistics, Sultan Qaboos University, Muscat, Oman

ABSTRACT: Mergers and acquisitions (M&As) are strategic decisions that enable consolidation of multiple firms in order to expand market share, gain access to new markets and enhance production capabilities. In this paper, we show that farms' mergers, as opposed to individual farms, can have a sizable contribution to sustainability. Using Inverse data envelopment analysis (InvDEA) as a methodological framework, we investigate the impact of mergers on enhancing sustainability in the agricultural production sector through optimizing energy requirements. The investigations are carried out on a sample of 15 cucumber GH farms in the Batinah region of Oman, Based on the data analysis, electricity and fertilizers emerged as major energy inputs, utilizing 81.84% and 15.61% of the total energy inputs, respectively. The application of the standard DEA model produced an average technical efficiency of 0.9635 with 80% of the GH farms declared efficient. The next major part of the study consists of implementing the inverse DEA approach to 435 pairwise consolidations of the GH farms, where only 18 mergers were identified as productive. In spite of the high level of efficiency among the GH farms, the results revealed that 50% of the productive post-merger GH farms comprise exclusively efficient GH farm, that is, energy savings are still conceivable even if the merging farms are seemingly efficient. The analysis of the energy outcomes indicated that the average proportions of potential energy gains per post-merger GH farm fall between 19.58% and 55.64%, with the highest figures reported for electricity. The post-mergers' energy savings for labor, machinery, fertilizers, chemicals, water and electricity reached proportions as high as 82.22%, 68.22%, 84.73%, 51.15%, 76.41% and 82.07%, respectively. Additionally, a synergy merge index (SMI) is developed to optimally pair GH farms. The selected plan contributed to improving the energy savings by a factor of more than 4, where the share of electricity represents alone 91.05%, followed by 7.88% for fertilizers. The application scope of the proposed methodology can be extended to other sectors where energy consumption might be a critical issue.

Keywords: Sustainability, Energy efficiency, Merger, Data envelopment analysis (DEA), Inverse DEA, Agriculture

1 INTRODUCTION

The world's population is expected to exceed 9.8 billion within the next three decades, which throws more burdens on the agricultural sector in meeting the food demand [1]. Under these prospects, more resources, including energy, are required for a sustainable agricultural production [2]. Therefore, developing robust tools for efficient energy use tends to be a major concern [3].

The most popular technique used in agricultural efficiency studies is undoubtedly data envelopment analysis (DEA) [4,5], which can deal with decision making units (DMUs)

 DOI: 10.1201/9781003496724-14

defined with several inputs and outputs [6,7] without necessitating functional forms on the data nor the usage of probability distributions [8,9]. Recent studies applied DEA to optimize energy consumption in open fields (e.g., [10–19]) as well as greenhouse (GH) agricultural systems (e.g., [20–25]). These studies focused on estimating energy savings for inefficient farms via slack analysis [26]. In this paper, we propose sharing resources (inputs) among the farms through mergers as a new way to optimize energy consumption. Mergers and acquisitions (M&As) are strategic decisions that enable consolidation of multiple firms in order to expand market share, gain access to new markets and enhance production capabilities [27]. In order to anticipate, at a pre-merger stage, the levels of inputs that enable a prospective post-merger farm to reach a predetermined efficiency target, we use Inverse DEA (InvDEA) [28].

To the authors' best knowledge, no previous study has ever investigated the impact of mergers on energy optimization.

The practical significance of the proposed approach is highlighted through a case study that consists of 15 cucumber GH farms from the Batinah region in Oman.

It is noteworthy that, though the discussion is concerned with GH production as a case study, the proposed methodology can be extended to any business where energy efficiency is of interest.

The remainder of the paper unfolds as follows. Section II discusses the methodological foundations of the proposed procedure. Section III presents the case study with an in-depth discussion of the results. Section IV is dedicated to developing a new index for an optimal pairing of GH farms. We finish with concluding remarks.

2 METHODS

2.1 *Standard DEA model*

Let G the number of GH farms to be evaluated. Each farm GH_g is defined with inputs and outputs x_{ig} and y_{ig}, $i = 1, ..., I$ and $j = 1, ..., J$. To estimate the efficiency e_o^* of $GH_o = (x_o y_o)$, the following BCC model [29] is used:

$$
\textbf{(BCC)}\quad
\begin{aligned}
e_o^* = &\min \pi \\
\text{s.t.} &\sum_{g=1}^{G} \alpha_g x_{ig} \le \pi x_{io} && i = 1, ..., I && (1)\\
&\sum_{g=1}^{G} \alpha_g y_{jg} \ge y_{jo} && j = 1, ..., J && (2)\\
&\sum_{g=1}^{G} \alpha_g = 1 && && (3)\\
&\alpha_g \ge 0 && g = 1, ..., G &&
\end{aligned}
$$

The optimal efficiency e_o^* of GH_o measures minimum reduction of inputs' for given levels of outputs [30–32]. Thus, GH_o is efficient if $e_o^* = 1$; otherwise, it is inefficient, i.e., it is not using its inputs optimally [33,34]. The intensity vector $\boldsymbol{\alpha} = (\alpha_1 \alpha_2 ... \alpha_G)$ provides the weights of the peers for the evaluation of GH_o [35]. Let $\alpha^* = (\alpha_1^* \alpha_2^* ... \alpha_G^*)$ the optimal vector. The ith input's savings required from GH_o to become efficient is estimated through the slack value s_{io}^* [36], where $s_{io}^* = e_o^* x_{io} - \sum_{g=1}^{G} \alpha_g^* x_{ig}$. If GH_o is strongly efficient, $s_{io}^* = 0$ for all $i = 1, ..., I$, otherwise, GH_o is weakly efficient if $e_o^* = 1$ and $s_{io}^* > 0$ for at least one input i [37].

2.2 *Inverse DEA model*

The inverse DEA model that formulates the merger GH farms GH_A and GH_B into the post-merger GH farm F_m writes as follows [38].

$$
\text{(InvDEA)} \quad
\begin{aligned}
& \min \sum_{i=1}^{I} (\partial_{iA} + \partial_{iB}) \\
& s.\,t. \sum_{k \in P} \alpha_k x_{ik} + \alpha_m (x_{iA} + x_{iB}) - (\partial_{iA} + \partial_{iB}) \times \overline{\pi} \le 0 \quad i = 1, \ldots, I \\
& \sum_{k \in P} \alpha_k y_{jk} + \alpha_m (y_{jA} + y_{jB}) \ge (y_{jA} + y_{jB}) \quad j = 1, \ldots, J \\
& \sum_{k \in P} \alpha_k + \alpha_m = 1 \\
& 0 \le \partial_{iA} \le x_{iA},\ 0 \le \partial_{iB} \le x_{iB} \quad i = 1, \ldots, I \\
& \alpha_k \ge 0,\ k \in P,\ \alpha_m \ge 0
\end{aligned}
$$

InvDEA is an inverse DEA model because the efficiency target $\overline{\pi}$ is set a priori for the post-merger GH farm F_m and the objective is to find the minimum levels of inputs ∂_{iA} and ∂_{iB} that enable GH_A and GH_B, respectively, to reach $\overline{\pi}$ while keeping their entire outputs. If $\alpha_m = 1$ and $\overline{\pi} = 1$ is the only solution of model (InvDEA), the merger will be a major consolidation [38].

3 APPLICATION

Oman is one of the world's hot, arid regions though part of the south of the country has a tropical climate [39]. With temperatures exceeding 50°C, scarce rainfall and persistent drought, Oman's agricultural sector is completely dependent on irrigation [40]. In the GH production, water resources are needed not only for irrigation but also for fan pad cooling systems, which consume about 67% of the total water demand [41]. The GH production is mostly located in Al-Batinah South, where cucumber cultivation is largely dominating with 89% of the total GH production facility [42].

3.1 *Data collection*

Data were collected over the winter season from 15 GH farmers for the inputs labor (x_1), machinery (x_2), fertilizers (x_3), chemicals (x_4), water (x_5), and electricity (x_6), and one output, yield (y).

Table 1 summarizes the data per hectare of land.

Table 1. Descriptive summary of the data sample.

Data	Unit	Quantity (per ha)	Energy equivalent (MJ ha^{-1})	Proportion (%)
Inputs				
Labor	h	3876.58	7598.10	1.07
Machinery	h	94.93	1239.81	0.17
Fertilizers	kg	16822.40	110890.79	15.61
Chemicals	kg	46.41	5568.76	0.78
Water	m^3	3700.22	3774.22	0.53
Electricity	kWh	48744.68	581524.03	81.84
Total inputs			710595.71	100.00
Output				
Cucumber	kg	116666.53	93333.22	

The energy-equivalents are computed using the energy-equivalent coefficients, presented in Table 2.

Table 2. Energy equivalent coefficients in cucumber GH production.

Data (Unit)	Energy equivalent (MJ Unit^{-1})
Human labor (h)	1.96
Machinery (h)	13.06
Fertilizers (kg):	
Nitrogen	66.14
Phosphate,P_2O_5	12.44
Potassium, K_2O	11.15
Micronutriments	120.00
Manure (kg)	0.30
Chemicals (kg)	120.00
Water (m3)	1.02
Electricity (kWh)	11.93
Cucumber (kg)	0.80

The highest proportions of energy consumption are 81.84% and 15.61% for electricity and fertilizers, respectively. Electricity is the prominent energy consumer due to the fact that all GH farms use electric pumps for water abstraction and fan-pad cooling systems.

3.2 *Performance evaluation*

The computation of the optimal efficiency e_g^* for each GH farm GH_g, $g = 1, ..., 15$, reveals a high level of efficiency among the GH farms, with 12 GH farms out of 15 (80%) efficient, for an average efficiency of 0.9635, comparable to previous results of cucumber GH production studies (e.g., [43,44]). The inefficient GH farms, GH_2, GH_8 and GH_{15}, need to benchmark the strongly efficient farms to improve performance [45].

Table 3. Energy savings per resource.

Inputs	Present use (MJ ha^{-1})	Target use (MJ ha^{-1})	Energy saving (MJ ha^{-1})	Contribution to savings (%)
Labor	113972	110693	3279	0.72
Machinery	18597	17639	958	0.21
Fertilizers	1663362	1588921	74441	16.46
Chemicals	83531	81507	2025	0.45
Water	56613	54157	2456	0.54
Electricity	8722860	8353762	369098	81.61
Total inputs	10658936	10206679	452256	100.00

Accordingly, the energy saving pattern in Table 3 suggests the highest share of 81.61% for electricity, followed by 16.46% for fertilizers. The resulting percentage of energy saving index is 4.24%, which is relatively low compared to the highest 24.46% [46] and the lowest 8.12% [44] reported for similar GH production systems.

In the next section, we propose GH farms merger as a strategic alternative for resource optimization and discuss its effect on energy savings.

3.3 *Mergers of GH farms*

Model InvDEA is solved for pairwise mergers $F_m = (GH_A, GH_B)$, $A \neq B$, and $\overline{\pi} = 1$. Out of 105 possible mergers F_m, 87 mergers are major consolidations, i.e., $\partial_{iA}^* = x_{iA}$, $\partial_{iB}^* = x_{iB}$, for all $i = 1, ..., 6$. The remaining 18 post-merger GH farms, presented in Table 4, are potentially productive.

Table 4. potential energy usages (GJ ha^{-1}) of Post-merger GH farms.

	Labor		Machinery		Fertilizers		Chemicals		Water		Electricity	
F_m	∂^*_{1A}	∂^*_{1B}	∂^*_{2A}	∂^*_{2B}	∂^*_{3A}	∂^*_{3B}	∂^*_{4A}	∂^*_{4B}	∂^*_{5A}	∂^*_{5B}	∂^*_{6A}	∂^*_{6B}
F_1	7.55	0	0.81	0.48	34.62	74.98	5.57	2.56	1.96	2.09	490.18	0
F_2	5.61	0	0.81	0.44	34.62	26.15	5.57	4.14	1.57	2.46	392.42	0
F_3	4.68	0	0.81	0.21	34.62	34.11	5.57	3.94	0.98	3.53	625.17	0
F_4	11.54	0	0.81	0.71	0	262.95	5.57	1.92	3.37	0	276.19	0
F_5	4.97	0	0.81	0.14	0	91.58	5.57	3.52	3.16	1.49	720.67	0
F_6	8.61	0	0.81	0.62	34.62	52.09	5.57	0.82	3.24	0.48	457.15	0
F_7	7.35	2.05	1.54	0	74.98	26.15	2.56	4.14	1.20	2.46	386.11	0
F_8	5.96	0	1.08	0	74.98	34.11	2.56	6.48	2.09	2.26	590.29	0
F_9	7.35	2.81	1.41	0	0	223.86	2.56	5.34	0	3.60	345.35	0
F_{10}	4.84	4.80	0.62	0.86	26.15	52.36	4.14	1.14	0.85	3.28	635.49	0
F_{11}	4.84	0.08	0	1.12	26.15	34.11	0	9.80	0.75	3.53	501.09	0
F_{12}	4.84	2.81	0.02	1.34	0	131.07	2.82	6.32	0	3.74	313.73	0
F_{13}	4.84	3.76	0	1.27	26.15	158.52	0	8.25	2.44	1.49	463.58	0
F_{14}	4.84	2.41	0.63	0.62	26.15	52.09	4.14	0.82	2.46	0.48	379.37	0
F_{15}	4.84	1.27	0	1.07	26.15	90.33	3.92	5.01	0	4.37	607.37	0
F_{16}	5.66	2.58	1.23	0	0	176.96	8.30	0	3.53	0.49	110.01	389.21
F_{17}	2.81	1.69	0.91	0	79.27	0	6.32	2.89	4.75	0	573.43	179.15
F_{18}	2.81	2.47	0.99	0	98.41	0	6.32	0.82	3.42	0	470.25	0

The most striking result is certainly that 50% of the productive mergers comprise exclusively efficient GH farms, namely $F_2 = (GH_{01}, GH_{03})$, $F_3 = (GH_{01}, GH_{05})$, $F_5 = (GH_{01}, GH_{11})$, $F_6 = (GH_{01}, GH_{13})$, $F_{10} = (GH_{03}, GH_{04})$, $F_{11} = (GH_{03}, GH_{05})$, $F_{13} = (GH_{03}, GH_{11})$, $F_{14} = (GH_{03}, GH_{13})$ and $F_{15} = (GH_{03}, GH_{14})$. Though the status efficient GH farm suggests that no further input gains are possible, these mergers repudiate such a statement. In other words, more input gains can still be achieved through mergers, irrespective of the efficiency status of the merging farms.

Table 5 presents the cumulated energy gains g_{im} with the corresponding proportions for each input $i = 1, ..., 6$. The highest proportions of energy gains recorded on the averages are 55.64% for electricity and 44.18% for machinery. The mergers' energy gains for labor, machinery, fertilizers, chemicals, water and electricity hit proportions as high as 82.22% (F_5), 68.22% (F_8), 84.73% (F_{17}), 51.15% (F_{16}), 76.41% (F_4) and 82.07% (F_4), respectively.

Table 5. Cumulated energy (GJ ha^{-1}) of Post-merger GH farms.

	Labor		Machinery		Fertilizers		Chemicals		Water		Electricity	
F_m	g_{1m}	%	g_{2m}	%	g_{3m}	%	g_{4m}	%	g_{5m}	%	g_{6m}	%
F_1	14.99	66.49	1.47	53.22	0	0	0	0	1.66	29.04	1245.56	71.76
F_2	14.42	72.01	0.46	27.19	0	0	0	0	2.04	33.66	1452.01	78.72
F_3	16.17	77.55	1.25	55.19	0	0	6.72	41.39	2.64	36.88	451.91	41.96
F_4	6.46	35.89	0.63	29.38	132.40	33.49	4.40	36.97	10.93	76.41	1264.31	82.07
F_5	22.96	82.22	1.31	58.11	101.56	52.59	6.90	43.13	0.45	8.88	549.39	43.26
F_6	9.06	51.27	0	0	0	0	0	0	0.38	9.23	640.57	58.35
F_7	2.79	22.86	1.32	46.13	0	0	0	0	0.88	19.45	1259.92	76.54
F_8	7.05	54.19	2.32	68.22	0	0	4.19	31.63	1.26	22.51	288.40	32.82

(continued)

Table 5. Continued

	Labor		Machinery		Fertilizers		Chemicals		Water		Electricity	
F_m	g_{1m}	%	g_{2m}	%	g_{3m}	%	g_{4m}	%	g_{5m}	%	g_{6m}	%
F_9	0	0	1.88	57.02	211.84	48.62	0.98	11.01	9.18	71.82	996.75	74.27
F_{10}	4.83	33.36	0.28	16.13	0	0	0	0	1.61	28.01	1111.90	63.63
F_{11}	5.58	53.19	1.23	52.47	0	0	5.01	33.83	1.71	28.57	486.28	49.25
F_{12}	0	0	0.88	39.39	255.80	66.12	1.31	12.56	9.41	71.57	1137.06	78.38
F_{13}	8.98	51.09	1.08	46.03	0	0	6.31	43.34	0.02	0.52	716.77	60.73
F_{14}	0.06	0.82	0.28	18.1	0	0	0	0	0	0	628.64	62.36
F_{15}	2.28	27.19	1.36	55.79	0	0	0.22	2.37	4.07	48.24	690.66	53.21
F_{16}	0.23	2.72	1.56	55.91	217.88	55.18	8.69	51.15	10.20	71.74	184.22	26.95
F_{17}	11.05	71.09	1.88	67.52	439.97	84.73	7.53	44.97	7.43	61.01	123.84	14.13
F_{18}	14.99	66.49	1.47	53.22	0	0	0	0	1.66	29.04	1245.56	71.76

Figure 1 shows the distribution of the proportions of energy gains among the post-merger GH farms.

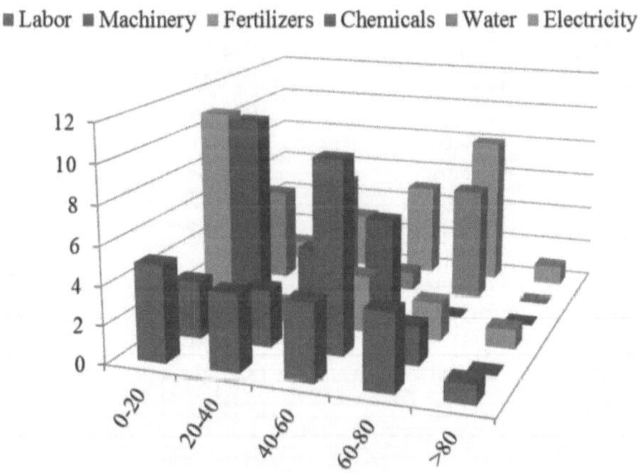

Figure 1. Distribution of energy gains' proportions of GH mergers.

There is a notable domination of electricity energy gains, with proportion caps of 60% and above for half of the post-merger GH farms. In other words, the majority of the mergers are able to achieve the efficiency target $\overline{\pi} = 1$ with less than 40% of the cumulated energy of the merging GH farms. The water's energy gains exhibit a slightly different pattern with only 6 post-merger GH farms recording gain proportions that vary within the range [60%, 80%] with a maximum of 76.41% reached for F_4. Though most of the energy gain proportions for fertilizers and chemicals fall on the bottom side of the picture, there are picks of 84.73% (F_{17}) and 51.15% (F_{116}) for these inputs, respectively, which reflect substantial energy reductions. The distribution of gain proportions for the labor energy is apparently more uniform.

4 SELECTING THE BEST PARTNERS OF A POST-MERGER

The synergistic impact of a merger cannot be perceived unless the post-merger outputs become far better than the pre-merger ones [47,58–61]. Selecting the best partners for each pair of merging GH farms is the best way to achieve an optimal synergy merge. Considering the cumulated energy gains $g_{im}(i = 1, ..., 6)$ are the most appropriate proxies for prospective mergers' outcomes, we develop the Synergy Merge Index (SMI) for the evaluation of alternative post-merger GH farms F_m. where:

$$SMI_m = \frac{Energy\ gains\ of\ F_m}{Energy\ usage\ of\ F_m} = 100 \times \frac{\sum_{i=1}^{6} g_{im}}{\sum_{i=1}^{6} x_{im}} \quad x_{im} = x_{iA} + x_{iB} \text{ for } i = 1, ..., 6.$$

As a ratio of the total energy gains of the post-merger GH farm F_m over its current energy consumption, SMI_m measures the impact of the merger on reducing energy usage. The SMIs have been computed for the productive post-merger GH farms and the results are shown in Table 6.

Table 6. Synergy merge Indices for post-merger GH farms.

F_m	GH_A	GH_B	Gains (GJ ha^{-1})	Usage (GJ ha^{-1})	SMI_m (%)	Rank
F_1	GH_{01}	GH_{02}	1263.68	1884.49	67.06	6
F_2	GH_{01}	GH_{03}	1468.94	1942.71	75.61	1
F_3	GH_{01}	GH_{05}	478.69	1192.31	40.15	16
F_4	GH_{01}	GH_{08}	1419.13	1982.18	71.59	3
F_5	GH_{01}	GH_{11}	682.58	1514.49	45.07	14
F_6	GH_{01}	GH_{13}	650.00	1214.01	53.54	9
F_7	GH_{02}	GH_{03}	1264.91	1773.45	71.32	4
F_8	GH_{02}	GH_{05}	303.23	1023.04	29.64	18
F_9	GH_{02}	GH_{08}	1220.63	1812.91	67.33	5
F_{10}	GH_{03}	GH_{04}	1118.62	1853.14	60.36	7
F_{11}	GH_{03}	GH_{05}	499.81	1081.27	46.22	13
F_{12}	GH_{03}	GH_{08}	1404.46	1871.14	75.06	2
F_{13}	GH_{03}	GH_{11}	733.16	1403.44	52.24	10
F_{14}	GH_{03}	GH_{13}	628.98	1102.97	57.03	8
F_{15}	GH_{03}	GH_{14}	698.58	1442.91	48.41	12
F_{16}	GH_{05}	GH_{08}	422.77	1120.74	37.72	17
F_{17}	GH_{08}	GH_{11}	591.70	1442.91	41.01	15
F_{18}	GH_{08}	GH_{13}	1263.68	1884.49	48.75	11

The best mergers are selected according to their ranking order while discarding subsequent pairs that overlap with previously selected ones. As such, $F_2 = (GH_{01}, GH_{03})$ and $F_9 = (GH_{02}, GH_{08})$, which are ranked 1st and 5th, are selected as best potential mergers. Here, $F_{12} = (GH_{03}, GH_{08})$, $F_4 = (GH_{01}, GH_{08})$ and $F_7 = (GH_{02}, GH_{03})$ are discarded, in spite of being ranked 2nd, 3rd and 4th, because of the overlaps on GH_{01} and GH_{03}. Therefore, out of 15 GH farms, the mergers are apparently most productive with only 4 farms, including GH_{01}, GH_{02}, GH_{03} and GH_{08}. Interestingly, both GH farms in F_2 are strongly efficient, whereas F_9 involves only inefficient GH farms.

With farms GH_{01}, GH_{02}, GH_{03} and GH_{08} reentering the market under the post-merger GH farms F_2 and F_9, the total usage of energy inputs becomes 7969.37 GJ ha^{-1} while the total energy output remains unchanged, i.e. 1400 GJ ha^{-1}. Accordingly, the post-merger percentage of energy savings will be 17.57%. Recalling that the pre-merger value of this index was 4.24% under the slack analysis, the latter result indicates a substantial improvement by a factor of more than 4. Figure 2 shows the distribution of these savings among the energy inputs.

Interestingly, the share of electricity consumption represents alone 91.05% of the post-merger energy gains, followed by 7.88% for fertilizers and only 0.54% and 0.42% for labor and water, respectively, besides relatively neglectable shares for chemicals and machinery

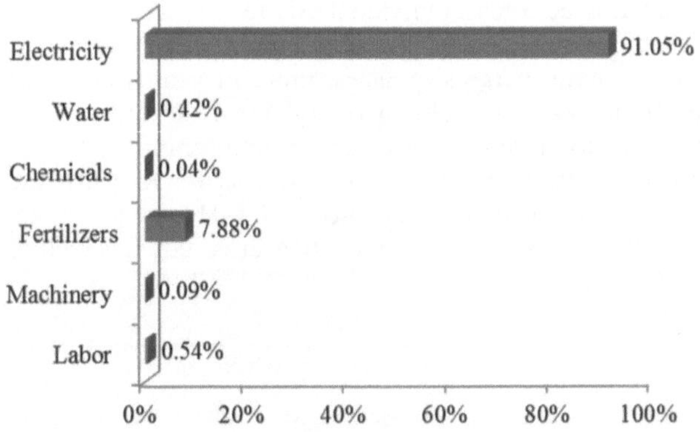

Figure 2. Distribution of the post-merger energy gains per input.

energy inputs. With a retrospective into the energy input contributions of each partner to pertaining post-merger GH farm, it is understood that the supply of labor and electricity is now the full responsibility of GH farmer GH_{01} in the case of F_2. Meanwhile, there is a better task sharing with F_9, where GH_{02} handles energy inputs for machinery and electricity whereas GH_{08} manages the energy needs for fertilizers and water. Such a collaborative pattern enables not only relieving a part of the managerial burden from GH farmers but also consolidating the partnership through delegating responsibilities and sharing experience among the mergers' partners.

5 CONCLUDING REMARKS

In this paper, we investigated possible contribution of M&As, as strategic decisions, towards optimizing energy usage and, as result, enhancing sustainability of the agricultural production. In order to measure the extent of such a contribution, the inverse DEA approach has been applied for the mergers of 15 cucumber GH farms from the Batinah region in Oman. With only 18 productive post-merger GH farms, the results confirmed that more input gains can be achieved through mergers, regardless of the efficiency status of the merging GH farms. More specifically, half of the mergers involve exclusively efficient GH farms, disclaiming as a fact that no further energy gains are possible with efficient GH farms. The average proportions of energy gains per post-merger GH farm range between 19.58% and 55.64%, with the highest figures recorded for electricity. The post-mergers' energy gains for labor, machinery, fertilizers, chemicals, water and electricity hit proportions as high as 82.22%, 68.22%, 84.73%, 51.15%, 76.41% and 82.07%, respectively.

Since the post-merger GH farms may include pairs with common pre-merger GH farms, it was crucial to select the best non-overlapping pairs. Therefore, we developed the Synergy Merge Index (SMI) for selecting the best partners of a merger based on the post-merger's expected outcomes. The implementation of the new approach enabled the selection of two most productive pairs of GH farms among the 18 candidate post-mergers. The inclusion of these GH post-merger farms into the GH farms' sample increased the percentage of energy saving from 4.24% to 17.57%, that is, more than 4 times improvement. In the meantime, it has been noted that the share of electricity consumption represents alone 91.05% of the post-merger energy gains, 7.88% for fertilizers and only 0.54% and 0.42% for labor and water, respectively, besides relatively neglectable shares for chemicals and machinery energy inputs.

Practically, these findings appear strategically important and may definitely support new policies that would promote mergers among GH farms. As long term decisions, mergers could ensure more rational consumption of scarce resources, such as water and electricity, in lieu of addressing related issues on an individual GH farm basis. The environmental impact of such decisions must also be stressed through mitigating the GHG emissions.

However, prior to framing mergers' policies, more investigations might be required to identify the market incentives that could encourage farmers to merge. Moreover, the data sample could be extended to more GH farms as a prerequisite for a better evaluation of the practical scope of the presented methodology. Future research may also focus on developing advanced techniques to select the best partners for each GH farm [47]. One alternative might consist in applying ranking techniques, such as DEA cross-efficiency [48–57].

REFERENCES

[1] Cortés, A., Feijoo, G., Fernández, M., and Moreira, M. T. (2021). Pursuing the route to eco-efficiency in dairy production: The case of Galician area. *Journal of Cleaner Production*, 285, 124861.

[2] Coscieme, L., Mortensen, L. F., Anderson, S., Ward, J., Donohue, I., and Sutton, P. C. (2020). Going beyond Gross Domestic Product as an indicator to bring coherence to the Sustainable Development Goals. *Journal of Cleaner Production*, 248, 119232.

[3] Unakıtan, G., and Aydın, B. (2018). A comparison of energy use efficiency and economic analysis of wheat and sunflower production in Turkey: A case study in Thrace Region. *Energy*, 149, 279–285.

[4] Oukil, A., and Zekri, S. (2021). Investigating farming efficiency through a two stage analytical approach: Application to the agricultural sector in Northern Oman. Preprint, *arXiv* preprint arXiv:2104.10943.

[5] Sow, O. Oukil, A. Ndiaye, B. M. Marcos, A. (2016). Efficiency analysis of public transportation subunits using data envelopment analysis (DEA) and bootstrap: Dakar Dem Dikk case study. *Journal of Mathematics Research*, 8, 114–127.

[6] Oukil, A. (2021). Building a consistent system for faculty appraisal. *Towards New Practices in Learning & Teaching for Higher Education*, 10–11 July 2018, Algiers, Algeria.

[7] Oukil, A. (2016). Assessing consistency of SMEs' performance measures from local to global production technology frameworks. International Colloquium on Governance and Performance of SMEs Between National and International Contexts, 16–17 October 2016, HEC Algiers, Algeria.

[8] Oukil, A. (2021). Data envelopment analysis–a flexible device to assessing SMEs performance. Preprint, *arXiv* preprint arXiv:2104.10444

[9] Oukil, A. (2023). A two-level Induced OWA procedure for ranking DMUs under a DEA cross-efficiency framework. Intelligent and Transformative Production in Pandemic Times: *Proceedings of the 26th International Conference on Production Research (ICPR)*, 19–21 July 2021, Taichung, Taiwan.

[10] Elhami, B., Akram, A., and Khanali, M. (2016). Optimization of energy consumption and environmental impacts of chickpea production using data envelopment analysis (DEA) and multi objective genetic algorithm (MOGA) approaches. *Information Processing in Agriculture*, 3(3), 190–205.

[11] Nabavi-Pelesaraei, A., Rafiee, S., Mohtasebi, S. S., Hosseinzadeh-Bandbafha, H., and Chau, K. W. (2017). Energy consumption enhancement and environmental life cycle assessment in paddy production using optimization techniques. *Journal of Cleaner Production*, 162, 571–586.

[12] Mohseni, P., Borghei, A. M., and Khanali, M. (2018). Coupled life cycle assessment and data envelopment analysis for mitigation of environmental impacts and enhancement of energy efficiency in grape production. *Journal of Cleaner Production*, 197, 937–947.

[13] Hosseinzadeh-Bandbafha, H., Nabavi-Pelesaraei, A., Khanali, M., Ghahderijani, M., and Chau, K. W. (2018). Application of data envelopment analysis approach for optimization of energy use and reduction of greenhouse gas emission in peanut production of Iran. *Journal of Cleaner Production*, 172, 1327–1335.

[14] Khoshroo, A., Izadikhah, M., and Emrouznejad, A. (2018). Improving energy efficiency considering reduction of CO2 emission of turnip production: A novel data envelopment analysis model with undesirable output approach. *Journal of Cleaner Production*, 187, 605–615.

[15] Ilahi, S., Wu, Y., Raza, M. A. A., Wei, W., Imran, M., and Bayasgalankhuu, L. (2019). Optimization approach for improving energy efficiency and evaluation of greenhouse gas emission of wheat crop using data envelopment analysis. *Sustainability*, 11(12), 3409.

[16] Ullah, A., Silalertruksa, T., Pongpat, P., and Gheewala, S. H. (2019). Efficiency analysis of sugarcane production systems in Thailand using data envelopment analysis. *Journal of Cleaner Production*, 238, 117877. https://doi.org/10.1016/j.jclepro.2019.117877

[17] Pishgar-Komleh, S. H., Zylowski, T., Rozakis, S., and Kozyra, J. (2020). Efficiency under different methods for incorporating undesirable outputs in an LCA + DEA framework: A case study of winter wheat production in Poland. *Journal of Environmental Management*, 260, 110138.
[18] Mushtaq, Z., Wei, W., Sharif, M., and Chandio, A. A. (2021). Evaluating energy consumption efficiency in tobacco production: Applying Data Envelopment Analysis. *E&M Economics and Management*, 24(3), 23–39.
[19] Zalaghi, A. H., Lotfalian Dehkordi, A., Abedi, A., and Taki, M. (2021). Applying data envelopment analysis (DEA) to improve energy efficiency of apple fruit, focusing on cumulative energy demand. *Energy Equipment and Systems*, 9(1), 37–52.
[20] Ebrahimi, R., and Salehi, M. (2015). Investigation of CO2 emission reduction and improving energy use efficiency of button mushroom production using Data Envelopment Analysis. *Journal of Cleaner Production*, 103, 112–119.
[21] Raheli, H., Rezaei, R. M., Jadidi, M. R., and Mobtaker, H. G. (2017). A two-stage DEA model to evaluate sustainability and energy efficiency of tomato production. *Information Processing in Agriculture*, 4(4), 342–350.
[22] Nourani, A., and Bencheikh, A. (2020). Energy requirement optimization of greenhouse vegetable production using Data Envelopment Analysis (DEA) Method in Algeria. *Acta Technologica Agriculturae*, 23(2), 60–66.
[23] Nourani, A., and Bencheikh, A. (2021). Applying the DEA approach to optimize the energy required and GHG emission for greenhouse vegetable production in Algeria. *Agricultural Engineering International: CIGR Journal*, 23(3), 156–164.
[24] Mardani Najafabadi, M., and Taki, M. (2020). Robust data envelopment analysis with Monte Carlo simulation model for optimization the energy consumption in agriculture. *Energy Sources, Part A: Recovery, Utilization, and Environmental Effects*, 1–15. https://doi.org/10.1080/15567036.2020.1777221
[25] Soheilifard, F., Taki, M., and van Zelm, R. (2021). Impact of energy flow optimization on the mitigation of environmental consequences and costs in greenhouse cucumber production. *Environmental Science and Pollution Research*, 28(7), 8421–8433.
[26] Soltani, A. A., Oukil, A., Boutaghane, H., Bermad, A., and Boulassel, M. R. (2021). A new methodology for assessing water quality, based on data envelopment analysis: Application to Algerian dams. *Ecological Indicators*, 121, 106952.
[27] Oukil, A. Zekri, S. (2015). Inverse optimization: An efficient tool for resource reallocation in the farming sector. *Operational Research Practice in Africa (ORPA)*, 20–22 April 2015, Algiers, Algeria.
[28] Amin, G. R. Oukil, A. (2019a). Flexible target setting in mergers using inverse data envelopment analysis. *International Journal of Operational Research*, 35 (3), 301–317.
[29] Banker, R. D., Charnes, A., and Cooper, W. W. (1984). Some models for estimating technical and scale inefficiencies in data envelopment analysis. *Management Science*, 9, 1078–1092.
[30] Oukil, A., and Govindaluri, S. M. (2017). A systematic approach for ranking football players within an integrated DEA-OWA framework. *Managerial & Decision Economics*, 38 (8), 1125–1136.
[31] Oukil, A., and Amin, G. R. (2015). Maximum appreciative cross-efficiency in DEA: A new ranking method. *Computers & Industrial Engineering*, 81, 14–21.
[32] Oukil, A. (2019). Embedding OWA under preference ranking for DEA cross-efficiency aggregation: Issues and procedures. *International Journal of Intelligent Systems*, 34(5), 947–965.
[33] Oukil, A., Channouf, N., and Al-Zidi, A. (2016). Performance evaluation of the hotel industry in an emerging tourism destination: Case of Oman. *Journal of Hospitality & Tourism Management*, 29, 60–68.
[34] Oukil, A., and El-Bouri, A. (2021). Ranking dispatching rules in multi-objective dynamic flow shop scheduling: a multi-faceted perspective. *International Journal of Production Research*, 59(2), 388–411.
[35] Oukil, A. (2024). Selecting material handling equipment through a market weight scheme based DEA cross-efficiency approach. *International Journal of Management Science and Engineering Management*, 19(1), 1–14.
[36] Oukil, A., Soltani, A. A., Boutaghane, H., Abdalla, O., Bermad, A., Hasbaia, M., and Boulassel, M. R. (2021). A Surrogate Water Quality Index to assess groundwater using a unified DEA-OWA framework. *Environmental Science and Pollution Research*, 28, 56658–56685.
[37] Oukil, A., Soltani, A. A., Zeroual, S., Boutaghane, H., Abdalla, O., Bermad, A., and Boulassel, M. R. (2022). A DEA cross-efficiency inclusive methodology for assessing water quality: A Composite Water Quality Index. *Journal of Hydrology*, 612, 128123.
[38] Amin, G. R., and Boamah, M. I. (2020) A new inverse DEA cost efficiency model for estimating potential merger gains: a case of Canadian banks. *Annals of Operations Research*, 295, 21–36.

[39] Oukil, A., and Al-Zidi, A. (2018). Benchmarking the hotel industry in Oman through a three-stage DEA-based procedure. *Journal of Arts and Social Sciences*, 9(2), 5–23.
[40] Al-Mezeini, N. K., Oukil, A., and Al-Ismaili, A. M. (2020). Investigating the efficiency of greenhouse production in Oman: A two-stage approach based on Data Envelopment Analysis and double bootstrapping. *Journal of Cleaner Production*, 247, 119160.
[41] Al-Mulla, Y. (2006). Cooling greenhouses in the Arabian Peninsula, *International Symposium on Greenhouse Cooling*, 719, 499–506.
[42] MAF (2014). *Agricultural Censes Report*, Oman. Ministry of Agriculture & Fisheries.
[43] Firoozi, S., Sheikhdavoodi, M. J., and Farani, S. M. (2014). Optimizing energy consumption efficiency for greenhouse cucumber production using the data envelopment analysis approach in Markazi Province of Iran. *Journal of Agricultural Technology*, 10(3), 543–558.
[44] Taki, M., Ajabshirchi, Y., and Mahmoudi, A. (2012). Application of parametric and non-parametric method to analyzing of energy consumption for cucumber production in Iran. *Modern Applied Science*, 6(1), 75.
[45] Oukil, A., and Govindaluri, S. M. (2020). A hybrid multi-attribute decision-making procedure for ranking project proposals: A historical data perspective. *Managerial and Decision Economics*, 41(3), 461–472.
[46] Khoshnevisan, B., Rafiee, S., Omid, M., and Mousazadeh, H. (2013). Reduction of CO2 emission by improving energy use efficiency of greenhouse cucumber production using DEA approach. *Energy*, 55, 676–682.
[47] Oukil, A. (2023). Investigating prospective gains from mergers in the agricultural sector through Inverse DEA. *IMA Journal of Management Mathematics*, 34(3), 465–490.
[48] Oukil, A., El-Bouri, A., and Emrouznejad, A. (2022). Energy-aware job scheduling in a multi-objective production environment–An integrated DEA-OWA model. *Computers & Industrial Engineering*, 168, 108065.
[49] Oral, M., Oukil, A., Malouin, J.-L., and Kettani, O. (2014). The appreciative democratic voice of DEA: A case of faculty academic performance evaluation. *Socio-Economic Planning Sciences*, 48, 20–28.
[50] Amin, G. R., and Oukil, A. (2019b). Gangless cross-evaluation in DEA: an application to stock selection. *RAIRO-Operations Research*, 53(2), 645–655.
[51] Oral, M., Amin, G. R., and Oukil, A. (2015). Cross-efficiency in DEA: A maximum resonated appreciative model. *Measurement*, 63, 159–167.
[52] Hassan, M. Oukil, A. (2021). Design of efficient systems of commercial material handling equipment for supply chain and logistics facilities using DEA. *International Journal of Logistics Systems & Management*, 39(2), 241–272.
[53] Oukil, A. (2020). Designing efficient material handling systems: a two-stage approach based on DEA cross-efficiency. *In 2020 5th International Conference on Logistics Operations Management (GOL)*, 1–5. https://doi.org/10.1109/GOL49479.2020.9314754
[54] Oukil, A. (2020). Exploiting value system multiplicity and preference voting for robust ranking. *Omega, International Journal of Management Science*, 94, 102048.
[55] Oukil, A. (2018). Ranking via composite weighting schemes under a DEA cross-evaluation framework. *Computers & Industrial Engineering*, 117, 217–224.
[56] Oukil, A. Amin, G. R. (2023). An Induced OWA aggregation operator with dual preference settings for DEA cross-efficiency ranking. *Soft Computing*, 27, 18419–18440.
[57] Oukil, A. Soltani, A. A. Al-Hajri, A. (2024). A novel stakeholder-driven approach for objective hotel quality rating. *Journal of Quality Assurance in Hospitality & Tourism*. In press
[58] Oukil, A. Soltani, A. A. Nourani, A. Benchikh, A. (2024). Improving agricultural sustainability through farm mergers: An energy efficiency perspective. *International Journal of Agricultural Sustainability*, 22(1), 2293598.
[59] Oukil, A. Kennedy, R. E. Al-Hajri, A. Soltani, A. A. (2024). Unveiling the potential of hotel mergers: a hybrid DEA approach for optimizing sector-wide performance in the hospitality industry. *International Journal of Hospitality Management*, 116, 103620.
[60] Oukil, A. Soltani, A. A. Al-Mezeini, N. Al-Ismaili, A. Nourani, A. (2023). Mergers as an alternative for energy use optimization: Evidence from the cucumber greenhouse production using the Inverse DEA approach. *Environment, Development and Sustainability*. https://doi.org/10.1007/s10668-023-04144-6
[61] Oukil, A. Nourani, A. Soltani, A. A. Benchikh, A. (2022). Using Inverse Data Envelopment Analysis to investigate potential impact of Mergers on Energy use optimization – Application in agricultural production. *Journal of Cleaner Production*. 381(1), 135199.

Innovation and Technological Advances for Sustainability – Al-Naemi et al. (Eds)

Smart produce prediction system for reducing food waste and improving food security in Qatar

Seifeddine Bouallegue & Salem Al-Naemi
College of Computing and Information Technology, University of Doha for Science and Technology, Doha, Qatar

ABSTRACT: Qatar is a small country with limited land resources for agriculture and thus it is essential to ensure food security by utilizing available resources efficiently. PPS provides a way to achieve this by using data from a variety of sources such as weather, population, economic indicators, hospitality needs, and consumption patterns to forecast demand for different types of produce. In this paper, we will explore how PPS can be used to improve food security and reduce food waste in Qatar by identifying potential benefits of PPS in several key areas. First, we will discuss how PPS can be used to optimize inventory management for farmers and retailers by forecasting future demand and adjusting production accordingly. Second, we will examine how PPS can improve the efficiency of the supply chain by identifying potential bottlenecks and inefficiencies in advance. Third, we will explore how PPS can be used to identify seasonal trends in produce consumption and assist farmers in planning for planting and harvesting times more effectively. Overall, the PPS is a powerful tool that can help to optimize agricultural production and reduce food waste in Qatar by providing insights and predictions based on data analysis. The use of PPS can help to improve food security by ensuring that the right amount of produce is produced and available for consumption in a timely manner.

Keywords: Produce prediction systems (PPS), Food waste reduction, Food security, Inventory management, Supply chain efficiency.

1 INTRODUCTION

From 2019 through 2022, there has been a continuous increase in global hunger, affecting about 10% of the worldwide population [1]. Meanwhile, the State of Food and Agriculture [2] report from the FAO estimates that around 14% of the world's food—worth $400 billion annually—continues to be lost after it is produced and before it is sold in stores [3]. Furthermore, an additional 17 percent of our food is wasted in retail and by customers [4]. With that being considered, the FAO estimates that the food lost and waste each year could feed 1.26 billion hungry people.

Qatar imports more than 90% of its food. In order to ensure food security, it is essential to optimize agricultural production and reduce food waste. Produce prediction systems (PPS) provide a way to do this by using data from a variety of sources to forecast demand for different types of produce.

2 PROBLEM STATEMENT

When the population's need in produce is not well predicted, it can lead to several problems related to food waste and food security. Therefore, it is important to use accurate data and

DOI: 10.1201/9781003496724-15

predictive analytics to better understand population needs and optimize food production and distribution.

2.1 *Food waste*

When population need in produce is not accurately predicted, farmers may end up producing more produce than is actually needed. This can lead to surplus produce that goes to waste, either by being thrown away or by not being able to be sold. This can result in a significant loss of resources, both in terms of the produce itself and the resources used to produce it, such as water and fertilizers.

2.2 *Food insecurity*

If population need in produce is not well predicted, there may be a shortage of produce available for consumption. This can lead to food insecurity, as people may not have access to enough nutritious and healthy food to meet their needs.

2.3 *Economic loss*

When population need in produce is not well predicted, there may be a mismatch between supply and demand, resulting in economic losses for farmers and businesses. Farmers may not be able to sell their produce at a profitable price, and businesses may not be able to meet customer demand.

2.4 *Environmental issues*

Food waste can lead to environmental issues, as discarded produce may end up in landfills where it releases methane, a potent greenhouse gas, and consumes valuable space.

2.5 *Dependence on imports*

If the population's need in produce is not well predicted, it may lead to over-reliance on imports, which can be affected by factors such as climate change, political instability and economic downturns, which can put the food security at risk.

3 PROPOSED SOLUTION

In this section, we propose a produce prediction system can help reduce food waste and improve food security in Qatar in several ways.

3.1 *Accurately forecasting demand*

PPS can use data from weather, population, economic indicators, and consumption patterns to accurately forecast demand for different types of produce. This can help farmers to plan their crop production more effectively and reduce the risk of surplus or wasted produce.

There are several predictive models that can be used for forecasting demand for different types of produce and optimizing crop production. Some of the most commonly used models for this problem are:

- Time Series Models are a set of statistical models that are used to analyze and make predictions based on historical data that has been collected over time. These models are particularly useful for forecasting demand for different types of produce and optimizing crop production, as they can be used to identify trends and patterns in the data and make predictions based on those patterns. [5]

There are several types of time series models that can be used for this problem, including:

3.1.1 *ARIMA (Auto-Regressive Integrated Moving Average)*

This model is used to analyze and forecast time series data that has a clear trend and/or seasonal pattern. The model uses the past values of the time series to make predictions about future values, taking into account the trend and seasonality components. [6]

3.1.2 *Holt-Winters*

This model is an extension of the exponential smoothing technique, which is used for forecasting time series data with a clear trend and/or seasonal pattern. The model takes into account both the trend and the seasonality components to make predictions about future values. [7]

3.1.3 *Exponential Smoothing*

This model is used for forecasting time series data with a clear trend but no clear seasonality. The model uses a weighted average of past values to make predictions about future values, taking into account the trend component. [8]

3.1.4 *State Space Model (SSM)*

This model is a general class of models that can be used to model a wide range of time series data, including data with complex patterns, irregular trends and outliers. It can be used with various likelihood functions and can be applied to univariate and multivariate time series.

3.1.5 *Prophet*

This model developed by Facebook is a procedure for forecasting time series data. It decomposes the time series into trend, seasonality and holidays. It is particularly useful for datasets with strong seasonal effects and multiple seasons patterns.

- Machine Learning Models are a set of algorithms that can learn from data, identify patterns and make predictions or decisions without explicit instructions. These models can be used to analyze large amounts of data and identify complex relationships between different factors. [9] There are several types of machine learning models that can be used for forecasting demand for different types of produce and optimizing crop production, such as:

3.1.6 *Regression models*

These models are used to predict a continuous output variable based on one or more input variables. For example, a linear regression model could be used to predict the demand for a certain type of produce based on weather data, population data and other relevant factors. [10]

3.1.7 *Classification models*

These models are used to predict a categorical output variable based on one or more input variables. For example, a decision tree model could be used to predict which type of produce is most likely to be in high demand based on weather data, population data and other relevant factors. [11]

3.1.8 *Clustering models*

These models are used to find natural grouping or patterns in the data. For example, a k-means clustering model could be used to group different types of produce based on their demand patterns and other relevant factors. [12]

- Hybrid Models are a combination of time series and machine learning models. These models use the strengths of both types of models to make predictions. For example, a hybrid model may use a time series model to identify trends and patterns in the data, and then use a machine learning model to make predictions based on those patterns. One example of a hybrid model is using an ARIMA model to decompose the time series data into trend, seasonality and residuals, then using the residuals as input to a machine learning model such as Random Forest or Neural Network to make predictions. Another example could be using a decision tree to identify the most important features in the data, and then using a time series model to make predictions based on those features. Hybrid models can often provide more accurate predictions than either time series or machine learning models alone.
- Artificial Neural Networks (ANNs) are a type of machine learning algorithm that are inspired by the structure and function of the human brain. They are particularly useful for modeling complex and non-linear relationships and have been used in a wide range of applications, including time series forecasting.

An ANN is composed of layers of interconnected nodes called neurons, which are organized in an input layer, one or more hidden layers, and an output layer. Each neuron in the input layer represents an input variable, and each neuron in the output layer represents an output variable. The neurons in the hidden layers are used to extract features from the input data and make predictions based on those features. [10]

There are several types of ANNs that can be used for time series forecasting, including:

3.1.9 *Feedforward Neural Networks (FFNNs)*

These are the most basic type of ANNs, which consist of an input layer, one or more hidden layers, and an output layer. The input data is passed through the network in a feedforward manner, from the input layer to the output layer, through the hidden layers.

3.1.10 *Recurrent Neural Networks (RNNs)*

These are a type of ANNs that are designed to handle sequential data, such as time series data. RNNs have a feedback loop, which allows them to remember previous inputs and use them to make predictions about future inputs. [14]

3.1.11 *Long Short-Term Memory (LSTM) networks*

These are a special kind of RNNs. LSTMs are particularly useful for modeling long-term dependencies in time series data. They have the ability to remember information over long periods of time and are less affected by the vanishing gradient problem compared to other recurrent neural networks. [15]

3.1.12 *Convolutional Neural Networks (CNNs)*

These are a type of ANNs that are designed to handle image and signal data. CNNs use a technique called convolution, which allows them to extract features from the input data and make predictions based on those features. They are particularly useful for time series forecasting problems when the data is periodic or has a repeating pattern. [16]

- Deep Learning models (DL) models are a subset of machine learning algorithms that are based on artificial neural networks (ANNs) but with multiple hidden layers. These models have a large number of parameters and are able to automatically learn representations of the data, which makes them particularly useful for complex and non-linear problems. [17]

There are several types of DL models that can be used for time series forecasting, including Deep Feedforward Networks (DFFNs). These are the most basic type of DL models, which consist of an input layer, multiple hidden layers, and an output layer. The input data is passed through the network in a feedforward manner, from the input layer to the output layer, through the hidden layers. DFFNs can be used for time series forecasting by treating each time step as a separate input.

3.2 *Optimizing inventory management*

In this section, we will discuss how the Produce Prediction System (PPS) can be used to optimize inventory management for farmers and retailers.

Effective inventory management is crucial in the fresh produce industry, as it helps to reduce food waste by ensuring that the right amount of produce is produced and available for consumption. To achieve this, the PPS uses data collected from the Ministry of Statistics and the hospitality industry to forecast future demand for different types of produce. By analyzing historical data on population demographics, economic indicators, weather patterns, and produce consumption patterns, the PPS can make predictions about which types of produce will be in high demand and when.

Once the predictions are made, the PPS can use this information to adjust production levels accordingly. For example, if the PPS predicts that demand for tomatoes will increase in the coming months, farmers can adjust their production levels to ensure that they have enough tomatoes available to meet that demand. This can help to avoid stockouts and ensure that there is a steady supply of fresh produce available for consumption.

Additionally, the PPS can also help farmers and retailers to manage their inventory levels by identifying when certain types of produce are likely to be in low demand. By knowing when demand for certain types of produce is likely to be low, farmers and retailers can adjust their production and inventory levels to avoid overproduction and reduce food waste.

To achieve this, the PPS can be integrated with other inventory management systems, such as warehouse management systems, so that farmers and retailers can easily access the predictions and adjust their inventory levels accordingly. This integration enables a real-time update of the inventory management and reduces the risk of stockouts and overproduction.

In conclusion, the PPS is a powerful tool for optimizing inventory management for farmers and retailers. By forecasting future demand and adjusting production accordingly, the PPS can help to reduce food waste by ensuring that the right amount of produce is produced and available for consumption. The integration of PPS with other inventory management systems enables a real-time update of the inventory management and reduces the risk of stockouts and overproduction.

3.3 *Improving supply chain efficiency*

In this section, we will discuss how the Produce Prediction System (PPS) can be used to improve the efficiency of the supply chain by identifying potential bottlenecks and inefficiencies in advance.

As described in the previous section, the produce prediction can be used to optimize the supply chain to ensure that produce is distributed and available for consumption in a timely manner. For example, if the PPS predicts that demand for strawberries will increase in the coming months, the system can notify suppliers and distributors to ensure that they have enough strawberries available to meet that demand. This can help to avoid stockouts and ensure that there is a steady supply of fresh produce available for consumption.

Additionally, the PPS can also be used to identify potential inefficiencies in the supply chain. For example, if the PPS predicts that demand for a certain type of produce will be low in a particular region, the system can notify suppliers and distributors to reduce their inventory levels in that region and avoid overproduction.

To achieve this, the PPS can be integrated with other supply chain management systems, such as logistics and transportation management systems, so that suppliers and distributors can easily access the predictions and adjust their inventory levels accordingly. This integration enables a real-time update of the supply chain and reduces the risk of stockouts and overproduction.

In conclusion, the PPS is a powerful tool for improving the efficiency of the supply chain by helping to identify potential bottlenecks and inefficiencies in advance and optimize the distribution of fresh produce. This can help to ensure that produce is available for consumption in a timely manner and reduce food waste. The integration of PPS with other supply chain management systems enables a real-time update of the supply chain and reduces the risk of stockouts and overproduction.

3.4 *Identifying seasonal trends*

In this section, we will discuss how the Produce Prediction System (PPS) can be used to identify seasonal trends in produce consumption.

Understanding seasonal trends in produce consumption is crucial for farmers as it enables them to plan for planting and harvesting times more effectively. This can help to ensure that the right crops are available at the right time, reducing food waste and improving food security. The PPS uses data collected from the Ministry of Statistics and the hospitality industry to identify seasonal trends in produce consumption. By analyzing historical data on population demographics, economic indicators, weather patterns, and produce consumption patterns, the PPS can make predictions about which types of produce will be in high demand and when.

Using the prediction information, the PPS can identify seasonal trends in produce consumption, such as which types of produce are typically in high demand during certain times of the year. For example, if the PPS predicts that demand for strawberries will increase in the coming summer months, farmers can plan to plant more strawberries during the spring to ensure that they have enough strawberries available to meet that demand.

Additionally, the PPS can also be used to identify potential opportunities for crop diversification. For example, if the PPS predicts that demand for a certain type of produce will be low in a particular region, farmers can consider planting other types of produce that are in high demand in that region.

To achieve this, the PPS can be integrated with other agricultural management systems, such as crop management systems, so that farmers can easily access the predictions and adjust their planting and harvesting times accordingly. This integration enables a real-time update of the crop management and reduces the risk of food waste.

The PPS is a powerful tool for identifying seasonal trends in produce consumption. By using data to identify seasonal trends, the PPS can help farmers to plan for planting and harvesting times more effectively, ensuring that the right crops are available at the right time. The integration of PPS with other agricultural management systems enables a real-time update of the crop management and reduces the risk of food waste.

4 SYSTEM MODEL

As shown in Figure 1, the produce prediction system can gather data from multiple sources, including the Ministry of Statistics and the hospitality industry (e.g. hotels, restaurants). The data can be collected through various means, such as web scraping, APIs, or manual data entry.

Once the data is collected, it can be preprocessed and cleaned to ensure that it is in the correct format for analysis. The cleaned data can then be used to train and test various predictive models.

The Ministry of Statistics data can include information on population demographics, economic indicators, and weather patterns. This data can be used to identify trends and patterns in produce consumption and to make predictions about future demand.

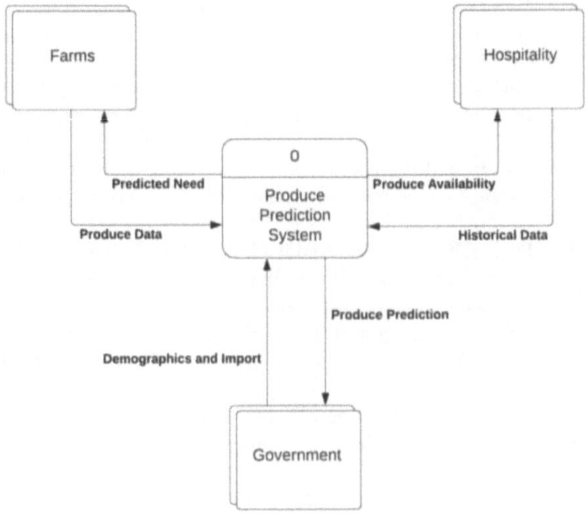

Figure 1. Produce prediction system context diagram.

The data from the hospitality industry can include information on the types and quantities of produce that are being consumed by hotels and restaurants. This data can be used to identify which types of produce are in high demand and to make predictions about future demand.

Both sets of data can be used in combination to make more accurate predictions about import needs, and the system can use the data to optimize crop production and reduce food waste. The system can also include tools for visualizing the data and for monitoring performance, so that adjustments can be made as needed.

5 CONCLUSION

In conclusion, this paper has discussed the potential benefits of using Produce Prediction Systems (PPS) for reducing food waste and improving food security in Qatar. By analyzing data from various sources such as weather, population, economic indicators, hospitality needs and consumption patterns, PPS can provide valuable insights and predictions for the agricultural sector.

PPS can help farmers to optimize crop production, retailers to manage inventory levels, and improve the efficiency of the supply chain. These benefits are crucial for Qatar, a country with limited land resources for agriculture, as it works towards food security and sustainability. Therefore, the implementation of PPS is a significant step towards achieving these goals.

REFERENCES

[1] Hunger A. A., *World Hunger Facts & Statistics — Action Against Hunger.*
[2] Food and A. O. of the United Nations, *Tackling Food Loss and Waste: A Triple Win Opportunity.*
[3] Food and A. O. of the UN, *Tackling Food Loss and Waste: A Triple Win Opportunity.*
[4] U. U. E. Programme, *Unep Food Waste Index Report 2021.*
[5] Hou C., Wu J., Cao B., and Fan J., (2021). A deep-learning prediction model for imbalanced time series data forecasting, *Big Data Mining and Analytics*, vol. 4, no. 4, pp. 266–278.

[6] Desai J. and Bohara M. H., (2021). Farmer connect – a step towards enabling machine learning based agriculture 4.0 efficiently, in *2021 6th International Conference on Communication and Electronics Systems (ICCES)*, pp. 1440–1445.
[7] ————, Farmer connect- a step towards enabling machine learning based agriculture 4.0 efficiently, in *2021 6th International Conference on Communication and Electronics Systems (ICCES)*, (2021), pp. 1440–1445.
[8] Zhao M. A. and Setyawan B., (2020). Sales forecasting for fresh foods: A study in indonesian fmcg, in *2020 International Conference on Information Science and Communications Technologies (ICISCT)*, pp. 1–9.
[9] Harshini K., Madhira P. K., Chaitra S., and Reddy G. P., (2021). Enhanced demand forecasting system for food and raw materials using ensemble learning, in *2021 International Conference on Artificial Intelligence and Machine Vision (AIMV)*, pp. 1–6.
[10] Jin M. and Jin C., (2008). Forecasting agricultural production via generalized regression neural network, in *2008 IEEE Symposium on Advanced Management ofInformation for Globalized Enterprises (AMIGE)*, pp. 1–3.
[11] Priyanka S. and Prema A., (2022). Real-time cold chain forecasting model for agricultural logistic transportation using feature centric deep neural classification for seasonal recommendation, in *2022 6th International Conference on Electronics, Communication and Aerospace Technology*, pp. 1222–1228.
[12] Fang Y., Min X., Zheng L., and Zhang D., (2019). Fruit market trend fore-cast using kmeans-based deep learning models, in *2019 IEEE 10th International Conference on Software Engineering and Service Science (ICSESS)*, pp. 628–633.
[13] DeepAI, *Feed Forward Neural Network.*
[14] IBM, *What is Recurrent Neural Networks?*
[15] Goel N., Kaur S., and Kumar Y., (2022). Chapter 23 – machine learning-based remote monitoring and predictive analytics system for crop and livestock, in *AI, Edge and IoT-based Smart Agriculture, ser. Intelligent Data-Centric Systems*, A. Abraham, S. Dash, J. J. Rodrigues, B. Acharya, and S. K. Pani, Eds. Academic Press, pp. 395–407.
[16] IBM, *Convolutional Neural Networks.*
[17] Santosh K., Das N., and Ghosh S., (2022). Chapter 3 – deep learning models, in *Deep Learning Models for Medical Imaging, ser. Primers in Biomedical Imaging Devices and Systems*, K. Santosh, N. Das, and S. Ghosh, Eds. Academic Press, pp. 65–97.

Renewable energy sources

Innovation and Technological Advances for Sustainability – Al-Naemi et al. (Eds)

An efficient extraction of the maximum power in a photovoltaic system based on enhanced P&O-Fuzzy logic controller

I.M. Elzein
Department of Telecommunication and Networking Engineering, University of Doha for Science and Technology, Doha, Qatar

M. Kurdi
Department of Computer Science, City University, Lebanon

ABSTRACT: The concept of solar cell is changing the world in obtaining a clean energy that is an environment user friendly source of energy. Solar arrays process and convert the irradiance which is representing the quantity of light that comes from sun to a usable energy. The maximum power point which is denoted by Maximum power (P_{max}) can be efficiently extracted from solar panels through the implementation of a tracking system denoted by the maximum power point tracking algorithm (MPPT). Such a tracking system is well known and utilized in detecting the ultimate maximum power that can be taken from the PV solar arrays under the circumstances of the various changes in both the temperature and irradiance. Through this analysis study, the Perturb and Observe (P&O) algorithm will be proposed to adjust the perturbation steps that will be controlled by an enhanced fuzzy logic controller. It will utilize the fuzzy concepts to extract the maximum power. In this research paper an analysis will be conducted to find the unknown parameters of the PV module along in selecting the appropriate set of fuzzy rules/fuzzy table as well as introducing a high gain dc-dc boost converter to improve the efficiency of power, response time, and reducing the oscillation around the P_{max} in the proposed photovoltaic system.

Keywords: Photovoltaic, maximum power point tracking (MPPT), Fuzzy logic controller (FLC), Perturb and Observe, DC-DC converter

1 INTRODUCTION

Renewable energy resources are getting more visible to humankind due to global warming concerns, and the gradual increase in the pollution that is caused by the conventional power-plants. During the past decades, solar energy proved to be one of the best alternative clean energy sources. As we look at the solar power benefits we may conclude many advantages which may be listed below in a concise form.

As mentioned above solar energy is a clean source of energy that has the effect in reducing the carbon footprint factor on the environment. Over the past years a noticeable decrease in the cost of solar panels that led to an increase on the demand of such a resource of energy. Conventional power-plants may heavily depend on fossil fuels; an example of that would be fuel or furnace oil, natural gas and coal which not only cause different types of pollution but makes the electricity bill more expensive since the market prices are volatile and continuously increasing their cost [1].

Another advantage of solar energy is that solar systems reduce the loss of electricity since it requires to be transported from power-plants to end users (consumers) via a set of multiple extensive networks. Feeding consumers by electricity through a long-distance transmission will result in a major power loss whereas if compared to a solar PV system those solar panels

DOI: 10.1201/9781003496724-16

are at a short proximity from the consumer house and thereafter would lead to increase the electricity efficiency.

From an economical growth point of view, solar power systems are generating more jobs in this domain through the fabrication, manufacturing, installation, and maintenance of them. The demand on solar systems and opting towards them globally is creating additional jobs in this booming sector.

The major drawback of a photovoltaic system is the non-linear characteristics through the current-voltage (I-V); represented as a group of graphical curves that serve the function of revealing the current and voltage output of a PV module [2,3].

Maximum power point tracker (MPPT) is represented by a dc-dc converter. Its role is to generate the optimal power at different environmental condition. The input current or voltage in a photovoltaic module is being regulated by MPPT. This is achieved through the utilization of a dc-dc converter which is responsible in making the output voltage /current of MPPT constantly at the maximum power point of PV system.

A dc-dc converter may be implemented in many varieties such as buck, boost or buck-boost for instance. In this paper we will use a high gain boost dc-dc converter

One of the well-known MPPT algorithms that has a great deal in detecting the maximum power point is the perturb and observe (P&O). This algorithm used to track the maximum power point (MPP) of the module. A slight perturbation is presented to vary the power at the photovoltaic system. The output power of a PV module will be periodically measured and compared to the previous power. If the power (at the output) increased the process is continued with higher perturbation steps otherwise the perturbation will be working in the reverse mode to hit the maximum power point (MPP).

A perturb and observe is known due to its ease of implementation and the ability to allow a high level of conversion. MPPT algorithms serve the PV module in extracting the maximum power (P_{max}) from a photovoltaic array. However, it cannot detect accurately the maximum power point due to the oscillations that take place around it. This is one of the disadvantages of the P&O where the MPP will never be reached accurately and such an algorithm will keep the operating point on a PV curve oscillating around the MPP which results in not extracting the maximum energy of the PV system.

An adaptive fuzzy logic controller will be introduced in the design of the PV system to enhance; the effectiveness and minimizing the oscillation around the maximum power point of a pv module and track it in a faster response time regardless of the changes of the environmental conditions (temperature and /or irradiance). The FLC will be designed to use different sets of membership functions and rules and compare the result of the Maximum power obtained from the implementation of the subsets, membership functions and overall rules. As a result, FLC controller will be targeting how to improve the maximum power point tracking (MPPT) algorithm through tuning membership functions in accordance with the design topology as well as the switching frequency (fs) of the high gain dc-dc boost converter. These factors will be covered in the upcoming sections [4,5].

Traditionally a boost converter increases the output voltages from a low input voltage module. A drawback in the design of a boost converter is that it cannot operate in a range of high ratio duty cycle. This is due to twofold; The transistor's switches off time and the limitation on the switching frequency and minimum off time of transistor switches. Further in this work we will present a non-classical boost converter, through a high gain boost converter. This converter, will provide two essential enhancements to the PV system; a high voltage gain and a lower switching stress at circuit components [6].

2 SYSTEM ARCHITECT AND COMPONENTS

The implementation illustration of the photovoltaic system is shown in Figure 1. In general, tracking the maximum power point (MPP) the impedance that is between the photovoltaic arrays and the load (under varying environmental conditions) must be matched through the employment of dc-dc boost converter.

On the other side, the MPPT controller has a main function in regulating the duty ratio cycle of the dc-dc converter by getting a real time operational parameters that are required by the MPPT controller algorithm. The block components of Figure 1 will be explained in the following sub-sections.

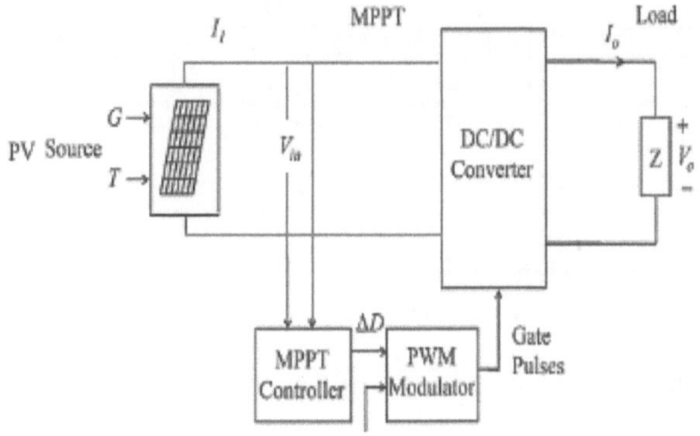

Figure 1. Electrical circuit of PV system.

The mathematical representation and modelling of a photovoltaic solar system is shown in the next section.

3 MATHMATICAL REPRESENTATION OF PV MODULE

Figure 2 represents the equivalent electrical circuit of a PV cell. This is considered a simple single diode which is mostly used during the modelling of PV systems. Five parameters of a PV module are required to be found out for the design of such scheme and those are the photocurrent source (I_{ph}), the saturation current (I_{sat}), the two resistances namely, the series and shut resistances (R_s and R_{sh}) and the ideality factor (n).

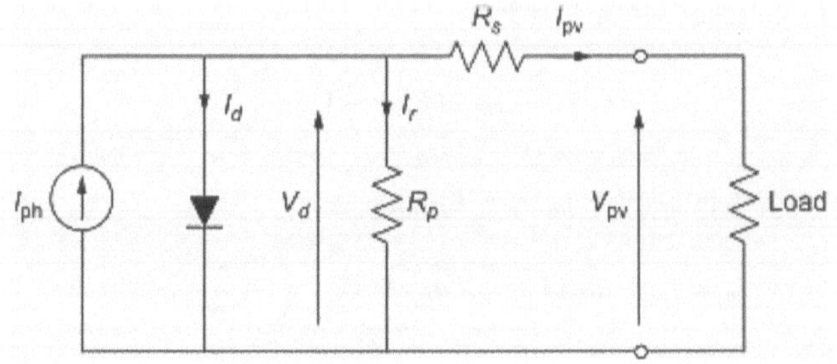

Figure 2. PV cell equivalent circuit.

There are two equivalent resistances as mentioned earlier the (R_p) and the (R_s). These resistances are representing the circuit's losses. Knowing that the output current of this system is also affected by losses within the circuit and this is denoted by [7];

$$Ipv = I_{ph} - I_d - I_{sh} \tag{1}$$

where;

Ipv : is the output PV current
I_d : is the diode current
I_{sh} : is the leakage current

The output current of a solar is expressed in the below equation.

$$I = Np \times Iph - Np \times Is\left[\exp\left(Vpv + \frac{\left(\frac{Ns}{Np}\right)RsI}{n_s\alpha v_t}\right) - 1\right] - \left(Vpv + \frac{\left(\frac{Ns}{Np}\right)RsI}{\left(\frac{Ns}{Np}\right)Rp}\right) \quad (2)$$

where;

I : output current
V_{pv} : output voltage
N_s : number of cells connected in series
N_p : number of cells connected in parallel
R_s : series resistance
R_p : parallel resistance
A : ideality factor of p-n junction

The current (I) and voltage (V) are interrelated through the following equation;

$$I = Iph - Isat\left[\exp\left(V + \frac{IRs}{nNsVth}\right) - 1\right] - \left(V + \frac{IRs}{Rsh}\right) \quad (3)$$

And the thermal equivalent voltage of PV is:

$$Vth = \frac{KT}{q} \quad (4)$$

The short-circuit current: $I = I_{sc} = I_{pv}$, $V = 0$

$$I_{sc} = I_{pv} - I_0[(AR_sI_{sc}) - 1] - \left(\frac{R_sI_{sc}}{R_{sh}}\right) \quad (5)$$

For open-circuit voltage: $I = 0$, $V = V_{oc}$

$$0 = I_{pv} - I_0[(AV_{oc}) - 1] - \frac{V_{oc}}{R_s} \quad (6)$$

The maximum power point: $I = I_p$, $V = V_p$

$$I_p = I_{pv} - I_0\left[\left(V_p + R_sI_p\right) - 1\right] - \left(\frac{V_p + R_sI_p}{R_{sh}}\right) \quad (7)$$

The reverse saturation current is:

$$Irs = Isc/\left(e^{\left(\frac{q \times V0c}{Ns \times K \times A \times Tc}\right)} - 1\right) \quad (8)$$

$$Is = Irs\left(\frac{Tc}{Tref}\right) \times e^{q \times Eg/A \times K\left(\frac{1}{Tc} - \frac{1}{Tref}\right)} \quad (9)$$

where;

Q: electronic charge (−1.602 × 10-19 C)
T_c: functional temperature of a cell
T_{ref}: reference temperature of a cell
E_g: band gap energy of semiconductors
Ns: number of the series-connected cells.
K: Boltzmann constant (1.381 × 10-23 J/K)

The shunt resistance current is:

$$Ish = V + \frac{I \times Rs}{Rp} \quad (10)$$

The diode (dark) current is:

$$Id = \left(e^{\frac{q(V+I\times Rs)}{K\times Tc\times Ns\times A}} - 1\right) \times Is \times Np \quad (11)$$

The electrical characteristics generated by PV module is a function of both a current (I) and a voltage (V).

Examining the relationship between the above parameters while varying the load connected to the system through a combination of open circuit/short circuit that lead to get characteristic current-voltage (I-V) and (P-V) curves as being generated in Figure 3.

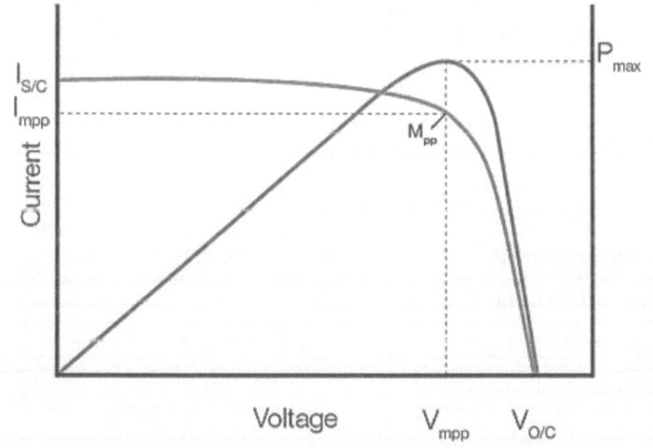

Figure 3. I-V and P-V curves of a general solar cell.

The I-V and P-V curves of a photovoltaic system are presenting the short circuit current at zero voltage (Imax) and the open circuit voltage which is the maximum voltage (Vmax) at a zero current as shown in Figure 3.

In this paper the testing of extracting the maximum power (P_{max}) using an intelligent tracking algorithm(P&O) fuzzy based controller will be based on selecting an MSX-60 PV module. The characteristics curves for both the I-V and P-V are shown in Figure 4 and those will be used as a reference for the upcoming simulations and obtained results.

In addition, the electrical characteristic table of MSX-60 is represented in Table 1. The below values will be compared with the simulation results.

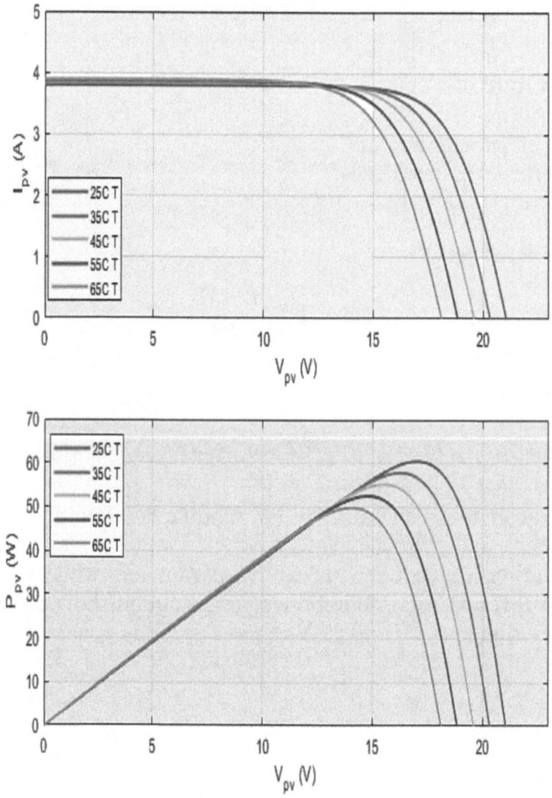

Figure 4. I-V and P-V characteristics curves (STC).

Table 1. Data-sheet of MSX-60 (STC).

Maximum power (P_{max})	60 W
Short-circuit current (I_{sc})	3.87 A
Voltage at MPP (V_{mp})	16.8 V
Open-circuit voltage (V_{oc})	21.0 V
Current at MPP (I_{mp})	3.56 A
Temperature coefficient of I_{sc}	(0.065 ± 0.015) %/°C
Temperature coefficient of V_{oc}	−(80 ± 10) mV/°C
Series-connected cells (Ns)	36

4 SIMULATION AND DISCUSSION OF RESULTS

In this section, we will support the theoretical analysis by validating the principle operation of the overall system and verifying the performance at the standard test conditions (STC). A combination of comparisons took place through the P&O and adaptive P&O fuzzy controller using a five by five fuzzy rule set that was adapted for a twenty five rules in total. The power with respect to time was examined in both techniques as well as a simulation of tracking the maximum output power versus the voltage was verified in regards to the two techniques.

In Figure 5 we ran a simulation at a different samples of time to test the systems 'response of tracking the maximum power P_{max} and extracting it at the shortest sample of time.

The efficient power was clearly obtained between the timing intervals of 0.03 sec. and 0.2 sec. However, the closet point to the MSX-60 power spec. was started to be reached at 0.05 sec. In contrast, when we examine the ultimate results obtained in Figure 6 for the P&O technique it was reached at 0.5 sec. and the maximum power P_{max} tracked did not reach the power matching the spec. Table of the MSX-60 which is expected to be 60 W.

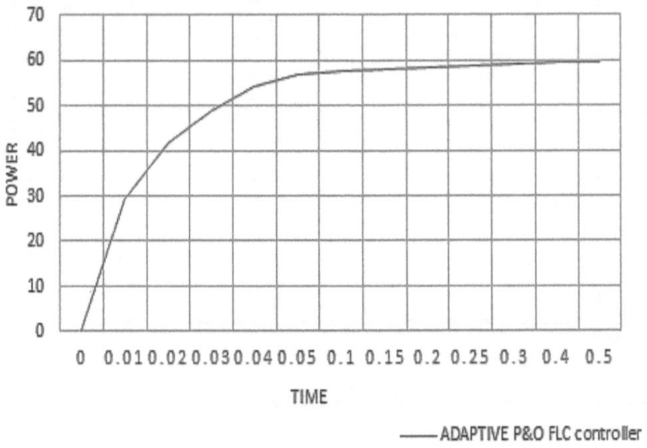

Figure 5. Adaptive P&O FLC tracking of maximum power.

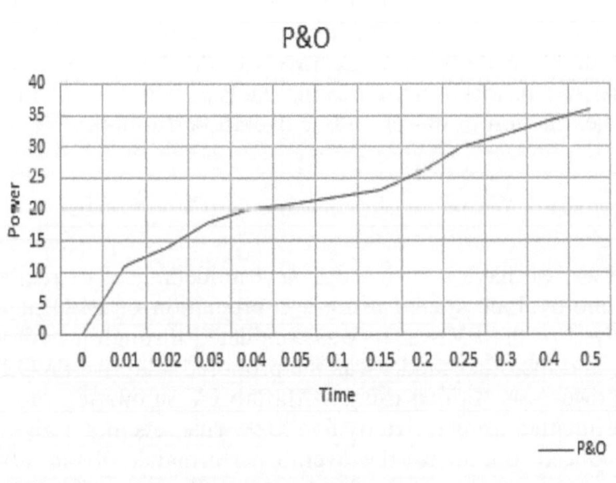

Figure 6. P&O tracking of maximum power.

Referring back to the graph in Figure 6, the P&O algorithm was able to tackle a maximum power up to 36 W at 0.5 sec.

Comparing the P&O tracking algorithm to the Adaptive P&O FLC controller, we can deduce the following:

The Adaptive P&O FLC controller was faster in detecting the maximum power point as compared to the P&O tracking algorithm, and was very close in reaching the P_{max} value which is 60W. In contrast the P&O algorithm was unable to extract the P_{max} and thus the efficiency in extracting the ultimate power was not achieved by using this algorithm. As a result, the proposed system was achieving better results in terms of faster tracking time response and more efficiently extracting the P_{max}.

In Figure 7 we analysed the P-V characteristics curve between our proposed system (Fuzzy P&O) and the (P&O).

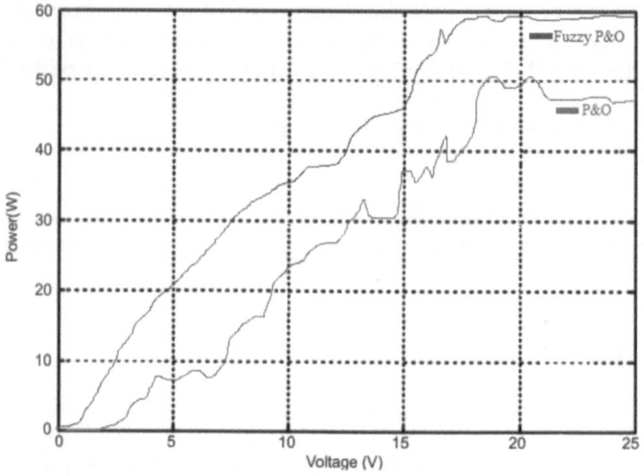

Figure 7. Comparative analysis of two tracking systems.

Our proposed systems' response time as compared to the P&O was faster in tracking the P_{max}, and smoothly ascending up to reach the P_{max}. The P&O was suffering from many oscillations that affected its overall performance in extracting the P_{max} and due to that its efficiency was declining in losing the power and hence a major loss of power was observed. Multi oscillations took place at different intervals with P&O whereas the proposed adaptive P&O FLC controller system was faster in tacking the P_{max} with no oscillations and thus, was more reliable in the extraction of power and its overall performance.

5 CONCLUSION

In this research paper, we have introduced a new modelling of extracting the maximum power point of a photovoltaic system using a combination of P&O algorithm and FLC controller. An analysis of the PV system was conducted through a twofold; one was conducted using P&O, and the other model was the proposed adaptive P&O fuzzy controller.

The maximum power was verified using a Matlab PV simulation model. A fuzzy logic controller was implemented using a five by five fuzzy rule sets and a high gain dc-dc boost converter was introduced to enhance the overall performance of the introduced adaptive P&O fuzzy controller. The results obtained proved the efficiency of the proposed modelling in terms of the overall performance, the least power loss during the tracking and extraction of the P_{max} and the least oscillation around the maximum power point.

REFERENCES

[1] Subudhi B. and Pradhan R. (2013). A comparative study on maximum power point tracking techniques for photovoltaic power systems [J]. *IEEE Transactions on Sustainable Energy*, 4(1): 89–98.

[2] Bendib B., Belmili H. and Krim F. A survey of the most used MPPT methods: Conventional and advanced algorithms applied for photovoltaic systems [J]. *Renewable and Sustainable Energy Reviews*, 2015, 45: 637–648.

[3] Reza Reisi, A., Hassan Moradi, M. and Jamasb, S., (2013). Classification and comparison of maximum power point tracking techniques for photovoltaic system: A review, *Renewable and Sustainable Energy Reviews*, 19: 433–443.

[4] Salam, Z., Ahmed, J. and Merugu, B., (2013). The application of soft computing methods for MPPT of PV system: A technological and status review, *Applied Energy*, 107(C): 135–148.

[5] Chen S. M., Liang T.J., Yang L. S., and Chen J. F., (Apr. 2013). Aboostcon verterwith capacitor multiplier and coupled inductor for Fuzzy Control based Quadruple Boost Converter (IJSRD/Vol. 3/ Issue 10/2015/180) AC module applications, *IEEE Trans. Ind. Electron.*, vol. 60, no. 4, pp. 1503–1511.

[6] Ismail E. H., Al-Saffar M. A., Sabzali A. J., and Fardoun A. A., (May 2008). A family of single-switch PWM converters with high step-up conversion ratio, *IEEE Trans. Circuits Syst. I, Reg. Papers*, vol. 55, no. 4, pp. 1159–1171.

[7] Elzein I. A. and Petrenko Y. N. (2014). A study of maximum power point tracking algorithm for photovoltaic system using a fuzzy logic controller. *2014 International Conference on Automation, Mechanical and Electrical Engineering (AMEE2014), China, WIT Transaction on Engineering Sciences*, Vol. 96, WIT Press.

[8] Ali, A., Almutairi, K. and Padmanaban, S. (2020). Investigation of MPPT Techniques Under Uniform and Non-Uniform Solar Irradiation Condition—A Retrospection. *IEEE Access*, 8, 127368–127392

[9] Shmilovitz D., (2005). On the control of photovoltaic maximum power point tracker via output parameters, *IEE Proc.-Electr. Power Appl.*, Vol. 152, No. 2, pp. 239–248.

[10] Ching-Tsai Pan, Chen-Feng Chuang, and Chia-Chi Chu, Member, (2014). A Novel Transformer-less Adaptable Voltage Quadrupler DC Converter with Low Switch Voltage Stress", *IEEE transactions on power electronics*, Vol. 29, NO. 9.

[11] João Bosco RF. Cabral, Tiago Lemes da Silva. (2013). A new high gain non-isolated dc-dc boost converter for photovoltaic application. *IEEE proc. Brazilian Power Electronics Conference*, Brazil.

[12] Zhao Y., Li W., Deng Y., He X., Lamber S. and Pickert V., (2010). High step-up boost converter with coupled inductor and switched capacitor, *International Conference on Power Electronics, Machines and Drives (PEM)*.

Innovation and Technological Advances for Sustainability – Al-Naemi et al. (Eds)
© 2025 The Author(s), ISBN 978-1-032-80372-2
Open Access: www.taylorfrancis.com, CC BY-NC-ND 4.0 license

Desalination and emission reduction techniques in a diesel engine powered with MWCNT blended emulsions

Sadhik Basha Jaffar, Jahad Al Jabri, Maryam Al Ghaithi, Maryam Al Saadi & Maitha Al Balushi
Department of Process Engineering, National University of Science & Technology (IMCO), Sohar, Oman

ABSTRACT: The two most crucial issues facing the world today are environmental pollution and water scarcity. In order to address these two crucial issues, the current investigation has made significant efforts to produce potable water from seawater using thermal and solar desalination methods in tandem with reducing hazardous emissions from diesel engines by blending diesel fuel with Multiwalled Carbon Nanotubes (MWCNT). Totally six stages were adopted to carry out the current research work. Both thermal and solar desalination experimental setups were designed, constructed, and eventually placed in the exhaust side of the diesel engine during the first stage. The reference readings were obtained in a diesel by using neat diesel along with thermal and solar desalination setups in the second stage. In the third stage, MWCNTs (50 ppm/litre and 100 ppm/litre) were combined with the diesel fuel (with or without water content) in the fourth stage. In the fifth stage, stability and fuel characteristics were assessed in accordance with ASTM standards for all prepared fuels. The performance and emission characteristics of each of the stable MWCNT blended additive fuels were examined in a diesel engine during the sixth stage. It was discovered that the fuels blended with MWCNT had better performance and lower emissions compared to that of neat diesel fuel. This was ascribed due to the MWCNT's presence in the fuel, which has led to improved fuel properties and played a vital role in enhancing performance and reduced emissions in diesel engine. On the other hand, on accustoming the thermal and solar desalination setups, sufficient amounts of 1.75 litres of potable water (from seawater) were also produced in addition to the performance and emission results of this investigation (on utilizing waste heat from exhaust gases from the diesel engine and solar energy convincingly).

Keywords: Emissions, Diesel engine, Desalination, Solar, Exhaust Gases, MWCNT

1 INTRODUCTION

Environmental pollution and water scarcity are the two main problems currently considered as big threats to the human community. The technical community has developed numerous ways to reduce harmful pollutants like nitrogen oxides, carbon monoxide, unburned hydrocarbons, carbon dioxide, soot particles, particulate matter (PM), etc. that are produced by various power generation sectors (such as diesel engine power plants, gas turbine power plants, the automotive and transportation sectors, etc.). The global environment has been getting worse recently, resulting in severe climatic changes, global warming, acid rain, ozone layer depletion, etc. [1–3]. This is because those sectors emit inordinately high amounts of harmful pollutants. This stance has led many researchers and scientists to test a wide range of solutions to the aforementioned global problems in an effort to uphold the ecological environment. Therefore, many researchers devised prominently three possible methods that can be used to reduce the pollutants from the

 DOI: 10.1201/9781003496724-17

power generation sectors: 1. Modification of engine design; 2. Exhaust gas treatment; and 3. Fuel modification. Due to the viability, accessibility, and simplicity attributes associated with the fuel modification method, many researchers and scientists [4–9] have tried out this method to reduce the pollution level. Considering this viewpoint, the main goal of the current investigation is to mix some non-hazardous nano-additives (such as Multiwalled Carbon Nanotubes – MWCNT) with diesel fuel considering the benefits of the fuel modification method. According to recent research findings [4–19], emulsified diesel fuel along with H2O + additives will significantly reduce harmful emissions without affecting the diesel engine's performance features. The use of MWCNT (Multi-Walled Carbon Nanotubes) as fuel additives in water-diesel emulsion has not yet been researched in-depth in diesel engines. Henceforth, in the current research work, MWCNT was blended with water and diesel systematically, and eventually, it was subjected for evaluation in a single cylinder air cooled constant speed diesel engine to evaluate the operating features (viz., performance and emission).

In addition, in the current research work, efforts were also taken to utilize waste heat gases from the diesel engine to boil seawater to produce fresh distilled water by incorporating thermal & solar desalination techniques together. Further, with the viewpoint of the research findings [20–25], 30% of the waste heat in exhaust gases (with a temperature range of 400–600°C) from the diesel engine goes as waste to the atmosphere. This exhaust gas contains a tremendous amount of heat energy, and it can also be used to implement thermal and solar desalination techniques to produce fresh water from seawater. From this point forward, two significant investigations were conducted concurrently in the current work:

(1) A diesel engine's performance and emission characteristics using MWCNT blended water-diesel emulsions.
(2) The production of fresh distilled water by using both thermal and solar desalination techniques on utilizing the waste heat exhaust gases from the diesel engine.

2 EXPERIMENTAL METHODOLOGY

Six steps of the research were conducted to complete the aforementioned responsibilities. To produce fresh distilled water from seawater, experimental setups for both thermal and solar desalination were devised and constructed in the first stage and installed near the diesel engine's immediate exhaust, as illustrated in Figure 1.

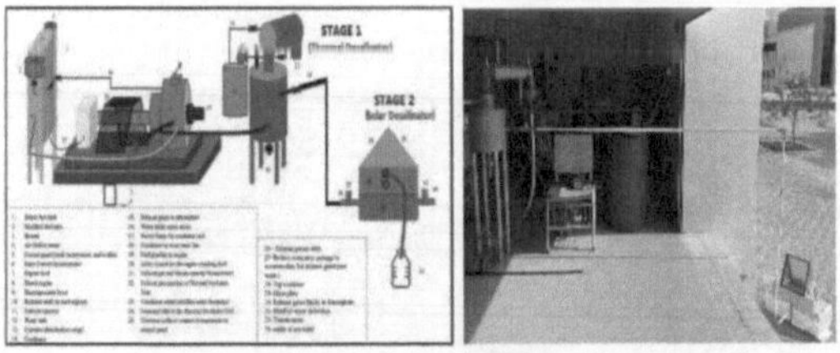

Figure 1. Engine setup with thermal & solar desalination setups.

To acquire the baseline readings, a single-cylinder diesel engine was operated on pure diesel as the only fuel in the second stage (both performance and emission elements). Six litres of seawater were filled into the respective containers in both thermal and solar desalination setups before the experimentation began. The investigation's experimental duration was fixed for 2 hours, and the same duration was fixed for other fuels for the freshwater production investigations. When the diesel engine was run from zero load to full load, an adequate amount of fresh distilled water was produced from seawater (using a thermal & solar desalination setup). In the third stage, MWCNT was systematically blended with diesel fuel in the amounts indicated (50 ppm of MWCNT + Diesel, 100 ppm of MWCNT + Diesel) using an ultrasonicator and a digital mechanical stirrer [9–14]. The results are shown in Figure 2. The characterization of MWCNT is shown in Figure 3.

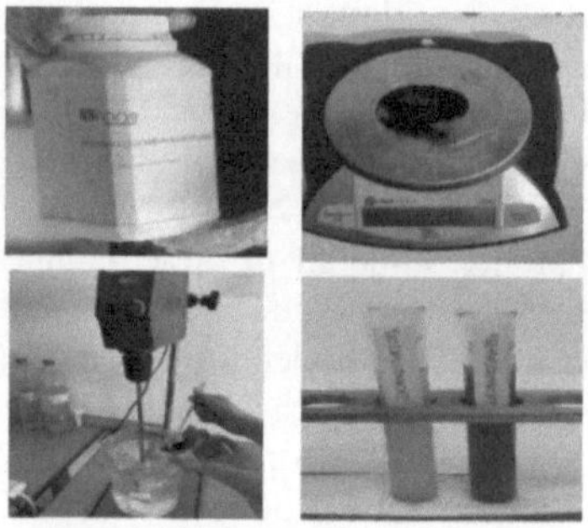

Figure 2. MWCNT blended diesel fuels.

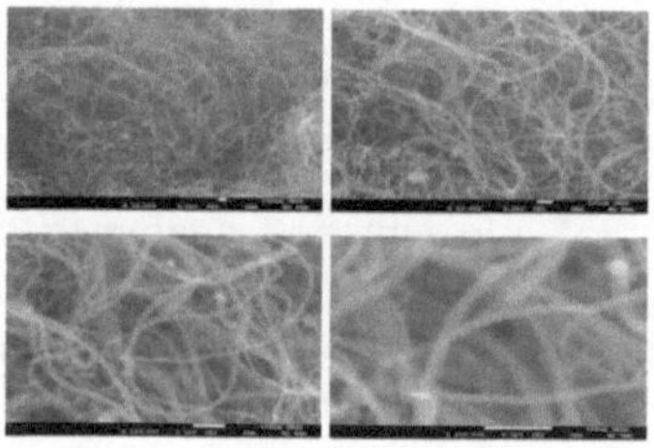

Figure 3. FESEM of MWCNT [26].

In the fourth stage, MWCNT blended water-diesel emulsions (98% diesel + 1% H_2O + 1% surfactants (Span80 + Tween80), (50 ppm & 100 ppm of MWCNT & 96% diesel + 2% H_2O + 2% surfactants) were prepared using a mechanical stirrer (as per the procedure adopted by the researchers [9–11]).

In the fifth stage, the stability (refer to Figure 4) and fuel properties of all prepared fuels were determined, as shown in Table 1.

Figure 4. MWCNT blended emulsions preparation.

Table 1. Fuel Properties.

No	Fuel Code	Density (g/cm^3)	Kinematic Viscosity @ 40°C ($\times 10^{-6}$ m^2/s)	Flash point (°C)	Fire point (°C)
1	D50MWCNT	0.865	4.68*	76	79
2	D100MWCNT	0.870	4.72*	74	78
3	98D2WS	0.872	4.85*	82	85
4	96D4WS	0.875	4.91*	85	87
5	98D2WS50MWCNT	0.875	4.94*	78	81
6	96D2WS100MWCNT	0.882	5.01*	76	80

In the sixth stage, a diesel engine was used to evaluate all of the stable MWCNT blended fuels to look at the performance and emissions characteristics. Furthermore, the experimental investigation's uncertainty error was calculated by the following equation:

$$\text{Error of Uncertainty}(\%) = \sqrt{\begin{array}{l}(NO_x)^2 + (CO)^2 + (CO_2)^2 + (HC)^2 + (Burette)^2 + (Stirrer)^2 + (Stopwatch)^2 + \\ (Smoke)^2 + (Load)^2 + (Thermocouples)^2 + (Speed)^2 + (Torque)^2 + (Ammeter)^2\end{array}}$$

$$= \sqrt{(1)^2 + (1)^2 + (0.01)^2 + (0.1)^2 + (1)^2 + (1)^2 + (1)^2 + (0.05)^2 + (0.1)^2 + +(1)^2 + (1)^2 + (0.1)^2 + (0.1)^2}$$

$$= 2.791\%$$

3 RESULTS AND DISCUSSIONS

The following section pertains to the outcomes of the current investigation and the behaviour of characteristic traits of the tested fuels of a diesel engine. The performance traits (i.e., engine load vs. brake thermal efficiency & brake specific fuel consumption), emission traits (engine load vs NOx, Smoke Opacity) and Quantity of fresh water produced in terms desalination techniques (viz., thermal and solar desalination) were discussed in the ensuing subsequent sections.

3.1 *Diesel engine performance features*

The correlation between brake power and brake thermal efficiency as well as brake-specific fuel consumption for the investigated fuels is shown in Figure 5. MWCNT blended diesel fuels, neat water-diesel emulsions, and MWCNT blended water-diesel emulsions outperformed pure diesel fuel in terms of brake-thermal efficiency. Thermal brake efficiency was high due to the lower brake-specific consumption values for the additive fuels while that of neat diesel fuel. This is because the thermal efficiency of the brake and the specific fuel consumption are inversely related. Furthermore, secondary atomization and micro-explosion effects improved the degree of thermal brake efficiency for MWCNT additive blended fuels when contrasted to pure diesel, as shown in Figure 6. Because water has a lower boiling point than diesel fuel, whereas, MWCNT additive fuels could have instantly absorbed energy to produce water vapours, resulting in the rapid evaporation of water droplets within the fuel droplet during combustion. Additionally, the degree of fuel-air mixing in the presence of MWCNT and the subsequent atomization of water droplets encasing MWCNT may have enhanced brake thermal efficiency and brake-specific fuel consumption [16].

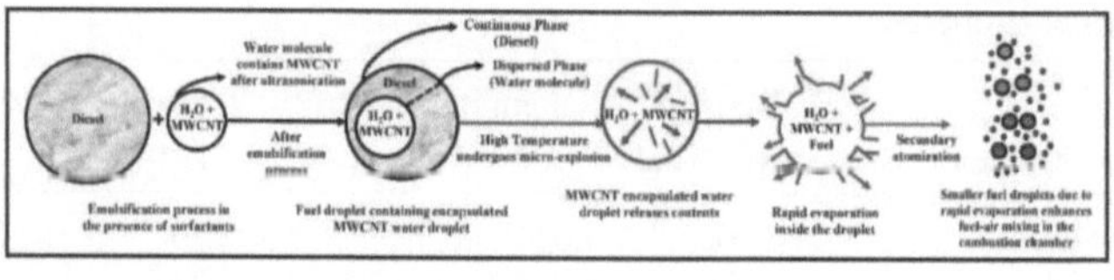

Figure 5. Schematic Illustration of the micro-explosion phenomenon in water-diesel emulsions [16].

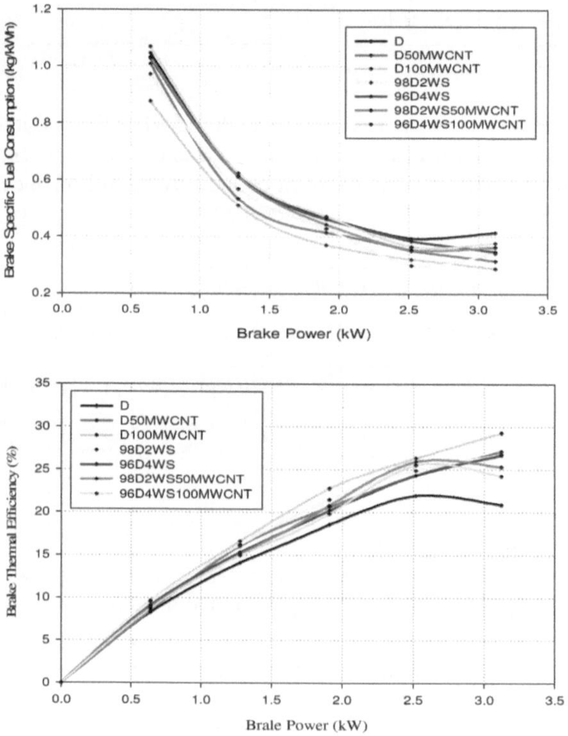

Figure 6. Performance (Brake specific fuel consumption & brake thermal efficiency) deviations.

The thermal efficiency of pure diesel at the maximum load observed was 20.9%, whereas that of additive-blended diesel fuels was 27.12%, 29.3%, 26.8%, 26.7%, 25.32%, and 24.21%. Similarly, at the maximum load, the fuel consumption for diesel fuel was 0.415 kg/kWh, whereas it was 0.289 kg/kWh for D100MWCNT.

3.2 *Diesel engine emissions characteristics*

In fact, as the engine load on the CI engine raises, the harmful emissions of the fuels also increased as a result of the fuel burning at a higher temperature inside the combustion chamber. However, due to additives like water and nanoparticles presence in the fuel, the fuel attributes (such as higher surface area/volume ratio) have headed to reduce emissions for the MWCNT blended fuels. The graph shows that the NOx percentage is distinctly lower for MWCNT-modified fuels than that of pure diesel fuel. The NOx emissions for the D100MWCNT fuel were 1182 ppm at full load, while the diesel fuel was 1586 ppm (refer Figure 7). Black smoke emissions were higher for diesel fuel when compared to MWCNT blended fuels. This is due to the incomplete combustion of diesel fuel, black smoke is primarily

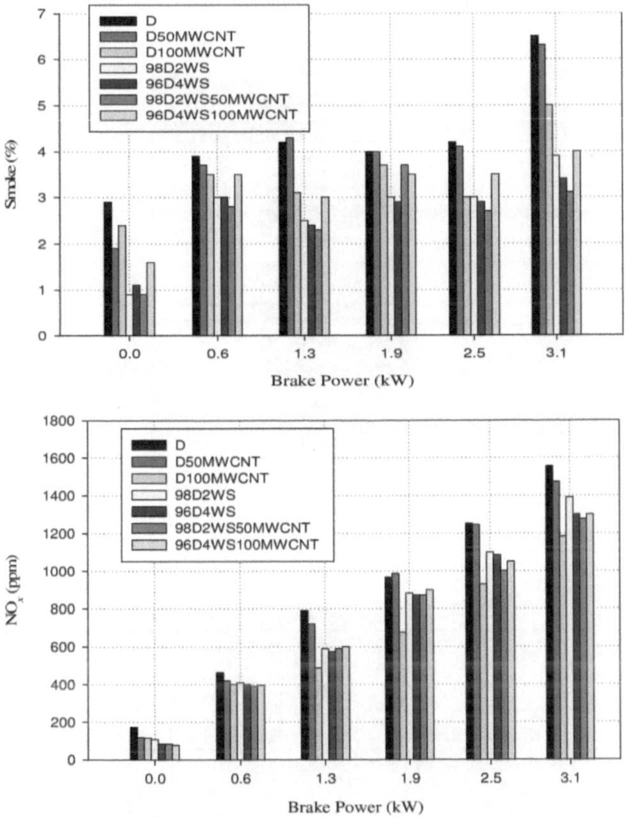

Figure 7. MWCNT blended diesel fuels characteristics with regards to emissions.

composed of elemental carbon with traces of burned-out engine oil components. As shown in Figure 7, at the engine's full load, the percentage of smoke opacity for 98D2WS50MWCNT fuel was 3.1%, whereas it was 6.5% for neat diesel fuel. This could be due to the improved fuel attributes of MWCNT existed in the fuel which have played an important role in emission reduction and performance enhancement features of compression ignition engine.

3.3 *Fresh water from seawater by thermal & solar desalination techniques*

In this current research work, the desalination process was carried out in two phases on utilizing the waste heat gases from the diesel engine. As illustrated in Figures 8 and 9, a thermal desalination and solar desalination setups were built unanimously in the first phase and placed near to the exhaust of the CI engine. The setup for the thermal desalination process comprises two cylinders (inner and outer). The seawater is contained in the inner cylinder as shown in the Figure 10, and the hot exhaust gases from the diesel engine are contained in the outer cylinder of thermal desalination setup. At the maximum load, the temperature of the diesel engine's exhaust fumes was approximately 300°C, which was sufficient to boil the seawater stored in the inner cylinder. When the boiling point of seawater was attained, vapors were produced, and it finally began to condense (through a condenser) and transformed into freshwater. After operating the diesel engine continuously for 2 hours (120 minutes), about 1 liter of fresh water was produced. After the exhaust of the thermal desalination setup, a solar desalination setup was kept in the second stage, as shown in Figures 9 and 11. Around, 150°C, the temperature of the exhaust gases leaving Phase-I. The

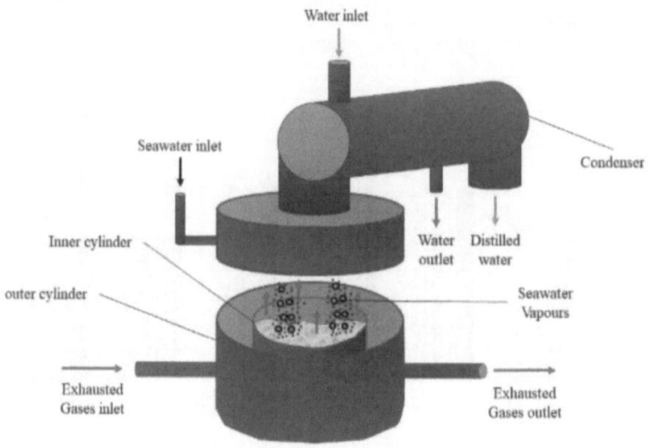

Figure 8. Stage 1 – Thermal desalination setup.

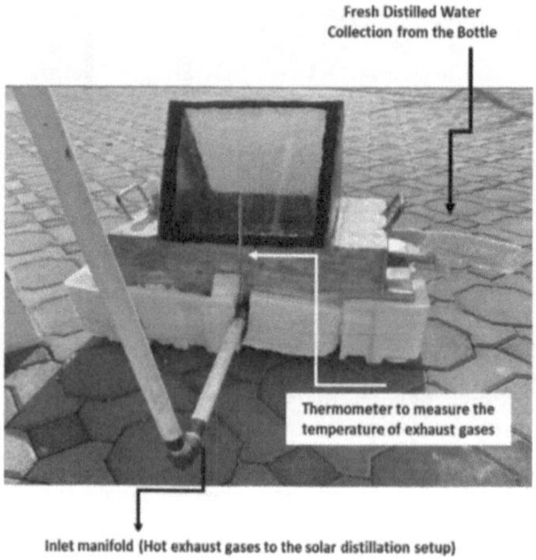

Figure 9. Stage 2 – Solar desalination setup.

Figure 10. Normal filtration and filling of seawater in thermal desalination setup.

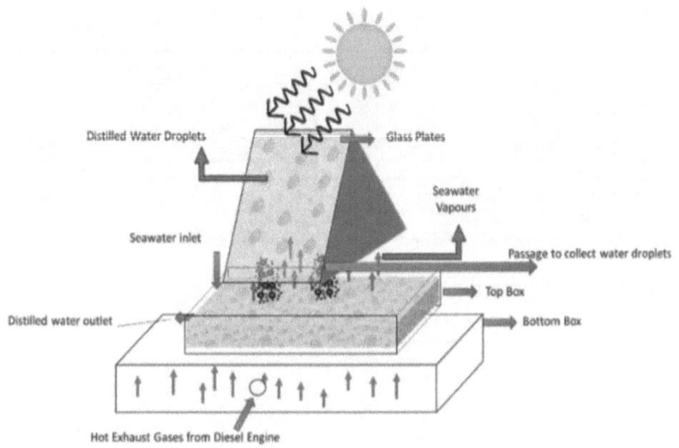

Figure 11. Solar desalination setup (3-D) view.

water in the solar desalination system boiled at this temperature of exhaust gases (with the additional heat energy from the solar radiations). In this solar setup, there are two containers in the solar desalination arrangement (top and bottom). The top container was filled with seawater and it's top was covered with slanting transparent glass plates (which receive solar energy), while the bottom container receives hot exhaust gases (about 150°C) from the phase 1 output. The upper container after receiving heat energy from waste hot gases and solar radiations, freshwater was produced. This is due to the fact that the enormous heat transfer from both sources. As a result, saltwater began to boil and subsequently began to condense on inclined glass plates. To collect the sliding condensed water vapor as fresh water in the top container, a provision was made inside the top container. The solar desalination setup produced about 750 ml of freshwater (when the diesel engine was running for 2 hours. Figure 12 showed a clear correlation between the amount of freshwater collected and the diesel engine's experimental duration.

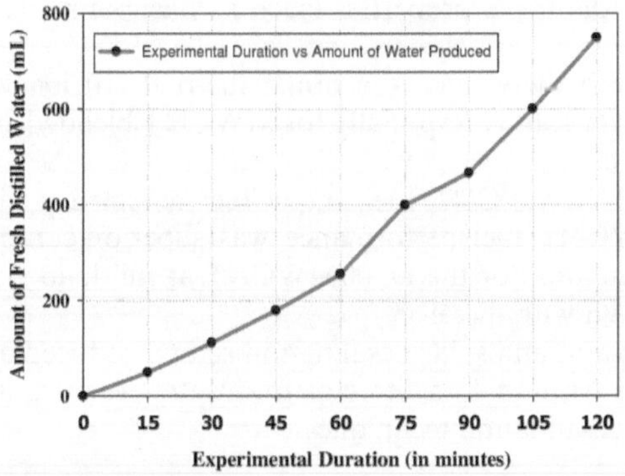

Figure 12. The amount of water produced vs experimental duration.

As shown below, the pH value, dissolved oxygen content (D.O.), and total dissolved solids (TDS) of the distilled water were measured using a calibrated typical water quality meter (as shown in the Figure 13) available in the research laboratory as shown in Table 2.

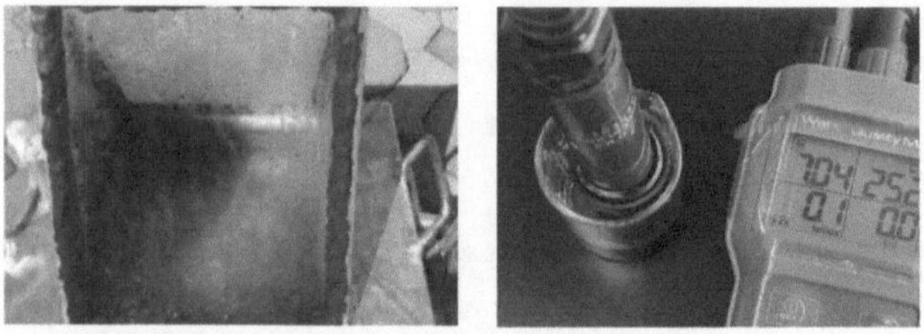

Figure 13. Condensed vapor droplets & quality of distilled water testing.

Table 2. Distilled water vs. natural water.

Parameter	Distilled water from Desalination Setups	Nominal values of drinkable water
TDS	220 μS/cm	250 μS/cm
D.O.	6.25 mg/L	5 mg/L
pH	6.99	6.5 – 8.5

4 CONCLUSIONS

After This research investigation examined the simultaneous effects of diesel engine performance, emission level, and desalination processes. The following are some of the most notable outcomes from the current investigation:

- A sufficient amount of fresh water was produced using the suggested thermal and solar desalination setups, which were installed successfully.
- The successful incorporation of MWCNT with diesel fuel in either the presence or absence of water revealed that the fuels can be stable for three weeks.
- By using MWCNT, the fuel's properties (such as flashpoint, burn point, and calorific values) were enhanced.
- For MWCNT blended diesel fuel, the brake thermal efficiency was heightened, and emissions lessened significantly (especially for MWCNT blended emulsions) compared to that of neat diesel fuel.
- In comparison to 98D2W50CNT fuel, diesel fuel reflected a higher amount of NOx emissions. D100MWCNT fuel performance was superior compared to normal diesel fuel. The fuel consumption for the D100MWCNT at full load was 0.289 kg/kWh compared to 0.415 kg/kWh with diesel.
- The thermal desalination and solar desalination setup produced a significant amount of fresh distilled water. Around 1.75 litres of freshwater was produced out of 6 litres of seawater from both desalination techniques.

Overall, the prominent benefits from the current investigation for the GCC are to achieve one of its major future sustainability goals (i.e., Preserving Environment Sustainability). The proposed research can be applied in gas turbine/diesel engine/steam power plants to produce fresh distilled water economically. Further, on using MWCNT additives with conventional fuels/biodiesel fuels in automotive and power generation sectors, the magnitude of emission levels can also be reduced on applying the proposed techniques.

NOMENCLATURE

D	–	Diesel
D50MWCNT	–	Diesel + 50 ppm MWCNT
D100MWCNT	–	Diesel + 100 ppm MWCNT
98D2WS50MWCNT	–	98% Diesel + 2% H_2O + Surfactant + 50 ppm MWCNT
96D4WS100MWCNT	–	96% Diesel+4% H_2O + Surfactant + 100 ppm MWCNT
98D2WS	–	98% Diesel + 2% Water + Surfactant
96D4WS	–	96% Diesel + 4% Water + Surfactant
FESEM	–	Field Emission Scanning Electron Microscope
MWCNT	–	Multi-Walled Carbon Nano Tubes
NO_x	–	Nitrogen Oxides

REFERENCES

[1] Ribeiro N. M. et al., (Jul. 2007). The role of additives for diesel and diesel blended (ethanol or biodiesel) fuels: A review, *Energy & Fuels*, vol. 21, no. 4, pp. 2433–2445, doi: 10.1021/ef070060r.

[2] Ullman T. L., Spreen K. B., and Mason R. L., (Mar. 1994). Effects of cetane number, cetane improver, aromatics, and oxygenates on 1994 heavy-duty diesel engine emissions, *SAE Technical Paper Series*, doi: 10.4271/941020.

[3] Basha J. S. and Anand R. B., (2012). Effects of nanoparticle additive in the water-diesel emulsion fuel on the performance, emission and combustion characteristics of a diesel engine, *International Journal of Vehicle Design*, vol. 59, no. 2/3, p. 164, doi: 10.1504/ijvd.2012.048692.

[4] Prakash R., Singh R. K., and Murugan S., (Jun. 2013). Experimental investigation on a diesel engine fueled with bio-oil derived from waste wood–biodiesel emulsions, *Energy*, vol. 55, pp. 610–618, doi: 10. 1016/j.energy.2013.03.085.

[5] Lapuerta M., Rodríguez-Fernández J., and García-Contreras R., (Dec. 2015). Effect of a glycerol-derived advanced biofuel –FAGE (fatty acid formal glycerol ester)– on the emissions of a diesel engine tested under the New European Driving Cycle, *Energy*, vol. 93, pp. 568–579, doi: 10.1016/j.energy.2015. 09.070.

[6] Rakopoulos D. C., Rakopoulos C. D., Giakoumis E. G., Papagiannakis R. G., and Kyritsis D. C., (Aug. 2014). Influence of properties of various common bio-fuels on the combustion and emission characteristics of high-speed DI (direct injection) diesel engine: Vegetable oil, bio-diesel, ethanol, n-butanol, diethyl ether, *Energy*, vol. 73, pp. 354–366, doi: 10.1016/j.energy.2014.06.032.

[7] Tsukahara M., Yoshimoto Y., and Murayama T., (Feb. 1989). W/O emulsion realizes low smoke and efficient operation of DI engines without high pressure injection, *SAE Technical Paper Series*, doi: 10. 4271/890449.

[8] Afify E. M., Korah N. S., and Dickey D. W., (Feb. 1987). The effect of air charge temprature on performance, ignition delay and exhaust emissions of diesel engines using W/O emulsions as fuel, *SAE Technical Paper Series*, doi: 10.4271/870555.

[9] Hountalas D. T., Mavropoulos G. C., and Zannis T. C., (Apr. 2007). Comparative evaluation of EGR, intake water injection and fuel/water emulsion as NOx reduction techniques for heavy duty diesel engines, *SAE Technical Paper Series*, doi: 10.4271/2007-01-0120.

[10] Tzirakis E. et al., (Oct. 2006). Diesel-water emulsion emissions and performance evaluation in public buses in attica basin, *SAE Technical Paper Series*, doi: 10.4271/2006-01-3398.

[11] Yoshimoto Y., (Oct. 2006). Performance of DI diesel engines fueled by water emulsions with equal proportions of gas oil-rapeseed oil blends and the characteristics of the combustion of single droplets, *SAE Technical Paper Series*, doi: 10.4271/2006-01-3364.

[12] Hall D., Thorne C., and Steven G., (Oct. 2003). An investigation into the effect of a diesel/water emulsion on the size and number distribution of the particulate emissions from a heavy-duty diesel engine, *SAE Technical Paper Series*, doi: 10.4271/2003-01-3168.

[13] Ithnin A. M., Ahmad M. A., Bakar M. A. A., Rajoo S., and Yahya W. J., (Jan. 2015). Combustion performance and emission analysis of diesel engine fuelled with water-in-diesel emulsion fuel made from low-grade diesel fuel, *Energy Conversion and Management*, vol. 90, pp. 375–382, doi: 10.1016/j.enconman.2014.11.025.

[14] Koc A. B. and Abdullah M., (May 2013). Performance and NOx emissions of a diesel engine fueled with biodiesel-diesel-water nanoemulsions, *Fuel Processing Technology*, vol. 109, pp. 70–77, doi: 10.1016/j.fuproc.2012.09.039.

[15] Srinivasa Rao M. and Anand R. B., (Apr. 2016). Performance and emission characteristics improvement studies on a biodiesel fuelled DICI engine using water and AlO(OH) nanoparticles, *Applied Thermal Engineering*, vol. 98, pp. 636–645, doi: 10.1016/j.applthermaleng.2015.12.090.

[16] Sadhik Basha J. and Anand R. B., (Mar. 2011), Role of nanoadditive blended biodiesel emulsion fuel on the working characteristics of a diesel engine, *Journal of Renewable and Sustainable Energy*, vol. 3, no. 2, p. 023106, doi: 10.1063/1.3575169.

[17] Sadhik Basha J., (Apr. 2015). Preparation of water-biodiesel emulsion fuels with CNT & alumina nano-additives and their impact on the diesel engine operation, *SAE Technical Paper Series*, doi: 10.4271/2015-01-0904.

[18] Sadhik Basha J. and Anand R. B., (Jun. 2014). Performance, emission and combustion characteristics of a diesel engine using carbon nanotubes blended jatropha methyl ester emulsions, *Alexandria Engineering Journal*, vol. 53, no. 2, pp. 259–273, doi: 10.1016/j.aej.2014.04.001.

[19] Basha J. S. and Anand R. B., (2010). Effects of nanoparticle-blended water biodiesel emulsion fuel on working characteristics of a diesel engine, *International Journal of Global Warming*, vol. 2, no. 4, p. 330, doi: 10.1504/ijgw.2010.037589.

[20] Ghojel J., Honnery D., and Al-Khaleefi K., (Dec. 2006). Performance, emissions and heat release characteristics of direct injection diesel engine operating on diesel oil emulsion, *Applied Thermal Engineering*, vol. 26, no. 17–18, pp.2132–2141, doi: 10.1016/j.applthermaleng.2006.04.014.

[21] Ha P. K., Kwak I., and Oh S., (2004). The effect of water emulsified fuel on a motorway-bus diesel engine, *Journal of Mechanical Science and Technology*, vol. 18, no. 11, pp. 2049–2057.

[22] Basha J. S. and Anand R. B., (Apr. 2011). An experimental study in a CI engine using nanoadditive blended water–diesel emulsion fuel, *International Journal of Green Energy*, vol. 8, no. 3, pp. 332–348, doi: 10.1080/15435075.2011.557844.

[23] Basha J. S. and Anand R. B., (May 2011). An experimental investigation in a diesel engine using carbon nanotubes blended water–diesel emulsion fuel, *Proceedings of the Institution of Mechanical Engineers, Part A: Journal of Power and Energy*, vol. 225, no. 3, pp. 279–288, doi: 10.1177/2041296710394247.

[24] Kadota T. and Yamasaki H., (Jan. 2002). Recent advances in the combustion of water fuel emulsion, *Progress in Energy and Combustion Science*, vol. 28, no. 5, pp. 385–404, doi: 10.1016/s0360-1285(02)00005-9.

[25] Qi D. H., Chen H., Lee C. F., Geng L. M., and Bian Y. Z., (Jan. 2010). Experimental studies of a naturally aspirated, DI diesel engine fuelled with ethanol−biodiesel−water microemulsions, *Energy & Fuels*, vol. 24, no. 1, pp. 652–663, doi: 10.1021/ef900814r.

[26] Nano Material Suppliers in India | Graphene Manufacturing in India, Ad Nano Technologies. https://ad-nanotech.com/. (accessed Jan. 14, 2023).

Innovation and Technological Advances for Sustainability – Al-Naemi et al. (Eds)

Wind energy production using novel HCS algorithm to reach MPPT for small-scale wind turbines under rapid change wind speed

Ahmed Badawi & Hassan Ali
College of Engineering Technology, University of Doha for Science and Technology, Doha, Qatar

Nasser A. Ismail & Alhareth Zyoud
Department of Electrical and Computer Engineering, Birzeit University Birzeit, Ramallah, Palestine

Siti Hajar Yusoff
Department of Electrical and Computer Engineering International Islamic University Malaysia, Kuala Lumpur, Malaysia

ABSTRACT: In this paper, a novel algorithm had been proposed to improve the efficiency performance of the wind energy conversion system (WECS) to reach the maximum possible captured power under fast varying wind speed. The permanent magnet synchronous generator (PMSG) had been selected because of its reliability and efficient performance, especially in low-speed conditions environment. The novel algorithm applied the power prediction mode to make optimization for the reached decision to the maximum peak point. The proposed algorithm shows better efficiency performance for different wind speed conditions. The novel algorithm accelerates convergence to detect the maximum peak point without sacrificing efficiency. The novel algorithm's simplicity reduces the overall time required by 33% when compared to the three-mode algorithm.

Keywords: Novel algorithm, Maximum power point tracking MPPT, Wind Turbine, Power Prediction mode, Hill Climbing Search HCS

1 INTRODUCTION

Renewable energy sources are cleaner and have low power generation costs when compared with traditional energy sources. Traditional sources are limited, emit CO2, and their prices keep increasing rapidly. Therefore, renewable resources gained a huge interest in the last two decades [1–6]. Given that for a wind energy system, the total output power varies instantaneously with the wind speed. It is a challenging process to keep maximizing the obtained energy. Using Maximum Power Point Tracking (MPPT) techniques will improve the efficiency of the renewable energy system from 39.99% to a minimum of 99.83% [4]. Therefore, several MPPT techniques have been utilized in the literature [7–13]. The best popular proposed techniques will be explained in brief in the next section. In terms of speed, wind turbines could be labeled into two main types, 1) variable speed, 2) fixed speed. The fixed speed turbines generate maximum power at the given speed and their efficiency degrades at other speeds.

The generated power normally fluctuate and produce disturbance to the grid. Moreover, in this type of turbine, the stress on the tower and the shaft is huge. While variable-speed wind turbines keep tracking the maximum efficiency for a large range of wind speeds. Once the speed is variable, the ratio of the wind speed to rotational speed keeps constant at an optimum value. Thus, the power is maximized, and the stress is minimized. This also extends the operation life of

DOI: 10.1201/9781003496724-18

the wind system and allows the manufacturers to build large-scale turbines that are cost-efficient. Therefore, this type of turbine is more attractive than fixed-speed turbines [10,14–16].

PMSGs are preferable due to many features such as light volume and weight combined with excessive power density and torque capability. Furthermore, the maintenance cost is low and such type of generator operates at low rotating speeds [17–22]. In the next section, the wind energy conversion system (WECS) components will be described.

2 WIND ENERGY CONVERSION SYSTEM

The universal electrical structure of the WECS applied in this paper is presented in Figure 1.

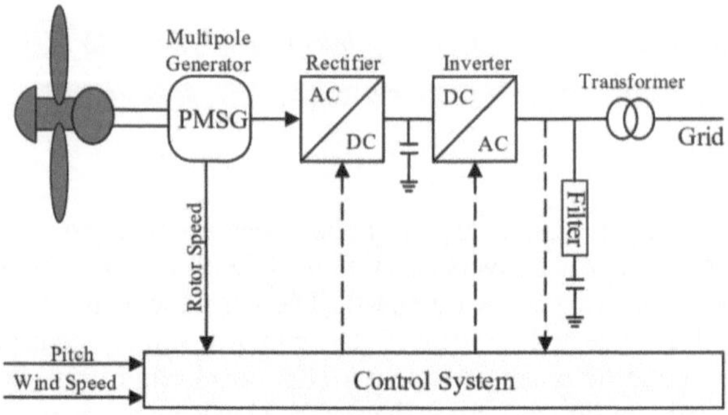

Figure 1. The PMSG-WECS connected to the control schemes [28].

Figure 1 shows the main parts of the system are the wind turbine, followed by PMSG which is connected to a DC to DC boost converter after that linked to the load. The load could be a stand-alone home or national grid. Two types of converters are normally applied. Either a converter based on a diode rectifier with DC chopper or a converter based on an Insulated-Gate Bipolar Transistor (IGBT) rectifier. This structure transfers mechanical energy to an electrical one to be used by the end-user [1,16,23–26]. In the WECS system, electricity can be produced by wind turbines. Wind turbines use the force of aerodynamics from the rotor blades. In this paper, PMSG is utilized due to the reliability particularly in low wind speed conditions of small-scale wind turbines. One of the most important challenges is very fast change in wind speed. PMSG-WECS had been used in this research for many reasons like high efficiency, reliability high power density, low rotor losses, and it can be used without gearbox [27].

A wind turbine converts the power of the wind into electricity utilizing the aerodynamic force from the swept area of the rotor blades [23]. Based on Betz theory, wind turbines extracted mechanical power from the instantaneous wind speed as follows [24,29]:

$$P = 0.5c_{P(\lambda,\beta)}\rho\pi R^2 V_w^3 \tag{1}$$

where, ρ is the density of air (kg/m^3), πR^2 is the swept area of the rotor blades (m^2), the wind speed is V_m (m/s), the power Coefficient symbol is Cp . Cp practically lies between 0.2 and 0.5 and it is considered a function of the tip speed ratio λ which can be expressed as:

$$\lambda = \frac{V_{tip}}{V_w} = \frac{\omega_r}{V_w} \tag{2}$$

Where R is the radius of rotor blade (m), ω is the angular speed of the rotor (rad/s).

3 MAXIMUM POWER POINT TRACKING CONTROL TYPES

There are two types of MPPT based on power maximization direct power controller and indirect power control. The first type is our scope in this research because the output power value can be maximized using the DPC.

$$P = \eta_g \eta_c P_{wind}$$

Equation (3) describes the relationship between the mechanical and the output electrical power, where ηg is the efficiency of the generator and ηc is the converter efficiency. [16,30–33], The Hill climbing search HCS algorithm is considered the most suitable for the small-scale wind turbine. Based on the step size classification there are two types of HCS, the fixed step size which is considered the conventional (traditional) and the adaptive or the intelligent developed HCS in [34–38].

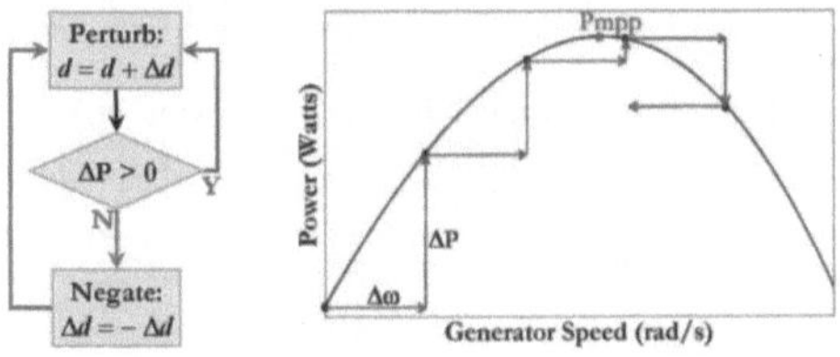

Figure 2. Principle of HCS algorithm [36].

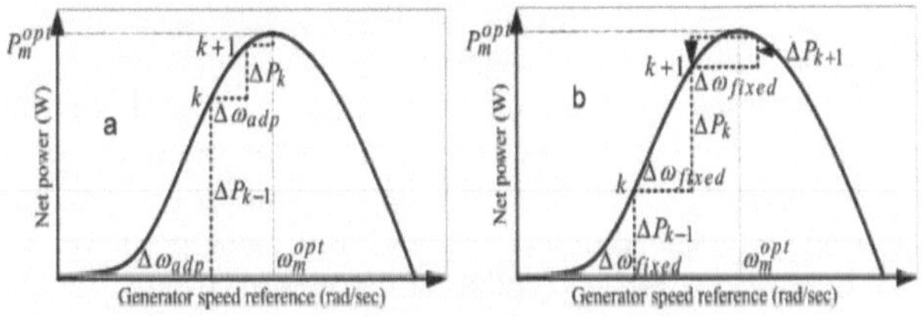

Figure 3. The adaptive algorithm with large step size and traditional HCS with small step size [37].

The HCS concept is illustrated in Figure 2 and it depends on perturbing the control variable in the same direction until the output power is reduced. The electric power is the input of HCS algorithm. In Figure 3 the perturbation in figure a is greater than perturbation in b, therefore the speed of tracking is faster. However, the oscillation around the peak point is much bigger. Therefore, there is a trade-off between the convergence speed and the accuracy [36]. The small step size is more efficient to reach the maximum peak point however, the control algorithm will be unable to reach the maximum peak point due to the rapid change in wind speed. Therefore, the conventional algorithm has always the trade-off between the efficiency performance and speed of tracking. To find a solution for the trade-off, knowledge about wind speed change should be used to avoid misleading to reach the highest peak point.

Figure 4 illustrates the relationship between the rotor speed and the output power for different wind speeds, the MPPT algorithm can be applied at the rated rotor speed (ω_{opt}) (optimum). In this case, the optimum tip speed ratio (λ_{opt}) can be achieved in order to capture the maximum possible power from the WECS. The following equation can be used to calculate the MPPT [37].

$$P_{\max} = K_{opt} \omega_{ropt}^3 \tag{4}$$

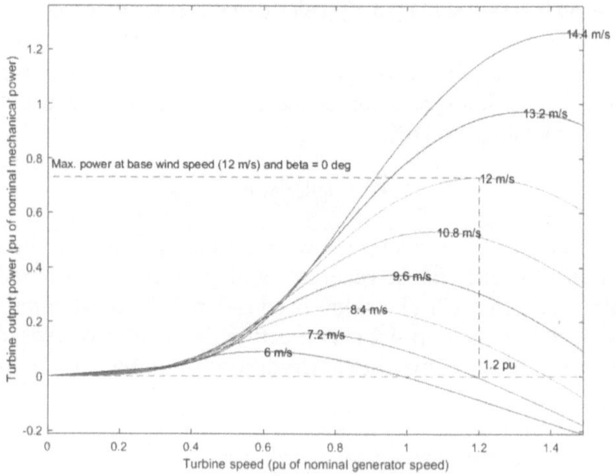

Figure 4. Turbine speed versus the output power for different wind speed profiles.

$$K_{opt} = \frac{0.5\pi\rho C_{pmax}R^5}{\lambda_{opt}^3} \tag{5}$$

where, K_{opt} is an optimum wind constant, c_{pmax} is the maximum power coefficient of the wind turbine, ω_{opt} is given by equation 6 [37].

$$\omega_{opt} = \frac{\lambda_{opt}V_\omega}{R} \tag{6}$$

Equations 4, 5 and 6 are the special case of equation 1, and this is only applicable at maximum power point tracking and PMSG operating at the rated speed.

That is why the controller of MPPT turns on the wind turbine at the optimum speed for every wind speed change. Thus, reaching the maximum peak point on the different wind speed profiles will be determined by the type of controller and the value of the parameters. Therefore, the tuning parameter can help the WECS to reach the maximum peak point. In addition, the speed detection to the peak point based on the fast decision plays an essential role to avoid misleading and delay due to the fast change of the wind speed.

In this paper, the authors used the following novel technique:

(A) Power prediction mode.
(B) Simplified three-mode adaptive algorithm.

The first stage of the novel algorithm used the power prediction mode to divide the data wind speed to five main intervals. Therefore, the new mode can calculate the amount of power energy based on the wind speed data. Moreover, the power prediction mode can exclude non efficient values. The second stage the HCS adaptive algorithm had been simplified from three-mode to two-main mode to increase the speed of tracking to reach the maximum peak point [8,38].

4 SIMULATION OF WIND TURBINE AND PERMANENT MAGNET SYNCHRONOUS GENERATOR (PMSG)

In order to reach the MPPT, the parameters must be tuned for the circuit model to reach the best efficiency performance. As a result, a unity power factor is desired to achieve the maximum power for the load resistor of the DC-to-DC boost converter. This can be accomplished by setting $X_L = X_C$.

Figure 5 represents five main blocks for model of WECS using the Simulink. Wind Turbine, Generator, MPPT Controller and 3-phase diode bridge (rectifier). In the model illustrated in the figure, the wind turbine and Generator linked with the MPPT control to reach to the maximum power point. The first block is a wind turbine, which converts the aerodynamic force of the rotor blades into mechanical torque. The mechanical torque is fed to the PMSG, which can be converted to electric power. The inverter converts the AC power to DC, which is then delivered to the DC-to-DC boost converter. The output of the Generator had been rectified using the-phase diode bridge. The converter is utilized to control the V_{dc} across c_1. The controller system delivers a reference voltage, the algorithm compares the reference value with the actual value of V_{dc}. The result after the comparison process is fed into a novel controller whose output is compared to the previous state $(k-1)$ to decide when to turn the dc-to-dc boost converter switch off or on. It operates the transistor (MOSFET) as a switch, causing the output voltage amplitude to change and reach the maximum power point.

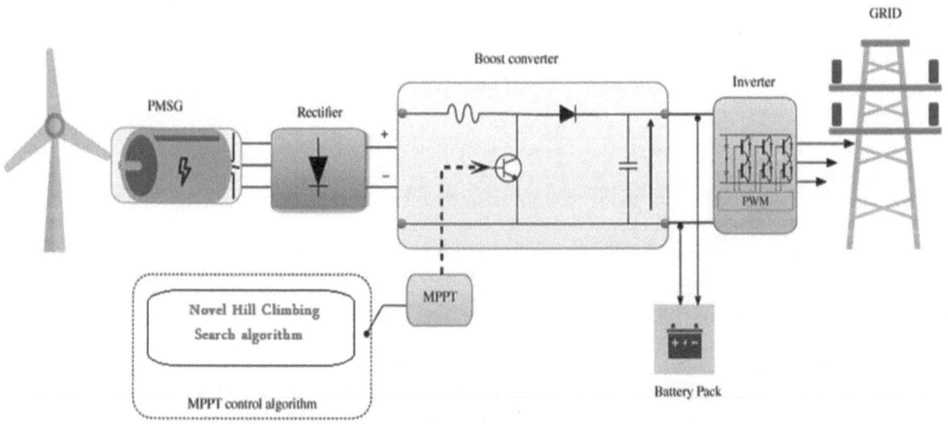

Figure 5. wind turbine connected to the PMSG and HSC control scheme.

Table 1. The parameters value for simulation models.

Parameter	Value
Rs stator resistance	0.18 Ω
Flux which induced by magnets	0.1716 V.s
β Pitch angle	0
L1	1.67 mH
R1	0.36 Ω
T the torque	60 N.m
J Inertia	4.5 kg. m2
NPP Number of poles pairs	4 poles
Constant Voltage	51.83 volt
Phase to phase nominal voltage Vrms	400 V
P Active power	650 W
Frequency	50 Hz
QL Reactive Power (inductive)	100 VA
QC Reactive Power (Capacitive)	−100 VA
C1 Capacitor	1000μF
R2	100 kΩ
Inductor L2	10 Mh

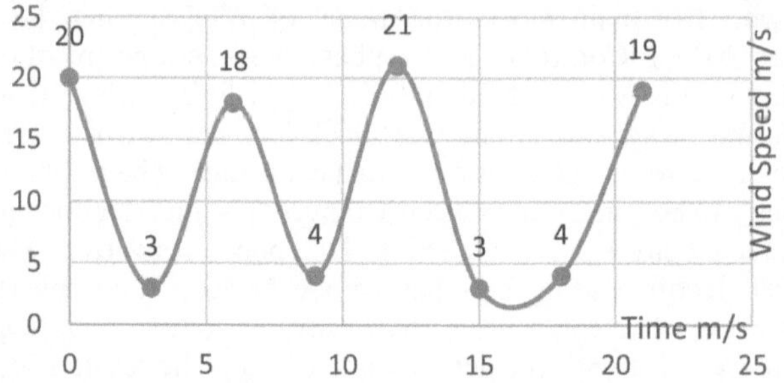

Figure 6. Different wind speed values.

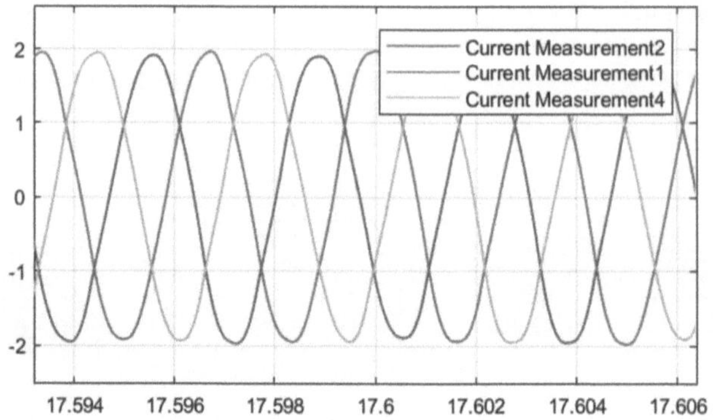

Figure 7. The three phase PMSG stator current.

Figure 6 shows nine different wind speed profiles applied to the WECS. The rapid change wind speed has been applied to check the efficiency performance for the two-mode Novel algorithm. Figure 7 shows the three-phase stator current is considered a pure sine wave. Because of the proportional relation between the speed value and power, the value of Amplitude increased dramatically as a result of to the wind speed value change. Figure 8 represents the AC voltage output for the PMSG. The graph shows the maximum output voltage due to the unity power factor. Figure 9 illustrates the value of output power for the PMSG. It can be seen the pure sine wave due reaches the rated power for the generator, due

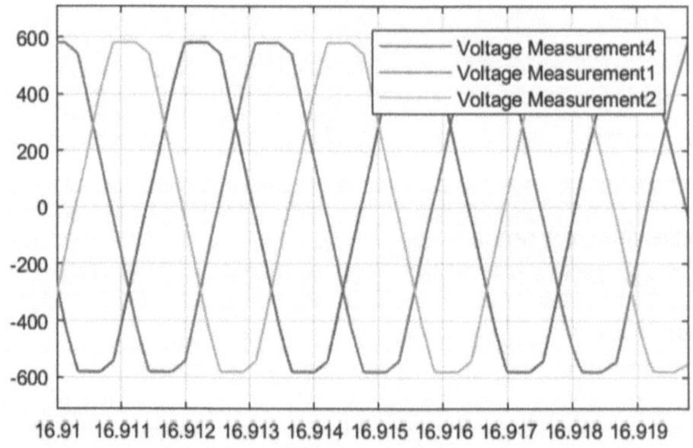

Figure 8. The three phase PMSG stator voltage.

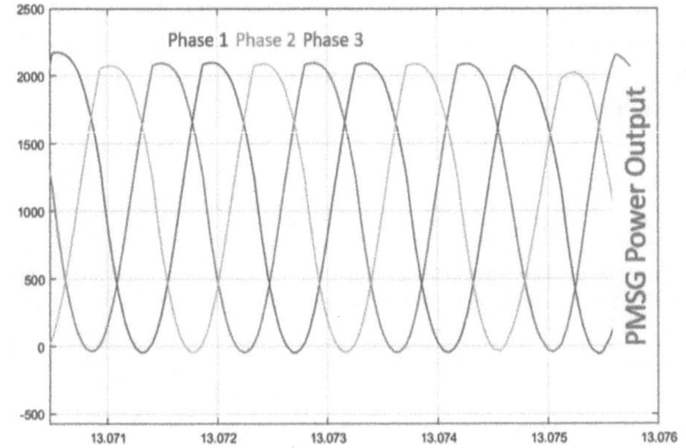

Figure 9. The three phase PMSG output power.

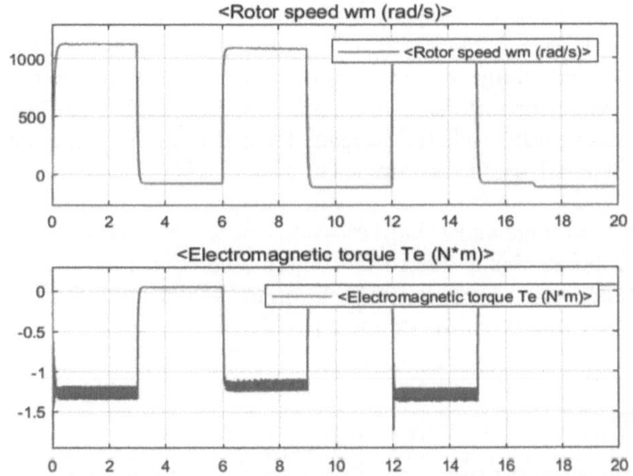

Figure 10. The three phase PMSG rotor speed and elecromagnetic torque.

to the unity power factor. In Figure 10 it can be seen that the rotor speed increased roughly this is due to the wind speed increasing. The rotor shaft speed and electromagnetic torque T_e also are increased dramatically because of the power of wind speed. It can be seen that the T_e fluctuates when wind speed changes; this is also considered because of the mechanical stress of the generator. As a result of the cubic proportionate relationship between power and wind speed, wind speed plays a crucial role in wind energy generation.

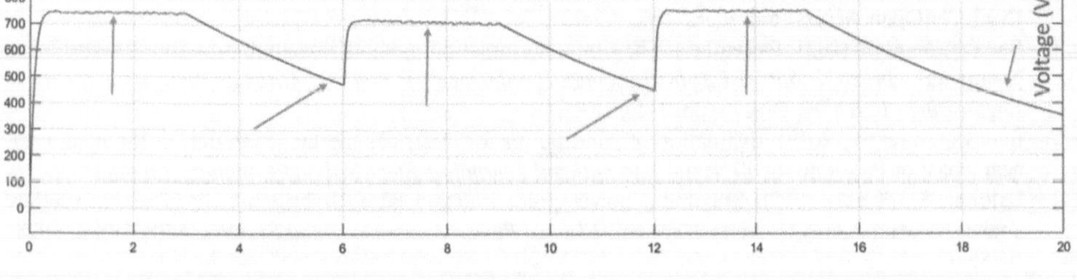

Figure 11. The DC voltage before controller.

In Figure 11 the DC link voltage for the conventional DC to DC boost converter before using the novel two-mode controller is shown. It can be seen that this DC voltage drops considerably due to the change in wind speed profile, this will definetly affect the ability of WECS in reaching MPP.

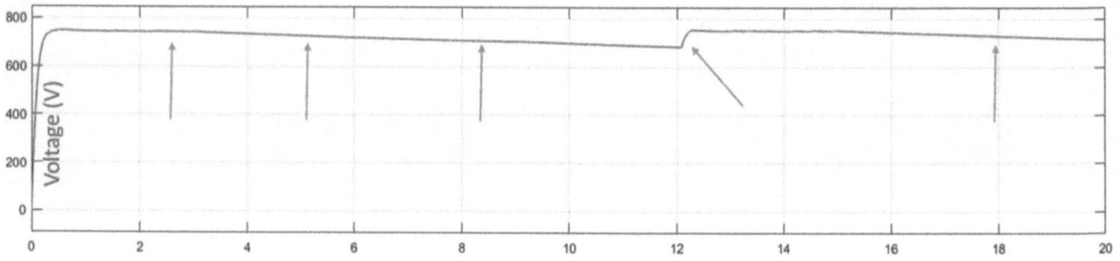

Figure 12. The DC voltage after the proposed novel controller.

In Figure 12 the influence of the proposed simplified adaptive algorithm (two-mode) is shown on this signal. It can detect the MPP in an a fast and efficient manner. The figure shows the DC link voltage value after the novel MPPT control to reach the maximum peak point on the curve for different wind speed conditions. The novel MPPT algorithm controlls the value of the duty cycle (D) to reach the maximum possible power. The value of $\beta = 0$, regardless of wind speed direction. The novel MPPT algorithm shows very quick tracking efficiency, there is no trade-off between the efficiency of tracking and the tracking speed. The optimization stage makes limitations for the total number of calculations to reach the maximum peak point, this is due to the power prediction mode. The simplicity of the novel algorithm makes a considerable contribution to the detecting decision due to the two-mode algorithm rather than the three-mode algorithm.

5 CONCLUSION

This paper represents a novel MPPT algorithm to increase the efficiency performance of the WECS-PMSG for wind speed power production on a small scale under different wind speed conditions. The proposed novel MPPT algorithm shows very fast-tracking potency, particularly at low wind speed conditions. The novel algorithm can reach the maximum peak point when the fast-changing of wind speed due to the new technique of power prediction mode intervals. The proposed algorithm is noteworthy faster by 33% compared to the three-mode algorithm in terms of overall time required.

REFERENCES

[1] Cámara, E. M., et al., (2010). *Environmental impact of modern wind power under LCA methodology.* INTECH Open Access Publisher.

[2] Badawi, A., et al. (2021, December). The simplest estimation method of weibull probability distribution parameters. *In 2021 6th IEEE International Conference on Recent Advances and Innovations in Engineering (ICRAIE)* (Vol. 6, pp. 1–5). IEEE.

[3] Badawi, A., et al. (2019). Evaluation of wind power for electrical energy generation in the mediterranean coast of Palestine for 14 years. *International Journal of Electrical and Computer Engineering.*

[4] Lalouni, S., et al., (2015). Maximum power point tracking based hybrid hill-climb search method applied to wind energy conversion system. *Electric Power Components and Systems*, 43(8–10): p. 1028–1038.

[5] Badawi, A., et al. (2018). Energy and power estimation for three different locations in Palestine. *Indonesian Journal of Electrical Engineering and Computer Science*, 10.
[6] Badawi, A., et al. (2019). Weibull probability distribution of wind speed for gaza strip for 10 years. In *Applied Mechanics and Materials* (Vol. 892, pp. 284–291). Trans Tech Publications Ltd.
[7] Hong, C.-M. and Chen C.-H., (2014). Intelligent control of a grid-connected wind-photovoltaic hybrid power systems. *International Journal of Electrical Power & Energy Systems*, 55: p. 554–561.
[8] Badawi, A. S. A., (2019). *Maximum Power Point Tracking Control Scheme For Small Scale Wind Turbine.*
[9] Badawi, A. S., et al. (2021). Maximum power point tracking controller technique using permanent magnet synchronous generator. in *2021 6th IEEE International Conference on Recent Advances and Innovations in Engineering (ICRAIE)*. IEEE.
[10] Dahbi, A., Nait-Said N., and Nait-Said M.-S., (2016). A novel combined MPPT-pitch angle control for wide range variable speed wind turbine based on neural network. *International Journal of Hydrogen Energy*, 41(22): p. 9427–9442.
[11] Abdullah, M., Yatim A., and Tan C. (2014). An online optimum-relation-based maximum power point tracking algorithm for wind energy conversion system. in *2014 Australasian Universities Power Engineering Conference (AUPEC)*. IEEE.
[12] Badawi, A. S., et al. (2020). Paper review: maximum power point tracking for wind energy conversion system. in *2020 2nd International Conference on Electrical, Control and Instrumentation Engineering (ICECIE)*. IEEE.
[13] Badawi, A. S., et al. (2020). Power prediction mode technique for Hill Climbing Search algorithm to reach the maximum power point tracking. in *2020 2nd International Conference on Electrical, Control and Instrumentation Engineering (ICECIE)*. IEEE.
[14] Wu, B., et al., (2011). *Power Conversion and Control of Wind Energy Systems*. John Wiley & Sons.
[15] Chiang, M.-H., A novel pitch control system for a wind turbine driven by a variable-speed pump-controlled hydraulic servo system. *Mechatronics*, 2011. 21(4): p. 753–761.
[16] Kumar, D. and K. Chatterjee, (2016). A review of conventional and advanced MPPT algorithms for wind energy systems. *Renewable and Sustainable Energy Reviews*, 55: p. 957–970.
[17] Li, H. and Z. Chen, (2008). Overview of different wind generator systems and their comparisons. *IET Renewable Power Generation*, 2(2): p. 123–138.
[18] Alnasir, Z. and M. Kazerani, An analytical literature review of stand-alone wind energy conversion systems from generator viewpoint. *Renewable and Sustainable Energy Reviews*, 2013. 28: p. 597–615.
[19] Geng, H., et al., Unified power control for PMSG-based WECS operating under different grid conditions. *IEEE Transactions on Energy Conversion*, 2011. 26(3): p. 822–830.
[20] Lin, W.-M., et al., Hybrid intelligent control of PMSG wind generation system using pitch angle control with RBFN. *Energy Conversion and Management*, 2011. 52(2): p. 1244–1251.
[21] Dehghan, S. M., Mohamadian M., and Varjani A. Y., A new variable-speed wind energy conversion system using permanent-magnet synchronous generator and $ Z $-source inverter. *IEEE Transactions on Energy Conversion*, 2009. 24(3): p. 714–724.
[22] Dahbi, A., et al. (2014). Analysis of different converters used in wind energy conversion system. in *2014 International Renewable and Sustainable Energy Conference (IRSEC)*. IEEE.
[23] Maroufi, O., A. Choucha, and L. Chaib, (2020). Hybrid fractional fuzzy PID design for MPPT-pitch control of wind turbine-based bat algorithm. *Electrical Engineering*, 102(4): p. 2149–2160.
[24] Abdullah, M. A., et al., (2012). A review of maximum power point tracking algorithms for wind energy systems. *Renewable and Sustainable Energy Reviews*, 16(5): p. 3220–3227.
[25] Chaudhuri, A., et al., (2022). Energy conversion strategies for wind energy system: Electrical, mechanical and material aspects. *Materials*, 15(3): p. 1232.
[26] Oğuz, Y., Güney İ., and Çalık H., (2013). Power quality control and design of power converter for variable-speed wind energy conversion system with permanent-magnet synchronous generator. *The Scientific World Journal*. 2013.
[27] Polinder, H., et al., Trends in wind turbine generator systems. *IEEE Journal of Emerging and Selected Topics in Power Electronics*, 2013. 1(3): p. 174–185.
[28] Honarbari, A., et al., MPPT Improvement for PMSG-Based Wind Turbines Using Extended Kalman Filter and Fuzzy Control System. *Energies*, 2021. 14(22): p. 7503.
[29] Abdelkafi, A. and L. Krichen, (2011). New strategy of pitch angle control for energy management of a wind farm. *Energy*. 36(3): p. 1470–1479.
[30] Pao, L. Y. and Johnson K. E. (2009). A tutorial on the dynamics and control of wind turbines and wind farms. *in American Control Conference*. ACC'09. 2009. IEEE.

[31] Badawi, A., Hasbullah, N. F., Yusoff, S. H., Aisha, H., and Zyoud, A. (2020). Novel technique for hill climbing search to reach maximum power point tracking. *International Journal of Power Electronics and Drive Systems (IJPEDS)*.

[32] Hua, G. and Geng Y. (2006). A novel control strategy of MPPT taking dynamics of wind turbine into account. in *Power Electronics Specialists Conference*, PESC'06. 37th IEEE. 2006.

[33] Shu, Z., Li Q., and Chan P., (2015). Statistical analysis of wind characteristics and wind energy potential in Hong Kong. *Energy Conversion and Management*. 101: p. 644–657.

[34] Buehring, I. and Freris L. (1981). Control policies for wind-energy conversion systems. in *IEE Proceedings C* (Generation, Transmission and Distribution). IET.

[35] Badawi, A., et al. (2021). Data bank: nine numerical methods for determining the parameters of weibull for wind energy generation tested by five statistical tools. *International Journal of Power Electronics and Drive Systems (IJPEDS)*.

[36] Kazmi, S. M. R., et al. (2010). Review and critical analysis of the research papers published till date on maximum power point tracking in wind energy conversion system. in *2010 IEEE Energy Conversion Congress and Exposition*. IEEE.

[37] Kazmi, S. M. R., et al., A novel algorithm for fast and efficient speed-sensorless maximum power point tracking in wind energy conversion systems. *IEEE Transactions on Industrial Electronics*, 2011. 58(1): p. 29–36.

[38] Echchhibat, M. E., Bahatti, L., Raihani, A., and Elmagri, A. (2017). An adapted method for a small wind turbine to extract the optimal power. *WSEAS Transactions on Systems*, 16, 261–270.

Innovation and Technological Advances for Sustainability – Al-Naemi et al. (Eds)

Weibull probability distribution based on four years wind speed data using nine numerical methods

Ahmed Badawi & Hassan Ali
College of Engineering Technology, University of Doha for Science and Technology, Doha, Qatar

Nasser A. Ismail & Alhareth Zyoud
Department of Electrical and Computer Engineering, Birzeit University Birzeit, Ramallah, Palestine

Siti Hajar Yusoff
Department of Electrical and Computer Engineering, International Islamic University Malaysia, Kuala Lumpur, Malaysia

Ashraf Al-Rimawi & Wasel Ghanem
Department of Electrical and Computer Engineering, Birzeit University Birzeit, Ramallah, Palestine

ABSTRACT: Wind energy potential can be estimated in a specific area using the Probability density function (PDF). PDF is used for the mean wind speed data on daily basis. This paper represents the analysis of the climatology of the mean wind speed data for Askalan city on the Mediterranean coastal plain of Palestine. A selection of the PDF had been implemented using nine numerical analysis methods to represent the cumulative distribution function and Weibull probability distribution. The two main parameters of the Weibull shape factor are k = 1.7848 and the scale factor parameter c = 4.3642 ms-1. The study assessment finds out the characteristics of wind energy conversion and results confirm that it can produce energy using small-scale wind turbines. Matlab software was used to implement the numerical methods and extract the wind characteristics.

Keywords: Weibull, PDF, CDF, shape factor, scale factor

1 INTRODUCTION

Due to universal energy crises, most countries nowadays have implemented and developed policies and strategies to employ renewable resources in producing energy. Based on a recent study by International Energy Agency (IEA), it is expected that the worldwide renewable energy capacity is increased by 50% for the next year and by 75% (2400 GW) in the coming five years [1]. IEA reported that PV and wind energy are considered the two most promising and growing sectors due to their commercial attractive cost. Figure 1 shows that the total installed capacity of wind power is increased to 837GW [2,3]. This increment reduced CO2 emissions by 1.2 billion tonnes. In addition, Figure 1 also shows that this increment of installation needs to quadruple by 2030 to meet Net Zero Scenarios (NZS) [1,3]. NZE is an approach to realize zero CO2 Emissions by 2050 proposed by the IEA. This also meets with the seventh, thirteenth and fourteenth of the United Nations Sustainable Development Goals (SDGs) [4]. The NZE and SDGs will not be achieved unless all countries worldwide participate and work together in order to increase renewable energy resources. This research considered an effort to investigate the feasibility of applying a small-scale wind turbine in our region [5].

DOI: 10.1201/9781003496724-19

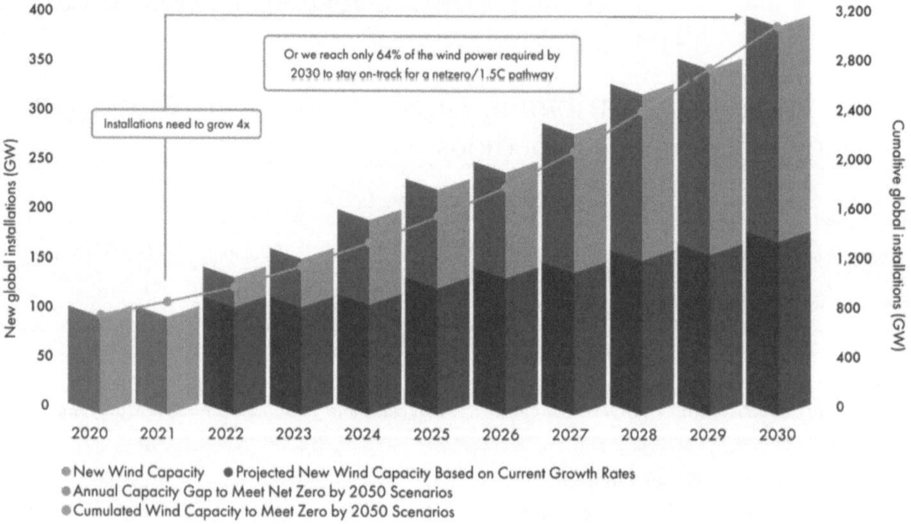

Figure 1. Lagging growth in this decade leads to wind energy shortfalls by 2030 [1].

The availability of meteorological data is crucial for an accurate wind power assessment of a specific region. The wind behaviour over time is normally investigated and presented [6]. Local weather data for six years from one station located on Hong Kong Island were discussed and studied by Lu et al [7]. The monthly mean wind speed data were analysed. It was concluded that the height of the turbine is an important factor that may affect wind distribution. Chang *et al.* [8] evaluated many stations located across Taiwan. The majority of the investigated data were collected for more than 39 years. Mean wind speed data based on hourly intervals were extracted and applied to estimate the Weibull parameters. It was found that most of the investigated locations have an average wind energy potential of 200 W/m2. The authors of [9] studied data of mean wind speed that were measured for four years from a single station located at Aksehir in Turkey to evaluate the Weibull parameters. It was concluded that an average of 8.67 W/m2 of wind energy potential was estimated. Himri *et al.* [10] investigated data of mean wind speed that were collected over 10 years from three different stations to investigate the wind power production ability in Algeria. Three hours interval mean wind speed data were used to examine the ability to install 30MW wind farm at the investigated locations. Three hours interval wind data for a minimum of three years were also studied by the authors of [11] at three stations in the Ionian Sea. An average wind energy potential of 197 W/m2 was estimated. Wind speed data was utilized for almost seven years in Cyprus [12]. Ten minutes of mean speed data were analysed. Wind speed is considered the most significant parameter that needs to be studied. The energy potential records are directly affected by the mean of the speed [13–16]. To reduce the generating cost, it is essential to optimize the turbine design according to the variation of the wind speed. It is found in the literature that wind speed followed two parameters of Weibull distribution [17]. The two parameters are extracted from the measured data and could be calculated and estimated by different numerical methods. Nine of them will be investigated in this paper.

2 WEIBULL PARAMETERS CALCULATION

To estimate the wind power potential for any site, the knowledge of the wind speed data is essential. Normally, the Weibull PDF curve can be applied to calculate two important factors named the shape factor (k) and the scale factor (c). These two factors could be evaluated from the measured data or may be estimated using numerical methods [8,12,15,18–35]. The Weibull probability distribution can be represented as Weibull PDF, f(v) or a cumulative distribution function (CDF) F(v) . The CDF can be calculated by determining the integral value of the Weibull PDF, f(v) [32,36–38], and the value of the CDF can be determined using equation (1) [32,33,36,37,39–41].

$$F(v) = 1 - \exp\left(-\left(\frac{v}{c}\right)^{k}\right) \tag{1}$$

The derivative of the CDF is considered the Weibull probability function:

$$f(v) = \frac{dF(v)}{d(v)} = k\frac{v^{(K-1)}}{c^{K}}e^{-\left(\frac{v}{c}\right)^{K}} k > 0, v > 0, c > 1 \tag{2}$$

where, v the mean wind speed and the parameters of Weibull c and k. The scale factor c unit is (m/s), the shape factor is k, where k is dimensionless. The shape factor c indicates the abscissa scale of the PDF, which represents the wind speed profile in a specific location [32,42,43]. The shape factor k parameter indicates the width of the PDF, which shows the wind PDF peak for the location study [14,32,44]. Weibull parameters value can be estimated using several numerical analysis technique in the literature such as Method of Moments (MM), where the Weibull parameters can be estimated after calculating the average and the standard deviation of measured wind speed data. The second method is Maximum Likelihood Method (MLM) in which extensive calculations are required. Empirical Method (EM), Standard Deviation Method (STDM), Least mean Square Method (LSM), Modified Maximum Likelihood Method (MMLM) and Second Modified Maximum Likelihood Method (SMMLM) which are similar to their reference method (MLM) in terms of extensive calculations, Graphical Method (GM) where double logarithm of Equation (1) is applied, and finally Energy Pattern Factor method (EPF), where the average wind speed is used to calculate the energy pattern factor. The efficiency performance of each method can be changed based on the wind speed condition for the case study site. Therefore, the appropriateness of the numerical method can be determined based on the goodness-of-fit test to check the percentage of error [18,45]. Using the Weibull PDF, wind power density can be calculated using Equation (3) therefore the required electrical output power can be simulated for the wind turbine model based on the PDF [32,46,47]:

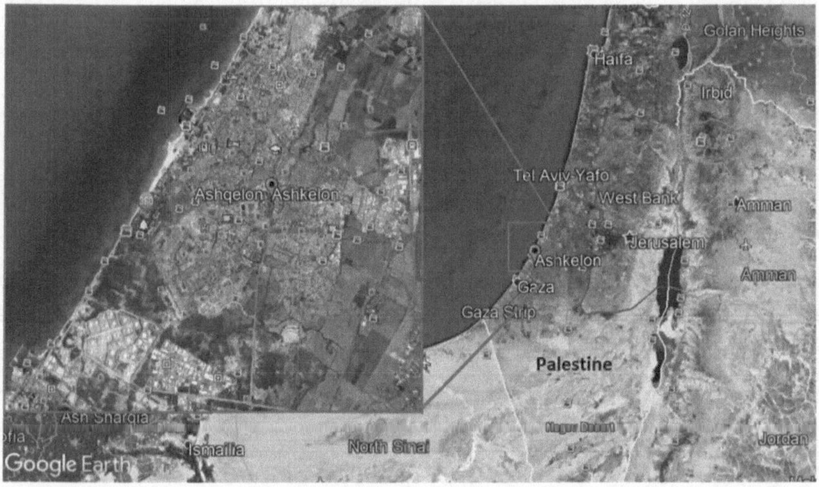

Figure 2. Askalan city on the South coastal plain of Palestine.

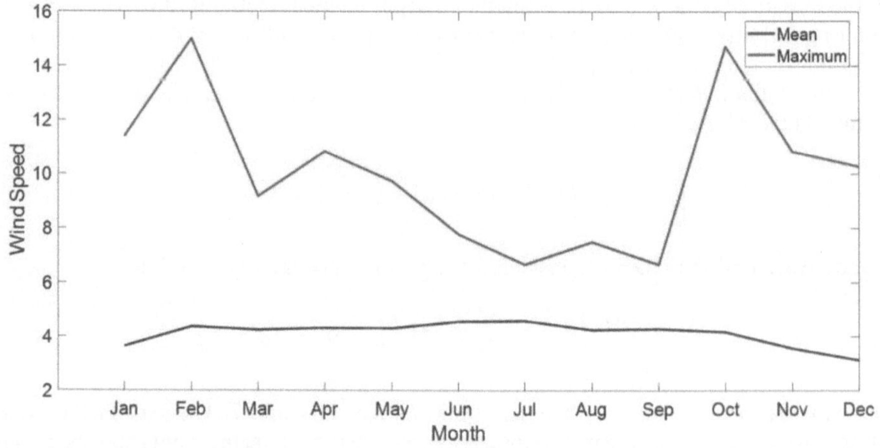

Figure 3. The maximum versus mean wind speed in Askalan of four years.

$$\overline{P} = \frac{1}{2}\rho \int_0^{\infty} v^3 f_w(v) dv = \frac{1}{2}\rho c^3 \Gamma\left(1 + \frac{3}{k}\right)\left(\frac{\text{W}}{\text{m}^2}\right) \tag{3}$$

The percentage of errors of each numerical analysis method can be estimated using the statistical tools tests such as RRMSE Root Mean Square Error (RMSE), Relative Root Mean Square Error (RRMSE), Chi-square test (X^2), Index of Agreement (IA), and Mean Absolute Percentage Error (MAPE).

3 DATA WIND SPEED AT ASKALAN CITY AS A CASE STUDY

The data of wind speed were extracted for Askalan city on the Mediterranean coastal plain of Palestine as shown in Figure 2. The weather for the coastal plain is dry and hot in Summer, whereas rainy and warm in Autumn. The mean wind speed is generally below 7 m/s, however, the maximum wind speed can be reached to 25 m/s [48]. Figure 3 represents the mean wind speed for a period of four years of collected wind speed data from January 2012 to December 2015. The mean average wind speed based on the Askalan data is around 4 m/s as shown in the blue line in Figure 3. Based on the red curve there is a fluctuation during March, April and August until September between 16m/s to 6 m/s to reach the minimum at the end of August. The red line represented the average maximum wind speed based on the recorded data. The wind speed increased dramatically in the winter season as shown in the red line in the figure, the maximum wind speed reach around 15 m/s in February and November due to the winter season where the wind speed reaches the maximum. Based on the mean wind speed it can be produced the power energy using a small-scale wind turbine. Figure 4 shows the number of iterations of the measured mean wind speed data. The bar graph shows Weibull probability distribution of the wind speed data. Most of the iterations' mean wind speed are around 3 to 5 m/s. These speed values considered suitable for small scale wind turbines.

4 PDF FOR ASKALAN CITY

The Weibull PDF had been illustrated using nine numerical analysis methods for the mean wind speed data for 4 years. The measured value was also represented using Matlab Simulink to compare the result with the numerical methods Figure 5 represents Weibull PDF for nine numerical analyses compared to the measured Weibull curve. The numerical analysis EPF, MLM, STDM, EM, FMLM and MM show identical Weibull curves. Whereas LMS and GM have the same Weibull. The unique Weibull curve was SMLM. Figure 6 represents the CDF which is considered the integral value of the Weibull Probability distribution function. The CDF numerical methods show the same behavior as the Weibull PDF. Table 1 illustrates the value of shape factor and scale factors for the mean wind speed for four years of data based on nine numerical analysis methods. The value of shape factors lies between 1.7607 and 2.1996, therefore the curve of the probability distribution function is considered Weibull Curve. Whereases the scale factors value varies between 4.1362 and 4.6449 m/s. In terms of scale factor, MLM method shows a good prediction since the estimated value is the closest to the observed. While Table 2 illustrates the error percentages for numerical analysis which are calculated out from different goodness of fit test such as RMSE, RRMSE, X^2, IA, and MAPE and. LSM, GM, and SMMLM show the best efficiency performance based on the RMSE test followed by MLM, MM, MMLM and STDM. The numerical methods MLM, MM, and EPF show average efficiency performance according to the RMSE, RRMSE and X^2. The lowest efficiency performance was STDM based on RRMSE, X2, RMSE, and IA.

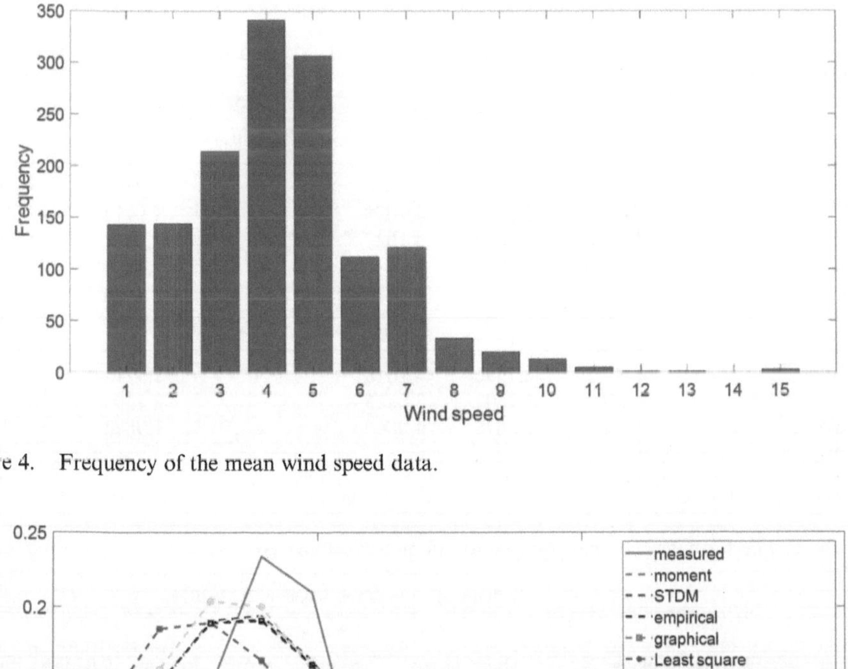

Figure 4. Frequency of the mean wind speed data.

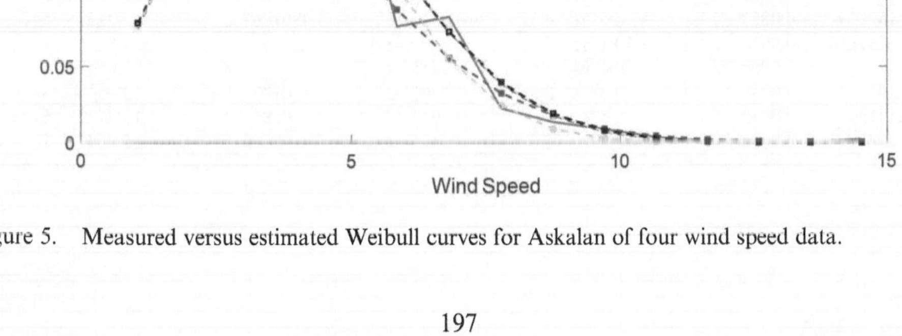

Figure 5. Measured versus estimated Weibull curves for Askalan of four wind speed data.

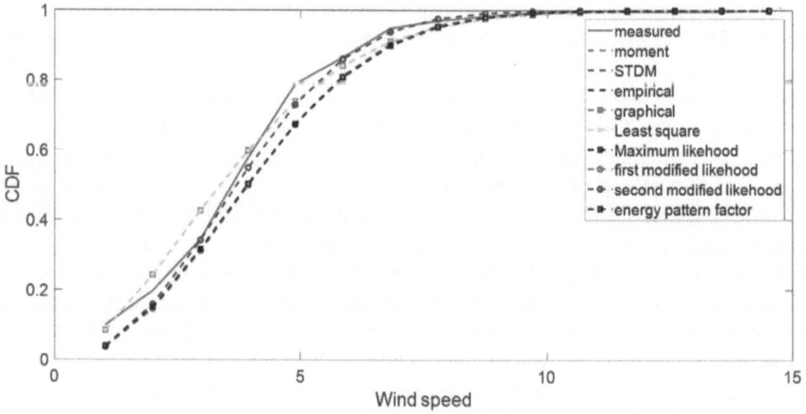

Figure 6. Comparison of measured and estimated CDF values for Askalan for four years of mesurments (2012–2015).

Table 1. Weibull parameters value, coefficient of variance (COV) and standard deviation.

	Shape and scale factor values				
Numerical Techniques	*c(m/s)*	*k*	*Mean wind speed(m/s)*	*σ(m/s)*	*COV (%)*
MM	4.6383	2.1886	4.1078	1.9802	48.2072
EM	4.6383	2.1996	4.1078	1.9714	47.9915
MLM	4.6338	2.1652	4.1037	1.9974	48.6739
MMLM	4.6449	2.1913	4.1136	1.9809	48.1540
SMMLM	4.3435	2.2728	3.8475	1.7930	46.6028
GM	4.1362	1.7607	3.6825	2.1598	48.3920
EPF	4.6384	2.1567	4.1078	2.0065	48.8467
STDM	4.6383	2.1996	4.1078	1.9714	47.9915
LSM	4.1362	1.7607	3.6825	2.1598	58.6486
Observed	4.6053	2.0616	4.0800	1.9873	51.1173

Table 2. The performance comparison rate for all numerical analysis.

Numerical Techniques		Goodness of fit tests for Askalan 2012–2015									
		Comparative analysis									
		RMSE	Ranking	X2	Ranking	IA	Ranking	MAPE	Ranking	RRMSE	Ranking
1	MM	0.0082	5	0.5134	5	0.7990	6	0.0186	2	12.2952	4
2	EM	0.0082	5	0.5769	6	0.7912	5	0.0185	1	12.3735	6
3	MLM	0.0081	4	0.4140	3	0.7895	4	0.0187	3	12.1149	3
4	MMLM	0.0082	5	0.5126	4	0.8316	6	0.0186	2	12.3227	5
5	SMMLM	0.0012	1	10.1575	7	0.7795	3	0.0196	4	1.8226	1
6	GM	0.0034	2	0.1969	1	0.7415	1	0.0252	5	5.0647	2
7	EPF	0.0080	3	0.3750	2	0.7705	2	0.0187	3	12.0544	3
8	STDM	0.0082	5	0.5769	5	0.7912	5	0.0185	1	12.3735	6
9	LSM	0.0034	2	0.1969	1	0.7415	1	0.0252	5	5.0647	2

5 CONCLUSION

This is a case study of Askalan city implementing the Weibull PDF and the CDF using nine numerical method analysis. The PDF represents the Weibull curve due to the value of the shape factor lying around 2. In addition, it was found that the scale factor parameter for the Weibull curve varies between 4.1362 and 4.6449 m/s. Furthermore, the mean wind speed for Askalan city from 2012 to 2014 is 4.10 m/s on daily basis. Based on the statistical tools LMS and GM show the best performance followed by EPF. the numerical analysis STDM, MM, MLM and EM show average parameters estimation and almost the same behaviour for the percentage of errors.

REFERENCES

[1] International Energy Agency, (2022). *Global Wind Report* [Online] [access 2023, January 31 Available: https://gwec.net/global-wind-report-2022/.

[2] Lai, J.-P., et al., (2020). A survey of machine learning models in renewable energy predictions. *Applied Sciences*. 10(17): p. 5975.

[3] International Energy Agency, Fuels and Technologies [Online] [access 2023, January 31 Available: https://www.iea.org/fuels-and-technologies/renewables, 2022.

[4] United Nations, The 17 Goals [Online] [access 2023, January 31 Available: https://sdgs.un.org/goals.

[5] Badawi, A. S. A., (2019). Maximum Power Point Tracking Control Scheme for Small Scale Wind Turbine.

[6] Guarienti, J. A., et al., (2020). Performance analysis of numerical methods for determining Weibull distribution parameters applied to wind speed in Mato Grosso do Sul, Brazil. *Sustainable Energy Technologies and Assessments*. 42: p. 100854.

[7] Lu, L., Yang H., and Burnett J., (2002). Investigation on wind power potential on Hong Kong islands—an analysis of wind power and wind turbine characteristics. *Renewable Energy*. 27(1): p. 1–12.

[8] Chang, T.-J., et al., (2003). Assessment of wind characteristics and wind turbine characteristics in Taiwan. *Renewable Energy*. 28(6): p. 851–871.

[9] Genc, A., et al., (2005). Estimation of wind power potential using Weibull distribution. *Energy Sources*. 27(9): p. 809–822.

[10] Himri, Y., et al., (2008). Wind power potential assessment for three locations in Algeria. *Renewable and Sustainable Energy Reviews*. 12(9): p. 2495–2504.

[11] Bagiorgas, H. S., et al., (2016). Wind power potential assessment for three buoys data collection stations in the Ionian Sea using Weibull distribution function. *International Journal of Green Energy*. 13(7): p. 703–714.

[12] Solyali, D., et al., (2016). Wind resource assessment of Northern Cyprus. *Renewable and Sustainable Energy Reviews*. 55: p. 180–187.

[13] Badawi, A. S., et al., (2018). Energy and power estimation for three different locations in Palestine. *Indonesian Journal of Electrical Engineering and Computer Science*. 10.

[14] Badawi, A. S., et al. (2019). Weibull Probability Distribution of Wind Speed for Gaza Strip for 10 Years. in *Applied Mechanics and Materials*. Trans Tech Publ.

[15] Badawi, A., et al., (2021). Data bank: nine numerical methods for determining the parameters of weibull for wind energy generation tested by five statistical tools. *International Journal of Power Electronics and Drive Systems (IJPEDS)*.

[16] Badawi, A. S., et al. (2021). The simplest estimation method of weibull probability distribution parameters. in *2021 6th IEEE International Conference on Recent Advances and Innovations in Engineering (ICRAIE)*. IEEE.

[17] Seguro, J. and Lambert T., (2000). Modern estimation of the parameters of the Weibull wind speed distribution for wind energy analysis. *Journal of Wind Engineering and Industrial Aerodynamics*. 85(1): p. 75–84.

[18] Akdağ, S. A. and A. Dinler, (2009). A new method to estimate Weibull parameters for wind energy applications. *Energy Conversion and Management*. 50(7): p. 1761–1766.

[19] Akpinar, E. K. and Akpinar S., (2004). Determination of the wind energy potential for Maden-Elazig, Turkey. *Energy Conversion and Management*. 45(18–19): p. 2901–2914.

[20] Ucar, A. and Balo F., (2009). Investigation of wind characteristics and assessment of wind-generation potentiality in Uludağ-Bursa, Turkey. *Applied Energy*. 86(3): p. 333–339.

[21] Chang, T. P., (2011), Performance comparison of six numerical methods in estimating Weibull parameters for wind energy application. *Applied Energy*. 88(1): p. 272–282.

[22] Oyedepo, S. O., Adaramola M. S., and Paul S. S., (2012). Analysis of wind speed data and wind energy potential in three selected locations in south-east Nigeria. *International Journal of Energy and Environmental Engineering*. 3(1): p. 1–11.

[23] Rocha, P. A. C., et al., (2012). Comparison of seven numerical methods for determining Weibull parameters for wind energy generation in the northeast region of Brazil. *Applied Energy*. 89(1): p. 395–400.

[24] Kitaneh, R., Alsamamra H., and Aljunaidi A., (2012). Modeling of wind energy in some areas of Palestine. *Energy conversion and Management*. 62: p. 64–69.

[25] Pishgar-Komleh, S., Keyhani A., and Sefeedpari P., (2015). Wind speed and power density analysis based on Weibull and Rayleigh distributions (a case study: Firouzkooh county of Iran). *Renewable and Sustainable Energy Reviews*. 42: p. 313–322.

[26] Gugliani, G., et al., (2018). New methods to assess wind resources in terms of wind speed, load, power and direction. *Renewable Energy*. 129: p. 168–182.

[27] Usta, I., et al., (2018). A new estimation approach based on moments for estimating Weibull parameters in wind power applications. *Energy Conversion and Management*. 164: p. 570–578.

[28] Emeksiz, C. and T. Cetin, (2019). In case study: Investigation of tower shadow disturbance and wind shear variations effects on energy production, wind speed and power characteristics. *Sustainable Energy Technologies and Assessments*. 35: p. 148–159.

[29] Li, Y., et al., (2020). Comparative study of onshore and offshore wind characteristics and wind energy potentials: A case study for southeast coastal region of China. *Sustainable Energy Technologies and Assessments*. 39: p. 100711.

[30] Saeed, M. K., et al., (2019). Comparison of six different methods of Weibull distribution for wind power assessment: A case study for a site in the Northern region of Pakistan. *Sustainable Energy Technologies and Assessments*. 36: p. 100541.

[31] Azad, A. K., Rasul M. G., and Yusaf T., (2014). Statistical diagnosis of the best Weibull methods for wind power assessment for agricultural applications. *Energies*. 7(5): p. 3056–3085.

[32] Mohammadi, K., et al., (2016). Assessing different parameters estimation methods of Weibull distribution to compute wind power density. *Energy Conversion and Management*. 108: p. 322–335.

[33] Costa Rocha, P. A., et al., (2012). Comparison of seven numerical methods for determining Weibull parameters for wind energy generation in the northeast region of Brazil. *Applied Energy*. 89(1): p. 395–400.

[34] Celik, A. N., (2004). A statistical analysis of wind power density based on the Weibull and Rayleigh models at the southern region of Turkey. *Renewable Energy*. 29(4): p. 593–604.

[35] Celik, A., Makkawi A., and Muneer T., (2010). Critical evaluation of wind speed frequency distribution functions. *Journal of Renewable and Sustainable Energy*. 2(1): p. 013102.

[36] Legates, D. R., (1999). Evaluating the use of "goodness-of-fit" measures in hydrologic and hydroclimatic model validation *Water Resources Research*.

[37] Manwell, (2002). Wind Energy Explained: Theory, Design and Application.

[38] Badawi, A. S., (2013). An Analytical Study For Establishment Of Wind Farms In Palestine To Reach The Optimum Electrical Energy.

[39] Ohunakin, O., Adaramola M. S., and Oyewola O. M., (2011). Wind energy evaluation for electricity generation using WECS in seven selected locations in Nigeria. *Applied Energy*. 88(9): p. 3197–3206.

[40] Kwon, S.-D., (2010). Uncertainty analysis of wind energy potential assessment. *Applied Energy*, 87(3): p. 856–865.

[41] Bhattacharya, P., (2011). Weibull distribution for estimating the parameters, in *Wind Energy Management*. InTech.

[42] Shu, Z., Li Q., and Chan P., (2015). Statistical analysis of wind characteristics and wind energy potential in Hong Kong. *Energy Conversion and Management*. 101: p. 644–657.

[43] Badawi, A. S., et al., (2019). Practical electrical energy production to solve the shortage in electricity in palestine and pay back period. *International Journal of Electrical and Computer Engineering*.

[44] Carrasco-Díaz, M., et al., (2015). An assessment of wind power potential along the coast of Tamaulipas, northeastern Mexico. *Renewable Energy*. 78: p. 295–305.

[45] Parajuli, A., (2016). A statistical analysis of wind speed and power density based on weibull and rayleigh models of Jumla, Nepal. *Energy and Power Engineering*. 08(07): p. 271–282.

[46] Tizpar, A., et al., (2014). Wind resource assessment and wind power potential of Mil-E Nader region in Sistan and Baluchestan Province, Iran–Part 1: Annual energy estimation. *Energy Conversion and Management*. 79: p. 273–280.

[47] Boudia, S. M. and O. Guerri, (2015). Investigation of wind power potential at Oran, northwest of Algeria. *Energy Conversion and Management*. 105: p. 81–92.

[48] (PCBS), P.C.B.o.S., Wind Speed Data2018.

Innovation and Technological Advances for Sustainability – Al-Naemi et al. (Eds)

Investigating the service quality factors that influence the use of Sahel service in Kuwait

Afnan Alkhaldi
Department of Management, Arab Open University, Ardiya, Kuwait

Huda Alrashidi
Assistant Professor, Department of Computer Science and Engineering, Doha, Kuwait

ABSTRACT: The purpose of this study is to investigate the factors that influence Sahel e-government service quality in Kuwait. The study is using a quantitative approach where a questionnaire was developed to collect data. The target is citizens and residents that use the e-government services. The required sample is 385. Findings will reveal the most important factors influencing the service quality of Sahel services, which will help in suggesting recommendations to enhance the use and dependency on Sahel services. The research findings could prove of value to the government of Kuwait in order to enhance the acceptability of Sahel services and enhance the adoption of Sahel e-government services.

Keywords: E-GovQual, Sahel Service, COVID 19, Kuwait

1 INTRODUCTION

E-government is about the employment of information technology in order to make avail services to the public. The public sector realized the advancement in the technology, which led a lot of people to be dependent and reliant on it [1]. It has been noticed that there is an escalating rate of use of technology in government services in the recent years, which is noticed all over the world. Thus, this has created a new and rich area for scholars to investigate and explore [2]. In Kuwait, the government has announced the launch of the Sahel Application in September 2021, which is a platform that brings all the government services in one location. In the first month of its launch, Sahel has seen more than 130 thousand subscribers to the application and more than 14 thousand successful transactions [3]. Currently, there are 750,000 subscribers and more than 300,000 transactions per month. The focus of this study is to examine the factors that influence the use of Sahel service in Kuwait.

2 LITERATURE REVIEW

Electronic government, usually referred to as egovernment, represents those public services that are provided online by means of information and communication technology (ICT). These services have been provided so that citizens can access information online as well as availing of various services online and to widen citizen participation in government activities [4]. This description emphasizes the special kind of interaction that e-government makes possible between the government itself, employees and the public at large. Fang [5] elaportes that e-government concept is simple and it involves using the available innovations to deliver the services electronically that were once delivered in a traditional manner. This has created the urge to develop portals that can facilitate these services to the public. Of the many

DOI: 10.1201/9781003496724-20

advantages that could be realized from the use of technology is reducing the time for completing the services and unify all the services to all individuals [6]. Moreover, another advantage is that it reduces the costs of operations and helps in creating a consistent experience for all individuals. Alasem [7] sees eGovernment as the application of ICT innovations, previously so successful in e-business where it has improved customer service, to the work of government in providing information and services. E-Government offers considerable benefits for both developing and developed countries, such as improving service quality and transparency in all procedures and accountability [6].

A. E-government in Kuwait before and during COVID 19 The Kuwaiti Government, as with most governments throughout the world, has been cognizant of the potential benefits to be gained by implementing e-government and decision-makers within the government alongside The Gulf Cooperation Council and the United Nations have been the driving force for change and the implantation of innovative technology. To underpin this development in Kuwait, more emphasis has been placed on IT in schools, IT training programmes for government employees and the expectation that management positions would be filled by personnel with IT certification [8]. However, the most striking innovation was the founding of the Central Agency for Information Technology (CAIT) in 2006 CAIT in accordance with the Amiri Decree 266/2006.

Under the wing of CAIT, the e-government project was launched in 2008, thus making way for the use of technology in the government services. The e-government portal (http://www.egov.kwwww.egov.kw) was the first attempt in Kuwait to introduce technology. Although the portal was first non-transactional, it started to grow by creating integration with the services of the other ministries, allowing individuals to find most of the services they want in one place [9]. However, the portals that were created at that time were primitive and did not rise to the name of being automated services but rather online services.

Lockdown of services due to the COVID-19 services online and for erstwhile reluctant citizens to now embrace the online provisions [10]. Because of the outbreak of Covid 19, government visits were difficult due to the prolonged lockdowns that were enforced in the country. Sahel was a solution to deliver to the citizens and residents in order to finish all their transactions from home and not to delay work due to lockdown.

1.1 *Advantages of e-government*

The advantage of an e-government system is that it permits citizens to interact directly with public organizations [11]. As this time in human development is usually referred to as the "knowledge era", the government is eager to offer an increased portfolio of public services to citizens efficiently and innovative public. In terms of policy, most governments make costeffectively. Thus, the main advantages of egovernments is the mistake of not having well-defined policies on e-gaining speed and accessibility to the services anytime and backup and security had adversely affected the adoption of anywhere [12]. online government. Papadomichelaki & Mentzas [31] broke down the barriers to using e-government services to four main

As previously discussed, other advantages of implementing categories; strategy, technology, policy and organization. The egovernment include enhanced efficient of work, easiness of authors listed a number of drivers to avoiding the use of accessing information, higher quality of services and technology from a transparency of the transactions [13]. Additionally, egovernment allows various ministries and departments to strategic point of view. Amongst these barriers are the lack engage more effectively with service users [14–16]. Thus, of shared government goals and objectives and being the online services or portals helps in strengthening the overambitous about what is aimed to be achieved. By doing so, confidence of the people in the government as well as their a government is dooming the use of e-government services to satisfaction level [17,18], delivery of the egovernment fail. In terms of technology, it is common to see various services, transparency of the processes and procedures and government entities using various systems where some are boosting trust of the individuals in the government [19–21]. legacy system, thus, making it hard to roll-out e-services to the.

1.2 *Challenges facing e-government*

One of the main hurdles for developing e-portals is the absence of having a well reformed and complete IT infrastructure in the country, which might be an issue especially in developing countries [22,23]. This results in a communication breakdown between the central and local government with citizens living in remote or mountainous areas where there is no Internet connection [24,25].

Bhuiyan [26] has drawn attention to the deleterious effects of the absence of genuine political support for the implementation of e-government. This lack of political support may frequently emanate from dishonorable practices, resistance rooted in traditional values and, more importantly, the lack of understanding of IT and poor skills in using technology resulting in a reluctance to recognise its benefits However, effective leadership at the local level which values a more participative approach, if supported by the central government, can provide a facilitative climate in which government employees can thrive in the provision of egovernment within communities [24].

However, the importance of a legal framework for egovernment is essential, as its absence can delay the development and implementation of e-government enterprises [25]. Frequently, such projects suffer owing to the poor systems of answerability and reporting which result in a lack of openness and fairness in procedures [27].

A framework incorporating an architectural e-government framework has been successfully applied by Ebrahim and Irani [17] within the public sector and resulted in the identification of obstacles that could hold back the development and the launch of e-government. These obstacles were categorised according to five broad classes: the backend infrastructure, government and not elaborating on issues such as privacy of data and information. Finally, some government agencies might not be ready for the change and this makes it difficult to make the transition.

1.3 *Background and framework for E-GovQual model*

EGovQual is a framework designed to measure citizens' perceptions of the quality of services provided by governments on websites or e-Government portals. An eGovernment portal is a gateway through which citizens can access information or can activate various services which they may require [32]. However, e-GovQual is much more dedicated to evaluating the quality of electronic services, more commonly referred to as eGovernment. Thus, eGovQual ismore focused on evaluating the quality of government websites and portals [31]. Papadomichelaki provides a scale for measuring e-service quality which incorporates four dimensions: efficiency, trust, reliability, and citizen support. Each of these four dimensions contains a number of attributes by which the respective dimension can be measured. Once these attributes have been identified, a method is employed for assessing the relative importance of each attribute in contributing to the overall measure of its related dimension [33].

3 THEORETICAL FRAMEWORK

According to the existing literature related to models and frameworks to measure the use of e-government services, the below framework in Figure 1 has been selected. It contains six independent variables, one dependent variable and all the demographics as moderating variables. The proposed independent variables will be studied to address their impact on Sahel service quality as the dependent variable. The privacy and security concerns, availability of technological skills, governing laws and the budget needed for its development. Utilising semi-structured interviews with 14 consultants with significant e-government experience in Singapore Lam [28] found a number of impediments to implementation by analysing the interview narratives. The barriers were broken into four main categories;

strategy, technology, policy and organisation. Lack of e-government shared objectives and funding issues were among the strategic barriers facing e-government. For technological barriers, the authors listed a number of them but the main ones are incompatible data standards and inflexibility of legacy systems. For policy-related barriers, the authors stated that concerns over citizens' privacy, data ownership and egovernment policy evaluation were very much impeding the development and evolution of e-government. Finally, lack of agency readiness and legacy government processes were among the top-rated barriers facing e-government. Elsheikh and Cullen [29] in a Jordanianbased study of e-government based on the country's published vision and strategy and concluded that serious delays existed in implementing the vision. In a study of e-government adoption in 49 urban regions in Italy Previtali and Bof [30], analysing narrative data from interviews targeting key personnel involved in its implementation found that deficiencies in infrastructure and other functionalities such as researcher has added demographic variables as a moderating factor in this relationship.

A. Reliability and its impact on service quality of Sahel Reliability refers to the level of confidence which a citizen has about the accuracy and timeliness of a government online site in delivering information or services [31]. This confidence extends to the technical functioning of the site in terms of accessibility, accuracy and availability. Accessibility is a general term that denotes how usable the site is to the majority of end-users without any adjustment is necessary. This is distinct from the concept of usability which is more concerned with ease of use for the majority of end-users. It is also important that the site has the capacity to have autonomy in running on the web browsers of all types [34]. Availability refers to the extent to which is resistance to degradation or interruption of service to citizens due to a breakdown in one or more of its constituent parts and can be measured in terms of the probability that the service will function and not fail [35]. The availability can be increased via various methods; one of which is to have the site available 24/7, and ensuring that the bandwidth is high to have consistency and speed for all the transactions [36]. Based on the above discussion, the following hypothesis is proposed:

Hypothesis H1: Reliability of Sahel Service has a positive significant relationship with Sahel service quality in Kuwait.

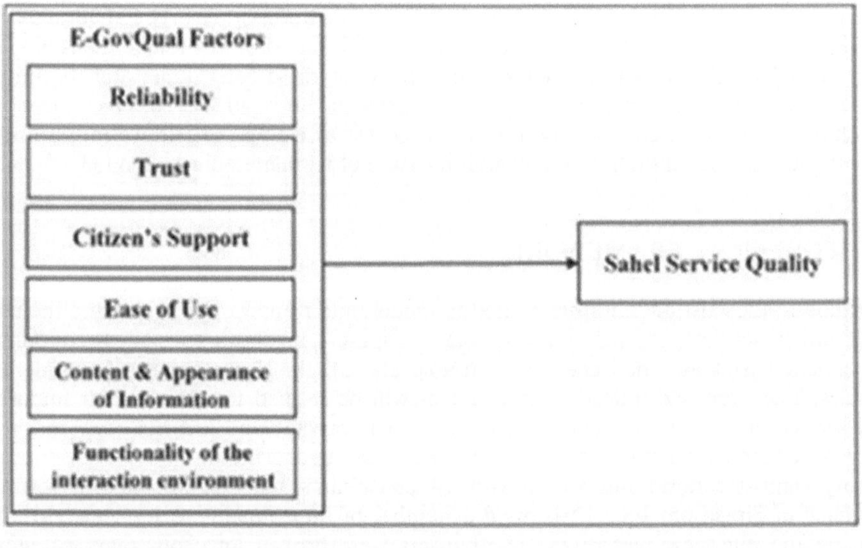

Figure 1. Research conceptual model.

3.1 *Trust (Privacy/Security) and its impact on service quality of Sahel*

Trust includes issues of privacy and security and can be described in terms of the confidence that an end-user has that a particular site presents no risk of hazard to them as they access the site. Many previous studies have seen trust as being a critical issue for e-service [37,38].

Privacy involves the protection of personal information and that it is not shared with other parties or agencies without the individual's permission. It implies that data required for statistical purposes is collected in an anonymous manner and stored securely. Informed consent is usually required for processing information [31].

Finally, security concerns the measures taken to protect end-users from being exposed to fraud or to the risk of having their credit cards or other financial information disclosed to other parties. Such breaches can usually be safeguarded by using encryption, restricting access to verified users, digital signatures, unique usernames and passwords [39]. Based on the above discussion, the following hypothesis is proposed:

Hypothesis H2: Trust in Sahel Service has a positive significant relationship with Sahel service quality in Kuwait.

3.2 *Citizens' support and its impact on service quality of Sahel*

Citizens' Support is intended to signify the role that the organization employees play in order to assist citizens in conducting their daily transactions on the e-government portals [31]. The support can take several shapes and forms, such as providing a full and detailed help menu, providing a Frequently Asked Questions (FAQs) page, or having a hotline ready to take calls. The provision of multiple alternative communication channels (i.e. phone, e-mail, whatsapp etc.) is also important where complete information is not possible on the site, tailored assistance can be offered either through email, phone or letter. In situations that involve the online person to person interaction, there can be quality dimensions attached to the nature of the Parasuraman et al. [40] including promptness of replies to customer inquiries, the breadth of knowledge of the employees interacting with the end user, the politeness of the employees, their ability to give the users the ability to trust them and to have confidence in them by establishing empathy and by their ability to efficaciously resolve problems. This implies the possession of personal qualities and not just technical knowledge or IT skills. Finally allowing the users to monitor the progress of their transactions reflects positively on the citizens' support from employees even though this type of interaction may only occur when citizens encounter difficulty with the website [41]. Based on the aforementioned discussion, the following hypothesis is proposed:

Hypothesis H3: Citizens' Support has a positive significant relationship with Sahel service quality in Kuwait.

D. Ease of use and its impact on service quality of Sahel Ease of use refers to the simplicity and clarity of the website for citizens wishing to interact with services provided on that site, a dimension which is frequently noted by researchers as a most important dimension [37,42].

Having an easy-to-access location and an address or a name that citizens can remember plays a significant role in allowing citizens to access the e-government services with ease. Additionally, consistency in how the site is logically structured can greatly facilitate navigation and a site map can also be useful to guide the end-user to the general area of interest and to avoid unwanted subsites. In this respect, customised search functions based on key terms are also helpful in guiding the end-user to the exact service which is desired. There are other useful navigation tools including menus, tabs, buttons, and various forms [43]. In a study by Zhang and Von Dran [44], it was determined that the availability of searching tools, the clarity of the layout and the easiness of the portal were most important elements in that particular study.

Additionally, the more personalised the information can be in its presentation is important for matching what the site offers to the individual needs and preferences of the end-user. This includes the choice of language and diction used which should be simple and straightforward avoiding technical jargon as much as possible. In particular, the site should be inclusive in its design and helpful for people with disabilities or people from ethnic minority groups. The application of artificial intelligence is useful for customising the site to the individual by presenting first those sites which the end-user has tended to visit more frequently in their history of using the site. Previous research has shown that governmental websites tend to be well-designed and easy to use [42]. Based on the above discussion, the following hypothesis is proposed:

Hypothesis H4: Ease of Use of Sahel Service has a positive significant relationship with Sahel service quality in Kuwait.

3.3 *Content and appearance of information and its influence on service quality of Sahel*

This dimension is about the quality of the information presented on the e-government portal in terms of the presentation, size, location, layout and colors of the fonts used [31].

The quality of the information depends on how accurate and comprehensive it is and whether it is presented concisely and without unnecessary detail. However, an important aspect is also how timely the information is and expired or dated information militates against the quality of the content of the site as has been established by earlier studies [42]. Linkage is a term used to refer to the number and quality of hyperlinks offered by a site. This requires constant monitoring of the site to establish that the links are still current or have not expired. Government documents can sometimes be presented with highly technical jargon – evidence of bad planning from the site designer. Another issue refers to the site aesthetics which involves the appealing use of color, graphics and animation to make the site attractive. The visual impact of a web page can affect the overall experience of using the site and also the effectiveness of communicating its intended message [45] and, in particular, the interaction that takes place between citizens and the e-government [46]. Hence, the following hypothesis is proposed:

Hypothesis H5: Content and Appearance of Information has a positive significant relationship with Sahel service quality in Kuwait.

3.4 *Functionality of the interaction environment and its impact on service quality of Sahel*

Very often, the format of an e-government site involves the filling out of a form online so that administrative departments collect the requisite information. This is mostly the case that is observed for e-government services that are not very developed and mature [4], where forms are used as the principal instrument for submitting information online. Thus, the quality of such online forms becomes an important issue for citizens during their interaction with an egovernment portal and this can impact the quality of the service being provided [31].

When filling out forms, having the ability to see previous forms and to connect with the database for general information to be filled [47]. the ability of the system to provide citizens and organizations with various related options such as submitting or printing a form, or the pop-up help menu that can facilitate the use of the e-government services to conduct the daily transactions [31]. Hence, the following hypothesis is proposed:

Hypothesis H6: Functionality of the environment has a positive relationship with Sahel service quality in Kuwait.

4 RESEARCH DESIGN AND METHODOLOGY

This research will adopt a quantitative methodology through the use of a structured questionnaire. The questionnaire will be formulated to collect data about the factors that influence individuals perception of Sahel service quality. The questionnaire is the primary source to collect the data of the targeted segment. The questionnaire will target all individuals that are above the age of 18 years old in Kuwait, where they will be asked to assess several factors leading to Sahel service quality. The first part of the questionnaire will include demographic questions about the participant while the second part will include statements that cover the factors affecting sahel service quality. The second part will be based on a five-point Likert response scale. The range will be from 1 (strongly disagree) to 5 (strongly agree).

The target population are all individuals in Kuwait that are above the age of 21 years old. As per Public Authority for Civil Information (PACI) statistics, the total number of targeted population is more than N = 3.35 million individual. According to Krejcie and Morgan (1970) who determined the required sample size for research activities, we can ascertain that we need a sample of n = 385 questionnaires. The sampling method is simple random sampling. An electronic survey was setup using survey monkey. A direct web link to the electronic questionnaire was developed to ease the participation of individuals in this study. The distribution of the questionnaire web link was carried out through social media such as Twitter, Instagram, Facebook, and WhatsApp groups. Data collection will take about three weeks. Data will be analyzed using smartpls v. 4.0.

5 EXPECTED FINDINGS

It is expected in this study that the researcher would be able to examine the factors that influence the use of sahel egovernment services in Kuwait. The service is gaining popularity in the media, however, there is still low transaction numbers and people are still using traditional methods to do their transactions. Hence, the findings will allow the researcher to come up with the recommendations to the government in order to enhance and highlight the role of Sahel services, which links perfectly with the future development plans of 2035 for Kuwait, which was announced towards the end of November 2022.

ACKNOWLEDGMENT

This research was supported and funded by the research sector, Arab Open University – Kuwait Branch under decision number "23004".

REFERENCES

[1] Meijer, A. (2015). E-governance innovation: Barriers and strategies. *Government Information Quarterly*, 32(2), 198–206.

[2] Deng, H., Waththage, K. K. T., and Xu, W. (2018). Evaluating the performance of e-government in developing countries: A public value perspective. *Internet Research*, 28(1), 169–190.

[3] Alrai newspaper. (2021). *The Launch of Sahel Services Soon.* Available from: https://www.alraimedia.com (Accessed 3 April 2022).

[4] United Nations. (2008). *Global E-Government Survey 2008*. Available from: https://www.un.org/en/development/desa/publications/global-e-government-survey-2008.html (Accessed 5 Nov 2022).

[5] Fang, Z. (2002). E-government in digital era: concept, practice, and development. *International Journal of the Computer, the Internet and Management*, 10(2), 1–22.

[6] Sabri, A., Sabri, O., and Al-Shargabi, B. (2012). A cultural e-government readiness model. *Intelligent Information Management*, 4(5), 212–216.

[7] Alasem, A. (2009). An overview of e-government metadata standards and initiatives based on Dublin Core. *Electronic Journal of e-Government*, 7(1), 1–10.

[8] CAIT. (2020). About Us. Available from: https://www.cait.gov.kw (Accessed 12 Nov 2022).

[9] Kuwait Government Online (2020). *E-Government.* Available from: https://www.e.gov.kw (Accessed 13 Nov 2022).

[10] Aljarida newspaper. (2020). *COVID 19: A Chance to Complete the E-government Services*. Available from: https://www.aljarida.com/articles/1595518918898243000/ (Accessed 5 Nov 2022).
[11] Almukhlifi, A., Deng, H., and Kam, B. (2019, April). Critical factors for the adoption of e-government in developing countries: validation of a measurement model. In *Proceedings of the 12th International Conference on Theory and Practice of Electronic Governance* (pp. 397–407).
[12] El-Haija, M. I. A., Kara, S. M. S., and Nasereddin, T. Y. (2019). E-Government legislative Constraints. *International Journal of Civil Engineering and Technology*, 10(6), 240–244.
[13] e-spincorp.com. (2018). The Advantages and Disadvantages of E-government: E-SPIN, Global themes and feature topics. Available from: https://www.e-spincorp.com/the-advantages-and-disadvantages-of-e-government/ (Accessed 13 Nov 2022).
[14] Helbig, N., Gil-García, J. R., and Ferro, E. (2009). Understanding the complexity of electronic government: Implications from the digital divide literature. *Government Information Quarterly*, 26(1), 89–97.
[15] Rana, N. P., and Dwivedi, Y. K. (2015). Citizen's adoption of an e-government system: Validating extended social cognitive theory (SCT). *Government Information Quarterly*, 32(2), 172–181.
[16] Zheng, L., Dawes, S., and Pardo, T. A. (2009, November). Leadership behaviors in cross-boundary information sharing and integration: comparing the US and China. In *Proceedings of the 3rd International Conference on Theory and Practice of Electronic Governance* (pp. 43–50).
[17] Ebrahim, Z., and Irani, Z. (2005). E-government adoption: architecture and barriers. *Business Process Management Journal*, 11(5), 589–611.
[18] Karunasena, K., and Deng, H. (2012). A citizen-oriented approach for evaluating the performance of e-government in Sri Lanka. *International Journal of Electronic Government Research (IJEGR)*, 8(1), 43–63.
[19] Ngulube, P. (2007). The nature and accessibility of e-government in Sub Saharan Africa. *International Review of Information Ethics*, 7, 155–167.
[20] Ciborra, C., and Navarra, D. D. (2005). Good governance, development theory, and aid policy: Risks and challenges of e-government in Jordan. *Information Technology for Development*, 11(2), 141–159.
[21] Eyob, E. (2004). E-government: breaking the frontiers of inefficiencies in the public sector. *Electronic Government, an International Journal*, 1(1), 107–114.
[22] Alshehri, M., and Drew, S. (2010). E-government fundamentals. In *IADIS International Conference ICT, Society and Human Beings*, 35–42.
[23] Chen, Y. N., Chen, H. M., Huang, W., and Ching, R. K. (2006). E-government strategies in developed and developing countries: An implementation framework and case study. *Journal of Global Information Management (JGIM)*, 14(1), 23–46.
[24] Janenova, S. (2010, May). E-Government in Kazakhstan: Challenges for a transitional country. In *18th NISPAcee Annual Conference "Public Administration in Times of Crisis* (pp. 12–14).
[25] Brimkulov, U., and Baryktabasov, K. (2018). E-government development in the Central Asian states: best practices, challenges and lessons learned. In *International E-Government Development* (pp. 121–154). Palgrave Macmillan, Cham.
[26] Bhuiyan, S. H. (2010). E-government in Kazakhstan: Challenges and its role to development. *Public Organization Review*, 10(1), 31–47.
[27] Ibrayeva, A., Seifullina, A., and Yessetova, S. (2012). Problems in the implementation of the transparency principle in the activities of the public authority bodies of Kazakhstan. In *ICMLG2016-4th International Conference on Management, Leadership and Governance: ICMLG2016* (p. 165). Academic Conferences and Publishing Limited.
[28] Lam, W. (2005). Barriers to e-government integration. *Journal of Enterprise Information Management*, 18(5), 511–530.
[29] Elsheikh, Y., Cullen, A., and Hobbs, D. (2008). E-Government in Jordan: Challenges and Opportunities. *Transforming Government: People, Process and Policy*, 2(2), 83–103.
[30] Previtali, P., and Bof, F. (2009). E-government adoption in small Italian municipalities. *International Journal of Public Sector Management*, 22(4), 338–348.
[31] Papadomichelaki, X., and Mentzas, G. (2012). e-GovQual: A multiple-item scale for assessing e-government service quality. *Government Information Quarterly*, 29(1), 98–109.
[32] Wijatmoko, T. E. (2020, April). E-Government Service Quality Using E-GovQual Dimensions: Case Study Ministry of Law and Human Rights DIY. In *Proceeding International Conference on Science and Engineering* (Vol. 3, pp. 213–219).
[33] Wijatmoko, T. E., and Siregar, M. U. (2020). Evaluation of IT Service Management (ITSM) Using E-GovQual Dimensions: Case Study Regional Office Ministry of Law and Human Rights DIY. *IJID (International Journal on Informatics for Development)*, 8(2), 55–63.

[34] Alateyah, S., Crowder, R. M., and Wills, G. B. (2013). Identified factors affecting the citizen's intention to adopt e-government in Saudi Arabia. *International Journal of Social, Behavioral, Educational, Economic, Business and Industrial Engineering*, 7(8), 244–252.
[35] Lin, Y. K., and Chang, P. C. (2012). A novel reliability evaluation technique for stochastic-flow manufacturing networks with multiple production lines. *IEEE Transactions on Reliability*, 62(1), 92–104.
[36] Bhattacharya, D., Gulla, U., and Gupta, M. P. (2012). Emerald Article: E-service quality model for Indian government portals: citizens' perspective. *Management*, 25(3), 246–271.
[37] Gefen, D., Karahanna, E., and Straub, D. W. (2003). Trust and TAM in online shopping: An integrated model. *MIS quarterly*, 51–90.
[38] Zhao, J. J., and Zhao, S. Y. (2010). Opportunities and threats: A security assessment of state e-government websites. *Government Information Quarterly*, 27(1), 49–56.
[39] Taherdoost, H. (2017). Understanding of e-service security dimensions and its effect on quality and intention to use. *Information & Computer Security*, 25(5), 535–559.
[40] Parasuraman, A., Zeithaml, V. A., and Berry, L. (1988). SERVQUAL: A multiple-item scale for measuring consumer perceptions of service quality. 1988, 64(1), 12–40.
[41] Zeithaml, V. A., Parasuraman, A., and Malhotra, A. (2002). Service quality delivery through web sites: a critical review of extant knowledge. *Journal of the academy of marketing science*, 30(4), 362–375.
[42] Santos, J. (2003). E-service quality: a model of virtual service quality dimensions. *Managing Service Quality*, 13(3), 233–246.
[43] Clyde, L. A. (2000). A strategic planning approach to Web site management. *The Electronic Library*, 18 (2), 97–108.
[44] Zhang, P., and von Dran, G. (2001, January). Expectations and rankings of Web site quality features: Results of two studies on user perceptions. *In Proceedings of the 34th Annual Hawaii International Conference on System Sciences* (pp. 10-pp). IEEE.
[45] Hoffmann, R., and Krauss, K. (2004, October). A critical evaluation of literature on visual aesthetics for the web. In Proceedings of the 2004 annual research conference of the South African institute of computer scientists and information technologists on IT research in developing countries (pp. 205–209).
[46] Schenkman, B. N., and Jönsson, F. U. (2000). Aesthetics and preferences of web pages. *Behaviour & information technology*, 19(5), 367–376.
[47] Alawneh, A., Al-Refai, H., and Batiha, K. (2013). Measuring user satisfaction from e-Government services: Lessons from Jordan. *Government Information Quarterly*, 30(3), 277–288.

Innovation and Technological Advances for Sustainability – Al-Naemi et al. (Eds)

Vertical-axis wind turbine for powering light gadgets in outdoor Qatar areas

Nasser Al-Marri
Distribution and Quality Department, Qatar General Electricity & Water Corporation, Doha, Qatar

Saleh Bahlool
Rotating Equipment Department, Qatar Gas, Doha, Qatar

Faysal Fayez Eliyan
Engineering Technology Department, Community College of Qatar, Doha, Qatar

ABSTRACT: A vertical wind turbine was designed with SOLIDWORKS® and constructed and tested in-house. Computational Fluid Dynamics (CFD) was carried out to validate the results of rotational speed and distribution of air flows. Theoretical efficiency in relation to the rotational speed in rounds per minute (rpm) was calculated. Charging capability of the turbine of a 12V battery was measured – achieved optimum performance at 120 rpm with a 0.2 Amps current. Mobile phones were connected to the battery to measure the battery discharging performance.

Keywords: wind turbine, VAWT, battery, charging, renewable

1 INTRODUCTION

The increase in demand for renewable energy in areas with limited access to regular supply of electricity has incentivized innovation and investment in wind turbines. Development of technology has made wind turbines more efficient, generating more power from wind, with virtually no negative environmental effects or harmful emissions. Renewable power from wind turbines are becoming also cheaper and more abundant and reliable in comparison to energy generated from fossil fuels [1]. G. Darrieus was first to introduce the concept of vertical axis wind turbines in 1931 [2]. Vertical axis wind turbines (VAWTs) have more merits over the horizontal wind turbines (HAWTs), including the ability to harness power from winds flowing in any direction [3], with good range of wind speeds, and the flexibility for installation at closer distances for multiple turbines within small area.

In this work, a preliminary design of an existing turbine was upgraded to the one outlined here, for the purpose of enhanced function and higher efficiency. The new design is of higher compactness, smaller diameter, and of better meeting design constraints. The turbine is for personal or small-scale use, not industrial use, and it serves as a prototype for efficiency studies. The upgraded turbine is of better portability, ease to assemble and to do maintenance, for which high power output at higher efficiency is achieved – at exposure to winds in rural areas, or in camps at sea sides to harness offshore winds.

The prototype in this paper is introduced as concept for the new sources of energy in the market, for renewable and reliable power especially in areas that depend on intermittent supplies of fossil fuels. Also, it serves to better understand the aspects of development of technology of vertical axis wind turbines, whilst experimenting new designs of lower cost and higher functionality. The turbine proved to be a quick energy supplier to charging mobile

DOI: 10.1201/9781003496724-21

phones and light gadgets in areas of high winds and limited supply of electricity from conventional sources.

Table 1. Specifications, constraints, & values.

Specification	Constraint	Value
Redesigning is costly and requires lots of resources [4]	Low cost	Efficient turbine reassembled from previous prototype
Prototype might not generate enough power from wind [5]	Functional	The system will be able to recharge mobile phones
Due to dust in flowing wind and other factors, regular maintenance is needed [6]	Maintainability	Maintenance possible in easy ways, and by users with no technical expertise
Low-cost, experimental, functional prototype serves as model to encourage utilization of renewable energy sources, with high efficiency that needs specifications [7]	Functionality, constructability, marketability	Upgraded turbine achieving all specified objectives – set as an ecofriendly solution of economic value as a marketable product

2 DESIGN CONSTRAINTS AND STANDARDS

2.1 *Design constraints and standards*

The main objectives of the paper include upgrading the design of a preliminary turbine to achieve higher efficiency, to model and characterize the performance of the turbine at different wind speeds, to investigate the significance of the number and orientation of the blades on the performance, and to determine the limit of charging of mobile phones in terms of their number and initial battery charging prior to plugging them into the turbine.

2.2 *Function of vertical axis wind turbine*

The VAWT typically provides a source of renewable energy for residential purposes. The working principle involves a rotor revolving around a vertically positioned shaft driven by vertical multiple blades [8]. The generator is positioned at the bottom of the turbine while the blades wrap around the shaft, thereby generating electricity by harnessing power from the wind. The spinning rotor enables the generator to effectively convert the mechanical energy to electricity. The components of the VAWT include the blades support plates, frame, bearings, shaft, and the vertical blade. The main advantage of the VAWT is it rotates by wind flowing from all directions and its orientation does not need to be controlled. As the gearbox, generator, and other turbine parts can be installed at "ground" low level, the design and construction of the wind tower is easily simplified, reducing the cost. It should be noted that the VAWT has limited maximum practical height, as the axis in modern designs is supported only on one end.

2.3 *Use of VAWTs, and distribution of winds in Qatar*

Gorgulu *et al.* studied utilization of VAWTs in Doha's Metro tunnels for renewable power generation from the massive airflow generated by trains traffic inside tunnels [9]. They studied critical speeds as high as 3 m/s for feasibility of power generation. In Qatar coastal regions, the wind speed was found to remain above the required 3 m/s speed at 80% of measurements time. Other researchers measured the wind speeds in Doha, Al Khor, Mesaieed, Dukhan, Ruwais, and Ras Laffan, in onshore and offshore areas, over 12-month period [10]. On average the wind speeds were between nearly 5 m/s (lowest in September, October), and 7 m/s (highest in February, June). While there might not be soon-to-execute plan for wide adoption of VAWTs, a memorandum of uunderstanding was signed between

Qatar and the International Renewable Energy Agency (IRENA), the organization tasked with coordinating The Global Atlas for Renewable Energy. Such initiative is for sharing expertise and relevant necessary data for countries that plan for adoption of renewable resources upon assessing the potential in renewable energy [11].

3 DESIGN, CONSTRUCTION, AND TEST APPARATUS

Definition of the problem was set as the need of designing a low-cost, efficient wind turbine that supplies enough energy for charging mobile phones. The design and construction were mapped in steps in a roadmap. The road map in a Gantt chart outlined the tasks of: 1) The literature review of existing test prototypes, 2) Free-hand sketching, and designing of the subject prototype, 3) specifying the parts to be ordered/assembled from suppliers, 4) Constructing the prototype, 5) Checking the functionality of the prototype, 6) Testing the prototype, 7) Refining the design, and 8) CFD simulation of the airflow by SOLIDWORKS®, for the purpose of optimizing power output in relation to orientation and number of the blades.

3.1 *Concept and design proposal*

A reliable cost-effective VAWT would take a wide variety of different shapes and designs. So it was a challenge to propose a design that meets the constrains previously mentioned. First, the impeller was designed to consist of three parts; the upper plate, the lower plate, and the blades. The impeller had three slots for orientation of blades for optimum angle of attack; *i.e.* the orientation of the blade is changeable, to correlate it with the efficiency at different rotation speeds. The alternator, other mechanical parts, and other accessories were acquired from commercial sources.

3.2 *Parts of the test VAWT*

A series of blades in between an upper plate and a lower plate are fitted. The plates were engraved in a way to accommodate the blades at three different orientations, as shown in Figure 1, for lower plate. One orientation goes in the direction of the outer arc of the plate, and the other two orientations are at 45° and −45° in reference to the first orientation. Four holes were drilled in the lower plate for attachment to the base and the alternator, directly, achieving velocity ratio of 1:1 between the plate and the alternator. The design of the blade, shown in Figure 1, is concave with a good radius of curvature that capture and utilize the force of air flow. The thickness of the blade, made of plastic, is 10 mm, of light weight for good overall rotation efficiency.

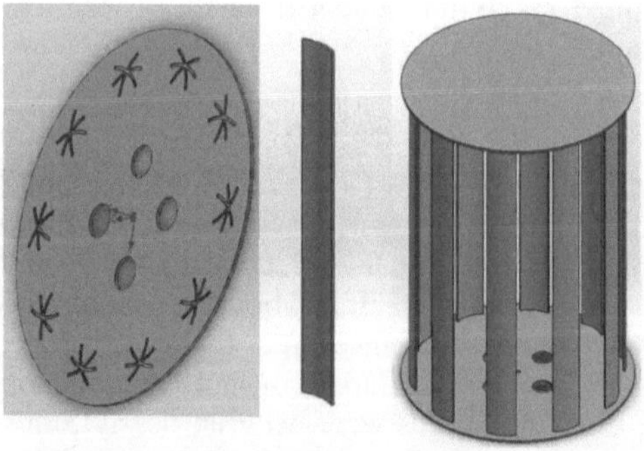

Figure 1. Lower plate of turbine, a blade, and assembly of plates and blades.

3.3 *Auxiliary parts*

- Alternator: An electrical generator utilizing a rotating magnetic field with a stationary armature for conversion of rotational mechanical energy into electrical energy in form of alternating current. The alternator connected to the lower plate of the assembly is shown in Figure 2, and the specifications are in Table 2.

Figure 2. Lower plate of turbine, a blade, and assembly of plates and blades.

Table 2. Alternator specifications.

Specification	Details
Model	S 100W-300W
Type	Gearless permanent magnet AC alternator
Current type	3 Phase
Voltage	12V–24V

- Battery: The battery is rechargeable, sealed, lead-acid type from Power Sonic, shown in Figure 3, of specifications in Table 3.

Figure 3. Power Sonic Battery, Model PS-1270 F2.

Table 3. Battery specifications.

Specification	Details
Model	PS-1270 F2
Type	Lead-Acid, DC power, constant voltage charge
Cycle use	14.4–15.0V (20°C)
Standby use	13.5–13.8V (20°C)
Initial current	2.1A max.

- Solar charge controller: It is used to display the voltage, current, and other parameters. It indicates the status of charging when the turbine rotates charging, with a signal of on an icon representing the battery. The charger is shown in Figure 4, of specifications in Table 4.

Table 4. Solar charge cotroller specifications.

Model	MX-20A
Battery voltage	12V/24V auto
Charge current	20A
Discharge current	10A
Max solar input	12V battery, the highest 23V; 24V battery when the highest 46V
Equalization	B1 sealed 14.3V
Float charge	13.7V (default, adjustable)
Discharge stop	10.7 V (default, adjustable)
Discharge reconnect	12.6 V (default, adjustable)
Charge reconnect	13V
Voltage of open light	Solar panel 8V
Voltage of close light	Solar panel 8V
USB output	2 way USB output, 5V/2.5A(MAX)
Self-consume	<10mA
Operating temp	−35~+60C
Size/Weight	133.5*70*35mm / 165g

- Bridge Rectifier Diode three phase: It is used to rectify the voltage from AC to DC. It is shown in Figure 5, of specifications in Table 5.

Figure 4. Bridge rectifier diode three phase.

Table 5. Bridge rectifier diode three phase specifications.

SQL-100A Type 3 phase Diode Heatsink Bridge Rectifier 100A 1600V	
Product	Bridge Rectifier
Model	SQL-100A-1000V
Output current (DC)	1000A
Repetitive peak reverse voltage (AC)	1600V
Material	Metal

- Wooden Base: It is used to mount the assembly on, shown in Figure 5.

Figure 5. Wooden base.

- Upper plate and lower plates: They were cut from wooden sheets of 5 mm thickness, with slots engraved for the blades, shown in Figure 6.

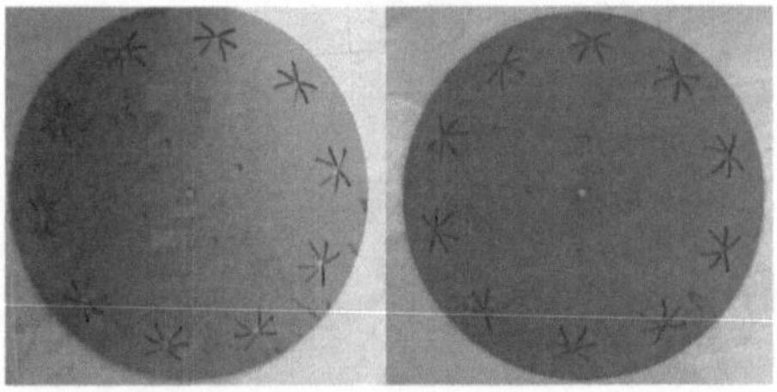

Figure 6. Upper plate and lower plate.

- Threaded rods: Four threaded Aluminum rods were used to connect the upper plate with the lower plate and to provide support for the assembly, shown in Figure 7.

Figure 7. Assembly of plates and four rods.

- Blades: They were cut from light PVC pipes, equal in length and radius of curvature, shown in Figure 8, together with the assembly.

Figure 8. Assembly with blades.

3.4 *Testing appartus*

- Photo/Contact Tachometer: It was used to measure the rotation speed in (rpm), shown in Figure 9, and specifications in Table 6.

Figure 9. Photo/contact tachometer.

Table 6. Alternator specifications.

Module Specifications	EXTECH 461895	
	Range	Max. Resolution
Photo tachometer	5 to 99,999rpm	0.1rpm < 1000rpm
Constant tachometer	0.5 to 20,000rpm	1rpm >1000 rpm
Surface speed		
m/min	0.05 to 1999.9m/min	0.01m/min < 100m/min; 0.1m/min > 100m/min

- Blower (Stanley STPT600): The blower was used to provide air flow at different speeds for rotational speed testing, shown in Figure 10, and specifications in Table 7.

Figure 10. Test blower.

Table 7. Blower specifications.

Module	STPT600
Power Input	600-Watt
No Load speed	16500 RPM
Manufacturer	Stanly
Weight	3.2 kg

- Watt meter and power analyzer: It was used to record the real-time data of the battery, and to measure wattage and analyze power, shown in Figure 11, and specifications in Table 8.

Figure 11. 200A watt meter and power analyzer.

Table 8. Blower specifications.

Specification	Details
Operation range	5V-60V
Measuring sensibility	0–200V, resolution 0.01A 0–60V, resolution 0.01V 0–6554W, resolution 0.1W 0–65Ah, resolution 0.001Ah 0–6554Wh, resolution 0.1Wh
Display	16 × 2 backlit lcd display
Size	85 × 42 × 25mm
Weight	82g

- Test mobile phones: The phones used to test discharge the turbine battery were Huawei Mate 20 X, Google Pixel, and Nubia Red Magic 6 Pro, with their battery specifications shown in Table 9.

Table 9. Test mobile phone specifications.

Specification	Huawei	Google	Nubia
Battery technology	Li-Polymer	Li-Ion	Li-Ion
Maximum capacity	4200mAh	2770mAh	5050mAh
Voltage	4.057V	4.165V	4.135V
Starting charge level	58%	18%	61%

- 3-outlet socket: It was used to connect the three test phones to the battery. The socket is from TravelMate® with an input of 12VDC, 6AMP and 72W, and an output of 12VDC, 2AMP and 24W for each socket.

4 RESULTS AND DISCUSSION

4.1 *Battery charging*

In a setup, outlined and pictured in Figure 12, the alternator was connected to the rectifier to rectify the voltage form AC to DC. The rectifier was connected to the solar charge controller for display of the voltage, for the battery connected to the controller.

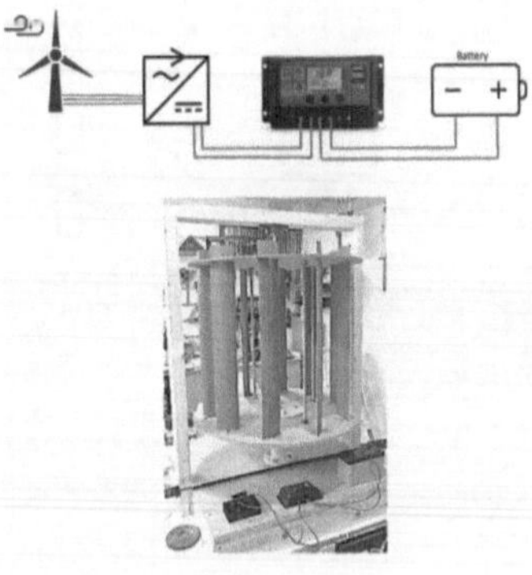

Figure 12. Battery charging connection setup.

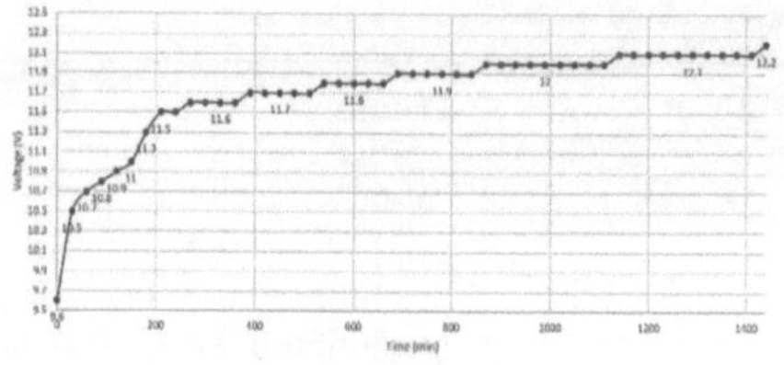

Figure 13. Voltage of battery vs. time (charging).

For a turbine rotating at 120 rpm, and an output current of 0.2 A, the voltage of the battery was recorded with respect to time, shown in Figure 13. In the first 200 minutes the charging rose quickly from nearly 9.5V to 11.5V, and increased slowly to nearly 12 V for 1200 minutes, indicating at no steady charging process of the battery at this specific rotation condition. The battery was fully charged in nearly 1440 minutes (1 day), which was a relatively long time, yet satisfactory for a test small-scale prototype. In another test, shown in Figure 14, at the same rotational speed at 0.2A, the charging started from a lower 7.8V for 40 minutes. The voltage rose to 9.6V, indicating at lower voltage the turbine charged the battery decently within short period of time.

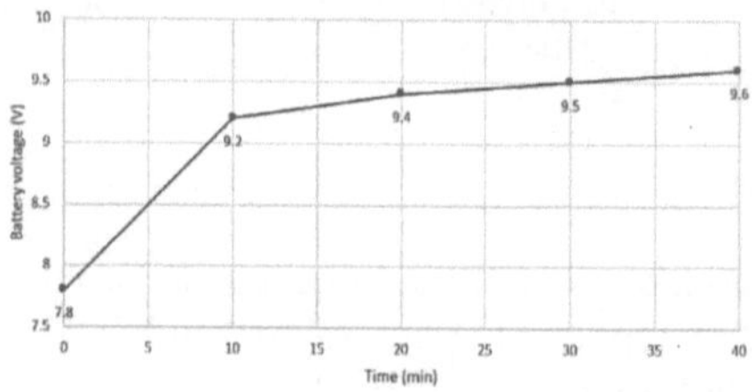

Figure 14. Voltage of battery vs. time (charging), from 7.8V for 40 minutes.

4.2 *Battery discharge*

The battery discharge performance was analyzed with respect to fully charging three mobile phones. The battery capacities were 4200mAh (Huawei), 2770mAh (Google) and 5050mAh (Nubia). The discharge schematic and setup are shown in Figure 15. The battery was connected to the watt meter for power analysis, which was connected to the 3DC socket. The phones were together connected to the outputs of the socket.

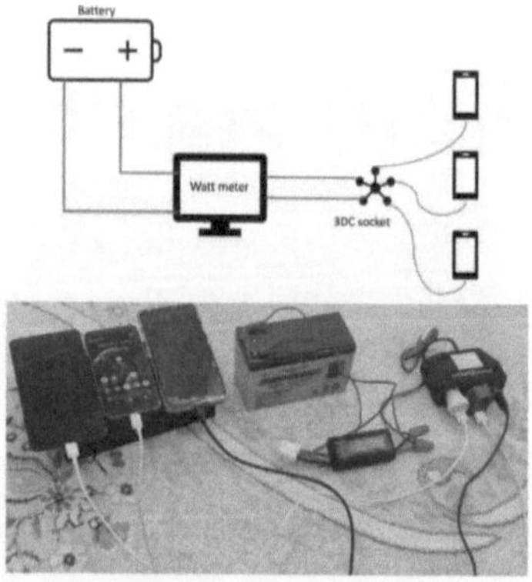

Figure 15. Battery discharge connection setup.

The data from the battery and the phones were collected at 10-minute periods for 90 minutes. They included the current and voltage output from the battery, and voltage and current of the phones. The battery voltage and current data while discharging are shown in Figure 16, and the voltage and current of the test mobile phones are shown in Figure 17. For the battery, the voltage drop in the first 10 minutes was followed by relatively stable 12V during discharge. The peak current was nearly 2.5A, sustained for 20 minutes, and then gradually steadily dropped towards discharge period. All test mobile phones showed similar rising battery voltage responses, characterized by gradual increase for the first 40 minutes prior to stability towards full charge, interestingly irrespective of their initial starting charge level. The current trends were similar, but there were discrepancy, attributed to the battery capacity [], for which the currents of Google mobile phone was lowest, and of Nubia was highest.

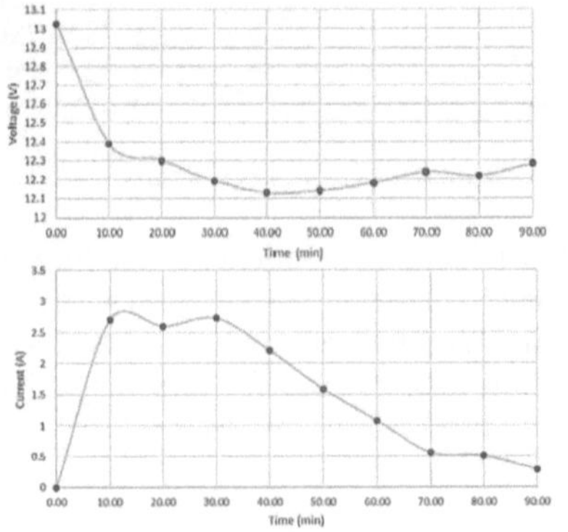

Figure 16. Battery discharge voltage and current.

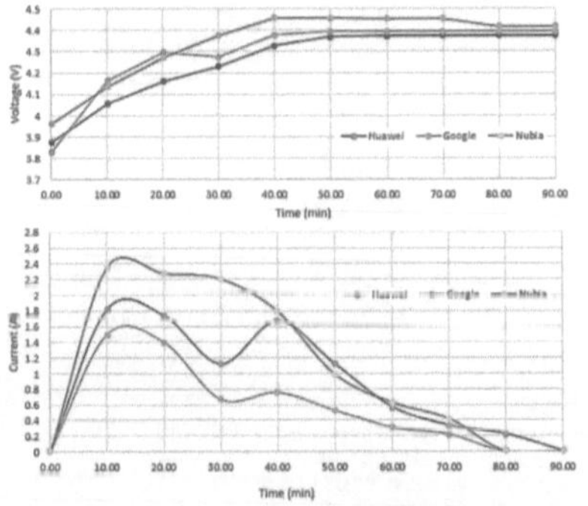

Figure 17. Test mobile phones voltage and current.

4.3 *CFD simulations*

CFD simulations were used effectively for VAWTs effectively to analyze performance, velocity, pressure, etc. []. In this study, the wind tunnel dimensions, as a virtual box that contains the wind turbine was specified and the turbine design by SOLIDWORKS® was in the tunnel for a perpendicular air flow, shown in Figure 18. The boundary conditions were specified with xyz reference axis, at atmospheric pressure of I 101.325 kPa, and the inlet fluid was specified as air flowing in an open space in a real-life turbine, with the properties shown in Table 10.

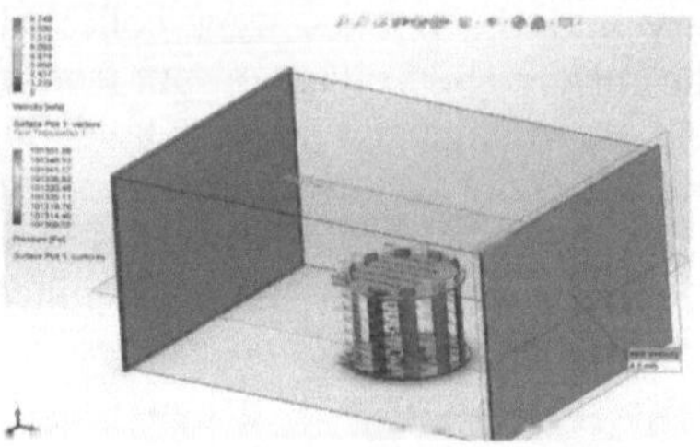

Figure 18. CFD simulation of wind turbine by SOLIDWORKS®.

Table 10. Properties of air for CFD simulations.

Density	1.225 kg/m^3
Specific volume	0.78 m^3/kg
Viscosity (dynamic)	17.22 μPa
Bulk modulus of elasticity	1.01325×10^5 N/m^2
Velocity	30 m/s (maximum real velocity)

The CFD simulations were run in correspondence to the three orientations of the blades of the rotating turbine, of the original orientation (that goes with the outer arc of the plate), and 45° and −45° in reference, and with different numbers of blades. For the original orientation arrangement, of 10 blades, the torque calculated from the software was 0.025 N.m, which was not apparently sufficient to meet power requirements.

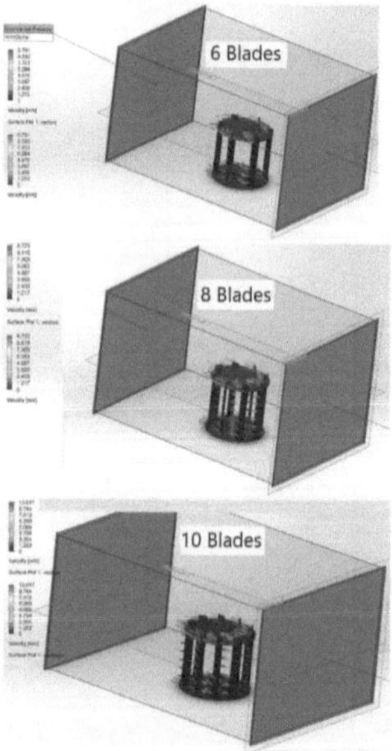

Figure 19. Velocity simulation of turbine of different number of blades (45° orientation).

In Figure 19, the velocity is shown simulated for different number of blades of 6, 8, and 10 at 45° orientation. Correspondingly, the velocity within the turbine increased as well as the torque. This was also the case for the −45° orientation.

4.4 *Calculation of turbine efficiency*

Turbine power output $\dot{W}_{turbine}$ was calculated with the following formula:

$$\dot{W}_{turbine} = \frac{\alpha\rho\dot{V}V^2}{2}$$

$\dot{V}$ is flow rate, V is the velocity, and ρ is the density of air (1.19 kg/m^3, at 23°C), and α is the correction factor that equals 1. The volume flow rate $\dot{V}$ = V A, for which A is the cross-sectional area of the blower, equals 4.52×10^{-4} m^2. The power drag was calculated from the following formula:

$$P = \frac{C_D A_\rho V^2 \omega}{2}$$

C_D is the coefficient of drag that was set to be 2.3, and ω is the angular velocity. Dividing P by $\dot{W}_{turbine}$ yields the efficiency. In the calculations, increasing wind velocity from 5, 8.5, to 11 m/s, corresponding to 63, 105, and 120 rpm, increased the drag power from 0.102, 0.49, and 0.94 kW, and the efficiency from 33, 34, and 38%.

5 CONCLUSION

The turbine design and constructed in-house proved reliable in providing power for charging a battery over a good range of wind velocity, and functionally in terms of orientation and number of blades, over relatively short period of time. The battery supplied power to three test mobile phones whose charging characteristics were measured. CFD simulations indicated that the greater the number of blades the greater the velocity and torque, also that the blade orientation of 45° is optimum for power generation. The efficiency was higher with greater wind speed, which induces greater rotational speed.

REFERENCES

[1] Leung D., and Yang Y., (2012). Wind energy development and its environmental impact: A review. *Renewable Sustainable Energy Rev.*, vol. 16, pp. 1031–1039.

[2] Alaimo A., Esposito A., Messineo A., Orlando C., Tumino D., (2015). 3D CFD analysis of a vertical axis wind turbine. *Energies*, vol. 8, pp. 3013–3033.

[3] Johari M., Jalil M., Shariff M., (2018). Comparison of horizontal axis wind turbine (HAWT) and vertical axis wind turbine (VAWT), *Int. J. Eng. Technol.*, 7, pp. 74–80.

[4] Gur M., Toklu E., (2022). Designing and modelling of vertical axis wind turbine with a new wing structure and investigation of electricity production, *J. Eng. Appl. Sci.*, vol. 11, pp. 1998–2006.

[5] Zhang L., Pei Y., Liang Y., Zhang F., Wang Y., Meng J. and Wang H., (2015). Design and implementation of straight-bladed vertical axis wind turbine with collective pitch control. In *2015 IEEE International Conference on Mechatronics and Automation (ICMA)* (pp. 1258–1263). IEEE.

[6] Jaen-Cuellar A., Elvira-Ortiz D., Osornio-Rios R., Antonino-Daviu J., (2022). Advances in fault condition monitoring for solar photovoltaic and wind turbine energy generation: A review, *Energies*, vol. 15, pp. 5404.

[7] Möllerström E., Gipe P., Beurskens J., Ottermo F., (2019). A historical review of vertical axis wind turbines rated 100 kW and above. *Renewable Sustainable Energy Rev.*, vol. 105, pp. 1–13.

[8] Sampath S., Shetty S. and Chithirai Pon Selvan M., (2015). Estimation of power in low velocity vertical axis wind turbine. *Front. Mech. Eng.*, vol. 10, pp. 211–218.

[9] Gorgulu Y., Furkan Y., Ozgur M., and Ramazan K. (2020). Wind Turbine Applications in Doha *Metro, Qatar. Int. J. Mech. Eng*, vol. 7, pp. 21–25.

[10] Mohammed A., Shanas P., Veerasingam S., Al-Ansari E., Sadooni F., Vethamony P. (2021). Long-term assessment of onshore and offshore wind energy potentials of Qatar. *Energies*, vol. 14, pp. 1178.

[11] Heubaum H. and Biermann F. (2015). Integrating global energy and climate governance: The changing role of the International Energy Agency. *Energy Policy*, vol. 87, pp. 229–239.

Innovation and Technological Advances for Sustainability – Al-Naemi et al. (Eds)

The design and analysis of an advanced hybrid renewable energy microgrid for Cyangugu Remote rural community in Rwanda using Homer Pro

Adimchinobi Asiegbu, Ali Almaktoof & Marco Adonis
Electrical Electronics and Computer Engineering Department, Cape Peninsula University of Technology, Cape Town, South Africa

ABSTRACT: In Rwanda, and in Cyangugu Remote community near the Kivu Lake, there is high solar sunshine energy, significant wind energy, abundant methane gas, appreciable geothermal energy, peat(biomass), hydro renewable energy sources, and grid power access. To ensure feasibility, accessibility, and sustainability, a balance need to be maintained to optimize this energy mix options, in terms of cost (NPV, COE etc.), maintenance, sensitivity, optimization, and other system parameters. The best combination of these Hybrid Renewable Energy Sources (HRES) considering the community's energy needs, is modelled using the Hybrid Optimization of Multiple Electric Renewables – Professional (HOMER Pro), innovatively, to design and find out the proposed optimum system requirements with environmental advantages or benefits, best payback duration, lower or zero net GHGs and radiation emissions, performance optimization with lest value of Net Present Cost (NPC) and Cost of Energy (COE), sensitivity analysis involving components supply changes like wind speed change based atmospheric conditions, variable sunshine frequencies based on time of the day, seasons, fuel price changes etc. These seven energy supply components is used to provide access to modern, clean and reliable energy to Cyangugu.

Keywords: hydrogen, renewable, storage, hybrid, energy mix, microgrid, modular: inverter, power quality, photo voltaic array, natural gas, biomass, peat

1 INTRODUCTION

Cyangugu is a small town in Rwanda, and a border town with the Democratic Republic of Congo (Congo DRC). Due to its proximity to lake Kivu, that has large natural gas deposit, it has large supply of methane gas. The hill topography of the environments present good hydro energy resources, the abundant sunshine ensures sufficient solar energy resources, the peat (decayed vegetation/organic matter accumulation in natural areas called peatlands), significant wind energy resources, Power grid access point [1,2], a standalone microgrid system will help provide clean and modern renewable energy to this remote tourist community separated from the Congo DR by the Kivu Lake. The microgrid analysis is done using homer pro, which needs only basic parameters like the load requirements, cost of components, etc. to perform complete power system and economic analysis of the microgrid. A Hybrid Renewable Energy System (HRES) is shown in Figure 1, the Optimization of Multiple Electric Renewables – Professional (HOMER Pro), innovatively and systematically evaluates and analyse off grid and grid connected power systems in a way that reduces the time and human effort required, when designing a power system or an advanced microgrid consisting of seven energy supply components.

 DOI: 10.1201/9781003496724-22

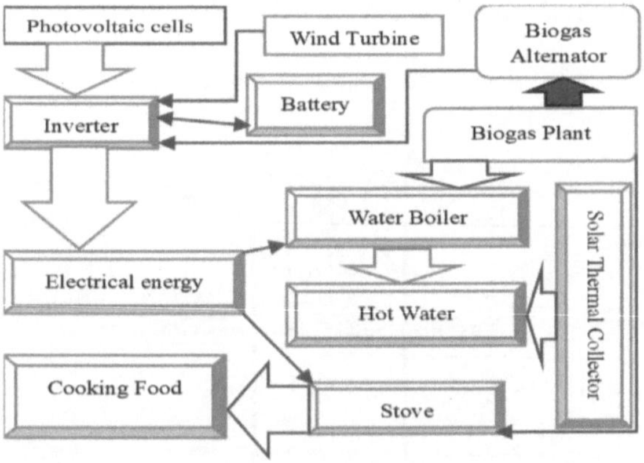

Figure 1. A hybrid renewable energy system (HRES) [3].

This helps in decisions making about the system configurations, selection of the best components, the size and number of components required, sensitivity analysis, most efficient components, cost considerations, optimum system conditions, technology, and energy renewable resources availability are also provided to make selection of the best system configuration easier [4].

2 PROBLEM STATEMENT

The About 80% of Rwandans are living in the rural communities [5], energy access through the national grid is limited, thus, there is a need for analyses of an advanced microgrid with access to the national grid that can supply a remote community sustainably and reliably. The potential geothermal renewable energy resource (temperature from 90°C to 140°C), and the geothermal fluid mass flow (from 20 kg/s to 80 kg/s) [6], the windier conditions from May to September each year [7], the solar energy from sufficient sunshine, and the river flow due to the gradients created by the hills and landscape of the Cyangugu community can provide renewable energy from an advanced microgrid utilizing this distributed generation.

3 LOAD DEMAND ASSESSMENT

The population of Cyangugu community is 63,883 persons [8,9], which translates to 12828 households of five persons, government health expenditure per person is 3 015,00 Rwandan franc. Using the minimum threshold electricity consumption level of 250 kWh/annum for a rural household, the Cyangugu community energy need is 3206900kWh/annum or 8786kWh/day translating to (8786kWh/24h) 366.1kW minimum power plant overall capacity. If power factor and other losses are factored in, then it should be 0.5MW power plant total capacity.

4 SOCIAL RESOURCE ASSESSMENT

The renewable energy resources assessed includes solar energy (5.2GWh/m^2/day), hydro-energy (0.35TWh), geothermal energy (0.01TWh), natural gas (0.22TWh), and wind energy (18.4W/m^2 to 24.4W/m^2, average = 21.4W/m^2, at 115.5 m^2, power delivered/day = 59.3208kWh/day/Wind turbine (2.5kW capacity per wind turbine) = 21652.092kWh/annum/Wind turbine) according to sustainable energy for all (SE4ALL) [10,11–13]. The average flow velocity of hydrokinetics is 6.92 m/s [14]. Biomass (peat) energy density = 15MJ/kg [15], inverter, and deep cycled Lead-Acid battery.

5 CASE STUDY DESIGN OVERVIEW

To The system configuration developed is show in Figure 2.

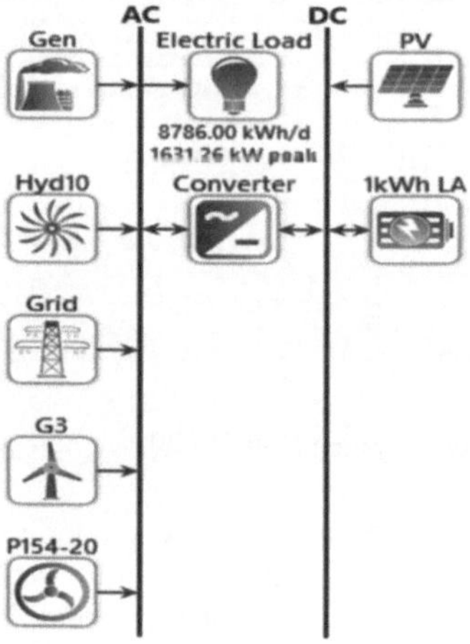

Figure 2. Advanced hybrid renewable energy scheme.

The motivation and reasons for choosing this system configuration is based on the various renewable energy resources available, accessible and the low cost associated with the installation, running, and maintaining these power supplies. The return of investment, quick payback period, low NPV, and COE, makes foreign direct investment in this energy sector [4], attractive and appealing to both local and foreign investors[16]. However, according to National Institute of Statistics of Rwanda (NISR), World Bank (WB), International Monetary Fund (IMF) and International Statistics (IS), and various data files, the discount rate is 16.2%, inflation rate is 10.78% (increasing at the rate of 2.7% per month going forward), GNI per Capital is $850, but Rwanda has experience infrastructural and economic advancement for a decade and managed to maintain this growth despite COVID-19 pandemics [17,18]. This a pitfall for investors because increase in inflation causes decrease in Gross Domestic Product (GDP) [19], since cost of getting loan from financial institutions is high based on Rwanda financial data [10]. Because of the well-crafted sensitivity analysis, optimization analysis, and guaranteed energy resource availability, financial institution will be willing to make their loan accessible in this sector, with or without minimal commitment deposit by the investors seeking financial loan [16]. The government will incentivize this energy sector to further strengthen the investment opportunities, and the job creation associated with it, for the betterment of this community [20]. These seven components emit low GHGs, and the carbon footprint is low. The money spent on the hospital by the government and members of this community will drop, and the quality of life will certainly improve [21]. Also, to account for unmet energy load needs in Rwanda, the Annual Capacity Shortage is set at 5% (at least all load energy needs is met 95% of the time) [4].

Geographic tool, like the geographic information system (GIS) can be used to model the supply chain managements systems like the Hydrogen Supply Chain (HDC) systems. Model for Optimization of Regional Hydrogen Supply is used to do geographic analysis (and economic feasibility) of the German HDC [22]. FCEVs will be sustainable, since it is an example of hydrogen to renewable energy technology, which is reliable (it can travel long distances without the need to recharge or refuel car batteries), can be integrated into the distributed generation system, and can capture wind energy, hydro energy, nuclear energy, and solar energy directly from a natural energy rich environment.

6 MATHEMATICAL MODELLING OF SYSTEM COMPONENTS

Detailed description of each of the components used in the design are described below.

Solar energy source. A 30kWh PV generic flat array (available supply during the day) consisting of solar panel rated 300W each with an area of 1.5329m^2 is used to generate electricity at ((300W X 12 hrs.) / (1.5329m^2)) 2.35kWh/m^2/day from the sun at 20% conversion efficiency. Solar PV contribution to the energy mix is 110kW.

Wind energy from the generic turbine (5kWh variable energy source, providing energy day and night), delivers power/day = 59.3208kWh/day/each standalone wind turbine at 40% conversion efficiency. The capacity and contribution to the energy mix is 2.5kW. Table 1 shows the distribution of wind speed from January to December across Rwanda.

Table 1. Wind speed across Rwanda.

Month	Wind Speed (m/s)
January	1.65
February	1.69
March	1.83
April	2.06
May	2.32
June	2.37
July	2.41
August	2.37
September	2.15
October	1.92
November	1.79
December	1.65

Hydro energy from runoff river (constant 10KWh energy supply), and hydrokinetic from runoff water sources from the hills and valleys (cheaper) as result of the Rwanda landscapes at a conversion efficiency of 90% [23]. The stream flow rate (Liter per second (L/s)) from January to December is: 280, 334, 321, 394, 327, 250, 197, 340, 378, 376, 372, 392 [15].

Storage/source. Lead-acid battery is used because of its deep cycled charging/discharging nature, and stability at various temperature variations (capacity for storage from solar PV = 120AH X 10 = 1200AH).

Transmission and distributions. The connecting cables is selected based on the current rating =300W/12V = 25A per solar panel connection to the inverter (each 300W capacity solar panels needs 300W inverter capacity). Losses from this component is considered in this research.

Biogas and natural (methane powered) power plant capacity is 92kW supplying 2196.5kWh/day to the energy mix.

Geothermal (conversion of earth internal heat energy to electricity) plant capacity is at 4% of total energy demand = 351.44kWh, which is 14.64kW capacity and contribution to the energy mix. The geothermal energy is used to heat water, and the hot water is a feedstock to hydrokinetic turbine, as shown in Homer Pro simulation software. It produces 20kW at 4m/s at cost of 299.43 Rwanda franc/kWh according to Guinard energies.

Peat power plant capacity = 10% of total energy requirement = 878.5kWh (3162600kJ) = 36.61kW. 0.1757 tons is required to supply this energy. The peat energy resources can last for 30 years, with heat value of 12999.69kJ/kg to 21999,47kJ/kg, (average value = 17999.57 kJ/kg). [24].

National grid component energy supply is variable, considered to supply an average of 100kW to the energy mix.

Total energy mix = 100 + 94 + 37 + 15 + 10 + 110 + 2.5 = 368.5kW. I recommend an increase in supplied power from the national grid or addition of more solar panels, which will result in topping up the total energy mix value to 500KW or 5MW due losses and power quality/power factor issues.

These energy sources are optimum and best suited to form a micro grid, integrated with the national grid, that can supply this community, forming a Distributed Generation (DG).

The mathematical models for each electrical supply component as follow:

Solar PV energy density Is = IbFb + IdFd + Fr(Ib + Id) [25]. Where **Is** is the total solar radiation (kWh/m2), **Ib** is the direct irradiation and **Id** is the diffuse solar radiation (kWh/m2), **Fb** is the tilt factors for the beam, **Fd** is the diffuse parts of solar radiation, and **Fr** is the reflected parts of solar radiation.

The Solar Power Generation (SPV) **PPV = Rpv X IsA** [26]. Where **PPV** is the power generation from the **PV** array (kWh), **Rpv** is the system efficiency, **A** is the available panel area.

The Wind Energy (WES) PWT = Pw X AwRW [27]. Where **PWT** is the electric power obtained by the wind turbine (kWh), **Pw** is the wind generator power, **Aw** is the summation of the swept area, **Rw** is the total gearbox, generator, and associated components efficiency.

The Biomass energy (BM) PBM = Rg(φw − φB) [28,29]. **PBM** is the biomass energy (kWh), **Rg** is the gasifier system efficiency, which includes the biomass efficiency of the producer gas conversion, internal combustion engine, and generator, **φw** is the biomass present now and instantly (kg), **φB** is the required kilogram per hour of the biomass present.

Biogas/natural gas: **PBG = Rb (BGtotal − BGC−H)** [30,31]. **PBG** is the generated power by the BG engine fired electrical machine (kWh), Rb is the biogas-fueled engine and generator system overall efficiency, **BGtotal** is the biogas available instantly (m3), **BGC−H** is the consumed biogas for clean cooking and modern generation boiling of water (m3).

Solar Thermal Collector: **PSTC = IbRtcAa − Ploss** [31]. **PSTC** is the generated power by solar thermal collectors (kWh), **Rtc** is the optical efficiency of the collector, **Aa** is the absorber surface area (m2), **Ploss** is the power lost in the solar collector (kWh).

Hydro power and hydrokinetics are mathematically modelled as Off-grid operation M = **J {(π/30) X(dn/dt)}** [32]. Where **J** is the moment of inertia of the turbine, **n** is the rotational speed, **t** is the time, **M** is the resistance moment of the water flow.

80% of the Cyangugu community are energy poor based on national statistics and demographics, (80% of 63,883 persons) = 51107 people, which translates to 10222 households of five persons. Using the minimum threshold electricity consumption level of 250 kWh/annum for a rural household, the Cyangugu rural community energy need is 2555500kWh/annum or 7002kWh/day. To meet this energy demand, 68% of the energy need will be supplied by the solar photo-voltaic array (cheaper, and the price keeps falling), 11% from wind turbine, and 21% from hydro sources (hydrokinetics, geothermal steam turbine and hydro-dam power plants) each based on the availability and accessible renewable energy resources in the rural community. Other available sources are not currently accessible (biogas, biomass, geothermal, etc.). GHGs, climate change, and ecological issues are some constraints of peat energy source. This mathematical modelling of the load is calculated from the percentage resource availability and accessibility considering the hills (river runoff), valleys (water runoff), rainfalls (change in water volume, flow rate and water flow velocity), sunshine (night and day) and wind speeds. Thus, the total of these contributions based on these percentages will fulfill the energy demand of 43.93kWh.

Total energy = 0.68 X SGP + 0.31 X HE + 0.11 X WE. WE = Wind Energy, HE = Hydro-Energy.

7 RESULT AND DISCUSSION

The result analysis procedure which involves the project input parameters, Sensitivity analysis, optimization, and simulation is shown in Figure 3 flow chart [15].

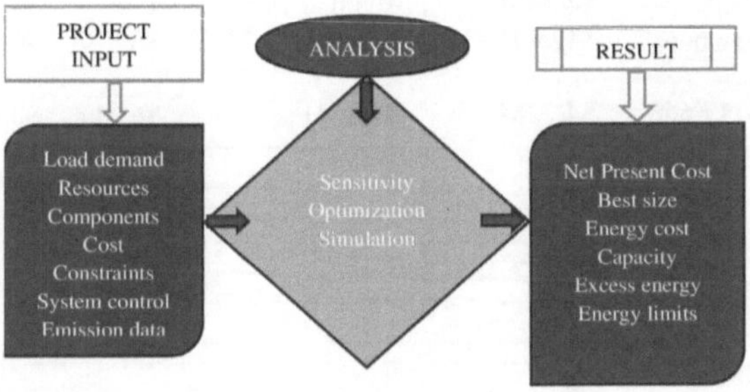

Figure 3. Result and discussion to solution implementation model.

The Homer generated results for the best-case scenario for Least Net Present cost (NPC) = Rwandan franc 3910000000, Initial Capital (IC) = Rwandan franc 117666000 Operation and Maintenance (O&M) = Rwandan franc 296623000/annum, Least Cost of Energy (LCOE) = Rwandan franc 9.54kWh, internal Rate of Return (IRR) = 38%, Return on investment (ROE) = 34%, Simple Payback Period (SPP) = 2.6 years is shown Table 4.

Table 2. Electrical summary (excess and unmet).

Component	Consumption (kWh/yr)	Percent
AC Primary Load	3,206,890	100
DC Primary Load	0	0
Deferrable Load	0	0
Grid Sales	75.9	0.00237
Total	3,206,966	100

The data retrieved according to the simulated results include the electrical production, Net Present Cost (NPC), and Annualized Cost, is shown in Tables 2 to 4, and electrical production and consumption is shown in Tables 5 and 6. Also, the grid rate, renewable energy summary, and compare economics is provided in Tables 7 to 9. The cost summary, cash flow charts, proposed annual nominal cash flows, base system annual nominal cash flows, and cumulative discounted cash flows of the Advanced Hybridge Renewable System (AHRES) is shown in Figures 4 to 8. This optimized value ensures reliability and continuous power supply of 250kWh (minimum energy threshold expected to increase to the average energy threshold of 500kWh) of modern and clean energy to the household with additional health and environmental gains, appealing payback period and low or net zero emissions. The sensitivity analyses shows that the variation in component parametric dimensions is within +/-10%, which is smooth and represents narrow wind speed variations, almost constant average sunshine during the day, and fairly water stream uniform velocity, with near zero acceleration. The hydrokinetics energy resource is not only affordable, but is made from simple technology, that can be transferred to the republic of Rwanda, for local skill content development in the power sector, and reducing the chances of importation of this vital component. This sensitivity includes wind speed, hydro speed, battery cost, sunshine frequency (Solar radiant energy frequency must be greater than Threshold frequency (constant) + work function threshold).

Table 3. Net Present Costs (NPCs).

Name	Capital Cost	Operating Cost	Replacement Cost	Salvage Value	Resource Cost	Total Cost
10kW	$80,000	$34,193	$0.00	$0.00	$0.00	$114,193
Grid	$0.00	$4.14M	$0.00	$0.00	$0.00	$4.14M
Guinard-P154 [20kW]	$37,666	$42,741	$11,541	-$6,817	$0.00	$85,131
System	$117,666	$4.22M	$11,541	-$6,817	$0.00	$4.34M

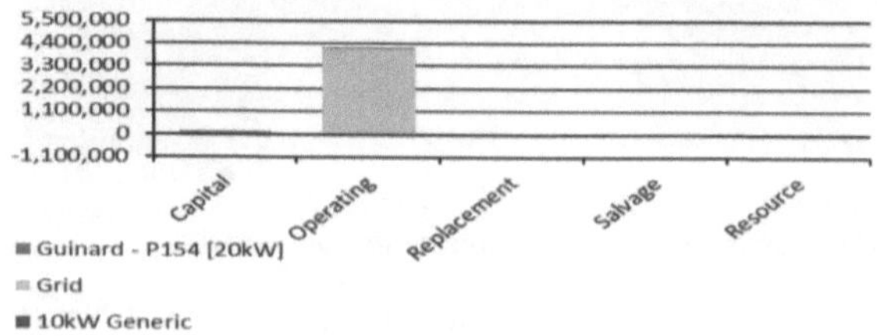

Figure 4. Cost summary.

Table 4. Annualized Costs (ACs).

Name	Capital	Operating	Replacement	Salvage	Resource	Total
10kW Generic	$5,615.00	$2,400.00	$0.00	$0.00	$0.00	$8,015.00
Grid	$0.00	$290,901.00	$0.00	$0.00	$0.00	$290,901.00
Guinard – P154	$2,644.00	$3,000.00	$810.04	-$478.46.00	$0.00	$5,975.00
System	$8,259.00	$296,301.00	$810.04	-$478.46.00	$0.00	$304,892.00

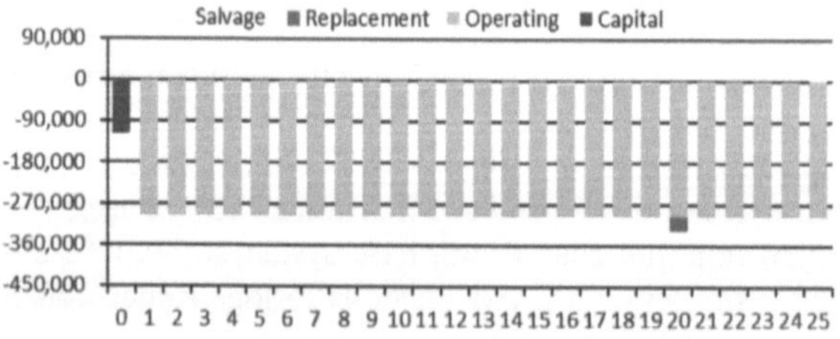

Figure 5. Cash flow chart.

Table 5. Electrical production summary.

Component	Production (kWh/yr)	Percent
Guinard – P154 [20kW]	175,200.00	5.46
Hydro	122,716.00	3.83
Grid Purchases	2,909,050.00	90.70
Total	3,206,966.00	100.00

Table 6. Electrical consumption summary.

Component	Consumption (kWh/yr)	Percent
AC Primary Load	3,206,890	100.00000
DC Primary Load	0.00	0.00000
Deferrable Load	0.00	0.00000
Grid Sales	75.90	0.00237
Total	3,206,966.00	100.00000

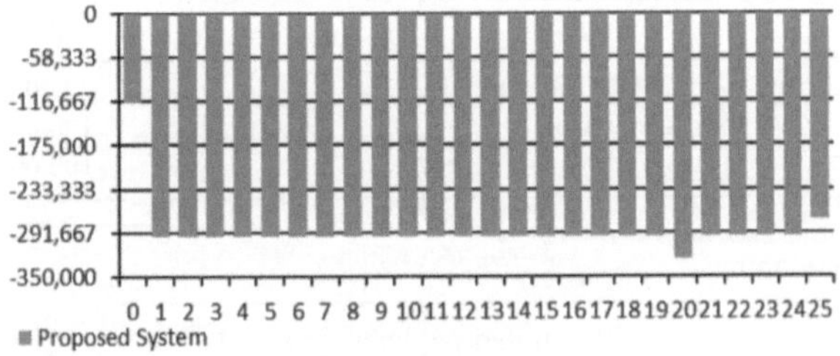

Figure 6. Proposed Annual Nominal Cash Flows (PANCF).

The BSANCF is $340,000 after 20 years, and the PANCF is $304,892 for 25 years.

Table 7. Grid rate: All.

Month	Energy Purchased (kWh)	Energy Sold (kWh)	Net Energy Purchased (kWh)	Peak Demand (kW)	Demand Charge	Energy Charge
January	243,967.00	6.1100	243,961	1,471	$24,396.00	$0.00
February	217,132.00	123.0000	217,118	1,281	$21,713.00	$0.00
March	253,777.00	0.0000	253,777	1,448	$25,378.00	$0.00
April	24.001,625.00	3.0000	241,622	1,480	$24,162.00	$0.00
May	240,978.00	15.2000	240,963	1,328	$24,097.00	$0.00
June	241,197.00	0.0000	241,197	1,236	$24,120.00	$0.00
July	244,985.00	0.0000	244,985	1,524	$24,499.00	$0.00
August	256,447.00	0.0000	256,447	1,416	$25,645.00	$0.00
September	243,087.00	0.0588	243,087	1,424	$24,309.00	$0.00
October	241,976.00	5.3500	241,970	1,360	$24,197.00	$0.00
November	236,966.00	32.9000	236,933	1,597	$23,695.00	$0.00
December	246,913.00	0.0000	246,913	1,326	$24,691.00	$0.00
Annual	2,909,050.00	75.9000	2,908,974	1,597	$290,901.00	$0.00

Annua l	2,909,05 0.00	75.90 00	2,908, 974	1,597	$290,90 1.00	$0.0 0

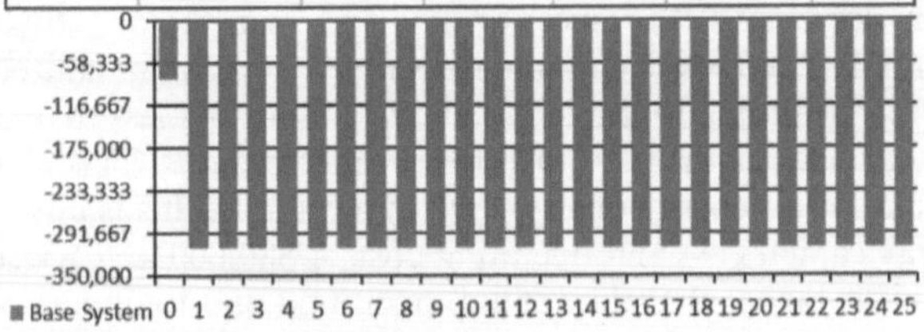

Figure 7. Base System Annual Nominal Cash Flows (BSANCF).

Table 8. Renewable summary.

Capacity-based metrics	Value	Unit
Nominal renewable capacity divided by total nominal capacity	100.00	%
Usable renewable capacity divided by total capacity	100.00	%
Energy-based metrics	**Value**	**Unit**
Total renewable production divided by load	9.29	%
Total renewable production divided by generation	9.29	%
One minus total nonrenewable production divided by load	100.00	%
Peak values	**Value**	**Unit**
Renewable output divided by load (HOMER standard)	82.40	%
Renewable output divided by total generation	100.00	%
One minus nonrenewable output divided by total load	100.00	%

Table 9. Compare economics.

IRR (%):38.5	Discounted payback (yr):2.85	Simple payback (yr):2.5
	Base System	Proposed System
Net Present Cost	$4.51M	$4.34M
CAPEX	$80,000.00	$117,666.00
OPEX	$310,817.00	$296,633.00
LCOE (per kWh)	$0.0987.00	$0.0951.00
CO_2 Emitted (kg/yr)	1,949,198.00	1,838,520.00
Fuel Consumption (L/yr)	0.00	0.00

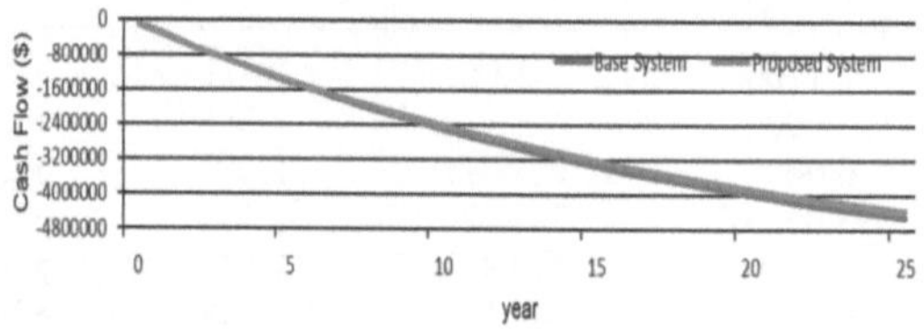

Figure 8. Cumulative discounted cash flows.

The top 3 system configurations, as determined by Homer, in a comparative analysis, and the homer software simulated results are clearly depicted in Figures 4 to 8. The top three is different due to different component combinations, but they are net zero emitters, except the national grid power. Another difference is in the initial cost capital ratio, which is 118: 119: 125. Thus, the first two configuration is better than the last third component configuration. The HOMER pro ensures that gas powered power plant is only used after all the renewable sources are fully engaged. Tables 3 to 9 shows that both the discounted and simple payback period is within 3 years, the compare economics for base and proposed system are similar, the discounted cash flow decreases with respect to time and spans across 25 years durations, the electrical consumption is 3,206,966 kWh/annum, and the system annualized cost is $304,892. The wind energy speed according to Table 1 is above 2.0 m/s but less than 2.5 m/s from April to September yearly and lower than 2.0 m/s in the other months. This coupled with uneven hills in Rwanda, makes wind energy unattractive in Rwanda. This is also shown in Table 2, depicting wind energy capital cost and operating cost of $80,000 and $34,193, giving a total cost of $114,193. This is higher that the capital, operating and total cost of other sources of renewable energy like the solar and hydro energy sources, yet these renewable energy sources have a low sensitivity, more reliable and higher yield in Rwanda than the wind energy sources.

8 CONCLUSION AND RECOMMENDATIONS

In the first case scenario, the Homer Pro recommended components are hydrokinetics, hydro-energy, and battery energy sources. This complies with the International Renewable Energy Agency (IREA) requirements, that recommends the selection of primary energy sources in the community, with the best renewable energy source potential. The second option presents all the renewable sources except wind energy sources, but at a high cost. Hence considering the low and unpredictable wind turbine output, it is not the optimum choice base on sensitivity analysis. The third choice, has a solar PV component, but based on optimization and sensitivity analysis, and considering the NPC and COE, wind energy sources is not recommended. The first two cases are different in terms of components implemented and initial cost of investment, which is a difference of 1,021,877.00 Rwandan franc. Also, the last two is within the same amount in terms of initial cost of investment but differ from the last two by almost 10,218,770.00 Rwandan franc. Hence, the first two component configuration is much affordable than the last two component configurations. The importance of the Homer results and the impact they have for the affected community is reflected in the revelation that wind energy source is not adequate and accessible. Indeed, in Cyangugu, the solar irradiation is available and reliable based on HOMER Pro results and offers the best energy source with the least carbon footprint. This is followed by the hydro resources, and the least renewable energy source is the wind because of the numerous hills blocking the wind. Thus, the result is a good reflection of what is best for this rural community, and it truly agrees with all the researched articles cited in this paper. These results will have a positive impact on this remote community of Rwanda because it assesses the energy demand of the vulnerable and poor households and proffer sustainable and reliable solution, with the least effect on the climate, environment, GHGs, and other far-reaching effects.

The quality of life, air quality, environment, and health of this remote community will be improved if this erudite and meticulously researched component configuration integrated with microgrid controller and national grid is implemented. Either first or second case should be implemented for increased energy access that is above the energy poverty line (250kWh/annum) at an affordable rate. Also, the time spent in fetching firewood, and the inconveniences and health problems associated with cooking with firewood and biomass will be mitigated, since 80% of Rwandans use Biomass as a primary energy source. Finally, economics, emissions, renewable energy penetration, Levelized COE, operating cost, HOMER cycle charging, energy-based metrics, capacity-based metrics, peak values, cash flow, cash summary, electrical performance, and efficiency is done by HOMER-Pro since conversion efficiency and high.

REFERENCES

[1] Safari B., (2010). A review of energy in Rwanda, *Renewable and Sustainable Energy Reviews*, vol. 14, no. 1. doi: 10.1016/j.rser.2009.07.009.

[2] Bolson N., Yutkin M., and Patzek T., (2021). Energy efficiency and sustainability assessment for methane harvesting from Lake Kivu, *Energy*, vol. 225, doi: 10.1016/j.energy.2021.120215.

[3] Masip Y., Gil A. F., Sanchez M. G., Castro C., and Gonzalez S. M. N., (2019). Optimization of a smart integrated renewable energy system for isolated rural villages using integer linear programming, *in Proceedings – 2019 7th International Engineering, Sciences and Technology Conference*, IESTEC 2019, doi: 10.1109/IESTEC46403.2019.00-83.

[4] Homer Energy, "*Homer Pro*," Manual Homer Energy, (2019).

[5] Nishimwe G. et al., (2020). Natural capital accounting for land in Rwanda, *Sustainability (Switzerland)*, vol. 12, no. 12, doi: 10.3390/su12125070.

[6] Wakana F., (2013). Preliminary study of binary power plant output comparing ORC and kalina for low-temperature resources in Rusizi Valley, Burundi, *Unu-Gtp*, no. 36.

[7] Langenberg V. T., Sarvala J., and Roijackers R.,(2003). Effect of wind induced water movements on nutrients, chlorophyll-a, and primary production in Lake Tanganyika, *in Aquatic Ecosystem Health and Management*, vol. 6, no. 3. doi: 10.1080/14634980301488.

[8] Deribe K. et al., (2019). Geographical distribution and prevalence of podoconiosis in Rwanda: a cross-sectional country-wide survey, *Lancet Glob Health*, vol. 7, no. 5, doi: 10.1016/S2214-109X (19)30072-5.
[9] National Institute of Statistics Rwanda, (2014). *RPHC4: Population Projections*, Population Projections.
[10] Padam G. et al., (2018). *Rwanda Beyond Connections – Energy Access Diagnostic Report Based on the Multi-Tier Framework*, World Bank Group.
[11] Bashahu M. and Nkundabakura P., (1994). Analysis of daily global irradiation data for five sites in Rwanda and one in Senegal, *Renew Energy*, vol. 4, no. 4, doi: 10.1016/0960-1481(94)90050-7.
[12] Hagumimana N. et al., (2021). Concentrated solar power and photovoltaic systems: a new approach to boost sustainable energy for all (Se4all) in Rwanda, *International Journal of Photoenergy*, vol. 2021, doi: 10.1155/2021/5515513.
[13] E. (Energy S. M. A. Program), (2019). Lifting the burden of electricity subsidies, while expanding access., *Energy Subsidy Reform Facility Rwanda Country Brief.*
[14] Munyaneza O W. U. G., Nzeyimana Y. K, (2013). Hydrolic structures for Flood control in the Nyabugogo Wetland, Rwanda, *Nile Basin Water Science&Engineering Journal*, vol. 6, no. 2.
[15] Niyonteze J. D. D., Zou F., Norense Osarumwense Asemota G., Bimenyimana S., and Shyirambere G., (2020). Key technology development needs and applicability analysis of renewable energy hybrid technologies in off-grid areas for the Rwanda power sector, *Heliyon*, vol. 6, no. 1, doi: 10.1016/j.heliyon.2020.e03300.
[16] E. Mutandwa, "Foreign Direct Investment and Rwanda's Economic Performance (1970-2011)," 2014.
[17] "Rwanda (Overview)," in World Encyclopedia of Contemporary Theatre, 2020. doi: 10.4324/9781315011172-35.
[18] Mustari K., Rusibana C., and Nzamalu A., (2020). Internal audit profession and financial performance in commercial banks in Rwanda, *Journal of Advance Research in Business Management and Accounting* (ISSN: 2456-3544), vol. 6, no. 10, doi: 10.53555/nnbma.v6i10.924.
[19] Nkikabahizi F., Ndagijimana J., and Musabanganji E., (2017). *Relationship Between Inflation and Real Economic Growth in Rwanda.* doi: 10.1007/978-981-10-4451-9_6.
[20] Okito M., (2019). Rwanda poverty debate: summarising the debate and estimating consistent historical trends, *Rev Afr Polit Econ*, vol. 46, no. 162, doi: 10.1080/03056244.2019.1678463.
[21] Thomas P., (2018). Whose Vision 2020? The World Bank's development and educational discourse in Rwanda, *Development Studies Research*, vol. 5, no. 1, doi: 10.1080/21665095.2018.1469422.
[22] Ball M., Wietschel M., and Rentz O., (2007). Integration of a hydrogen economy into the German energy system: an optimising modelling approach, *International Journal of Hydrogen Energy*, vol. 32, no. 10–11, doi: 10.1016/j.ijhydene.2006.10.016.
[23] Silvennoinen P., Vieno T., and Vira J., (1980). Multigoal fuel cycle optimization including non-proliferation objectives, *Nucl Technol*, vol. 48, no. 1, doi: 10.13182/NT80-A32445.
[24] Mugerwa T., Ehinola O. A., Oladosu I. A., and Rwabuhungu D., (2020). Assessment of peat deposits in Rwanda, *Journal of Geoscience and Environment Protection*, vol. 08, no. 10, doi: 10.4236/gep.2020.810001.
[25] Li T. et al., (2019). Optimized operation of hybrid system integrated with MHP, PV and PHS considering generation/load similarity, *IEEE Access*, vol. 7, doi: 10.1109/ACCESS.2019.2932851.
[26] Huang H., Nie S., Lin J., Wang Y., and Dong J., (2020). Optimization of peer-to-peer power trading in a microgrid with distributed PV and battery energy storage systems, *Sustainability (Switzerland)*, vol. 12, no. 3, doi: 10.3390/su12030923.
[27] Babatunde O. M., Munda J. L., and Hamam Y., (2019). Selection of a hybrid renewable energy systems for a low-income household, *Sustainability (Switzerland)*, vol. 11, no. 16, doi: 10.3390/su11164282.
[28] Hamilton J., Negnevitsky M., Wang X., and Lyden S., (2019). High penetration renewable generation within Australian isolated and remote power systems, *Energy*, vol. 168, doi: 10.1016/j.energy.2018.11.118.
[29] Kanase-Patil A. B., Saini R. P., and Sharma M. P., (2011). Sizing of integrated renewable energy system based on load profiles and reliability index for the state of Uttarakhand in India, *Renew Energy*, vol. 36, no. 11, doi: 10.1016/j.renene.2011.04.022.
[30] Chang C. C. et al., (2019). A case study on the electricity generation using a micro gas turbine fuelled by biogas from a sewage treatment plant, *Energies (Basel)*, vol. 12, no. 12, doi: 10.3390/en12122424.
[31] Masip Y., Gutierrez A., Morales J., Campo A., and Valín M., (2019). Integrated renewable energy system based on IREOM model and spatial-temporal series for isolated rural areas in the region of Valparaiso, Chile, *Energies (Basel)*, vol. 12, no. 6, doi: 10.3390/en12061110.
[32] Kishor N., Saini R. P., and Singh S. P., (2007). A review on hydropower plant models and control, *Renewable and Sustainable Energy Reviews*, vol. 11, no. 5. doi: 10.1016/j.rser.2005.06.003.

Innovation and Technological Advances for Sustainability – Al-Naemi et al. (Eds)

Integrated production and maintenance planning optimization of a photovoltaic system: A case study of Nigeria

Aisha Sa'ad, Aimé C. Nyoungue & Zied Hajej
LGIPM, Université de Lorraine, Metz, France

ABSTRACT: Global energy crises has been a huge challenge which causes the need for diversification in order to be able to satisfy the increasing demand. In this work, we developed optimal production and maintenance planning strategies on two (2) techniques to; the support vector regression (SVR) and the theoretical techniques for a photovoltaic (PV) system. Optimization of the production planning was performed to determine the minimal number of PV panels and battery storage required to satisfy the demand. After successful production planning, an integrated maintenance model was the developed to determine the cost for optimal number of maintenance actions to be performed on the system over a production horizon. A case study of Sokoto power plant in Nigeria was considered to validate the algorithm. The two results were compared to determine the best technique for the PV system management. The results obtained presents the machine learning (SVR) approach as the most suitable technique of the two considered.

Keywords: Optimization, integrated maintenance, photovoltaic, minimal repair, Support vector regression (SVR)

1 INTRODUCTION

The utilization of renewable energy has gained recognition since the early 1970s after the oil crisis and rapid depletion of fossil fuel. These fossil fuels are observed to be harmful by increasing carbon footprint in the atmosphere which in-turn affects the ecosystem. Because of this, climate change and global warming have been serious topic of discussion globally over the recent decades. Nations signed and got into treaties such as Rio de Janeiro conference (1992), and the most recent COP 26, Glasgow (2021) which centered on ways to restrict global carbon emission (Global Wind Energy Council Report, 2019). This caused a tremendous shift and diversification of energy generation sources to renewable energies such as Solar, Wind, Tidal, Ocean current and hydro. Renewable energy has then become more attractive and cost-effective. PV energy is one of the promising and one of the most harnessed sources of the renewable energies globally. According to [1], global solar energy has a cumulative installed capacity of about 633.7 GW by the end of 2019 and expected to reach 1TW by 2023 which increases the solar penetration capacity by 23%.

This means that electricity generated from solar energy has grown by nearly 400% from its inception since early time of the century which occurred when Germany launched feed-in tariff law on grid-connected solar plants [1]. In Figure 1, the cumulative global installed solar PV capacity from 2000 to 2020 is shown.

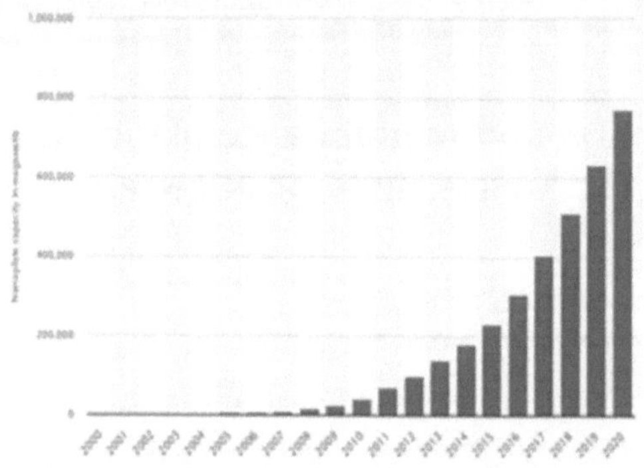

Figure 1. Cumulative installed solar PV capacity worldwide from 2000 to 2020 (statistica.com).

Harnessing renewable energy sources requires appropriate and efficient planning. This is to guarantee that energy can be made available even when the source produces less than the required demand since these sources depend on weather conditions for their availability. To do so, many researchers have studied and developed algorithms to estimate the capacity of renewable generating units capable of supplying energy reliably at low cost by different approaches [2]. Ref [3] presented an iterative method for determining the optimal number of generating as well as the storage units in an autonomous PV/wind hybrid energy system in their model. Wind turbines of different sizes and capacities were considered, and for each of the turbine, a combination of the number of PV panels and storage units required to satisfy the demand at minimal cost. A model to estimate the levelized cost of energy was developed by [4] which properly sizes the dimensions of generating units in a stand-alone PV/wind hybrid energy system. In their work, the authors believed that appropriate selection of the battery storage system plays an important role in minimizing the cost.

The above cited works were performed by the conventional methodology (empirical, analytical, numerical, hybrid, etc.). For sizing PV-systems generally, data were used for sites where the required weather data (irradiation, temperature, humidity, clearness index, wind speed, etc.) and information concerning the selected site are available. However, these techniques could not be used for sizing PV systems in remote areas, where the required data are not available. In all of these, accuracy is achieved by using data from daily global irradiation series. Majority of these conventional approaches need long term meteorological data such as total solar irradiation, air temperature, wind speed, etc. for their operation. Therefore, when the relevant meteorological data are not available, these methods cannot be used, especially in remote areas. In order to overcome this situation, methods that are more recent have been developed for sizing parameters for PV-systems based on artificial intelligence (AI) techniques [5]. Also, to improve the system performance, maintenance is encouraged to be periodically performed.

Maintenance is an important factor for sustainable production in industrial systems to ensure continuous operations considering reliability, plant and machines/components failures [6]. Development of maintenance strategies has gained recognition since early 1950s [7]. A Preventive Maintenance model with imperfect continuous condition monitoring of a wind turbine components was presented by [8] for calculating the expected cost of maintenance per unit of time and the average lifetime maintenance cost which are derived for an arbitrary distribution of time to degradation failure. Another maintenance strategy model was formulated using neural network in [9] where optimization of turbine system was performed by clustering and mapping similar turbines to obtain maximum power while minimizing cost. Ref [10] integrated the deterioration of quality in the integration of production and maintenance problem for an imperfect maintenance process. An opportunistic preventive maintenance (PM) strategy was developed for a 2-unit system where inspection was considered both periodically every time T and when failure occurs on the first unit [11]. In [12], a mathematical model for maintenance based on the effective age of the components, determined after periodic inspections also considering replacement during PM intervention.

The main goal of this work is to develop an algorithm that jointly minimizes the production and maintenance costs of a PV system in a solar power plant dealing with identical panels arranged in series to satisfactorily supply an unsteady demand over a finite horizon. The strategy adopted in this work is to first perform production planning by applying two (2) techniques; Support vector regression and theoretical techniques for the power production forecasting/estimation. Significant time for preventive maintenance is found to cause shortage (loss of production). Therefore, an integrated maintenance plan minimizing cost of loss of production due to the non-negligible PM time is developed. Another goal is to determine the optimal number of preventive maintenance sessions to be performed for each of the power production technique.

2 GENERAL DESCRIPTION

This work consists of developing a joint production and maintenance planning optimization algorithm which deals with a varying load demand for a solar energy power plant. Solar energy is dealt with as a random variable due to its fluctuating nature. We therefore carry out power production estimation by two different approaches; forecast from support vector as well as estimation using the theoretical method in order to compare and determine the best of the two methods The production planning entails determining the power generated, number of panels required and the quantity of energy stored during each period. However, all solar panels are assumed to be the same with same ratings, therefore, the power output is expected to be the same. Subsequently, a preventive maintenance schedule is developed consisting of perfect maintenance where minimal repair is considered at failure. More so, the maintenance is planned over a finite horizon which helps specify the total economic implication associated with solar energy power plant production.

3 MATHEMATICAL MODELLING

This section deals with problem formulation for production planning and maintenance planning. Before we proceed with the modelling, first we present the notation as follows:

Notations

a	Weibull shape parameters
AEC	Available energy capacity of battery
c	Weibull scale parameters
C_b	Battery capacity
C_p	Cost of production
C_s	Cost of storage
C_l	Cost of shortage
C_{bat}	Cost of purchasing battery
C_{pv}	Cost of purchasing PV panel
C_{sell}	Cost of selling 1 kW
DOD	Depth of discharge
H	Operation horizon

(continued)

Continued

H_{system}	Cost index of the system
H_{cap}	Capital cost of the system
H_{ope}	Operational cost of the system
k	Production period index (k = 1, 2, ... H)
N	Number of preventive maintenance actions
N_{PV}	Number of PV panels
P_b	Power stored in the battery
P_g	Power generated
P_k	Power generation during period k
P_L	Power load demand
P_S	Power from solar energy
$P_{\max}$	Maximum power production level over the horizon
SOC	State of charge of battery
TC_b	Total battery capacity of the system
V_b	Voltage capacity of battery
Δt	Duration of each production period
$\varphi(N)$	Number of Average failures
θ	Service level rate
λ_n	Nominal Failure rate
λ_{pv}	Panel Failure rate
μ_P	Preventive maintenance time
μ_c	Corrective maintenance time

3.1 *Power estimation using SVR*

A Support Vector (SV) is a nonlinear kernel-based machine learning technique used for regression as well as classification. It consists of creating or mapping training data into hyper-planes to gaudily discriminate predictions from training data in the feature space as shown in Figure 2 which deals efficiently with nonlinear regression problem.

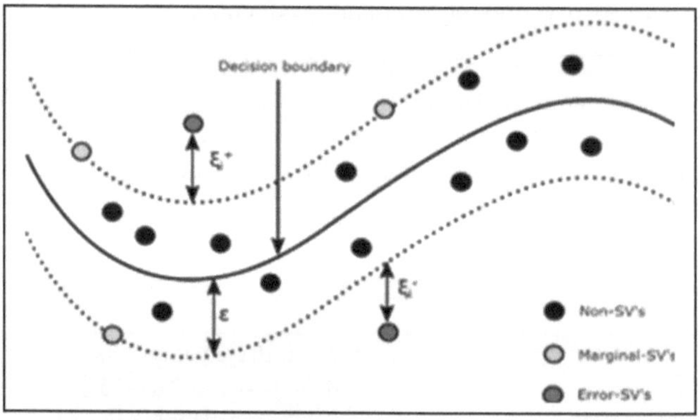

Figure 2. An SVR hyperplane representation.

Support Vector Regression approximates the predictions using a function f that has a fixed deviation threshold from actual targets [13]. The function is in the following form:

$$f(x) = w\varphi(x) + b \tag{1}$$

Where φ is the activation function, x is an input data point.

However, the solution to this function may not be feasible sometimes, therefore, to accommodate for errors, slack variables $(\xi_i + \xi_i^*)$ are introduced. This helps to cope with infeasible constraints of the optimization problem [14]. The optimization problem is then remodeled as

$$\text{minimize } P_{Pv} = \frac{1}{2}\|w\|^2 + C\sum_{i=1}^{l}(\xi_i + \xi_i^*) \tag{2}$$

$$\text{Subject to}\begin{cases} y_i - \varphi(x_i) - b \leq \epsilon + \xi_i \\ \varphi(x_i) + b - y_i \leq \epsilon + \xi_i \\ \xi_i, \xi_i^* \geq 0 \end{cases}$$

Where y_i is the ith target, n is the of training samples, ϵ is the deviation threshold of the function f, C is the penalty parameter and ξ_i is the slack error that guarantees the solution by coping the in-feasible constraints [15]. The kernel acts as activation power which makes the model exhibit good prediction ability. There are four types of activation (kernel) functions used by SVR and these are:

(i) Linear function

$$k(x_i.x_j) = x_i^T, x_j + c$$

(ii) Radial Basic Function (RBF)

$$k(x_i.x_j) = e^{\left(-\gamma\|x_i - x_j\|^2\right)}$$

(iii) Hyperbolic Tangent Function

$$k(x_i.x_j) = \tanh\left(\alpha x_i^T x_j + c\right)$$

(iv) Polynomial Function

$$k(x_i.x_j) = (x_i.x_j + c)^d$$

However, for the purpose of this work, the Radial Basic Function (RBF) is selected because it fits well with the data.

3.2 *Theoretical technique of power generation*

Due to the intermittent nature of solar PV and its dependent on climatic condition for power generation, the solar power output from a panel is modelled based on temperature and solar irradiance, thus obtained by the following equation;

$$P_S(G, \Delta T) = k_1 \times A_S \times G \times (1 - k_T \Delta T) \tag{3}$$

Nonetheless, due to other insolation phenomena such as dust and cloud cover, solar irradiance is represented as random variable [16] characterized by beta distribution function with α *and* β parameters expressed in the following form;

$$f(G) = \frac{\Gamma(\alpha + \beta)}{(\Gamma\alpha)\Gamma(\beta)}\left(\frac{G}{G_{\max}}\right)^{\alpha - 1}\left(1 - \frac{G}{G_{\max}}\right)^{\beta - 1} \tag{4}$$

In a PV system, a PV panel consists of several solar cells as described connected in series, therefore, the total obtainable power output from a panel could be expressed as

$$P(S) = \int_0^{G_{max}} \left(k_1 \times A_S \times G \times (1 - k_T \Delta T) \times \frac{\Gamma(\alpha+\beta)}{(\Gamma\alpha)\Gamma(\beta)} \times \left(\frac{G}{G_{max}}\right)^{\alpha-1} \times \left(1 - \frac{G}{G_{max}}\right)^{\beta-1} \right) d(G) \times \Delta t \quad (5)$$

G is the solar radiation for each generating period, G_{max} is the maximum solar radiation over the production horizon in W/m^2, k_1 is the PV panel generating efficiency, A_S is the panel area in m^2, k_T is the cell temperature coefficient, $\Delta T = T_c - T_{cref}$ representing the error between the cell temperature and the chosen reference cell temperature in °C, α *and* β are the beta function parameters.

3.3 *Battery sizing and service rate modelling*

The battery storage behavior is recognized by its state of charge (SOC). The SOC depends on the past data and the power in the battery system (the difference between the generated power $P_G(t)$ and the load power demand $P_L(t)$) at every time instance t. Therefore,

$$SOC(t) = SOC(t-1) + \frac{P_G(t) - P_L(t)}{V_b.TC_b} \quad (6)$$

Given that

$$P_G(t) = \sum_{t=1}^{H} N_{pv} \times P_S(t) \quad (7)$$

Where V_b *and* C_b are the battery voltage and capacity respectively.

To ensure good performance of the battery bank and availability of energy in the battery at all time, a cycle invariance criterion is introduced such that, the battery SOC returns to its initial state $SOC(t_0)$ at the end of each production period. The problem therefore is modelled as in equation 8 and 9 below such that the $AEC(t)$ at each period is identified by the SOC variation

$$SOC(t) = SOC(t-1) \quad (8)$$

$$AEC(t) = SOC(t).C_b \quad (9)$$

And in order to determine the total battery bank capacity (TC_b) over a finite time horizon H, the TC_b must satisfy the following condition in order not that the battery capacity does not become negative

$$TC_b \geq AEC_{max} - AEC_{min} \quad (10)$$

AEC_{max} *and* AEC_{min} are the maximum and minimum available energy capacity in the battery.

To ensure power security regardless of the solar irradiance fluctuations, a service level $S(k)$ is imposed for each period. This service level is the probability that failure of the panel does not occur and must be greater than a minimum threshold (θ). The probability is often represented as:

$$prob(S(k) \geq 0) \geq \theta \quad (11)$$

3.4 *Optimization problem*

First, we formulate the production problem to minimize the production cost of the system by determining the optimal number of PV panels and the storage capacity sufficient enough to satisfy the demand. The optimization problem is presented thus:

$$\text{Min } H_{system}(N_{PV}, C_b)$$

Subject to:

$$N_{pv} * Ppv(t) + V_b * SOC(t) \geq P_L(t)$$

$$AEC_{\min} \leq AEC(t) \leq AEC_{\max}$$

$$N_{pvm} \leq N_{pv}(t) \leq N_{pvM}$$

The system cost H_{system} is expressed as the sum of the initial capital cost and the operational cost represented in equation 12.

$$H_{system} = H_{cap} + H_{ope} \tag{12}$$

H_{cap} *and* H_{ope} are the initial capital cost and operational and maintenance cost respectively. They are mathematically evaluated in the following equations:

$$H_{cap} = \sum_{t=1}^{H} N_{pv}(t) * C_{pv} + C_b(t) * C_{bat} \tag{13}$$

$$H_{ope} = C_p + C_s + C_l \tag{14}$$

Where C_p, C_s *and* C_l are the cost of production, cost of storage and cost of shortage respectively

On another hand, a preventive maintenance (PM) planning aimed to determine the optimal number of PM actions required. The maintenance planning of the system is formulated taking into account the cost of loss of production (C_{post}) evaluated as follows:

$$\min C = \sum_{k=1}^{H} (C_p.P_k + C_s.S_k + C_l.S_l) + C_{post} \tag{15}$$

with

$$C_{post} = C_{sell} \times \frac{\sum_{k=1}^{H} P_k \times (\mu_c \times \bar{Y}(P_k, N) + \mu_P \times N)}{H} \tag{16}$$

And

$$\bar{Y}(P_k, N) = \sum_{j=1}^{\frac{H/T}{\Delta t}} \left[\sum_{i=1}^{\frac{T}{\Delta t}} \int_0^{\Delta t} \frac{P_{i,T}}{P_{\max}}.\lambda_n(t).dt + \sum_{i=2}^{\frac{T}{\Delta t}} \left(\frac{T}{\Delta t} - (i-1) \right).\left(\frac{P_{i-1,T}}{P_{\max}}.\lambda_n(\Delta t).\Delta t \right) \right]$$
$$+ \left(\sum_{i=\frac{H/T}{\Delta t}.T+1}^{H} \int_0^{\Delta t} \frac{P_i}{P_{\max}}.\lambda_n(t).dt + \sum_{i=\frac{H/T}{\Delta t}.T+2}^{H} \left(H - \frac{H/T}{\Delta t}.T \right) - \left(i - \frac{H/T}{\Delta t}.T - 1 \right).\frac{P_i}{P_{\max}}.\lambda_n(\Delta t).\Delta t \right) \tag{17}$$

subject to

$$0 < P_k < P_{\max}$$

$$Prob(S(k) \geq 0) \geq \theta$$

The priority of this work is to obtain the optimal preventive maintenance planning presenting the optimal number of PM actions (N*) and optimal periodicity (T*) between two (2) successive PM times with minimal cost. The relationship between N* and T* is expressed in the equation below:

$$T^* = \frac{H}{N^*} \tag{18}$$

The preventive maintenance is planned to consider minimal repair at failure, which returns the machine to a condition same as its previous condition before breakdown. Hence, substantial time resources are allocated to this event. Therefore, failure rate and degradation factor of the PV system is considered as parametric relationship between nominal failure rate ($\lambda_n(t)$) and energy production function $\left(f(P_k) = \frac{P_k}{P_{\max}}\right)$ as expressed in equation 17. The nominal failure rate $\lambda_n(t)$ is modelled with Weibull distribution as follows

$$\lambda_n(t) = \frac{a}{c}\left(\frac{t}{c}\right)^{a-1} \tag{19}$$

4 NUMERICAL EXAMPLE

An algorithm was built on the above formulated problem for production and maintenance planning optimization for a PV power plant. A case study of Sokoto solar power plant was chosen for this study concentrating on Yabo town supply. Sokoto is a state in northern Nigeria on $13° \, 44' N$ and $5° \, 02' E$. It has an average daytime temperature of 40°C and average daily sunshine hours of 12 hours 53 minutes. It also has an average daily solar irradiance of 258.51W/m^2 all year round. We also obtained a one-year (2020) data from the location and implemented it for the proposed developed algorithm. Table 2 presents the production and maintenance planning parameters for this example. It is worthy of mention that the power generation inputs are the temperature and solar irradiance. The average load demand for this case study is presented on Figure 3 and the average solar irradiance is presented on Figure 4.

Considering the generated excess quantity of energy from the analysis, a 125 Amp-hour (Ah) Lithium-Iron (Li-Fe) battery was selected for this work with the characteristic of PV panel presented on Table 1. Accordingly, the parameters associated with the production and maintenance as well as the Weibull are also presented on Table 2.

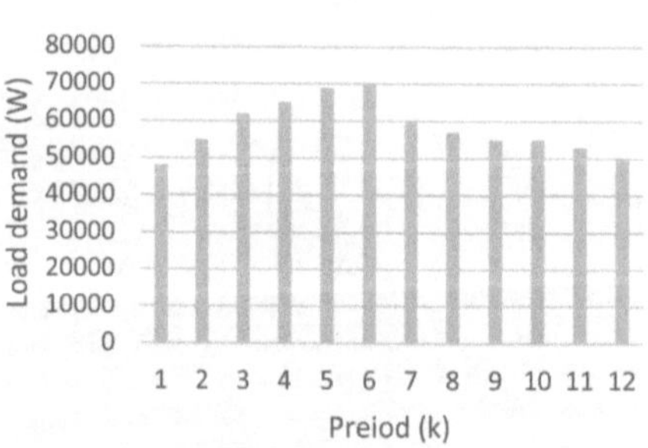

Figure 3. Average load demand of Yabo town.

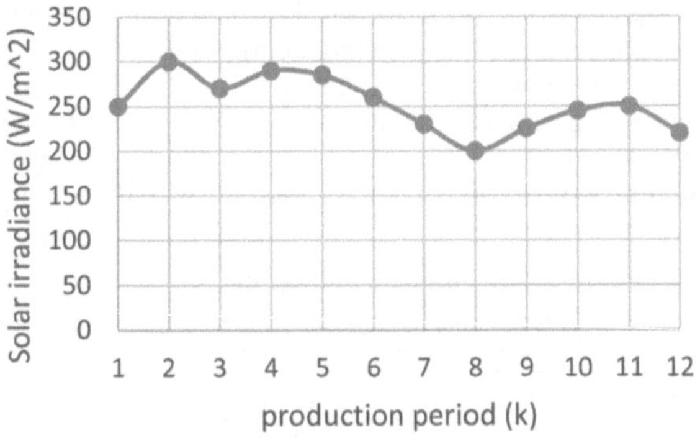

Figure 4. Average solar irradiance for Sokoto State, Nigeria.

Table 1. PV panel characteristics.

Parameter	Value
Nominal Power (W)	215
Max power voltage (V)	37.18
Max power current (A)	9.28
Module efficiency (%)	17.78
Operating temperature (°C)	−40 to +85
NOCT (°C)	45 ± 2
Module dimensions (mm^2)	1956 x 992

Table 2. Production and maintenance parameters.

Parameter	Value
H	12
Δt	1
Cp (NGN/kW)	12
Cs (NGN/kW)	15
C_l (NGN/kW)	25
C_{sell} (NGN/kW)	50
μ_c (Time-units)	10
μ_P (Time-units)	0.3
θ (%)	90
a	2
c	10

5 RESULTS AND DISCUSSION

The power production plan is presented on Tables 3 and 4 for SVR and theoretical techniques respectively. Optimal production cost for each period together with the optimal number of panels and battery capacity required are also presented. For each of the periods, the quantity of power generated is also stated. This helps the manager plan for forecasted uncertainties such as cloud cover and dusts on the solar panels. From Table 3, high power generation is estimated for periods 3 to 5 with the SVR technique. With the theoretical technique, the power estimation is high in period 6 to 9 with average of 89 kW.

The cost of generating electricity is assessed to be 15.59 million Nigerian Naira with the SVR method. The theoretical technique presents about 18.05 million Nigerian Naira. This shows that estimation using the machine learning yields optimal cost and reduces material and equipment wastage by minimizing the number of panels and the number of batteries to be used. It is also able to minimize the shortage so as to curtail sourcing from the grid. Table 5 present the quantity of energy stored per period for both techniques.

Table 3. Optimal production plan using the SVR approach.

Total cost = 15.5924 x10^6 (NGN)				
Period	*Optimal cost x10^6 (NGN)*	*Number of panels (PV)*	*Battery capacity (x10^4Ah)*	*Power generated (x10^5W)*
1	1.3005	517	0.5378	5.2562
2	1.3006	522	0.6378	5.9915
3	1.3177	672	1.4780	6.4794
4	1.3329	621	1.9200	6.3193
5	1.3444	601	2.5800	6.0545
6	1.3743	599	4.2800	5.9110
7	1.3740	558	3.8300	5.3620
8	1.3812	503	3.6500	4.9972
9	1.3282	582	2.2800	5.4541
10	1.3809	618	3.3700	6.1011
11	1.3783	583	3.5200	5.9129
12	1.3805	529	3.6300	5.6607

Table 4. Optimal production plan with the theoretical technique.

Total cost = 18.05 x10^6 (NGN)				
Period (k)	*Optimal cost x10^6 (NGN)*	*Number of panels (PV)*	*Battery capacity (x10^4Ah)*	*Generated power (x10^5W)*
1	1.4880	634	0.5183	1.174
2	1.4930	660	2.8951	2.277
3	1.4931	710	2.9496	3.263
4	1.4931	770	2.9496	3.274
5	1.4949	741	2.7882	3.089
6	1.5374	897	2.8843	9.869
7	1.5005	856	3.7292	8.961
8	1.5119	837	3.3224	8.213
9	1.5006	752	3.2679	1.031
10	1.5050	787	3.9722	6.232
11	1.5223	767	3.7625	7.723
12	1.5102	647	2.7383	1.184

After successfully planning the production, an integrated joint maintenance optimization was then developed to determine the optimal number of PM actions (N*) and the optimal periodicity (T*). The minimal total production cost yielding the optimal number of PM actions for both techniques are presented on Table 6. It is worthy of mention that all panels are similar and identical. The SVR has the best optimal lifecycle cost at 15.9186 x10^6 NGN requiring 2 maintenance sessions every 6 months. For the theoretical techniques, 3 PM sessions are required expected to be performed after every 4 months which costs about

Table 5. Quantity of energy in the battery.

	Quantity of energy stored	
Period (k)	*SVM technique (W)*	*Theoretical Technique (W)*
1	4562	892
2	4914	2829
3	2793	4976
4	1806	6437
5	4454	6052
6	1089	7223
7	6380	5450
8	5027	3450
9	4582	3159
10	4011	1099
11	6129	1817
12	6607	1270

Table 6. Optimal number of PM sessions for solar energy.

	SVM	Theoretical
N*	2	3
Cost (NGN)	15.9186×10^6	18.4406×10^6

18.4406×10^6 NGN. It was also observed that the processing time for the SVR is lesser than that of the theoretical method.

6 CONCLUSION

We have successfully developed an integrated production and maintenance model in MATLAB 2021b. Power estimation was performed by both machine learning (SVR) and theoretical techniques to assess the power output for the selected site. Result from these models were integrated into maintenance model in order to obtain the minimal number of PM actions (N*) and cost. A perfect maintenance with minimal repair was proposed at the time for PM session. Optimal number of preventive maintenances with the machine learning proved to be more economical having minimal cost at a less processing time. Unlike the theoretical method which is based on random value generation for the input data, the machine learning technique is better because it works with real-life, real-time data. For further study, we will study the influence of the environment on the system by considering other meteorological factors as the input parameters while also developing other strategies of maintenance.

REFERENCES

[1] Solar Power Europe, (2020). Global Market Outlook for Solar Power 2020–2024.

[2] Zhou W., Lou C., Li Z., Lu L., and Yang H., (Feb. 2010). Current status of research on optimum sizing of stand-alone hybrid solar–wind power generation systems, *Applied Energy*, vol. 87, no. 2, pp. 380–389, doi: 10.1016/j.apenergy.2009.08.012.

[3] Borowy B. S. and Salameh Z. M., (Jun. 1996). Methodology for optimally sizing the combination of a battery bank and PV array in a wind/PV hybrid system, *IEEE Trans. On energy Conversion*, vol. 11, no. 2, pp. 367–375, doi: 10.1109/60.507648.

[4] Diaf S., Diaf D., Belhamel M., Haddadi M., and Louche A., (Nov. 2007). A methodology for optimal sizing of autonomous hybrid PV/wind system, *Energy Policy*, vol. 35, no. 11, pp. 5708–5718, doi: 10.1016/j.enpol.2007.06.020.

[5] Mellit A., Kalogirou S. A., Hontoria L., and Shaari S., (Feb. 2009). Artificial intelligence techniques for sizing photovoltaic systems: A review', *Renewable and Sustainable Energy Reviews*, vol. 13, no. 2, pp. 406–419, doi: 10.1016/j.rser.2008.01.006.

[6] Sa'ad A., Nyoungue A. C., and Hajej Z., (Sep. 2021). Improved preventive maintenance scheduling for a photovoltaic plant under environmental constraints, *Sustainability*, vol. 13, no. 18, p. 10472, doi: 10.3390/su131810472.

[7] Barlow R. and Hunter L., (Feb. 1960). Optimum preventive maintenance policies, *Operations Research*, vol. 8, no. 1, pp. 90–100, doi: 10.1287/opre.8.1.90.

[8] Raza A. and Ulansky V., (Oct. 2019). Optimal Preventive Maintenance of Wind Turbine Components with Imperfect Continuous Condition Monitoring, *Energies*, vol. 12, no. 19, p. 3801, doi: 10.3390/en12193801.

[9] Hameed Z. and Wang K., (Jun. 2012). Development of optimal maintenance strategies for offshore wind turbine by using artificial neural network, *Wind Engineering*, vol. 36, no. 3, pp. 353–364, doi: 10.1260/0309-524X.36.3.353.

[10] Nourelfath M., Nahas N., and Ben-Daya M., (Apr. 2016). Integrated preventive maintenance and production decisions for imperfect processes, *Reliability Engineering & System Safety*, vol. 148, pp. 21–31, doi: 10.1016/j.ress.2015.11.015.

[11] Yang L., Zhao Y., Ma X., and Qiu Q., (Dec. 2018). An optimal inspection and replacement policy for a two-unit system, *Proceedings of the Institution of Mechanical Engineers, Part O: Journal of Risk and Reliability*, vol. 232, no. 6, pp. 766–776, doi: 10.1177/1748006X18761488.

[12] Merizalde Y., Hernández-Callejo L., Duque-Perez O., and Alonso-Gómez V., (Jan. 2019). Maintenance models applied to wind turbines. A comprehensive overview, *Energies*, vol. 12, no. 2, p. 225, doi: 10.3390/en12020225.

[13] Cortes C. and Vapnik V., (Sep. 1995). Support-vector networks, *Mach Learn*, vol. 20, no. 3, pp. 273–297, doi: 10.1007/BF00994018.

[14] Antonanzas J., Osorio N., Escobar R., Urraca R., Martinez-de-Pison F. J., and Antonanzas-Torres F., (Oct. 2016). Review of photovoltaic power forecasting, *Solar Energy*, vol. 136, pp. 78–111, doi: 10.1016/j.solener.2016.06.069.

[15] Vapnik V. N., (1995). *The Nature of Statistical Learning Theory*. New York, NY: Springer New York. doi: 10.1007/978-1-4757-2440-0.

[16] Li J., Wei W., and Xiang J., (Dec. 2012). A simple sizing algorithm for stand-alone pv/wind/battery hybrid microgrids, *Energies*, vol. 5, no. 12, pp. 5307–5323, doi: 10.3390/en5125307.

Digital technologies for sustainable development

Innovation and Technological Advances for Sustainability – Al-Naemi et al. (Eds)

Backstepping control based on neural network estimation

Walid Alqaisi
School of Engineering Technology, The University of Doha for Science and Technology, Jelaiah St., Doha, Qatar

Claude Ziad El-Bayeh
Canada Excellence Research Chairs Team, Concordia University, Montreal, QC, Canada

ABSTRACT: This paper is implementing a backstepping controller to a quadrotor that is subject to unknown disturbances and uncertainty in dynamics. Estimation and approximation of unkowns and uncertainties are performed by using Radial Base Function Neural Network (RBFNN). Along with a backstepping controller, they provide robustness to the robotic systemdespite of the presence od uncertainties. The RBFNN's output layer is utilized as an estimator then compensation eliminates the undesired effect occurs due to uncertainties. As a result, faster error convergence is achievable. A Lyapunov function was used to analyze the closed loop system, and Matlab/Simulink was used to evaluate the system performance. The demonstrated results prove the efficiency of the proposed approach.

Keywords: Radial Base Function, neural network, Backstepping control, quadrotor, Unmanned Arial vehicles. Adaptive control system

1 INTRODUCTION

In the last decades, robotics technology has been developed remarkably sin terms of control systems and intelligence. Robot systems suffer from Nonlinearity, uncertain parameters, unmodeled and coupled dynamics in addition to external disturbances. Such challenges in robotic control systems are large areas of research. Many researchers have developed nonlinear controllers to overcome nonlinearity problems. For example, a terminal sliding mode is used to stabilize a control system as in [1], sliding mode control is utilized in many robotics applications [2], a sliding mode with second order is implemented [3], feedback linearization controllers were introduced in [4], super-twisting is used in [5,6]. Although the aforementioned controllers are successful, their accuracy depends on how accurate are the mathematical system model and system parameters. Additionally, changes in mass and inertia also reduce the efficiency of the controller's performance. To attain more robustness, researchers created adaptive controllers. To tackle unmodeled dynamics, inaccurate parameters, and external disturbances [7]. An adaptive backstepping controller is used [8], and a nonlinear time-varying systems was controlled by an adaptive control [9]. Adaptive controllers requires a predefined model structure which his not the case in most robots. To tackle this problem researchers designed observers as in [10,11]. An observer of three loops was designed to estimate and compensate unknowns [12]. Disturbance compensation is utilized in [13].

Radial Base Function Neural Networks (RBFNNs) are a different strategy for providing estimation of nonlinearity and uncertainty of robotic systems. The development of a controller incorporating neural network feedback was performed [14,15]. On the other hand, to improve the effectiveness of the controllers, the singularity issue was studied in many research projects. Super-Twisting Non-Singular Fast Terminal Control is built in [16].

DOI: 10.1201/9781003496724-24

The backstepping control is robust for robotic structures with multiple coupled dynamics and unkowns [17,18]. Backstepping controllers provide a systematic framework for error tracking. The effectiveness of the backstepping depends on the exact modeling of the mechanical system.

A promossing solution to such problems is the use of Neural Networks (NNs) to estimate uncertain nonlinear functions where the neural network weights are tuned online without requiring a learning phase. However, further research on RBFNNs is required for the high computational cost and approximation lag.

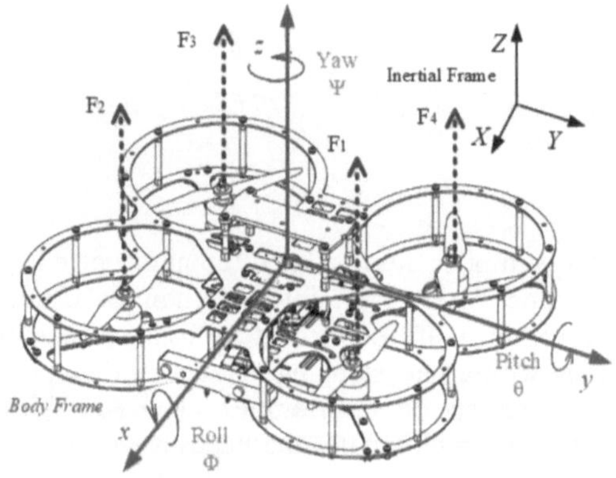

Figure 1. Quadrotor configuration.

In this article, the estimation ability of RBF Neural Networks and the effectiveness of backstepping control motivate us to implement such techniques to a quadrotor drone. In order to ensure the stability of the overall system, Lyapunov stability was examined. The RBFNN's response lag and high computational cost, both are eliminated by using the node number as a tuning variable, then selecting the minimum possible node number that achieves the desired stability. We used a quadrotor drone model to test the backstepping controller design with RBF approximation.

The main important highlight of the proposed design can be summarized as:

- *The RBF system provides an accurate estimatiom for imperfect modeling and varying parameters without the need for exact system modeling.*
- *The RBF system performs constant estimation, upper bound of model uncertainty is not necessary.*
- *The entire design of the Radial Base Functin and the Backstepping controller guarantees finite time error convergence and tracking.*

This paper is prepared as follows, introduction in Part I, the quadrotor model in Part II. The problem formulation and the design objective are in Part III, Section IV explains the RBFNN. Part V describes control design, Part VI shows simulation results. Part VII briefed the conclusion.

2 QUADROTOR DYNAMICS

The proposed controller is to be applied to a quadrotor drone. The quadrotor has four rotors, and variations in rotor velocities move the quadrotor in different directions. Opposite rotors direction is similar to cancel the gyroscopic effect. The quadrotor goes through many forces, moments, and gyroscopic effects, Figure 1, which have to be considered for developing the dynamic model [19]. Based upon Newton Euler formulas, the quadrotor model is given as [20,21]:

$$
\begin{aligned}
\ddot{x}_1 &= \ddot{\phi} = \dot{\theta}\dot{\psi}\frac{(i_{yy}-i_{zz})}{i_{xx}} - \dot{\theta}\omega_r\frac{j_r}{i_{xx}} - \frac{k_{fax}}{i_{xx}}\dot{\phi}^2 + \frac{1}{i_{xx}}u_2 \\
\ddot{x}_2 &= \ddot{\theta} = \dot{\phi}\dot{\psi}\frac{(i_{zz}-i_{xx})}{i_{yy}} + \dot{\phi}\omega_r\frac{j_r}{i_{yy}} - \frac{k_{fay}}{i_{yy}}\dot{\theta}^2 + \frac{1}{i_{yy}}u_3 \\
\ddot{x}_3 &= \ddot{\psi} = \dot{\theta}\dot{\phi}\frac{(i_{xx}-i_{yy})}{i_{zz}} - \frac{k_{faz}}{i_{zz}}\dot{\psi}^2 + \frac{1}{i_{zz}}u_4 \\
\ddot{x}_4 &= \ddot{x} = \frac{1}{m}(\sin\theta\cos\phi\cos\psi + \sin\phi\sin\psi)u_1 - \frac{k_{ftx}}{m}\dot{x} \\
\ddot{x}_5 &= \ddot{y} = \frac{1}{m}(\sin\theta\cos\phi\sin\psi - \sin\phi\cos\psi)u_1 - \frac{k_{fty}}{m}\dot{y} \\
\ddot{x}_6 &= \ddot{z} = \frac{-1}{m}(\cos\phi\,\cos\,\theta)u_1 + g - \frac{k_{ftz}}{m}\dot{z}
\end{aligned} \tag{1}
$$

The variables are described as follows:

u_1, u_2, u_3 *and* u_4	: *Actual control input.*
u_x, u_y	: *Axillary control inputs:* $u_x = (\sin\theta\cos\phi\cos\psi + \sin\phi\sin\psi)$ $u_y = (\sin\theta\cos\phi\sin\psi - \sin\phi\cos\psi)$
x, y, z	: *The position in three coordinates respectively.*
ϕ, θ, *and* ψ	: *Rolling, pitching, and yawing angles.*
ϕ_d, θ_d *and* ψ_d	: *The desired angles.*
i_{xx}, i_{yy} *and* i_{zz}	: *Moments of inertia.*
$k_{fax}, k_{fay}, k_{faz}$	: *Aerodynamic friction coefficients.*
$k_{ftx}, k_{fty}, k_{ftz}$	: *The translation drag coefficients.*
m	: *Quadrotor mass.*
g	: *Gravity force.*
ω_i	: *Rotors' angular velocity* $(i = 1, 2, 3, 4)$.
ω_r	: *Angular equilibrium* $\omega_r = -\omega_1 + \omega_2 - \omega_3 + \omega_4$
d_c	: *Drag coefficient.*
B	: *Thrust coefficient.*
j_r	: *Rotor inertia.*
l_e	: *Length of the moment arm.*

We use axillary control inputs u_x, u_y to find the desired roll and pitch ϕ_d and θ_d which will be controlled by u_2 and u_3. Desired angles, roll and pitch [19,22]:

$$
\begin{aligned}
\phi_d &= \sin^{-1}(u_x\sin\psi_d - u_y\cos\psi_d) \\
\theta_d &= \sin^{-1}\left(\frac{u_x\cos\psi_d + u_y\sin\psi_d}{\cos\phi_d}\right)
\end{aligned} \tag{2}
$$

3 PROBLEM FORMULATION

Considering the following nonlinear second-order robotic system described as:

$$
\begin{cases}
\dot{x}_1 = x_2 \\
\dot{x}_2 = f(x,t) + +g(x,t)u + d(t)
\end{cases} \tag{3}
$$

Where x is the state variable, f(x, t) and $g(x,t)$ are nonlinear and uncertain functions and $g(x,t) \neq 0$, $d(t)$ is the external disturbances. In real-time robotic systems, $f(x,t), g(x,t)$ and $d(t)$ cannot be precisely identified. Accurate approximation and estimation of the unknowns helps the controller to improve the entire performance.

3.1 *Objective*

- *Using the RBF-neural network to estimate uncertain dynamics and unknown disturbance to design a robust backstepping tracking control system.*

4 RADIAL BASIS FUNCTION NEURAL NETWORK

The Radial Base Function (RBF) is used to enhance control performance in various robot applications [23]. RBFNN can be used to learn the unknown upper bound of model uncertainty. They can also be used to estimate nonlinear continuous functions. Their objective is to provide estimation for uncertain modelling [14]. Three layers, Figure 2, make up the RBF: an input layer, an output layer, and a hidden layer. The input to the network is chosen as [24]:

$$\Gamma = \left[e^T \dot{e}^T\right]^T \tag{4}$$

The error and its derivative are denoted as $e = x_d - x$ and $\dot{e} = \dot{x}_d - \dot{x}$. Where, $x_d, \dot{x}_d$ are the desired trajectory and velocity respectively. $\phi\ (\Gamma(t))$ is the Gaussian radial basis function used for each node.

The hidden layer, is a linear combiner that maps nonlinearity into a new space. It performs a nonlinear transform of the input and output layer. RBF algorithm is given as [24]:

$$\begin{gathered}\varphi_i = \exp\left(-\frac{\|\Gamma - R_i\|^2}{b_i^2}\right) i = 1, 2, ..n \\ Q = \boldsymbol{W}^T\phi + \varepsilon\end{gathered} \tag{5}$$

Where,

$\phi = [\varphi_1\varphi_2\cdots\varphi_n]^T$	*Gaussian function*
n	*Hidden layer number of nodes*
R_i	*ith node center vector*
b_i	*Smoothness control of interpolating function.*
E	*Approximation error*
Γ	*Input vector of neural network.*
Q	*Desired estimation*
$\boldsymbol{W}$	*Weight matrix*

The RBF neural network's approximation function is:

$$\widehat{Q}(x(t)) = \widehat{\boldsymbol{W}}^T\phi \tag{6}$$

Selecting an arbitrary positive initial value for the weight matrix $\widehat{\boldsymbol{W}}$. The network output can provide accurate approximation; thus, error tracking can converge to zero asymptotically.

The desired estimation of system (1) is the uncertain functions and the disturbance f, g, and d which can be approximated using the NNRBF as:

$$\begin{cases} f_\delta = \boldsymbol{W}_1^T\phi_1 + \varepsilon_1 \\ g = \boldsymbol{W}_2^T\phi_2 + \varepsilon_2 \end{cases} \tag{7}$$

Where $f_\delta = f + d$, ε_1 and ε_2 are approximation errors.

<u>*Assumption:*</u> approximation errors ε_1 and ε_2 are assumed to be bounded by a positive number ε_N, i.e. $\|\varepsilon_1 + \varepsilon_2\| \le \varepsilon_N$.

The output estimation of the RBF-NN is given as:

$$\begin{cases} \tilde{f}_\delta = \tilde{W}_1^T \phi_1 \\ \tilde{g} \ = \tilde{W}_2^T \phi_2 \end{cases} \tag{8}$$

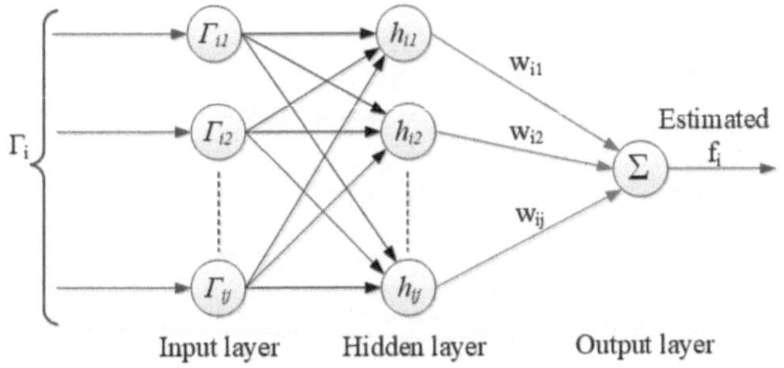

Figure 2. Neural network.

5 CONTROL DESIGN

Consider the nonlinear second-order robotic model described in (3). Backstepping controller design for the given system is achieved as described below. At first, we define the error and its derivative as e_1 and $\dot{e}_1$:

$$\begin{aligned} e_1 &= x_1 - x_{1d} \\ \dot{e}_1 &= \dot{x}_1 - \dot{x}_{1d} \end{aligned} \tag{9}$$

Where x_{1d} is the desired trajectory. To achieve $e_1 \to 0$, a Lyapunov function is sutilised as:

$$v_1 = \frac{1}{2} e_1^2 \tag{10}$$

Then,

$$\dot{v}_1 = e_1 \dot{e}_1 = e_1(\dot{x}_1 - \dot{x}_{1d}) = e_1(x_2 - \dot{x}_{1d}) \tag{11}$$

We choose $(x_2 - \dot{x}_{1d}) = -k_1 e_1$ where $k_1 > 0$, then we have:

$$\dot{v}_1 = -k_1 e_1^2 \tag{12}$$

The stabilization is obtained by introducing a virtual control input x_2 chosen as:

$$x_{2d} = \dot{x}_{1d} - k_1 e_1 \tag{13}$$

A second error is introduced for x_2 and x_{2d}:

$$e_2 = x_2 - x_{2d} \tag{14}$$

Then,

$$\dot{e}_2 = \dot{x}_2 - \dot{x}_{2d} = f(x,t) + g(x,t)u - \dot{x}_{2d} \tag{15}$$

To have a finite time errors convergence, a second Lyapunov analysis is required:

$$V_2 = V_1 + \frac{1}{2}e_2^2 = \frac{1}{2}\left(e_1^2 + e_2^2\right) \tag{16}$$

Then,

$$\begin{aligned} \dot{V}_2 &= e_1\dot{e}_1 + e_2\dot{e}_2 \\ e_1\dot{e}_1 &= e_1(\dot{x}_1 - \dot{x}_{1d}) = e_1(x2 - \dot{x}_{1d}) \\ &= e_1(x2d + e2 - \dot{x}_{1d}) = e_1(x2d + e2 - \dot{x}_{1d}) \\ &= -k_1 e_1^2 + e_1 e_2 \\ e_2\dot{e}_2 &= e_2(f(x,t) + g(x,t)u + d(t) - \dot{x}_{2d}) \end{aligned} \tag{17}$$

Then

$$\dot{V}_2 = -k_1 e_1^2 + e_1 e_2 + e_2(f(x,t) + g(x,t)u + d(t) - \dot{x}_{2d}) \tag{18}$$

If the nonlinear functions $f(x,t)$ and $g(x,t)$ are known exactly, it would be a routine nonlinear controller. The controller is designed to use the estimation provided by the RBF neural network, the control u is selected as:

$$u = \frac{1}{\widehat{g}}\left[-e_1 - \widehat{f}_\delta + \dot{x}_{2d} - k_2 e_2\right] \tag{19}$$

Where, $f(x,t)$ and $g(x,t)$ are denoted as f and g respectively, $\widehat{f}_\delta$ is the estimation of $(f + d(t))$ and $\widehat{g}$ is the estimation of g. Substituting (19) in (18) we find,

$$\begin{aligned} \dot{V}_2 &= -k_1 e_1^2 + e_1 e_2 + e_2\left(\left(f - \widehat{f}_\delta\right) + e_2\,(g - \widehat{g})u\right. \\ \dot{V}_2 &= -k_1 e_1^2 - k_2 e_2^2 + e_2 \tilde{f}_d + e_2 \tilde{g} u \end{aligned} \tag{20}$$

Where $\tilde{f}_\delta = f + d(t) - \widehat{f}_\delta$ and $\tilde{g} = (g - \widehat{g})$ are the estimation errors in the dynamic system. Let the adaptive law as:

$$\dot{\widehat{Z}} = \Gamma h \xi^T - n\Gamma\|\eta\|\widehat{Z} \tag{21}$$

Where Z and $\tilde{Z}$ are given as:

$$Z = \begin{bmatrix} W_1 & 0 \\ 0 & W_2 \end{bmatrix} \tilde{Z} = \begin{bmatrix} \tilde{W}_1 & 0 \\ 0 & \tilde{W}_2 \end{bmatrix}, \tilde{Z} = Z - \widehat{Z}$$

And n is a positive number, Γ is a positive-definite matrix,

$$\Gamma = \begin{bmatrix} \Gamma_1 & 0 \\ 0 & \Gamma_2 \end{bmatrix} \mathit{using}\ \xi = [e_1 e_2]^T \text{ and } V_2 = \frac{1}{2}\xi^T\xi.$$

Lyapunov stability for the backstepping control is selected as follows [25]:

$$V = \frac{1}{2}\boldsymbol{\xi}^T\boldsymbol{\xi} + \frac{1}{2}\mathrm{tr}\left(\tilde{\mathbf{Z}}^{\mathrm{T}}\mathrm{B}^{-1}\tilde{\mathbf{Z}}\right) \tag{22}$$

$$\dot{V} = \boldsymbol{\xi}^T\dot{\boldsymbol{\xi}} + \mathrm{tr}\left(\tilde{\mathbf{Z}}^{\mathrm{T}}\mathrm{B}^{-1}\dot{\tilde{\mathbf{Z}}}\right) \tag{23}$$

Using (7), (8), (18), and (23), it can be found that

$$\dot{V} = -k_1 e_1^2 - k_2 e_2^2 + e_2\left(\tilde{W}_1^T h_1 + \varepsilon_1\right) + e_2\left(\tilde{W}_2^T h_2 + \varepsilon_2\right) + \mathrm{tr}\left(\tilde{\mathbf{Z}}^{\mathrm{T}} \mathrm{B}^{-1} \dot{\tilde{\mathbf{Z}}}\right) \tag{24}$$

Let $\lambda_{\min}$ be the minimum eigenvalues of $K_e = [k_1 k_2]$. Substituting $\dot{\tilde{Z}} = -\dot{\hat{Z}}$ in (21) we find that $\dot{V} < 0$ if the following conditions are met:

$$\|\xi\| > \frac{\varepsilon_N + \frac{n}{4} Z_M^2}{\lambda_{\min}} \; or \; \|\tilde{Z}\|_F > \frac{Z_M}{2} + \sqrt{\frac{Z_M^2}{4} + \frac{\varepsilon_N}{n}} \tag{25}$$

The proof of adaptive law design using RBF networks is further detailed in [26].

6 SIMULATION

In this part, the entire system is tested to verify the accuracy of trajectory tracking. Matlab/Simulink simulation is used for this purpose, parameter values are taken from a real quadrotor, and the trajectory is given as follows:

$$\begin{aligned} x_d &= \cos(t) \\ y_d &= \sin(t) \\ z_d &= 1 \\ \psi_d &= 0 \end{aligned} \tag{26}$$

The RBF estimation function (8) and the backstepping control (19) are used to create control and compensate for unknown dynamics and parameters in (3). Figure 3 displays the desired trajectory as given in (26) and the measured trajectory in three dimentions. Figures 4 and 5 displays the translational and orientational trajectories. The given graphs show the performance of the trajectory tracking system. The control's ability to provide precise estimation, bring the quadrotor to stability, and track the desired position is indicated by the fast convergence and the low values of errors. Figure 6 displays control signals. It is obvious that during the flight, the signals are continuous and never achieved saturation which indicate the capability to apply to a practical system to verify controller's quality and stability.

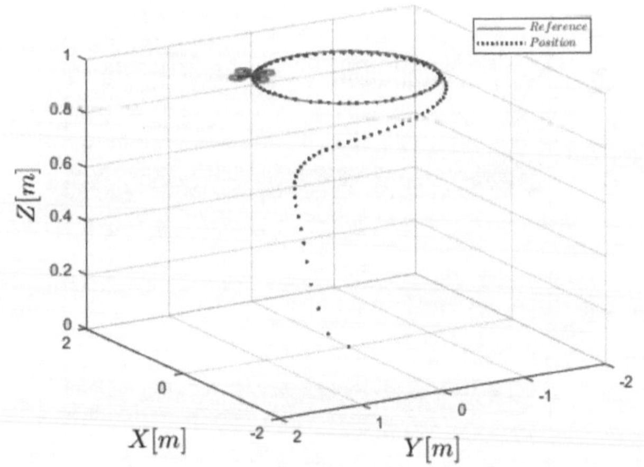

Figure 3. Trajectory tracking.

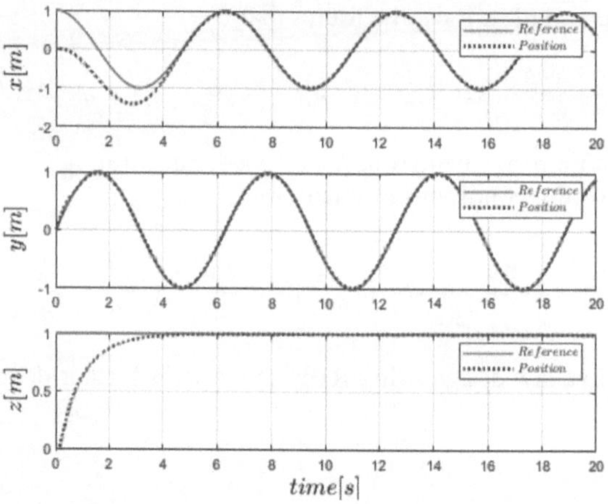

Figure 4. Trajectory tracking.

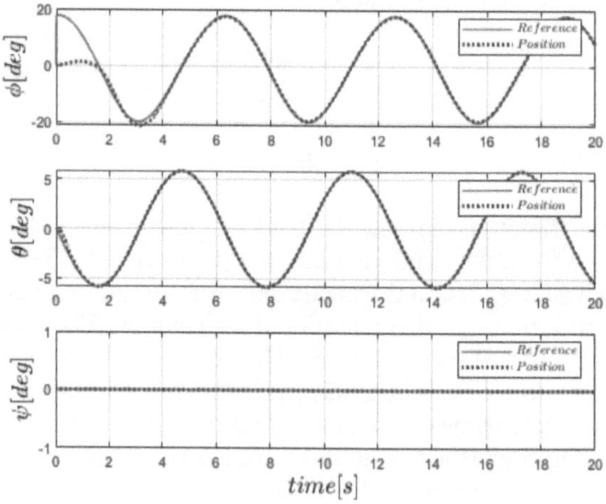

Figure 5. Response of roll, pitch, and yaw angles.

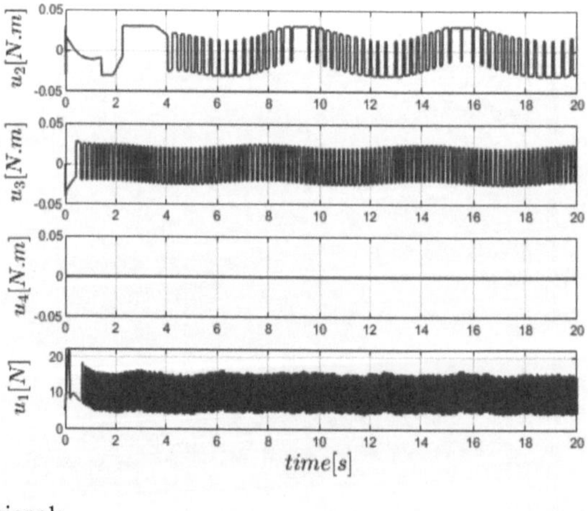

Figure 6. Controller signals.

7 CONCLUSION

This article proposed a design of a backstepping controller with a neural network radial base function, applied to a quadrotor. The coupled dynamics and the hard nonlinearities of the quadrotor make it ideal to test and prove the effective performance of a proposed system. Continuous estimation and compensation of uncertainties lead to quick convergence of errors. Performance and trajectory tracking are improved as a result. Matlab simulation and Lyapunov analysis demonstrated the system's effectiveness as well.

REFERENCES

[1] Mu C. and He H., (2017). Dynamic behavior of terminal sliding mode control, *IEEE Trans. Ind. Electron.*, vol. 65, no. 4, pp. 3480–3490.

[2] Utkin V., Guldner J., and Shi J., (2017). *Sliding Mode Control in Electro-mechanical Systems*. CRC press.

[3] Ding S., Park J. H., and Chen C.-C., (2020). Second-order sliding mode controller design with output constraint, *Automatica*, vol. 112, p. 108704.

[4] Alqaisi W., Brahmi B., Ghommam J., Saad M., and Nerguizian V., (2019). Vision-based leader-follower approach for uncertain quadrotor dynamics using feedback linearisation sliding mode control, *Int. J. Model. Identif. Control*, vol. 33, no. 1, pp. 9–19, doi: 10.1504/IJMIC.2019.103980.

[5] Alqaisi W., Kali Y., and El-Bayeh C. Z., (2021). Modified fast terminal super-twisting control for uncertain robot manipulators, *18th IEEE Int. Multi-Conference Syst. Signals Devices, SSD 2021*, no. 1, pp. 1142–1147, doi: 10.1109/SSD52085.2021.9429461.

[6] Alqaisi W., Brahmi B., Ghommam J., Saad M., and Nerguizian V., (2018). Multivariable super-twisting control in a vision-based quadrotor utilized in agricultural application, *CIVEMSA 2018 – 2018 IEEE Int. Conf. Comput. Intell. Virtual Environ. Meas. Syst. Appl. Proc.*, doi: 10.1109/CIVEMSA.2018.8439964.

[7] Xu W., Qu S., Zhao L., H. Z.-I. Transactions on power, and undefined 2020, "An improved adaptive sliding mode observer for middle-and high-speed rotor tracking," *ieeexplore.ieee.org*, Accessed: Nov. 03, 2021. [Online]. Available: https://ieeexplore.ieee.org/abstract/document/9119877/.

[8] Li Y., Qiang S., Zhuang X., and Kaynak O., (2004). Robust and adaptive backstepping control for nonlinear systems using RBF neural networks, *IEEE Trans. Neural Networks*, vol. 15, no. 3, pp. 693–701.

[9] Zelina J. K., Dogan K. M., Prazenica R. J., and Henderson T., (2022). Adaptive control for time varying systems with actuator dynamics and sensor noise," in *AIAA SCITECH 2022 Forum*, p. 2029.

[10] Chen F., Jiang R., Zhang K., Jiang B., and Tao G., (2016). Robust backstepping sliding-mode control and observer-based fault estimation for a quadrotor UAV, *IEEE Trans. Ind. Electron.*, vol. 63, no. 8, pp. 5044–5056.

[11] Madani T. and Benallegue A., (2007). Sliding mode observer and backstepping control for a quadrotor unmanned aerial vehicles, in *2007 American control conference*, pp. 5887–5892.

[12] Alqaisi W., Ghommam J., Alazzam A., Saad M., and Nerguizian V., (Jan. 2020). Three-loop uncertainties compensator and sliding mode quadrotor control, *Comput. Electr. Eng.*, vol. 81, p. 106507, doi: 10.1016/j.compeleceng.2019.106507.

[13] Alqaisi W., Brahmi B., Ghommam J., Saad M., and Nerguizian V., (2018). Sliding mode controller and hierarchical perturbation compensator in a UAV quadrotor, doi: 10.1109/CIVEMSA.2018.8440003.

[14] Alqaisi W. K., Brahmi B., Ghommam J., Saad M., and Nerguizian V., (2019). Adaptive sliding mode control based on RBF neural network approximation for quadrotor, doi: 10.1109/ROSE.2019.8790423.

[15] Wu J., Peng H., and Chen Q., (2010). RBF-ARX model-based modeling and control of quadrotor, in *2010 IEEE International Conference on Control Applications*, pp. 1731–1736.

[16] Alqaisi W., Kali Y., and Lucia W., (2020). Finite-time flight control of uncertain quadrotor UAV based on modified non-singular fast terminal super-twisting control, in *2020 IEEE Conference on Control Technology and Applications (CCTA)*, pp. 37–42.

[17] Wei Y., Chen Y., Liang S., and Wang Y., (2015). A novel algorithm on adaptive backstepping control of fractional order systems, *Neurocomputing*, vol. 165, pp. 395–402.

[18] Cai J., Wen C., Su H., Liu Z., and Xing L., (2016). Adaptive backstepping control for a class of nonlinear systems with non-triangular structural uncertainties, *IEEE Trans. Automat. Contr.*, vol. 62, no. 10, pp. 5220–5226.

[19] Khebbache H., (2018). Tolérance aux défauts via la méthode backstepping des systèmes non linéaires: application système UAV de type quadrirotor, Accessed: Nov. 02, 2021. [Online]. Available: http://dspace.univ-setif.dz:8888/jspui/handle/123456789/1300.

[20] Sabatino F., (2015). Quadrotor Control: Modeling, Nonlinearcontrol Design, and Simulation.

[21] Nakamura M., Takaya K., Ohta H., Shibayama K., and Kroumov V., (2019). Quadrotor modeling and simulation for industrial application, in *2019 23rd International Conference on System Theory, Control and Computing (ICSTCC)*, pp. 37–42.

[22] Gupta N., Kothari M., and Abhishek, (Sep. 19, 2017). Modeling and control of inverted flight of a variable-pitch quadrotor," *arXiv*. arXiv.

[23] Alqaisi W. and El-Bayeh C. Z., (2022). Adaptive control based on radial base function neural network approximation for quadrotor, in *2022 17th Annual System of Systems Engineering Conference (SOSE)*, pp. 214–219.

[24] Liu Q., Li D., Ge S. S., Ji R., Ouyang Z., and Tee K. P., (2021). Adaptive bias RBF neural network control for a robotic manipulator, *Neurocomputing*, vol. 447, pp. 213–223.

[25] Karar M. E., (2018). Robust RBF neural network–based backstepping controller for implantable cardiac pacemakers, *Int. J. Adapt. Control Signal Process.*, vol. 32, no. 7, pp. 1040–1051.

[26] Liu J., (2013). *Radial Basis Function (RBF) Neural Network Control for Mechanical Systems: Design, Analysis and Matlab Simulation*. Springer Science & Business Media.

Innovation and Technological Advances for Sustainability – Al-Naemi et al. (Eds)

Temperature aware energy efficient delay tolerant routing protocol networks

Rachid Selt
Laboratoire des Méthodes de Conception des Systèmes, Ecole nationale Supérieure d'Informatique, Oued-Smar, Algeria

Yacine Challal
College of Computing and IT University of Doha for Science and Technology, Doha, Qatar

Abdelmalik Bachir
University of Biskra, Algiers, Algeria

ABSTRACT: Wireless transmission is one of the most energy consuming tasks in nowadays communicating devices and the Internet of Things in general. This generates huge amount of dangerous and harmful waste for nature such as batteries. Therefore, reducing wireless transmissions has been one of the most investigated levers to optimize energy consumption in the Internet of Things and hence design more nature-friendly IT landscape. However, few of those works considered the impact of temperature and other environmental factors on the quality of radio transmission and leverage that impact to improve network lifetime through optimized energy usage. In this work, we design a new routing protocol that takes into consideration temperature impact on the quality of transmissions. Indeed, many studies showed that high temperatures negatively influence the quality of transmissions which requires higher transmission powers and hence more energy consumption and rapid battery depletion. We consider Delay Tolerant Networks and leverage atmosphere temperature sensing of devices to decide when and where to send data in an energy-effective way. We show, through simulations, that our approach allows to improve energy consumption at the cost of extra-latency that should not harm applications running over Delay-Tolerant Networks.

Keywords: delay tolerant network, Temperature Aware wireless transmission, energy-aware routing protocol, wireless sensor networks, Sustainable Development

1 INTRODUCTION

Wireless sensor networks are a technological revolution in measuring tools created by the convergence of microelectronics and communication systems. They represent a group of small electronic devices powered with Lithium-ion batteries that measure certain physical phenomena from their deployment environment .

Due to micro-factors and manufacturing costs, sensor nodes usually have limited resources for energy intensity, computation capacity, data storage space and transmission speed. They must comply with certain constraints in order to ensure optimal functioning: the existence of an end-to-end path, sufficient bandwidth and permanent electrical supply.

Wireless sensor networks have numerous prominent applications in various fields such as agriculture, pollution monitoring, fleet management, remote health, etc.

DOI: 10.1201/9781003496724-25

However, a large scale deployment of such tiny devices equipped with Lithium batteries would be a disaster for nature. Indeed, the environmental damage [1] caused by the valuable chemical components such as cobalt, copper, lithium, a mixture of organic electrolyte and salts, which are dumped into the environment, requires responsive recovery technologies and system optimisation. Also from [18] the Environmental Services Association estimates that between April 2019 and March 2020, there were 250 fires caused by small Lithium Ion Batteries at waste treatment facilities in the UK.

One of the levers to minimize those harmful wastes is to optimize the life length of batteries and prolong it as far as possible. Optimizing wireless transmission would contribute to prolong the life time of batteries, since wireless transmission is one of the most consuming functions in wireless devices in general and wireless sensor networks in particular.

In this work, we consider Delay Tolerant Networks and applications over wireless sensor networks. We explore new factors that are rarely considered in optimizing energy consumption in wireless networks: atmospheric impact on radio transmission quality. Indeed, many studies showed that high temperatures negatively influence the quality of transmissions which requires higher transmission powers and hence more energy consumption and rapid battery depletion. We design a routing protocol that leverages atmosphere temperature sensing of devices to decide when and where to send data in an energy-effective way. The main idea is to route the packets through the coldest routes and coldest periods of the day. Of course, not all applications could tolerate such delays, and this is why we limit the application of our approach to delay-tolerant networks (DTNs).

We show, through simulations, that our approach allows to improve energy consumption at the cost of extra-latency that should not harm applications running over DTN.

In this paper, we first present related literature research on DTN routing protocols, then present the temperature effects on wireless communications. Then, we present our approach to optimizing energy consumption in delay-tolerant networks: we present two variants of our protocol, one considers temperature change over time and the second variation considers temperature variation in time and location at the end, we present performance evaluation results through simulation and we conclude this paper with future works.

2 RELATED WORKS

Various routing protocols have been proposed in the DTN community, each with their own characteristics and features. However, current DTN routing techniques do not consider atmospheric impact on radio transmission quality and power constraints when choosing routes. only a few research papers have investigated energy-aware routing either for conventional DTN protocols [5–7,20] or for specific proposed DTN protocols [8–10,19]. In the following, we present a brief overview of these related works.

In [5], authors simulated the energy consumption issues of the PROPHET [14], Bubble Rap [21], Epidemic [14] and Spray and Wait [12] protocols. Their simulation showed how the different forwarding strategies of the five DTN routing protocols can affect the energy usage in the mobile node. These studies only took into account energy consumption. They did not use energy data for DTN routing decisions. Patel, et al. [7] Improved spray and wait protocol energy efficiency with a new scheme named Vibrant Energy-aware Spray and Wait (VESW). In [6], authors proposed an ordinary Differential Equations (ODEs) model to solve the issue of energy-efficiency in Epidemic and Two-Hop routing protocols. They proposed a routing strategy on the basis of requirements for message lifetime and probability of delivery. In [20] Authors propose a load-balanced and energy-efficient PRoPHET routing algorithm (ELPR), where the forwarding of messages depends on the residual energy and the buffer space available at a node. ELPR improves the lifetime of a network by distributing the load of nodes with high traffic that would otherwise die early due to overload, resulting in a disconnected network. Their simulation results have shown that the number of dead nodes is reduced by more than 60% compared to the basic PRoPHET protocol, improving the performance of several other network parameters such as residual energy, number of message drops and probability of delivery. In [7] and [6], authors proposed energy-efficient approaches, but these were only applicable to the Epidemic and Spray and Wait solutions, and cannot be adapted to other protocols.

In [8,10,19] and [9] papers, new specific energy-aware strategies have been proposed for DTN protocols. In [10], Authors proposed an energy-aware Epidemic Routing (EAER) protocol which is an extension of the Epidemic routing protocol. EAER targets high performance in terms of energy consumption and packet delivery ratio. For that, the strategy of n-parameter is proposed to optimize the ability of a node to transmit packets to its neighbors when they are within transmission range of each other. Sghaier et al. [8] Proposed a routing protocol for mobile WSN named EXLIOSE (Extending network Lifetime in Opportunistic Sensor Networks) which is built on a new measure named Energy Shortage Cost (ESC). The purpose of this strategy is to extend the lifespan of the network by selecting subsequent hops through ESC. This process evaluates both the energy consumed and the energy available to a sensor node and provides an optimal method for delivering messages from one host to another. By calculating each host with its own ESC value. Authors in [19] developed an energy saving approach that enables energy savings by controlling the wake-up period at regular periods, while keeping the node idle for the rest of the time. The results obtained show that their energy saving scheme has improved battery life time by a factor of five compared to constant access. Also, Cabacas and Ra [9] They present a new context-aware queuing method, termed CEAMS, designed to maximise the probability of delivery while giving priority to messages and optimising the energy consumption of nodes during transmission. EAER, CEAMS and EXLIOSE techniques guarantee superior data rates and performance compared to protocols that ignore the energy factor. But the remaining energy and the consumption of the devices are computed without taking into account the energy consumed for message generation and without taking into account energy harvesting.

Previous research into energy-efficient DTN protocols has primarily examined the energy performance of protocols. Few studies have been conducted on energy conscious DTN routing protocols for wireless networks that take into consideration environmental impact of transmission quality and the required energy.The aim of this work is to propose a temperature-aware energy model to enhance existing DTN routing protocols and propose a new routing algorithm to improve energy consumption and increase the lifetime of the network.

3 TEMPERATURE IMPACTS ON WIRELESS COMMUNICATIONS

Wireless communications may be affected by changes in weather conditions such as rain, snow, winds, temperature and humidity in particular. Temperature and humidity have little effect on the propagation of radio waves in the air, but they have a major impact on radio equipment [2,3]. High temperatures affect radio frequency circuits, resulting in deterioration of the quality of radio link connections.

It is well known that temperature affects radio transmitters on both receivers and transmitters. The information provided in widely used low-power radio data sheets (such as CC2400s) shows that sensitivity and output power decrease with temperature increase as shown in Figure 1 [4].

Research has also increased evidence that high temperatures affect the transmission signal on the transmitter side. According to [4], high temperatures affect transmitters and lead to distortion of output waves during the final phase of amplitude, leading to bad signal transmission. In addition, [2] authors showed that link quality was reduced to 8dB during the 40°C temperature range.

The key challenges under consideration are the limited energy resources available and the limited connectivity of nodes in long-term variant temperature deployment applications. The

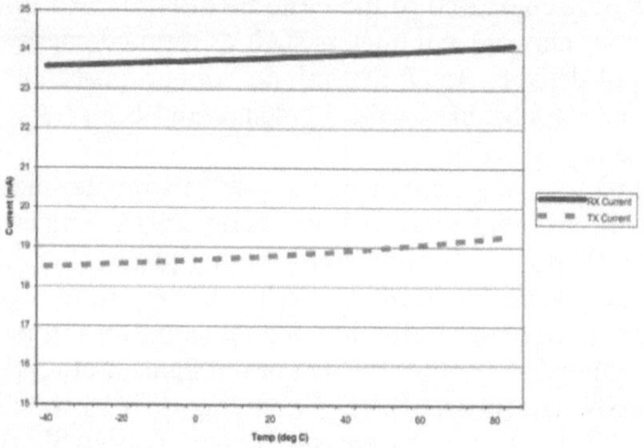

Figure 1. Typical RX and TX current vs. temperature [4].

main idea of our work is to design a routing protocol that allows forwarding packets through the coldest path and hence save the extra-energy that would have been spent to allow connectivity in hot areas and/or during hot periods.

One category of networks that allows monitoring the best opportunity, in terms of connectivity, to forward packets is Delay Tolerant Networks (DTN).

Our aim is then to design a thin control layer over DTN to leverage temperature impact on the quality of wireless transmissions for energy saving through forwarding packets via the coldest paths.

In our work, we consider the energy model used in a previous work [4]. In this system model, the quality of a wireless communication link between two nodes, a transmitter and a receiver, is represented by the observed PER (Packet Error Rate). Sensitivity is defined as the lowest level of signal strength measured at the receiver that is required to achieve a PER below 1%. The power Pt emitted by the emitter and its attenuation along the communication channel determine the signal strength measured at the receiver. We also consider that a connection exists if the received power PR is equal to or greater than the sensitivity P_s.

We use the expression $P_{t,\tau}$ (I) to describe the transmitted power as a function of the temperature τ sensed at the transmitter when an input current I is used.

$$P_{t,\tau}(I) = P_{t,\tau}(I^*) + P_{t,\tau^*}(I) \tag{1}$$

In what follows, we will recall definitions and design of Delay Tolerant Networks. Then, we will introduce our temperature-aware routing protocol.

4 DELAY TOLERANT ROUTING PROTOCOLS

DTN routing protocols can be categorized based on the information gathered by their nodes to make forwarding decisions. The search for a target requires two different characteristics: Replication and Knowledge. Replication creates multiple copies of each message to ensure on-time delivery and maximum possibility of successful transfer. Knowledge takes a decision from network information, based on these properties, DTN routing protocols are divided into two categories [11]. Forwarding Routing Protocols and Flooding Routing Protocols.

The forwarding routing protocols include Location-Based Routing, Source Routing, Per-Hop Routing and Per-Contact Routing. The forwarding routing is a knowledge-based approach that is not replicated. In this strategy, routing protocols require network knowledge, nodes gather information from other nodes of the network to select the best way to forward the message to the destination node.

The flooding routing protocols include Epidemic Routing, Two-Hop Relaying, Tree-Based Flooding and Prioritized Epidemic Routing. In flood protocols, each node sends a number of messages to other nodes with unknown network structure. Nodes store messages in buffers and store them until they find the next node or destination. Replication of messages increases the likelihood of sending messages, as more nodes are responsible for transmitting messages to the base station.

The following paragraph provides more information on the commonly used DTN routing protocol considered in this study, in particular: Direct Delivery.

4.1 *Direct delivery*

Direct delivery routing is a single copy routing protocol [12], in which nodes transport only one copy of the message to the end without relaying any other nodes.

This strategy does not require network information, and only when the source node is in contact with the destination allows the message to be transmitted by one hop. This simple protocol has a minimum overhead, but suffers from a low delivery rate. Normally, it is used as a baseline for other protocols.

As we have proposed in our approach, we recommend using this protocol to ensure that mobile sink nodes (e.g., drones) communicate directly to fixed nodes to exchange packets.

4.2 *Epidemic*

A flooding-based forwarding routing protocol: Epidemic, was proposed by Vahdat and Becker [13]. In the Epidemic Protocol, nodes send a copy of the message to all nodes they contact without knowing the network. Identified nodes save messages to buffers by store-carry-forward until they reach other nodes or destinations. When two nodes are in contact, they exchange their own list of messages, called a summary vector. This is done to compare what each node is holding and to decide which new messages need to be downloaded from the nodes they encounter.

4.3 *Prophet*

PROPHET [14], unlike the Epidemic, uses information derived from the last encounters of nodes such as the number of packet copies, the mobility pattern, the location, the time of contact and the social behaviour of nodes to optimize the delivery of the packet.

When two nodes meet in PROPHET, they share their data, which contains all the delivery information that each node knows. Once this information has been exchanged and updated, the node can make a decision about which messages to send based on its delivery strategy. If node a needs to send its message to destination b, it will pass through node c if the probability of reaching b from point a is lower than that of passing through node c.

5 TEMPERATURE AWARE ENERGY EFFICIENT DELAY TOLERANT ROUTING PROTOCOL

In this section we present an effective routing protocol called TAEEF for the delay tolerant networks (DTN) which takes into consideration the ambient temperature to optimize the energy of network components. A lot of research been conducted to propose Routing protocols for DTN, but this is the first work that uses ambient temperature information to decide when to store and forward data.

In our proposal, we assume that each node in the network has a state table and collects the ambient temperature on a periodic basis. When two nodes connect to each other, they exchange their state table

After state table exchange, a node considers the temperature and positions. There are two variants:

5.1 *Temperature variation over time only*

in this variant we consider that temperature is the same in the whole network deployment area and varies in time. This could be the case of a small-scale network where temperature does not vary too much from a location to another. In This case, the main idea is to store the packets during hot temperatures and forward them during the coldest periods. For the sake of simplicity, we considered the operation of CC2400 antenna [4]. According to that data-sheet, the optimal operation of the antenna is met at 25C. Therefore, in our approach as shown in Figure 2 each node senses temperature ambient periodically and if temperature is > 25 the node stores the packet until the weather get cold (< 25).

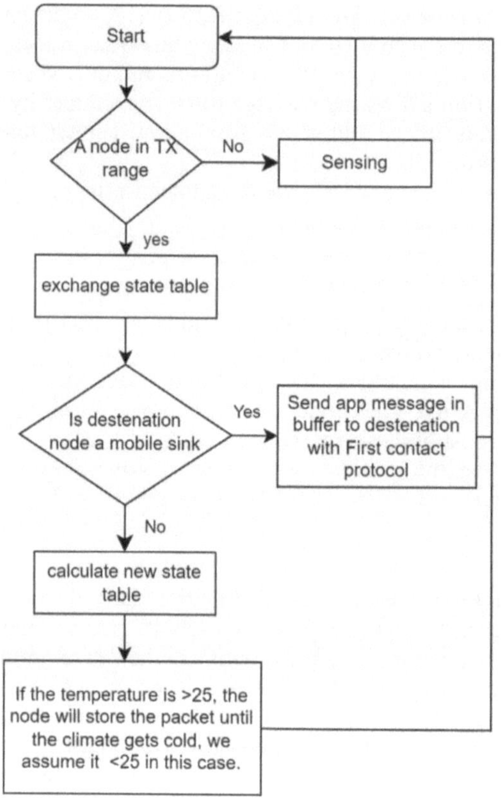

Figure 2. First variant (Temperature variation over time only) routing protocol flowchart.

5.2 *Temperature variation over time and space*

in this variant we consider that temperature varies not only in time but also from a location to another. This could be the case of large-scale networks covering a wide area where temperature varies from a location to another. The main idea of this variant is to forward the packets through the coldest routes. To do this, a node calculates the shortest coldest path to the sink node as shown in Figure 3 and sends the state table to the next hope and adds timestamp as the time of last calculation so we update the calculation if it is too old. We based on Dijkstra algorithm to calculate the shortest path considering ambient temperature as a metric. Our protocol takes the decision to forward packets based on temperature and chooses the coldest neighbor as next hope. We assume that we already segmented the map in sub areas where there is a sink in every coldest sub area.

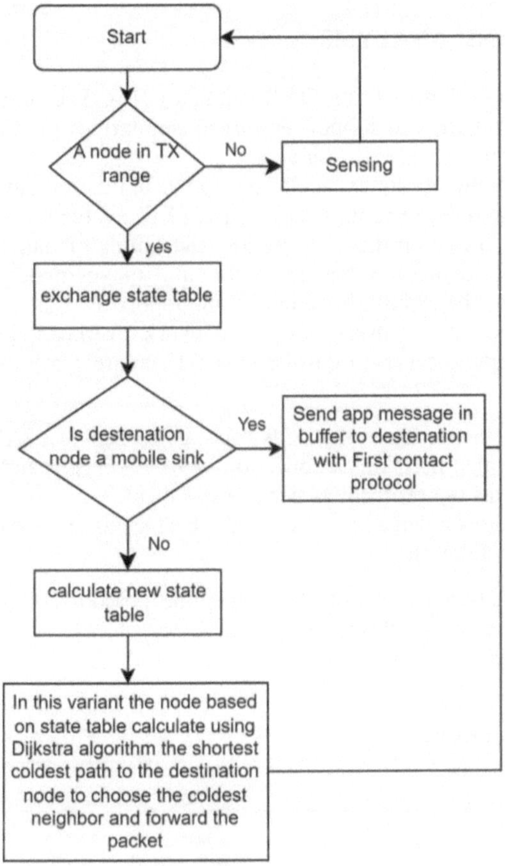

Figure 3. Second variant routing protocol flowshart.

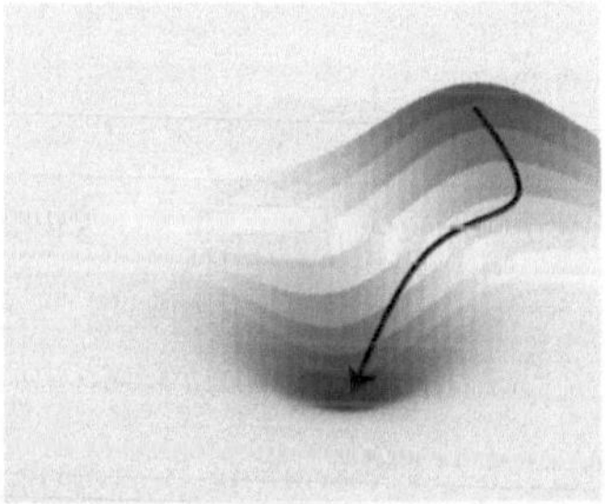

Figure 4. TAEEF routing protocol coldest path route illustration.

Table 1. State table.

field	description
ID	Unique ID
Type of node	Mobile or fixed sink or fixed node
Temp	Ambientes temperature
Neighbours	Neighbours' temperature and position
position	Position of node

6 EVALUATION AND ANALYSIS

The proposed work is evaluated using ONE simulator [15]. The simulation scenarios and extensions of ONE simulators to support empirical simulations are described, followed by discussion of evaluation metrics and results.

The simulator reads the setting text file to set up the simulation environment such as the number of nodes, the number of groups, time to live (TTL) value for packets, buffer size, or speed of node groups. Based on this information, the simulator builds the simulation map which is called the world in the ONE simulator, and makes nodes on the world. After running the simulation, the simulator generates one report file that contains many metrics such as delivery probability, delivery latency, network overhead, packet drop ratio, or average number of hops to reach to destinations. The entire system is developed in three stages, as below.

(1) We wrote the implementation of the DTN routing protocol in a JAVA class provided by ONE Simulator, which itself was a collection of JAVA classes and packages
(2) Then ONE simulates the protocol as per our specifications
(3) Therefore, it generates a detailed report on the performance assessment of the protocol according to the simulation

In the simulation, we based on temperature model proposed in [16] to calculate the energy consumption at temperature T. We considered a network of 200 nodes distributed randomly in area of 4500m * 3400m

Table 2. Simulation parameters.

Parameters	Values
Simulation Time (In Seconds)	43200
Update Interval	0.1 second
Transmit Speed	250 kbps
Transmit Range	15 m
Routing Protocols	TAEEF, prophet, epedimic
Buffer Size	5 MB
Message Generation Rate (message/min.)	2 (One new message in every 25 to 35 seconds)
Message TTL	30.60.110.150
Message Size	500 KB
Simulation area	4500m * 3400m

We also compared three different DTN routing protocols: TAEEF, epidemic and prophet. To assess performance, we have checked the performance ratio of each routing protocol. The delivery ratio is equal to the total number of delivered packages divided by the total number of packages. Since it takes into account only the number of packages delivered, the number of copies is not important in this case.

We set a simple scenario in which all simulation regions assume the same temperature, then run simulations at different temperatures, and then run simulations at the same network at different temperatures, in the same execution.

The first impact studied is the delivery ratio in three different scenarios at 30, 18 and variable temperature. The delivery ratio equals to the total number of delivered packets divided by the total number of packets. Since it only counts the number of delivered packets, the number of copies does not matter in this case.

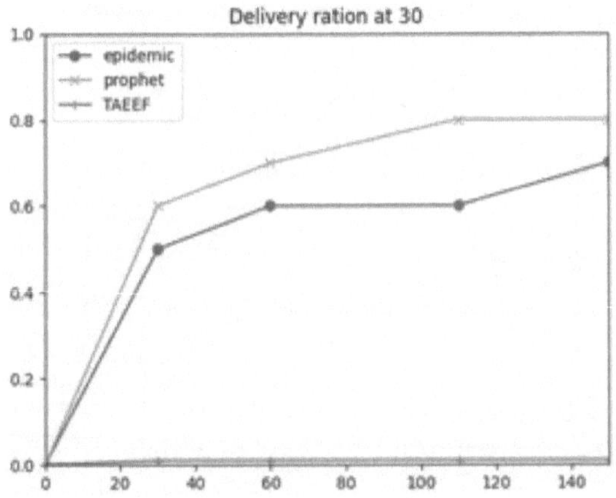

Figure 5. Delivery ration over TTL changes in 30°.

Figure 4 shows the delivery ratio of the three routing protocols at temperature 30°. When the TTL value is more than 30 minutes, the transmission rate of the epidemic routing protocol is the highest among the three protocols. However, it decreases when the TTL value increases. Because the epidemic generates too many copies on the network, the packets may not be eliminated from time out, but will be eliminated because of the overflow of the buffer at each node. However, our proposed protocol, is near 0 because of our algorithm to store packet if temperature is bigger than 25.

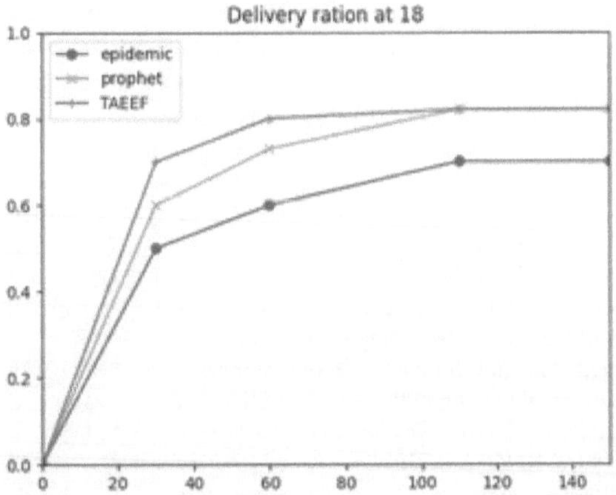

Figure 6. Delivery ratio over TTL changes at 18°.

Figure 5 shows the delivery ratio of the three routing protocols at temperature 18°. The results are in line with previous simulation except for TAEEF that perform near best of all after TTL 30.

Figure 6 shows the delivery ratio of the three routing protocols at variant temperatures. The results are in line with previous simulation except for TAEEF that performs best of all after TTL 110. Our approach requires a long TTL because of its policy of storing the packet in hot weather and delivered in cold weather.

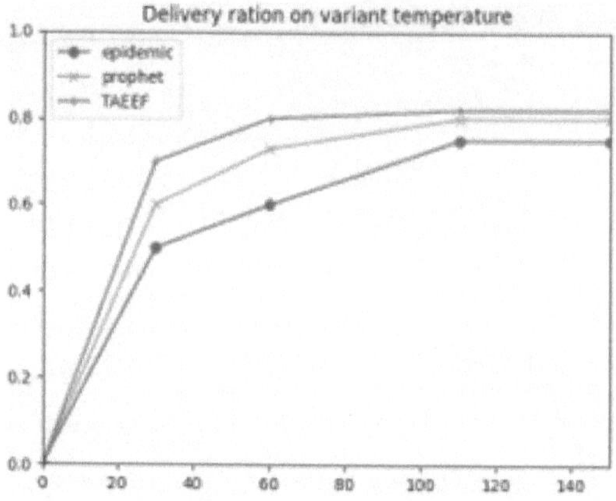

Figure 7. Delivery ration over TTL changes in variant temperature.

7 CONCLUSION AND FUTURE WORK

In this paper we presented a temperature aware energy efficient routing protocol for DTN, which uses the temperature ambiant factor to forward message to nodes. First, we defined the flow diagram of our approach, then we simulated the proposed TAEEF routing protocol in which nodes send packets when cold and store packets when hot.

We believe that this paper represents a first step in establishing energy-efficient routing protocols using ambient temperature data. In the future, we plan to develop and test our approach in real scenario where temperature varies in the same area over time and space. We consider to use temperature forecasting model to predict temperature changes and determine when a node is likely to send a packet in the near future.

REFERENCES

[1] Bankole, Oluwatosin Emmanuel, Chunxia Gong, and Lixu Lei. (2013). Battery recycling technologies: Recycling waste lithium ion batteries with the impact on the environment in-view. *Journal of Environment and Ecology* 4.1: 14–28.

[2] Otero J., Yalamanchili P., and Braun H.-W., (2001). *High Performance Wireless Networking and Weather*, University of California, San Diego, Tech. Rep..

[3] Bannister K., Giorgetti G., and Gupta S., (2008). Wireless sensor networking for hot applications: Effects of temperature on signal strength, data collection and localization, *in Proceedings of the 5nd IEEE workshop on Embedded Networked Sensors*, Charlotteville, VA, USA.

[4] Corporation C., (2008). *CC2400 2.4 GHz Low-Power RF Transceiver, Data Sheet*, [Online]. Available: http://www.ti.com/lit/gpn/cc2400

[5] Socievole A. and Marano S., (2012). Evaluating the impact of energy consumption on routing performance in delay tolerant networks, *in Wireless Communications and Mobile Computing Conference (IWCMC)*, pp. 481–486.

[6] Manam V., Gurav G., and Murthy C. (2013). Performance modeling of messagedriven based energy-efficient routing in delay-tolerant networks with individual node selfishness, *in Fifth International Conference on Communication Systems and Networks (COMSNETS)*, pp. 1–6.

[7] Patel V. G., Oza T. K, and Gohil D. M., (2013). Vibrant energy aware spray and wait routing in delay tolerant network, *Journal of Telematics and Informatics*, vol. 1, pp. 43–47.

[8] Sghaier N., Augustin B., and Mellouk A., (2014). On enhancing network-lifetime in opportunistic Wireless Sensor Networks, in *Wireless Communications and Networking Conference (WCNC)*, IEEE, pp. 2617–2622.

[9] Cabacas R. and Ra I.-H., (2014). A novel energy-aware priority transmission scheme based on context-metric queuing for delay tolerant networks, in *International Conference on Information Science, Electronics and Electrical Engineering (ISEEE)*, pp. 1095–1099.
[10] De Rango F., Amelio S., and Fazio P., (2013). Enhancements of epidemic routing in delay tolerant networks from an energy perspective, in *Wireless Communications and Mobile Computing Conference (IWCMC)*, pp. 731–735.
[11] Jones, E. P. and Ward, P. A. (2006). Routing Strategies for Delay-Tolerant Networks.
[12] Spyropoulos, T.; Psounis, K. and Raghavendra, C. S. (4–7 October 2004). Single-copy routing in intermittently connected mobile networks. In *Proceedings of the Conference on Sensor and Ad Hoc Communications and Networks (SECON 2004)*, Santa Clara, CA, USA, pp. 235–244.
[13] Vahdat, A. and Becker, D. (2000). Epidemic Routing for Partially Connected Ad Hoc Networks; Technical Report CS-200006; Duke University: Durham, NC, USA.
[14] Lindgren, A., Doria, A. and Schelén, O. (2003). Probabilistic routing in intermittently connected networks. *ACM SIGMOBILE Mob. Comput. Commun. Rev.* 7, 19–20.
[15] Keranen, A. (2008). *Opportunistic Network Environment Simulator; Special Assignment report*; Helsinki University of Technology, Department of Communications and Networking: Helsinki, Finland.
[16] Bachir A., Bechkit W., Challal Y. and Bouabdallah A., (1 July 2015). Joint connectivity-coverage temperature-aware algorithms for wireless sensor networks, *in IEEE Transactions on Parallel and Distributed Systems*, vol. 26, no. 7, pp. 1923–1936
[17] Cerf, V., Burleigh, S., Hooke, A., Torgerson, L., Durst, R., Scott, K., Fall, K. and Weiss, H. *Delay-Tolerant Networking Architecture*. RFC 4838. April 2007. Available online: https://tools.ietf.org/html/rfc4838 (accessed on 01/12/2022).
[18] Mrozik, Wojciech, et al. (2021). Environmental impacts, pollution sources and pathways of spent lithium-ion batteries. *Energy & Environmental Science* 14.12: 6099–6121.
[19] Singh, Amit Kumar, Rajendra Pamula, and Gautam Srivastava. An adaptive energy aware DTN-based communication layer for cyber-physical systems. *Sustainable Computing: Informatics and Systems* 35 (2022): 100657.
[20] Malik, Amit, Amita Rani, and Milan Kumar. ELPR: Energy Efficient Load Balanced PRoPHET Routing for Delay Tolerant Networks. *Advances in Micro-Electronics, Embedded Systems and IoT*. Springer, Singapore, 2022. 365–375.
[21] Hui, Pan, Crowcroft, Jon and Yoneki, Eiko. (2011). BUBBLE Rap: Social-based forwarding in delay-tolerant networks. *IEEE Trans. Mob. Comput.* 10. 1576–1589. 10.1145/1374618.1374652.

Innovation and Technological Advances for Sustainability – Al-Naemi et al. (Eds)

Smart and sustainable threat intelligence

Duaa Shoukat
Riphah Institute of Systems Engineering, Islamabad, Pakistan

Muhammad Taimoor Khan
Riphah Institute of Science and Engineering, Islamabad, Pakistan

Syed Muhammad Sajjad
Riphah Institute of Systems Engineering Islamabad, Pakistan

Adnan Akhunzada (Senior Member; IEEE)
College of Computing & IT, University of Doha for Science & Technology (UDST), Doha, Qatar

ABSTRACT: Distributed Denial of Service (DDoS) attack is known to be the most dangerous attack in cyber-threats which is identically collective in the world of today's interconnected networks, as it mainly disrupts and makes vital services unavailable. Hackers are skilled in using multi- classification DDoS attacks to evade the detection for exploitation of the targeted networks. A proactive and efficient detection mechanism is much needed to secure the networks from multiclass DDoS attacks. However, implementing DDoS detection schemes is very difficult due to the factors such as cost, complexities, and inflexibility. Most of the work has been done to tackle these attacks; still, the open question is to find out an efficient, smart and sustainable model amongst available choices. Therefore, in this research, the author proposes a novel, lightweight Cuda- powered Deep Neural Network Long Short-Term Memory (Cu-DNNLSTM) enabled smart and sustainable threat intelligence system for large and distributed enterprise networks and is simply termed as Multiclass DDoS Detection mechanism (MDDM). The proposed approach is implemented using the current state of the art Canadian Institute of Cybersecurity (CIC)-DDoS2019 dataset that is publicly available and has been thoroughly evaluated. The preliminary results achieves 99.60% detection accuracy with a relatively low ratio of False Positives (FP) (i.e., 0.0003). Additionally, the proposed approach has also been compared with two other DL models to show the promising performance of the proposed approach. Finally, the proposed approach is highly scalable, cost effective, flexible and sustainable to be customized for any emerging computational and communication paradigm.

Keywords: DDoS, Denial of service attacks, multiclass, Detection, Deep Learning, hybrid technique, threat intellignce

1 INTRODUCTION

Most key infrastructures and sectors are based on computer networks, which play a significant role; the areas may incorporate government, military, and other organisations. Computer Networks have become a strength of many institutions nowadays, including management and exploration. The environments containing multiple networks and interactions with the public inherit various cyber threats and attacks. One of the significant threats to the organisational servers and databases is an emerging.

DOI: 10.1201/9781003496724-26

Distributed Denial of Service attack. The danger of this attack is not just limited to making the computer resources unavailable but also disturbs the available bandwidth, thereby bringing the network to a halt. Therefore, it is vital to prevent cyber attackers from attacking and compromising systems and networked computers operating in a safe and normal environment [1]. Thus, it is an essential and urgent task for current network administrators to study how to prevent malicious attacks by attackers on network systems successfully.

Various restrictive and open-source arrangements exist to detect and mitigate DDoS attacks. These threats keep developing recurrence, modernity, and seriousness [2], [3]. Quickly classifying DDoS attacks has become a serious concern as cyber-attackers utilise novel methods to dispatch DDoS threats [4]. The rising number of DDoS trends, combined with developing decent variety in their sorts, causing appalling effects, has made DDoS threat discovery and anticipation the top need for example, Arbor Networks Inc. One of the primary DDoS danger insurance arrangements suppliers revealed 334 Gbps attacks focusing on a system administrator in Asia.

Moreover, it revealed numerous threats more significant than 100 Gbps in 2018 [6]. Numerous occurrences show that new ways are required to address the DDoS issues. These new methodologies must be intended to meet the presentation and adaptability necessities of current server centres and give the most significant degrees of security against rising and unpredictable attacks.

Emerging cyber-attacks are evolving progressively and becoming more sophisticated. Attackers use multiclass DoS/Distributed Denial of Service attacks to harm the target organisation's network. A simple DoS attack tries to contradict access to authentic alternative users and disable the resources or services. Attackers sometimes use various distributed computing services to initiate a coordinated DoS attack on one or many objectives. Therefore, the main concern is to come up with a solution to tackle evolving nature of DDoS attacks. Enormous contributions of efforts from different researchers are studied and discussed. Two mechanisms are developed for detecting malicious events, usually called signature and behaviour-based techniques. Solutions based on signature-based classification are inefficient for the identification of zero-day threats. However, techniques based on Machine Learning (ML) algorithms are still valuable and majorly secure the environments; however, ML has many challenges, such as feature engineering problems highlighted by researchers [4]. So, due to the inefficiency in the solutions of ML techniques, the trend is shifted towards Deep Learning (DL) based security measures. DL techniques provide efficient solutions and high accuracy in cyber-attack mitigation and detection mechanisms[5]. The adoption of DL models is mainly due to the resolved feature engineering problem done by the algorithm in DL.

Therefore, in this article, the authors propose an innovative, robust, and state-of-the-Art DL-enabled architecture for efficiently detecting multiclass DDoS threats in open organisations. The main contributions of the paper are explicitly given as under:

- A comprehensive literature analysis of DDoS attacks and their impact on big organisations is carried out.
- The authors studied existing solutions for recognising and avoiding DDoS attacks established using DL/ML solutions and highlighted their limitations.
- A practical, robust, and proficient DDoS detection system has been developed by applying efficient deep learning algorithms to detect evolving DDoS attacks.
- The CICDDoS2019 is a Canadian current state of the art dataset containing 13 varied sophisticated DDoS attacks that is employed for our proposed Multiclass DDoS Detection mechanism (MDDM).
- Standard evaluation metrics are used for an exhaustive assessment of the proposed technique, i.e., precision, recall, detection accuracy, and F1-score.

- The GPU accelerated hybrid technique Cu-DNNLSTM has been adopted for the training and analysis of the proposed architecture.
- Finally, the authors compare the performance of the proposed scheme with the latest techniques for showing the promising results of a proposed model.

The rest of the paper is organised as follows. Section II includes the closest research contributions to the proposed model. The problem statement and proposed model description are given in section III. Implementation of the proposed model and the experimental setup along with the dataset and proposed model architecture and structure is explicitly discussed in Section IV. Then it is followed by Section V, which demonstrates the experimental results and graphs of the model performance. Finally, the proffered work is concluded in Section VI.

2 RELATED WORK

Several methods and mechanisms were suggested, and various systems were developed to detect intrusions. Researchers have already highlighted their research with some ML and DL algorithms in the same area to correctly use vast amounts of data and their security. This section elaborates on the available solutions with their applied technique and performance. For instance, the researchers in [7] proffer a DL-based methodology for detecting malicious intents using the DL-based Deep Neural Networks (DNN) technique for model construction. The proposed model has employed two old datasets named UNSW-NB15 and NSL-KDD99 to perform training and testing steps. The given approach has achieved an accuracy of 99.24%, which is low in the case of efficient detection. The accuracy in the given research paper is low. The DDoS detection scheme proposed in [8] applied EABC elite for classifying malicious network traffic. The authors have employed well known and majorly adopted KDDCUP99 dataset for model training. But the inefficient learning technique has reached low accuracy with a high rate of false positives (i.e., 99.30% and 2.43). The dataset used provides low accuracy as compared to the proposed model. DDoSNet has been proposed in [9] as an intrusion detection system against DDoS attacks in SDN environments. This method was based on Deep Learning (DL) technique, merging the Recurrent Neural Network (RNN) with an autoencoder. The model used in [9] is the newly released dataset CICDDoS2019, which consisted of a wide variety of DDoS attacks and addressed the gaps in the existing current datasets. Hence, this model provided great confidence in securing these networks. The dataset used in [9] is the old one as compared to the proposed model. A mechanism for detection by Berral *et al.* in [10] proposed malicious activities. The nodes in the network are collaborated within the network and share their local traffic observations that aid in the improvement of their overall traffic, and each node use the naive Bayes algorithm independently for acquiring data. The detection of DDoS attacks using a neural network is done after obtaining information from each packet on the network. To effectively secure networks against DDoS attacks, an efficient DL-based defence mechanism is introduced in [11], consisting of DL constructed layered model that reaches high accuracy of 99.8% by effectively classifying malicious instances. However, the training of the proposed model has been done using old network traffic patterns from the ISCX2012 dataset, which can miss the unknown and evolving threats. The accuracy of the given dataset is low and the dataset used is old. GRU-RNN [12] proffers a Deep Recurrent Neural Network for Intrusion Detection system in Software Defined Networks (SDN). Network traffic characterisation is done with a robust and hybrid DL technique. GRU-enabled RNN performs fast learning and efficiently remembers the critical states of learning gates.

The significance of the proposed scheme is that it is constructed using a model constructed using GRU layers. However, the simulation results demonstrate that the model achieves the lowest accuracy of 89%, with the highest rate of false alarms, which is 10. The accuracy of the proposed scheme in [12] is very low as compared to the proposed scheme. In [13], the authors proposed a scheme for effective DDoS detection for the most extensive networks using Deep layers. The model employed the ISCX2012 dataset neither outdated nor up to date. The researchers used an efficient LSTM model with RNN layers for model construction. Thus, it achieves an overall 97% evaluation accuracy in classifying DDoS attacks. The proposed scheme results in low accuracy in comparison to the proposed model. Another security system for detecting cyber threats in networks using DL- enabled techniques is proposed in [14]. It provides a solution to tackle different attacks coming from the digital era. Like other schemes, the model uses an older KDDCUP99 dataset for model implementation. Performance-wise, the proposed scheme has provided fair results for metrics like accuracy (i.e., 99.20%). The authors have evaluated the old traffic dataset, due to which the model's efficiency could be reduced while working on the latest traffic patterns.

This research focuses on mitigating drawbacks and issues by choosing the latest openly available CICDDoS2019 dataset, which contains novel traffic patterns of multiclass attacks for classifying emerging DDoS threats. Therefore, a robust and state-of-the-art DL-based hybrid technique is proposed to effectively mitigate the current challenges and improve the performance of standard metrics such as accuracy and False Positive Rate (FPR).

3 PROBLEM STATEMENT AND PROPOSED METHODOLOGY

The infrastructure of an enterprise comes at the cost of sophisticated DDoS threats and attacks. DDoS attacks achieve effectiveness by utilising multiple compromised systems as sources of attack traffic. It can completely take down the availability of a server/system in a network. The existing DDoS detection schemes lack in handling emerging and evolving DDoS threats. Further, limited mechanisms are available for multiclass DDoS detection. Therefore, an efficient, state-of-the-art, and robust detection mechanism is needed against multi-classification DDoS attacks that are evolving terrifically in the organisational network. Therefore, this study proposes a highly efficient and scalable Cu-DNNLSTM-enabled framework for detecting sophisticated and evolving multi-vector DDoS attacks.

3.1 *GPU enabled model (Cu-DNNLSTM)*

DNNLSTM is a predefined hybrid model of deep learning (DL) which is picked and proposed as a best classifier for DDoS detection in this study. The DNNLSTM model is the efficient algorithm, which by default combines the two state of the art algorithms for achieving the maximum level of accurate detections. We have trained the model using DNNLSTM layers with dense layers and then tested the system for determining the attack type. The model of DNN and LSTM alone is defined below.

3.2 *Significance*

The organizations that are using computer networks as their backbone, poses a high threat of DDoS attacks. Therefore, the security of such environments is a ceaseless as well as significant concern; hence, an advanced comprehension of the security weaknesses present in the network infrastructures and the advancement of relating DDoS detection methodologies are mandatory. In big organizations, the security vulnerabilities and open public ports create security dangers by which the server maybe got affected and crashed. As the network, security is a genuine and constant issue; there is a crucial need to come up with best DDoS detection mechanism, which will protect the networked systems and the availability of servers during peak or average demand.

4 IMPLEMENTATION AND EXPERIMENTS

This section provides a brief description of the deployment and implementation of the proposed model. Three state-of-the-art DL models have been used to evaluate the proposed system. The simple DNN and GRU are first evaluated for mapping the efficiency, and then the evaluation of GPU accelerated Cu-DNNLSTM is proposed because of its superiority and efficiency. The section carries subsections providing descriptions of the practical environment, dataset, proposed hybrid Cu-DNNLSTM architecture, and the structure of the constructed algorithm for the proposed scheme.

4.1 *Practical environment*

The experiments have been conducted on 16 GB RAM and 7 core processors., Python language is used with the Keras and Tensorflow packages using the anaconda environment to implement DL functions. The training and testing phases of the chosen models have used supervised schemes to perform feature learning. The same features extracted and used for the model training phase are later used with the test features for the detection phase. Before embedding the models for the fulfilment of several tasks, python provides numerous tools with many libraries. The presence of many libraries makes python's greatest strength. The author of this research utilises numerous Python libraries such as Keras, Numpy, Matplotlib, Sklearn, OS, and Pandas, for the model execution and also for the performance of many essential tasks, perhaps, Sklearn and Numpy libraries are used for data reorganisation, splitting, and for formulation as well as analysis of test and train sets. Moreover, the authors imported libraries such as Matloib and Pandas to display the results after performing some numerical activities using these two packages.

4.2 *Description of dataset*

This research uses the publicly available dataset for model implementation, called CICDDoS2019. Two types of traffic have been contained in it, benign and the most updated and commonly found DDoS attacks, looks like (PCAPS), the correct and real-world data. The results of network traffic analysis are also included by it, which are captured using CICFlowMeter-V3 along with labelled and categorised traffic flows. The author aimed to generate realistic background traffic as the top priority while building the dataset. The authors have constructed this dataset's theoretical behaviour of 25 users based on the HTTP, HTTPS, FTP, SSH, and email protocols. The author in [15] proposed B-Profile system dataset testbed is used to abstract human behaviour summary and naturalistic benign traffic. The dataset contains 86 of features in total, including class labels. According to the authors, the 86 best features are extracted to get the best possible detection results. The dataset contains the traffic of multiclass DDoS attacks, which are of two types, the reflection and the exploitation attacks. Reflection attacks are based on TCP, UDP, and the standard TCP/UDP protocols such as DNS, MSSQL, NTP, and TFTP. While the exploitation attacks are TCP and UDP based, such as flooding attacks (SYN flood, UDP flood). One of the initial and foremost steps in Artificial Intelligence or Deep Learning is preprocessing data before using it. The captured traffic data must be prepared before the classification algorithm, or proposed model is implemented. The dataset file usually contains some categorical, Nan, and numeric values. In most cases, the author does the initial preprocessing of data, excluding attack and benign class labels. The system training and learning of models must be done on numeric values, so all absolute values are firstly converted while the data preprocessing phase. The time, date, IPs, and all the data which doesn't contain numeric values would be preprocessed first using python libraries and code. The Nan values and empty fields contained in the dataset may result in inadequate training if it remains unhandled. Therefore, python functions would be used to remove and drop Nan values and empty fields.

Afterwards, data scaling using python MinMax/Standard scalar is done. The following step would be the splitting of the data by range. The data is split into the ratio of 80:20 and saved in variables as the train and test sets. Training data would later be used in the model implementation phase for training purposes. The testing data would remain untouched and used later for model testing and predictions. Figure 1 highlights some preprocessing steps.

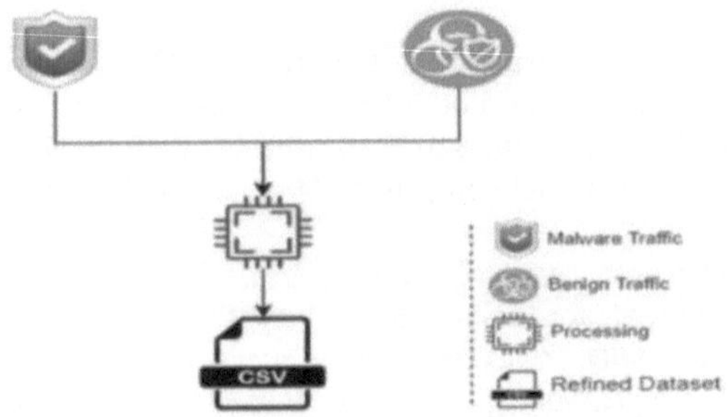

Figure 1. Dataset preprocessing.

4.3 *Architecture of algorithm*

The DNNLSTM model is the efficient algorithm, which by default, combines the two states of the art algorithms for achieving the maximum level of accurate detection. The model has been trained using DNNLSTM layers with dense layers and then tested the system to determine the attack type. The architecture of the model is shown in Figure 2. The rare contributions of the chosen dataset are employed for the embedding layer of DL-based algorithms; the dataset contains 80 features of benign and malicious traffic in total. Conceptually, the system operates from start to end in the following way, the preprocessed data in CSV with extracted features of network traffic is taken as input and given to the layered model of neurons. The first layer is an input layer that applies the activation function RELU and then uses the data for model training. Multilayer perceptrons (MLP) are used as the hidden layers arranged in descending order for learning of a system to achieve improved and highest levels of detection accuracy. The output layer with the activation function SOFTMAX contains 4 neurons to provide multiclass threat detection. For regulating the learning process, the attack categories and features are provided by the detection phase.

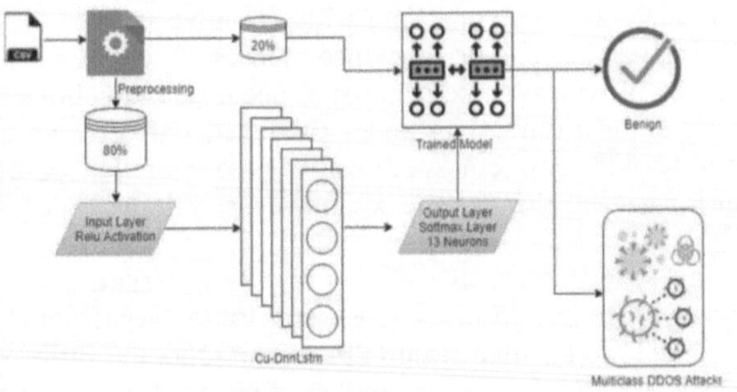

Figure 2. Proposed CuDNNLSTM architecture.

It has two phases, training and testing. First, the model is trained using the training data split into 80% of the whole dataset. Once the feature normalisation and data preprocessing phase are completed, the DL model is applied over normalised data to process the training phase. The iterations are set accordingly for pattern learning. When the model completed the training process, the trained would be used for the testing phase, where 20% of data was stored untouched while splitting for the prediction of results and performance evaluation.

4.4 *Structure of algorithm*

This subsection demonstrates the structure of the proposed DL model illustrated in Table 1. The layers and neurons used for the model training phase are structured as follows. Four layers of DNNLSTM with 4 dense layers and one output layer are employed. The first layer of GPU enabled DNNLSTM model comprises 700 neurons. In the second layer, 600 neurons are contained.

Further, the third and fourth layers of the model are composed of 500 and 200 neurons, respectively. The return sequence for the first three layers is valid; for the last DNNLSTM layer, the sequence is false. Additionally, the exact sequence uses four dense layers with the neurons of 500, 400, 200, and 50. The activation function (AF) RELU is used for all these layers, and the categorical cross-entropy is used for the loss function (LF) for DL models. The optimiser ADAM is employed with a batch size of 64 for the iterations of 70 epochs. Finally, the last layer of the model contains 14 neurons, one for benign class and 13 for the prediction of 13 attack classes.

Table 1. Proposed architecture of the algorithm.

Algorithm	Neurons	Layers	AF	LF	Model Optimiser	Epochs	Batch Size
Cu-DNNLSTM	(700,600,500,200)	CuDNNLSTM Layers (4)	Relu		Adam	70	64
	-	Dropout layer					
	(500,400,200,50)	Dense Layers (4)					
	14	Output Layer	Softmax				
	Activation Function: AF, Loss Function : LF						

5 RESULTS AND DISCUSSION

This section explains the overall results of all three DL models with relevant figures and graphs. Some standard and well-known metrics have been chosen for analysing each model's performance, such as Accuracy, Precision, Recall, FPR, and F1-score. The values or outcomes using these metrics are discussed and compared to each other to prove the proposed scheme's superiority. Moreover, as most of the research has been conducted in the same area, the proposed model has been compared with available state-of-the-art detection systems to show promising results. The performance metrics are defined for the evaluation of DL classifiers. Accuracy provides the percentage of accurate predictions made by a trained model. Recall describes the ratio of instances predicted correctly as positive. Precision evaluates the probability of the correctness of positive prediction. F1-score works to bend the recall and precision for getting the average weight. This is done to acquire relative values such as 1 and 0.

Further, TP such as True Positive as well as (TN) True Negative denotes the correctly predicted values, whereas False Positives (FP) and False Negatives (FN) indicate misclassified incidents. The True Positive Rate (TPR) represents the ratio of accurate predictions. False Positive Rate (FPR) depicts the number of illustrations classified in class X while belonging to a distinct class, besides the instances not present in class X. However, the experimentation results of the GPU-enabled Cu-DNNLSTM are discussed.

DDoS detection scheme. The proposed model has been compared with two DL models to show promising results. The overall detection accuracy of each model is illustrated in Figure 3.

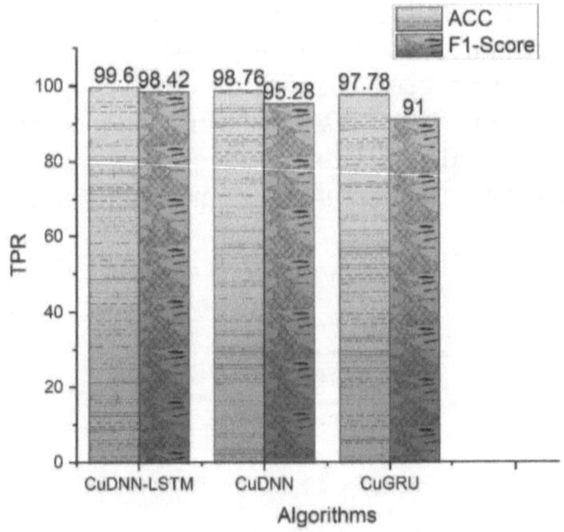

Figure 3. Obtained accuracy of DL models.

The detailed analysis of this system is illustrated in Figure 4, where all metrics results of compared models are presented. The Cu-DNNLSTM obtained a high True Positive Rate (TPR) of 98.25% than the other two models. Further, the proposed scheme outperformed the other DL models by achieving high values of precision, recall, and f1-score of 98.65%, 98.36%, and 98.43%, respectively.

On the other hand, the DNN and GRU are less efficient and achieved values of less than 96%. Additionally, the assessment of True Positive Rate (TPR) as depicted in the figure, has also been considered for enhanced evaluation of the offered mechanism.

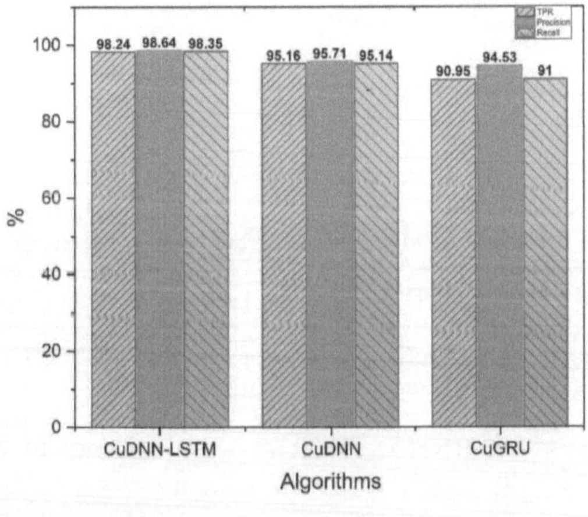

Figure 4. Precision, recall, F1-score, and TPR of DL models.

For further evaluation of the proposed model, other metrics are also analysed. For instance, FPR, FNR, and False Discovery Rate (FDR) are illustrated in Figure 5. The proportion of incorrectly classified positive samples gives the False Negative Rate (FNR). The False Positive Rate (FPR), also called the false alarms rate or FAR, characterises the ratio between the negative samples that are incorrectly classified and the total amount of negative samples. False Discovery Rate (FDR) measures to supplement the NPV and PPV, correspondingly.

As illustrated in Figure 5, the FPR, FNR, and FDR values of the proposed scheme are in the range of 0 and 0.02, which is appropriate for DDoS detection in organisational environments. The experimental results for all algorithms on the DDoS dataset state the efficiency and promising performance of the detection model by all metrics. The best results for the Cu-DNNLSTM model have been ensured by comparing it with the other two DL models. Lastly, the comparison of the research work with other current state-of- the-art detection schemes has been provided in Table 2, which displays the efficiency and scalability of the Cu- DNNLSTM enabled architecture by comparing it with following other existing solutions.

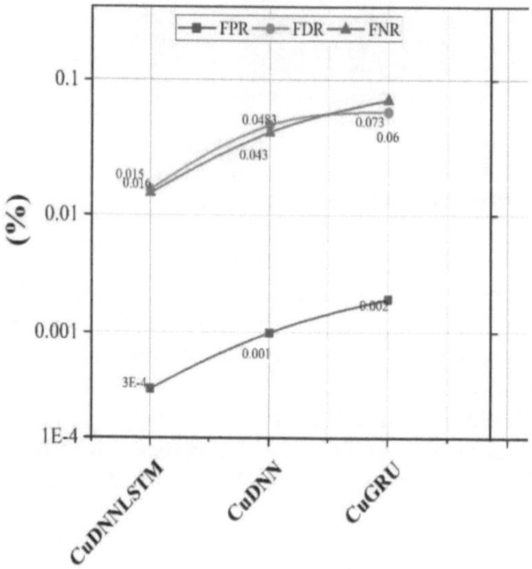

Figure 5. FDR, FPR, and FNR of DL-models.

Table 2. Comparison of the proposed model with current state-of-the-art detection mechanisms.

Parameters	Hussain [16]	Elsayed [17]	Ujjan [18]	Novaes [19]	Our model
Dataset	CICDDoS2019	CICDDoS2019	Author's own generated dataset	CICDDoS2019	CICDDoS2019
Technique	ML	AE-RNN	SAE	LSTM- FUZZY	Cu- DNNLSTM
Binary	u·	u·	u·	–	–
Multiclass	–	–	–	u·	u·
Accuracy	–	99%	91%	–	99.60%
Precision	96.9%	99%	95%	97.89%	98.64%
Recall	96.4%	99%	83%	93.13%	98.36%
F1-score	96.2%	99%	88.1%	–	98.43%
FPR	3.65%	–	4%	–	0.0003
TPR	96.35%	–	95%	96.22%	98.25%
Epochs	500	50	–	–	70

6 CONCLUSION

The article proposed a DL based approach for smart and sustainable threat intelligence in enterprise networks. The proposed approach efficiently detects multi-class sophisticated DDoS attacks. The major shift from classical machine learning to deep learning has motivated us to propose a novel DL-based technique for smart and sustainable threat intelligence in emerging networks. Our primary goal is achieve high detection threat accuracy with low false positives and that is the main goal of every smart and sustainable threat intelligence system. Besides, the proposed solution is cost effective and scalable. Consequently, the proposed solution can be easily customized and implemented for any emerging network. Therefore, the authors employed a Cu-DNNLSTM deep model to develop a multi-DDoS detection mechanism (MDDM) for large and distributed enterprise networks. The experimental analysis of the approach is conducted on the CICDDoS2019 state of the art dataset by importing Keras on Tensorflow using the Anaconda environment.

Further, the proposed MDDM is compared with two state-of-the-art DL algorithms with various evaluation metrics to confirm the efficiency of the offered scheme. To compare, the author has used standard metrics such as accuracy, FPR, precision, TPR/recall, FDR, and F1-score. The evaluation outcomes prove that the Cu-DNNLSTM-enabled MDDM scheme outperformed the other benchmark DL schemes for detecting varied DDoS attacks. The proposed scheme reaches high detection accuracy of 99.6% with a significantly low rate of false positives. Therefore, the MDDM evidences the appropriateness of DDoS detection in organisational networks.

In the future work, the proposed model will be evaluated with the different state-of-the-art ememrging learning models. Besides, the proposed technique can also be tested and validated for the emerging computational and communication paradigms such as Internet of Things (IoT), Industrial Internet of Things (IIoT), Mobile Edge Comuting, FoG computing and Volunteer Computing.

REFERENCES

[1] Peng, W., Kong, X., Peng, G., Li, X., and Wang, Z. (2019, July). Network Intrusion Detection Based on Deep Learning. *International Conference on Communications, Information System and Computer Engineering (CISCE)* 2019 (pp. 431–435). IEEE.

[2] Samta, R., and Sood, M. Analysis and Mitigation of DDoS Flooding Attacks in Software Defined *Networks. International Conference on Innovative Computing and Communications (2020)* (pp. 337–355). Springer, Singapore.

[3] Dhayanidhi, G. (2022). *Research on IoT Threats & Implementation of AI/ML to Address Emerging Cybersecurity Issues in IoT with Cloud Computing.*

[4] Gao, D., Liu, Z., Liu, Y., Foh, C. H., Zhi, T., and Chao, H. C. (2018). Defending against Packet-In messages flooding attack under SDN context. *Soft Computing*, 22(20), 6797–6809.

[5] Khan, Muhammad Taimoor, Adnan Akhunzada, and Sherali Zeadally. (2022). Proactive defense for Fog-to-Things critical infrastructure. *IEEE Communications Magazine.*

[6] Bawany, Narmeen Zakaria, Jawwad A. Shamsi, and Khaled Salah. (2017). DDoS attack detection and mitigation using SDN: methods, practices, and solutions. *Arabian Journal for Science and Engineering* 42.2 : 425–441.

[7] Nagisetty, A., and Gupta, G. P. (2019, March). Framework for detection of malicious activities in iot networks using keras deep learning library. *In 2019 3rd International Conference on Computing Methodologies and Communication (ICCMC)* (pp. 633–637). IEEE.

[8] Yu, Xueshan, et al. (2019). Design of DDoS attack detection system based on intelligent bee colony algorithm. *International Journal of Computational Science and Engineering* 19.2: 223–232.

[9] Elsayed, M. S., Le-Khac, N. A., Dev, S., and Jurcut, A. D. (2020, August). Ddosnet: A deep-learning model for detecting network attacks. In *2020 IEEE 21st International Symposium on "A World of Wireless, Mobile and Multimedia Networks" (WoWMoM)* (pp. 391–396). IEEE.

[10] Berral, J.L., Poggi, N., Alonso, J., Gavalda, R., Torres, J., and Parashar, M. (2008). Adaptive distributed mechanism against flooding network attacks based on machine learning. *In Proceedings of the 1st ACM Workshop on AISec ACM*, 43–50.
[11] Li, Chuanhuang, et al. (2018). Detection and defence of DDoS attack-based on deep learning in OpenFlow-based SDN. *International Journal of Communication Systems* 31.5: e3497.
[12] Tuan A Tang, L. M. (2018). Deep recurrent neural network for intrusion detection in SDN-based networks. *IEEE International Conference on Network Softwarization (NetSo. 2018)*. Montreal, QC, Canada: IEEE.
[13] Yuan, X., Li, C., and Li, X. (2017, May). DeepDefense: identifying DDoS attack via deep learning. *In 2017 IEEE International Conference on Smart Computing (SMARTCOMP)* (pp. 1–8). IEEE.
[14] Diro, A. A., and Chilamkurti, N. (2018). Distributed attack detection scheme using deep learning approach for Internet of Things. *Future Generation Computer Systems*, 82, 761–768.
[15] Sharafaldin, I., Lashkari, A. H., and Ghorbani, A. A. (2018, January). Toward generating a new intrusion detection dataset and intrusion traffic characterisation. *In ICISSP* (pp. 108–116).
[16] Hussain, Y. S. (2020). *Network Intrusion Detection for Distributed Denial-of-Service (DDoS) Attacks using Machine Learning Classification Techniques.*
[17] Elsayed, M. S., Le-Khac, N. A., Dev, S., and Jurcut, A. D. (2020). DDoSNet: A Deep-Learning Model for Detecting Network Attacks arXiv preprint arXiv:2006.13981
[18] Ujjan, R. M. A., Pervez, Z., Dahal, K., Bashir, A. K., Mumtaz, R., and Gonzalez, J. (2020). Towards sFlow and adaptive polling sampling' for deep learning based DDoS detection in SDN. *Future Generation Computer Systems*, 111, 763–779.
[19] Novaes, M. P., Carvalho, L. F., Lloret, J., and Proenc¸a, M. L. (2020). *Long Short-Term Memory and Fuzzy Logic for Anomaly Detection and Mitigation in Software-Defined Network Environment*. IEEE Access, 8, 83765–83781.

Innovation and Technological Advances for Sustainability – Al-Naemi et al. (Eds)

Mitigating energy poverty in sub-Saharan Africa through smart metering and sustainable energy access programs

Bennour Bacar
Department of Electrical, Electronic and Computer Engineering, Cape Peninsula University of Technology, Cape Town, South Africa

Ali Almaktoof
Department of Electrical, Electronic and Computer Engineering, Cape Peninsula University of Technology, Cape Town, South Africa
ORCID ID: 0000-0002-1134-5587

ABSTRACT: Electricity access is a challenge faced globally, especially in sub-Saharan Africa (SSA), where over half of the world's energy-poor population is located. Energy data provided by evolving smart metering (SM) technologies, applied to energy access frameworks such as Sustainable Energy Access Planning (SEAP), can enable the planning, implementation, and maintenance of energy systems for energy-poor areas. This study aims to analyse such applicability of SM to the SEAP framework, where household (HH) energy data based on an SSA country's criteria for energy access is studied and defined into an energy profile, which then serves as input to a smart metering experiment comprising a household's electrical load, a smart meter that measures electricity usage, and a Meter Data Management System (MDMS) software that remotely collects the data through General Packet Radio Service (GPRS) communications. An analysis is performed on the applicability of the acquired energy data to the SEAP framework and over three exercises that explore the analysis and calculation of energy access indicators, demand forecasting via machine learning (ML), and the optimization and cost analysis of energy systems. The household's measured energy data not only applies to all the assessments in the SEAP framework but also reveals the prospect of generating data that can be further used on applications requiring a particular range of datasets. With these capabilities enabled, policymakers and energy planners can use the enhanced data to determine energy access indicators within a program, unlock data forecasting elements, and optimize energy systems when aspects like renewables, low emissions, sizing, and cost are concerned.

Keywords: demand prediction, energy access, energy poverty, renewable systems optimization, smart metering, Sub-Saharan Africa

1 INTRODUCTION

Worldwide, nearly 761 million people have no access to electricity, with over 53% in SSA alone; furthermore, while on one end only 28% can use electricity, on the other end the rural population is the most affected by its lack of access, despite the region's economic growth [1]. In the context of basic access to electricity for human and economic growth, the UN has established global access to modern energy services as a universal goal for 2030, defining access to electricity as the "availability of an electricity connection at home or the use of electricity as the primary source for lighting" [2,3]; particularly, the SEAP addresses this

DOI: 10.1201/9781003496724-27

challenge by supporting the design of clean, affordable, and accessible energy systems for both the poor and non-poor [4]. Through this framework, household and energy aspects are identified based on primary (surveys, interviews, etc.) and secondary (records from government and health institutions, etc.) data collection to estimate future energy needs [5,6]; these bring challenges though, from planning input data that could be estimated missing, or old, to no output data on monitoring and tracking of the energy systems implemented. By monitoring, logging, collecting, and providing accurate data, SM technologies address these limitations, allowing energy planners to correctly measure and acquire information that, in combination with other sources (e.g., data from HH surveys), assist in motivating and making informed decisions for an energy access program; moreover, the measured HH energy data collection and monitoring renders insights, forecasting, better planning, effective implementation, tracking, and improved results.

1.1 *Smart metering systems*

Reinforced by technology growth, SM has evolved from just basic manual meter readings to incorporating advanced smart grids, including elements of prosuming, distributed generation, and charging stations for electric vehicles (EVs) while playing a more active role in reducing carbon footprint [7]. A typical SM system is defined by [8] as one that comprises smart meters (metering gateway), control devices, appliances, a communication link, and a control centre; furthermore, it may include energy measured at a distribution point for an area [9]. These elements are shown in Figure 1.

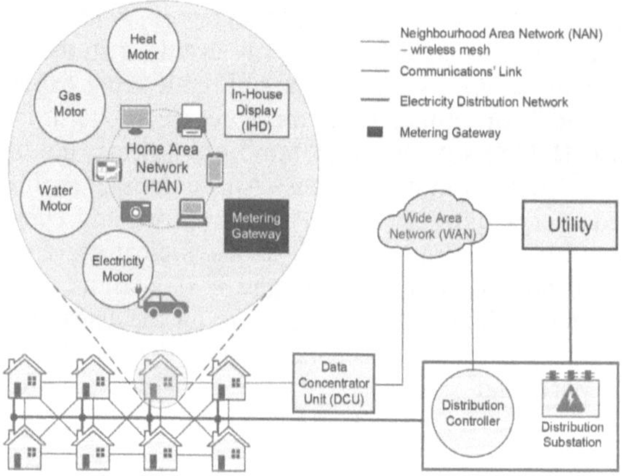

Figure 1. Elements of a smart metering system, based on [7].

As stated by [5], SM systems contribute to increased efficiency in utilities' operations and quality of supply (QoS), as well as improved consumer satisfaction and involvement, with key benefits that include reduced carbon footprint from reduced physical presence, more accurate bills, reduced aggregated technical and commercial (AT&C) losses, integration with renewable energies, electricity usage monitoring and profiling, among others.

The smart meter is the main device of an SM system, capable of both energy metering and intelligence [8]. From an endpoint (such as a household), it records energy data and remotely sends it to the software application (middleware) using communication mediums such as Global Systems for Mobile Communications (GSM), GPRS, Radio Frequency (RF), Power Line Carrier (PLC), etc [10]; moreover, it can be accessed locally for both meter programming and data collection, through interfaces such as an infra-red (IR) optical probe [11]. While communication networks are an important element of SM systems, they also bring disadvantages to security and privacy, such as Internet attacks and exposure to personal data

[9]. The United States is the most targeted country by cyber threats, while sector-wise the energy and public sectors as the least and most targeted respectively [12]. This indicates that while SSA and SM systems might not be a favourite target for cyber threats, attention should be given to the security of SM data stored by public utilities and NGOs.

Smart meter deployments are rising across the world, expected to double by 2024 [13]. With Asia on the lead (Figure 2), countries like China are at the forefront, while projections for emerging market countries indicate 178 million smart meters' projected deployment; within SSA, some of those emerging countries such as Kenya, Ghana, Nigeria and South Africa plan on taking the same path to address AT&C losses, however with financing as the main challenge to overcome [14].

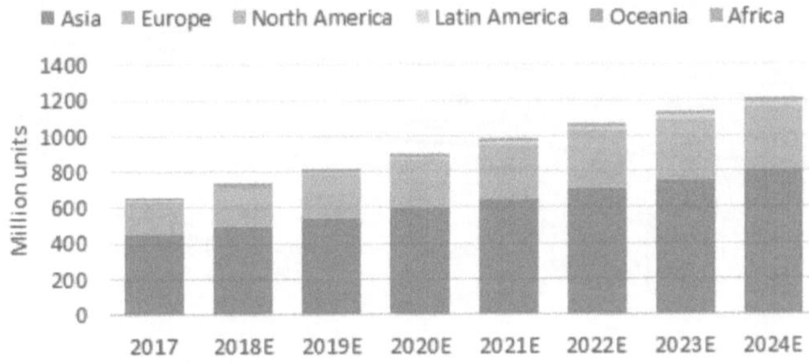

Figure 2. Smart meter deployment, based on [13].

Along with such growth, the emergence of COVID-19 in 2020 has highlighted the importance of SM systems, with benefits that included reduced physical presence to collect meter readings since these could be read remotely, and analysis of energy usage data collected from the meters, as shown by [15] in Figure 3, where during strict lockdown energy demand had decreased for non-residential consumers (businesses closing, employees working from home, etc.).

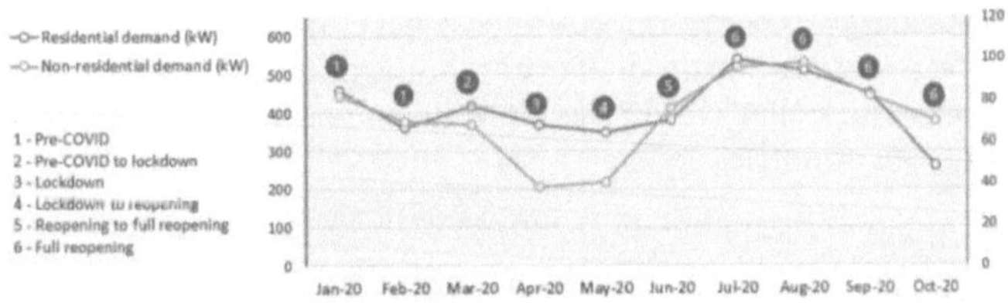

Figure 3. Demand profile during COVID-19, based on [15].

1.2 *Energy access through the SEAP framework*

A key element to planning the ideal solution for an energy access program, the SEAP framework identifies sustainable and cost-effective technologies and resources, aiming at "providing universal access to basic energy services and to assess the affordability of cleaner-energy service options to energy-poor households" [3]; furthermore, its features, in particular for developing countries in the Middle East and SSA regions, distinguish this framework from other conventional energy and electricity planning frameworks, through: (i) social inclusiveness, considering the poorest HHs when access to electricity and other cleaner energies is concerned; (ii) affordability, quality and sustainability by locally analysing acceptable, sustainable, cleaner, and reliable energy options; (v) environment and social well-being on generation of key data for investment requirements, along with benefits/improvements in aspects such as energy inequality, greenhouse gases (GHGs), environmental quality, and social well-being; continuity of the energy access program by examining the adequacy, reliability, and sustainability of clean-energy services' options.

The SEAP framework encompasses seven assessments; Figure 4 shows those assessments and their relationship (vertical, sequential, and horizontal), e.g., the resource, cost, benefit, and affordability assessments provide inputs required for the sustainability assessment.

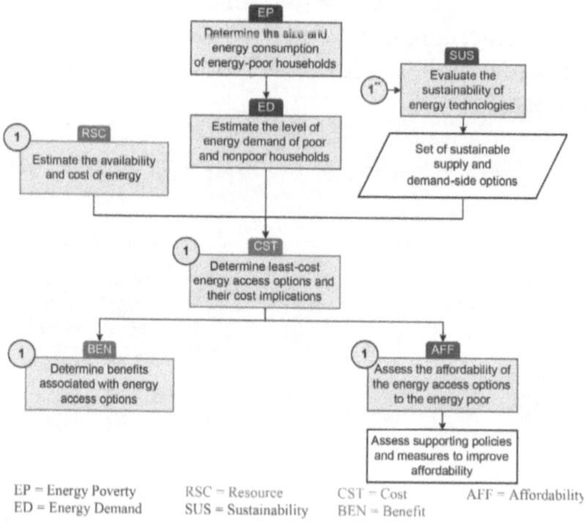

Figure 4. SEAP assessments' flow diagram [3].

1.3 *Considerations*

Smart metering and energy access have been addressed separately in the literature reviewed; in particular, a research by [16] analysed the use of smart meters at HHs to address energy poverty in a developed world context, considering data given by the utility. This study, instead, approaches energy poverty in SSA's least developed and developing countries by directly using the SM system (smart meter, communications, software) to record low-tier energy consumption data, collect it, and explore its suitability to an energy access methodology relevant to those countries.

2 METHODOLOGY

Primary and secondary data are used in this research, collected from smart meter and existing SSA country databases respectively, among others; therefore, a quantitative research method is used. The data comprises energy consumption over time primarily, complemented by other related variables. Simple random probability sampling is applied in this research, based on the criteria described by [17], with the resulting random numbering removing the data bias factor; the data is sampled at 278 – 370, from a 1000 – 10 000 population range and within an error margin of 5% [17]. The specific size depends on the EAP's availability and detail, as well as other data relevant to the energy profile. The collected secondary data is based on area and time series, and derives from different sources that include the country's website, International Renewable Energy Agency, The World Bank, and others [17, p. 249]. For a period of 1–3 months, an SM experiment provides the primary data, by analyzing energy profile from secondary data, the latter including energy data and demographic data. Figure 5 shows the process for the collection and use of data for the SEAP framework.

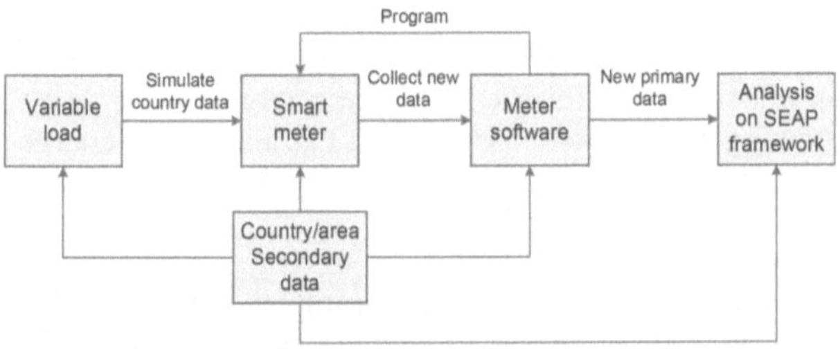

Figure 5. Method for data collection and analysis against SEAP.

3 EXPERIMENT, SIMULATION, AND ANALYSIS

3.1 *Analysis of country data*

Niger was the selected Western SSA country for the secondary data. It is placed third as one of the countries with the lowest per capita energy consumption in the world [18], with a population of nearly 24 million [19], mostly rural, as shown in the secondary data from Table 1. Considering the multi-tier framework (MTF) [20], an average Niger HH would be on tier two. But since most of the population in the country, which is rural, has no access to electricity (Table 3), lower tiers must be taken into account; what validates this is that on one end rural areas have no access to modern energy services, while on the other, urban areas having some connectivity also have challenges around poor access and unreliability [21]. At this point, SM technology would plug in to collect such data and provide an energy profile that is accurate.

Table 1. Niger energy and household data For 2014.

Indicator	Value	Unit
Electricity access: total population	15.77	% population
Electricity access: rural population	7.72	% population
Rural population	84.00	% population
Average HH size	6.00	Members per HH
Number of HHs	4.03	Million HHs
Electricity consumption per HH	307.14	kWh

3.2 *SM experiment*

From the acquired country profile, the SM system shown in Figure 6 was built to record energy data for a typical MTF tier-1 HH in Niger, with elements that include:

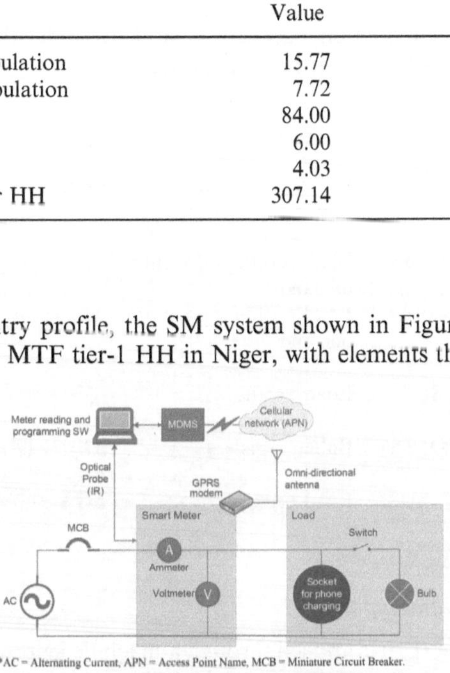

Figure 6. Designed system to simulate MTF tier-1 data for a household. *AC = Alternating Current, APN = Access Point Name, MCB = Miniature Circuit Breaker.

(1) The smart meter: recording bi-directional active and reactive power and energy, billing and demand registers, load profile (LP), events and network monitoring, acquired locally using an optical probe (infrared port), or remotely.
(2) The load: considering MTF's tier 1 for a HH, it has a socket to charge a cell phone, and a bulb; the latter has two stages, an initial 60W incandescent (stage-1), afterwards replaced by a 9W LED bulb (stage-2).
(3) The MDMS: with a daily schedule configured to remotely acquire meter data.
(4) The computer (PC): to access energy data using the MDMS, as well as to locally/remotely read and program the meter using meter configuration software (SW).
(5) The energy data collected: LP (also known as interval usage); billing registers (BR), and maximum demand (MD) registers.

3.3 *Data applicability to the SEAP framework*

Linking the energy data to the SEAP assessments identifies which indicators can be determined through the equations presented in Table 2; while to some assessments the calculations based on energy data are applicable, to others, it is a direct/indirect benefit. The equations consider the following variables:

- $\overline{E_{HE}}$ = average electricity consumption per household h, for the intended area/region and a given period, or if applicable the HHs minimum electricity threshold specific to the same area, in kWh.
- ΔE_{DP} = additional electricity required, in kWh, to raise a region's number of households h_P, from tier-n to tier-$(n+1)$, based on a HH's average electricity consumption E_n (given by $\overline{E_{HE}}$) and the minimum electricity threshold $E_{(n+1)}$ (less than E_n).
- $\overline{t_S}$ = the average daily hours of supply for a given number of households.
- ΔE_{de} = period-based difference in electricity consumption given by the start and end registers at stage 1 $(R_{12} - R_{11})$ and stage 2 $(R_{22} - R_{21})$.
- ΔC = cost difference between stage 1 C_1 and stage-2 C_2, based on the MDMS's simulated tariff and provisional bill.
- ΔG = GHG emission reduction between stage 1 G_1 and stage 2 G_2, based on the MDMS's consumption-derived environmental data.

Table 2. Applicable SEAP indicators and assessments.

Indicator and assessment	HH energy data input	Equations
Basic minimum energy requirement: EP, ED	Load profile, maximum demand	$\overline{E_{HE}} = \frac{\sum_{i=1}^{h} E_{Hi}}{h}$ (1)
Energy demand: EP, ED	Load profile, maximum demand	$\Delta E_{DP} = h_P \times \left(E_{(n+1)} - E_n\right)$ (2)
Hours of electricity supply: SUS, AFF	Load profile	$\overline{t_S} = \frac{\sum_{i=1}^{h} t_{Si}}{h}$ (3)
End-use device efficiency: ED, CST, SUS, BEN	Billing registers	$\Delta E_{pe} = (R_{12} - R_{11}) - (R_{22} - R_{21})$ (4) $\Delta C = C_1 - C_2$ (5)
GHG emissions: RSC, CST, SUS, BEN	Load profile	$\Delta G = G_1 - G_2$ (6)

3.4 *Load forecasting: Data to predict*

Electrical load forecasting (LF) is applied through a machine learning model to: (i) complete a year's worth of LP data (primary data from the SM experiment is of 3 months) to use for the optimization and cost analysis of energy systems, described further in this paper;

(ii) within the context of EP, ED, CST, and SUS, highlight SM energy data integration into ML and load prediction for future demand planning, preventive actions/maintenance, financial and tariff planning [22], especially when, according to [23], close to 90% of the nine models mostly used in LF are based on artificial intelligence (AI).

The small village of Gorou, in Niger, was selected for the exercise; according to [24] and [25], the village had no access to electricity, i.e., on MTF tier-0 [20] and was far from the national electric grid, reinforcing the need for micro-grid renewable energy options. The data set was built as shown in Figure 7; the smart meter's LP data is exported from the MDMS, and assuming it as the average demand per interval per HH, the total demand for the village is calculated and allotted to Gorou's historical 2021 weather data [26], as indicated in the sample from Table 3. A regression model is then created to predict demand. A total of five models is run, with the Neural Network Regression (NNR) model being selected based on, among other metrics, the lowest root-mean-square-error (RMSE) and highest coefficient of determination, or R-squared (R2), as shown in Table 4.

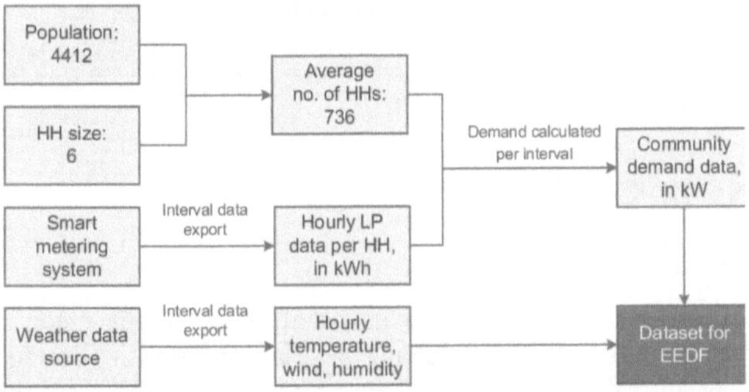

Figure 7. Dataset flow.

Table 3. A sample from the 12-month dataset.

Date	Time	Temp (°C)	Wind (km/h)	Humidity (%)	Pwr _cut	Demand (kW)
2021/01/01	00:00	24.00	6.00	19.00	NO	0.00
2021/01/01	01:00	23.00	2.00	19.00	NO	0.00
2021/01/01	02:00	23.00	6.00	20.00	NO	44.16
2021/01/01	03:00	21.00	0.00	21.00	NO	44.16

Table 4. Regression models and metrics

Metric	Linear	Boosted Decision Tree	Neural Network	Bayesian Linear	Decision Forest
RMSE	6.04	6.03	5.74	6.04	6.07
R^2	0.07	0.07	0.16	0.07	0.06

3.5 *Energy potential and renewables: Data to optimize*

An analysis of RSC, CST, BEN, AFF, and SUS is performed with techno-economic simulation software, using 1 year of measured (SM experiment) and predicted (ML model) datasets as input to evaluate and optimize design, costs, load profiles, components, and environmental variables [27]. While covering all the elements of the assessments, it allows the use of raw energy data collected from a HH, by using LP as direct input for the load setup, instead of the pre-defined synthetic load; the latter being the main benefit to be shown with this approach. Besides the LP dataset, other inputs include the solar and wind resources acquired directly from the software based on the location input (Gorou), as well as economic variables.The hybrid renewable energy system (HRES) comprises six electrical supply elements that include the renewable solar and wind options, the related storage (battery), and current conversion components (Table 5), with the respective input values of both capacity and costs.

Table 5. HRES components

Component	Capital (R)	Replacement (R)	O&M* (R/year)	Lifetime (years)
Electric load	–	–	–	–
Genset (diesel)	7 830.00	7 830.00	4 115.45	1.71
PV (solar)	39 150.00	39 150.00	156.60	25.00
G1 (wind)	109 620.00	109 620.00	1 096.20	20.00
1kWh lead-acid (battery)	4 698.00	4 698.00	156.60	10.00
Converters	4 698.00	4 698.00	–	15.00
Grid	125 280 (per km)	–	2 505.6 (per km)	–

*O&M = Operations and maintenance

4 RESULTS AND DISCUSSION

4.1 *Meter data collection*

With the configuration SW and IR optical probe attached to the meter's optical sensor, energy data from the load as well as device and installation data is collected from the meter. Real-time power, energy values, and event logs provide QoS power cut events, accurately logged from the time when the alarm was raised for a phase-cut, then a power cut, to restoration; one of the power cuts lasted for exactly 6 minutes and 58 seconds. These measures are significant to the SEAP framework, quantifying and tracking the aspects of adequacy of supply, reliability, and convenience for a household. From an MDMS perspective, different types of views and reports are provided for the data analysis. The daily usage graph in Figure 8 shows insights from the acquired LP data, where the measurements in different scenarios underline, in the context of the SEAP framework, the implementation of efficient options for demand-side technology [4]: the LED lamp, consuming 6 times less energy, proves more efficient and cost-effective [28] while providing more light, addressing the

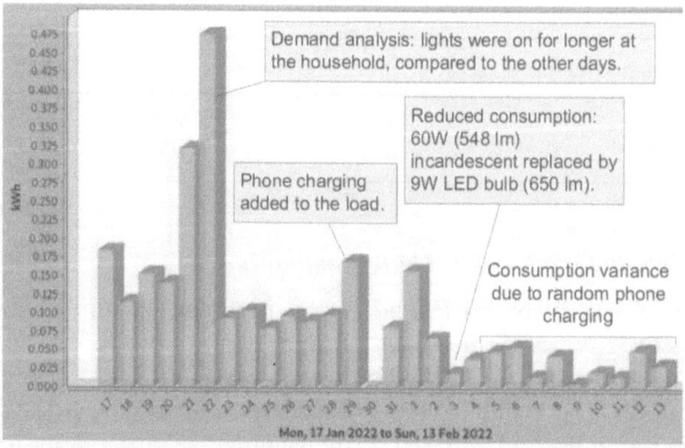

Figure 8. Daily usage analysis (kWh per day).

affordability (for both lamp and electricity bill) and adequacy (low power rating) factors, while at the same time reducing electricity access costs and power generation capacity requirements. Now further analysing LP to the level from every few hours to an hour, to half an hour, more usage patterns are observed as indicated in Figure 9, with higher energy demand at night for both charging and lighting, and stable voltage levels (only one phase used); such granularity would address affordability challenges via improved electricity tariffs [4] for the energy-poor, optimized tariffs for the non-energy poor, and enhanced monitoring to improve QoS (reliability). The United Nations states the importance of providing, where feasible, energy access from sources with low GHG emissions [3]. With the energy data collected, environmental impact analysis data (Table 6) is generated from the MDMS, with the compounds' emission/usage values over the specified consumption period, based on constants as defined by the local utility supply.

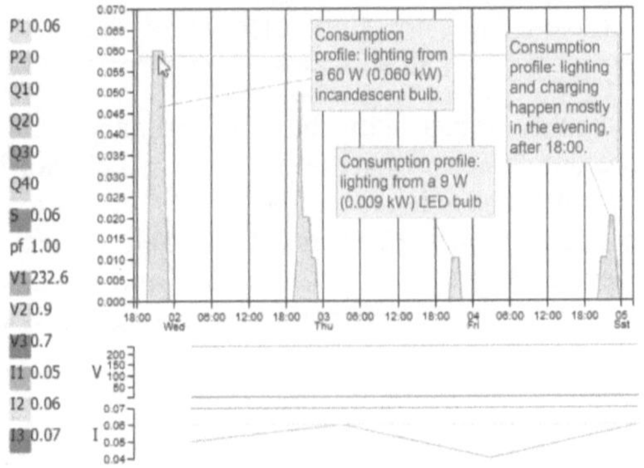

Figure 9. Half-hourly consumption analysis (kW).

4.2 *Assessment calculations*

The following aspects are discussed from the results, shown in Table 6: (i) based on the 3-month data from the SM experiment, the average from (1) is applied for a year (16 kWh), falling within MTF tier-1 as per load used for the experiment; (ii) Niger's number of rural HHs with access to electricity (Table 3) is considered as the sample for the energy demand calculation; moving these HHs to MTF tier-2 requires nearly 15 GWh of additional energy (2); (iii) from (3), electricity is available for 97% of the considered supply period; (iv) the consumption difference, based on the change from an incandescent to a more efficient LED bulb (Figure 8), indicates a monthly reduction by 246 Wh (4); from an MDMS-simulated tariff and bill, a derived saving of 8.37 R on the HH bill (5), and decrease in carbon emissions of 228.29 $kgCO_2$ (6), considering South Africa's grid emission factor of 0.928 $kgCO_2$ per kWh [29].

Table 6. Environmental impact (based on South Africa).

Implication	17/01/2022 16:00 to 13/02/2022 20:30	Daily Average	Monthly Average
Ash produced	0.4280 kg	0.0157 kg	0.4791 kg
CO_2 emissions	2.7334 kg	0.1005 kg	3.0601 kg
Coal use	1.4633 kg	0.0538 kg	1.6383 kg
NO_x emissions	0.0115 kg	0.0004 kg	0.0129 kg
Particulate emissions	9.1113 kg	0.3351 kg	10.200 kg
SO_2 emissions	0.0214 kg	0.0008 kg	0.0240 kg
Water use	0.0039 kl	0.0001 kl	0.0043 kl

4.3 *Predicted demand*

The ML exercise by NNR model is applied to predict the demand values based on temperature, wind, and humidity. Table 7 shows a sample of the predicted data, not only with the applicable transformations to improve predictive power, but also negative predicted data (scored labels); as there can be no negative demand, these were replaced by zeros for the final dataset.

Table 7. Sample of predicted data (scored labels).

Temperature	Sqrt (Wind)	LnPlus1 (Humidity)	Scored Labels
−0.855093474	0.293972368	0.498921986	0.669818521
−1.049094918	0.293972368	0.498921986	1.927718997
−1.049094918	0.333333333	0.476092568	−0.176963925
−1.243096362	0.368513866	0.476092568	0.534485459

The predicted data is then combined with the input primary/measured data (January to March 2021) for the final dataset covering the full year, which can be seen graphically in Figure 12. The initial spike in January relates to the use of a higher load (incandescent bulb), then changed to a lower load in early February (LED bulb); these stages are part of the experiment on end-use device efficiency for affordability and sustainability. Also observed with the predicted data is the increase in consumption during the Winter and Autumn seasons (July to November), typical for annual usage patterns.

4.4 *Energy system optimization and cost analysis*

The full dataset from the meter and prediction exercises is then used as load input to the techno-economic simulation software to perform a resource and cost analysis of an energy system to implement for energy access, in line with SEAP frameworks' resource, cost, benefit, affordability and sustainability assessments. The results in Table 8 compare the base

Table 8. Base system against HRES and full PV systems.

			Best-case system (optimized)	
Metrics	Unit	Base system: Genset	*Renewable-only: PV + 1 kWh LA + Converter*	*HRES: PV + Gen + 1 kWh LA + Converter*
NPC	R	9.04M	5.69M	3.02M
Capital	R	446 310.00	4.28M	1.11M
O&M	R/year	1.01M	167 058	225 317.00
LCOE	R/kWh	38.18	24.78	12.77
IRR	%	–	22.20	109
Discounted payback	Year	–	4.65	1.05
Simple payback	Year	–	4.28	1.01
Renewable fraction	%	–	100.00	31.10
CO_2 emission	kg/year	108 665.00	0.00	20 805.00
Fuel Consumption	L/year	41 513.00	0.00	7 948.00

*NPC = Net Present Cost, LCOE = Levelized Cost of Electricity, IRR = Internal rate of return.

energy system supplying electricity via diesel fuel-based generator, with the optimized systems; a carbon-free, renewable system is a clean, off-grid, but costly option, with a payback period of nearly 4 years (4 times longer than an HRES). Furthermore, on the HRES there were periods of the year with near-zero carbon emissions.

5 CONCLUSION

In this study, data-acquisition features of smart metering technologies have been coupled with energy access (SEAP) methods through energy data. Using input data acquired from a selected country in SSA, a smart metering system has been assembled, with a smart meter measuring electricity usage data and an MDMS wirelessly collecting it. The applicability of this data to the energy access framework has then been analysed against its assessments and explored by applying it to different, but interdependent applications; these included the calculation of indicators using the energy access framework as the baseline, data prediction (demand forecasting) via machine learning exercise to build a full year dataset from the available meter data, and energy systems' optimization and cost sensitivity analysis by applying the full dataset (measured and predicted) to simulation software. With the contribution provided in this study, energy planners can effectively extract, interpret, and use real energy data toward implementing and sustaining an optimal energy system for the energy poor. It is a ripple effect that will be valuable to the EAP, implementation, and tracking, but ultimately and most importantly, to those who need access to modern energy. Future work on aspects such as reliable SM communication technologies fit for the reality and area, local empowerment to maintain such systems, and improving predictive power for energy data, will be beneficial to the solution proposed.

REFERENCES

[1] The World Bank, Access to electricity (% of population), (2021). https://data.worldbank.org/indicator/EG.ELC.ACCS.ZS?locations=ZG (accessed Jan. 24, 2022).

[2] International Energy Agency & The World Bank, *Sustainable Energy for All 2013–2014: Global Tracking Framework Report*, Washington, DC, (2014).

[3] Shrestha R. M. and Acharya J. S., (2015). *Sustainable Energy Access Planning: A Framework.* Manila: Asian Development Bank.

[4] Asian Development Bank, *Sustainable Energy Access Planning: A Case Study, no.* August. Manila: Asian Development Bank, (2018).

[5] Asian Development Bank, *Smart Metering Roadmap For Nepal*, Asian Development Bank, Manila, (2018). doi: 10.1201/b11300-4.

[6] Benedictine University Library, Public Health Research Guide: Primary & Secondary Data Definitions, (2021). https://researchguides.ben.edu/c.php?g=282050&p=4036581 (accessed Apr. 14, 2021).

[7] Živi N. S., Ur-rehman O., Ruland C., Member S., Meter A. M., and Mmr R., (2015). *Evolution of Smart Metering Systems*, pp. 635–638.

[8] Weranga K. S. K., Kumarawadu S., and Chandima D. P., (2014). *Smart Metering Design and Applications.* Singapore: Springer.

[9] Sun H., Hatziargyriou N., Poor H. V., Carpanini L., and Fornié M. A. S., (2016). *Smarter Energy: From Smart Metering to the Smart Grid.* London: The Institution of Engineering and Technology.

[10] Toledo F., (2013). *Smart Metering Handbook.* Tulsa: PennWell.

[11] Bajer M., (2019). Building advanced metering infrastructure using elasticsearch database and IEC 62056-21 protocol, [Online].

[12] Lambert J., (2021). *Microsoft Digital Defense Report Shares New Insights on Nation-state Attacks.*

[13] Wood Mackenzie, Global smart meter total to double by 2024 with Asia in the lead, (2019).

[14] Chakerian S., (2021). *Emerging Markets Poised to Achieve Smart Metering Despite COVID-19.*

[15] García S., Parejo A., Personal E., Guerrero J. I., Biscarri F., and Léon C., (2021). A retrospective analysis of the impact of the COVID-19 restrictions on energy consumption at a disaggregated level, [Online].

[16] Casals L. C., Herrero S. T., Barbero M., and Corchero C., (2020). Smart meters tackling energy poverty mitigation: Uses, risks and approaches, *2020 IEEE Electr. Power Energy Conf. EPEC 2020*, vol. 3.
[17] Saunders M., Lewis P., and Thornhill A., (2007). *Research Methods for Business Students*, 4th ed. Harlow: Pearson.
[18] The World Bank, Access to Electricity, Rural (% of rural population) – Niger, (2022).
[19] Central Intelligence Agency, *The World Factbook*, (2022). https://www.cia.gov/the-world-factbook/countries/niger/#geography (accessed Feb. 13, 2022).
[20] The World Bank, *Beyond Connections: Energy Access Redefined*, pp. 1–244, (2015).
[21] The World Bank, *International Development Association Project Paper on a Proposed Additional Credit in The Amount of Eur 48.0 Million (US$56 Million Equivalent) and a Proposed Additional Grant in The Amount of SDR 9.9 Million (US$14 Million Equivalent) to the Republic of*, (2018).
[22] Malik H. and Iqbal A., (2021). *Intelligent Data Analytics for Time-Series Load Forecasting Using Fuzzy Rein- forcement Learning (FRL)*.
[23] Nti I. K., Teimeh M., Boateng O. N., and Adekoya A. F., (2020). Electricity load forecasting: a systematic review, *J. Electr. Syst. Inf. Technol.*, vol. 8.
[24] Plan International España, *Access to Electricity in Gorou*, Niger, (2017).
[25] ECOWAS Regional Centre for Renewable Energy and Energy Efficiency, Inauguration of a Solar Mini-Grid Project in Niger, (2018).
[26] Freemeteo, *Gorou – Daily Weather History*, (2022). https://ke.freemeteo.com/weather/ (accessed Apr. 30, 2022).
[27] HOMER Energy, *HOMER Pro (Version 3.14) [Computer Program]*. (2017).
[28] Desai S., *Handbook of Energy Audit*. New Delhi: McGraw Hill Education, (2015).
[29] Carbon Footprint, *Country Specific Electricity Grid Greenhouse Gas Emission Factors*, (2020). [Online]. Available: https://www.carbonfootprint.com/docs/2020_06_emissions_factors_sources_for_2020_electricity_v1_1.pdf
[30] The World Bank, *Access to Electricity (% of population) – Sub-Saharan Africa*, (2021).
[31] The United Nations, *Household Size and Composition*, (2019). https://population.un.org/Household/index.html#/countries/562 (accessed Feb. 13, 2022).

Innovation and Technological Advances for Sustainability – Al-Naemi et al. (Eds)

Assessment of the GHG performance of a gas plant through an interactive business intelligence dashboard

A.A. Al-Mohannadi, M.A. Masoud & B.N. Al-Naemi
Engineering Management Program, Qatar University, Doha, Qatar

M. Kucukvar
Department of Mechanical and Industrial Engineering, Qatar University, Doha, Qatar

ABSTRACT: This study investigated the greenhouse gas performance of an operating plant from a business intelligence perspective. 2017-2021 GHG data was collected from the plant, transformed, and loaded into Microsoft Power BI, which was used to build the data model. A dynamic visual dashboard was created for data analysis. The analysis was categorized into areas based on the requirements of the functional dashboard. These categories were GHG emissions per time period, top emitting process units, distribution of GHG emissions by pre-defined categories, and energy intensity. The dashboard showed stable emissions and energy intensity from 2018-2021. Dashboard analysis also demonstrated major emissions contributions from gas turbine generators and fired heaters, mainly from fuel *I*, highlighting key focus areas for future GHG reduction initiatives. GHG abatement strategies were proposed to provide direction for improvement. Although the dashboard was intended to aid strategic decision making, it can be adapted for greater insights and for operational and tactical decision making by using more data in the model and closer cooperation with data and process owners and operators.

Keywords: Data Visualization, Business Intelligence, Data Analytics, Descriptive Analytics, Big Data, Greenhouse Gases, Dashboard, Power BI, Energy Efficiency, Energy Intensity, Gas Plant

1 INTRODUCTION

1.1 *Sustainability and greenhouse gases*

Breakthrough developments in technology since the late 20th century allowed humanity to transition into the Information Age [1,2]. In parallel, industries are shifting from improving production processes through automation to a greater reliance on artificial intelligence and machine learning to bring optimization to the next level [3]. Such an industrial framework, "Industry 4.0", is characterized by smarter production processes, decentralization, human centricity, digitalization, and an emphasis on corporate social responsibility [4].

From a business perspective, companies should strive to demonstrate an active concern for the environment and society at large to remain competitive while contributing towards sustainable development [5]. Sustainable development refers to the achievement of society's needs without negatively impacting the ability of future generations to meet their needs [6], [7]. Sustainability is founded on three pillars: economic, environmental, and societal aspects [7]. Despite the minor improvement in the fight against climate change during the pandemic, the outcome is still bleak. Research demonstrates that unmitigated climate change could

DOI: 10.1201/9781003496724-28

result in the same death rates as the pandemic within the next 40 years [8]. Hence, there is more weight on tackling the environmental pillar of sustainability with greater urgency.

One of the major actors in the global climate crisis is the energy industry. Globally, operations based on fossil fuel extraction and operations represent 40% of CO_2 emissions [9]. Although the world is gradually shifting away from fossil fuels, emissions from the use of fossil fuels continues to increase, especially as the need for affordable, accessible energy increases with higher levels of economic growth [10,11]. To align with shareholder and stakeholder concerns for the environment, key players such as Shell, BP, and ExxonMobil, have set targets to have a net-zero carbon footprint by 2050 [12]–14]. Investments in renewables, carbon offsets, and infrastructure electrification indicate that the energy transition is under way.

However, operations cannot be improved if they are not measured using relevant metrics. The industry standard for reporting emissions is in tCO_2e, where greenhouse gases (GHGs) are normalized based on their global warming potential relative to a CO_2 molecule [15]. For operating assets, energy intensity is a common indicator used to represent energy consumption per unit of production. These are key metrics in the reporting of an asset's GHG performance, but collecting the required data and organizing it in a structured manner is not always a simple task.

1.2 *The role of data analytics*

Ease of access to data since the dawn of the information revolution in the 1990s introduced a wealth of potential knowledge sources that has driven data-based decisions [16]. Large-scale analysis of vast amounts of data to identify trends and extract value is the core of big data analytics [16]. Subsets of analytics include descriptive, predictive, and prescriptive, depending on the underlying intention of gathering insights from the data [16]. One such framework that can be applied is that of business intelligence (BI), which covers the end-to-end process of converting raw data into information, followed by useful knowledge, and ultimately into action plans to achieve operational, tactical, or strategic objectives [17,18]. Performance dashboards play a role by enabling process monitoring, data analysis, and visual reporting in a dynamic way to support the managerial decision-making process [17].

BI dashboards have been used to visually quantify the sustainability of a multitude of processes. On the global scale, a global sustainable tourism dashboard was developed, using indicators pertaining to energy, water usage, and waste generation by hotel guests and CO_2 emissions by flight cabin class, to aid policy makers in assessing the carbon footprint of their tourism industry and how it changes over time [19]. More recently, a dashboard was developed to visualize carbon hotspots of cities as a result of pandemic lockdowns [20]. On a more local scale, dashboards were developed to track the GHG performance of a university campus, also assessing its recycling, transport, and waste management initiatives [21]. To support urban planning, a dashboard was developed to track electric vehicle adoption across several US states [22]. BI dashboards were applied in identifying the distribution of emissions across a dairy food supply chain [23]. In the energy industry, a GHG performance dashboard was developed to track the heat rates and GHG intensities from a power plant [24]. However, the authors found limited literature on GHG dashboards specifically for gas plants.

1.3 *Objectives and context*

The objective of this study is to analyze GHG data for a gas plant through the BI framework. An overview of the natural gas value change is shown in Figure 1 for perspective. It includes upstream natural gas extraction to LNG, CNG, GTL, and methanol processing routes, as well as direct export as sales gas.

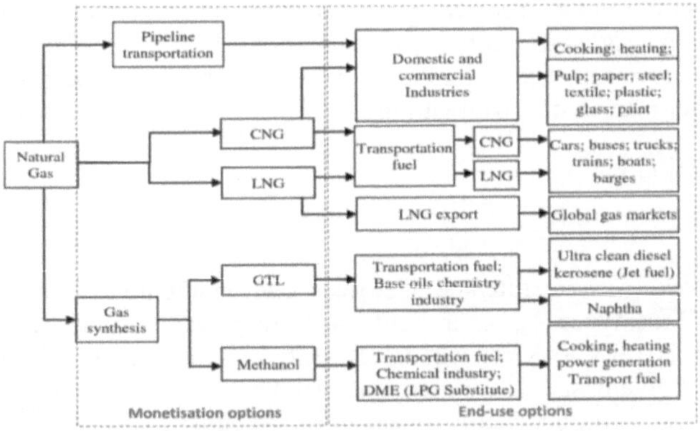

Figure 1. Natural gas value chain in context. Source: Reproduced from [25].

The aim is to identify patterns in the data, possible reasons for that behavior, and provide strategic-level guidance to improve the plant's GHG performance. Specifically, the primary means of applying BI is to generate a data visualization dashboard. Visualizations on the dashboard are driven by the type of GHG data available and the end-goal desired by plant management. Hence, requirements include the need to assess performance over time (internal benchmarking), identify the main emissions contributors (to drive optimization initiatives), and a breakdown of GHG distribution by pre-defined categories (to understand nature of emissions for deeper insights).

The main limitation of this study is data accessibility, as the nature of the data source requires retaining confidentiality, limiting the amount of available data. The amount of data was limited to scope 1 emissions (sources controlled by the plant), with no access to scope 2 (from purchased energy consumed by the plant) and scope 3 (use of products exported by the plant) emissions data. Furthermore, entities and attributes have been generalized.

2 METHODOLOGY

The research methodology is illustrated in Figure 2.

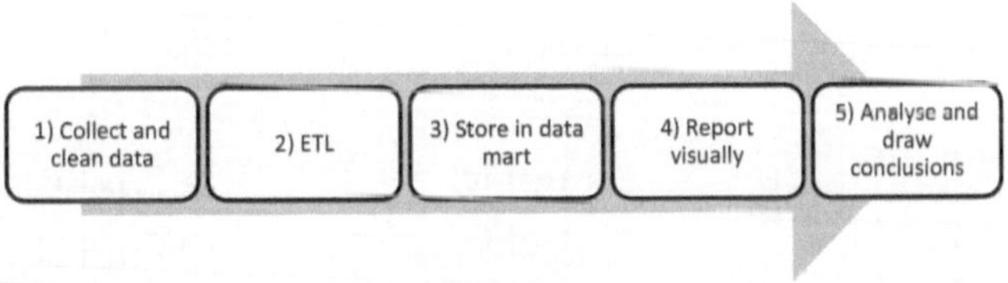

Figure 2. Research methodology.

Data from 2017-2021 was collected from the plant. The time series data includes total GHG emissions, process units, production units, production numbers, energy intensity, and distribution of emissions by category. For energy intensity, 2017 data was unavailable. The source data is then cleaned and re-organized to ensure consistency in the data structure. It is then extracted, transformed, and loaded (ETL) into Microsoft Power BI. The transformation step is conducted using the Query Editor in Power BI, which requires renaming source tables for ease of identification, appending the individual tables into a single query, and creating the relationships between the entities in the data model. Due to the complexity and scope of data warehouses, the data was stored instead in a data mart which is specific for the purpose of this study [17]. The dashboard elements are then generated ensuring each visual achieves one or more of the specified objectives. Finally, the visualized information was analyzed to develop insights to justify recommendations. In this case, the steps from ETL to analysis can all be accomplished using Power BI. Data analysis will be based on descriptive analytics since the focus is on investigating past operational data to understand historical performance.

3 RESULTS

The final dashboard (without filters applied) is shown in Figure 3. The full legend of the anonymized GHG fuel types is given in Figure 4. As displayed, the data represents the entire period of 2017-2021. A year slicer was implemented to select a specific year or multiple years, and production unit slicer to analyze data from a specific production unit. Cumulative emissions and average product energy intensity for the entire period are also provided.

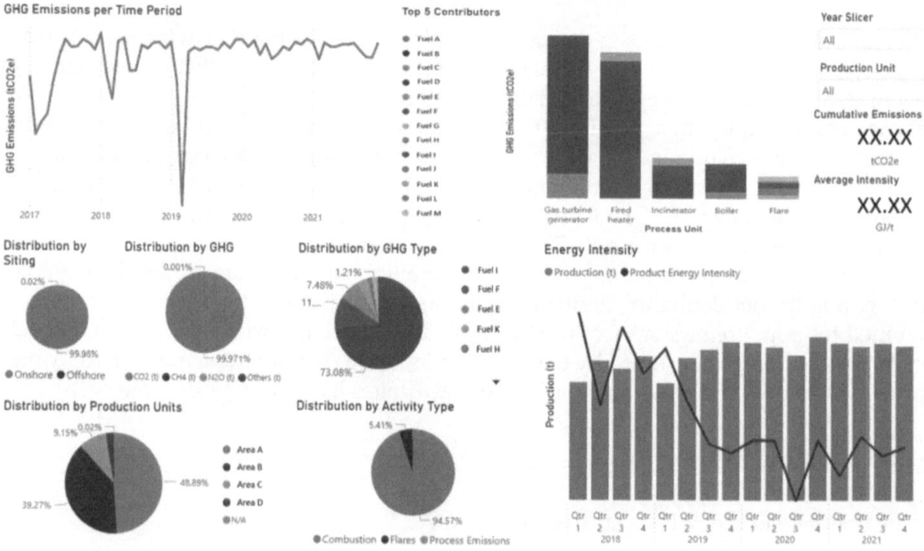

Figure 3. Power BI GHG performance dashboard.

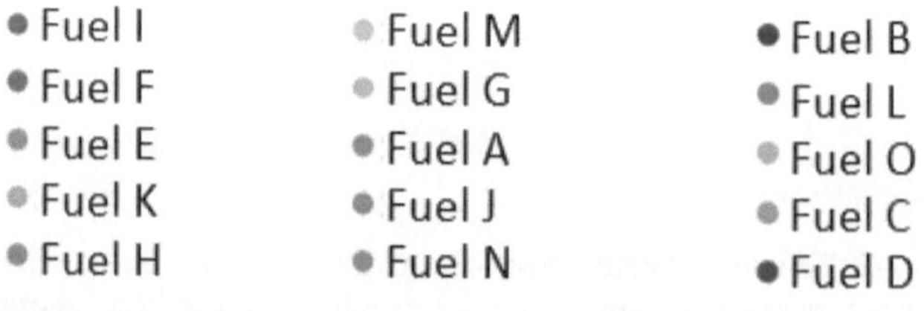

Figure 4. Complete GHG fuel type legend used in visualization.

3.1 *GHG emissions per time period*

Emissions are relatively stable throughout the dataset. A trough is observed in Q1 2017 and a sharp dip to zero in March 2019. The lowest emissions periods were generally in Q1 for 2017-2019 and 2021, with 2020 deviating from the norm. The lowest emissions period occurred in June for 2020.

3.2 *Top contributors*

Overall, the gas turbine generators (GTGs) were identified as the largest emitters at the plant. The majority of GTG emissions are due to combustion of reactor fuel *I* relative to fuel *E*. When analyzing specific years, it was observed that fuel *I* was the largest contributor for every process unit. The GTGs were the primary emitters until 2020-2021, where fired heaters took the top spot. The remaining 3 process units appear each year in the same order except for 2017, where the flare came in 3rd place with more fuel *H* flaring compared to future years, and 2018 where boiler emissions surpassed incinerator emissions with increased fuel *E* firing.

3.3 *GHG emissions distributions*

The quantities of offshore emissions and those caused by non-CO_2 GHGs were insignificant ($< 1\%$) relative to onshore flaring caused by direct CO_2 emissions.

Overall, Area *A* generated the most emissions (49%), followed by Area *B* (39%), Area *C* (9%), and the residual Area *D* (2%). The distributions remain relatively consistent each year, except for 2017 where Area *A* units accounted for more than half (54%) of emissions.

The main emissions type is through combustion-related activities (94%), followed by flaring (5%), and the rest process emissions. In this context, flaring is the release of gases to the atmosphere post-ignition (gas is converted into a less environmentally hazardous form) whereas process emissions include direct release of unignited gas. A trend of decreasing emissions from flaring and process emissions over time was observed.

In terms of GHG fuel sources, fuel *I* remains the majority (73%), followed by fuels *F* and *E* (18%), fuel *K* (4%), with the remaining fuels balancing. The data showed that the aforementioned sources (with the exception of fuels *I*, *E*, *F*, and *K*) are unique to flaring activities.

3.4 *Energy intensity*

The energy intensity visual shows a reduction in product energy intensity over time, peaking in 2018 and reaching its lowest value of in 2020. Production increased from. It was observed that 2019 was the only year with a continuous decrease in energy intensity.

4 DISCUSSION

4.1 *GHG emissions per time period*

Stability of the emissions trend (excluding the sharp dips) indicates there is no appreciable change in GHG performance since 2017, which is worth investigating by plant management. The first data point at the start of 2017 was lower than the same point in any of the future years, hence the steady state emissions value could have been lower prior to 2017. This observation is logical considering the increase throughout 2017, indicating the commissioning of a major project at the plant that contributed to its emissions. The dips in Q1 for 2017-2019 indicate likely plant maintenance periods where part of the plant is shutdown, reducing overall emissions sources. This is aligned with the reduced production numbers in Q1 2018-2019 and a lack of production downtime in Q1 2020. The sharp decline in March 2019 is unique and is likely due to a complex outage. Hence, it is important to understand the reason for peaks and dips such that unexplained observations are investigated further.

4.2 *Top contributors*

As GTGs rely on gas combustion to generate electricity for the rest of the plant (versus importing electricity), it is understandable why it is the largest emitter. However, it is also a positive sign that GTGs combust fuel *I* a majority of the time, versus flaring fuel *I* or reducing production due to unavailability of an consumer of fuel *I*. This requires the GTG to be designed to run on fuel *I*, as otherwise it would have led to expensive retrofitting. It also reduces the operating cost of using fuel *E* as a fuel rather than for other purposes that add value to the process. As fuel *I* emissions were predominant in every process unit, it implies significant amounts of fuel *I* are generated. From a systems perspective, improving plant performance in the future by minimizing fuel *I* production may have a knock-on effect on GTGs, heaters, incinerators, and other process units that rely on it as a fuel (i.e., more value-added production at the cost of substituting fuel *I* for more expensive fuels).

Although emissions from GTGs increased from 2019-2020, emissions from fired heaters increased by a larger amount over the same period, leading to slightly more emissions from fired heaters in 2020 compared to GTGs. The increase for fired heaters includes an increase for each fuel type, implying the result of an increase in plant throughput rather than the introduction of a new type of fuel. This observation is aligned with the large increase in production numbers from 2019 to 2020.

The increase in emissions contribution by flaring in 2017 is likely due to unplanned operational upsets which did not require immediate complex shutdown, since a larger portion of fuel *H* (typically a feed) flaring was observed compared to future years. In 2018, the increase in boiler emissions may be due to start-up of new boilers or frequent boiler outages that lead to greater reliance on fuel *E* as the start-up fuel.

4.3 *GHG emissions distributions*

Distribution by siting and by GHG showed that emissions are predominantly from onshore emissions and are attributed to CO_2, hence these should be the focus areas for abatement. Similarly, Area *A* units remain the major contributors, which aligns with the observation that two of the top 5 process emitters, GHGs and flares, are exclusive to Area *A*. Hence, the GHG abatement strategy should include a focus towards improving efficiency of Area *A* units and improving reliability to minimize unplanned flaring. Once improvements in Area *A* are exhausted and it reaches the point where further improvements are grossly disproportionate to the gain in GHG performance, the focus can shift to the next largest emitting production units in order.

The majority contribution of Area *A* to overall emissions in 2017 compared to future years is due to much lower emissions from Areas *B* and *C*, potentially implying that Area *A* units were operating normally while others were operating at lower throughput.

In terms of activity type, combustion is the most energy intensive relative to flaring and process emissions since it involves the direct combustion of fuels to produce CO_2. Compared to flaring, combustion is part of the process and hence large amounts of combustion take place, whereas flaring is utilized in case of operational upsets and is infrequent. Venting is even rarer since it wastes energy but is required in case there is a need to vent gases with high oxygen concentration which would otherwise lead to explosive mixtures if routed to the flare.

As with the top contributors visual, fuel *I* is the main participant in GHG emissions. It is crucial that initiatives implemented to manage fuel *I* production consider how it may impact other units in the plant, particularly its consumers. Although there are several GHG sources being flared, the focus should start with reducing emissions from fuels *I*, *F*, *E*, and *K* as the main sources from combustion-related activities, since combustion accounts for nearly 90% of overall emissions. Fuel *E* is used in reactor burners, boilers, and GTGs, while fuel *F* is used in fired process heaters. Hence, reducing fuel *E* usage may be achieved by modifying fuel sources in GTGs and boilers and improving combustion efficiency, but it would be more difficult to mitigate fuel *F* emissions since their usage in fired process heaters is directly reliant on process throughput, and retrofitting to accommodate different fuels would require extended outages and incur retrofitting costs.

4.4 *Energy intensity*

The energy intensity visual is not explicitly representing emissions, but rather the energy consumption per ton of product. Since production and energy intensity data was unavailable prior to 2018, it cannot be determined whether the decrease in intensity started in 2018 or prior. Overall, the decreasing energy intensity, when coupled with increasing production levels, implies an improvement in how energy is utilized. The continuous decline in energy intensity in 2019 is likely due to the implementation of energy efficiency initiatives which had a large impact on energy usage, although this should be investigated further such that other contributing factors can be identified and replicated in the future. After the 2019 decline, energy intensity reached a steady-state value in future years.

4.5 *Emissions abatement*

Several strategies exist which can improve the GHG performance of an industrial asset. The highest level is that which applies at the legislative level, and examples have already been adopted in China. China adopts a range of "command and control" policies, such as emissions standards on SO_2 and NO_x and levying fines on exceedances, both of which requiring mandatory compliance [26]. Market-based regulations also exist, which dynamically adjust based on market conditions [26]. This includes the tax credit system, whereby portions of portions of tax liabilities are offset for more environmentally sustainable operations, and the Clean Development Mechanism where the government provides financial aid to projects which promote environmental sustainability [26,27]. Focusing more on the lifecycle emissions of a product, the Low Carbon Fuel Standard (LCFS) was introduced by California in 2007 to limit the carbon intensity of commercial fuels [28]. Policies similar to LCFS may potentially find success when being applied on the end-products of a gas plant. However, more practical projects at the plant level can also be undertaken to reduce emissions on a shorter-term timeframe.

It was already identified that greater in-house consumption of fuel *I* would significantly reduce the plant's emissions, since it is a major emissions source. In addition, plants can be retrofitted with carbon capture units to sequester or utilize surplus CO_2 streams; a favorable direction considering Qatar Energy's commitment to construct CO_2 capture facilities with capacities in excess of 7 MTPA by 2030 [29]. Qatar is well suited to capitalize on carbon storage due to the availability of existing depleted reservoirs [29]. Clean fuel or process streams may also be utilized in the plant, particularly for high energy intensity combustion processes (e.g., blue, or green hydrogen firing), although it entails a complex techno-economic feasibility study [30]. If carbon capture gains significance in the country, it can potentially provide economic advantages in case a future carbon cap and trade system is developed, as Qatar could sell any excess emissions rights [29,31]. A plant may also reduce scope 1 emissions through increasing import of renewably sourced electricity (scope 2). In Qatar, new solar farms launching in 2024 are expected to reduce carbon intensity of natural gas processing by up to 35%, with room to increase the efficiency of solar currently limited by sandstorms [29,32]. However, GHG management initiatives can be as simple as having plans to clean air-cooled heat exchangers to improve their efficiency and having a robust maintenance plan in place to improve equipment reliability and minimize unplanned downtime.

5 CONCLUSION

By assessing GHG data of an operating plant through the BI-based interactive dashboard, this study has surfaced trends highlighting various aspects of the plant's GHG performance across multiple dimensions. The results showed relatively stable emissions and energy intensity from 2018 onwards, highlighting potential stagnation of the ongoing GHG abatement initiatives at the plant.

Hypotheses were generated to explain specific observations, such as maintenance, downtime, capacity changes, and project commissioning. However, this highlights the value of BI in democratizing access to information such that dashboards can be shared and analyzed by a wider team who may draw different insights by bringing alternative perspectives.

Analysis also revealed large emissions contributions from combustion activities, and specifically of fuel *I* from Area *A* units. Several GHG abatement strategies were recommended at different intensity levels, from government legislation to plant operational level. In this case, a systems perspective should be applied to understand the causes and how these emissions can be mitigated without creating unforeseen problems elsewhere in the plant.

It is recommended to further enhance the dashboard by linking it to a data source which can pull live plant data, to avoid excessive manual data collection and reorganization. Expanding the dashboard with data since plant commissioning would highlight changes over a longer period, which is vital in establishing a baseline for comparison. Greater insights could be drawn from future analysis if the dashboard is interpreted by those familiar with the plant, without restrictions on data confidentiality. This includes expanding the dashboard to also account for scope 2 and scope 3 emissions.

Overall, the dashboard generated demonstrates the potential for BI in fueling the energy transition by providing businesses with a means of visualizing their carbon footprint such that informed, data-driven decisions can be made in achieving sustainability targets.

REFERENCES

[1] Hughes G. C., (Apr. 1991). The information age, *Information Development*, vol. 7, no. 2, pp. 72–74, doi: 10.1177/026666699100700202.

[2] Spira J. B., Ed., (2015). *Welcome to the information age, in Overload!*, Hoboken, NJ, USA: John Wiley & Sons, Inc., pp. 43–65. doi: 10.1002/9781119200642.ch5.

[3] Khanpara P. and Tanwar S., (2020). Additive manufacturing: concepts and technologies, in *A Roadmap to Industry 4.0: Smart Production, Sharp Business and Sustainable Development*, A. Nayyar and A. Kumar, Eds., in Advances in Science, Technology & Innovation. , Cham: Springer International Publishing, pp. 171–185. doi: 10.1007/978-3-030-14544-6_10.

[4] Lasi H., Fettke P., Kemper H.-G., Feld T., and Hoffmann M., (Aug. 2014). Industry 4.0, *Bus Inf Syst Eng*, vol. 6, no. 4, pp. 239–242, doi: 10.1007/s12599-014-0334-4.

[5] Coombs W. T. and Holladay S. J., (2011). *Managing Corporate Social Responsibility: A Communication Approach.* Wiley. doi: 10.1002/9781118106686.

[6] Institution of Chemical Engineers, The Sustainability Metrics. (2010). [Online]. Available: https://www.icheme.org/media/16021/sustainability-metrics.pdf

[7] Elliott J. A., (2013). *An Introduction to Sustainable Development*, 4th Edition. in Routledge perspectives on development. London; New York: Routledge, Taylor & Francis Group.

[8] Gates B., *COVID-19 is awful. Climate Change could be Worse.*, GatesNotes. Accessed: Oct. 22, 2022. [Online]. Available: https://www.gatesnotes.com/Energy/Climate-and-COVID-19

[9] Foster V. and Bedrosyan D., (May 2014). *Understanding CO2 Emissions from the Global Energy Sector.* The World Bank. [Online]. Available: https://openknowledge.worldbank.org/bitstream/handle/10986/17143/851260BRI0Live00Box382147B00PUBLIC0.pdf?sequence=6

[10] Lamb W. F. et al., (Jul. 2021). A review of trends and drivers of greenhouse gas emissions by sector from 1990 to 2018, *Environ. Res. Lett.*, vol. 16, no. 7, p. 073005, doi: 10.1088/1748-9326/abee4e.

[11] Jackson R. B., Friedlingstein P., Andrew R. M., Canadell J. G., Le Quéré C., and Peters G. P., (Dec. 2019). Persistent fossil fuel growth threatens the Paris Agreement and planetary health, *Environ. Res. Lett.*, vol. 14, no. 12, p. 121001, doi: 10.1088/1748-9326/ab57b3.

[12] BP p.l.c., *Our sustainability aims*, bp global. Accessed: Oct. 22, 2022. [Online]. Available: https://www.bp.com/en/global/corporate/sustainability/our-aims.html

[13] Exxon Mobil Corporation, *Net-zero Ambition*, ExxonMobil. Accessed: Oct. 22, 2022. [Online]. Available: https://corporate.exxonmobil.com:443/climate-solutions/advancing-climate-solutions-progress-report/net-zero-ambition

[14] Shell International B. V., *Achieving Net-zero Emissions*, Shell Global. Accessed: Oct. 22, 2022. [Online]. Available: https://www.shell.com/powering-progress/achieving-net-zero-emissions.html

[15] Jones C. S. and Mayfield S. P., (2016). The Impact of Energy Usage on Climate Change, in *Our Energy Future: Introduction to Renewable Energy and Biofuels*, Oakland, California: University of California Press.

[16] Pusala M. K., Amini Salehi M., Katukuri J. R., Xie Y., and Raghavan V., (2016). Massive data analysis: Tasks, tools, applications, and challenges, in *Big Data Analytics*, S. Pyne, B. L. S. P. Rao, and S. B. Rao, Eds., New Delhi: Springer India, pp. 11–40. doi: 10.1007/978-81-322-3628-3_2.

[17] Eckerson W. W., Ed., (2012). *Performance Dashboards: Measuring, Monitoring, and Managing Your Business*, 1st ed. Wiley, doi: 10.1002/9781119199984.

[18] Bordeleau F.-E., Mosconi E., and Santa-Eulalia L. A. D., (Jan. 2018). *Business Intelligence in Industry 4.0: State of the Art and Research Opportunities*, presented at the Hawaii International Conference on System Sciences, Hilton Waikoloa Village, Hawaii, p. 10. [Online]. Available: http://hdl.handle.net/10125/50383

[19] Miller P. G., *Global Sustainable Tourism Dashboard*, p. 30.

[20] Kuze A. et al., (Jul. 2021). Covid-19 impact monitoring for climate environment (Greenhouse Gases), in *2021 IEEE International Geoscience and Remote Sensing Symposium IGARSS*, Brussels, Belgium: IEEE, pp. 1560–1562. doi: 10.1109/IGARSS47720.2021.9554328.

[21] Zin P. P., Heggen S., and Pauly J., (2017). Sustainability dashboard management system, *Journal of Technologies in Society*, vol. 12, no. 3–4, pp. 1–9, doi: 10.18848/2381-9251/CGP/v12i3-4/1-9.

[22] Developing an interactive data visualization platform to present the adaption of electrical vehicles in Washington, California and New York, in *2020 IEEE 7th International Conference on Industrial Engineering and Applications (ICIEA)*, Bangkok, Thailand: IEEE, (Apr. 2020), pp. 1–1. doi: 10.1109/ICIEA49774.2020.9101986.

[23] Al-Obadi M., (Jun. 2021). *Dairy Wastage Footprint Analysis: A Farm-to-Fork Life Cycle Approach Across Dairy Supply Chain*, presented at the First Central American and Caribbean International Conference on Industrial Engineering and Operations Management, Port-au-Prince, Haiti.

[24] Plant N., Cabrero E., and Saha T. K., (Jul. 2012). Power plant efficiency reporting and online auditing system for carbon pollution, in *2012 IEEE Power and Energy Society General Meeting*, San Diego, CA: IEEE, pp. 1–6. doi: 10.1109/PESGM.2012.6344776.

[25] Yusuf N. and Al-Ansari T., (Jan. 2023). Current and future role of natural gas supply chains in the transition to a low-carbon hydrogen economy: A comprehensive review on integrated natural gas supply chain optimisation models, *Energies*, vol. 16, no. 22, Art. no. 22, doi: 10.3390/en16227672.

[26] Zhao X., Yin H., and Zhao Y., (Jul. 2015). Impact of environmental regulations on the efficiency and CO2 emissions of power plants in China, *Applied Energy*, vol. 149, pp. 238–247, doi: 10.1016/j.apenergy.2015.03.112.

[27] The China Clean Development Mechanism Fund (CCDMF) | UNFCCC. Accessed: Nov. 20, 2022. [Online]. Available: https://unfccc.int/climate-action/momentum-for-change/financing-for-climate-friendly/china-clean-development-mechanism-fund

[28] Lepitzki J. and Axsen J., (Aug. 2018). The role of a low carbon fuel standard in achieving long-term GHG reduction targets, *Energy Policy*, vol. 119, pp. 423–440, doi: 10.1016/j.enpol.2018.03.067.

[29] Babonneau F., Benlahrech M., and Haurie A., (Nov. 2022). Transition to zero net emissions for Qatar: A policy based on Hydrogen and CO2 capture & storage development, *Energy Policy*, vol. 170, p. 113256, doi: 10.1016/j.enpol.2022.113256.

[30] Kazi M.-K., Eljack F., El-Halwagi M. M., and Haouari M., (Feb. 2021). Green hydrogen for industrial sector decarbonization: Costs and impacts on hydrogen economy in qatar, *Computers & Chemical Engineering*, vol. 145, p. 107144, doi: 10.1016/j.compchemeng.2020.107144.

[31] Babonneau F., Badran A., Benlahrech M., Haurie A., Schenckery M., and Vielle M., (Apr. 2021). Economic assessment of the development of CO2 direct reduction technologies in long-term climate strategies of the Gulf countries, *Climatic Change*, vol. 165, no. 3–4, p. 64, doi: 10.1007/s10584-021-03058-4.

[32] *Qatar to Double its Renewable Energy Output with Two New Solar Power Plants*, Fast Company Middle East | The future of tech, business and innovation. Accessed: Nov. 20, 2022. [Online]. Available: https://fastcompanyme.com/news/qatar-to-double-its-renewable-energy-output-with-two-new-solar-power-plants/

Innovation and Technological Advances for Sustainability – Al-Naemi et al. (Eds)

Enhancing energy management system using Internet-of-Things sensors and energy simulation – A case study of industrial ventilation

Da-Chun Wu, Paria Movahed, Ali Razban & Jie Chen
Department of Mechanical & Energy Engineering, Indiana University-Purdue University, Indianapolis, USA

ABSTRACT: There is a need for energy conservation in order to achieve the 2030 Sustainable Development Goals. Many studies have demonstrated that delivering real-time assessment and energy saving goals are some of the most effective practices to reduce energy consumption and can be achieved by utilizing energy management. However, there remains a sizable gap in the development of the energy management and energy-efficiency evaluation for complicated manufacturing processes. With the current development of new technologies such as Internet of Things, big-data analytics, and machine learning, improvements can be made to further enhance the effectiveness of the energy management system. In this study, a real-time monitoring system that utilizes IoT sensors, cloud-based big data, and grey-box modeling is proposed. This paper describes the ongoing study of using the proposed energy management system in an electroplating manufacturer to reduce its ventilation energy consumption.

Keywords: IoT-based sensors, energy management system, grey-box modeling, fault detection, ventilation

1 INTRODUCTION

According to the united nation perspective, energy usage accounts for 60% of total global greenhouse gas emissions and play a pivot role in the climatic change. In this regard, reducing carbon emission is an important objective in long-term climatic targets. The 7th goal of the 2030 Sustainable Development Goals (SDGs) is to ensure access to affordable, reliable, and sustainable modern energy for all. Affordable, reliable, and sustainable energy are crucial for achieving many of the sustainable development goals, from poverty eradication to progress in health, water supply, education, and industrialization, to mitigating climate change. However, energy access is not distributed evenly across countries, and the current rate of advancement is not enough for reaching this milestone [1].

Energy plays a vital role in manufacturing industry as one of the main resources. On the other hand, it is essential to provide the uninterrupted energy supply in the future due to the limitation of fossil fuel resources and increasing global population. According to the 2012 UNESCO report, the food production and supply chain account for 30% of the global energy consumption [2]. To reduce energy costs, negative environmental impacts, and carbon emissions, the food manufacturing industry has investigated various methods to improve its energy efficiency. One of the most convenient and cost-effective approach for energy consumption reduction is to mitigate the energy intensive activities without affecting the profitability of the manufacture or better energy management and energy recovery [3]. Energy management system (EMS) is widely used in manufacturing sector to achieve energy efficiency [4]. However, due to the complexity and existence of various parameters including

DOI: 10.1201/9781003496724-29

energy conversion, energy production, energy import/export, energy storage, and energy transmission and uncertain parameters such as interval possibility and probabilistic distribution, the implementation of the EMS is complicated [5]. Therefore, to make EMS more accessible, a system is required to provide a detailed real-time energy usage breakdown of their production facilities.

The Internet of Things (IoT) provide us with valuable information to be used for successful energy management by having access to real time data on energy consumption. In general, IoT can be employed as an approach to improve operational performance, safety, and profits by monitoring the quality and maintenance prediction [6]. IoT is spreading swiftly across various operations within manufacturing sector, and energy, as one of the most influential and significant factors in the sector, has drawn special attention. IoT has enabled manufacturing industry to monitor real-time energy consumption and provide the decision makers with valuable data of equipment, operation and the product lines [7].

2 LITERATURE REVIEWS

Various studies have been conducted to investigate the feasibility of using IoT-based EMS coupled with different machine learning techniques in manufacturing industry. Rathor and Saxena [8] conducted a review research on integration of Energy Smart Grid and EMS to optimize the performance of the system. Reference [9] presented a comprehensive EMS to reduce the operations cost of the micro-CHP system and increase the reliability considering uncertainties in demand, energy resources, and market price. Ahmad *et al.* [10] analyzed four supervised based machine learning models which are Binary Decision Tree, Compact Regression Gaussian Process, Stepwise Gaussian Processes Regression, and Generalized Linear Regression Model to forecast electricity requirement in the distinct building environment with the limited data information. Moreover, [11] discussed the IoT application in the food industry, comprising various processes from food production to distribution, consumers, and disposal. A meta-analysis conducted by [12] reveals a comprehensive view on energy data analytics and their the greatest effects on energy conversation.

Although the usage of IoT in modern EMS is getting traction, many issues remain to be solved. Reference [12] indicated that higher energy efficiency can be achieved by comparing customers' historical values. Combining energy analytics with goal setting also may increase the energy conservation effectiveness. Finally, providing a more energy report to the customer also seems to be useful. Consequently, an IoT-based EMS needs to accomplish more than just real-time monitoring and displaying, as the effect on energy conservation diminishes over time [13]. Many scholars have proposed solutions to increase the efficacy of EMS [14–18] and we found the following issues and areas of improvement:

- The IoT-based EMS can be more effective if it provides evaluations and predictions.
- Black-box modeling requires additional data collection period to train the models sufficiently.
- White-box modeling requires detailed knowledge of the operation, as well as many setup parameters, which sometimes can be unrealistic to obtain.

In this study, instead of using black-box models which involves large amount of data collection and time-consuming processes, a gray box model is proposed for improving energy consumption in manufacturing systems. This approach is not merely based on physical characteristics or data-driven approaches, but the combination of both. This paper proposed a real time energy management based on IoT data information. The data extracted from IoT application is transferred to EnergyPlus for further energy assessment. Together with AI-assisted fault detection algorithms, energy managers can use the enhanced system to evaluate the company's energy consumption status and make decisions with ease.

3 PROPOSED METHODOLOGY

The real-time energy management system proposed here is to help companies more effectively monitor their manufacturing process, assess energy consumption, and receive alerts about potential problems. The proposed system utilizes IoT-based sensors, big-data analytics, and grey-box energy models incorporated with EnergyPlus.

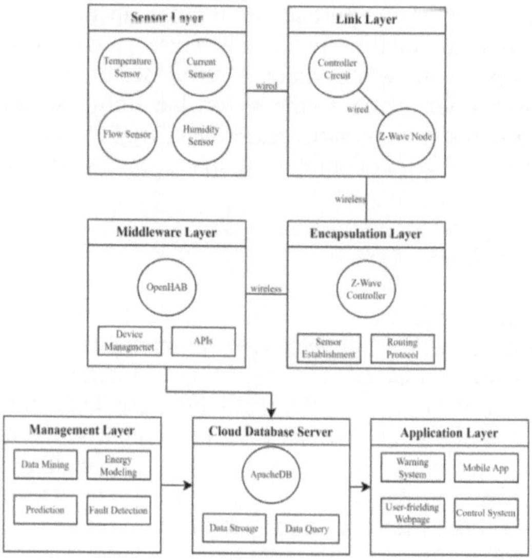

Figure 1. Proposed IoT real-time energy management.

3.1 *System design*

Figure 1 demonstrates Open Systems Interconnection (OSI) model for the proposed the real-time energy management system. The system is connected to various sensors and HVAC equipment as well as a cloud server for EnergyPlus simulation. The proposed energy management system has many functions such as data acquisition, AI-assisted data mining, and energy management, as well as smart control.

This system's data acquisition capability enables the prompt and precise gathering of energy usage data from multiple connection node of a large building. The nodes transmit the energy consumption data, the node's serial number, and other information data to the Z-Wave controller using 800–900 MHz radio transmission.

The data mining function involves using gaussian regression and decision tree to analyze the energy data, and thus finding correlations between decision variables and evaluations on the effects of any changes regarding to the process.

The energy management is made possible by reading the processing energy data stored in the cloud database. Managers have the ability to pull up the dynamic web page on their web browsers or use a mobile app to access information from any place at any time. They can access the data stored in the database and modify the settings instantly. The system also has features like system logs and fault detection in addition to the three core functions.

3.2 *EnergyPlus modeling*

The system proposed in this paper can not only achieve real-time monitoring but also connect to EnergyPlus energy equations in order to simulate and optimize the equipment's overall energy consumption. Also, by smart controlling the equipment, it can optimize operation and reduce energy consumption.

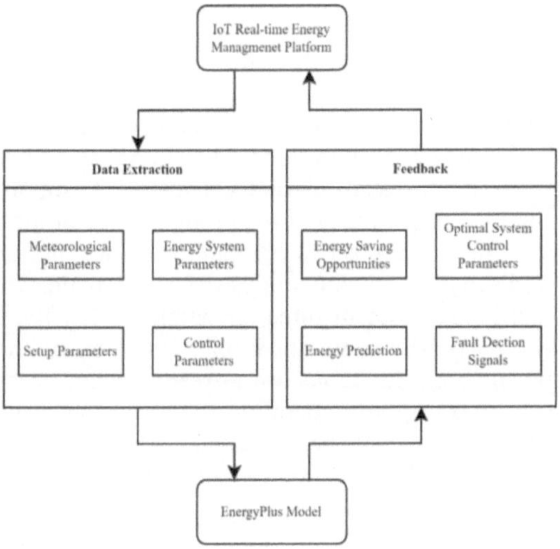

Figure 2. Proposed grey-box energy modeling.

The proposed model is used to coordinate the system with EnergyPlus in the cloud server is shown in Fig*ure* 2. The IoT system collects in real-time like the outside temperature and weather conditions, target parameters, operating status of the HVAC, control variables such as changes in production volume, and through the interface, converts them into data that EnergyPlus can recognize and process. The model does not presume and use any default values for the systems and the building envelop. The setup parameters can be extracted by using machine learning, then fed to the EnergyPlus simulation, whose results show the real-time load and energy consumption of the existing system and the correlation between room temperature, outdoor temperature, and operating conditions. Finally, the model generates and transmits back decisions and predictions to the IoT-enhanced EMS, thus achieving better performance in energy conservation.

3.3 *Test cases*

The research object of this paper is a electroplating manufacturer in the USA. Electroplating is a process of using electricity to coat a metal with a thin layer of another metal. It is done by passing an electric current through a solution that contains the metal to be deposited. The anode is purchased by the company to coat the parts (the cathode). The parts are shipped to the electroplating company by clients that need to be plated with various types of coating (e.g., nickel, copper, gold, and silver). The company consumed 3,383,378 kWh of electric energy and 5,769,934 MJ of natural gas in 2021. The total building area is about 3,665 m^2. The building is approximately 85 m long and 43 m wide. The facility has one production floor with a single thermal zone. The following test cases are what we want to experiment with the IoT-enhanced EMS.

(1) *How effective can IoT bring to enhance EMS:* Adding real-time evaluation and assessment, as well as future consumption predictions can improve the company's awareness of furthering energy conservation.
(2) *How effective can a IoT-enhanced grey-box modeling comparing to black box modeling:* IoT-enhanced grey-box modeling can achieve the same degree of accuracy using less data points and processing time compared to black-box modeling.

(3) *How effective can a IoT-enhanced grey-box modeling comparing to white box modeling*: IoT-enhanced grey-box modeling can achieve better accuracy compared to white-box modeling without the knowledge of high level of details.

4 CASE STUDY

The company has an intricate heating, ventilation, and air conditioning (HVAC) system that maintains the air quality on the production floor. The HVAC system serves three purposes: 1) ventilate the off-gas from the electroplating chemistry, 2) supply conditioned air to maintain the thermal quality (temperature) of the production floor air so that the chemistry does not crystalize, and 3) maintain positive pressure to ensure that the ventilation system can function properly and safeguard the building's structural against implosion. In order to serve the purposes of the HVAC system three air handling unit types are utilized by the company: exhaust air systems (EAS), make-up air units (MAUs), and roof top units (RTUs). The equipment's components and functions are intricate, each system currently functions independently to exhaust, make-up and condition the air. Each of these systems and the details related to current practices will be discussed individually to best understand the purpose, requirements, and procedures related to their operation.

4.1 *Exhaust Air Systems (EAS)*

The first purpose of the EAS is to ventilate the off-gas from the chemistry. The company operates chemical related process which requires electrolyte solutions containing ionic compounds that produce off-gases. The off-gases are a byproduct of the electroplating chemical process which can contain harmful substances that have the potential to endanger human health. Therefore, it is necessary to scrub these chemicals and exhaust them to the outdoors where they can be diluted into the atmosphere. To do this the various electroplating process lines have exhaust air unit systems which work to draw the off-gases from the process tanks and move them to the atmosphere. The exhaust air equipment that is used to ventilate the off-gassing from the process is shown in Table 1. We took the measurement of the power demand in kilowatt and volumetric flowrate in cubic meter per second of each EAS. The EAS have power demands ranging between 0.62 kW and 29.7 kW and volumetric flowrates ranging between 0.10 m^3/s to 9.15 m^3/s. Note that the system does not operate at the full design load, rather some EASs are controlled by the variable frequency drive (VFD). The company has regulated the EAS using VFDs for two reasons: 1) to reduce energy usage and 2) regulate the amount of air being exhausted from the facility.

Table 1. EAS operating information.

Name	Controlled by VFD?	Rated Power (kW)	Motor Setting (RPM)	Power Demand (kW)	Flowrate (m^3/s)
EAS01	Yes	37	1760	29.7	9.15
EAS02	Yes	19	1350	10.2	3.84
EAS03	Yes	37	1200	5.74	1.16
EAS04	Yes	5	1100	0.62	0.1
EAS05	No	4	N/A	3.08	1.88
EAS06	No	4	N/A	3.65	1.29

4.2 *Make-up Air Units (MAUs)*

The MAU information is shown in Table 2. The MAUs serve two purposes within the production facility 1) provide a surplus of supply air to maintain a positive pressure within the facility and 2) temper (pre-condition) the make-up air. Each MAU consist of three subsystems: the fan which supplies fresh air from the atmosphere to the facility, the burner which acts as a heat exchanger heating the air, and the cooling coil which transfers energy from the air to cool it. The MAUs have measured average power demands ranging between 13.84 kW and 19.97 kW and volumetric flowrates ranging between 10.21 m^3/s to 21.45 m^3/s.

Table 2. MAU fans operating information.

Name	Rated Power (kW)	Power Demand (kW)	Flowrate (m^3/s)
MAU01-FAN	15	13.84	10.21
MAU02-FAN	15	19.97	21.45
MAU03-FAN	15	16.61	10.58

4.3 *Rooftop Units (RTUs)*

The company uses rooftop air units (RTUs) to provide the final conditioning of the ventilation air on the production floor. The main difference between the RTUs and the MAUs is the RTUs return air from the production area for conditioning. This diagram shows that the ventilation within RTUs consists of four parts the return air, the exhaust air, the outside air, and the supply air. The facility does not utilize exhaust air on their RTUs as the have a dedicated EAS for this purpose. Furthermore, the facility does not utilize outside air as the MAUs are dedicated to serving this purpose. Therefore, the main function of the RTUs is to return the air from the production floor, condition it, and supply it back to the production area. The RTUs provide the final conditioning of the air which allows plant personnel to maintain an acceptable thermal quality for the air. This is necessary to preserve the chemistry in the electroplating tanks.

4.4 *Implementation of sensors*

IoT sensors were installed on each EAS, MAU, and RTU of interest, as well as both inside and outside. Figure 3 displays the locations and types of these IoT sensors. We decided on Z-Wave compatible current sensors, which are able to measure electrical current between

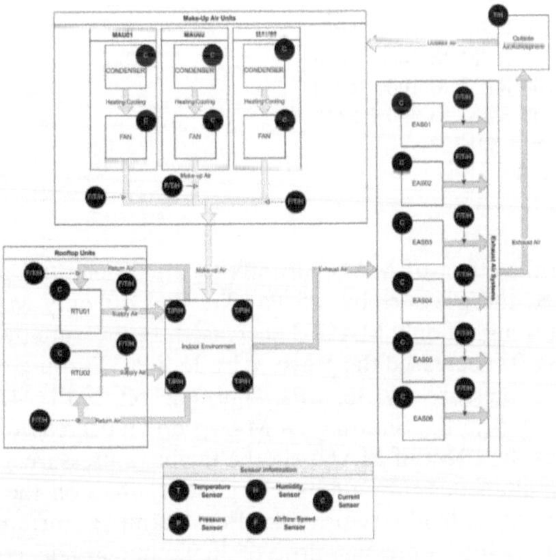

Figure 3. Implementation of sensors on the ventilation system.

0 A and 200 A. These sensors were attached to each EAS's VFD control panel to measure the incoming current. Other current sensors were installed on their repsective equipment's electric switchboard. Additionally, we chose temperature and humidity sensors that can record between -40°C to 120°C and 0% RH to 100% RH respectively, as well as airflow velocity sensors that can measure 0 m/s to 16.55 m/s. These sensors allow us to accurately determine the psychrometric properties of the make-up and exhaust air.

5 PRELIMINARY FINDINGS

We performed serveral expierments prior to the fully installed IoT system with a intend to study the effect on the facility pressure by each exhaust's and MAU's setting. It started with finding correlation models connecting the variables that are both energy and operational requirement related. We will build relationships using 4 measured variables that include the electric current of the HVAC equipment, the flow velocity, the differential pressure at the production floor, and the inside air temperature. The objective of the study is to create a data centric energy monitoring system that allows user to monitor, analyze, and optimize energy consumption of the facility.

The first experiment intended to study the effect on the facility pressure by each exhaust. Each exhaust was manipulated one at a time during the experiment while every other equipment remained at their operating conditions. All three MAUs, other five exhausts and RTU 1 & 2 were their on normal operation condition, respectively. Table 3 shows the differential pressure change when we changed the opertional conditions of each EAS.

Table 3. Effects on facility's differential pressure under different EAS operational conditions.

Name	Change in Operations	Change in Differential Pressures (Pa)
EAS01	1760 RPM to 1473 RPM	3.7
	1760 RPM to 1188 RPM	5.6
	1760 RPM to 950 RPM	12.8
EAS02	1350 RPM to 1200 RPM	0.0
	1350 RPM to 1050 RPM	2.5
	1350 RPM to 950 RPM	2.5
EAS03	1200 RPM to 1760 RPM	0.0
	1200 RPM to 1300 RPM	0.0
	1200 RPM to 1050 RPM	1.2
EAS04	1100 RPM to 1400 RPM	0.0
	1100 RPM to 1000 RPM	0.0
	1100 RPM to 950 RPM	0.0
EAS05	ON to OFF	3.3
EAS06	ON to OFF	2.5

Then, we examined the effect of MAU turnoffs to the differential pressure. Turning off only MAU01 decreased the pressure by 7.5 Pa. Turning off only MAU02 decreased the pressure by 13.3 Pa. Turning off only MAU03 decreased the pressure by 11.2 Pa. Turning off both MAU01 and MAU02 decreased the pressure by 36.9 Pa. Turning off both MAU01 and MAU03 decreased the pressure by 36.1 Pa. Turning off both MAU02 and MAU03 decreased the pressure by 41.5 Pa. Shutting all MAUs down decreased the pressure by 68.5 Pa. Table 4 summerizes the effect of MAUs on the facility's pressure.

It is evident that simultaneous turnoffs have a greater effect on the differential pressure compared to the sum of individual turnoffs. For example, turning off MAU01 and MAU02 simultaneously yielded a pressure drop of 36.9in magnitude, compared to only 20.8

by summing up effect by MAU01 and MAU02 individually. This additive effect is critical when developing the energy saving strategies concerning of turning off MAUs.

Table 4. Effects on facility's differential pressures by different MAU turn-off combinations.

Name	Change in Operations	Change in Differential Pressures (Pa)
MAU01	ON to OFF	-7.5
MAU02	ON to OFF	-13.3
MAU03	ON to OFF	-11.2
MAU01 & MAU02	ON to OFF	-36.9
MAU01 & MAU03	ON to OFF	-36.1
MAU02 & MAU03	ON to OFF	-41.5
MAU01 & MAU02 & MAU03	ON to OFF	-68.5

6 CONCLUSION

In this research we are proposing a real-time monitoring system that uses IoT-based sensors, big data analytics, and grey-box energy models. The proposed framework is expected to help managers monitor the status of the energy consumption and to identify faults in the process, thus reducing energy wastes and preventing losses caused by faults. Integration of IoT-based sensors with a big data processing system is effective for processing and analyzing large amounts of sensor data in real-time, as our study showed. The energy monitoring system proposed in this study utilizes Z-Wave, OpenHAB, ApacheDB. Experimental results showed there are many potential energy saving opportunities existing for the company. Once the proposed system is online more data can be processed and analyzed to provide a better view of energy saving strategies.

ACKNOWLEDGMENT

This work was supported by the United States Department of Energy (grant number DOE-DE-439-5534).

REFERENCES

[1] Goal 7 | *Department of Economic and Social Affairs*. https://sdgs.un.org/goals/goal7 (accessed Nov. 27, 2022).

[2] UN-Water, water, food and energy, *UN Water*. https://www.unwater.org/water-facts/water-food-and-energy (accessed Nov. 24, 2022).

[3] Woolley E., Luo Y., and Simeone A., (Oct. 2018). Industrial waste heat recovery: A systematic approach, *Sustainable Energy Technologies and Assessments*, vol. 29, pp. 50–59, doi: 10.1016/j.seta.2018.07.001.

[4] Menghi R., Papetti A., Germani M., and Marconi M., (Dec. 2019). Energy efficiency of manufacturing systems: A review of energy assessment methods and tools, *Journal of Cleaner Production*, vol. 240, p. 118276, doi: 10.1016/j.jclepro.2019.118276.

[5] Zhou S., Zou F., Wu Z., Gu W., Hong Q., and Booth C., (Jan. 2020). A smart community energy management scheme considering user dominated demand side response and P2P trading, *International Journal of Electrical Power & Energy Systems*, vol. 114, p. 105378, doi: 10.1016/j.ijepes.2019.105378.

[6] Luo Z. et al., (Aug. 2022). Application of the IoT in the food supply chain from the perspective of carbon mitigation, *Environ. Sci. Technol.*, vol. 56, no. 15, pp. 10567–10576, doi: 10.1021/acs.est.2c02117.

[7] Arunsankar G. and Parameswari G. A., (Aug. 2022). Energy management of smart grid equipped by an Internet of Things (IoT) configuration: A DDFAT methodology, *Energy & Environment*, p. 0958305X221117519, doi: 10.1177/0958305X221117519.

[8] Rathor S. K. and Saxena D., (2020). Energy management system for smart grid: An overview and key issues, *International Journal of Energy Research*, vol. 44, no. 6, pp. 4067–4109, doi: 10.1002/er.4883.

[9] Firouzmakan P., Hooshmand R.-A., Bornapour M., and Khodabakhshian A., (Jul. 2019). A comprehensive stochastic energy management system of micro-CHP units, renewable energy sources and storage systems in microgrids considering demand response programs, *Renewable and Sustainable Energy Reviews*, vol. 108, pp. 355–368, doi: 10.1016/j.rser.2019.04.001.

[10] Ahmad T. et al., (Sep. 2018). Supervised based machine learning models for short, medium and long-term energy prediction in distinct building environment, *Energy*, vol. 158, pp. 17–32, doi: 10.1016/j.energy.2018.05.169.

[11] Ben-Daya M., Hassini E., Bahroun Z., and Banimfreg B. H., (Jan. 2021). The role of internet of things in food supply chain quality management: A review, *Quality Management Journal*, vol. 28, no. 1, pp. 17–40, doi: 10.1080/10686967.2020.1838978.

[12] Ahir R. K. and Chakraborty B., (Sep. 2021). A meta-analytic approach for determining the success factors for energy conservation, *Energy*, vol. 230, p. 120821, doi: 10.1016/j.energy.2021.120821.

[13] Shen M., Lu Y., Wei K. H., and Cui Q., (Jul. 2020). Prediction of household electricity consumption and effectiveness of concerted intervention strategies based on occupant behaviour and personality traits, *Renewable and Sustainable Energy Reviews*, vol. 127, p. 109839, doi: 10.1016/j.rser.2020.109839.

[14] Zhang H., Li L., Li L., Cai W., Liu J., and Sutherland J. W., (Oct. 2022). An integrated energy efficiency evaluation method for forging workshop based on IoT and data-driven, *Journal of Manufacturing Systems*, vol. 65, pp. 510–527, doi: 10.1016/j.jmsy.2022.10.010.

[15] Al Dakheel J., Del Pero C., Aste N., and Leonforte F., (Oct. 2020). Smart buildings features and key performance indicators: A review, *Sustainable Cities and Society*, vol. 61, p. 102328, doi: 10.1016/j.scs.2020.102328.

[16] Ngarambe J., Yun G. Y., and Santamouris M., (Mar. 2020). The use of artificial intelligence (AI) methods in the prediction of thermal comfort in buildings: energy implications of AI-based thermal comfort controls, *Energy and Buildings*, vol. 211, p. 109807, doi: 10.1016/j.enbuild.2020.109807.

[17] Xianru W., Xinmin Z., Han Z., and Xuyun Y., (May 2021). Optimal control of boiler system based on IoT and energy consumption simulation, *J. Phys.: Conf. Ser.*, vol. 1904, no. 1, p. 012005, doi: 10.1088/1742-6596/1904/1/012005.

[18] Chang S., Castro-Lacouture D., and Yamagata Y., (Dec. 2020). Estimating building electricity performance gaps with Internet of Things data using bayesian multilevel additive modeling, *Journal of Construction Engineering and Management*, vol. 146, no. 12, p. 05020017, doi: 10.1061/(ASCE)CO.1943-7862.0001930.

Innovation and Technological Advances for Sustainability – Al-Naemi et al. (Eds)
© 2025 The Author(s), ISBN 978-1-032-80372-2
Open Access: www.taylorfrancis.com, CC BY-NC-ND 4.0 license

A reliable energy-efficient routing algorithm for WBAN using centrality measure

Shayan Ameen, Ghufran Ahmed, Rauf Shams Malik, Shahbaz Siddiqui & Muhammad Kariz Kamal
National University of Computer and Emerging Sciences (NUCES), Karachi, Pakistan

Adnan Akhunzada
College of Computing and Information Technology, University of Doha for Science and Technology, Doha, Qatar

ABSTRACT: WBAN (Wireless body area networks) have enabled continuous real-time monitoring of health professionals and patients. For example, monitoring blood pressure and other biological indicators can help predict heart attacks. Real-time applications require constant communication between source and destination sensor nodes. Developing real-time applications while maintaining the energy of the critical sensor nodes is difficult due to the dynamic network architecture of WBAN. To improve the network lifetime, throughput in terms of energy-efficient routing protocol in WBAN will play a vital role with the help of identifying the critical nodes in WBAN. In this study, we proposed an energy-efficient SD-WBAN (Software define-Wireless body area network) routing algorithm with the help of identifying the critical nodes and comparing them with the existing solutions. To determine the essential nodes of WBAN, we used the centrality measure.

Keywords: WBAN, SDN, network topology, centrality measure, SD-WBAN

1 INTRODUCTION

The Internet of Things (IoT) is one of the fastest expanding technologies in the computing industry. It assists people in becoming more critical thinkers in their everyday lives and obtaining control over their lives. People's everyday jobs are being automated by the IoT, which saves time and enhances processes and performance [1]. The use of IoT to connect health care systems increases diagnostic accuracy and treatment outcomes. It allows clinicians to act more quickly, which leads to more responsible healthcare that patients like [2]. Medical wearable devices can monitor a patient's physical and mental status, which can be monitored not only by professionals but also by the patient's guardian, who can look after the patient in an emergency circumstance. In addition, medical devices may offer bodily readings such as temperature, pulse rate, and blood pressure, which can play an important part in assessing data sets that can save a patient's life in an emergency [3]. In the IoT healthcare ecosystem, the performance requirements for apps and devices will change, necessitating different data-processing approaches [4]. Energy consumption has always been a matter of study in IoT health ecosystem more specifically to the scenario where medical IoT devices had no external power is attached except built-in battery power, which needs to be saved from being drained out [5]. The combination software-defined network with WBAN (Wireless body area network) has the potential to provide better control in terms of traffic forwarding with different routing algorithms such as Dijkstra's, RPL algorithm [6]. SDN has the ability to run data plane and control plane independently, which helps to tackle heterogeneous IoT devices in a healthcare ecosystem [7]. In this research, we present an energy-efficient solution for WBAN IoT nodes with the help of an efficient routing algorithm by identifying the critical nodes. The main contribution of this paper are as follow

1) Identify the critical IoT sensor nodes in an IoT health-care ecosystem that are highly engaged during any medical-related task such as IoT sensor are engaging to getting information of heart related information form the human body, for that, and we use the concept of social networking centrality measure algorithms such as degree centrality, closeness centrality, betweenness centrality
2) We proposed an energy-efficient routing algorithm based on centrality measure information. We create WBAN topology for IoT health Ecosystem with the help of SDN-Wise integrated IoT node environment

The following is a breakdown of how this article is structured. Section II discusses the importance of centrality measures in an IoT network. Section III discusses the literature review that is relevant to the problem statement. Sections IV and VI explain the methodology of the proposed algorithm and discussion of result.

2 IMPORTANCE OF CENTRALITY MEASURES IN WBAN IOT NETWORKS

Critical nodes make WSNs less reliable by making it more likely that the network will be cut off. So, figuring out (and fixing) the critical nodes is essential for setting up a safe WSN. Centrality measures are an excellent way to look at dipper to WBAN networks. Through centrality measure, the identification of the critical node in the WBAN network will help to create an energy-efficient WBAN network [8].

2.1 *Degree centrality*

Degree Centrality begins by looking at a WBAN node's how many direct, "one-hop" links each WBAN node has to other WBAN nodes in the network is defined as the degree of a WBAN node in a non-directed graph. Degree centrality measures a node's importance in a WBAN by counting the number of linkages of WBAN nodes. The more links a node has, the greater its degree. Mathematically, the Degree of Centrality is defined as equation-1. Degree Centrality is defined as equation 1. Figure 2A is a way to figure out how important a node is in a network that plays a "mediation" role. If the node in the WBAN network is where all other nodes have to go through, such as communication or connecting with other nodes, that means this node should be essential and very likely have a high degree of centrality [8]

$$C_d(N_i) = d(n_i)/(g-1) \tag{1}$$

$C_d(N_i)$ = Degree Centrality
$d(n_i)$ = Number of Connected Nodes
$(g-1)$ Size of Network

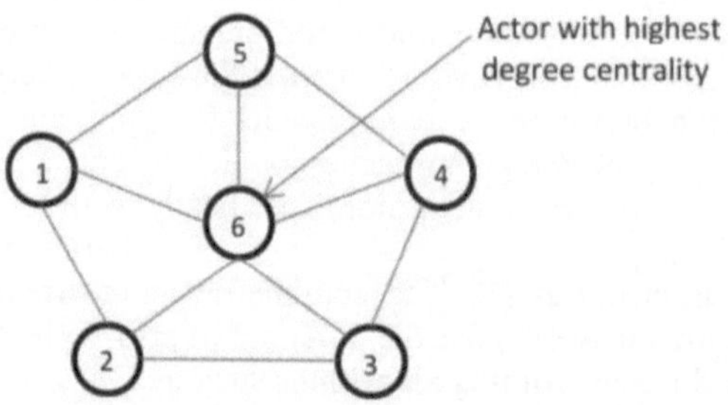

Figure 1. Degree centrality.

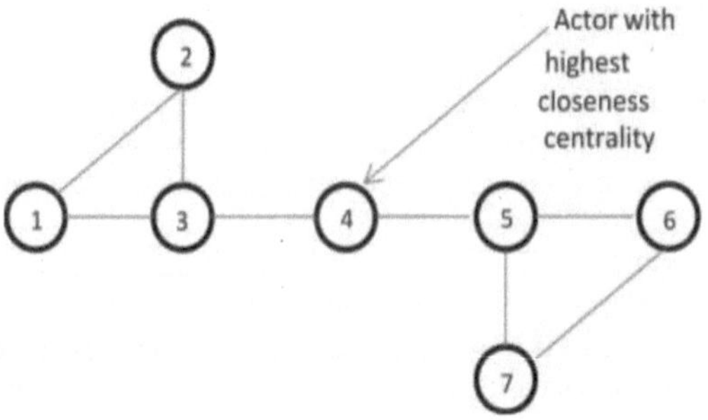

Figure 2. Closeness centrality.

2.3 *Betweenness centrality*

Betweenness centrality is define in terms of geodesic distances. A node's role as a 'bridge' is determined by the number of times it occurs on one of the network's shortest paths (known as the geodesic distances). This technique counts the number of times a node appears in between the most straightforward path rather than summing up the geodesic distance from one node to all other nodes.The higher the value of betweeenness the nodes importance in the WBAN network is high as shown in Figure 2C.

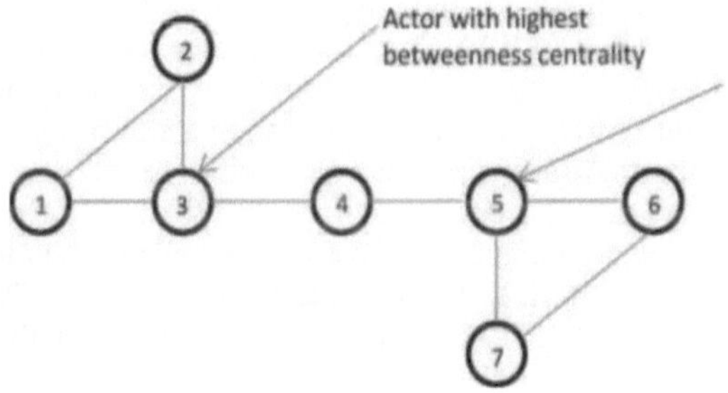

Figure 3. Betweenness centrality.

2.2 *Closeness centrality*

The significance of proximity. The closeness centrality is calculated using the normalized average of all of its geodesic distances (or closeness). To calculate an overall score for each node, the shortest paths connecting it to other nodes are added together. The shortest path between two WBAN nodes, i and j, is used to calculate the Geodesic distance d between them (the one with the fewest edges). The Geodesic distance, as defined by mathematics, is as follows in Equation-2:

$$C_c(N_i) = g - 1/X_d(n_i, n_j) \quad (2)$$

$C(N)$ = Degree closeness
$(g - 1)/L_d(n_i, n_j)$ = distance d between two nodes i j

$$C_b(N_i) = Xg_{ij}(n_{ij})/g_{ij} \quad (3)$$

$C(N)$ = Degree betweenness
$g_{ij}(n\text{ij})$ = shortest path between two nodes i j

3 LITERATURE REVIEW

Numerous related state-of-the-art works have been reviewed in detail, and a comprehensive summary table has been included to offer a complete overview of this work's current status. The bulk of efforts have been directed on the development of resource-constrained routing algorithms for WBAN and IoT frameworks. The provided works are compared based on the protocol offered, the methodologies utilized, the performance parameters used for evaluation, and the simulation or implementation environment. Rashed et al. [9] presented a low-energy routing technique for extending the stability time of wireless heterogeneous networks. This proposed method combines a cluster-based approach with a chain routing algorithm to ensure the wireless sensor network's energy and stability in a diverse environment.

Ivanov et al. [10] described the WBAN in their work Virtual Groups for Patient WBAN Monitoring in Medical Environments. They encountered issues with the WBAN and monitoring procedures, resulting in inadequate data quality. The authors created an architecture in which personal digital assistants and patients formed a virtual group. Additionally, the author created a new measure, named Quality of health monitoring, that provides medical specialists with information about the quality of data they get. Monowar et al.

In [11] used the term TMQoS to refer to a WBAN routing system that is thermally aware, multiconstrained, and QoS-aware. This protocol's principal purpose is to maintain good multi-constraint qualities such as latency and dependability. Additionally, the sensors' temperatures are kept within a safe range by using a hotspot avoidance method to avoid sensor node damage. The proposed TMQoS algorithm prioritizes data transmission between forwarder nodes according to the predicted QoS requirements of the intended traffic types.

Ahmad et al. [12] proposed a WBAN-compatible thermally aware routing scheme. The placement of fixed wireless sensing nodes is done in this case based on the nodes' energy level. In this protocol, data is transported from source to sink in a single hop, i.e., directly from the source to sink, whereas data is transferred under a multi-hop, i.e. via a forwarder node, in normal conditions. Javaid et al. [13] proposed an improved routing system for the WBAN. The author developed a multi-hop communication protocol that allows data to be sent between the deployment sensor node and the sink through intermediate nodes. This protocol used less energy, extended the network's stability period, and boosted its throughput. Tanvir et al. [14] have proposed a body coordinator node placement strategy, which can increase the network lifetime. Here authors proposed three different algorithms for placement of body sensor node coordinator. The output result shows that the algorithms along with an appropriate routing protocol are used for energy-saving and an increased network lifetime of WBAN up to 47.45

Adhikary et al. [15] proposed a unique routing protocol for WBAN with the objective of lowering energy consumption and prolonging the network's life. We add fixed nodes to WBAN in this article, which act as forwarder nodes. A fixed node is a subset of sensor nodes that are selected based on the cost function metrics associated with them. The measurements are of the nodes' transmission power, receiver velocity, residual energy, and sensor node location in respect to the coordinator. This multi-hop routing system enhances packet delivery and extends the life of the network. Smail et al. [15] demonstrated a reliable, trustworthy, and energy efficient WBAN routing system. In this situation, the author used an objective model to determine the routing path. Furthermore, the author discusses the state of the art in terms of energy consumption and overhead routing.

4 METHODOLOGY

A wireless body area network (WBAN) is a collection of microscopic sensors implanted on the human body, as shown in Figure 1. Battery life is one of the most crucial elements to consider because most devices are powered by batteries. If the sensor's battery dies, the node will likewise die. As a result, energy-efficient technologies are necessary to extend the node's lifetime.

The system concept based on SDN represented in Figure 4 consists of a sensing device and a controller. The sensing units are a system that uses sensors to detect required values, such as the nervous system sensor, which is used to assess the sensory nervous system properties of the human body. The controller is in charge of sending and receiving data from numerous sensor devices. A sensing unit consists of several distinct sensors.

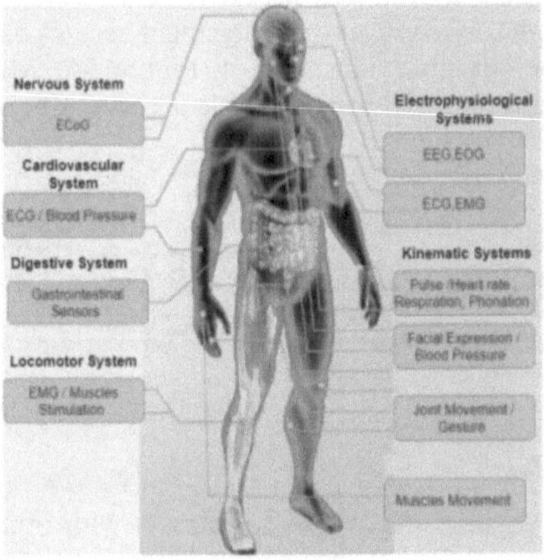

Figure 4. Wireless body area network.

4.1 *SD-WBAN structure*

Software-Defined Networking (SDN) is a networking architecture that is dynamic, manageable, cost-effective, and adaptive enough for today's high-bandwidth, dynamic applications [1]. This separation of network control and forwarding operations enables direct programmability of network control while hiding the underlying infrastructure from applications and network services [16]. SDN is gaining attraction as a means of replacing traditional networks with more dynamic and flexible network management. It has piqued the curiosity of academic and corporate researchers alike. In particular, SN integration in Wireless Body Area Network (WBAN) applications shows promise in terms of addressing challenges such as traffic management, authentication, and energy efficiency while also boosting administrative control [17].

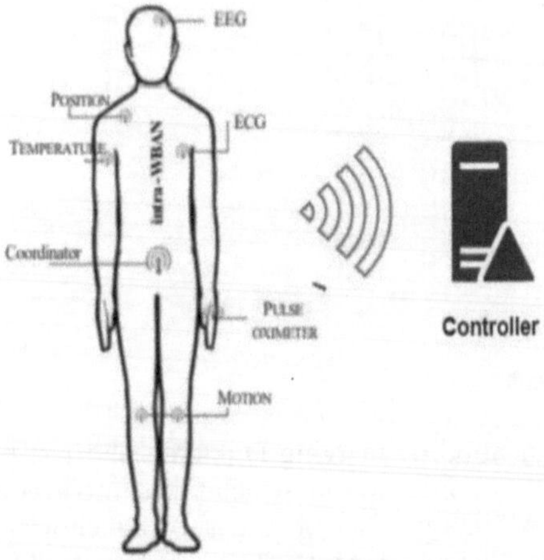

Figure 5. SD-WBAN structure.

We used the IEEE 802.5.6 Wireless Body Area Network (WBAN) standard for communication in the proposed system, which distributes nodes on or inside the human body. 802.15.6 provides a standard frame structure as well as physical and data link layer elements for communication between systems. It supports three physical layer standards: narrowband, ultra-wideband, and human body communication band. It also offers Beacon mode, which has access phases such as random access (RAP), exclusive access phase (EAP1, EAP2), contention access phase (CAP), and management access phase (MAN), and employs a superframe as the frame structure, as shown in Figure 5A (MAP). This standard also supports the CSMA/CA and Slotted Aloha multiple access (MA) protocols.

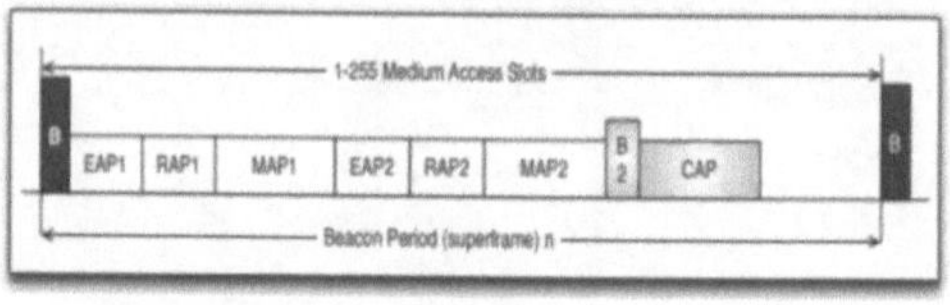

Figure 6. Beacon packets.

4.2 *System to system interaction*

In the system model, the sensing unit wishes to interact with other sensing units that are experiencing various difficulties. One of the identified challenges, for example, is the energy-efficient routing algorithm. A unit is required to provide vital information to other units as rapidly as possible. A software-defined network (SDN) controller is used. The controller's default routing technique is the Dijkstra shortest path, with the Radio Signal Strength Indicator as the cost function (RSS). In this standard implementation, the g reactive routing technique is also used to install routes in the network. A sensor node checks a packet against all flow entries in the flow table as it analyses it. If a flow entry is found that matches, the packet is processed according to the flow entry's instructions. When a packet does not match any flow entry, a request packet is sent to the controller. The controller determines the path from the requesting sensor node to the packet's destination and responds to it.

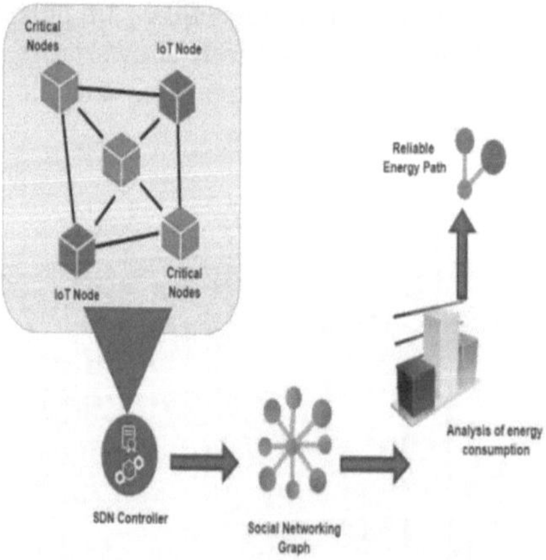

Figure 7. System interaction design.

We refined this routing method by merging Dijkstra's shortest path logic with criticality measures for maximum energy consumption, which takes into account the node's remaining energy and selects the most energy-efficient and shortest choice. In the proposed architecture, we use SD as a controller and T-Mote as a sensing unit. SD-defined network packets are a vital component in the development of new network applications. There are seven different types of network packets in the framework. However, we will only talk about the ones that are relevant to our research.

As shown in Figure 4B, packet length (LEN), network id (ID), source node address (SRC), destination node address (DST), packet type (TYP), Time to Live (TTL), and Next Hop Address are some of the packets that are included in each other network packet.

LEN	ID	SRC	DST	TYP	TTL	NX HOP
1 Bytes	1 Bytes	2 Bytes	2 Bytes	1 Bytes	1 Bytes	2 Bytes

Figure 8. Packet length.

Sensor nodes that work together as a system send data to the controller in the form of Report Packets with priority values and neighbor table information, as discussed above. The Report Packets frame structure is shown in Figure 5. The report packet contains the header packets, the sensing unit's distance attribute from the controller, the battery value of sensing units, and the number of connected Sensing units, as well as its neighbor attribute information, which includes the Address, RSSI, LQI, and Data Priority characteristics.

Header	Distance	Battery	$ Neighbour	Address	RSSI	...	Address	RSSI
10 Bytes	1 Bytes	1 Bytes	1 Bytes	1 Bytes	1 Bytes	...	2 Bytes	1 Byte

Figure 9. Packet length.

5 PROPOSED ROUTING ALGORITHM

At the start of the system of the WBAN network, the SDN controller is responsible for calling the network graph of the existing network topology. The sink is the interface library between WBAN nodes and the SDN controller. The WBAN node is responsible for providing detailed information on connected neighbours and RSSI, battery consumption, and centrality measure to the SDN controller. SDN controller passes this information to the default Dijkstra algorithm in the SDN-Wise library in order to find the energy-efficient path based on the default parameter with a new parameter that is a centrality measure.

Algorithm 1 Centrality measure based Routing Alogorithm

Find Engery Efficient routing path Controller Code
Initialization: Node Id, Controller Id
Node ID = Get Node ID of node;
Controller ID = Get Controller ID; NetWork Graph()
CM = Centerlity measure parameter
Array List [Rssi,LQI, Nodes Battery Level] = Report Packet()
Dijkstra(Network Packets) //Finding Shortest Path From Sink Array List [] = Sort (Rssi, LQI,Battery Level,CM)
Beacon Rules (Rssi,LQI, Nodes Battery Level,CM); Send Beacon Rules to Sensor Nodes
Controler Send Messages According to Energy Efficient Nodes
while Node Id ! = Sink **do**
  Nodes Id++
  Array List[i]
  Send Messages to High Energy Reserve IoT
end while

5.1 *Identification of critical nodes*

For the identification of critical nodes, a complex network analysis will be performed. Complex network analysis will help study the description and visualization of network topology. It will help to find the critical nodes in the subnet and avoid using them for data transmission within the subnet. The critical nodes will only be used in the subnet to communicate with the other subnet's node. The criteria for selecting the critical node are based on three centrality measurements: degree centrality, closeness centrality, and Betweenness centrality. We consider network topology with a probability of edge creation parameter that is the representation of the interaction of each node; that is, if the parameter is 0.3, a mean of 30percent of nodes are connected in the network.

5.1.1 *Identification of critical nodes*

The degree of centrality of a WBAN node is defined as the number of linkages that pass through the particular WBAN nodes.The degree value might be understood in terms of how important a node is for catching whatever is now going through the network. Our Efficient routing algorithms require the input value of degree centrality. We first calculate the degree of centrality of each node from the network topology. As defined above, the probability of edge creation parameter represents the interaction of connected WBAN nodes. We consider the 60percent WBAN nodes connected, as shown in Figure 10. To calculate the degree of centrality, we use the networkx library in python. Algorithm-2 find out the degree of centrality, closeness centrality, betweenness centrality of WBAN nodes. Figure 11 shows the WBAN node graph with high degree centrality based on Equation-1.

Algorithm 2 Algorithm for Centrality measure of WBAN Nodes

Centrality measure of WBAN Nodes
Initialization:WBAN Node Id NetWork Graph()
CM:Centerlity measure parameter using
Networkx d = nx.degree$_{c}entrality(G)$
$b = nx.betweenness_{c}entrality(G)$
print(d)

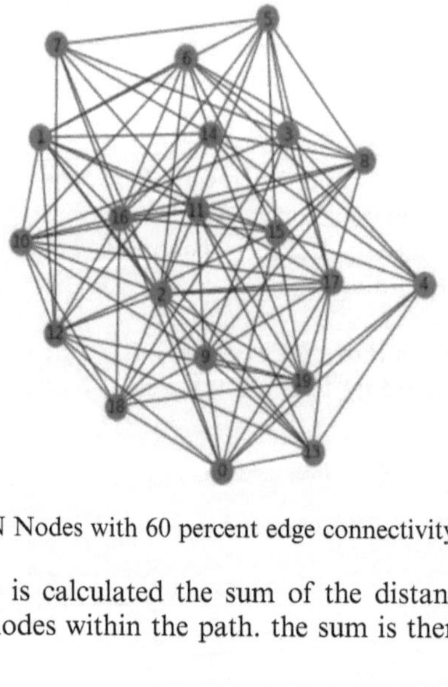

Figure 10. Simulated WBAN Nodes with 60 percent edge connectivity.

The Closeness centrality is calculated the sum of the distance from node k to node n travelling between all the nodes within the path. the sum is then inverted to determine the closeness centrality score for that node. Distances between nodes in an undirected graph are meaningless; yet, in a directed graph, these distances might have entirely different effects (e.g. a website can have a high closeness centrality from outgoing link, but low closeness centrality from incoming links). Figure 12 shows the result of closeness centrality of WBAN nodes from the simulated WBAN nodes Betweenness centrality is a measure of centrality in a graph that is based on the shortest paths. For every pair of vertices in a connected graph, there exists at least one shortest path between the vertices that minimizes either the number of edges that the path passes through (in the case of unweighted graphs) or the sum of the weights of the edges (in the case of weighted graphs). The number of these shortest paths that pass through a given vertex is known as the betweenness centrality of that vertex. Figure 13 Shows the WBAN nodes Betweenness centrality. Figure 13 shows the result of betweenness centrality of WBAN nodes from the simulated WBAN nodes.

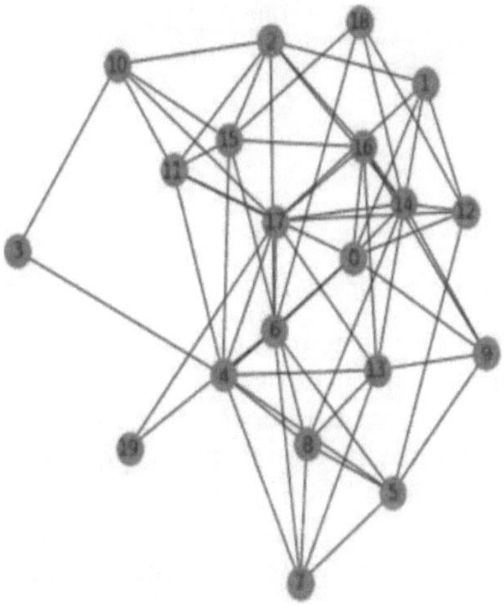

Figure 11. High degree centrality nodes.

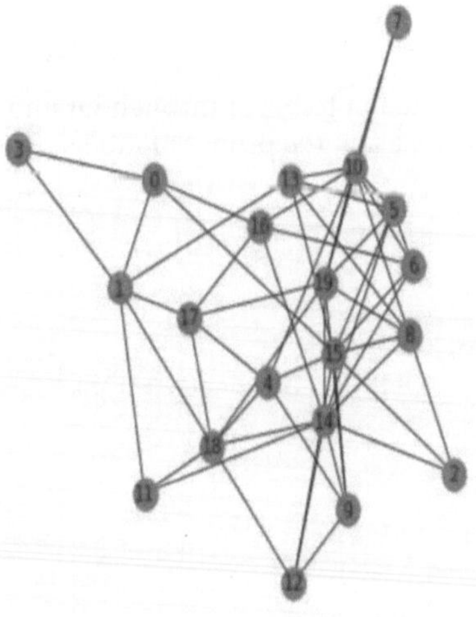

Figure 12. High closeness centrality nodes.

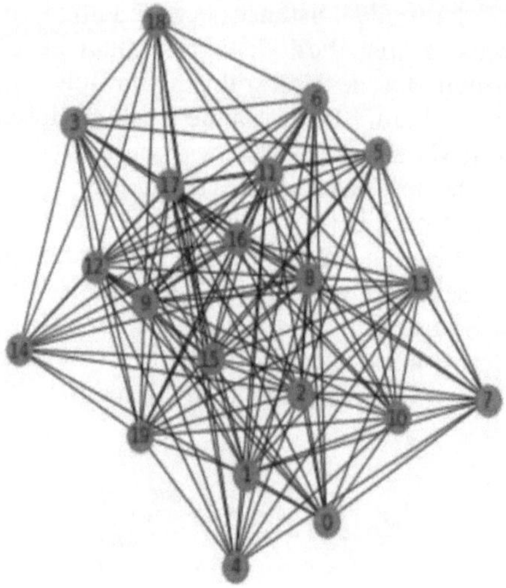

Figure 13. High closeness centrality nodes.

5.2 *Data communication using routing paths*

After collecting all the result of degree centrality. We will collect all the routing possibility considering the above centrality major from point A (source node) to point B (destination node). Based on the data gather of critical nodes and battery health, we will select the best match path in the network to traverse the data from one point to another. This will help us to maintain balance in the battery life and from getting critical node failure.

For Example, As shown in Figure 14, if node A in the subnet 1 wants to transmit data to the node C in the subnet 2 the algorithm will choose the critical node for making the tunnel. However, if node A wants to transmit data to node B which lies within the same subnet. It will select the alternate path to avoid using critical nodes for saving the energy of the node.

6 EXPERIMENTAL RESULT

6.1 *Testbeds environment*

The proposed approach is evaluated in terms of throughput and energy consumption. To do so, we undertake two experiments using the proposed model. The initial configuration of the system is shown in the table.

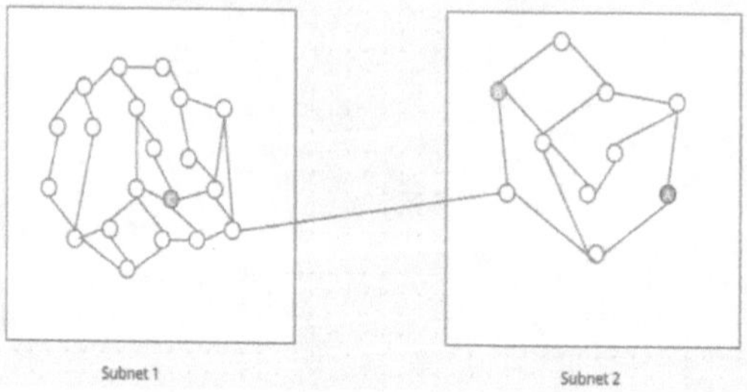

Figure 14. Data transmission between two clusters.

Parameter	Values
Simulation Time	2400s
Simulating Tool	Contiki, Cooja
Mac Protocol	IEEE 802.15.6
Sensing Unit	T-motes
Number of Sensing Units	40
Number of Controllers	4

6.2 *Design*

One of the most important design goals of the proposed algorithm is to ensure that packets reach the gateway by selecting the most energy-efficient acceptable pathways depending on priority levels. We created a scenario in which priority data were successfully transmitted to other sensing units by requesting an energy-efficient path from the controller when the sensing unit was outside the controller's coverage area, as shown in Figure 16, to assess the proposed model's throughput based on energy-efficient path selection.

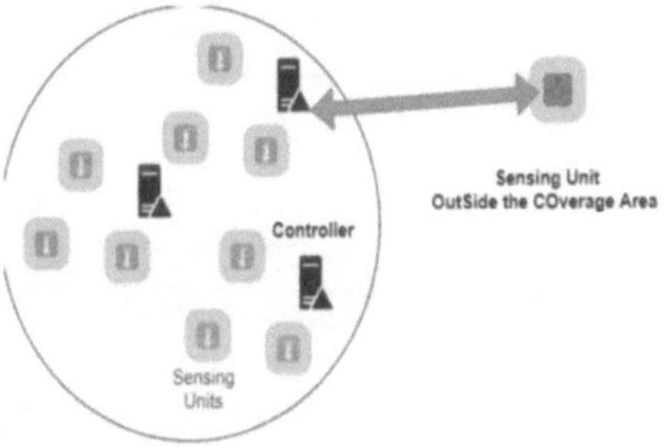

Figuer 15. System environment.

6.3 *Simulation results*

We placed the sensor nodes outside of the gateway or sink nodes coverage region. All priority packets were successfully transferred to the gateway. The simulation takes 4800 seconds, or 40 minutes, to complete. In this Test Bed, we consider four priority data packets [UP7, UP4, UP2, UP0] in order of greatest to lowest priority, with a delay of [0.25,0.5,1,2] second. The results reveal that outside of the coverage region, the sensing units correctly transport all distinct priority packets to the controller. The simulation takes 4800 seconds, or 40 minutes, to complete.

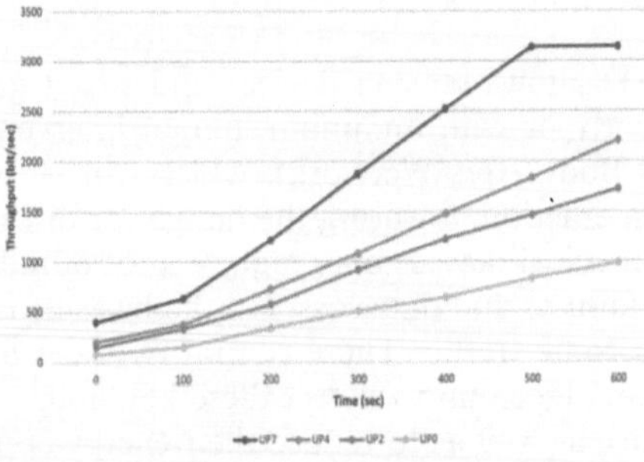

Figure 16. Throughput of WDAN network.

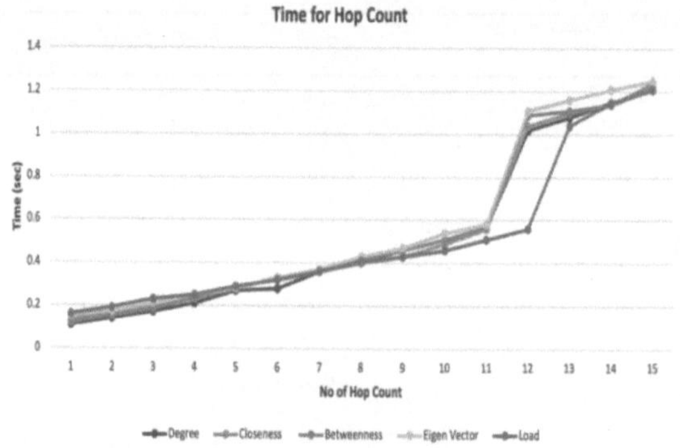

Figure 17. Time for Hop count.

In Figure 17, it express that the time will grow as the hop count increases, because it will take longer to reach the destination. So, we tried to get the shortest path possible for the by searching for the path avoiding the critical nodes presented in the path which will help us to keep the network alive.

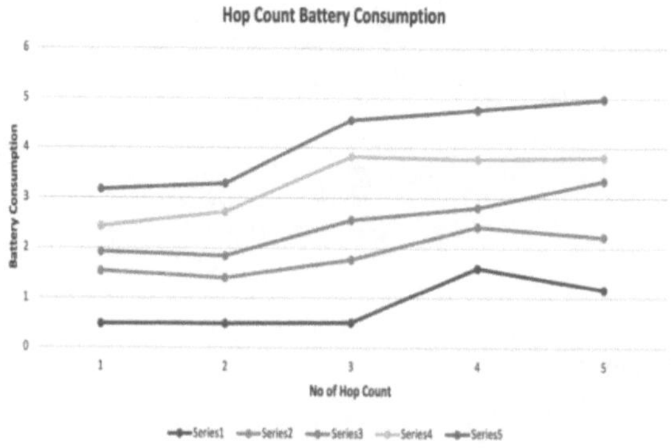

Figure 18. Battery consumption on hop count.

In Figure 18, it shows that with increase in the number of the hop will also effect the battery consumption, as it will increase node present in one path. We tried to find the shortest way feasible for the by avoiding the crucial nodes that were present in the path, which would help us keep the network alive.

7 CONCLUSION

The Internet of Things (IoT) allows sensing devices and smart gadgets to connect to the internet and run independently or with minimum human intervention. This technology is useful in WBAN (Wireless Body Area Network), which can be useful in medical science because it reduces human interaction, lowering the danger of disease transfer. The data set transactions between these medical devices may require a lot of memory, CPU, and battery power, resulting in a lot of failures and data loss. On the basis of these centrality measures, we created an algorithm to route traffic. These centrality nodes aided in the reduction of battery usage. We devised an algorithm to detect these key nodes based on their condition criticality and alternate paths, as well as to reduce their energy consumption. We proposed criteria for identifying crucial nodes that can cause major disconnection or disjoint in an Ad-hoc network. These five nodes, known as Edge Betweenness Centrality (EBC) nodes, had a significant impact on the network. Closeness centrality nodes are those that are close to the data transmitting nodes. The Eigen Vector centrality nodes are linked to the network's most important nodes. Stress centrality refers to nodes that have a lot of data travelling through or between them, while degree centrality refers to nodes with the most connected degrees.

REFERENCES

[1] Souri A., Hussien A., Hoseyninezhad M., and Norouzi M., (2022). A systematic review of iot communication strategies for an efficient smart environment, *Transactions on Emerging Telecommunications Technologies*, vol. 33, no. 3, p. e3736.

[2] Koren A. and Prasad R., (2022). Iot health data in electronic health records (ehr): Security and privacy issues in era of 6g, *Journal of ICT Standardization*, pp. 63–84.

[3] Saravanan M., Ajayan J., Maheswar R., and Parthasarathy E., (2022). An overview of architecture and applications of iot-based health care systems, *IoT-Enabled Smart Healthcare Systems, Services and Applications*, pp. 87–121.

[4] Alkhayyat A., Thabit A. A., Al-Mayali F. A., and Abbasi Q. H., (2019). Wbsn in iot health-based application: toward delay and energy consumption minimization, *Journal of Sensors*, vol. 2019.

[5] Kumar S., Buckley J. L., Barton J., Pigeon M., Newberry R., Roden-cal M., Hajzeraj A., Hannon T., Rogers K., Casey D. et al., (2020). A wristwatch-based wireless sensor platform for iot health monitoring applications, *Sensors*, vol. 20, no. 6, p. 1675.

[6] Cicioğlu M. and Çalhan A., (2020). Energy-efficient and sdn-enabled routing algorithm for wireless body area networks, *Computer Communications*, vol. 160, pp. 228–239.

[7] Ja'afreh M. A., Adhami H., Alchalabi A. E., Hoda M., and El Saddik A., (2021). Toward integrating software defined networks with the internet of things: a review, *Cluster Computing*, pp. 1–18.

[8] Zhang J. and Luo Y., (2017). Degree centrality, betweenness centrality, and closeness centrality in social network, in *Proceedings of the 2017 2nd International Conference on Modelling, Simulation and Applied Mathematics (MSAM2017)*, vol. 132, pp. 300–303.

[9] Gupta P., Tripathi S., and Singh S., (2021). Energy-efficient routing protocols for cluster-based heterogeneous wireless sensor network (het-wwsn)—strategies and challenges: A review, *Data Analytics and Management*, pp. 853–878.

[10] Ivanov S., Foley C., Balasubramaniam S., and Botvich D., (2012). Virtual groups for patient wban monitoring in medical environments, *IEEE transactions on biomedical engineering*, vol. 59, no. 11, pp. 3238–3246.

[11] Monowar M. M., Mehedi Hassan M., Bajaber F., Hamid M. A., and Alamri A., (2014). Thermal-aware multiconstrained intrabody qos routing for wireless body area networks, *International Journal of Distributed Sensor Networks*, vol. 10, no. 3, p. 676312.

[12] Ahmad A., Javaid N., Qasim U., Ishfaq M., Khan Z. A., and Alghamdi T. A., (2014). Re-attempt: a new energy-efficient routing protocol for wireless body area sensor networks, *International Journal of Distributed Sensor Networks*, vol. 10, no. 4, p. 464010.

[13] Javaid N., Ahmad A., Nadeem Q., Imran M., and Haider N., (2015). im simple: improved stable increased-throughput multi-hop link efficient routing protocol for wireless body area networks, *Computers in Human Behavior*, vol. 51, pp. 1003–1011.

[14] ul Huque M. T. I., Munasinghe K. S., and Jamalipour A., (2014). Body node coordinator placement algorithms for wireless body area networks, *IEEE Internet of Things Journal*, vol. 2, no. 1, pp. 94–102.

[15] Adhikary S., Choudhury S., and Chattopadhyay S., (2016). A new routing protocol for wban to enhance energy consumption and network lifetime, in *Proceedings of the 17th International Conference on Distributed Computing and Networking*, pp. 1–6.

[16] Cicioğlu M. and Çalhan A., (2018). Sdn-enabled wireless body area networks, in *2018 6th International Conference on Control Engineering & Information Technology (CEIT)*. IEEE, pp. 1–5.

[17] Hasan K., Wu X.-W., Biswas K., and Ahmed K., (2018). A novel framework for software defined wireless body area network, in *2018 8th International Conference on Intelligent Systems, Modelling and Simulation (ISMS)*. IEEE, pp. 114–119.

Innovation and Technological Advances for Sustainability – Al-Naemi et al. (Eds)

Classification of nuclei in whole slide imaging based on ensemble deep learning

Khaled Al-Thelaya, Mahmood Alzubaidi, Fahad Majeed, Gilal Nauman Ullah, Marco Agus & Jens Schneider
Division of Information and Computing Technology, College of Science and Engineering, Hamad Bin Khalifa University, Qatar Foundation, Doha, Qatar

ABSTRACT: Recently, several innovative digital imaging technologies led to the development of several medical imaging data representations. Whole slide imaging (WSI) refers to capturing and storing tissue glass slides and biopsy samples in a digital form. This newly emerging modality is increasingly gaining huge interest in pathology departments worldwide due to the unique advantages introduced by this medical imaging modality for diagnostic, educational, and research purposes. This work provides a comparative analysis of multiple classification approaches of nuclei in WSI where multiple nuclei context-aware manipulation strategies are evaluated by our nuclei classification experiments. We propose a context-aware deep ensemble approach based on ResNet deep architecture using different forms of input images pre-processed using multiple manual feature extraction techniques. Results show that our approach attains higher accuracy performance compared to traditional context-ware strategies in the literature.

Keywords: Whole Slide Images, Digital Pathology, Deep Ensemble Learning, Handcrafted Features, Nuclei Classification

1 INTRODUCTION

Digital pathology technologies have significantly improved modern clinical practices. The impact of these technologies on the development of modern healthcare systems is rapidly increasing by improving disease detection and tissue sample analysis in modern laboratory environments (Niazi *et al.* 2019a). Influenced by the progress of digital image processing, recent digital pathology technologies led to the development of new imaging technology named whole slide images (WSI) captured by digital slide scanners. They are mainly used to store scans of tissue samples to facilitate storing, viewing, processing, and sharing digital tissue samples (Al-Janabi *et al.* 2012). This new imaging technology motivates simultaneous case reviews and enables the international sharing of samples with experts. It represents a rich source of information with a high level of complexity and the presence of color information with multiple resolutions. WSI opens the door for the development of various solutions for complex problems followed from the highly sophisticated content of these images revealing a set of unique imaging features and enabling several artificial intelligence solutions in digital pathology applications. However, the resolution of these images is large, with sizes up to a few gigabytes (Al-Thelaya *et al.* 2023). This leads to management, transmission, and processing challenges. Research in this field actively continues to raise and solve different information extraction and decision-making issues and challenges (Niazi *et al.* 2019a).

Automation in digital pathology provides numerous solutions to extract meaningful information from WSIs resulting in fast and accurate clinical reasoning. The application of

DOI: 10.1201/9781003496724-31

automated image analyses to WSI ensures higher productivity, accuracy, and reproducibility. It also helps to validate the standardization of study outcomes and inclusion criteria, extracting a useful set of features and measurements that can be used to derive inferences and make decisions (Pell *et al.* 2019). Novel features and diagnosis information in histological samples can be extracted using digital image analysis and deep learning techniques. These features can be used in different hematological trials (Pell *et al.* 2019). The texture of tissue samples is one of the main features commonly used to analyze histopathology images. These features are widely used to determine regions of interest with unique clinical properties showing the spread of abnormal cells in tissue samples (Abels *et al.* 2019).

The biological structures and topologies of human tissue are sophisticated and dense. WSIs absorb this information in a complex static form represented by color and texture. Therefore, the information embedded in these images is huge, which raises many issues concerning information extraction and manipulation. Moreover, WSIs are very large and characterized by high variations in color and texture, in addition to variations in scanning parameters and technologies. Therefore, traditional image analysis solutions may not provide the best analysis mechanism for these images. Recently, a new kind of image feature extracted using deep learning provides an accurate representation of WSI contents. These deep features proved superior in terms of accuracy performance in many problems related to histology image analysis, including image segmentation, classification, and description (Faust *et al.* 2018; Khened *et al.* 2021; Xu *et al.* 2017). Convolutional neural networks (CNNs) are one of the deep architectures commonly used to process WSI. Some deep customized architectures are proposed in the literature based on CNNs (Gao *et al.* 2022; J. Zhang *et al.* 2021). Other studies adopt pre-trained, well-established, publicly available deep architectures based on CNNs, such as ResNet (Al-Thelaya *et al.* 2022; Wang *et al.* 2021). Our model is developed based on the ResNet (He *et al.* 2016), one of the most widely used deep architectures for image processing and classification. ResNet is leveraged by our study based on an ensemble deep learning architecture using three vertically stacked encoders to extract features from different forms of input images that we pre-process using manual feature processing techniques. The idea is to combine manual feature analysis and automatic deep learning feature extraction methods to provide a better representation of WSI images for nuclei classification.

The rest of this paper is organized as follows. First, we discuss some of the most related work in Section 2. Section 3 provides further discussion of our proposed approach. Section 4 provides a description of the data set used to conduct evaluation experiments. It also presents experiments and results. Finally, we summarize our conclusions in Section 5.

2 RELATED WORK

Some clinical histopathology procedures rely on the measurement and counting of cell nuclei in tissue samples (Graham *et al.* 2019). For instance, the prognosis of cancer is the identification of biomarkers and T-cells to measure tumor immune infiltrates. The manual visual estimation of nuclei is a cumbersome and tedious inaccurate process. It is a subjective process that can result in many observations across and between histopathologists (Shakya *et al.* 2020). Assessment reliability and productivity can be increased by automatic nuclei detection and quantification. The nuclei detection, segmentation, and classification are sophisticated tasks due to variations in human biological tissue structures and different staining and tissue preparation parameters (Niazi *et al.* 2019b).

Since the public release of the PanNuke data set (Gamper *et al.* 2019), many solutions have been developed for nuclei segmentation, detection, and classification. The data set provides a comprehensive source of nuclei labeling for 19 different tissue types. It creates a large pan-cancer data set for segmenting and classifying nuclei instances containing around 455 visual fields, 312 of which were randomly selected from over 20K whole slide images at various magnifications from multiple data sources. This data set is used in our work for training and evaluating classification models. It was also used by Al-Biltaji (2021) for the classification of inflammatory cells in pediatrics based on an improved deep SqueezNet architecture. The PanNuke data set was also used to evaluate the knowledge-based representation learning approach developed by W. Zhang *et al.* (2022) for nuclei classification in WSI. The study used a ResNet encoder to extract features from image patches.

Chen *et al.* (2021) trained a ResNet-50 model to describe WSI images divided into patches. These features are then used to form graph embeddings for the whole slide image, which is then used as input to a GCN model to perform survival prediction of cancer patients. ResNet-50 is also compared with other deep architectures in Arvaniti *et al.* (2018) for Gleason grading in prostate cancer; the deep learning architectures produce comparable accuracy performance compared to human experts. ResNet-152 and ResNet-101 are used by Dodington *et al.* (2021) for tumor region identification and nuclei segmentation, where a set of texture and shape features are extracted from nuclei to analyze the impact of neoadjuvant chemotherapy treatment on breast cancer patients.

In our work, we propose an ensemble deep learning approach based on ResNet encoders for the classification of nuclei in WSI. We compare multiple classification approaches based on three context-aware strategies. Several per-pixel manual feature analysis techniques are adopted in our study to extract useful information from image patches and generate new input images for our ensemble deep learning approach.

3 METHODOLOGY

The application of deep learning methods in digital pathology is rapidly increasing to assist medical diagnosis. The output of these applications proved to be superior, with a high impact on digital pathology progression. Deep learning methods are used for the segmentation and detection of nuclei and regions of interest within WSI images. They are also used for cancer diagnosis, grading, and nuclei classification (Pantanowitz *et al.* 2018). CNNs are used in the most popular classification architectures for various image classification problems, including nuclei classification. Features learned by CNNs provide an effective description of WSI images. These features can be used as input to other deep learning models or fully connected layers to achieve many tasks such as classification, segmentation, and detection. Many well-known deep network architectures are developed based on CNNs, such as Inception, DenseNet, ResNet, etc. Due to their high accuracy performance, numerous solutions are developed based on these architectures to process WSI images. These models are pre-trained on well-known huge data sets, such as ImageNet (Krizhevsky *et al.* 2012).

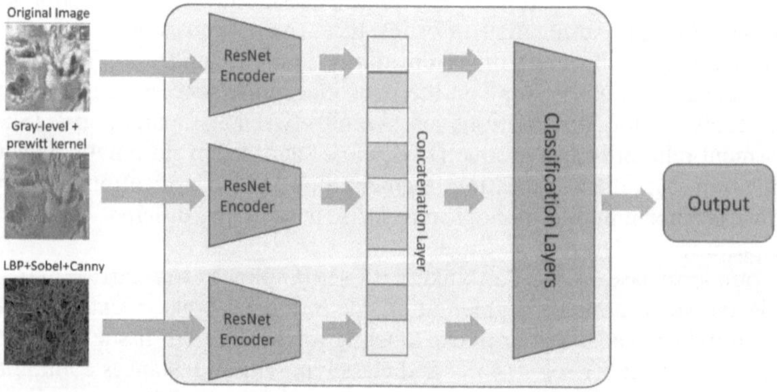

Figure 1. The proposed ensemble deep architecture.

We propose a deep learning approach for nuclei classification in histopathology images based on transfer learning and ensemble deep learning. In ensemble deep learning, the predictions of multiple models are usually combined to exploit the accuracy performance of these models using different aggregation methods such as voting or averaging. In our approach, we propose to combine features generated by the ensemble instead of combining the final predictions of classification models. The combined features are then used as input to the classification layers attached to the ensemble deep feature learning architecture. We utilize three ResNet networks for deep feature learning and encoding, as shown in Figure 1. Input images are first transformed into several other forms based on various image processing techniques where each encoder receives a different version of the same input image. Color channels in the new images represent features extracted using manual feature extraction methods. Manually engineered feature analysis has a long tradition for extracting information from images; it is an explainable process often producing significant results. In contrast, machine learning models are trained to perform different tasks, including nuclei classification, freeing humans from the complex feature engineering task. Leveraging the benefits of both approaches, we thus propose to perform pre-processing of original images using manual feature extraction techniques, such as the Sobel operator, Prewitt operator, Canny edge detection, and local binary patterns (LBP) to provide different representation forms of input images and extract meaningful information. In the following, we discuss these manual feature extraction methods and explain how we use them as input to our proposed deep learning architecture.

3.1 *Sobel operator*

The Sobel operator (Dim and Takamura 2013), developed and named in 1968 after Irwin Sobel, emphasizes edges in the image. It is a discrete differentiation operator. It computes an approximation of the gradient of the image intensity function. The Sobel operator produces either the corresponding gradient vector or the norm of this vector at each point in the image. It is based on convolving the image in the horizontal and vertical directions with a small, separable, integer-valued filter and is thus computationally inexpensive. The main difference between Sobel and Prewitt operators is that in the Sobel operator, the mask coefficients are not fixed and can be adjusted to meet different objectives as long as they do not violate the properties of derivative masks (Sobel and Feldman 1968).

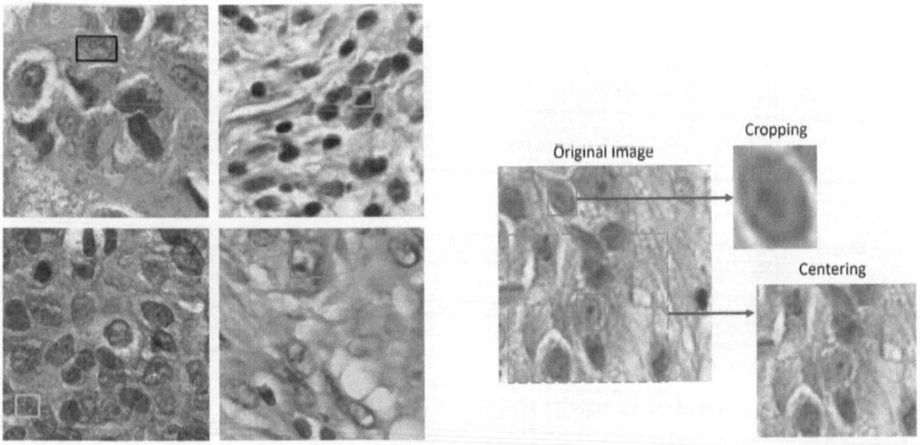

Figure 2. Left: images with bounding boxes around target nuclei. Right: cropping and centering based on the position of target nuclei in original images.

3.2 *Prewitt operator*

The Prewitt operator is a discrete differentiation operator in the sense that it computes an approximation of the gradient of the image intensity function. It is an edge detection algorithm developed by Prewitt in 1970 (Prewitt *et al.* 1970). The Prewitt operator produces either the corresponding gradient vector or the norm of this vector at each point in the image. Like the Sobel and Kayyali operators, it is based on convolving the image with a small, separable, and integer-valued filter in horizontal and vertical directions and is thus relatively computationally inexpensive. However, the gradient approximation is relatively crude, especially for high-frequency variations in the image.

3.3 *Canny algorithm*

John F. Canny developed the Canny edge detection algorithm in 1986. It uses the calculus of variations to find an optimal function described by the sum of four exponential terms. However, the first derivative of a Gaussian is often used to approximate it. The algorithm begins by smoothing the image with a Gaussian filter in order to remove noise, followed by discrete differentiation to find the image's intensity gradients. Gradient magnitude thresholding is then used to eliminate spurious responses of edge detection. To identify potential edges, double thresholding is used. Edges are finally detected by suppressing all weak edges that are not connected to strong edges (Canny 1986).

3.4 *The Local Binary Pattern (LBP)*

The LBP method can be used to obtain visual texture descriptors. LBP features are derived from pixel intensity values, and neighboring pixels within a fixed radius are assigned new binary values. Pixels with intensities greater than or equal to the center pixel are assigned a value of one, while the rest are assigned a value of zero. The generated binary representation are then normalized and binned to describe patterns within images. The resulting features can be used for image segmentation, classification, and registration (Brahnam *et al.* 2014).

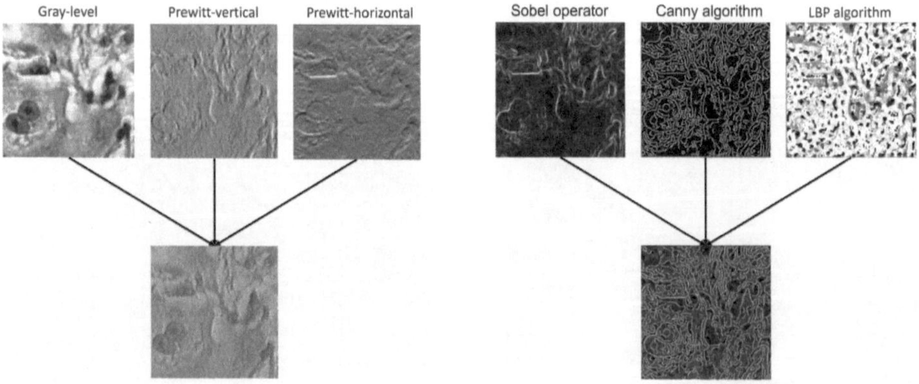

Figure 3. Images generated using manual feature techniques.

Table 1. Classification results of different models trained using PanNuke data set based on color-coded tissue frames. Compared to (Gamper *et al.* 2020). ($^{+}$) Input images are plain nuclei cropped and extracted from original image patches as shown in Figure 2 (right). ($^{-}$) Input images are plain image patches with nuclei at the center of each image. (*) original images with tissue color-coded bounding boxes around target nuclei as shown in Figure 2 (left).

	Connective/Soft			Dead			Epithelial			Inflammatory			Neoplastic			
Model	precision	recall	f1-score	precision	recall	f1-score	precision	recall	f1-score	precision	recall	f1-score	precision	recall	f1-score	Weighted Accuracy
$^{+}$Single-Encoder ResNet34	0.68051	0.61246	0.64470	0.51104	0.28477	0.36574	0.65770	0.59334	0.62387	0.69056	0.69413	0.69234	0.86503	0.90018	0.78442	0.70923
$^{+}$Single-Encoder ResNet50	0.68514	0.63993	0.66177	0.49243	0.33869	0.40135	0.6937	0.58691	0.63585	0.70397	0.7043	0.70413	0.75641	0.83186	0.79235	0.72054
$^{+}$Single-Encoder ResNet101	0.69151	0.63081	0.65977	0.54817	0.3122	0.39783	0.68207	0.63149	0.6558	0.70107	0.69783	0.69944	0.76274	0.83488	0.79718	0.7239
$^{+}$Muli-Encoder ResNet34	0.679	0.6054	0.6401	0.4992	0.2942	0.3702	0.6845	0.5756	0.6253	0.6611	0.7094	0.6844	0.745	0.8221	0.7817	0.7059
$^{+}$Muli-Encoder ResNet50	0.6876	0.63511	0.66032	0.47668	0.34815	0.4024	0.69934	0.58623	0.6378	0.71119	0.68997	0.70042	0.75282	0.842	0.79492	0.72135
$^{+}$Muli-Encoder ResNet101	0.70699	0.7126	0.70979	0.54126	0.3226	0.40426	0.70775	0.63386	0.66877	0.70133	0.6468	0.67297	0.77155	0.84261	0.80552	0.73426
$^{-}$Single-Encoder ResNet34	0.75447	0.75512	0.75479	0.73828	0.46168	0.56810	0.87653	0.83894	0.85732	0.76345	0.73552	0.69234	0.74678	0.82607	0.88225	0.82039
$^{-}$Single-Encoder ResNet50	0.78349	0.77123	0.77731	0.71981	0.56386	0.63236	0.89108	0.87257	0.88173	0.78706	0.75187	0.76906	0.88141	0.91764	0.89916	0.8408
$^{-}$Single-Encoder ResNet101	0.77269	0.7882	0.78037	0.77517	0.43709	0.55898	0.8956	0.87336	0.88434	0.78208	0.75954	0.77064	0.88971	0.9098	0.89964	0.84124
$^{-}$Muli-Encoder ResNet34	0.7559	0.7192	0.7371	0.5837	0.4191	0.4879	0.8595	0.8354	0.8473	0.751	0.7222	0.7363	0.8525	0.9067	0.8788	0.8105
$^{-}$Muli-Encoder ResNet50	0.7941	0.7564	0.7748	0.667	0.5516	0.6038	0.8989	0.9027	0.9008	0.7883	0.7576	0.7727	0.8789	0.9219	0.8999	0.8435
$^{-}$Muli-Encoder ResNet101	0.79	0.756	0.7726	0.6621	0.5043	0.5725	0.8999	0.8946	0.8972	0.7862	0.7555	0.7705	0.8775	0.923	0.8997	0.8417
*Single-Encoder ResNet34	0.76733	0.76395	0.76564	0.69433	0.47493	0.56404	0.89762	0.87777	0.88758	0.77559	0.73718	0.7559	0.88012	0.91546	0.89744	0.83486
*Single-Encoder ResNet50	0.77873	0.78252	0.78062	0.75349	0.45979	0.57109	0.90905	0.88894	0.89888	0.77197	0.75778	0.74922	0.89453	0.91721	0.90573	0.84507
*Single-Encoder ResNet101	0.79357	0.76572	0.7794	0.67488	0.54399	0.60241	0.90768	0.9	0.90383	0.77046	0.75751	0.76393	0.88814	0.92164	0.90458	0.84532
*Muli Encoder ResNet34	0.79598	0.78854	0.79224	0.72257	0.43614	0.54395	0.90914	0.90462	0.90687	0.79805	0.75057	0.77358	0.89042	0.93013	0.90984	0.85270
*Muli-Encoder ResNet50	0.80399	0.77398	0.78870	0.69202	0.53358	0.60256	0.90747	0.91986	0.91362	0.78799	0.76535	0.77651	0.89480	0.92883	0.91150	0.85430
*Muli-Encoder ResNet101	0.81145	0.78395	0.79746	0.75278	0.57615	0.65273	0.92914	0.93544	0.93228	0.78249	0.77903	0.78076	0.90446	0.93073	0.91741	0.86269

3.5 *Input data*

In addition to the original image, we generate two other images. The first image is generated by the concatenation of gray-level features stored in the first channel and vertical and horizontal Prewitt kernel features in the second and third channels, respectively. The second image is a combination of three channels storing LBP, Sobel kernel, and Canny's algorithm features. We feed images as patches rather than plain single nuclei to preserve the context around each nucleus, which we found to be useful for nuclei classification since nuclei tend to exist within clusters of the same classes. Yet, there are many cases where nuclei clusters can overlap, such as in cancerous tissue samples. We add bounding boxes around each target nucleus before feeding input images to leak information about the nucleus' position to the model. Since the tissue type is essential information that medical practitioners should know before tissue examination, we also add color-coded tissue types using bounding boxes to feed this information to the model during the learning process. Figure 3 depicts our methodology to generate input images based on manual feature extraction methods.

3.6 *Architecture*

Our proposed architecture comprises three vertically stacked ResNet encoders, each simultaneously receiving a different input image as described above. A concatenation layer combines feature vectors to generate the final feature vector, which is then input to the classification layer. The latter is composed of an average pooling layer. Then a flatten layer, followed by two blocks of batch normalization, dropout, and linear layers. The output is a softmax probability vector, whereby nuclei labels are identified based on the maximum probability value.

4 EXPERIMENTS AND RESULTS

To evaluate our proposed approach, we use the PanNuke data set (Gamper *et al.* 2020), one of the most comprehensive data sets available. PanNuke is based on several publicly available data sets. It combines a variety of visual fields, making it one of the best choices for training localization and classification models that can be generalized to process data from different visual fields. PanNuke uses five non-background classes. Unlike the binary (neoplastic, non-neoplastic) labels sometimes used in the literature, our classifiers are designed to predict all five classes. Neoplasms include different benign or malignant tumors, such as melanomas, carcinomas, etc. Non-neoplastic cells are split into epithelial, connective/soft tissue cells, inflammatory, and dead cells. The data set consists of three folds. We use the first and second folds for the training, whereas the third fold is used for testing. All the results reported in this work are based on the testing data.

Our evaluation experiments are conducted using Python, utilizing PyTorch and FastAi to develop and train the deep learning models, whereas the OpenCV package is used for image pre-processing and manual feature extraction. We evaluate the proposed ensemble architecture using ResNet34, ResNet50, and ResNet101. We performed the inference and data processing on a workstation running Ubuntu 20.04 Linux. The machine is equipped with a 16-core Intel(R) Core(TM) i9-9900KS CPU with 4.00GHz frequency and GPU Nvidia Geforce RTX 2080 with 11 GB RAM. We use weighted accuracy, F_1 score, precision, and recall to evaluate and compare the classification models defined as follows.

$$\text{precision} = \frac{TP}{TP + FP} \tag{1}$$

$$recall = \frac{TP}{TP + FN} \tag{2}$$

$$F_1 = 2 * \frac{\text{precision} * \text{recall}}{\text{precision} + \text{recall}}. \tag{3}$$

$$Weighted\ Accuracy = \sum_i \frac{n_i}{N} * F_{1_i} \tag{4}$$

where TP is the true positives, FP is the false positives, FN is the false negatives, n_i is the number of samples in class i, and N is the total number of samples.

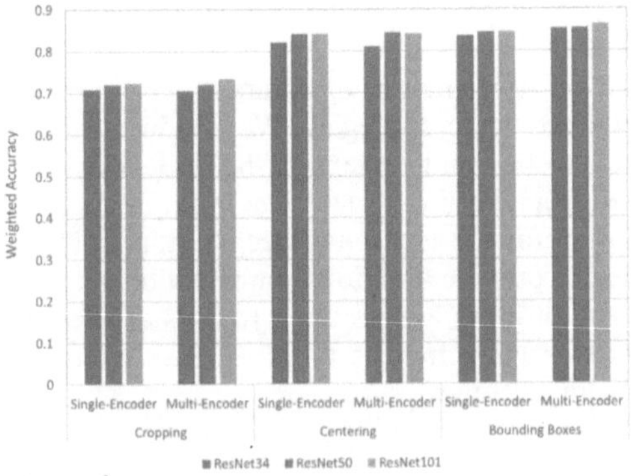

Figure 4. Comparison between the trained models based on weighted accuracy.

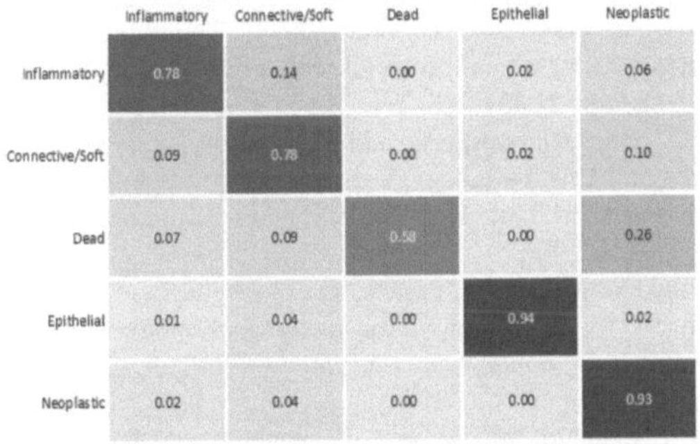

Figure 5. Confusion matrix of the classification results of our proposed model based on the multi-encoder ResNet101 with bounding boxes approach.

In our approach, we use plain original image patches with tissue color-coded bounding boxes around target nuclei, as shown in Figure 2 (left). To show the impact of the bounding boxes on the accuracy of the model, we train single- and multi-encoder models based on three ResNet architectures using nuclei centering and cropping strategies. The results of our evaluation experiments are shown in Table 1. The nuclei cropping strategy omits very important contextual information, as shown in Figure 2 (right). Our experiments indicate that cropping the nucleus out of its context has a negative impact on the classification accuracy performance, which might be due to the lack of information provided to the model. Our findings are corroborated by prior studies (Gamper *et al.* 2020). On the other hand, the nuclei-centering strategy preserves this information, but images have to be resized, cropped, and padded. Besides, the position of the nuclei is left to be figured out by the model. This strategy adds extra unnecessary overhead, reducing the accuracy of the classification models, as shown by the results of our evaluation experiments. To avoid this limitation, we use the bounding boxes to keep the image context and emphasize the nuclei position at the same time.

Results show that feeding images as patches with tissue color-coded bounding boxes around target nuclei preserves the context, which is very useful for classification performance. The model gives more attention to the classification of the target nucleus than to deciding its position. Moreover, combining original images with the pre-processed images based on manual feature analysis further improves the accuracy performance. It enables the models to extract useful information from input images, increasing their classification accuracy. Comparisons between different trained models based on the weighted accuracy results are illustrated in Figure 4 and the confusion matrix of the best model is shown in Figure 5.

5 CONCLUSION

This work provides a comparative analysis of different nuclei-oriented image manipulation approaches based on context-aware strategies. We conducted evaluation experiments to show the impact of nuclei position and context on the nuclei classification performance. We proposed ensemble-based deep learning to leverage multiple manual feature analysis techniques as a pre-analysis step for the automatic deep feature extraction step. Results show that context-aware strategies provide additional information that is vital for the classification process. Pre-processing of input images also provides an extra level of abstraction that enables the deep feature extraction models to gain better feature representation, thereby improving their classification accuracy.

REFERENCES

Abels, E., Pantanowitz, L., Aeffner, F., Zarella, M. D., Van Der Laak, J., Bui, M. M., Vemuri, V. N., Parwani, A. V., Gibbs, J., Agosto-Arroyo, E., et al. (2019). Computational pathology definitions, best practices, and recommendations for regulatory guidance: A white paper from the digital pathology association. *The Journal of Pathology*, 249(3), 286–294.

Al-Biltaji, H. R. H. (2021). Nuclei Classification of Inflammatory Cells in Pediatrics: A Novel Transfer Learning Approach for Digital Histopathology. https://www.proquest.com/docview/2605622755

Al-Janabi, S., Huisman, A., and Van Diest, P. J. (2012). Digital pathology: Current status and future perspectives. *Histopathology*, 61(1), 1–9.

Al-Thelaya, K., Joad, F., Gilal, N., Mifsud,W., Pintore, G., Gobbetti, E., Agus, M., and Schneider, J. (2022). Histocontours: A framework for visual annotation of histopathology whole slide images. *Eurographics Workshop on Visual Computing for Biology and Medicine (VCBM)*, 99–109. https://doi.org/10.2312/vcbm.20221192

Al-Thelaya, K., Gilal, N. U., Alzubaidi, M., Majeed, F., Agus, M., Schneider, J., and Househ, M. (2023). Applications of discriminative and deep learning feature extraction methods for whole slide image analysis: A survey. *Journal of Pathology Informatics*, 14, 100335.

Arvaniti, E., Fricker, K. S., Moret, M., Rupp, N., Hermanns, T., Fankhauser, C., Wey, N., Wild, P. J., Rueschoff, J. H., and Claassen, M. (2018). Automated gleason grading of prostate cancer tissue microarrays via deep learning. *Scientific Reports*, 8(1), 1–11.

Brahnam, S., Jain, L. C., Nanni, L., Lumini, A., et al. (2014). *Local Binary Patterns: New Variants and Applications* (Vol. 506). Springer.

Canny, J. (1986). A computational approach to edge detection. *IEEE Transactions on Pattern Analysis and Machine Intelligence*, (6), 679–698.

Chen, R. J., Lu, M. Y., Shaban, M., Chen, C., Chen, T. Y., Williamson, D. F., and Mahmood, F. (2021). Whole slide images are 2d point clouds: Context-aware survival prediction using patch-based graph convolutional networks. *International Conference on Medical Image Computing and Computer-Assisted Intervention*, 339–349.

Dim, J. R., and Takamura, T. (2013). Alternative approach for satellite cloud classification: Edge gradient application. *Advances in Meteorology*, 2013.

Dodington, D. W., Lagree, A., Tabbarah, S., Mohebpour, M., Sadeghi-Naini, A., Tran, W. T., and Lu, F.-I. (2021). Analysis of tumor nuclear features using artificial intelligence to predict response to neoadjuvant chemotherapy in high-risk breast cancer patients. *Breast Cancer Research and Treatment*, 186(2), 379–389.

Faust, K., Xie, Q., Han, D., Goyle, K., Volynskaya, Z., Djuric, U., and Diamandis, P. (2018). Visualizing histopathologic deep learning classification and anomaly detection using nonlinear feature space dimensionality reduction. *BMC Bioinformatics*, 19(1), 1–15.

Gamper, J., Alemi Koohbanani, N., Benet, K., Khuram, A., and Rajpoot, N. M. (2019). Pan-Nuke: An open pan-cancer histology dataset for nuclei instance segmentation and classification. *European Congress on Digital Pathology*, 11–19.

Gamper, J., Koohabani, N. A., Benes, K., Graham, S., Jahanifar, M., Khurram, S. A., Azam, A., Hewitt, K., and M., R. N. (2020). *PanNuke Dataset Extension, Insights and Baselines*. arXiv:2003.10778v7. https://doi.org/10.48550/arXiv.2003.10778

Gao, Z., Lu, Z., Wang, J., Ying, S., and Shi, J. (2022). A convolutional neural network and graph convolutional network based framework for classification of breast histopathological images. *IEEE Journal of Biomedical and Health Informatics*.

Graham, S., Vu, Q. D., Raza, S. E. A., Azam, A., Tsang, Y. W., Kwak, J. T., and Rajpoot, N. (2019). Hover-net: Simultaneous segmentation and classification of nuclei in multitissue histology images. *Medical Image Analysis*, 58, 101563.

He, K., Zhang, X., Ren, S., and Sun, J. (2016). Deep residual learning for image recognition. *Proceedings of the IEEE Conference on Computer Vision and Pattern Recognition*, 770–778.

Khened, M., Kori, A., Rajkumar, H., Krishnamurthi, G., and Srinivasan, B. (2021). A generalized deep learning framework for whole-slide image segmentation and analysis. *Scientific Reports*, 11(1), 1–14.

Krizhevsky, A., Sutskever, I., and Hinton, G. E. (2012). Imagenet classification with deep convolutional neural networks. *Advances in Neural Information Processing Systems*, 25.

Niazi, M. K. K., Parwani, A. V., and Gurcan, M. N. (2019a). Digital pathology and artificial intelligence. *The Lancet Oncology*, 20(5), e253–e261. https://doi.org/https://doi.org/10.1016/S1470-2045(19)30154-8

Niazi, M. K. K., Parwani, A. V., and Gurcan, M. N. (2019b). Digital pathology and artificial intelligence. *The Lancet Oncology*, 20(5), e253–e261.

Pantanowitz, L., Sharma, A., Carter, A. B., Kurc, T., Sussman, A., and Saltz, J. (2018). Twenty years of digital pathology: An overview of the road travelled, what is on the horizon, and the emergence of vendor-neutral archives. *Journal of Pathology Informatics*, 9.

Pell, R., Oien, K., Robinson, M., Pitman, H., Rajpoot, N., Rittscher, J., Snead, D., Verrill, C., quality assurance working group, U. N. C. R. I. (C.-M. P. (.-P., Driskell, O. J., *et al.* (2019). The use of digital pathology and image analysis in clinical trials. *The Journal of Pathology: Clinical Research*, 5(2), 81–90.

Prewitt, J. M., et al. (1970). Object enhancement and extraction. *Picture Processing and Psychopictorics*, 10(1), 15–19.

Shakya, R., Nguyen, T. H., Waterhouse, N., and Khanna, R. (2020). Immune contexture analysis in immuno-oncology: Applications and challenges of multiplex fluorescent immunohistochemistry. *Clinical & Translational Immunology*, 9(10), e1183.

Sobel, I., and Feldman, G. (1968). A 3×3 Isotropic Gradient Operator for Image Processing.

Wang, K.-S., Yu, G., Xu, C., Meng, X.-H., Zhou, J., Zheng, C., Deng, Z., Shang, L., Liu, R., Su, S., et al. (2021). Accurate diagnosis of colorectal cancer based on histopathology images using artificial intelligence. *BMC Medicine*, 19(1), 1–12.

Xu, Y., Jia, Z., Wang, L.-B., Ai, Y., Zhang, F., Lai, M., Eric, I., and Chang, C. (2017). Large scale tissue histopathology image classification, segmentation, and visualization via deep convolutional activation features. *BMC Bioinformatics*, 18(1), 1–17.

Zhang, J., Hua, Z., Yan, K., Tian, K., Yao, J., Liu, E., Liu, M., and Han, X. (2021). Joint fully convolutional and graph convolutional networks for weakly-supervised segmentation of pathology images. *Medical Image Analysis*, 73, 102183.

Zhang, W., Zhang, J., Yang, S., Wang, X., Yang, W., Huang, J., Wang, W., and Han, X. (2022). Knowledge-based representation learning for nucleus instance classification from histopathological images. *IEEE Transactions on Medical Imaging*.

Innovation and Technological Advances for Sustainability – Al-Naemi et al. (Eds)

Leveraging artificial intelligence and data science for addressing climate change (SDG 13)

Shrouq Al-Fuqaha'a, Mohammad Abu-Soud, Ghayd'a Al-Hyasat & Bushra Alhijawi
Princess Sumaya University for Technology, Jordan

ABSTRACT: Changes in the global climate attract the attention of the whole world. Climate change results in severe impacts on humans and the environment, such as extremely high temperatures, freezing, floods, hurricanes, and a shorter winter. The United Nations took the initiative to lead the world to a stable climate with its 2030 Agenda, SDG 13. Artificial intelligence (AI) and data science technologies are enablers of sustainable development, including addressing climate change. This article reviews and classifies the research efforts on climate change solutions using AI and data science technologies. We categorized the research efforts into four groups: classification-based, regression-based, data analysis-based, and transfer learning-based solutions. Furthermore, the article surveys the applications and datasets considered by the reviewed articles. Also, we categorized the applications into causes and impacts of climate change prediction. Finally, the article provides promising future directions for developing the research field.

Keywords: Climate change, SDGs, SDG 13, artificial intelligence, data science, machine learning, data analysis

1 INTRODUCTION

The United Nations Member States adopted the 2030 Agenda for Sustainable Development that includes 17 Sustainable Development Goals (SDGs) in September 2015[1]. The 17 SDGs are an urgent call for action by all countries to achieve peace and prosperity for people and the planet, considering major challenges, such as poverty, hunger, inequality, environmental issues, and climate change. The 17 SDGs have been defined focus on different aspects of the economy, society, and the environment and fundamentally targeting People, Prosperity, Peace, Partnership, and Planet [1].

Currently, a rapidly changing climate presents enormous challenges worldwide [2]. Climate change is causing more severe and frequent disasters, negatively impacting social and economic stability and development [3]. Rising global temperatures continue unabated, leading to more extreme weather. According to the 2030 Agenda for Sustainable Development, climate change will result in sea level rises 30-60cm by 2100, drought displacing 700 million people by 2030, and medium-to-large-scale disasters increase 40% from 2015 to 2030 (e.g., forest fire, flood, and storms). For example, the number of flood and storm disasters has increased by 7.4% yearly in recent decades [2]. Therefore, Goal 13 of the SDGs aims to take urgent actions to combat climate change and its impact.

Data science and artificial intelligence (AI) are changing a variety of industries and sectors. Currently, AI reached a sufficient level of maturity as an umbrella scientific discipline and a technology, having spread to the entire society [4]. Thus, AI and data science offer

[1]2030 Agenda for Sustainable Development: https://sustainabledevelopment.un.org/post2015/transformingourworld

 DOI: 10.1201/9781003496724-32

great opportunities to help achieving the SDGs, including climate change and its impact. Several solutions integrated AI with the Internet of Things (IoT) to address climate change by predicting natural disasters and energy resource needs, optimizing the exploitation of green and renewable energy, and improving traffic [4].

Several reviews have considered the research and progress on AI solutions for achieving the SDGs [4–8]. However, these articles emphasized different perspectives other than the applications, datasets, and events for addressing climate change and its impact. For instance, Palomares *et al.* [4] provided a general view of the AI impact in attaining the SDGs and performed a SWOT analysis. Goralski and Tan [5] analyzed the AI impact on SDGs from the business strategy and public policy perspectives. The AI applications for sustainability solutions and related challenges are discussed by Nishant *et al.* [6]. Vinuesa *et al.* [7] investigated the positive and negative impacts of AI on the SDGs. Cowls *et al.* [8] investigated the opportunities offered by AI for global climate change.

In that context, this survey aims to review the recent AI and data science-based solutions that address climate change during 2018–2023. Climate change research articles using AI and data science are categorized based on the application and techniques. This survey reported the following:

- Most countries are interested in working in AI and data science-based climate change solutions.
- Techniques used to build the solutions.
- Application of AI and data science-based climate change solutions.
- Datasets used to build and evaluate the solutions.
- Events focused on awareness-raising, supporting, and developing innovative solutions for climate change and its impact.

The rest of this paper is organized as follows. Section 2 presents the methodology followed in this survey. Section 3 classifies the climate change solutions using AI and data science according to the techniques used. Section 4 shows the applications of AI and data science-based climate change solutions. Section 5 presents the datasets used in the selected articles and climate change events. Finally, Section 6 concludes the survey and summarizes the future directions.

Table 1. Distribution of articles based on publishing year.

Year	Articles(%)
2018	10%
2019	3%
2020	42%
2021	39%
2022	6%

2 METHODOLGY

We have collected 32 articles focused on employing AI and data science to address climate change published during 2018-2022. Table 1 presents the distribution of papers according to the publishing year. Searches are conducted on IEEE Explorer, Google Scholar, Science Direct, and ACM Digital Library to identify the relevant literature. The collected research works are sought using a combination of keywords such as Data Science, Artificial Intelligence, Climate Change, Prediction, Machine Learning, and Data Analysis. Initially, we collected more than 40 articles and excluded 18 papers based on predefined criteria where the selected article should be: (1) written in English. (2) relevant to climate change and its impact. (3) employ an AI or a data science technique. (4) evaluated with a clear methodology.

Overall, the reviewed researches are full research articles. 35% of the research papers have between eight and fifteen pages, as shown in Table 2. Most of the research works are completed in Indian universities, with 31%. In the second place, 19% of the researchers are conducted in Chinese universities. Table 3 shows the distribution of papers based on country.

Table 2. Distribution of articles based on page count.

Page count	Articles(%)
1-3	6%
4-7	44%
8-11	16%
12-15	19%
16-19	9%
20-23	6%

Table 3. Distribution of research based on country.

Country	Articles(%)
India	31%
China	19%
Italy	6%
others	44%

3 AI AND DATA SCIENCE AGAINST CLIMATE CHANGE

Data science helps organizations extract useful knowledge and insights from sizable data to create effective climate change solutions. The advancement of AI, including machine learning technology, has dramatically accelerated modern data science. AI techniques are able to discover hidden patterns in vast amounts of data more quickly. Data scientists can then use these patterns to identify workable solutions to typical problems. We classified the collected articles into four groups based on the used AI/Data Science technique:

- Classification-based solutions. Classification problems involve identifying/predicting an observation (i.e., data point/record) to a predefined set of classes. The target column (i.e., output) in a classification problem is a discrete value that refers to a particular class. Several classification algorithms are adopted in the literature to handle climate change, such as k-nearest neighbors [9–12], random forest [9–11,13–16], gradient-boosting [17], and neural networks and deep learning [12,18–28]. Different deep learning architectures are used in the reviewed articles, such as long short-term memory (LSTM) [21,23], convolutional neural network (CNN) [19–21,23], and Radial Basis Function (RBF) [24].
- Regression-based solutions. Regression problems involve the prediction of continuous values. A variety of regression algorithms are used to develop solutions for climate change, including linear regression [12,13,16,19,29], multiple linear regression [11,30], support vector regression [11,13,20,21,31].
- Data analysis-based solutions. Analyzing the data to identify patterns and extract useful knowledge and insights is a common solution for a wide range of applications. Different measures and techniques are used to address climate change and its impacts, such as correlation, causation, time series, and principal component analysis. Correlation analysis describes how variables are associated (e.g., analyzing and identifying the relationship between climate change and abnormal weather) [32,33]. Causation analysis focuses on analyzing how a variable might affect another (e.g., study how greenhouse gases affect climate change) [34,35]. Meteorologists use time series analysis to understand and predict future weather and how

climate will change [32,36,37]. The principal component analysis is a technique for analyzing huge data with a high number of features to reduce the data dimensionality and increase the interpretability of data while preserving the maximum amount of information [32].

- Transfer learning-based solutions. Transfer learning is a machine learning method where a model developed for a particular task (pre-trained model) is reused on a new problem. Kadow *et al.* [38] benefits of a pre-trained deep learning model to reconstruct missing climate information of global temperature patterns.

Table 4 summarizes and classifies the collected research articles. Clearly, most of the authors handle the climate change challenge using classification algorithms.

Table 4. AI and data science techniques used for SDG 13.

Group	Method	Ref.
Data Analysis	Correlation Analysis	[32,33]
	Time Series Analysis	[32,36,37]
	Principal Component Analysis	[32]
	Causal Analysis	[33–35]
Classification	K-Nearest Neighbors	[9–12]
	Random Forest	[9–11,13–16]
	Artificial Neural Network	[12,25–28,39]
	LSTM / DNN /CNN / RBF	[18–24]
	Gradient-Boosting	[17]
Regression	Linear Regression	[12,13,16,19,29]
	Multiple Linear Regression	[11,30]
	Support Vector Regression	[11,13,20,21,31]
Transfer Learning	20CR / CMIP5	[38]

4 APPLICATIONS OF AI AND DATA SCIENCE FOR CLIMATE CHANGE

AI and data science are adopted to address climate change for several purposes and applications. We divided the applications into two categories: predicting the impact and cause of climate change. Table 4 shows the classification of the selected papers based on the application.

Table 5. Summary of applications.

Application	Type	Ref.
Rainfall Prediction	Impact	[11,27,36]
Climate/Weather Prediction	Impact	[29,32,37]
Power Distribution Network	Impact	[33]
Flood/Storm Prediction	Impact	[14,17]
Sea Surface Temperature and Level Prediction	Impact	[12,19,25]
Water Demand Prediction	Impact	[26]
Temperature Prediction	Impact	[9,22,39]
Humidity Prediction	Impact	[39]
Indoor Building Temperature Prediction	Impact	[24]
Agriculture Applications	Impact	[15,18]
Evaporation Prediction	Impact	[31]
Gases (CO, CO2, O3) Emissions Prediction	Cause	[23,30,34,35]
Air Pollution Prediction	Cause	[13,16]
Reconstruct Missing Climate Information	Impact	[38]
Safeguarding of the Wetlands	Impact	[10]
Deforestation	Cause/Impact	[20]
Health	Impact	[21,28]

Climate change has several impacts, and considerable research efforts have focused on predicting these impacts. Climate change results in extreme weather events that can cause devastating impacts on communities and agricultural and natural ecosystems. Many authors developed solutions to predict the impact of weather factors on crops [10,15,18,20]. Extreme weather-related events are predicted using AI and data science-based solutions, including heat waves [9,24,37,39], heavy rainfall [27,33,36], storms [17], humidity [37,39], evaporation [31], and floods [14].

In addition, climate change and extreme weather events negatively impact human health, where the change in weather may increase diseases [21,23,29]. Also, weather-related events may damage infrastructure and ecosystems [17,27]. Climate change raises sea levels, thus, decreasing abundant water in coastal areas. Rising sea levels may negatively affect groundwater quality through saltwater intrusion [12,19,25,31].

Climate change results from many causes, including the industrial revolution. The industrial revolution increased the use of fossil fuels, including gas, oil, and coal, which led to global warming; thus, increasing tropospheric ozone (O3) and carbon dioxide (CO2) [23,30,34,35]. Air pollution is the deterioration of air quality that severely impacts humans and the environment. The air quality is negatively affected due to natural processes, and harmful human activity [13,16]. Air pollution leads to climate change and affects food, plant productivity, crop yields, and human health.

Deforestation is one of the significant contributors to climate change[20]. We classified Deforestation as a cause and an impact of climate change. Deforestation has many forms, including wildfires, clearing trees for agriculture and livestock, and logging for wood. Forest burning releases carbon dioxide into the atmosphere, one of the leading greenhouse gases contributing significantly to global warming and climate change. Removing trees negatively affects air quality, increasing CO2 and reducing oxygen.

5 DATASETS AND EVENTS

All the reviewed articles proposed a data-driven solution, which depends mainly on the data used to develop and evaluate the suggested solutions. Table 6 presents the distribution of the reviewed research works based on datasets. We classified the datasets into three groups based on availability, public, private, and data that can be requested from authors.

Table 6. Summary of datasets.

Dataset Availability	Ref.
Public	[9,10,12,14,18–20,25,28,29,35,38]
Private	[11,13,15–17,21–24,26,30–33,36,37,39]
Can be requested	[34]

Two kinds of data are used in the literature: (1) sensory data [9,16,18,23,28]. (2) imagery data [10,12,14,18–20,29,35,38]. The sensory data are collected using sensors, such as temperature sensors and humidity sensors. Authors collected sensory data from various sources, e.g., Kaggle, U.K. National Centre for Earth Observation, Copernicus Climate Change Services, USDA's National Agricultural Statistical Service, Soil data from SSURGO. The most used imagery dataset is called Landsat [10,12,14,20], a public dataset. Other imagery data are used in the literature, such as Sentinel-5P satellite imagery using Google Earth Engine [35], Met Office SST [38], HadCRUT4 [38], NASA dataset [29], Laboratory for Satellite Altimetry / Sea Level Rise data [19], Gridded surface meteorological data of the Northwest Knowledge Consortium [18].

Recently, several events focus on awareness-raising, supporting, and developing innovative solutions for climate change are organized. We classified the climate change events into three categories:

- Events encourage developing innovative solutions [40–43]. For example, the University of Doha for Science and Technology will launch ITAS Arab Youth Competition 2023 for university students across the Arab world to develop Sustainable and Practical Solutions using technology [40]. In September 2022, Egypt University of Informatics (EUI) launched a Climate change Hackathon for all the university students in Egypt, Arab Countries, and Africa to find effective solutions to climate change challenges [41]. Also, the Women in Data Science (WiDS) Datathon 2023 challenge is "Adapting to Climate Change by Improving Extreme Weather Forecasts." [42]. WiDS Datathon is a platform for data scientists to learn and apply their data science skills through the social impact challenges presented to them.
- Academic conferences and workshops [44]. For instance, the 5th international conference on climate change: impacts & responses will be held in Canada in April 2023. The conference focuses on research works that propose scalable solutions for the climate-nature intersects.
- High-level, discussion, negotiations, press conferences [45–48]. The 27th conference of the parties to the United Nations framework convention on climate change (COP27) was held in Sharm El-Sheikh, Egypt included high-level events, negotiations, and press conferences, hosting more than 100 Heads of State and Governments, over 35,000 participants [45]. Women in climate change conference 2022 (#WiCC) was held in Uganda to recognize women's role in addressing climate change [47]. In addition, the global conference on climate change 2022 was announced by the International Astronautical Federation and the Norwegian Space Agency [48]. In October 2023, the 2nd world conference on climate change & sustainability (Climate Week 2023) will be held in Italy. Climate Week 2023 is an annual event where people from around the world gather to discuss climate change challenges and developed technologies to overcome them.

6 CONCLUSION AND FUTURE DIRECTIONS

This survey article aimed to review the research efforts for addressing climate change and its impact using AI and data science methods from 2018 to 2022. Three main contributions are reported in the article. First, the paper presented a new classification of AI and data science solutions for climate change based on the technique used. The classification included classification-based, regression-based, data analysis-based, and transfer learning-based methods. Second, the paper analyzed and classified the applications of AI and data science for climate change into causes and impact prediction. Third, the paper reviewed and summarized the datasets used to develop and evaluate the proposed methods and worldwide climate change events.

There are several promising future research directions. First, a promising future direction would be to develop additional transfer learning-based solutions against climate change. Second, more contributions can be made to collect and publish public climate data where data-driven solutions rely on the data and its quality. Third, a promising future direction would be to unify the evaluation methodology where authors used different datasets, baselines, and evaluation methodology, and their selections were seemingly arbitrary, which led to inconsistent and unreliable reporting of scores. Also, many evaluation parameters need unified tuning to ensure reliable results. Fifth, designing and developing data-driven solutions for climate change using other kinds of machine learning and data analysis methods.

REFERENCES

[1] Wu J., Guo S., Huang H., Liu W., and Xiang Y., (2018), Information and communications technologies for sustainable development goals: state-of-the-art, needs and perspectives, *IEEE Communications Surveys & Tutorials*, vol. 20, no. 3, pp. 2389–2406.

[2] Balogun A.-L., Marks D., Sharma R., Shekhar H., Balmes C., Maheng D., Arshad A., and Salehi P., (2020). Assessing the potentials of digitalization as a tool for climate change adaptation and sustainable development in urban centres, *Sustainable Cities and Society*, vol. 53, p. 101888.

[3] I. Adopted, (2014). Climate change 2014 synthesis report, *IPCC: Geneva, Szwitzerland.*

[4] Palomares I., Martnez-Cámara E., Montes R., Garca-Moral P., Chiachio M., Chiachio J., Alonso S., Melero F. J., Molina D., Fernández B., et al., (2021). A panoramic view and swot analysis of artificial intelligence for achieving the sustainable development goals by 2030: Progress and prospects, *Applied Intelligence*, vol. 51, no. 9, pp. 6497–6527.

[5] Goralski M. A. and Tan T. K., (2020). Artificial intelligence and sustainable development, *The International Journal of Management Education*, vol. 18, no. 1, p. 100330.

[6] Nishant R., Kennedy M., and Corbett J., (2020). Artificial intelligence for sustainability: Challenges, opportunities, and a research agenda, *International Journal of Information Management*, vol. 53, p. 102104.

[7] Vinuesa R., Azizpour H., Leite I., Balaam M., Dignum V., Domisch S., Felländer A., Langhans S. D., Tegmark M., and Fuso Nerini F., (2020). The role of artificial intelligence in achieving the sustainable development goals, *Nature communications*, vol. 11, no. 1, pp. 1–10.

[8] Cowls J., Tsamados A., Taddeo M., and Floridi L., (2021). The AI gambit: leveraging artificial intelligence to combat climate change–opportunities, challenges, and recommendations, *Ai & Society*, pp. 1–25.

[9] Pavuluri B. L., Vejendla R. S., Jithendra P., Deepika T., and Bano S., (2020). Forecasting meteorological analysis using machine learning algorithms, in *2020 International Conference on Smart Electronics and Communication (ICOSEC)*, pp. 456–461.

[10] Granata F., Gargano R., and de Marinis G., (2020). Artificial intelligence based approaches to evaluate actual evapotranspiration in wetlands, *Science of The Total Environment*, vol. 703, p. 135653.

[11] Premachandra J. S. A. N. W. and Kumara P. P. N. V., (2021). A novel approach for weather prediction for agriculture in sri lanka using machine learning techniques, in *2021 International Research Conference on Smart Computing and Systems Engineering (SCSE)*, vol. 4, pp. 182–189.

[12] Wei-Hao S. and Tseng S.-H., (2021). Comparisons of classification models on coastsat, in *2021 International Symposium on Intelligent Signal Processing and Communication Systems (ISPACS)*, pp. 1–2.

[13] Saikiran K., Lithesh G., Srinivas B., and Ashok S., (2021). Prediction of air quality index using supervised machine learning algorithms, in *2021 2nd International Conference on Advances in Computing, Communication, Embedded and Secure Systems (ACCESS)*, pp. 1–4.

[14] Chen H., Liqiang T., Zhaocheng G., Jienan T., Hua W., and Peng H., (2021). A dynamics trend analysis method of thermokarst lakes based on the machine learning algorithm, in *2021 IEEE International Geoscience and Remote Sensing Symposium IGARSS*, pp. 6484–6487.

[15] Suresh N., Ramesh N., Inthiyaz S., Priya P. P., Nagasowmika K., Kumar K., Shaik M., and Reddy B. N. K., (2021). Crop yield prediction using random forest algorithm, in *2021 7th International Conference on Advanced Computing and Communication Systems (ICACCS)*, vol. 1, pp. 279–282.

[16] Pal S., Pramanik D., and Jain E., (2021). Effectiveness of machine learning algorithms in forecasting aqi, in *2021 International Conference on Technological Advancements and Innovations (ICTAI)*, pp. 492–495.

[17] Wu Z., Zhou Y., and Wang H., (2020). Real-time prediction of the water accumulation process of urban stormy accumulation points based on deep learning, *IEEE Access*, vol. 8, pp. 151938–151951.

[18] Crane-Droesch A., (oct 2018). Machine learning methods for crop yield prediction and climate change impact assessment in agriculture, *Environmental Research Letters*, vol. 13, p. 114003.

[19] Hassan K. M. A., Haque M. A., and Ahmed S., (2021). Comparative study of forecasting global mean sea level rising using machine learning, in *2021 International Conference on Electronics, Communications and Information Technology (ICECIT)*, pp. 1–4.

[20] Nichols K. and Hosein P., Estimating deforestation using machine learning algorithms, in *2021 Second International Conference on Intelligent Data Science Technologies and Applications (IDSTA)*, pp. 82–87, 2021.

[21] Abdullahi T. and Nitschke G., (2021). Predicting disease outbreaks with climate data, in *2021 IEEE Congress on Evolutionary Computation (CEC)*, pp. 989–996.

[22] Song W. and Ma X., (2021). Research on climate prediction based on deep learning, in *2021 3rd International Symposium on Smart and Healthy Cities (ISHC)*, pp. 107–109.

[23] Ibrahim S. K., Ziedan I. E., and Ahmed A., (2021). Study of climate change detection in north-east africa using machine learning and satellite data, *IEEE Journal of Selected Topics in Applied Earth Observations and Remote Sensing*, vol. 14, pp. 11080–11094.

[24] May Tzuc O., Livas-Garca A., Jiménez Torres M., Cruz May E., López-Manrique L. M., and Bassam A., (2020). Artificial intelligence techniques for modeling indoor building temperature under tropical climate using outdoor environmental monitoring, *Journal of Energy Engineering*, vol. 146, no. 2, p. 04020004.

[25] Wie L., Guan L., Qu L., and Li L., (2019). Prediction of sea surface temperature in the south china sea by artificial neural networks, in *IGARSS 2019 - 2019 IEEE International Geoscience and Remote Sensing Symposium*, pp. 8158–8161.

[26] Zubaidi S. L., Ortega-Martorell S., Al-Bugharbee H., Olier I., Hashim K. S., Gharghan S. K., Kot P., and Al-Khaddar R., (2020). Urban water demand prediction for a city that suffers from climate change and population growth: Gauteng province case study, *Water*, vol. 12, no. 7, p. 1885.

[27] Mishra N., Soni H., Sharma S., and Upadhyay A., (01 2018). Development and analysis of artificial neural network models for rainfall prediction by using time-series data, *International Journal of Intelligent Systems and Applications*, vol. 10, pp. 16–23.

[28] Shaffiee Haghshenas S., Pirouz B., Shaffiee Haghshenas S., Pirouz B., Piro P., Na K.-S., Cho S.-E., and Geem Z. W., (2020). Prioritizing and analyzing the role of climate and urban parameters in the confirmed cases of covid-19 based on artificial intelligence applications, *International journal of environmental research and public health*, vol. 17, no. 10, p. 3730.

[29] Debdas S., Roy K., Saha A., Kundu S., Dasray S., and Chouhan S., (2020). Analysis and prediction of climate change in post-covid19 india, in *2020 4th International Conference on Electronics, Communication and Aerospace Technology (ICECA)*, pp. 1530–1536.

[30] Tanania V., Shukla S., and Singh S., (2020). Time series data analysis and prediction of co2 emissions, in *2020 10th International Conference on Cloud Computing, Data Science& Engineering (Confluence)*, pp. 665–669.

[31] Ahmadi F., Mehdizadeh S., Mohammadi B., Pham Q. B., Doan T. N. C., and Vo N. D., (2021). Application of an artificial intelligence technique enhanced with intelligent water drops for monthly reference evapotranspiration estimation, *Agricultural Water Management*, vol. 244, p. 106622.

[32] Guo S., Qiao W., Chen B., and Wang B., (2020). Prediction and abnormality analysis of climate change based on pca-arma and pcc, in *2020 IEEE International Conference on Networking, Sensing and Control (ICNSC)*, pp. 1–6.

[33] GU Y. Z., WU Z. R., ZHAO J. G., HAN L. Q., YUAN L. L., and HUANG W. T., (2020). Research on intelligent early warning algorithm for distribution network considering extreme climate conditions, in *2020 15th IEEE Conference on Industrial Electronics and Applications (ICIEA)*, pp. 412–417.

[34] Leung F., Sitch S., Tai A. P. K., Wiltshire A. J., Gornall J. L., Folberth G. A., and Unger N., Co0rw1s34rfesdcfkexd09rt421rw1s34rfesdcfkexd09rt4 fertilization of crops offsets yield losses due to future surface ozone damage and climate change, *Environmental Research Letters*, vol. 17, p. 074007, 07 2022. Copyright - Â© 2022 The Author(s). Published by IOP Publishing Ltd. This work is published under http://creativecommons.org/licenses/by/4.0 (the "License"). Notwithstanding the ProQuest Terms and Conditions, you may use this content in accordance with the terms of the License; Last updated - 2022-08-06.

[35] Ikram N. M., Afifah L., Arthatia B. S., Wicaksono S. J., Maharani M., Ihsanudin T., and Apriyanti D., (07 2022). Monitoring co and o0rw1s34rfesdcfkexd09rt431rw1s34rfesdcfkexd09rt4 concentration that caused climate change periodically using google earth engine (study case: Java island), *IOP Conference Series.Earth and Environmental Science*, vol. 1047, p. 012021. Copyright - Published under licence by IOP Publishing Ltd. This work is published under http://creativecommons.org/licenses/by/3.0/ (the "License"). Notwithstanding the ProQuest Terms and Conditions, you may use this content in accordance with the terms of the License; Last updated - 2022-11-16; SubjectsTermNotLitGenreText - Indonesia; Java (island.

[36] Varghese L. R. and Vanitha K., (2020). A time-series based prediction analysis of rainfall detection, in *2020 International Conference on Inventive Computation Technologies (ICICT)*, pp. 513–518.

[37] Teggi P. P., Natarajan S., and Malakreddy B., (2020). Intelligent forecasting model for climate variations (inform): An urban climate case study, in *2020 7th International Conference on Computing for Sustainable Global Development (INDIACom)*, pp. 130–137.

[38] Kadow C., Hall D. M., and Ulbrich U., (2020). Artificial intelligence reconstructs missing climate information, *Nature Geoscience*, vol. 13, no. 6, pp. 408–413.
[39] Al-Shawwa M., Al-Absi A., Hassanein S., Baraka K., and Abu-Naser S., Predicting temperature and humidity in the surrounding environment using artificial neural network, *International Journal for Academic Development*, vol. 2, pp. 1–6, 09 2018.
[40] Innovation competition - itas, *University of Doha for Science and Technology. Access Date: 29/11/2022.*
[41] Climate change hackathon (climathon), *Egypt University of Informatics. Access Date: 29/11/2022.*
[42] Wids datathon 2023, *Women in Data Science. Access Date: 29/11/2022.*
[43] The climate change-makers challenge, *Devpost. Access Date: 29/11/2022.*
[44] 5th international conference on climate change: Impacts & responses, *Access Date: 29/11/2022.*
[45] Cop27:delivering for people and the planet, *UnitedNations COP27. Access Date: 29/11/2022.*
[46] 2nd world conference on climate change & sustainability (climate week 2023), *Access Date: 29/11/2022.*
[47] Women in climate change conference 2022 (#wicc), *Access Date: 29/11/2022.*
[48] The global conference on climate change 2022, *Access Date: 29/11/2022.*

Innovation and Technological Advances for Sustainability – Al-Naemi et al. (Eds)

Multi-Criteria Decision-Making (MCDM) methods for micro-grids: An overview

Walid Alqaisi
School of Engineering Technology, The University of Doha for Science and Technology, Jelaiah St., Doha, Qatar

Claude Ziad El-Bayeh
Canada Excellence Research Chair Team, Concordia University, Montreal, Canada

Khaled Alzaareer
Electrical Engineering Department, Philadelphia University, Amman, Jordan

Mohamed Zellagui
Department of Electrical Engineering, Faculty of Technology, University of Batna 2, Fesdis, Batna, Algeria

Aymen Flah
Processes, Energy, Environment and Electrical Systems, National Engineering School of Gabès, University of Gabès, Gabès, Tunisia

Brahim Brahmi
Electrical Engineering Department, Collège Ahuntsic, Montreal, QC, Canada

ABSTRACT: Multi-Criteria Decision-Making (MCDM) methods are great tools for solving complex problems with a huge number of alternatives, conflicting criteria and objectives. The main goal of using MCDMs in engineering, especially in the design of micro-grids (MGs), is to find the best solution for a specific complex problem among many possibilities and alternatives, such as selecting the best solar panel from many alternatives considering contradicting criteria. This paper presents a brief overview of the usage of MCDM methods in MGs. It shows, (a) the most used MCDM algorithms for MG design and components' selection, (b) the main objective functions in modeling MCDM problems, (c) the main criteria (*e.g., technical, technological, economic, social, and environmental*), and (d) the prons and cons of each MCDM method. In addition, it points out the available MCDM software in the market and categorizes them.

Keywords: multi-criteria decision-making, micro-grid, optimization, solar photovoltaic system

1 INTRODUCTION

In the last few decades, energy demand and electricity consumption have witnessed tremendous growth in developed countries as well as developing countries. With this growing need, the exploitation of fossil fuels for conventional energy supply is nearing its end [1]. Recently, renewable energy systems are being deployed all over the world [2], to tackle both increasing energy demand and climate change. Renewable Energy Sources (RES) presented 181 GW power generation in 2018, and in particular solar PV systems have the highest

DOI: 10.1201/9781003496724-33

generation capacity of 55% between all RES [3]. Simultaneously, the integration of such Distributed Energy Resources (DERs) into the utility grid has paved the way for the creation of micro-grids (MGs). The MG concept was introduced to have a self-sustained system consisting of DERs that can operate either in an islanded-mode or while connected to the grid. MGs present many benefits, such as but not limited to, decentralizing energy supply and energy demand, reducing GHG emissions, reducing the dependency on the power systems and distribution network, and providing ancillary services to the power grid [2]. In the context of implementing an MG project, it is important to establish an optimal techno-economic approach. This means selecting the best electrical components (i.e., PV system, storage system, etc.) based on energy demand and weather data [4], and maintaining an optimal energy management system [5].

All these constraints of different nature and importance can lead to a complex situation where solving it, requires advanced computing tools as well as expert opinions on a managerial level [6–8]. From this place, using operations research, such as, optimization techniques, and multi-criteria decision-making methods help to displace and size many MG components (*e.g., PV systems, battery storage systems, displacing electric charging stations, etc.*), and manage the energy in a way to increase the independency from the power grid, improve the network's stability, and reduce losses. Hence, it is necessary to write a brief literature review on the importance of using MCDM methods in studying and designing MGs' components.

2 LITERATURE REVIEW

2.1 *Brief review on MCDM methods used for micro-grids*

Operations research is a field in pure and applied mathematics that deals with the development and application of advanced analytical methods to improve decision-making. Operations research is used in many fields including, but not limited to, energy management, MGs, renewable energy, etc., in which the main goal is to minimize or maximize certain objectives while considering many constraints. To solve optimization problems, advanced optimization algorithms are used, such as exact algorithms (i.e., linear and non-linear programming, dynamic programming, etc.), heuristic algorithms (i.e., GA, PSO, etc.), and many others. For example, Ref. [6] used several multi-objective optimization techniques to optimize the power flow of many MGs connected to the same distribution network. Likewise, authors in [9] proposed an interactive three-level multi-energy management strategy for heterogeneous multi-MGs and coherently applied many case studies to back up their proposed solution. As a result, they effectively minimized MGs' operation costs by sharing surplus energy among many MGs. Despite the effectiveness of such algorithms in solving complex problems, advanced tools are required to solve realistic and more complex problems where there are lots of conflicting objectives and criteria for decision-makers, as for MGs. This kind of complexity can be solved using MCDM.

MCDM algorithms are advanced tools that are designed specifically to quickly solve multi-objective optimization problems with conflicting criteria and objectives. In the literature, extensive applications of different MCDM methods were presented. The most notable and widely used methods are, but are not limited to, TOPSIS, ELECTRE, AHP, and PROMETHEE, [2,3], [7,8], [10–13].

They concluded that their proposed RWR method was faster than TOPSIS in finding the best solution with good accuracy. Other papers [11–14] incorporated fuzzy hybrid methods in an effort to improve the outputs of their studies. It should be noted that by applying fuzzy logic to MCDM methods, better results were yielded compared to conventional MCDMs. However, the complexity of solving problems with fuzzy logic becomes higher [14]. On the other hand, Ref. [13] extensively detailed and compared different MCDM methods by listing their prons and cons. They concluded that no MCDM method is the worst or the best, since it all depends on the application at hand. Ref. [12] determined weight coefficient factors of different criteria that can arise in MCDM models by using the Full Consistency Method (FUCOM). On a larger scale, paper [15] assessed different MCDM methods for energy planning problems. Article [16] thoroughly studied smart power management using SWOT-AHP (Strength–Weakness–Opportunity–Threat – AHP) and hybrid MCDM analysis to emphasize the importance of developing advanced metering infrastructure and smart meters for better grid energy management. Figure 1 shows the usage of MCDM methods for MGs, (*e.g., selecting the best solar panel for a specific application*).

Other characteristics of MGs were being considered and deployed including Vehicle-to-Grid (V2G) and Virtual Power Plant (VPP) concepts. With V2G technology, Electric Vehicles (EVs) support MGs by providing ancillary services when needed [17], thus, making use of existing resources and reducing concerns for grid overload. In [18], authors analyzed the economic benefits offered by the V2G technology to a VPP structure, and then, experimentally concluded that the LiFePO4 battery technology was one of the best candidates for EVs. Similarly, authors in [19] provided an in-depth case study of smart charging of EVs with PV and V2G technology in an MG. They simulated three EV charging control algorithms and demonstrated that EVs can contribute significantly to a well-balanced demand and supply.

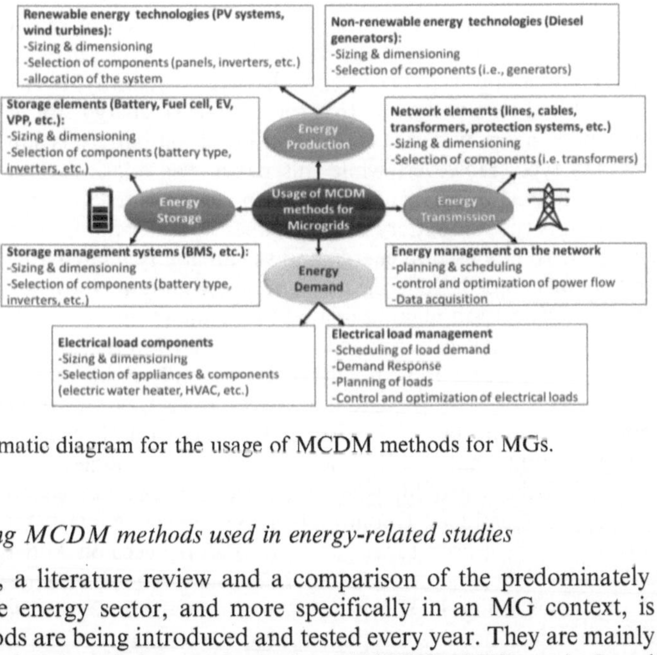

Figure 1. Schematic diagram for the usage of MCDM methods for MGs.

2.2 *Comparing MCDM methods used in energy-related studies*

In this section, a literature review and a comparison of the predominately used MCDM methods in the energy sector, and more specifically in an MG context, is made. Novel MCDM methods are being introduced and tested every year. They are mainly iterations and extensions of already existing methods (e.g., TOPSIS, AHP, etc.). In addition, hybrid MCDM methods (two or more) are incorporated together to solve certain problems more efficiently. As well, the fuzzy logic approach is integrated into MCDM methods, effectively handling the uncertainty in linguistic terms in some applications, yet also presenting complexity limitations. Lastly, MCDM approaches can be associated with other computational methods, such as grey sets, machine learning, and Geographic Information Systems [20]. Regarding the use of conventional MCDM applications, a literature review [21] presented the advantages and drawbacks of each method. The authors argued that the CRiteria Importance Through Intercriteria Correlation (CRITIC) is one of the most objective methods because it did not require the intervention of decision-makers – yet it did not consider the type of criteria at hand, thus, cannot be applied to our case study. While the authors successfully pinpointed the best uses, as well as, the restrictions of each method, they only limited the review to assess renewable energy technologies in small-scale applications (households). Conversely, Ref. [22] treated a medium to large-scale MG case, which is relevant to our study. In fact, authors, proposed an optimal MG setting specified for a rural remote village in India. They started by using AHP in order to weight each criterion, such as total Net Present Cost (having the highest weight) and total cost of Operations and Maintenance (O&M) (having the lowest weight) among others. They then used TOPSIS to present a final ranking for each MG alternative.

On the other hand, fuzzy MCDM methods can be similarly applied to the same decision-making problems as the conventional ones. Their major issue, however, is that they significantly raise the complexity of solving the problem. Paper [23] exhibited a comprehensive literature review on different fuzzy MCDM applications for energy policy-making problems. It mainly aimed to showcase the differences between them but more importantly revealed gaps in the fuzzy-MCDM literature related to energy problems. For example, the fuzzy-AHP has been the most popular in the last decade compared to other fuzzy-MCDM applications and was mostly used for solving problems in the renewable energy sector. This could infer that the fuzzy-AHP is a better medium than others in the areas of performance-type, environmental, and energy policy problems. Authors in [7], proposed an integrated fuzzy MCDM model combining: fuzzy Delphi, Best Worst Method (BWM) (to determine optimal criteria weights), and fuzzy-Cumulative Prospect Theory (CPT) for the assessment of Battery Storage System (BSS). In this case, the fuzzy theory is employed to maximize the use of objective data information. Here again, like the majority of other papers, the criteria and sub-criteria for decision-making were based on technology, economy, environment, performance, and sociality. Final results showed that for large-scale energy storage, regardless of different risk preferences given by decision-makers, the Li-ion battery is still the most optimal, followed by NaS type. However, while this paper was relatively recently published (2018) even when the LifePO4 batteries have long been implemented for large-scale use, they still failed to mention this type of battery in their paper. Furthermore, hybrid MCDM frameworks are being increasingly used in energy and sustainability related applications. Paper [24], reviewed and summarized those hybrid techniques, aiming to showcase those approaches as improvements for decision-making related to sustainability issues. According to the authors, a whopping 42% of articles employed hybrid MCDM in their studies, compared to 36% for conventional methods. Hybrid MCDM methods can be associations of two or more different MCDMs, an MCDM method, and a fuzzy or grey set, or an MCDM and another method that identifies the importance of criteria. This technique is especially helpful and even essential for applications that involve numerous phases of studies.

One relevant example is the case of [8], where they evaluated the 3E (*economy-environment-energy*) performance of a Combined Cooling Heat and Power (CCHP)-MG system. This evaluation included two modules: one to weight the indicators based on Anti-Entropy Weight (AEW) and grey-DEMATEL (*Decision Making Trial and Evaluation Laboratory*), and the other is to rank the cases using DQGRA (*Difference Quotient Grey Relation Analysis*)-TOPSIS. Therefore, on one hand, this solution ensured the reliability of the weighting results (by *reducing subjectivity*). On the other hand, DQGRA-TOPSIS overcomes the limitation of the conventional TOPSIS, by dealing with both the geometric and numerical proximities between the sequences. Nonetheless, the authors did admit that the applicability and effectiveness of the proposed model still needed testing in other fields. One negative aspect of hybrid MCDM methods can be outlined which is the increased complexity of the solution, and the needed time spent on solving the problem. Of all the reviewed papers, rarely did the authors take into account this complexity angle – which can sometimes outweigh the complexity and importance of the case study, decreasing the proposed solution's pertinence. Last but not least, there is an emergence of new methods that often derive from already established original methods. Authors in [25], defined a newly developed MCDM method called

Plithogenic Hypersoft Sets (PHSS) which is in turn based on the conventional TOPSIS. The authors applied it to a parking problem with different choices of a set of alternatives and compared it to the fuzzy TOPSIS version to prove its validity and efficiency. The PHSS indeed excelled in decomposing attributes into several attribute values, such that each alternative is then required to be evaluated against each attribute value in order to perform a detailed analysis. This distinguishing feature should be further explored in MG situations. Indeed, typically in an MG, a solar PV system can be installed. In this case, a set of attributes of solar inverters (brand, cost, etc …) for example are under consideration, with each one having a set of attribute values (popularity of the brand, cheap or costly, etc …). PHSS handles this problem by choosing the best alternative out of the whole alternative set, hence solving the attribute value problem fairly well. It is important to note that the main criteria used in the decision-making process creates an essential part of the decision-making framework. In our case study, for MGs, the main criteria are but are not limited to, technical, economic, environmental, and social. Indeed, an overwhelming majority of research articles treating this type of study employed the aforementioned criteria. Consequently, the main criteria are divided into several sub-criteria. Article [26], provided a summary of many papers that treat renewable energy and energy management problems using MCDM methods while defining their own criteria and sub-criteria. This article reviewed papers from 2002 till 2018. In our study, some recently published papers (2018–2022) are summarized in Table 1, which presents the main criteria and sub-criteria used in MCDMs related to renewable energy sectors. Table 2 provides an extensive summary of different types of MCDM methods (mainly in energy-related applications) and their respective characteristics. It can be concluded that there is no best or worst application of an MCDM method; rather depending on the type of application, the adequate selection of the method should be made.

Table 1. Criteria and sub-criteria of main MCDM methods.

					Main criteria				
Method	Area of application	Technological	Technical	Economic	Environmental	Social	Performance	Risk	Ref.
					Sub-criteria				
Fuzzy-Delphi	BSS sizing and selection	Cell voltage, storage capacity, lifecycle	*Data not provided*	Battery cost, operation cost, energy storage system profit	Energy saving, CO2 emissions reduction	Reduction of power grid loss	Energy efficiency, energy intensity	*Data not provided*	[7]
Fuzzy AHP	Ranking of renewable energy alternatives in India	Technology maturity	capacity factor, reliability	Capital cost, O&M cost, Levelized Cost Of Electricity (LCOE)	Land requirement, impact on ecosystem, climate change	Job creation, social acceptance, impact on human health	*Data not provided*	*Data not provided*	[30]
AHP	Solar power plant location selection in Vietnam	Distance from power network, potential demand	*Data not provided*	Construction cost, O&M cost	Sunshine hours, humidity	Protection laws, legal compliance	*Data not provided*	*Data not provided*	[28]
BWM	Sustainable evaluation of potential solar sites in east of Iran	*Data not provided*	Distance to substation, land availability, solar radiation intensity	Construction cost, initial investment, payback period	Protected areas, ecosystem destruction	Social acceptance, job creation	*Data not provided*	Time delay risk, investment risk	[7]
VIKOR, AHP, Fuzzy DEMATEL	Literature review: location of EV battery swapping station	*Data not provided*	Impact on load levels of power grid, harmonic pollution to power grid	Equipment purchasing, O&M cost	*Data not provided*	Coordination with the transportation network	*Data not provided*	*Data not provided*	[32]
AHP PROMETHEE, TOPSIS, ELECTRE	Sustainability evaluation of Concentrated Solar Power projects	Operation temperature, capacity factor, exergy efficiency	Technology maturity, easy of decentralization, MG suitability	LCOE, costs, fuel saving, availability of funds	Land use, CO2 emissions reduction	Job creation, impact on local economy, impact of quality of life	*Data not provided*	*Data not provided*	[20]

Table 2. Extensive summary table of different MCDM methods.

Types of MCDM	Name	Areas of Application	Summarized Steps	Advantages	Disadvantages	Notes	Ref., year
Conventional	TOPSIS	Engineering, energy management, renewable energy, human resources	(a) Calculate the matrices (b) Normalize (c) Calculate positive & negative ideal solutions (d) Calculate the separation & the relative closeness.	Easy to use; simple process; does not require complex computations; results obtained quickly	Strong deviation of one indicator from the ideal solution; strongly influences results; only suitable when indicators do not vary a lot	Rarely used in its raw form (when solving real life issues), often applying fuzzy sets or other extensions of the original TOPSIS	[8], 2022 [14], 2020 [21], 2020
	DEMATEL	Risk assessment, management studies, energy, IT sector	(a) Assess relationships between factors in systems (b) Establish a normalized matrix (c) Construct the total influence matrix (d) Produce the Influential Relation Map	Ability to achieve possible results using the least amount of data; helps to draw a structural output using matrices & graphs	Requires a large set of experts' opinions, subjective results		[27], 2021
	VIKOR	Supply chain, energy planning, risk management	(a) Determine the best & worst values (b) Calculate weighted & normalized Manhatten distance, & Chebyshev distance (c) Rank & sort alternatives	Appropriate to use when dealing with quantitative or objective data, with many alternatives and attributes	Lacks provision to weigh elicitation of the decision-makers' preferences		[7], 2018
	AHP	Performance-type problems, energy policy, energy management, technology selection	(a) Define objectives into hierarchical models. (b) Determine weights for each criteria. (c) Calculate the score of each alternative considering criteria. (d) Calculate overall score of each alternative.	Easy to use, computation process is simple compared to other methods, results obtained quickly, has a better focus on each criterion in the calculations	Interdependence between alternatives and objectives can lead to inaccurate results; higher complex weights for many decision-makers	Most used method in the last decade for solving energy policy related problems	[21], 2020 [28], 2021
	DEA	Economics, retail, business problems	(a) Define Decision Making Units (DMU) (b) Settle input & output factors relevant to DMUs (c) Apply DEA model and analysis	Handles well multiple inputs and outputs	Does not deal with imprecise data		[11], 2018
	ELECTRE	Energy, economics, environmental problems	(a) Determinate of threshold function (b) Concordance index & Discordance index. (c) Outrank degree. (d) Assign ranking based on above calculation.	Takes uncertainty into account	Its outcome can be difficult to explain in layman's terms	Advantage to be used in heuristic situations	[14],2020
	PROMETHEE	Business, financial management, energy, environmental	(a) Calculate & compare the evaluation matrices (b) Assign preference function with values from 0–1 depending on the difference between pairs (c) Calculate the global matrix & determine the rank	Easy to use, does not require assumption that criteria are proportionate	Does not provide a clear method by which to assign weights		[14], 2020
	FUCOM	Performance-type problems, technology selection	Same as AHP	Significantly smaller number of pairwise comparisons, calculation of reliable values of criteria weight coefficients	Still needs more implementations on real-world applications, in other domains	Compared with BWM and AHP	[12], 2018

(continued)

Table 2. Continued

Types of MCDM	Name	Areas of Application	Summarized Steps	Advantages	Disadvantages	Notes	Ref., year
CRITIC	Technology comparison, evaluation of hybrid energy systems	a. Estimate the Objective Weights of Decision Criteria	Simple calculations, one of the most objective methods, no need for decision-makers intervention	Does not express the relative importance of achieving decision makers' goals, does not consider criteria types		[21], 2020	
Fuzzy-based	Fuzzy TOPSIS	Same as TOPSIS	Both criteria and weights are normalized into fuzzy sets	Applying Fuzzy Set Theory to conventional	Require higher data volume than conventional	According to [38], fuzzy MCDM usage in	[11], 2018
	Fuzzy VIKOR	Same as VIKOR	Both criteria and weights are normalized into fuzzy sets	MCDM methods helps to get more sensitive, concrete & realistic results	MCDMs	energy related publications represented 76% of the total	[29], 2019
	Fuzzy AHP	Same as AHP	Both criteria and weights are normalized into fuzzy sets	Effectively handles the vagueness in linguistic statements		papers in 2019	[30], 2018
	Fuzzy DEMATEL	Same as DEMATEL	Both criteria and weights are normalized into fuzzy sets	found in conventional MCDM methods			[31], 2019 [32], 2018
Hybrid	FAHP – TOPSIS	Same as AHP, TOPSIS	Same as FAHP	Effectively handles the vagueness in	Selection results among other		[12], 2018
	AHP – PROMETHEE	Same as AHP, PROMETHEE	Same as FAHP	linguistic statements found in conventional MCDM methods	MCDM techniques should be compared		[12], 2018
	Fuzzy Delphi-BWM-CPT	Engineering, risk assessment	Three-step method, using the three methods relative to Delphi, BWM, and CPT	Establishes an evaluation indicator system, decides the optimal weights of all sub-criteria		3-step computation for BSS ranking and selection considering risk preferences	[7], 2018
	FAHP-VIKOR	Same as AHP	Same as FAHP and VIKOR	FAHP deals with decision makers' subjectivity encountered in AHP; VIKOR complements it by achieving the positive-ideal arrangement	Very complex, worth using in niche applications only	Compared with AHP	[33], 2020
	DEMATEL – DQGRA TOPSIS	Economy, environment, energy performance evaluation	Two-step method, using methods relative to DEMATEL then TOPSIS	Better robustness and sample discrimination than other mentioned hybrid methods	Applicability in other fields still needs analysis	Compared with other hybrid methods	[8], 2021
Novel	PIV	Engineering, manufacturing systems, environmental, human resources	(a) Identify alternatives & decision criteria (b) Determine the weighted normalized decision matrix (c) Evaluate the Weighted Proximity Index (WPI) and rank the Overall Proximity Value (d) Rank alternatives	Minimizes the rank reversal problem better than in conventional methods	More applicability should be done to further prove it	Compared with TOPSIS, AHP, VIKOR	[34], 2019
	PHSS	Same as TOPSIS	Integration of the Plithogenic Set into an algorithm with similar steps of TOPSIS	Same as TOPSIS	Still under development, yet to be applied to a wider range of applications	Fuzzy TOPSIS-based	[25], 2020

2.3 *MCDM based software*

In the market, there are lots of software that help the user or the decision-maker to solve complex problems, such as, conflicting criteria and multi-objective functions. The selection of the software depends on the usage, the frequency of the utilization, the complexity of the problem, type of clients (students, professional engineers, researchers, etc.), and the budget. Figure 2 presents the top available MCDM software in the market. They are categorized based on their cost (free, partially free, open source, web-based, and paid). Figure 3 shows the methods used in MCDM software development. It can be seen that AHP is the most popular and useful method.

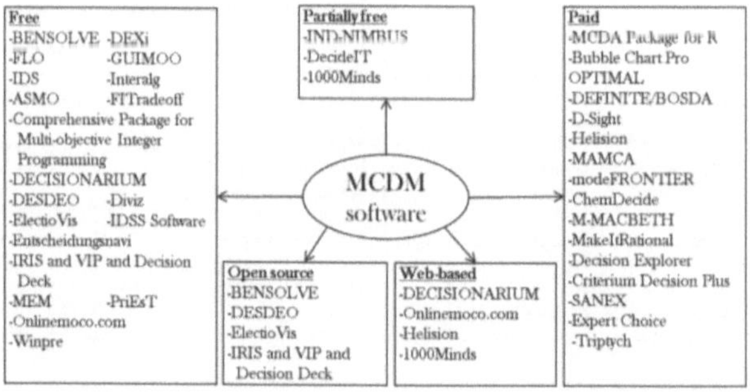

Figure 2. Schematic diagram for the usage of MCDM methods for MGs.

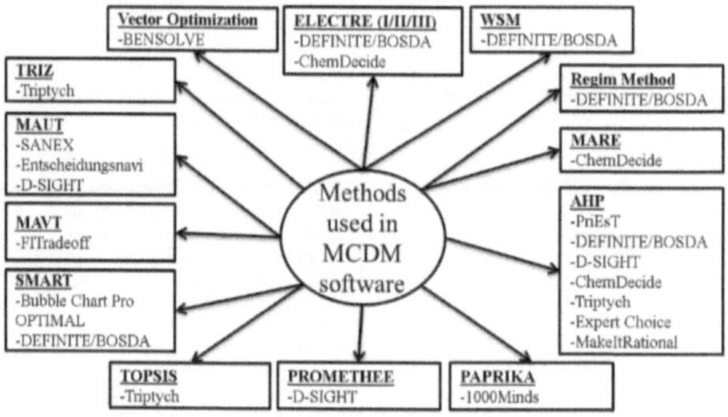

Figure 3. Methods used in developing the MCDM software.

2.4 *Particular case: Using MCDM for solar panels selection*

One of the main components of an MG is the photovoltaic (PV) system, in which the selection of solar panels from many alternatives in the market, has a high impact on the total efficiency, investment cost, power output, payback period, and many other important factors of the system. In presence of conflicting criteria, such as selecting the cheapest with the highest efficiency solar panels needs advanced tools to suggest the best solar panels for specific applications. The most important considerations in selecting the best and optimal SPs are but are not limited to, brand name and popularity, cost, availability, dimensions, efficiency, warranty, quality, photovoltaic technology type. The four most important criteria that need to be taken into consideration while selecting SPs are: technical, economic, environmental and social. Indeed, of the reviewed papers, a handful of them prioritized the technical aspect first, followed by economic, environmental and finally social criteria. Article [35], tackled the "unstructured problem" of the SP selection by employing the fuzzy-AHP. In this case, the primary criterion was the technical nature of SPs, followed by the financial and social aspects, respectively. Figure 4 presents the main criteria and sub-criteria used in the SP selection process. Evidently, there are many other sub-criteria that are not as important (in terms of criteria weights) as those already mentioned, but they can still play a meaningful part in the SP selection process. The usage of MCDM is not limited to selecting solar panels. In this paper, only a simple example is given in order to show the importance of using MCDM methods in different field including Electric vehicles [36–39], energy management [40–42], renewable energy allocation [43], blockchain [44], and many others.

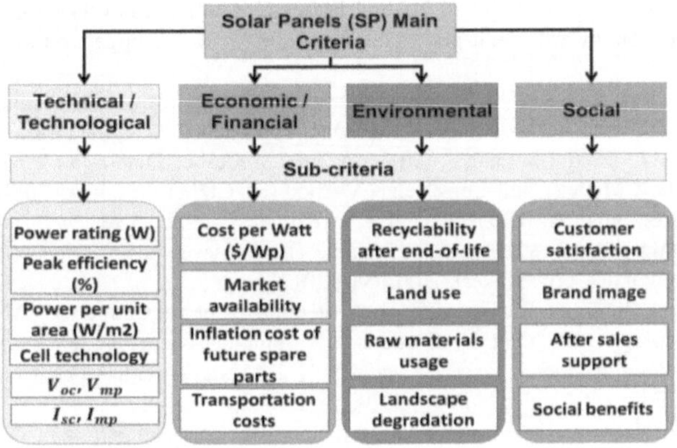

Figure 4. The main criteria and sub-criteria in the SP selection process.

3 CONCLUSION

This paper presents an overview of the usage of the Multi-Criteria Decision-Making (MCDM) methods in designing a micro-grid (MG) and selecting its main components. The paper discusses the most used MCDM algorithm in the field, the main objective functions and criteria (e.g., technical, technological, economic, environmental, and social), the advantages and drawbacks of each MCDM method, and finally the most used software to model an MCDM problem and solve it. In conclusion, MCDM methods are powerful tools to solve complex and conflicting problems and select the best items from a list of many alternatives considering many criteria.

REFERENCES

[1] Dawoud S. M., Lin X., and Okba M. I., Hybrid renewable microgrid optimization techniques: A review, *Renewable and Sustainable Energy Reviews*, vol. 82, pp. 2039–2052, 2018/02/01 2018.

[2] Zia M. F., Elbouchikhi E., and Benbouzid M., Microgrids energy management systems: A critical review on methods, solutions, and prospects, *Applied Energy*, vol. 222, pp. 1033–1055, 2018/07/15 2018.

[3] Ridha H. M., Gomes C., Hizam H., Ahmadipour M., Heidari A. A., and Chen H., Multi-objective optimization and multi-criteria decision-making methods for optimal design of standalone photovoltaic system: A comprehensive review, *Renewable and Sustainable Energy Reviews*, vol. 135, p. 110202, 2021/01/01 2021.

[4] El-Bayeh C. Z., Alzaareer K., Brahmi B., Zellagui M., and Eicker U., An original multi-criteria decision-making algorithm for solar panels selection in buildings, *Energy*, vol. 217, p. 119396, 2021/02/15 2021.

[5] Ferreira C. C., Vieira E., Gouvea da Gosta S. E., Loures E. R., and de Lima E. P., Application of MCDM method for technologies selection to support energy management, *Procedia Manufacturing*, vol. 39, pp. 1289–1296, 2019/01/01 2019.

[6] Ahmad Khan A., Naeem M., Iqbal M., Qaisar S., and Anpalagan A., A compendium of optimization objectives, constraints, tools and algorithms for energy management in microgrids, *Renewable and Sustainable Energy Reviews*, vol. 58, pp. 1664–1683, 2016/05/01 2016.

[7] Zhao H., Guo S., and Zhao H., Comprehensive assessment for battery energy storage systems based on fuzzy-MCDM considering risk preferences, *Energy*, vol. 168, pp. 450–461, 2019/02/01 2019.

[8] Zhao H., Li B., Lu H., Wang X., Li H., Guo S., et al., Economy-environment-energy performance evaluation of CCHP microgrid system: A hybrid multi-criteria decision-making method, *Energy*, vol. 240, p. 122830, 2022/02/01 2022.

[9] Cao Y., Zhou B., Or S. W., Chan K. W., Liu N., and Zhang K., An interactive tri-level multi-energy management strategy for heterogeneous multi-microgrids, *Journal of Cleaner Production*, vol. 319, p. 128716, 2021/10/15 2021.

[10] El-Bayeh C. Z., Zellagui M., Shirzadi N., and Eicker U. (2021), A novel optimization algorithm for solar panels selection towards a self-powered EV parking lot and its impact on the distribution system. *Energies* 14(15).

[11] Wang T.-C. and Tsai S.-Y. (2018), Solar panel supplier selection for the photovoltaic system design by using fuzzy multi-criteria decision making (MCDM) approaches. *Energies* 11(8).

[12] Pamučar D., Stević Ž., and Sremac S. (2018), A New Model for Determining Weight Coefficients of Criteria in MCDM Models: Full Consistency Method (FUCOM). *Symmetry* 10(9).

[13] Kumar A., Sah B., Singh A. R., Deng Y., He X., Kumar P., et al., A review of multi criteria decision making (MCDM) towards sustainable renewable energy development, *Renewable and Sustainable Energy Reviews*, vol. 69, pp. 596–609, 2017/03/01 2017.

[14] Kumar A., Sah B., Singh A. R., Deng Y., He X., Kumar P., et al., Chapter 1 – Multicriteria decision-making methodologies and their applications in sustainable energy system/microgrids, in *Decision Making Applications in Modern Power Systems*, S. H. E. Abdel Aleem, A. Y. Abdelaziz, A. F. Zobaa, and R. Bansal, Eds., ed: Academic Press, 2020, pp. 1–40.

[15] Løken E., Use of multicriteria decision analysis methods for energy planning problems, *Renewable and Sustainable Energy Reviews*, vol. 11, pp. 1584–1595, 2007/09/01 2007.

[16] Kabir K. H., Aurko S. Y., and Rahman M. S. (2021), Smart Power Management in OIC Countries: A Critical Overview Using SWOT-AHP and Hybrid MCDM Analysis. *Energies* 14(20).

[17] Yilmaz M. and Krein P. T., (2013). Review of the impact of vehicle-to-grid technologies on distribution systems and utility interfaces, *IEEE Transactions on Power Electronics*, vol. 28, pp. 5673–5689.

[18] Musio M., Lombardi P., and Damiano A., (2010). Vehicles to grid (V2G) concept applied to a virtual power plant structure, in *The XIX International Conference on Electrical Machines – ICEM* 2010, pp. 1–6.

[19] van der Kam M. and van Sark W., Smart charging of electric vehicles with photovoltaic power and vehicle-to-grid technology in a microgrid; a case study, *Applied Energy*, vol. 152, pp. 20–30, 2015/08/15 2015.

[20] Ghasempour R., Alhuyi Nazari M., Ebrahimi M., Ahmadi M. H., and Hadiyanto H., (2019). Multi-criteria decision making (MCDM) approach for selecting solar plants site and technology: A review, vol. 8, p. 11, 2019-02-02 2019.

[21] Siksnelyte-Butkiene I., Zavadskas E. K., and Streimikiene D. (2020), Multi-criteria decision-making (MCDM) for the assessment of renewable energy technologies in a household: A review. *Energies* 13(5).

[22] Kumar A., Deng Y., He X., Kumar P., Rayudu R., and Bansal R., (2020). An integrated design approach for rural electrification based on community microgrids, *Energy Sources*, vol. 12, pp. 14–19.

[23] Kaya İ., Çolak M., and Terzi F., A comprehensive review of fuzzy multi criteria decision making methodologies for energy policy making, *Energy Strategy Reviews*, vol. 24, pp. 207–228, 2019/04/01 2019.

[24] Zavadskas E. K., Govindan K., Antucheviciene J., and Turskis Z., Hybrid multiple criteria decision-making methods: a review of applications for sustainability issues, *Economic Research-Ekonomska Istraživanja*, vol. 29, pp. 857–887, 2016/01/01 2016.

[25] Ahmad M. R., Saeed M., Afzal U., and Yang M.-S. (2020, A Novel MCDM method based on plithogenic hypersoft sets under fuzzy neutrosophic environment. *Symmetry* 12(11).

[26] Kaya İ., Çolak M., and Terzi F., Use of MCDM techniques for energy policy and decision-making problems: A review, *International Journal of Energy Research*, vol. 42, pp. 2344–2372, 2018/06/10 2018.

[27] Peleckis K. (2021), Application of the DEMATEL model for assessing IT sector’s sustainability. *Sustainability* 13(24).

[28] Kannan D., Moazzeni S., Darmian S. m., and Afrasiabi A., A hybrid approach based on MCDM methods and Monte Carlo simulation for sustainable evaluation of potential solar sites in east of Iran, *Journal of Cleaner Production*, vol. 279, p. 122368, 2021/01/10 2021.

[29] Saraswat S. K. and Digalwar A. K., (2021). Evaluation of energy sources based on sustainability factors using integrated fuzzy MCDM approach, *International Journal of Energy Sector Management*, vol. 15, pp. 246–266.

[30] Wang C.-N., Nguyen V. T., Thai H. T., and Duong D. H. (2018), Multi-criteria decision making (MCDM) approaches for solar power plant location selection in Viet Nam. *Energies* 11(6).

[31] Du Y.-W. and Zhou W., New improved DEMATEL method based on both subjective experience and objective data, *Engineering Applications of Artificial Intelligence*, vol. 83, pp. 57–71, 2019/08/01 2019.

[32] Simsek Y., Watts D., and Escobar R., Sustainability evaluation of concentrated solar power (CSP) projects under clean development mechanism (CDM) by using multi criteria decision method (MCDM), *Renewable and Sustainable Energy Reviews*, vol. 93, pp. 421–438, 2018/10/01 2018.

[33] Ishak A., Asfriyati, and Nainggolan B., Integration of Fuzzy AHP-VIKOR methods in multi criteria decision making: Literature review, *IOP Conference Series: Materials Science and Engineering*, vol. 1003, p. 012160, 2020/12/01 2020.

[34] Khan N. Z., Ansari T. S. A., Siddiquee A. N., and Khan Z. A., Selection of E-learning websites using a novel proximity indexed value (PIV) MCDM method, *Journal of Computers in Education*, vol. 6, pp. 241–256, 2019/06/01 2019.

[35] Sasikumar G. and Ayyappan S., Multi-criteria decision making for solar panel selection using fuzzy analytical hierarchy process and technique for order preference by similarity to ideal solution (TOPSIS): An empirical study, *Journal of The Institution of Engineers (India): Series C*, vol. 100, pp. 707–715, 2019/08/01 2019.

[36] El-Bayeh C. Z., Mougharbel I., Saad M., Chandra A., Asber D., and Lefebvre S., (2018). Impact of considering variable battery power profile of electric vehicles on the distribution network, *in 2018 4th International Conference on Renewable Energies for Developing Countries (REDEC)*, pp. 1–8.

[37] El-Bayeh C. Z., Mougharbel I., Saad M., Chandra A., Lefebvre S., Asber D., et al., A detailed review on the parameters to be considered for an accurate estimation on the Plug-in Electric Vehicle's final State of Charge, in *2016 3rd International Conference on Renewable Energies for Developing Countries (REDEC)*, 2016, pp. 1–6.

[38] El-Bayeh C. Z., Alzaareer K., Aldaoudeyeh A.-M. I., Brahmi B., and Zellagui M. (2021), charging and discharging strategies of electric vehicles: A survey. *World Electric Vehicle Journal* 12(1).

[39] El-Bayeh C. Z., Mougharbel I., Saad M., Chandra A., Lefebvre S., Asber D., et al., (2016). A novel approach for sizing electric vehicles Parking Lot located at any bus on a network, in *2016 IEEE Power and Energy Society General Meeting (PESGM)*, pp. 1–5.

[40] El-Bayeh C. Z., Eicker U., Alzaareer K., Brahmi B., and Zellagui M. (2020), A novel data-energy management algorithm for smart transformers to optimize the total load demand in smart homes. *Energies* 13(18).

[41] El-Bayeh C. Z., Mougharbel I., Saad M., Chandra A., Asber D., Lenoir L., et al., Novel soft-constrained distributed strategy to meet high penetration trend of PEVs at homes, *Energy and Buildings*, vol. 178, pp. 331–346, 2018/11/01/ 2018.

[42] El-Bayeh C. Z., Alzaareer K., Brahmi B., and Zellagui M. (2020), A Novel algorithm for controlling active and reactive power flows of electric vehicles in buildings and its impact on the distribution network. *World Electric Vehicle Journal* 11(2).

[43] Zellagui M., Settoul S., Lasmari A., El-Bayeh C. Z., Chenni R., and Hassan H. A., (2021). Optimal allocation of renewable energy source integrated-smart distribution systems based on technical-economic analysis considering load demand and DG uncertainties, in *Artificial Intelligence and Renewables Towards an Energy Transition*, Cham, pp. 391–404.

[44] Ilieva-Obretenova M., (2022). Blockchain in smart grid for sustainable management of environmental resources usage, *Management*, vol. 10, pp. 275–283.

Innovation and Technological Advances for Sustainability – Al-Naemi et al. (Eds)

Towards real-time image mining for waste detection in deserts

Souheir Mehanna
Université Paris-Saclay (UVSQ), ECE Engineering School, Versailles, France

Nazim Berchiche, Assitan B. Niare & Abderrahmane Maaradji
ECE Engineering School, Paris, France

Hakim Hacid
Technology Innovation Institute, Abu Dabhi, UAE

Assia Soukane
ECE Engineering School, Paris, France

ABSTRACT: As the world's population has grown and mobility has become more accessible, human activities have expanded into remote areas and deserts. These activities (typically desert safaris) generate much left-behind waste. Untreated left-out wastes contribute directly to gas emissions that exacerbate the global warming, especially litter such as plastics, cans, or any other resistant material. We present a drone-based solution to find waste in extra-urban environments (mainly deserts). In this paper, we use an autonomous unmanned aerial vehicle (drone) that flies over a desert area to capture a video stream to locate and identify waste objects such as plastic bottles. Our solution is still a work in progress; hence, in this paper, we have explored an object detection algorithm using transfer learning (few-shot-learning), where a pre-trained classifier (YOLO) is retrained with our relatively small dataset DroneTrashNet. Our goal is to benchmark existing models and approaches for object detection and classification in order to provide a baseline for selecting the most appropriate models for detection and classification tasks in extra-urban contexts. Our DroneTrashNet dataset corresponds to three minutes of annotated video captured by a drone at a low altitude in a controlled experiment in a desert environment, where waste objects were manually placed, with some objects half-buried in the sand. Preliminary results show an achieved mAP of 42% in specific conditions that could be subject to a number of improvements. The first results are promising compared to the results obtained with the initial classifier. While the resulting model has been validated in a controlled desert environment, new challenges and opportunities remain to be addressed, such as full-scale validation and generalization to extra-urban environments.

Keywords: Desert Environment, Drone, Image Mining, Waste Detection

1 INTRODUCTION

Environmental issues and waste pollution have significantly developed over the years as global warming worsens. One of the causes of global warming is the gas emissions from undetected wastes left out in nature. Waste is an object or substance that has undergone physical or chemical changes at the end of its life and no longer has a destination or intended disposal. Today, protecting the environment is a priority, and managing waste in urban and extra-urban areas is a primary concern of communities. Several solutions tackling the issue have been implemented, ranging from collection and sorting combined in a circular economy. However, no systematic solution has been proposed for isolated extra-urban areas.

 DOI: 10.1201/9781003496724-34

As the world's population continues to grow and mobility becomes more accessible, people can easily access areas deep in nature or in the desert, resulting in more waste being littered in these areas. For example, the 2022 FIFA World Cup, hosted by Qatar, and the 2022 Saudi Games, hosted by Saudi Arabia, are attracting many tourists from around the world. As a result, the deserts in these two countries are subject to massive littering. Man-made waste causes significant environmental damage, especially in remote areas that are difficult to access.

This paper presents a solution based on drones to locate waste in these extra-urban environments. An autonomous drone flying over these spaces embeds a specially trained object classification algorithm to locate and identify waste objects such as plastic bottles. The tremendous volume of data generated from images and video footage captured by Unmanned aerial vehicles (UAV)s requires a lot of time to process in order to produce insightful knowledge. Using the technique of *transfer learning*, a pre-trained classifier (*YOLO*) is re-trained with our dataset, which is made public. The first results show that the performances of the new re-trained classifier are superior to those of the initial classifier.

In the remaining of this paper, Section 2 discusses existing works on object detection and image classification in different urban contexts. Section 3 presents our current work and approach. Section 4 shows identified challenges and future perspectives. Preliminary results are then presented in Section 5. Finally, we conclude the paper in Section 6.

2 RELATED WORKS

This section reviews several studies investigating image processing techniques for waste object detection and classification. These studies cover a wide range of topics, from convolutional neural networks and deep learning techniques to more traditional methods such as feature extraction and segmentation. They also explore different datasets and applications for waste object detection and examine ways to enhance the accuracy and efficiency of the methods. Overall, this research provides invaluable insight for information for further investigations in the field of waste object detection.

In [1], the authors propose a system that takes municipal solid waste (MSW) and then classifies plastic wastes into their appropriate subclass of plastics to help improve the effectiveness of the reprocessing systems of waste materials. The authors tested two convolutional neural networks (CNN) for the classification with varying underlying architectures on 122 × 122, and 277 × 277 input images. The first network consists of 23 layers, 9 × 9 convolution filters in the first convolution layer, six weave layers responsible for encoding the input image, and a three-layer fully connected layer. The second network consists of 15 layers, 64 convolution filters with 9 × 9 dimension in the first convolution layer, three layers of convolution to encode the input images, and a two-layer fully connected layer. The authors of [2] propose a waste detection system in urban cities with a camera embedded in a vehicle. The objectives of this work are first to localize all objects in an image frame and then to classify all detected objects into their correct litter categories. The authors used the Overfeat model [3] for the localization task where its classification layer is replaced by GoogLeNet [4] which will perform the classification task. The work of [5] presents a solution with a drone to detect anthropogenic marine debris waste on a beach. To tackle this problem, the authors proposed combining an unmanned aerial vehicle (Drone) equipped with a high-definition RGB camera and software based on deep-learning convolutional neural networks. The software is a commercial software called PlasticFinder. According to the article, the algorithm achieved a sensitivity of 67%. However, this software is limited by climatic conditions (for example, on a sunny day, the sand shines too brightly, which prevents the detection), which may also be due to a camera that is not sufficiently developed for this style of use. To the best of our knowledge, this work was the first to confirm the effectiveness of drones for automatically detecting anthropogenic marine debris on the beach and is the closest study to our project.

You Only Look Once (YOLO) [6] is one of the most popular model architectures and object detection algorithms. It uses one of the best neural network architectures to produce high accuracy and overall processing speed, which is the main reason for its popularity. The YOLO algorithm aims to predict a class of an object and the bounding box that defines the location of the object on the input image. In addition to increased accuracy in predictions and better *Intersection on Union* in bounding boxes (compared to real-time object detectors), YOLO has the inherent advantage of speed. YOLO is a much faster algorithm than its counterparts, running at up to 45 FPS. The authors of [7] present a deep learning-based two-stage detector of waste localization and classification in natural and urban environments. The authors also benchmark existing waste littering open-source datasets with unified annotations across the datasets. The work of [8] proposes a construction waste semantic segmentation model in the context of automated waste composition recognition. Compared to image classification and object detection approaches, semantic segmentation works at a finer granularity in a pixel-wise manner. The authors adopted the DeepLabv3+ as the main architecture of the proposed construction waste segmentation model.

However, as investigated in a literature review proposed in [7], the current studies are hardly comparable to ours since they focus on waste objects in well-controlled environments (such as waste management factories). This case is different from our target use case where we randomize the scattering of waste objects in a real desert, and we emulate a real-life situation where the objects are partially buried in the sand. Also, none of the existing works benchmark object detection and classification frameworks systematically and comprehensively for the waste pollution and littering context.

3 OUR APPROACH

This project aims to detect the presence of waste (bottles, glass, paper products, and food waste) using a drone equipped with a smart camera. The project aims to design an object detection algorithm embedded in a drone, specializing in detecting extra-urban waste. The real-time identification of litter is necessary to produce insights and statistics that will be sent to an operational center in order to decide on the intervention of the collection teams.

Figure 1 shows our general approach, which mainly consists of drones that will roam over the deserts to capture video footage of the desert from which we extract images covering the treatment area in the desert. And contrary to existing works that consider datasets taken in well-controlled environments where the images demonstrate a clear separation from the background and an angle facing the objects, our approach considers a practical, real-life situation where the images are taken from above using drones and in real-life circumstances where waste objects are actually in the desert and sometimes are partially covered with sand due to sandstorms. After the images are transferred to our data centers, our inference approach using YoloV4 is applied to localize and classify the waste objects to finally alert the local garbage collection entities of the computed insights about the existence of waste objects for their collection and treatment.

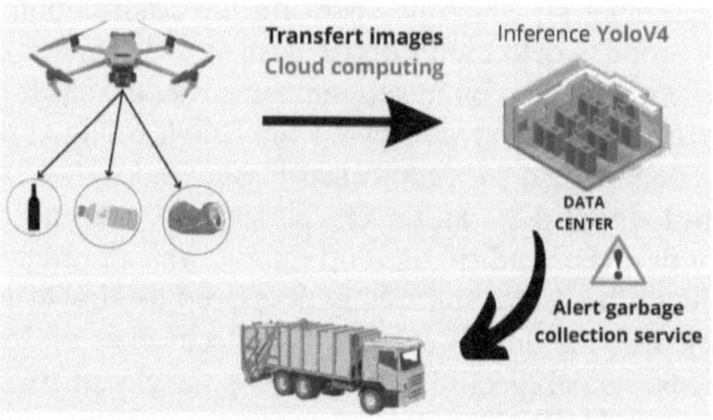

Figure 1. Our general approach for waste detection and treatment.

We started our studies with the pre-trained classifier (YOLO) [6]. Figure 2 shows an overview of our proposed approach. First, we train the pre-trained YoloV4 model on the DroneTrashNet dataset we have created, in order for the model to learn more about the different classes of waste in our context and to improve its detection accuracy on the specific images from the desert environment. The pre-trained model is trained on a set of manually annotated images captured by our drone in the desert. This training of the model in different settings, and specifically to images from our context, significantly improved the model's detection accuracy.

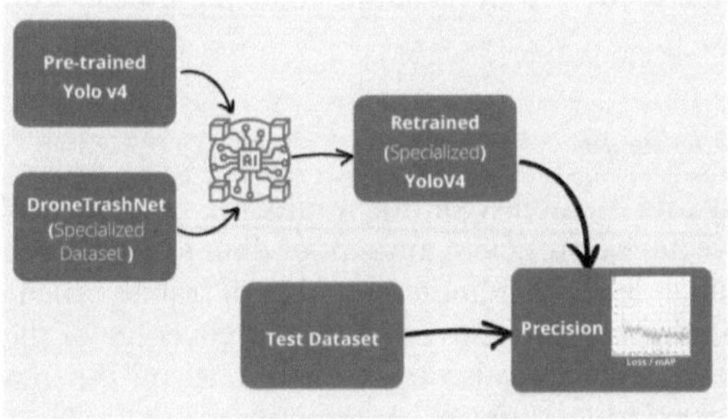

Figure 2. Structure of our approach using YOLO v4 to detect waste objects taken in the desert by drones.

4 CHALLENGES AND FUTURE PERSPECTIVES

Many existing works address waste pollution and littering using computer vision frameworks to localize, detect, and learn the classes of the objects detected. The work we described in Section 3 is only the beginning of our work on this subject, and it is still ongoing work that we are continuously working on enriching and improving. This section presents the research challenges we are currently addressing in our work and our future perspectives.

4.1 *Benchmark the existing models*

This is one of our primary goals in this work, as none of the existing works propose a benchmark that serves as a start point for an object detection and classification problem in the context of extra-urban environments and waste pollution. We are currently working on benchmarking existing algorithms and frameworks such as the Fast R-CNN (Detectron) by [9], the Single Shot multibox Detector (SSD) [10], the spatial pyramid pooling (SPP-net) [11], and others to have a solid basis for choosing the most suitable algorithm both to localize and classify the waste objects in real-life images and contexts. Also, it is in our future perspectives to test the Yolov6 algorithm with the best interference speed across the entire FPS spectrum. Currently, YOLOV4 is the model we trained and tested our data on. Our objective here is to explore the best-performing approaches of the proposed methods in terms of effectiveness, precision, and performance in order to provide a benchmark that allows readers to choose the most suitable object detection framework for their context.

4.2 *Real-time detection*

One of our primary goals is to develop a real-time detection approach that allows garbage collection services to act as quickly as possible after images have been taken. It takes a lot of time to take the videos, unload the drone and pre-process the data into frames to eventually detect and classify objects if each step is done manually and separately. However, if the detection and classification are done directly on the frames right after the drone has taken them, the whole process becomes faster. Then we are able to scan more areas, and collection services can act more quickly and efficiently. There are several challenges in designing a real-time approach to drones, as it is known that drones have limited capacities in terms of storage and computation abilities. The objective here is to propose a model that is light enough to be embedded and deployed on the drone and can compute the localization and classification of objects live while it is roaming over the area fast and efficiently. Also, it is crucial that the model can report its findings and communicate them to the appropriate authorities for action in real time. This is still a challenge because of the heavy weight of the detection models and the communication tasks, and the limited storage and computation capacities of the drones.

4.3 *Data collection and sharing*

To ensure the reliability and credibility of our results, we intend to collect more data in the near future, such that we have a sufficient amount of data to train and test our model. At the moment, we have a small video with a limited number of frames taken in the desert at a fixed altitude. The goal is to have more frames and broader coverage of the possible scenarios of waste objects in the desert. We will also train our model on other waste pollution public datasets to avoid the overfitting problem. Moreover, we will collect data with drones at different altitudes, such as 5, 10, and 20, above the surface of the desert and at different speeds of the drone and angles. Currently, we only have the frames taking footage of the waste objects from one angle, the top.

And finally, for the post-processing algorithm perspectives. We plan to contact waste collection companies to create a partnership. Then develop a system for sending signals to the collection companies indicating the location of the waste.

4.4 *Model development*

The objective is to elaborate new methods for supporting litter identification in wild areas. While developing the models, we should consider that we deal with real-time and heterogeneous data taken under different constraints. We will build on top of the knowledge learned from benchmarking existing approaches and the real-life constraints we will have in the wild environments. Also, we will secure dedicated machines to train and test the benchmark models on. Moreover, our current experiments solely relied on the mean average precision mAP of the bounding box detecting the objects for evaluation, while in the future, we will also add another evaluation metric which is the accuracy of the classification of the detected objects to facilitate the recycling process for the garbage treatment services. This project extensively uses experiments as a mean of validation of the results and the findings.

5 PRELIMINARY EXPERIMENTS AND EVALUATIONS

We present in this section the preliminary experiments that we conducted for our approach. The results of the experiments we present here were performed manually and in a non-systematic manner because this is an still ongoing work. However, conducting more systematic experiments with more data and a systematic design of the settings of the experiments is in our next steps for the project.

5.1 *Dataset*

Our DroneTrashNet dataset that was used for post-training the YOLOv4 model corresponds to 3 minutes of video footage taken by a manually piloted drone at a low altitude in a controlled desert environment. In this controlled experiment, the waste items were manually and randomly dispersed around the study area in the desert as depicted in Figure 3. To address other real-life use cases and to partially emulate realistic conditions, a part of these wastes was placed in a way that only 50% of the waste appears to the roaming drone recording footage while the rest half is buried in the sand. The reason behind this procedure is that, in reality, the longer time passes on dumped garbage in the desert, the more buried in the sand they will be due to the wind factor, for instance.

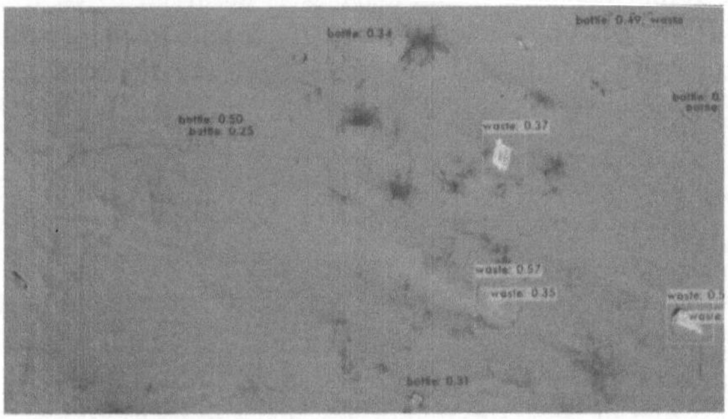

Figure 3. An example of the inference of the specialized-YOLO-model on a drone image capture.

We first extracted 200 frames (images) from the 3-minutes captured video. In order to improve the performance of the model, we augment the size of our dataset by applying a 90Â° rotation to these 200 frames. The resulting dataset contained 400 frames. We also manually annotated our dataset with 5 specific classes/labels and a generic one. The specific classes are categories of waste, they are glass, plastic, bottle, cardboard, and metal. And the generic class is just waste. Our dataset is available on a public repository on GitHub[1].

5.2 *Experiments settings*

For the experiments, we are using python on Jupyter Notebook on Google Colaboratory which is a cloud service offered by Google for training and research in machine learning. This platform makes it possible to train Machine Learning models directly on the cloud. We re-trained YOLO on our augmented dataset in order to achieve decent results that are comparable to the results achieved in [6].

As discussed earlier, our dataset comprises 200 images that are augmented to 400 images after applying data augmentation. Some of the experiments are done on the full dataset and others on a subset of the data that comprises only the initial 200 images extracted from the video.

5.3 *Evaluation metrics*

In this section, we present the evaluation metrics we are using to assess the performance of the approach on our data. Our approach detects and classifies wastes in the desert context. This means that we will need metrics to assess the appropriate trade-off between precision and recall. Hence, we chose the mean average precision (mAP), commonly used by object detection algorithms for the evaluations such as COCO[2]. The mean average precision (mAP) metric is used to analyze object detection. It compares the ground truth bounding box of detected objects to the actual detected box by our approach and returns a score. The higher

[1]https://github.com/ECE-StP/waste-desert-dataset
[2]Details available at https://cocodataset.org/#detection-eval

the score, the more accurate the model is in detecting. The mAP is the mean of the average precision (AP); it is a trade-off between precision and recall. It is computed as follows: $mAP = \frac{1}{N}\sum_{i=1}^{N} AP_i$, where N is the number of classes of the classification. AP is the average precision of each class.

Average precision (AP) values are averaged over recall values from 0 to 1. The mAP formula is based on precision and recall which are on the measures from the confusion matrix: the true positives (TP), true negatives (TN), false positives (FP), and false negatives (FN). The precision is computed as: $Precision = \frac{TP}{TP+FP}$, and the recall is computed as: $Recall = \frac{TP}{TP+FP}$. An actual detected bounding box is assumed to be a true positive if the detected part of the frame is above a certain threshold called intersection on Union (IoU). Intersection on Union (IoU) indicates the overlap of the coordinates of the predicted bounding box with the ground truth box. A higher IoU indicates that the predicted bounding box coordinates closely resembles the ground truth box coordinates.

5.4 *Preliminary results*

In this section, we report our preliminary results obtained from a limited evaluation as an exploratory step towards a more systematic and exhaustive evaluation in the future.

Before training the classifier on our dataset, we ran the pre-trained YOLO classifier during the exploratory process on our test dataset and obtained a *mAP* of about 13%. This first result indicates the irrelevance of the pre-trained YOLO classifier for the task at hand. This measure provides a baseline to evaluate the usefulness of our *few-shot-learning* approach.

We then retrained the YOLO classifier with our training dataset to test the retrained *specialized* YOLO classifier on our test data. This retraining operation was performed with different settings. At times we included the augmented data obtained from 90Â° rotation, and at other times we didn't; also, by varying the number of waste object classes. For example, the YOLO classifier was retained on the training dataset annotated with only two annotation labels, i.e., *Bottle* and *Waste* (for any other waste object). The resulting *specialized* YOLO classifier achieved a higher *mAP* of 42%. Figure 3 shows an example of waste detection using the specialized YOLO classifier that was re-trained on our data on a photo taken by a drone in our controlled stage.

Despite being only a preliminary step, we believe these results are encouraging on several aspects. Not only this specialized classifier could learn only with relatively few annotated examples, but it could also keep the same precision order of magnitude when tested in the very specific use case it was retrained for. Indeed, in [6], the authors reported that YOLO classifier trained and tested on the COCO dataset scored a 44% of precision. This same classifier performed poorly when tested on our specific use case of waste detection in a desert environment. However, when it was first retrained with relatively few annotated images from our controlled stage, this same classier performed almost as well as its original reported performance. These preliminary results could be further validated in a systematic and comprehensive evaluation.

6 CONCLUSION AND DISCUSSION

In this paper, we presented our ongoing work and the results we have achieved so far. We also elaborated on some relevant approaches from the literature. We presented the current challenges we are facing and our future perspectives and goals for our project. One of our primary goals is to find the model and settings that work best in the waste pollution context. We made our data publicly available on Github[3]. We managed to identify and implement an object detection algorithm based on YOLOV4 for waste localization. We showed our

[3]https://github.com/ECE-StP/waste-desert-dataset

preliminary experiments and results with 42% mAP, that is comparable to the results achieved by [6]; which proves that our approach can deliver promising results.

Currently, we have limitations such as the amount of data for the accuracy of the results because our experiments would be more reliable with a bigger dataset for training and testing. Also, in climatic conditions that are not favorable to the flight of the drone (heat), sandstorms can quickly hide the waste. The lack of networks and antennas for better control can affect the detection capacity of the drone. We also lack a setting that makes the detection and classification happen in real-time.

While our proposed solution might address the problem of waste management in desert areas and more generally in extra-urban areas, it is important to study the economical [12] and the carbon footprint of such a solution [13]. On the one hand, the economic value of a drone-based waste detection system would depend on the cost of its implementation compared to the potential cost savings and improved waste management it can provide. For example, by using drones to efficiently and effectively survey landfills and bodies of water for waste, the system can help reduce cleanup costs and prevent illegal dumping. On the other hand, it is also important to consider ways to minimize the carbon footprint of this technology, such as by using energy-efficient hardware, implementing green energy in the data centers, and optimizing the frequency and duration of drone flights to minimize energy consumption.

REFERENCES

[1] Bobulski J. and Kubanek M., (2019). Waste classification system using image processing and convolutional neural networks, in *International work-conference on artificial neural networks*. Springer, pp. 350–361.

[2] Rad M. S., Kaenel A. v., Droux A., Tieche F., Ouerhani N., Ekenel H. K., and Thiran J.-P., (2017). A computer vision system to localize and classify wastes on the streets, in *International Conference on computer vision systems*. Springer, pp. 195–204.

[3] Sermanet P., Eigen D., Zhang X., Mathieu M., Fergus R., and LeCun Y., ((2014)). Overfeat: Integrated recognition, localization and detection using convolutional networks, in *2nd International Conference on Learning Representations, ICLR 2014, Banff, AB, Canada, April 14-16, 2014, Conference Track Proceedings*, Y. Bengio and Y. LeCun, Eds., [Online]. Available: http://arxiv.org/abs/1312.6229=0pt

[4] Szegedy C., Liu W., Jia Y., Sermanet P., Reed S. E., Anguelov D., Erhan D., Vanhoucke V., and Rabinovich A., (2014). Going deeper with convolutions, *CoRR*, vol. abs/1409.4842, [Online]. Available: http://arxiv.org/abs/1409.4842=0pt

[5] Fallati L., Polidori A., Salvatore C., Saponari L., Savini A., and Galli P., (2019). Anthropogenic marine debris assessment with unmanned aerial vehicle imagery and deep learning: A case study along the beaches of the republic of maldives, *Science of The Total Environment*.

[6] Bochkovskiy A., Wang C.-Y., and Liao H.-Y. M., (2020). Yolov4: Optimal speed and accuracy of object detection, *arXiv preprint arXiv*:2004.10934.

[7] Majchrowska S., Mikoajczyk A., Ferlin M., Klawikowska Z., Plantykow M. A., Kwasigroch A., and Majek K., (2022). Deep learning-based waste detection in natural and urban environments, *Waste Management*, vol. 138, pp. 274–284.

[8] Lu W., Chen J., and Xue F., (2022). Using computer vision to recognize composition of construction waste mixtures: A semantic segmentation approach, *Resources, Conservation and Recycling*, vol. 178.

[9] Girshick R., (2015). Fast r-cnn, in *Proceedings of the IEEE international conference on computer vision*, pp. 1440–1448.

[10] Liu W., Anguelov D., Erhan D., Szegedy C., Reed S., Fu C.-Y., and Berg A. C., (2016). Ssd: Single shot multibox detector, in *European conference on computer vision*. Springer, pp. 21–37.

[11] He K., Zhang X., Ren S., and Sun J., (2015). Spatial pyramid pooling in deep convolutional networks for visual recognition, *IEEE Transactions on Pattern Analysis and Machine Intelligence*, vol. 37, no. 9.

[12] Wilts H., Garcia B. R., Garlito R. G., Gómez L. S., and Prieto E. G., (2021). Artificial intelligence in the sorting of municipal waste as an enabler of the circular economy, *Resources*, vol. 10, no. 4, p. 28.

[13] Busch J., Foxon T. J., and Taylor P. G., (2018). Designing industrial strategy for a low carbon transformation, *Environmental Innovation and Societal Transitions*, vol. 29, pp. 114–125.

Innovation and Technological Advances for Sustainability – Al-Naemi et al. (Eds)

Crypto-stenographic technique for safe transmission of images in Stego-audio in public networks

Shafi'i M. Abdulhamid, Aliya M. Almarri & Abdullah Hussein Al-Ghushami
Information Technology Department, Community College of Qatar, Doha, Qatar

ABSTRACT: There are many ways that deliver images through the use of the public network, as this network is considered one of the most vulnerable because it is open to anyone to use. The risks and threats are high when sending confidential images over the public network, where the messages sent can be exposed to the risks of theft, modification or deletion and other harmful risks. The solution lies in the use of masking methods by hiding secret images through the audio file and sending them over the network. This process takes place through several main steps, starting with image compression using Geometric Mean of Pairwise Ratio (GMPR) technology, followed by encryption using the Advanced Encryption Standard (AES) algorithm, and then the process of hiding the audio data by substitution and fitting for Least Significant Bits (LSB). The embedding process of the encoded image bits is done by inserting them into the various variable LSB layers according to the format adopted for the adopted method. Through the results obtained, we find that there is no significant difference between the stego-audio that contains the encoded images and the original audio file.

1 INTRODUCTION

The main concern in public networks is that any device can communicate with all other devices and herein lies the danger if it is considered an open network. According to [1], any public network, whether wired or wireless, is considered insecure, and there are many types of public networks, including: LAN, WAN, and others. When encryption is used for concealment, this process does not change the content of the secret message, but this message is hidden inside the audio file, making it difficult to discover and see its content. That is, the method of hiding information will not make anyone suspect the existence of this hidden data. Another advantage of the information concealment system is that it maintains the confidentiality of the data encoding method, but when the encryption system is known, the data masking system is easily penetrated [2].

Data masking hides data that we don't want to show inside the cover file. Stego also includes additional features that help with the masking process, although it looks identical to the cover file. The data masking system exploits human cognition; this is because human senses are not able to see the contents of the files being sent and know what's inside, although there are many applications that help detect steganography called steganalysis [3].

The main threat to companies and individuals is the process of security of information that is sent over private and public networks and the Internet. Therefore, it must contain a system that combines more than one technology in order to ensure the confidentiality of the transmitted data and its safe access to the recipient. Individuals and companies are working to find a secure transmission method for sending their confidential data, and we will learn about the method of data transmission by including it within the audio files [4]. So, the main problem is the process of maintaining the confidentiality of the message during the transmission process in order to reach the recipient safely and thus achieve secure communication.

 DOI: 10.1201/9781003496724-35

We outline the main contributions of this research paper as follows.

- To critically review related work in audio steganography.
- To determine how steganography can help overcome the problem of securely transmitting confidential images over a public network.
- To design an audio crypto-stenographic technique using Advance Encryption Standard (AES) encrypting and Least Significant Bits (LSB) scheme.
- To evaluate the proposed steganography scheme with other literatures.

The aim of this research is to propose techniques of hiding data in audio files to reach the second party (the recipient) safely. The process of combining cryptographic and steganographic algorithms is to be used in secure data transmission. The remaining sections of the papers are structured as follows: Section II presents previous related works, Section III details the methodology of the research and Section IV presents the results obtained. Section V shows the experimental set ups and Section VI draws major conclusions.

2 RELATED WORKS

In this section, there are many studies in this field that discussed about different techniques for hiding information over the public network using audio files, and these studies as shown in Table 1.

Table 1. Related works.

Reference	Method	Metric Parameters	Result	Limitations
[5]	AES-128 and LSB Data Hiding	standard hearing test, MSE, PSNR and waveforms	high results	The AES algorithm is complicate d to implement.
[6]	LSB + 1, LSB + 2 and LSB + 3	(PSNR) and bit error rate (BER)	good results	The algorithms LSB + 1, LSB + 2 did not give excellent results.
[7]	multiple layers of LSB	MSE, PSNR and BER	good results	bypassed by inserting three bits into the target bits.
[8]	The primary key is responsible for encryption and decryption, and a secondary key at the end of the encryption process.	SNR and BER	SNR 139 dB to 142 dB for 32- bit and 67 dB to 85 dB for 16-bit stereo files. The error rate is at Bit (BER) is in the range 0.23 to 0.32 percent for 32-bit and 0.018 to 0.028 percent for 16- bit files.	This method is useful because there are two keys used in this method.
[9]	Transparent LSB	MSE, PSNR and BER	good results	The signal width is not large enough to transmit large amounts of data.
[10]	LSB technique	SNR	good results	Its results are inaccurate.
[11]	AES-128 and LSB Data Hiding	standard hearing test, MSE, PSNR and waveforms	high results	The AES algorithm is complicate d to implement.

The study presented by [5], which works on the use of the AES-128 algorithm to encrypt confidential information. After the encryption is performed, the encrypted LSB secret information is entered into the voice envelope by cascading through a tan logistics plan. In this method, standard hearing test, MSE, PSNR and waveforms are used in order to know the effectiveness and efficiency of this method.

According to [6] Least Significant Bit (LSB) is one of the most widely used methods for hiding information within audio files. And because of the ease of work in it, it has attracted the attention of many researchers in order to develop and improve it. It focused on how to take full advantage by adding bits to the audio file and how the quality of the audio file is affected. There is more than one version of LSB technology, where it has been modified to version LSB + 1, LSB + 2 and LSB + 3. Through the results, we found that LSB + 3 gave the best results compared to the rest.

Based on [7] LSB technologies work by entering data at the same LSB location for adjacent parts, providing a great service for hackers to easily obtain confidential data. This study works to bypass this problem by entering confidential data within several layers of LSB that are chosen randomly and in non-sequential parts. But there was a flaw in this technology, which is the secret data in the non-sequential parts lose the ability of the stego sound. This problem was bypassed by inserting three bits into the target bits. This proposed method can introduce various payloads into the casing sound and is better than many other techniques.

The researchers [8], proposed to implement a great security for messages, two different schemes that contain two layers, where the first layer represents the encryption method, which implements AES-128 encryption, and the second layer represents the data hiding process, which works through LSB substitution technology. This technology got high ratings by using the AES encryption algorithm along with LSB technology.

According to [9], in various media in order to achieve secure communication, emphasis has been placed on achieving security for confidential information that is transmitted through a secure transmission channel which is always a difficult task in this field. With the existence of various technologies in order to send data secretly, hiding information through audio files appears to us, which offers many solutions in this field. A method has been devised to perform this process which is random key indexing in order to change the LSBs of the sound of the wave that carries the secret message. The primary key is responsible for making a change through TTP (a trusted third party) and for decryption a secondary key is generated at the end of the encryption process and at the same time as entering the secret data. This method is also supported by the use of the message recovery code, which adds another layer of protection to this method. This method, when tested, provided excellent results in payloads, SNR values, and BER modulation.

Based on [10], The process of hiding information within images is more widespread and widely known than hiding information within audio files. In this proposed method, a new scheme was designed to hide confidential information across the audio file through transparent LSB technology, which was tested through various audio tools. A method of hiding data, which is very similar to the method of LSB, was used to hide data through images, as this scheme provided high transmission rates and transparency. In addition, in this method, no indication of secret information hidden within the audio file was shown, and no noise was detected within the audio signal.

A study presented by [11], indicates that audio signals have an excellent repetitive nature which makes them one of the best techniques for concealment and transmission of confidential information. Steganographic method for inserting confidential information into audio using an algorithm based on LSB was adopted in this paper. This method works by taking each audio sample, changing it into bits, and then fusing it into the text data.

Upon input, each message character is changed to its equivalent binary. With this method and based on LSB, the Stego system increases its ability to hide confidential information. The performance results of this proposed algorithm are calculated by finding the SNR values which represent various audio inputs. In this section, the review found out that there were more than one way to hiding data within audio files which has been identified. The LSB algorithm has been adopted, which is characterized by high durability and efficiency in the process of including confidential data within the audio file. It was also discovered that to send an image through the audio file, where we used an algorithm to compress and encrypt the data and enter it within the audio file and send it to the recipient, where we identified it and the work steps. The technique to be used in this study was identified and what are its features that help us achieve the research objectives.

3 RESEARCH METHODOLOGY

In this section, we describe the research phases that have been followed to implement the research procedure. Figure 1. shows the research process framework.

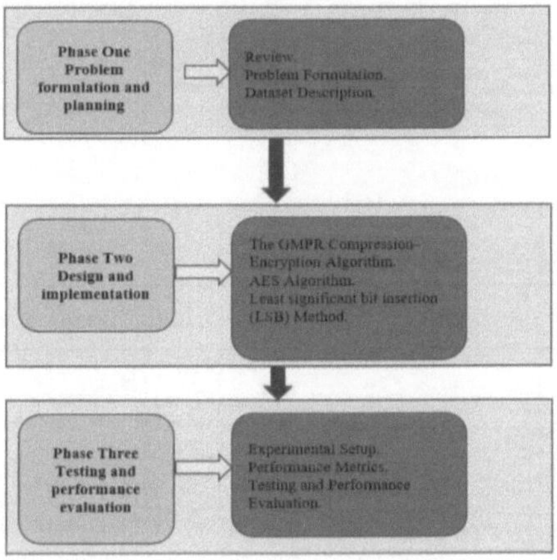

Figure 1. Research process framework.

3.1 *Problem formulation*

In the process of hiding confidential information via the audio file (in our research, hiding an image), we must compress and encrypt the image, the image will be compressed through the GMPR algorithm, and the encryption will be via the AES- 128 algorithm, and then both the audio and image are converted into a stream of bits in order to include the secret image within the audio file via LSB.

To formulate the image is segmented using the GMPR compression with n×n blocks, then DCT, is applied on each block independently. The result from DCT gives all A block of DC value (first position in block), while all other values are placed as AC parameters. It is done through Equation 1:

$$C(i,\ v) = a(u)a\sum_{x=0}^{n-1}\sum_{y=0}^{n-1} f(x,\ y)\cos\left[\frac{(2x+1)u\pi}{2n}\right]\left[\frac{(2y+1)v\pi}{2n}\right] \tag{1}$$

where $a(u) = \sqrt{\frac{1}{n}}$, for $u = 0$, $a(u) = \sqrt{\frac{2}{n}}$, for $u \neq 0$. where $i, j = 1, 2, \ldots, n$.

Next step, GMPR algorithm separates DC parts of each block is kept in a DCArray. Then the difference between each two adjacent values is calculated for increased pressure levels. To get the values in DC-Array are done by the Equation 2.

$$D_i = D_i - D_{(i+1)} \tag{2}$$

3.2 *Dataset description*

Four images were used for the purpose of using them in applying the process of compression and masking inside the audio file, as well as an audio file with a length of 0.42 seconds, the images are shown in Figure 2, and Table 2 shows the details of the sizes of these images and the audio file.

Figure 2. Images used in the search.

Table 2. Images details.

Image name	Original image size
Apples	1.37 MB
Lena	1.0 MB
Card	1.37 MB
Guitar	1.37 MB
Audio File	672 KB

3.3 *Design and implementation*

The proposed crypto-stenographic technique was implemented in three main steps as shown in the flowchart presented in Figure 3.

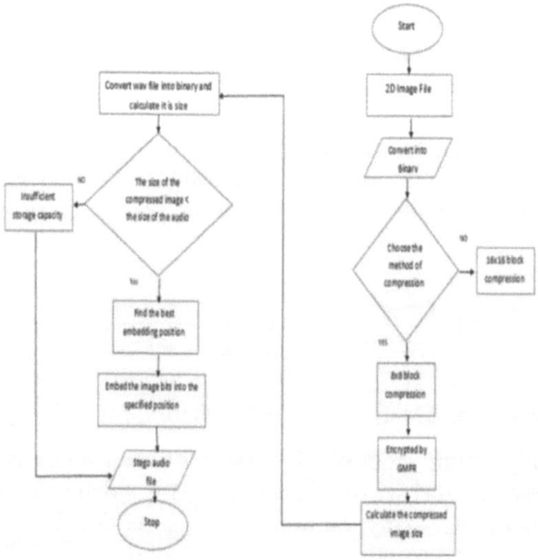

Figure 3. Flowchart of proposed system.

3.4 *LSB method*

In computer science, the Least Significant Bit (LSB) refers to a binary integer that represents the value of the number of units, in other words, whether the number of units is even or odd in a byte. The LSB is in the far right of the byte.

To hide part of the secret message by using LSB technology has been presented in [3]. The confidential information in this figure is in the middle of the Figure 6 and the cover file is an audio file. In this figure, the audio file feed has been converted into a stream of zeros and ones, and the secret message is obtained from the output of the GMPR algorithm. The method works simply by replacing the LSB column with the Secret Message column. The final output file is called a Stego-audio file.

3.5 *AES algorithm*

The AES algorithm is characterized by a very high level of security, and it is considered one of the algorithms least vulnerable to attacks and is considered very suitable in image encryption, because there is always concern about integrity and privacy, a unique combination of LSB and AES encryption appears Picture by using LSB technology (Least Significant Bits) Selects the LSB bit and replaces it with image bits. Image security is enhanced by encryption with AES because masking without encryption is considered weak, so the image is encrypted with AES to provide durability.

The phased implementation of AES is as follows:

(i) Convert user data to ASCII from int64.
(ii) Build an encryption key.
(iii) Work on converting data to ASCII integer and square matrix.
(iv) Include a key to the data.
(v) Repeat the following process. Class conversion.

Creates a matrix based on the column and multiplies it. Include a key to the data.

(i) Turn the first row one last time.
(ii) Add a key to the data for the last time.

4 EXPERIMENTAL SETUP

In this study, the main tool in the process of masking sound within the audio file is DeepSound, as this tool contains the encryption feature using the AES algorithm. either for the image compression process, which is done by using Matlab R2012a with Visual C++ 2008. Through the tool DeepSound, a password was specified when sending, and this word is required when decrypting the message. This proposed method has been applied for this research and tested by using a computer with Windows 10 installed with the following specifications: Processor: Intel(R) Core (TM) i5-3230M CPU @ 2.60GHz. Fixed Memory (RAM): 8.00 GB and System type: 64-bit operating system. This should considered as a minimum specification for better performance.

4.1 *Performance metrics*

The performance of the proposed system was evaluated using performance measures:

- Root Measure Square Error (RMSE) Determines how good the image is after compression. It is calculated by the Equation 3:

$$RMSE = \sqrt{\frac{\sum_{i=1}^{N} (Predicted_i - Actual_i)^2}{N}} \quad (3)$$

where actual is the image values after compression. Predicate represents the values of the image before compression.

N is the number of studied points.

Another metric that has been adopted is the Signal-to-Noise Ratio (SNR) It is used to compare the value of the desired signal with the value of the background noise signal. It mainly denotes the ratio of signal strength to noise strength, which is often expressed in decibels. This scale was applied through the Equation 4:

$$SNR = 10log\left(\frac{RMS^2_{signanl}}{RMS^2_{noise}}\right) \tag{4}$$

where RMS_signal is the RMS value of signal and RMS_noise is that of noise. log is the logarithm of 10.

As for the criteria that determine the quality of the audio file, the peak signal-to-noise ratio (PSNR) was used. This scale indicates the highest possible value of the signal and the strength of the damaged noise that affects the accuracy of its representation. PSNR is usually expressed as a logarithmic quantity using the decibel scale. This scale was applied through the Equation 5:

$$PSNR = 10 \cdot \log_{10}\left(\frac{MAX_I^2}{MSE}\right) \tag{5}$$

Here, MAXI is the maximum possible value of the audio file.

5 RESULTS AND DISCUSSION

The results obtained from the three stages are presented as follows.

5.1 *Images compression result*

After performing the compression process on the selected images, we check the quality of the compressed images by calculating the RMSE, where the root mean is used in order to measure the difference between the source image and the compressed image, the lower the RMSE value, the better the segmentation performance. The RMSE results were obtained through the compression process and are shown in Table 3. From the results obtained from the compression process, we found that the apple image achieved the best results in terms of quality among the other images, because it achieved the lowest value for RMSE, followed by Guitar image that scored RMSE with a score of 1.3, then Lena that achieved a score of 1.7 for the RMSE scale and card The last photos had a high RMSE score compared to the other photos, with a score of 3.5.

Table 3. Image compression results.

Image name	Original image size (MB)	Block size	Compressed image size (KB)	RMSE
Apples	1.37	8x8	152	1.06
Lena	1.0		105	1.7
Card	1.37		360	3.5
Guitar	1.37		95	1.3

After applying the process of hiding the secret message within the audio file, we now calculate the SNR and PSNR of the audio file. The results in Figure 8. are high for SNR and

PSNR, and high values mean that the hidden message remains undetected or it is difficult to notice its presence within the audio file.

Figure 4 shows the RMSE results for the images that were used.

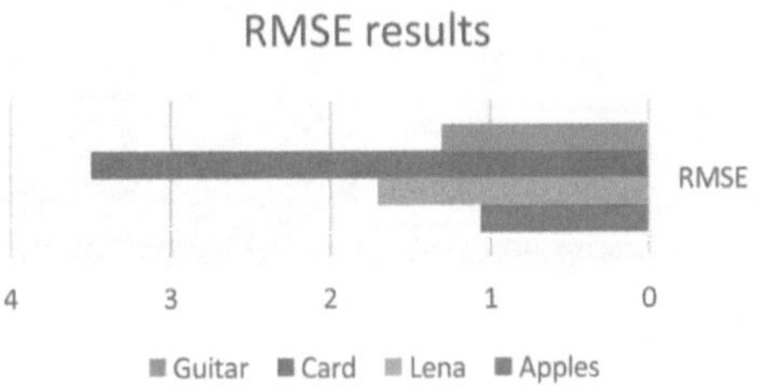

Figure 4. RMSE results.

5.2 *Results of using LSB technology*

After completing the process of embedding the image into the audio file, we can now see the difference between the original audio file format before and after the embedding process. Figure 5(a-b).

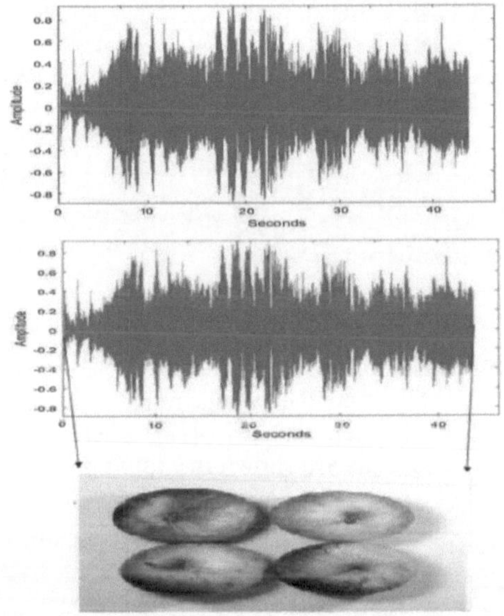

Figure 5(a-b). Audio file before and after embedding.

After executing the operation completely and getting the output file that contains the secret message, Figure 6 shows the details of the audio file size before and after the masking operation.

Table 4. SNR and PSNR results.

Image name	Original image size (MB)	Compressed image size (KB)	Audio SNR	Audio PSNR
Apples	1.37	152	61.18	74.08
Lena	1.0	105	67.22	79.70
Card	1.37	360	56.17	69.55
Guitar	1.37	95	70.08	83.98

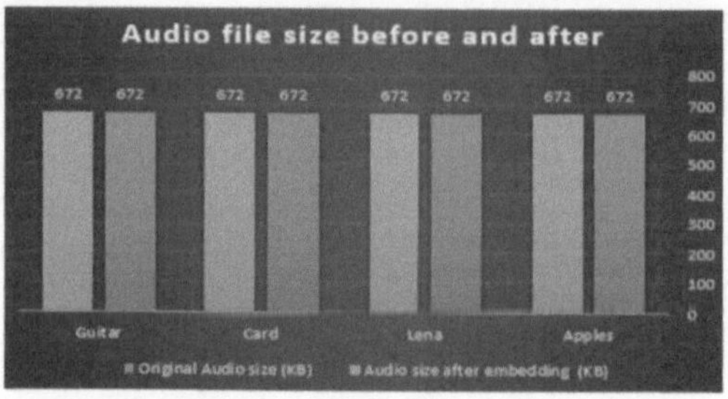

Figure 6. Audio file size before and after.

Through the results of SNR and PSNR, we find that the image Guitar is considered the best among the other images because it achieved the highest results, Figure 7, a chart showing the results of SNR and PSNR.

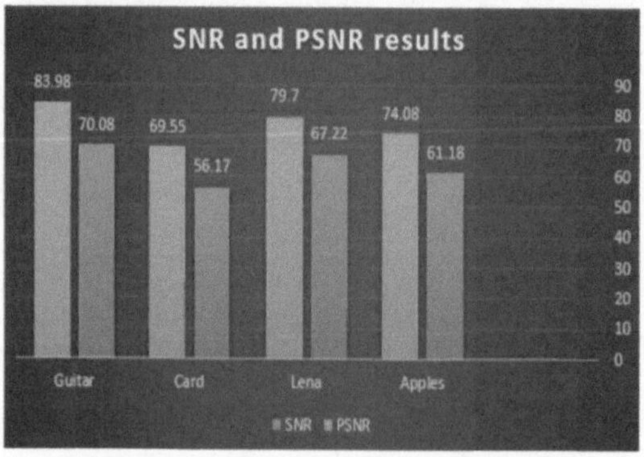

Figure 7. SNR and PSNR results.

Previous studies were identified in order to calculate the RMSE for the four compressed images, through the study Abood (2021) [7], where the average result of his study for RMSE was 5.41.

The second study was through Yuantao, 2019 [12], where the average result for compressed images was 2.71. These results are shown in Figure 8.

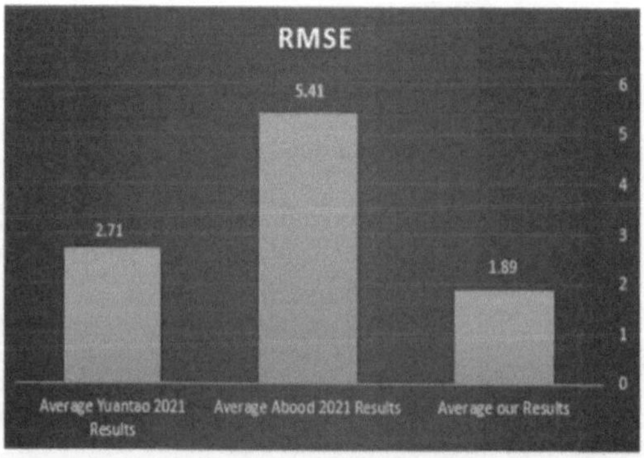

Figure 8. RMSE results comparison.

The SNR results were presented and compared with ours through Figure 9.

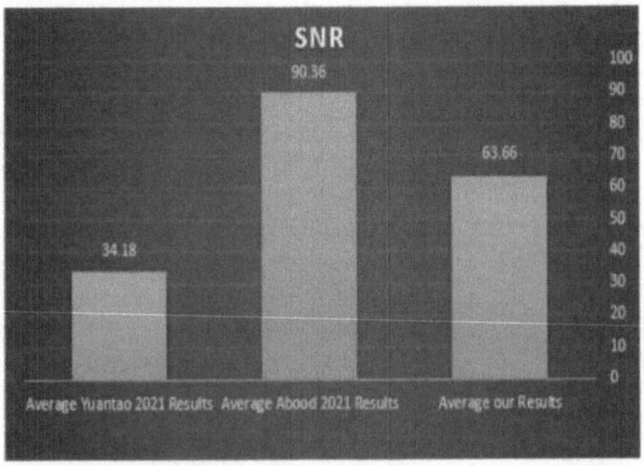

Figure 9. SNR results comparison.

6 RECOMMENDATIONS

After completing the application of this study and identifying the method of concealment using LSB technology, we can develop and expand this method by identifying other means and tools that achieve the same goal. In addition, we can study other technologies similar to LSB technology and work on their application and development.

7 CONCLUSION

This is the method that was followed in order to include secret messages of images within the audio files by masking the sound LSB, where the results showed that excellent values were obtained. The method used gave a large capacity for placing secret messages, and offset by no decrease in the quality of the embedded images and not affected by noise or distortions that can be exposed to other roads.

The proposed method is considered one of the most powerful methods and methods to counter stego analysis attacks. By analyzing PSNR and SNR shows that the cover generated after the embedding process is very similar to the audio file before the embedding process, so we conclude that the masking process for the audio message is imperceptible.

And by performing the analysis process for the algorithms used, which provided a high concealment of the secret message, we find the result, which was to obtain the image within the receiving party with high quality, after implementing the process of high pressure on it.

Finally, the method that was used in this study provided high levels of security, as no one can easily detect it except by knowing how LSB and AES decryption work and then using the GMPR algorithm in order to decompress the image. At the present time, the methods and methods of concealing confidential information have greatly evolved, and it has also been matched by an evolution in the process of its disclosure. In this study, we showed the means that make the process of hiding information easy and simple, and the focus is now on continuing to develop to achieve better results in the future.

ACKNOWLEDGMENT

Special appreciation also goes to Prof. Dr. Ahmad Al Omari, Acting Dean of Sciences and Technology Division, Community College of Qatar (CCQ) for all his support and encouragements.

REFERENCES

[1] Wahab F., (Apr. 24, 2018). *What Is a Public Network and Why Is It Dangerous?*, AddictiveTips. https://www.addictivetips.com/windows-tips/public-network-dangerous/ (accessed Jan. 30, 2023).

[2] Geeksforgeeks A., (Dec. 13, 2019). *Data Concealment Methods*, GeeksforGeeks. https://www.geeksforgeeks.org/data-concealment-methods/ (accessed Jan. 30, 2023).

[3] El_Rahman S. A., (Aug. 2018). A comparative analysis of image steganography based on DCT algorithm and steganography tool to hide nuclear reactors confidential information, *Computers & Electrical Engineering*, vol. 70, pp. 380–399, doi: 10.1016/j.compeleceng.2016.09.001.

[4] Prabhu R. D., *What Is Network Security? Definition, Types, and Best Practices*, Spiceworks. https://www.spiceworks.com/it-security/network-security/articles/what-is-network-security/ (accessed Jan. 30, 2023).

[5] Alexan E. A., (2019). Logistic tan map based audio steganography, in *2019 International Conference on Electrical and Computing Technologies and Applications* (ICECTA).

[6] Abdulrazzaq S. T., Siddeq M. M., and Rodrigues M. A., (Mar. 2020). A novel steganography approach for audio files, *SN Computer Science*, vol. 1, no. 2, doi: 10.1007/s42979-020-0080-2.

[7] Abood E. W., Abduljabbar Z. A., Al Sibahee M. A., Hussain M. A., and Hussien Z. A., (Jun. 2021). Securing audio transmission based on encoding and steganography, *Indonesian Journal of Electrical Engineering and Computer Science*, vol. 22, no. 3, p. 1777, doi: 10.11591/ijeecs.v22.i3.pp1777-1786.

[8] AlSabhany A. A., Ali A. H., Ridzuan F., Azni A. H., and Mokhtar M. R., (Nov. 2020). Digital audio steganography: Systematic review, classification, and analysis of the current state of the art, *Computer Science Review*, vol. 38, p. 100316, doi: 10.1016/j.cosrev.2020.100316.

[9] Farhani W. S. and Dwiharzandis A., (Jun. 2022). Steganografi Metode Least Significant BIT (LSB) pada mpeg spatial audio object coding, *Rang Teknik Journal*, vol. 5, no. 2, pp. 364–368, doi: 10.31869/rtj.v5i2.3260.

[10] Liu J., Yang C., Wang J., and Shi Y., (Sep. 2020). Stego key recovery method for F5 steganography with matrix encoding, *EURASIP Journal on Image and Video Processing*, vol. 2020, no. 1, doi:10.1186/s13640-020-00526-2.

[11] Anon A. What are audio files? (with picture), *Easy Tech Junkie*. https://www.easytechjunkie.com/what-are-audio-files.htm (accessed Jan. 30, 2023).

[12] Mahmou A., (Feb. 2020). Single image super-resolution algorithm using PSNR in the wavelet domain, *Journal of Advanced Research in Dynamical and Control Systems*, vol. 12, no. 01-Special Issue, pp. 677–691, doi:10.5373/jardcs/v12sp1/20201117.

Innovation and Technological Advances for Sustainability – Al-Naemi et al. (Eds)

Classification of visual stimuli using brain-computer interfaces by synchronously decoding reactive EEG signals: A novel approach

Nawfel Belgasium, Ramzi A. Haraty & Sergio Thoumi
Department of Computer Science and Mathematics, Lebanese American University, Beirut, Lebanon

ABSTRACT: Artificial intelligence and machine learning have been on a sharp rise in the last few years. One reason behind their popularity, is their usefulness as they can be seen in multiple fields including medicine and biology. Brain-computer interfaces are one of the many systems that have been developed and enhanced due to machine learning. In this work, we classify visual stimuli using an electroencephalogram-based brain-computer interface. The experimental results, applied on small data sets, are promising.

Keywords: Brain Computer Interface, Classification, Deep Learning, Electroencephalography, Neural Networks

1 INTRODUCTION

Analyzing neural activity and its underlying drivers require a thorough understanding of the convolutions of the human brain in its structure and functionality. The human brain consists of several discrete but highly interconnected parts: the cerebellum, the brain stem and the cerebrum. The cerebrum contains two hemispheres that in turn split into four lobes: the frontal, parietal, occipital, and the temporal lobes [1]. The principal indicators that signify neural activity in the brain are communication signals omitted by the neurons of the brain. The signals' intensity, type, and position are used in analyzing brain activity.

Brain Machine Interfaces (BMIs), or more commonly known as Brain Computer Interfaces (BCIs), is a system that takes signals emitted by the brain and translate them into commands that can be fed to a computer. Some BMIs are based on Electroencephalography (EEG) which records electrical brain signals (as opposed to chemical brain signals) emitted by neurons which corresponds to various positions from the scalp area [2].

There are three main types of BCIs: non-invasive, invasive, and partially invasive. This work deals with non-invasive BCIs where electrodes are placed on the scalp, this type is the least accurate of the three but is generally accepted by everyone. Moreover, it can be easily implemented. The second type is partially-invasive BCIs where electrodes are placed inside the skull but remain outside the brain. This type gives more accurate data but is harder to implement. The last type is invasive BCIs where the electrodes are placed under the scalp via surgery, this type gives the most accurate data but comes with its own downsides and side-effects related to the surgery performed to be able to do it. Both partially invasive and invasive BCIs have been controversial and are out of scope of this work.

Getting more familiar with object–induced EEG has been the main goal that BCIs set out to achieve. At their earliest stages, BCIs were set to function when signals pass a certain threshold or when a very specific pattern would be noticed, but thanks to deep learning, most modern EEG-based systems rely on machine learning, image classification, and other fields [3]. BCIs at their beginning stages were regarded as too complex as they had limited resolution, the reliability of the information they produced was too variable and the task they

DOI: 10.1201/9781003496724-36

were programmed to do would need to be very specific in the functional sense because of the challenges that come with trying to generalize a function on data that is very unpredictable [4].

Nevertheless, EEG is relatively easy to use and has a low cost when compared to other methods. However, despite it having great temporal resolution (i.e., being able to capture signals at millisecond timescales), it suffers from low spatial resolution (i.e., signals will contain a considerable amount of noise) since EEG electrodes are placed on the scalp area where many layers of tissue separate the receiver and the source of the signals, which is usually the case with non-invasive BCIs [4] [5] as opposed to other types. The most pressing hurdle with EEG data is the high subject inter-variability, which if not taken into consideration can highly limit your model's performance. Not only can it make the process of deriving a function that successfully maps an input to the intended output very difficult, but it can also lead the model to over-fitting at a very early stage [5] [6].

When it comes to visual stimuli and the distinct brain states that can be decoded from EEG recordings, two scenarios are often undertaken [6]:

1) The synchronous BCI approach in which discrete selections are being made based on external event registration.
2) The asynchronous BCI approach in which joint or vaguely discrete selections are being made based on the concentration or de-concentration of specific sub-mental states. In this approach, more weight is put on the voluntarily user-led state than the state an external stimulus can result in.

In this work, we set out to explore the classification of EEG signals induced by using pictures as stimulus, ten numbers ranging from 0 to 9 were displayed to each subject. We explored different approaches to decoding the EEG signals and made the application work in real-time which was one of the main challenges given the extensive pre-processing that EEG signals go through.

2 LITERATURE REVIEW

The technology and literature around BCI research has witnessed great development within the past 20 years [7]. Earlier research was restricted to around three topics. Today, BCI research includes more than a 100 topics.

Using deep learning techniques to recognize EEG patterns is a relatively recent field. Despite some authors claiming that a limited number of methods has been developed in interpreting EEG data (none of them exploiting deep learning) [8] and that their research was one of the first to deal with more than just a binary-class EEG classifier. According to another study [9], there was previous endeavors. One study [11], attempted to generate, record and interpret with deep learning three different motor imagery signals using EEG in order to control a video game character. The authors discussed the different classes or outputs the player should produce in order to complete the game. They used a 32-electrode setup and achieved an accuracy of 80% with their neural network.

Spampinato *et al.* [8] mentioned that high performance was recorded with another study [11], where they had attempted to distinguish twelve different classes worth of EEG signals and had achieved an accuracy of about 29% in which they used 124 electrodes as facilitators of the signal acquisition phase.

Due to its recent popularity and low cost of application, many different methods have been undertaken when approaching the interpretation of EEG data. For example, a study [12] attempted to fit a Convolutional Neural Network (CNN) model to be able to classify audio, over 12 classes, using 64+ channel electrodes. They have recorded an accuracy of 28%.

Another research [3], tackled EEG visual image classification with 40 classes and a setup of 64 electrodes, however, in this work, a more condensed review of the results was given. Even though they had reached an 80% accuracy on the best model they achieved, when they reviewed their accuracy with respect to each and every class, they found that multiple classes had less than a 50% accuracy making it indistinguishable from other EEG signal classes. At the same time, there existed classes with over 70% accuracy. This occurrence can be attributed to many reasons, some of them being:

1) Unbalanced data-set
2) Placing more weight on training the class with higher accuracy
3) The low accuracy class samples contained data that was not as easy to generalize as other classes

In one review [13], the authors extended BCI sub-categories to include the different states a user is in during a live human-machine system recording. They divided BCI applications into three types: active, reactive, and passive. All previously mentioned literature can be classified into one of these three categories. In Walker's research, the user directly and consciously attempted to control a robot using his generated EEG signal [10], this experiment was intended to be independent of external events. It only relied on the user to initiate communication and hence is regarded as active BCI communication. Passive BCI communication is the recording of brain activity without intention of enriching the EEG recording with any voluntary control from the user's side. An example of this is the recording of EEG sleep signals or more commonly known as "polysomnography". Lastly, we have the reactive BCI communication, which is carried out in reaction to external stimulations. The produced signals are open to modulation from the user's end in order to cause a specific outcome to arise. This type of communication is performed in most of the previously mentioned literature [3] [11] [12]. Stober *et al.* [12] recorded the human response to 12 classes of audio and attempted to classify them based on the assumption that the human reaction to every audio clip would be different. Visual image classification done in other studies [3] [11] is also regarded as reactive BCI communication, as it is built on the same assumption that there should exist a difference between the recordings that are perceptible and learnable by a function of some sorts.

In this work, we investigate an EEG-based BCI approach. We implement a reactive BCI communication paradigm to classify the visual stimulations that are evoked when a subject observes a number from 0 to 9 using state-of-the-art methods. Along with classifications, this model predicts samples in real-time, using different methods to make the simultaneous streaming and the recording of data as efficient as possible.

3 METHODOLOGY

This work could be divided into four parts:

1) Signal acquisition
2) Signal pre-processing and enhancement
3) Feature extraction
4) Classification and output

3.1 *Signal acquisition*

Participants in this work were of similar age and level of education, moreover, none of them reported any medical or health issues that may hinder their participation. During the recording, the participant were asked to monitor a screen around 0.8 to 1m away from them that displayed an image of a random number between 0 and 9 as shown in Figure 2. The signal evoked by each image was recorded separately. Each image was displayed for 6 to

7 seconds. The first 1 to 2 seconds were truncated from every recording to avoid the noise that is generated when a streaming session is initiated on the EEG hardware.

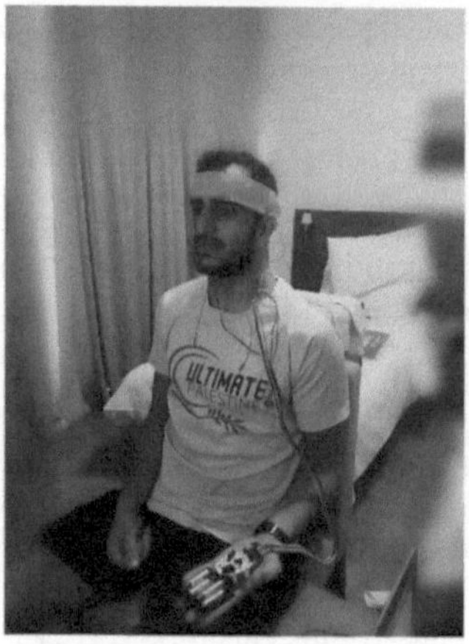

Figure 1. Subject undergoing an EEG recording session.

To record EEG data, we used a five channel setup with two other pins that enhance the functionality of the hardware. In total we had five EXG channels, one BIAS channel, and one SRB channel.

The BIAS channel implements the concept of a ground pin/channel, and also inserts interference waves that aid in cancelling out unwanted noise. The SRB channel acts as the Normal or Default that all other channels would use as reference. The remaining five channels are the pins that will transfer the data that is generated in the positions in which they were placed on the scalp area. In short, the data used for classification was obtained from five channels. The recording was done using OpenBCI Cyton which is a commonly used neural interface with a 32-bit processor.

3.2 *Signal pre-processing and enhancement*

3.2.1 *Sample rate*

Our sampling rate was 250Hz. Sample rates in professionally used EEG headsets have an average sampling rate of 500Hz. A low sampling rate can result in a jagged pattern in the recorded data, hence, we applied a basic smoothing function to help with the transitions from one sample to another.

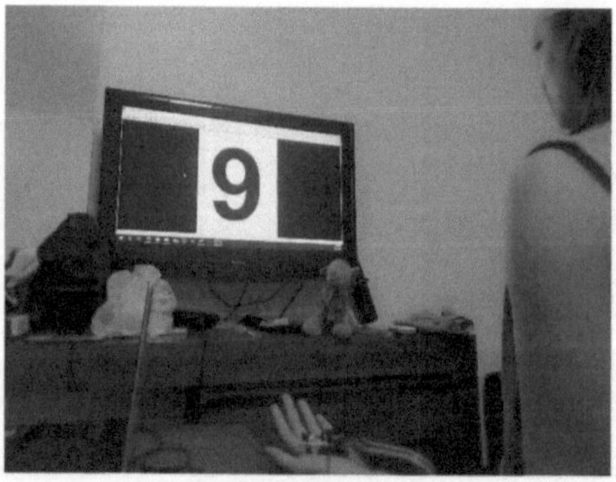

Figure 2. Another subject undergoing an EEG recording session.

3.2.2 *Notch and pass filters*

Typically, in EEG data recordings, high-pass and low-pass filters are used to filter out slow frequencies (ex: less than 0.1Hz) and fast frequencies (ex: above 40 or 50 Hz) respectively. If we consider that all brain electrical activity of interest lies within a contained bandwidth domain, we can use a combination of high and low-pass filters, and the resulting domain would contain our data of interest. This setup is usually referred to as a bandpass filter setup.

It is assumed by laypeople that a principle to work by when deciding what sort of filter to use, is that most brain electrical activity lies between 1Hz and 35Hz and that the more a signal deviates from this bandwidth, the more likely it is to consist of mostly noise or signal artifacts. Although this assumption has some truth to it, it is not completely true.

Out of all frequency ranges that adhere to the distribution on the scalp, Gamma rhythms is the least commonly used in EEG brain-computer interfaces. The most pressing issue that comes along with including Gamma rhythms is that they contain more artifacts than other frequencies [4]. In this work, we used a band-pass filter that spans the range 1-50Hz. Despite the drawbacks mentioned earlier, Gamma frequencies have some advantages compared to alpha, beta and other traditional frequencies. Namely, they have more spatial specificity and contain a bigger bandwidth range; thus, they may increase not only the information transfer quantity but also quality if pre-processed correctly.

The purpose of notch filters is to choose a specific frequency (as opposed to a range of frequencies), and to filter it out. In this work, we chose to implement a notch filter of 50Hz (49Hz – 51Hz), to filter out the alternating current emitted by electrical sockets in Lebanon which oscillates at 50Hz with a standard voltage of 220V. If this was not done, our data would be filled with noise and we would need to implement extra removal techniques to get rid of the noise.

3.2.3 *Distribution*

The data set was turned into a Gaussian distribution by using a normalization process in which all the data recorded from a single channel would have a mean of 0 and would be divided by its standard deviation.

3.2.4 *Feature representation*

Since the time-series data represented by EEG signals is highly multi-dimensional, we did not put all recorded data into one feature vector. Instead, we dealt with the recordings as a low-dimensional feature vector that encapsulates all the data needed which keeps its multi-dimensional characteristic.

3.2.5 *Neural network structure*

The core idea of this work revolves around the neural network structure and its implementation in classifying the EEG signal into the correct class. Considering the high temporal levels of EEG data, we adopted a Long Short-Term Memory (LSTM) based neural network. LSTMs are known for their great performance in Natural Language Processing (NLP) applications. In one time-series sample, there can a noise or lag of some sort that shouldn't included in the processing, LSTMs take care of that. A LSTM network will make predictions in a time-series sample based on a pre-defined and fixed context.

Dropout layers also play a significant role in the structure to prevent over-fitting. Since in each dropout layer, deactivated neurons are randomly sub-sampled. This prevents weights of layers to be concentrated or dependent on a certain section. This is most noticeable in networks with many layers. Moreover, it is able to prevent or at least minimize exploding and vanishing gradients which is an issue that is most evident in large neural networks.

The input EEG sample is sent into a sequential model, where the output of the nth LSTM layer will serve as the input layer to the nth Dropout layer, and the nth Dropout layer output will serve as input for the activation layer, which in this case we chose to be a "Leaky-Relu" function, for the following reason:

When $X < 0$ the slope is not a static zero. This prevents the "Dying Relu" problem, that takes place when too many RELU output neurons become inactive and only output zero for non-zero intended cases. This, in extreme cases can lead to a whole section of a neural network becoming inactive and not getting updated by back-propagation. However, this is exactly why optimizers such as the "Adam" optimizer exist. As long as not all inputs push the RELU function to be negative, the optimizer will work to keep the neural network regulated. Considering the inter-subject variability mentioned earlier, we thought it was better to use a "Leaky RELU" function.

3.3 *Classification and output*

The classifier used in this context is a SoftMax output layer, which can output the analog classification accuracy of the 10 classes used in this work (digits from 0 to 9). The overall neural network architecture is shown clearly in Figure 3.

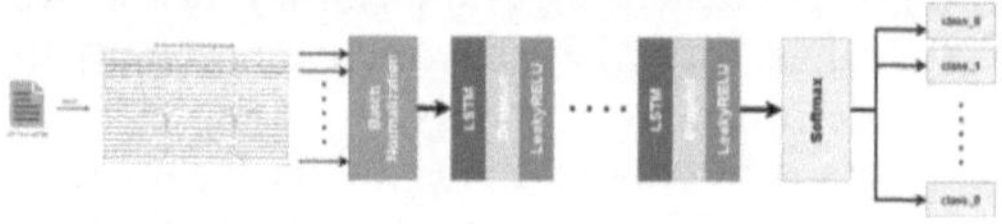

Figure 3. Neural network architecture.

3.4 *Discussion and results*

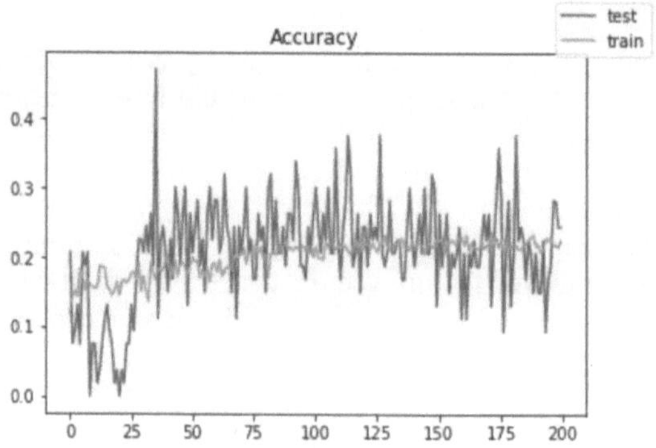

Figure 4. Accuracy score of the LSTM based model after 200 epochs.

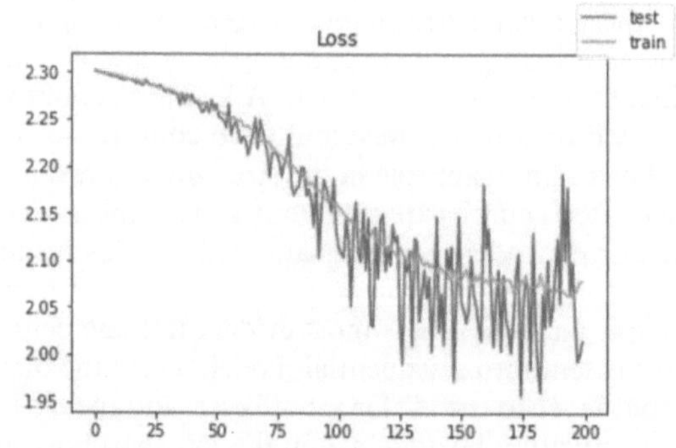

Figure 5. Loss score of the LSTM based model after 200 epochs.

The current recorded EEG data-set was divided into a training set and a testing set, with a 90% and a 10% split, respectively. However due to data-set size limitations, we did not implement the traditional cross-validation procedure when it came to testing the model. Instead, we decided to carry out a bootstrap validation scheme, where in each epoch the input sample-set is partitioned into two subsets, one for training and the other for testing, and with each epoch another split is randomly chosen and trained on (repetition is allowed).

Given the small data-set size and the type of validation we chose; we could not run our training for a long time to ensure that the model would be able to generalize after fitting. After 200 epochs, the model had reached an average accuracy of 20% and had minimized the loss to an average score of 2.1 as can be seen in Figures 4 and 5. At first, these result may seem discouraging. However, when compared to previous results, such as what was obtained by Kaneshiro *et al.* [11] where they had utilized 124 electrodes to reach an accuracy 29%, our results are encouraging given the amount of data at hand. Our model has reached 70% of the accuracy that some studies [11] [12] reached in their model with less than tenth the number of dimensions.

Given more recorded data, we believe that this model would be capable of reaching a global maximum at a faster rate than average models.

3.5 *Limitations*

This work suffered from multiple limitations. First, the number of channels that were available was low. Standard deep learning EEG related research would usually utilize 32 or more electrodes, in order to have more data with better quality in terms of spatial resolution. Electrodes were also placed on the designated scalp positions (AF3, AF4, T7, T8, Pz) manually, and not with a firm headset which hence can be prone to errors.

We had five participants and, we recorded around 10s worth of data for each class (two recordings per class), which accumulates to around 90s worth of data for each person at a sampling rate of 250Hz. This is considered a very small data-set to work on given the amount of variability in EEG data. Hence, we decided that we would train our model on the open-source data-set "the MNIST of brain digits", and then apply transfer learning to our created data-set. The "MNIST" data-set may have impeded or deterred the learning of the model if the data coming from it was different which could happen in many ways such as:

- Data was recorded on different devices using different materials
- Different pre-processing techniques were implemented on the MNIST dataset (even though this wasn't mentioned in the description)
- Since no information was documented regarding the recording process and how the electrodes were placed other than their general positions (AF3, AF4, T7, T8, Pz), we couldn't account for differences in the electrode placement.

4 CONCLUSION

In this work, we explored the ability to decode EEG signals generated after visual stimulation of certain information. We used five electrodes to record EEG data in different regions of the brain. We reached what we think an optimal pre-processing scheme and a neural network architecture that would optimize the mapping function of input EEG to output classification in a stable manner.

In future work, we plan to record more data, to firmly confirm our findings, and to design a new experimental paradigm in which we explore the translation of raw EEG signals to spectrograms and feed the outputted image into a CNN. We would also like to add on to the real-time streaming feature and optimize it by using parallel so streamed information gets lost while the previous is being evaluated by the model. Finally, we want to discover the application potential of Gamma rhythms.

ACKNOWLEDGMENT

This work was funded by the Lebanese American University, Beirut, Lebanon.

REFERENCES

[1] *A Beginner's Guide to Neuroscience - iMotions*, Imotions Publish, Jun. 30, 2020. https://imotions.com/blog/beginners-guide-neuroscience/ (accessed May 29, 2021).

[2] Kumar J. S. and Bhuvaneswari P., (2012). Analysis of electroencephalography (EEG) signals and its categorization–A study, *Procedia Engineering*, volume 38, pp. 2525–2536, doi:10.1016/j.proeng.2012.06.298.

[3] Zhou Q., Liu Y., Zhong S., Tang J., and Liu Y., (Oct. 2019). Decoding of visual information based on EEG data, *J. Physics Conference Series.*, volume 1325, p. 012092, doi: 10.1088/1742-6596/1325/1/012092.

[4] Nicolas-Alonso L. F. and Gomez-Gil J., (Jan. 2012). Brain computer interfaces, a review, *Sensors*, volume 12, no. 2, pp. 1211–1279, doi:10.3390/s120201211.

[5] Roy Y., Banville H., Albuquerque I., Gramfort A., Falk T. H., and Faubert J., (Aug. 2019). Deep learning-based electroencephalography analysis: a systematic review, *J. Neural Engineering*, volume 16, no. 5, p. 051001, doi:10.1088/1741-2552/ab260c.

[6] Rashkov G., Bobe A., Fastovets D., and Komarova M., (Oct. 2019). Natural image reconstruction from brain waves: a novel visual BCI system with native feedback, *Neuroscience, preprint.* doi: 10.1101/787101.

[7] Vidal J. J., (Jun. 1973). Toward direct brain-computer communication, *Annual Review Biophysics Bioengineering*, volume 2, no. 1, pp. 157–180, doi:10.1146/annurev.bb.02.060173.001105.

[8] Spampinato C., Palazzo S., Kavasidis I., Giordano D., Shah M., and Souly N., (Oct. 2019). *Deep Learning Human Mind for Automated Visual Classification*, ArXiv160900344 Cs, Accessed: May 24, 2021. [Online]. Available: http://arxiv.org/abs/1609.00344

[9] Wei X., Ortega P., and Faisal A. A., (Mar. 2021). *Inter-subject Deep Transfer Learning for Motor Imagery EEG Decoding*, ArXiv210305351 Cs, Accessed: May 29, 2021. [Online]. Available: http://arxiv.org/abs/2103.05351

[10] Walker I., *Deep Convolutional Neural Networks for Brain Computer Interface using Motor Imagery*, p. 75.

[11] Kaneshiro B., Perreau Guimaraes M., Kim H.-S., Norcia A. M., and Suppes P., (Aug. 2015). A Representational similarity analysis of the dynamics of object processing using single-trial EEG classification, *PLOS ONE*, volume 10, no. 8, p. e0135697, doi: 10.1371/journal.pone.0135697.

[12] Stober S., Sternin A., Owen A. M., and Grahn J. A., (2016). *Deep Feature Learning for EEG Recordings*, p. 7.

[13] Zander T. O. and Kothe C., (Apr. 2011). Towards passive brain–computer interfaces: applying brain–computer interface technology to human–machine systems in general, *J. Neural Engineering*, volume 8, no. 2, p. 025005, doi:10.1088/1741-2560/8/2/025005.

Innovation and Technological Advances for Sustainability – Al-Naemi et al. (Eds)

Web Content Management Systems (WCMS) vulnerabilities detection approaches: A comparative analysis

Shafi'i M. Abdulhamid, Jawza Alotaibi, Noora Alshamari & Tahani Abu Musa
Information Technology Department, Community College of Qatar, Doha, Qatar

ABSTRACT: Web applications are among the main targets of malicious attacks. Attacks are increasing intensely on websites, especially on Web Content Management Systems (WCMS), as the currently used web applications include a number of different security vulnerabilities that have been identified and discovered in recent times. The main problem in CMSs is the exploitation of weaknesses. This work was done on web CMS, where the focus was on WordPress, Drupal and Joomal systems. The work was done by examining the some sites for each tool, and demonstrated the vulnerabilities in each website. The Zmap recorded an accuracy of 95% in detecting the vulnerabilities. Other parameters of detection were used; total number of weaknesses, total number of open ports found and average treatment time. The information obtained from this study provide the necessary need to implement and widely disseminate updates on CMS to help stop these vulnerabilities.

1 INTRODUCTION

Content management system (CMS) often used with the WordPress, Drupal and Joomal blogging service. WCMS is built on the PHP and MySQL databases [1]. On the other hand, WCMS can also be set up on a web server so that it may be used to create a regular website or a network of websites. As of May 2019, the most popular website management system currently in use is WordPress, which is utilized by 33.6% of the top 10 million websites according to measurements taken by Alexa Internet [2]. Additionally, WCMS has been used in various other application fields, including ubiquitous display systems (PDS).

A comparison study of several online vulnerability detection programs will be carried out here in this research. Mainly, it will be concentrating on using WCMS. Millions of websites are run using WordPress, Drupal and Joomal content management system (CMS). Free software is distributed under the GNU General Public License [3]. WordPress is susceptible to various security flaws, the most common being cross-site scripting (also known as XSS), SQL injection, and remote code execution. Attackers have the potential to take advantage of these flaws to obtain access to sensitive data, deface websites, or even seize control of the server on which the websites are hosted [4]. Checking WordPress, Drupal and Joomal websites for security flaws may be done using one of a number of tools that are now available [5]. This article will examine and evaluate Zmap, Hackertarget and Sucuri, three of the most widely used WCMS security plugins currently available.

CMS websites are often attacked by cybercriminals who exploit vulnerabilities to take over the website or introduce malicious code [6]. Several WCMS security plugins have been built that can identify and defend against common security vulnerabilities in response to widespread attacks of this nature by evaluated each plugin through its steps based on criteria including features, accuracy, false positives, and false negatives. According to this research findings, even if each plugin does an excellent job of identifying common WCMS security

DOI: 10.1201/9781003496724-37

flaws, there are huge disparities between them in terms of their accuracy and the number of false positives they produce. This research led to the conclusion that WCMS users should choose a security plugin according to the trade-offs they are willing to make between the level of accuracy the plugin provides and the number of false positives it generates.

We outline the main contributions of this research paper as follows.

- To critically review the existing literature on web vulnerability detection tools.
- To identify the key features of web vulnerability detection tools.
- To evaluate the effectiveness of web vulnerability detection tools in detecting CMS vulnerabilities.
- To recommend improvements to web vulnerability detection tools and CMS.

The main aim of the research is to conduct a comparative analysis of CMS web vulnerability analysis tools. The remaining sections of the papers are structured as follows: Section II presents previous related works, Section III details the methodology of the research and Section IV presents the results obtained. Section V shows the experimental set ups and Section VI draws major conclusions.

2 RELATED WORKS

In this section previous related works were reviewed in the field of vulnerabilities in WCMS. Table 1 shows some related literatures and their results. There are many studies in this field that discussed about different WCMS weaknesses as shown in Table 1.

Table 1. Related works.

Reference	Method	WCMS tool	Result	Limitation
[7]	Security rules in dynamic web content managem ent systems applications	Mambo and vBulletin tools.	All security features have been implemente d on WCSM Content Management Systems.	Some features require a third party in order to apply them properly. There is no possibility to apply automatic update of these systems.
[8]	Security in Open Source Web Content Management Systems	TamperData and OWASP WebScarab	It gave good results in identifying and detecting attack methods.	Non-professionals cannot work on older versions of this system.
[9]	Access-Control Metamodel for Web Content Management Systems	Drupal system	The idea of meta data was successfuly presented and fully controlled by WCMS systems.	Access control overrides the permissions and permissions of the users, that is, there is no privacy for the users.
[10]	Web Vulnerability Scanner	SecuBat tool	It gave accurate results in the process of identifying security vulnerabilities.	Web developers are not aware of the security issue in web applications.
[11]	black-box web application n vulnerability testing	Rational AppScan (IBM), HailStorm Pro (Cenzic), WVS (Acunetix), McAfee Secure (McAfee), WebInspect (HP), QA Edition (N-Stalker), QualysGuard PCI (Qualys) and NeXpose (Rapid7) tools.	The scanners gave a high detection rate, which means the ability to control.	Finding vulnerabilities in scanners It was less than 50%.

Laitinen (2018)[7], puts forward that when talking about WCMS there are five features of information security. The first is confidentiality, meaning that information cannot be accessed except by authorized persons. The second is integration, that is, the inability to modify application data by unauthorized persons. The third is authentication is to ensure that the content of the information is correct and that there is no possibility of forgery in it. The fourth is availability, meaning that data and systems should be available only when authorized people need them. Fifth, there is an assurance that information can be transmitted between sender and receiver, Through the Five Attributes, plans are made for the security of WCMS Web Content Management Systems. In this study, two tools were used, Mambo and vBulletin, which helped in the implementation of the study.

The authors emphasize [8] that WCMS is open source and therefore excellent targets for hackers because of the code available to them, which makes them search for security holes and exploit them. By working on these vulnerabilities, they can access confidential and important information, carry out phishing operations, change code and others, through which the authors stated that web applications operate hostilely. Hackers can use different methods to carry out attacks, by setting plans and the possibilities they expect. There are several stages to carry out the attack, starting from studying the target and exploring vulnerabilities to exploitation.

The researchers adopted [9], In their study to adopt a method for access control (AC) within (WCMS.Writers) This method aims to maintain the confidentiality and integrity of information. In other words, access control manages content, procedures, and permissions. This method works by building an integrated approach, where this method or approach is based on a complete analysis of the access control information, without regard to the security features of the WCMS Web Content Management System. The idea of access control was applied to the Drupal system, which provided high results in controlling and reducing risks.

The researchers [10], in this method demonstrated the ease in the process of capturing and exploiting vulnerabilities Calles and others. Website vulnerabilities appear due to validation processes such as cross-site scripting and SQL injection. There are two ways to check for vulnerabilities, the first is by accessing the source code to check for security vulnerabilities. The second is through the ability to access the source code directly, in order to enter specific instructions to find security holes. The code is searched using the SecuBat tool and works by scanning websites for SQL injections that can be used as a weakness in cross-site scripting (XSS). The results showed that many problems were found that could be exploited by attackers.

The researchers [11], in this study aim to identify web application scanners in order to know the vulnerabilities that these scanners test, and their success rate in the test. The scanners used by the researchers in this study were Rational AppScan (IBM), HailStorm Pro (Cenzic), WVS (Acunetix), McAfee Secure (McAfee), WebInspect (HP), QA Edition (N-Stalker), QualysGuard PCI (Qualys) and NeXpose (Rapid7). These tools are used to compare identified security vulnerabilities. Through the results, it was revealed that there are security holes such as SQL injection, cross-site scripting, and intersection forms.

2.1 *Review of WCMS vulnerability points*

This section contains a breaf review of some of the WCMS attacks and vulnerabilities.

2.1.1 *Brute force attack*

This type of attack is used by trying to enter the names of people using this site or trying passwords or using many combinations of them and they try it a lot until a specific formula is found that secures them with successful entry. As per the design of WCMS sites, it does not prevent the possibility of repeated login attempts to the sites. Therefore, the hacker can set up customized software that attempts to log in to websites and initiate the attack using brute force. This attack is more successful if the passwords are weak, and as a result, the attacker will get the login, which means he gets the data. It can easily get login pages because most WCMS sites offer login pages [14].

2.1.2 *SQL injection*

SQL injection is performed if input fields are not sanitized with added values in an optimal and correct manner which means that SQL queries are allowed to be applied to them.

When SQL queries are executed correctly and successfully by the attacker, then he can gain many advantages such as creating a new account through which he can access and control the entire site. The attacker can also use SQL injection to control the site's database by entering new data, modifying the data stored in the database, or deleting data [14].

2.1.3 *Cross-site scripting*

This type of attack is carried out by the client by inserting malicious scripts, which leads to downloading or re-sending them to other dangerous sites.

2.1.4 *DDoS attacks*

This type of attack boils down to sending too many requests to the web server causing it to be out of service and unavailable to users.

2.1.5 *Exploit embed file*

When the codec is bad, weaknesses appear in the included files. This vulnerability becomes effective when the site allows the user to upload files to the site or add data entries, then the PHP code cannot verify the correctness of the operations performed by the user.

2.1.6 *Guide traversal*

It is considered an HTTP attack and the hacker can access directories, files, and commands that are outside the server's home directory

After a critical literature review in this section, the following were discovered:

- clear ideas were formed about the research and its contents, where the concept of Web Content Management Systems WCMS was identified, and the weaknesses that could be exposed to content management systems were presented.
- In addition to that, the tools that help in the process of scanning and discovering weaknesses were presented. and security holes.
- Then made a literature review of previous studies and research in this field, which greatly helps in a deeper understanding of this research through a review of the most important results obtained in this field.
- The research method was presented, which is through several steps to collect information about the sites that are being examined and give results about the security holes that were found.

3 RESEARCH METHODOLOGY

In this section, we describe the research phases that have been followed to implement the research procedure. Figure 1 shows the research process framework.

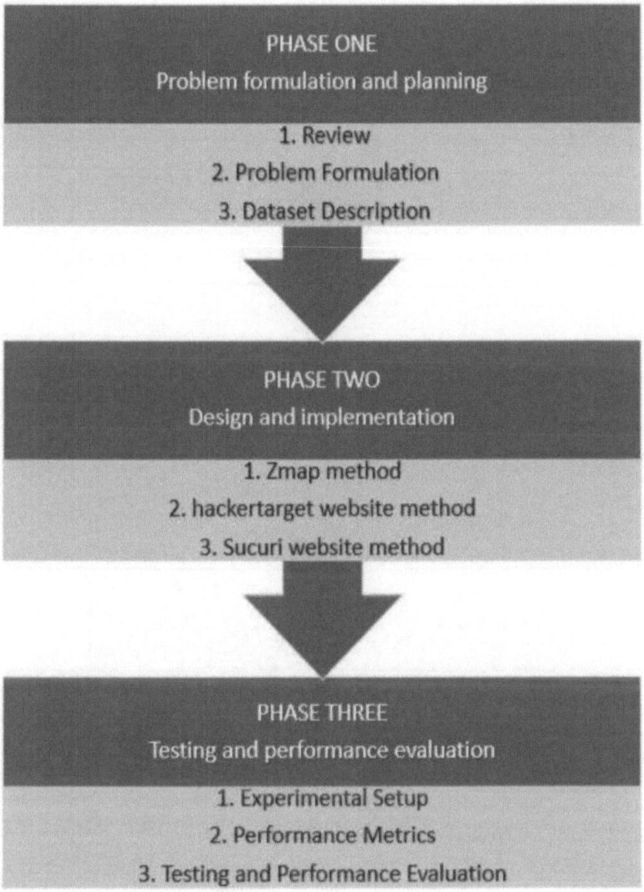

Figure 1. Research process framework.

3.1 *Problem formulation*

In order to detect and identify vulnerabilities use the tools Zmap, hackertarget and Sucuri website. These tools greatly help in detecting weaknesses, as the results obtained between these tools will be compared to see which tool has the best performance.

3.2 *Dataset description*

This comparative analysis will be done by examining three websites for each of the three tools.

Table 2. Dataset of web scan.

No.	Website	URL	WCMS
1	thegreenbayet	http://thegreenbayet.flywheelsites.com/	WordPress
2	programiz	http://programiz.com	Drupal
3	mindphp	http://mindphp.com	Joomla

3.3 *Design and implementation*

The proposed comparative analysis was implemented in the main steps as shown in the flowchart presented in Figure 2.

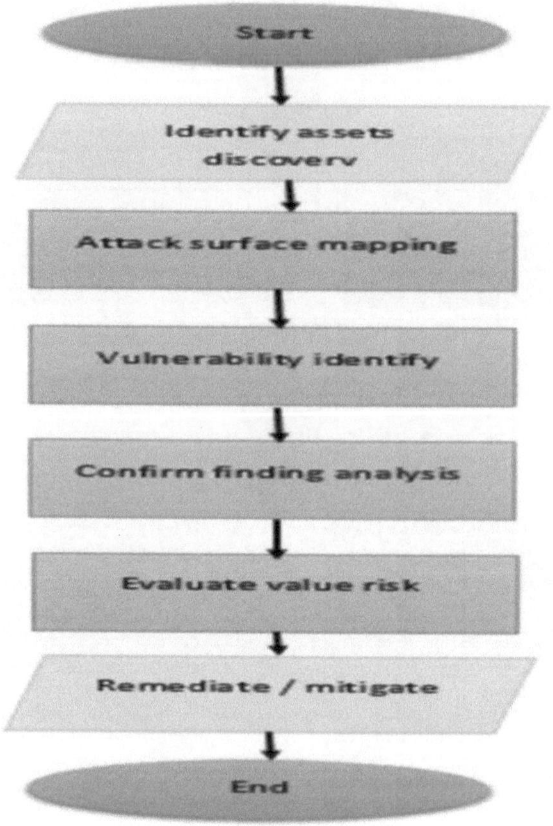

Figure 2. The assessment cycle.

3.3.1 *Zmap method*

Zmap was designed by a team of researchers at the University of Michigan in 2013. In order to help scan the entire Internet, the goal is to secure the information, as it scans a large group of addresses instead of a single address. The expected time for the scanning process does not exceed 45 seconds. It helps greatly in checking websites because of its speed which gives more possibilities in order to detect weaknesses.

Zmap increases its speed by using multiplication cycles. ZMap can scan the same IPv4 address space about 1,300 times faster than Nmap. ZMap works by an iterative syntax that ensures that each of the 32-bit IP address numbers is visited once in a pseudo-random order. The Figure 3 shows the working principle of the Zmap tool in the scanning process [12].

3.3.2 *Hackertarget method*

It is considered one of the most important sites for WCMS security scan online to detect weaknesses in WCMS settings. The checks contain security checks for the application, WCMS plugins, hosting environment, and web server as shown in *Algorithm 1.*

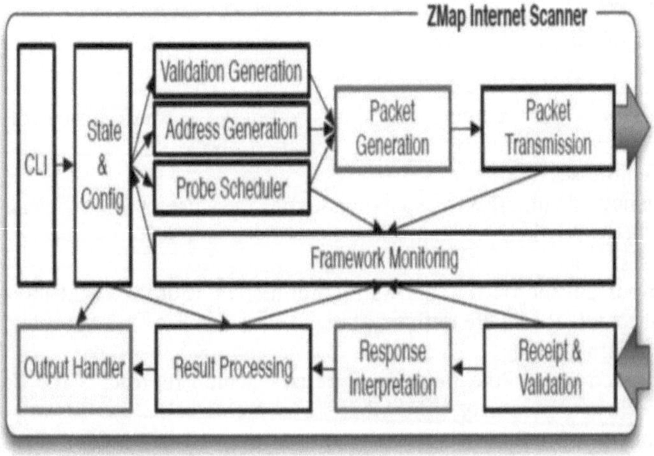

Figure 3. Zmap internet scanner.

Algorithm 1. Hackertarget

---

1. Start
2. Work to reveal the version of WCMS Core.
3. Find plugins in the HTML response.
4. Determine the topic in use.
5. Trying to detect the first 2 WCMS users.
6. List of page resources including js & iframes.
7. Activate the directory indexing test on the main sites.
8. Check Google Safe Browsing for popularity
9. End

3.3.3 *Sucuri method*

There are many tools available for WordPress Security Scanner and one of the most important of these is the use of Sucuri, which is one of the most prominent sites that provide services for anti-malware that poses a threat to WordPress sites, and at the moment it has become a part of Godaddy. It offers many features that help in identifying and detecting malware, general website errors, outdated software, or blacklisting [13].

4 EXPERIMENTAL SETUP

By working with the tools specified in this study, to see how efficient these tools are in giving accurate results on the sites that have been examined. After worked with Windows 10 and Linux, so a computer with no less than the following minimum specifications is needed: Processor: The type of processor is either Intel (Core i3 or i5) or AMD processor of the third generation.

Main memory (RAM): It must be at least 4 GB Systems: 64-bit operating system.

4.1 *Performance metrics*

The performance of the proposed system was evaluated using performance measures:

(i) Total number of weaknesses
In order to know how secure a site is, it is to know the total number of vulnerabilities that have been discovered by the tools specified above.

(ii) Repeat weaknesses
The frequency of vulnerabilities is an essential measure of the efficiency of the remediation process for these vulnerabilities.

(iii) IP addresses
It is an important metric that shows the number of addresses that have been checked through the tools that were defined above.

(iv) Average treatment time
It is considered one of the important metrics, which represents the time required in general to close the security vulnerability.

(v) Number of open ports
It shows the number of open ports on the device that could pose a risk to the device.

5 RESULTS AND DISCUSSION

After working through the methodology, in which the working method of each tool was explained, now in this section we will show the details obtained and compare the results of the three tools with each other to see which tool has the highest results in detecting and identifying weaknesses in web content management systems.

5.1 *Results obtained from Zmap tool*

The results obtained from the Zmap was presented in Figure 4. It shows the accuracy of detection, average treatment time, IP addresses extracted and total number of open ports discovered.

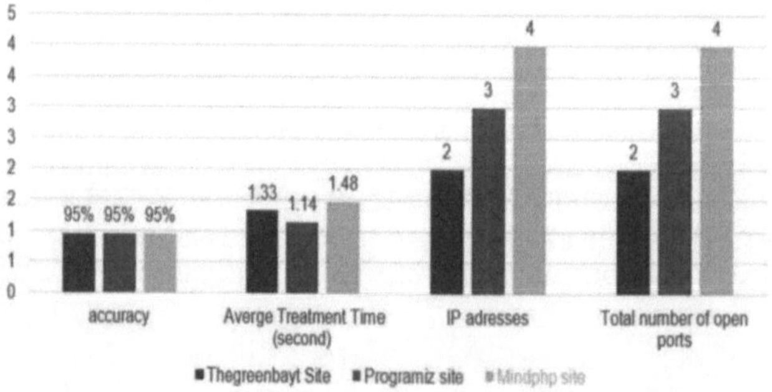

Figure 4. Results of Zmap.

5.2 *Thegreenbayet site check results*

The template has been found by examining the path /templates/ *template name* /. The results obtained by examining the three sites were extracted by Hackertarget and presented in Figure 5.

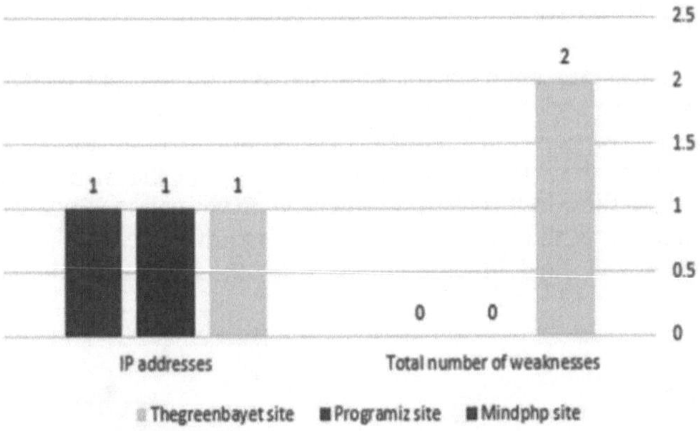

Figure 5. Results of hackertarget site.

5.3 *Sucuri site check results*

The results obtained by examining the three sites were extracted by Sucuri site and presented in Figure 6.

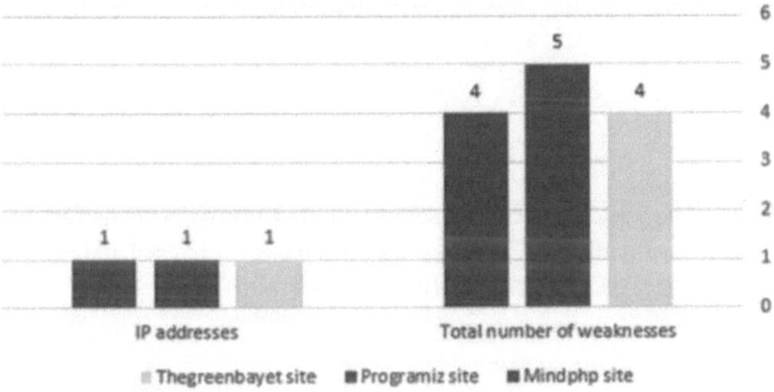

Figure 6. Result of sucuri site.

5.4 *Comparative results*

After completing the application and showing the results of the three tools, the comparison results for Zmap and Hackertarget and Sucuri will be presented in Figure 7.

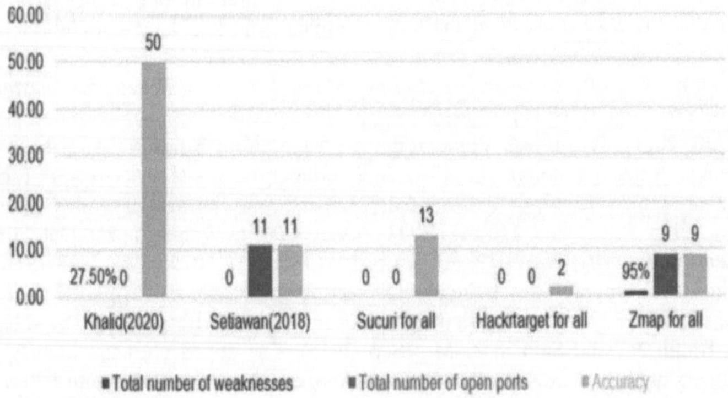

Figure 7. Camparative results.

6 RECOMMENDATIONS

On a daily basis, hacks are obtained for sites that rely on WCMS content management systems. This study gave a clear idea of how the tools and sites that identify and discover weaknesses work, as specialists in this field develop and update the necessary means to combat these intrusions through continuous research. About the ways and means used by hackers to access these sites. In this research will keep up with all the updates that take place in this field, which increases the scientific and academic knowledge, and helps in the future to continue studying in this field in order to provide the services to the community and the world.

7 CONCLUSION

Internet sites have recently spread greatly, so there is no company or institution, whether governmental or private, small or large, but it has become a website to carry out its work. This great development was accompanied by the emergence and great risks on these sites by hackers, who are working to access these sites in illegal ways. This type of hack is not a static type and there is no specific way to discover it, so it works by searching for weaknesses in its CMS platforms such as WordPress, Drupal, Joomal and other platforms. Therefore, these platforms must be constantly updated to gain the trust of their customers and address the weaknesses that may affect them in the future.

ACKNOWLEDGMENT

Special appreciation also goes to Prof. Dr. Ahmad Al Omari, Acting Dean of Sciences and Technology Division, Community College of Qatar (CCQ) for all his support and encouragements and also to the CCQ for supporting this research.

REFERENCES

[1] Baran K., Kopniak P., and Kozieł G., (2021). Using the WORDPRESS system for the implementation of student projects during the course: Fundamentals and technologies of the electronic economy, *INTED2021 Proceedings.*

[2] Maraga A., Awuor F., and Ogalo J., *Journal of Internet and Information Systems Model for Security Controls in Web Content Management System*, vol. 11, no. 1, pp. 1–12, 2022, doi: 10.5897/JIIS2021.0120.

[3] Walsh S., (Jun. 29, 2022). CMS market share trends: Top content management systems in 2022, *Search Engine Journal.* https://www.searchenginejournal.com/cms-market-share/454039/#close (accessed Jan. 25, 2023).

[4] Shah M. P., (Jan. 08, 2020). *Comparative Analysis of the Automated Penetration Testing Tools*, norma.ncirl.ie. https://norma.ncirl.ie/id/eprint/4165

[5] Billie, (Oct. 20, 2021). WordPress Vulnerabilities | 5 Common Security Issues to Avoid, Passionate Design Agency. https://passionates.com/wordpress-vulnerabilities-security-issues/ (accessed Jan. 25, 2023).

[6] Nguyen V.-L., Lin P.-C., and Hwang R.-H., (May 2019). Web attacks: Defeating monetisation attempts, *Network Security*, vol. 2019, no. 5, pp. 11–19, doi: 10.1016/s1353-4858(19)30061-3.

[7] Laitinen P., (2018). Vulnerabilities in the wild: Detecting vulnerable web applications at scale, jyx.jyu.fi, Accessed: Jan. 25, 2023. [Online]. Available: http://urn.fi/URN:NBN:fi:jyu-201804242355

[8] Seng L. K., Ithnin N., and Said S. Z. M., (Sep. 2018). The approaches to quantify web application security scanners quality: A review, *International Journal of Advanced Computer Research*, vol. 8, no. 38, pp. 285–312, doi: 10.19101/ijacr.2018.838012.

[9] Mackey T. K., Miyachi K., Fung D., Qian S., and Short J., (Sep. 2020). Combating health care fraud and abuse: Conceptualization and prototyping study of a blockchain antifraud framework, *Journal of Medical Internet Research*, vol. 22, no. 9, p. e18623, doi: 10.2196/18623.
[10] Khalid M. N., Iqbal M., Rasheed K., and Abid M. M., (Aug. 2020). web vulnerability finder (WVF): Automated black-box web vulnerability scanner, *International Journal of Information Technology and Computer Science*, vol. 12, no. 4, pp. 38–46, doi: 10.5815/ijitcs.2020.04.05.
[11] Anagandula K. and Zavarsky P., (2020). An analysis of effectiveness of black-box web application scanners in detection of stored SQL injection and stored XSS vulnerabilities, *2020 3rd International Conference on Data Intelligence and Security (ICDIS)*, South Padre Island, TX, USA, pp. 40–48, doi: 10.1109/ICDIS50059.2020.00012.
[12] otw, "*Zmap for Scanning the Internet: Scan the Entire Internet in 45 minutes*," hackers-arise, Oct. 06, 2021. https://www.hackers-arise.com/post/zmap-for-scanning-the-internet-scan-the-entire-internet-in-45-minutes (accessed Jan. 25, 2023).
[13] Borges E., *Top 5 WordPress Vulnerability Scanners*. Retrieved from securitytrails: https://securitytrails.com/blog/top-5-wordpress-vulnerability-scanners. 2018. (accessed Jan. 25, 2023).
[14] Nair S. V., *CMS Vulnerabilities: Why are CMS Platforms Common Hacking Targets?*, https://beagle-security.com/blog/article/cms-vulnerabilities.html (accessed Jan. 25, 2023).

Innovation and Technological Advances for Sustainability – Al-Naemi et al. (Eds)

Joint power allocation and splitting ratio for SWIPT-enabled UAV assisted time-shared cooperative NOMA network

Saniya Tariq
Department of Electrical Engineering Institute of Space Technology, Islamabad, Pakistan

Umber Saleem
Department of Electrical Engineering, Tsinghua University, Beijing, China

Sobia Jangsher, Ashfaq Ahmed & Arafat Al-Dweik
Department of Electrical Engineering, Khalifa University, Abu Dhabi, UAE

ABSTRACT: Unmanned aerial vehicles (UAVs) can support higher data rates and enhance network coverage by leveraging air-to-ground (A2G) line-of-sight (LOS) channels, which makes them ideal to act as relays in wireless communications networks. However, the limited battery constraint and severe interference are the fundamental challenges that may degrade the energy efficiency (EE) of UAV-aided communications. Therefore, this work considers UAV-assisted cooperative communications networks with multiple users in a shared spectrum. Taking into account the aforementioned challenges, we propose radio frequency (RF) energy harvesting at the UAV and adopt non-orthogonal multiple access (NOMA) scheme with simultaneous wireless information and power transfer (SWIPT) protocol to improve the performance of the considered network. In particular, NOMA allows spectral reuse and separates the users in the power domain while mitigating interference through successive interference cancellation (SIC). With the aim to maximize the total EE of the network, we formulate joint power allocation and power splitting ratio control while considering user pairing and quality of service (QoS) constraints in time-shared full duplex mode. The resulting optimization is a linear fractional programming problem that is non-convex. We relax the problem and propose a low-complexity algorithm based on subgradient to obtain the near-optimal solution. The obtained results show that the proposed algorithm may improve the EE of the system by more than 300% for several cases of interest.

Keywords: Unmanned aerial vehicle (UAV), non-orthogonal multiple access (NOMA), energy efficiency (EE), simultaneous wireless information and power transfer (SWIPT), power splitting (PS)

1 INTRODUCTION

Unmanned aerial vehicles (UAVs) assisted communications have gained significant attraction to enable massive connectivity required by the internet of things (IoT) paradigm, due to its potential to support ubiquitous coverage, high throughput and reliability [1–4]. Additionally, the instant deployment, cost-effectiveness, and agility of UAVs provide an edge over conventional terrestrial systems [5]. To enhance the performance of existing communications networks, integrating UAVs into cellular networks as flying base stations or relays has been recently considered [6]. Especially, the high altitude and line-of-sight (LoS) communications channels of UAVs enable long-range communications by improving the coverage and quality of service (QoS) for non-line-of-sight (NLoS) users. Although

 DOI: 10.1201/9781003496724-38

UAVs LoS links suffer from lower path-loss, shadowing, and fading, the resulting interference can degrade the communications performance of UAVs in downlink and ground users in uplink [7]. Meanwhile, the energy limitation due to finite battery capacity can also hinder the perpetual operation of UAV-assisted communications networks.

For the integration of UAVs in the fifth generation (5G) wireless networks, it is imperative to overcome the aforementioned challenges. In this context, multiple access schemes and renewable energy can play a vital role. In order to overcome the interference and improve the quality of communications in UAV-aided networks, non-orthogonal multiple access (NOMA) has been widely considered in the literature as a multiple access scheme. In NOMA, the users are paired according to their channel conditions or some other criterion [8], where the far user is assumed to have poor channel conditions and is paired with the near user which is assumed to be the strong user with good channel conditions [9]. The weak user in NOMA is assigned a higher level of power whereas, the strong user is assigned a relatively lower power level, and this difference in power level is used in order to perform successive interference cancellation (SIC) at the receiver side. SIC is the most important block of NOMA transmission for the mitigation of interference caused by sharing the same resource block by multiple users. The strong user after receiving a signal from the base station (BS) performs SIC i.e, it detects the high-power signal and subtracts it from the received signal, and it is left with its own signal. [10,11].

However, NOMA along with its numerous advantages also has some drawbacks, as it favors the users with poor channel conditions by assigning them almost all the power available for the transmission. On the other hand, users with good channel conditions are left with no power which may lead to QoS degradation of individual users. Keeping that in view, cooperative NOMA (C-NOMA) was introduced to overcome the challenges faced by NOMA in 5G [12]. In particular, C-NOMA utilizes cooperative relaying to ensure fairness among users and meanwhile enhances reception reliability. In C-NOMA systems, the relay is used to assist the far user in absence of a direct link from BS to the far user. The relay can either be an in-band relay (where a near user acts as a relay) or an out-band relay (a dedicated relay) [13]. Nevertheless, a fundamental challenge in C-NOMA systems in 5G is the energy-constrained relay which fails to support the perpetual operation of the network. In this regard simultaneous wireless information and power transfer (SWIPT) protocol is introduced [14–16] which enables energy harvesting at the relay and, meanwhile decodes the information using the power received from RF signal, as shown in Figure 1. Hence, the cooperative relay employees SWIPT to harvest the energy and consequently the lifetime of relay increases. The capability of relay to transmit information is completely powered up by the use of this protocol resulting in improved energy efficiency (EE) which is also the most essential requirement of 5G wireless communications networks [17].

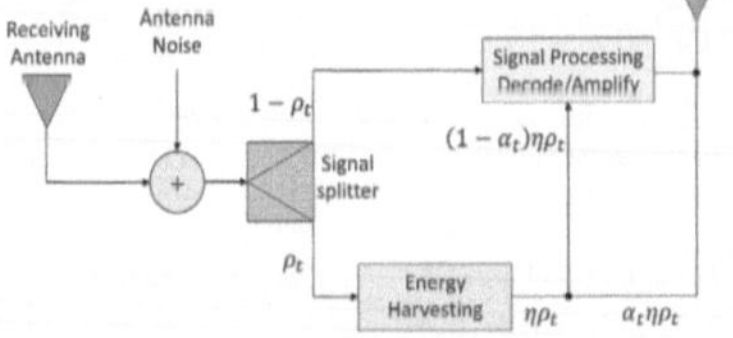

Figure 1. PS architecture for the SWIPT enabled UAV relaying.

In this context, numerous works investigated EE maximization for SWIPT-based C-NOMA systems. In [18], a C-NOMA based design was proposed using SWIPT to achieve high EE. The formulated problem jointly optimizes using beamforming vectors and power splitting (PS) ratio. The numerical results proved that the introduced scheme enhanced the EE of the system. In [19], the author also considered the SWIPT protocol for the strong user which assisted the weak user by harvesting energy. The impact of SWIPT on the outage performance of strong and weak users was studied. Also in [20], the energy-efficient C-NOMA model based on SWIPT is investigated where the power consumption of the system is a constraint. In order to get the maximum network energy efficiency, a three-stage strategy was proposed and results showed that using SWIPT with C-NOMA works far better than conventional NOMA and orthogonal multiple access (OMA) schemes.

In most of the existing works on C-NOMA and SWIPT, static relays for the transfer of information to the far users have been considered [18,21]. UAV combined with C-NOMA can provide us with further increase of spectral efficiency (SE), high reliability and increased transmission rates [22]. Since UAVs also suffer from severe uplink/downlink interference because of the LoS links, using C-NOMA scheme with UAV-based communications can inherently solve this extreme interference of UAV networks [7,23,24]. In this paper, we consider a downlink communications scenario where a UAV serves only the far users while the near user is served by the BS directly, which is not considered in the existing literature. UAV acts as a relay using full-duplex (FD) mode with SWIPT protocol. More precisely, we take into account the problem of far users which were being left unpaired due to the limited number of near users. We consider the UAV to use decode and forward (DF) protocol [25].

The main contributions of this paper are summarized as:

1) We propose a UAV-assisted C-NOMA framework to serve the multiple unpaired weak users by pairing them with a single strong user based on time sharing. Meanwhile, we employ the SWIPT protocol to enhance the EE of the UAV relay and ensure QoS.
2) To maximize the EE of the proposed framework, we jointly formulate power allocation to the users and power splitting at the UAV while considering QoS constraints of the users and mutual interference due to the pairing of near-to-far users in the shared spectrum.
3) Being a linear fractional programming problem, our formulation is nonconvex. We propose a low-complexity iterative approach by employing CVX tool where a near-optimal solution is obtained. Our proposed approach outperforms the other baseline schemes with ground relays.

This paper is organized as follows: Section 2 describes the system model. Section 3 discusses the problem formulation along with the proposed algorithm for joint optimization of EE maximization. Section 4 discusses the numerical results, and finally, the work is concluded in Section 5.

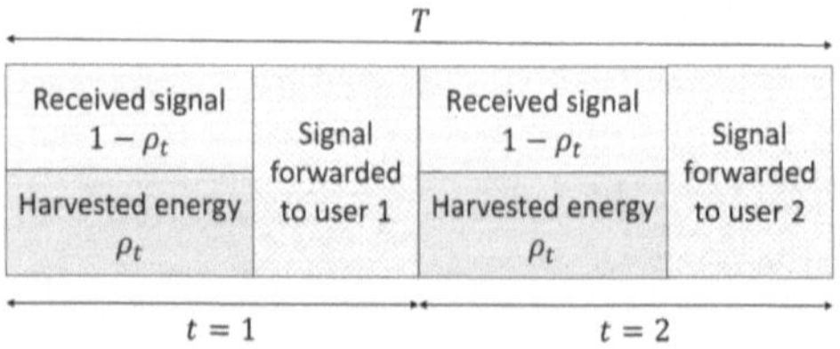

Figure 2. Block diagram of SWIPT PS protocol.

2 SYSTEM MODEL AND PROBLEM FORMULATION

We consider the downlink transmission scenario depicted in Figure 3, with user A as the near user and users B and C as the far users.

It is assumed that because of large scale fading and shadowing, direct communications between the BS and the far users is not possible. Consequently, the BS communicates directly with user A, while a UAV assists the BS in establishing communications with far users, where the UAV functions as an FD relay. We further assume that the BS has a single transmitting antenna and that all other nodes have single receiving antennas. The UAV has one receiving antenna and one transmitting antenna. Transmission is performed on a time-sharing basis, i.e., the duration of a single time-slot is divided into 2 sub-slots, as depicted in Figure 2. In time-slot 1, the BS broadcasts the NOMA signal for user A and user B, which is concurrently received by user A and the relaying UAV. Similarly, during time-slot 2, the BS transmits a fresh NOMA signal for user A and user C, which is again received by the relaying UAV and user A. In the second time slot, the relaying UAV receives the NOMA signal for user C and simultaneously decodes and forwards the signal for user B, which was received in the first time slot. Consequently, user A and user B are paired in the first time slot, and user A and user C are paired in the second time slot. This work considers a model of a three-dimensional Cartesian coordinate system, where BS is located at $(-d_{BS}, 0, 0)$, user A is positioned at $(-d_A, 0, 0)$, and users B and C are located at $(-d_B, 0, 0)$ and $(-d_C, 0, 0)$, respectively. The UAV is supposed to follow a circular path with a radius of r at a fixed altitude of H. The UAV's center location is specified as $(0, 0, H)$. The azimuth angle of a UAV flying in a circular trajectory is $0 \leq \theta \leq 180$, and the UAV's flight trajectory is $(r \cos \theta, r \sin \theta, H)$.

Table 1. Literature review.

Ref.	Objective	Power alloc.	QoS	Multi-User	SWIPT	FD	Time-Sharing	UAV Placement
[6]	**max EE**	√	√		√	√		
[7]	**max sum rate**	√	√	√				
[14]	**max sum rate**	√	√	√	√			
[15]	**QoS fairness**		√		√	√		
[17]	**max EE**		√		√	√		
[18]	**max EE**	√	√		√	√		
[26]	**max sum rate**			√		√	√	
[23]	**UAV position Optimization**			√		√		√
This work	**max EE**	√	√	√	√	√	√	√

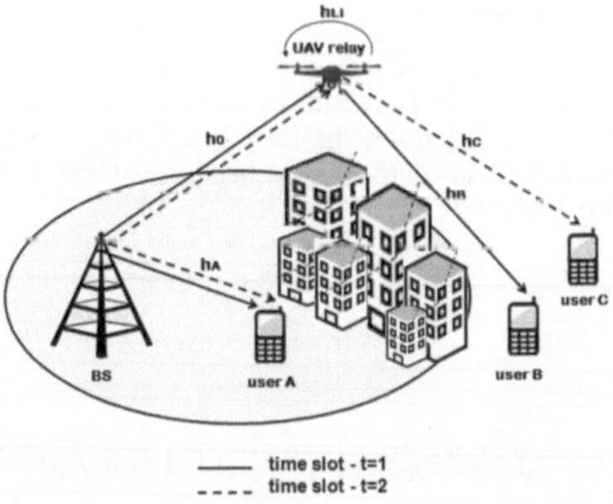

Figure 3. System model of UAV aided downlink transmission in C-NOMA network.

2.1 *Channel model*

Channels from BS-user A, BS-UAV, UAV-user B and UAV-user C are denoted by channel coefficients as h_A, h_0, h_B and h_C. The ground-to-ground channel that is the channel from

BS-user A is modeled as,

$$h_A = g_A d_A^{-\frac{a}{2}}, \tag{1}$$

where $g_A \sim CN(0,I)$ represents the normalized small scale fading coefficient, d_A is the distance from BS-user A and a is the path-loss exponent.

The ground-to-air channel between BS to UAV and the air-to-ground channel between the UAV and the far users consists of two components LoS and NLoS [27]. In general, the LoS component is dominant in UAV communications. We can model the channels from BS to UAV and UAV to far users as,

$$h_j = \frac{\beta \bar{h}_j}{d_j^{\tau_j}}, \; j \in \{B, C, 0\}, \tag{2}$$

where j denotes the far users or the UAV and β denotes the channel power gain at a distance of 1m. The channel coefficients $\bar{h}_0, \bar{h}_B$ and $\bar{h}_C$ are calculated as,

$$\bar{h}_j = \sqrt{\frac{K_j}{K_j + 1}} \bar{h}_{j,LoS} + \sqrt{\frac{1}{K_j + 1}} \bar{h}_{j,NLoS}, \tag{3}$$

where $\bar{h}_{j,LoS}$ denotes the LoS component of the channel, $\bar{h}_{j,NLoS}$ denotes fading component. The Rayleigh fading channel components are modeled as independent circularly symmetric complex Gaussian (CSCG) variables, where $\bar{h}_j^{NLoS} \sim CN(0, 1)$. Similarly, the Rician factor is given as $K_j = l_1 e^{\mu_1 w_j}$, where l_1, μ_1 and w_j denote the constant coefficients and flight angle of UAV. Here $\bar{w}_j = \arcsin \frac{H}{d_j}$, where d_j is the distance between BS-UAV and UAV-far users. The distances are calculated as,

$$d_0 = \sqrt{H^2 + (d_0 + r\cos\theta)^2}, \tag{4}$$

$$d_B = \sqrt{H^2 + (d_B - r\cos\theta)^2}, \tag{5}$$

$$d_C = \sqrt{H^2 + (d_C - r\cos\theta)^2}, \tag{6}$$

where H denotes the altitude of the UAV, which is assumed to be fixed in this work. The path-loss exponent between the BS-UAV, UAV-user B, and UAV-user C is represented as τ_j which can be written as $\tau_j = l_2 P_{LoS}(w_j + l_3)$, with LoS probability given as $P_{LoS}(w_j) = (1 + l_4 e^{-\mu_2 w_j})^{-1}$, where l_2, l_3, l_4 and μ_2 are the environment constants. In addition, $h_{LI} \sim CN(0,\theta_{LI})$ in Figure 3 denotes the loop interference channel between the transmitting and the receiving antennas of the UAV because of the FD mode. We assume a perfect Channel State Information (CSI) at the receiver side.

2.2 *Signal model*

The superimposed signal transmitted from the BS to users A and B or user A and user C is given as,

$$x^t = \sqrt{w_A^t P}\, x_A^t + \sqrt{w_j^t P}\, x_j^t, \; j \in \{J\}, j \neq A, \tag{7}$$

where $x_A^{(t)}$ and $x_j^{(t)}$ are message signals for near and far users, respectively, and $w_A^{(t)}$ and $w_j^{(t)}$ are power allocation coefficients for corresponding users during time-slots $t \in T = \{1, 2\}$. The power allocation constraint is expressed as $(w_A^{(t)} + w_j^{(t)})P \leq P$, where P represents the power budget at the BS.

Table 2. Time sharing Cooperative NOMA.

Time	Events
$t = 1$	(i) BS sends NOMA signal for user A and user B. (ii) User A employs SIC. (iii) UAV harvests energy from RF signal. (iv) UAV decodes user B signal. (v) UAV forwards decoded signal to user B.
$t = 2$	(i) BS sends NOMA signal for user A and user C. (ii) User A employs SIC. (iii) UAV harvests energy from RF signal. (iv) UAV decodes user C signal. (v) UAV forwards decoded signal to user C.

In this work, we assume an overall strong signal to noise ratio (SNR) between the communicating nodes. The harvested energy at the UAV during time slots 1 and 2 is written as,

$$P_0^t = \zeta(1 - \alpha^{(t)})\left(\sum_{T \in t}\sum_{j \in J} |h_0|^2 w_j^{(t)} P\right), \; t \in T \tag{8}$$

where $\zeta \in [0, 1]$ denotes the energy conversion efficiency at the UAV. Since the UAV is in FD mode, it can receive $x^{(t)}$ from BS and transmit the decoded $\tilde{x}_j^{(t)}$ to the user $j \in \{B, C\}$ in the same time-slot. Thereby, the signal received by the UAV is then given as,

$$y_0^{(t)} = h_0 x^{(t)} + \sqrt{w_0^{(t)} P_0^{(t)}}\, h_{LI} \tilde{x}_j^{(t)} + \eta_0^{(t)} \tag{9}$$

where $\tilde{x}_j^{(t)}$ is the decoded signal intended for user B or user C after performing SIC at the UAV. η_0 represents the additive white Gaussian noise (AWGN) at the UAV. We have considered PS architecture, with the splitting ratio $\alpha^{(t)} \in (0, 1)$. After self-interference cancellation, the signal at the UAV is given as,

$$\tilde{y}_0^{(t)} = \sqrt{\alpha^{(t)}}(h_0 x^{(t)} + \sqrt{w_0^{(t)} P_0^{(t)}}\, \tilde{h}_{LI} \tilde{x}_j^{(t)} + \eta_0^{(t)}) + z_0^{(t)}, \tag{10}$$

where $z_0 \sim CN(0, \delta_A^{(t)2})$ denotes noise caused by information decoding and $\tilde{h}_{LI} = \sqrt{\rho} h_{LI}$, where $\rho \in [0, 1]$ denotes the self residual interference level.

The SINR of user B or user C at UAV is given as,

$$\gamma_{0j}^{(t)} = \frac{\alpha^{(t)} |h_0|^2 w_j^{(t)} P}{\alpha^{(t)}(|h_0|^2 w_A^{(t)} P + \rho |h_{LI}|^2 w_0^{(t)} P_0^{(t)} + \upsilon_0^{(t)2}) + \delta_0^{(t)2}} \tag{11}$$

The signal received by user A is as follows:

$$y_A^{(t)} = h_A x^{(t)} + \eta_A^{(t)}, \tag{12}$$

where η_A represents the AWGN at user A. The observed SINR of user B or user C at user A can be written as,

$$\gamma_{Aj}^{(t)} = \frac{|h_A|^2 w_j^{(t)} P}{|h_A|^2 w_A^{(t)} P + \sigma_A^{(t)2}}. \tag{13}$$

Furthermore, the SNR at user A is given as,

$$\gamma_{AA}^{(t)} = \frac{|h_A|^2 w_A^{(t)} P}{\sigma_A^{(t)2}} \tag{14}$$

The signal received at user B or user C is denoted as,

$$y_j^{(t)} = h_j \sqrt{w_0^{(t)} P_0^{(t)}} \widetilde{x}_j^{(t)} + \eta_j^{(t)},\ j \in \{B, C\}. \tag{15}$$

Similarly, the SNR at user B or user C can be written as,

$$\gamma_{jj}^{(t)} = \frac{|h_j|^2 w_0^{(t)} P_0^{(t)}}{\sigma_j^{(t)2}} \tag{16}$$

The rate achieved by user A is given as,

$$R_A^{(t)} = \log(1 + \gamma_{AA}^{(t)}) \tag{17}$$

Likewise, the rate achieved by users B and C is given as,

$$R_j^{(t)} = \min\{\log(1 + \gamma_{Aj}^{(t)}),\ \log(1 + \gamma_{jj}^{(t)}),\ \log(1 + \gamma_{0j}^{(t)})\} \tag{18}$$

3 PROBLEM FORMULATION AND PROPOSED JOINT OPTIMIZATION OF ENERGY EFFICIENCY MAXIMIZATION

The EE is defined as the ratio of the sum data rate to the total power consumed in the system [28]. The EE is denoted as η_{EE}, and is mathematically expressed as,

$$\eta_{EE} = \frac{\text{Sum data rate}}{\text{Total consumed power}} = \frac{R_S}{P_S} \tag{19}$$

3.1 *Sum data rate*

The sum data rate of the system under consideration is expressed as,

$$R_S = \sum_{t \in T} \sum_{j \in J} R_j^{(t)},\ \ J \in \{A, B, C\} \tag{20}$$

3.2 *Power consumption*

The UAV harvests energy from the RF signal received from the BS. This harvesting eventually improves the EE of the system. The power consumption of the whole system is expressed as,

$$R_S = \sum_{t \in T} \sum_{j \in J} \left(w_j^{(t)} P + w_0^{(t)} P_0^{(t)} - P_0^{(t)} \right) + P_c,\ T = \{1, 2\},\ \ J = \{A, B, C\} \tag{21}$$

where P_c denotes the circuit power and $P_0^{(t)}$ represents the energy harvested at the UAV during time-slot $T = \{1, 2\}$.

3.3 *EE maximization*

The objective of this work is to find the optimal PS ratio and power allocation coefficients to maximize EE while taking the users' QoS requirements into account. The problem also considers the UAV and BS power constraints. The system model is expressed mathematically as,

$$\begin{aligned}
\textbf{OP1}: \quad & \max_{\substack{w_A^{(t)}, w_j^{(t)} \\ \alpha^{(t)}, w_0^{(t)}}} \quad \eta_{EE} \\
& \text{such that} \\
& C1: \; R_j^{(t)} \geq \bar{R}_j^{(t)}, \; \forall t \in T, j \in \{A, B, C\} \\
& C2: \; \sum_{t \in T} \sum_{j \in J} w_j^{(t)} P \leq P \\
& C3: \; w_0^{(t)} P_0^{(t)} \leq P_0^{(t)}, \; \forall t \in T \\
& C4: \; 0 \leq \alpha^{(t)} \leq 1, \forall t \in T.
\end{aligned} \tag{22}$$

where $\bar{R}_j^{(t)}$ denotes the minimum data rate required by each user. P in represents the total power budget at the BS, $P_0^{(t)}$ is the total harvested power by the UAV in time-slot t for simultaneous transmission of the received signal to far users. The constraint $C1$ ensures that the minimum data rate for each user is satisfied. Constraints $C2$ and $C3$ ensure that the transmission power of the base station cannot exceed the maximum available power and that the transmission power of the UAV cannot exceed the power harvested by the UAV in time slot t. Constraint $C4$ defines the PS ratio bounds for full duplex simultaneous transmission.

Due to the non-convex objective function and non-convex constraints, **OP1** is a non-convex optimization problem. Due to this non-convexity, it is challenging to determine the optimal solution for the stated problem. In order to handle this optimization challenge, we present a heuristic algorithm, as illustrated in Algorithm 1, which optimizes the power allocation $w_j^{(t)}$ for each node and PS ratio α to maximize the EE. We assume that interference from other users in R_S is small compared to the noise. This assumption enables the conversion to a concave function, which can be conveniently transformed to a convex function through trivial manipulation, i.e., by converting to $\min\{-R_S\}$. Finally, the optimal values of power allocation coefficients are obtained using CVX, a Matlab-based convex optimization modeling tool [29]. P_S is convex, and the optimal values for $w_0^{(t)}$ and $\alpha^{(t)}$ are determined to minimize P_S. The EE is calculated using the optimal values obtained from each iteration. The Algorithm 1 is executed in both time slots. For each time-slot t, it continues execution until the gap between successive obtained EE values falls below the set threshold tolerance ϵ.

4 NUMERICAL RESULTS

In this section, the numerical evaluation of the proposed EE scheme is presented and discussed. To confirm the effectiveness of the proposed scheme, it is compared to two other schemes: non-cooperative NOMA (NC-NOMA) and C-NOMA with a ground relay. In the considered NC-NOMA scheme, users are served directly by the BS without the use of any relay, whereas in the considered C-NOMA scheme, the BS uses a ground relay in FD mode to serve users in multiple time-slots, where users are serviced in each sub-slot.

The simulations are conducted in MATLAB, while CVX is utilized as an optimization tool. The path-loss of the loop interference channel is set as $\theta_{LI} = -10$ dB. The circuit power is assumed to remain constant at $P_c = 20$W, and is kept constant. The channel power gain at the reference distance of 1m is considered to be $\beta = 65$dB, and the other channel model related environmental constants are $l_1 = 5$, $l_2 = -1$, $l_3 = -1$, $l_4 = 42$, $\mu_1 = \frac{2}{\pi}\ln 3$, and $\mu_2 = 9$. The other simulation parameters are summarized in Table 2.

Algorithm 1. Power Allocation and Splitting Algorithm (PASA).

1. **Input:** BW, Channel gain, P_c, $\bar{R}_j^{(t)}$, $\varepsilon \leftarrow$ Tolerance
2. **Output:** $\alpha^{(t)*}$ $w_0^{(t)*}$, $w_j^{(t)*}$, $\forall j \in J$
3. **Initialize:** Set initial values of $w_j^{(t)}$, $i = 1$
4. **For** each time-slot t 1:2 do
5.   **While** $|\eta_{EE}[i] - \eta_{EE}[i-1]| \geq \varepsilon$
6.   Minimize $P_S[i]$ for known $w_j^{(t)}[i]$ from (21)
7.   Maximize $R_S[i]$ for known $a^{(t)}[i]$ and $w_0^{(t)}[i]$ from (20) using cvx
8.   Calculate $\eta_{EE}[i]$ from (19)
9.   Get the optimal values of $w_j^{(t)}[i]$, $\alpha^{t(i)}$, and $w_0^{(t)}[i]$
10.   update $w_j^{(t)}[i]$, $w_0^{(t)}[i]$, $\alpha^{(t)}[i]$
11. $i = i + 1$
12.   **End while**
13. **End for**

Table 3. Simulation parameters.

Parameter	Notation	Value
Bandwidth of the system		1 Mz
Noise variances	$\sigma_0, \delta_0, \sigma_j$	0.001
SI cancellation level	ρ	0.001
Minimum required data rate	$\bar{R}_j^{(t)}$	1 Mbps
Energy conversion efficiency	ζ	0.8
Power budget at the BS	P	20 W
Azimuth angle	θ	45°
Altitude of UAV	H	30 m
Radius of UAV	r	40 m

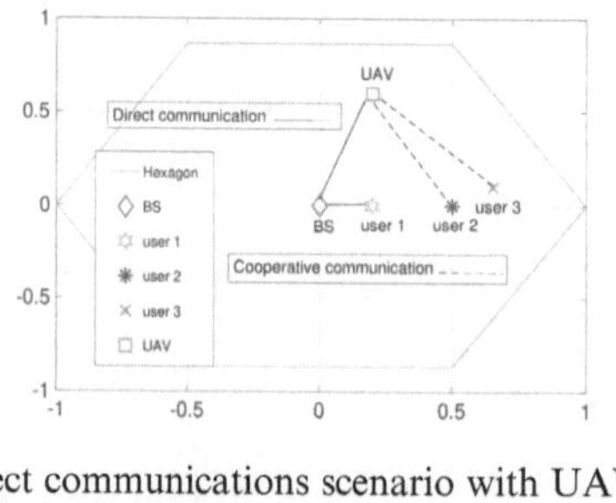

Figure 4. Cooperative and direct communications scenario with UAV as a relay.

Figure 4 depicts a C-NOMA communications scenario with a UAV acting as a relay, three ground users, and a BS located at the origin of the considered area of interest (AoI). The dotted lines in this figure represent cooperative communications between UAV and far users, whereas solid lines represent direct information transmission from BS to UAV and near users.

The average EE of the system with different iterations converges after the first iteration, indicating that the optimal performance is achieved immediately after the first iteration when using the proposed Algorithm 1. Figure 5 shows the average EE vs the BS's transmit power $P = 20, 25, 30, \ldots, 45$W. It is evident that as the BS power budget P increases so does the system's average EE. Though the EE of the other schemes increases as the power budget increases, the proposed scheme has a substantially larger improvement in average EE. When the power at the BS is increased, the system's sum data rate increases, which directly improves EE. However, due to the limited bandwidth, the EE of the system saturates after a certain point. According to Figure 5, the proposed scheme achieves an average increase in EE of 68.38% and 53.55% when compared to the NC-NOMA scheme and the ground relay C-NOMA scheme, respectively.

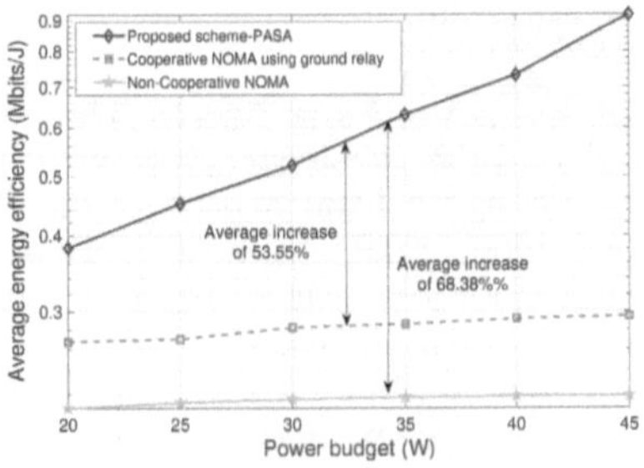

Figure 5. Average EE vs. the power budget at the BS.

Figure 6 demonstrates the impact of circuit power P_c on average EE. This experiment is simulated for p_c= 20, 25, 30,. . ., 45W. As can be seen, the average EE of all schemes considered decreases as P_c increases. This is because an increase in P_c necessitates the BS allocating more power to users in order to meet their QoS requirements. Furthermore, an increase in P_c increases the overall power consumption of the system, which has a detrimental effect on the EE of the system. The proposed scheme, as shown in Figure 6, has an average increase of approximately 65.55% and 61.89% when compared to the NC-NOMA scheme and the ground relay scheme, respectively. As a result, the proposed scheme outperforms the other schemes in terms of EE under the same parameters.

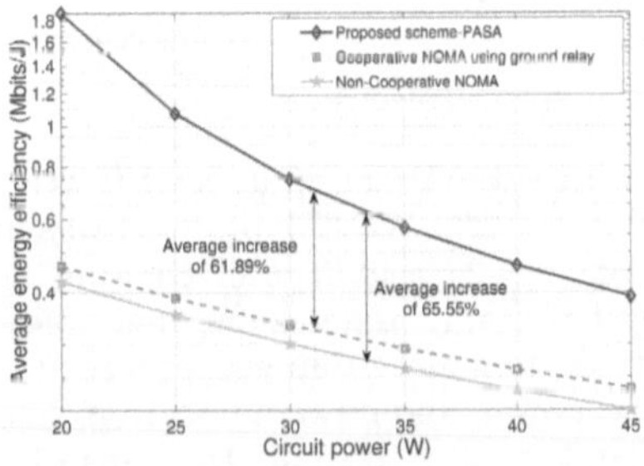

Figure 6. Average EE versus the circuit power consumption at P= 20W.

The bandwidth has a significant impact on the performance of the system. Figure 7 depicts the effect of various bandwidth values on the average EE. With increased bandwidth, the system's EE improves. This is because bandwidth is directly proportional to the achievable data rate. Therefore, an increase in the system's bandwidth increases the sum data rate, resulting in an increase in EE. Moreover, it is observed that, as bandwidth increases, the proposed scheme outperforms the alternatives in terms of average EE. Compared to the NC-NOMA scheme and the C-NOMA with ground relay, an increase in bandwidth results in an average EE increase of 76.44% and 75.23%, respectively.

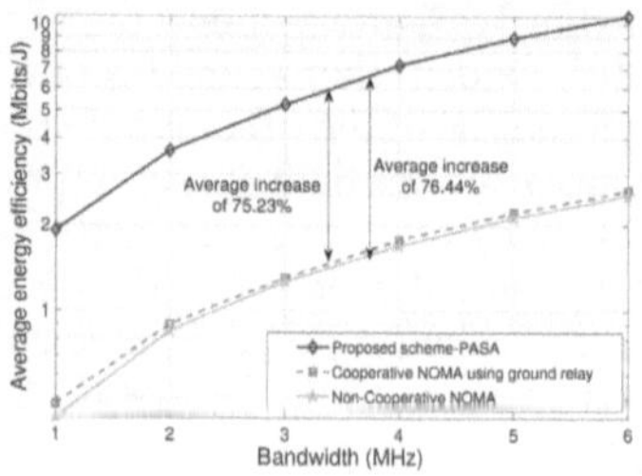

Figure 7. Average EE versus the bandwidth.

Finally, the effect of BS distance on EE is investigated. The average EE versus distance of BS from UAV and near user A with the power budget $P = 20\text{W}$ is shown in Figure 8. The average EE of all schemes decreases as the distance between the BS and the users increases. It is because the path-loss increases exponentially as the distance between the BS and the UAV and the user increases. Increased path-loss reduces the user data rate, so the BS allocates more power to the users to meet QoS requirements. It can be seen that the proposed scheme achieves an average increase of 71.66% when compared to the NC-NOMA scheme and an average increase of about 65.74% when compared to the C-NOMA with a ground relay. As a result, when the distance between the BS and the users/relay increases, the proposed scheme outperforms the other considered schemes in terms of EE.

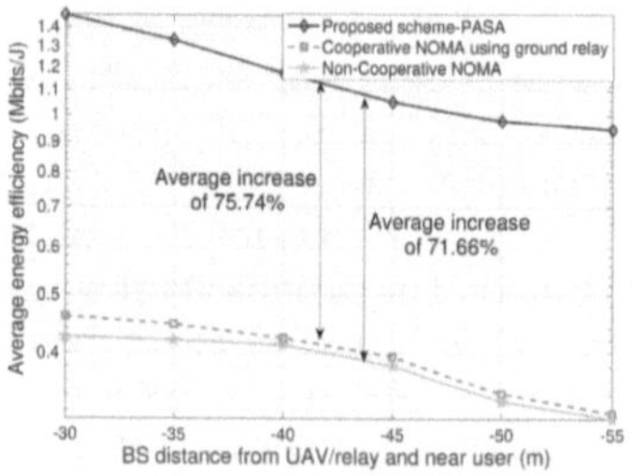

Figure 8: Average EE versus the distance of BS from UAV.

5 CONCLUSION

In this paper, a C-NOMA with a UAV acting as an FD relay was considered. Because UAVs are typically power constrained, a SWIPT protocol was implemented at the UAV to allow it to harvest energy from the RF signal received from the BS. Using the harvested energy, the UAV receives and transmits data to far users during the same time slot. To maximize the EE, a joint optimization algorithm was proposed to obtain the optimal values for the PS ratio, UAV transmit power, and power allocation. The practical constraints of all users' QoS requirements are taken into account. In terms of achieved EE, performance evaluation shows that the proposed scheme outperforms the other baseline schemes.

In future work, we intend to improve the proposed scheme by incorporating the varying channel conditions of the mobile UAV. Furthermore, we aim to consider minimizing the energy consumption of the mobile UAV in such a scenario.

REFERENCES

[1] Y. Zeng and R. Zhang, "Energy-efficient UAV communication with trajectory optimization," *IEEE Trans. Wireless Commun.*, vol. 16, no. 6, pp. 3747–3760, 2017.

[2] M. Al-Jarrah, A. Al-Dweik, E. Alsusa, Y. Iraqi, and M.-S. Alouini, "On the performance of IRS-assisted multi-layer UAV communications with imperfect phase compensation," *IEEE Trans. Commun.*, vol. 69, no. 12, pp. 8551–8568, 2021.

[3] S. Jangsher, M. Al-Jarrah, A. Al-Dweik, E. Alsusa, and P.-Y. Kong, "Energy constrained sum-rate maximization in irs-assisted UAV networks with imperfect channel information," IEEE Trans. Aerosp. Electron. Syst., 2022.

[4] S. Jangsher, M. Al-Jarrah, A. Al-Dweik, E. Alsusa, and M.-S. Alouini, "Ber reduction using partial-elements selection in IRS-UAV communications with imperfect phase compensation," IEEE Trans. Aerosp. Electron. Syst., 2022.

[5] H.Wang, J.Wang, J. Chen, Y. Gong, and G. Ding, "Network-connected UAV communications: Potentials and challenges," *China Commun.*, vol. 15, no. 12, pp. 111–121, 2018.

[6] G. Mu, "Joint beamforming and power allocation for wireless powered UAV-assisted cooperative NOMA systems," *EURASIP J. Wireless Commun. Netw.*, vol. 2020, no. 1, pp. 1–14, 2020.

[7] W. Mei and R. Zhang, "Uplink cooperative NOMA for cellularconnected UAV," *IEEE J. Sel. Topics Signal Process.*, vol. 13, no. 3, pp. 644–656, 2019.

[8] H. Yahya, A. Al-Dweik, E. Alsusa, and M. Debbah, "Priority-based dynamic IoT-downlink communications with blind transmission-mode recognition," *IEEE Internet Things Mag.*, vol. 5, no. 3, pp. 106–112, 2022.

[9] G. Li, D. Mishra, and H. Jiang, "Cooperative NOMA with incremental relaying: Performance analysis and optimization," *IEEE Trans. Veh. Technol.*, vol. 67, no. 11, pp. 11291–11295, 2018.

[10] M. R. Usman, A. Khan, M. A. Usman, Y. S. Jang, and S. Y. Shin, "On the performance of perfect and imperfect SIC in downlink non orthogonal multiple access (NOMA)," in Intl. conf. smart green technol. electr. inf. syst. (ICSGTEIS), pp. 102–106, 2016.

[11] Z. Ding, R. Schober, and H. V. Poor, "Unveiling the importance of SIC in NOMA systems—part 1: State of the art and recent findings," *IEEE Commun. Lett.*, vol. 24, no. 11, pp. 2373–2377, 2020.

[12] J. Zhang, X. Tao, H. Wu, and X. Zhang, "Performance analysis of user pairing in cooperative NOMA networks," *IEEE Access*, vol. 6, pp. 74288–74302, 2018.

[13] M. Zeng, W. Hao, O. A. Dobre, and Z. Ding, "Cooperative NOMA: State of the art, key techniques, and open challenges," *IEEE Netw.*, vol. 34, no. 5, pp. 205–211, 2020.

[14] M. Wu, Q. Song, L. Guo, and A. Jamalipour, "Joint user pairing and resource allocation in a SWIPT-enabled cooperative NOMA system," IEEE Trans. Veh. Technol., 2021.

[15] T.-N. Tran, T. P. Vo, P. Fazio, and M. Voznak, "SWIPT model adopting a PS framework to aid IoT networks inspired by the emerging cooperative NOMA technique," *IEEE Access*, vol. 9, pp. 61489–61512, 2021.

[16] L. Bariah, S. Muhaidat, and A. Al-Dweik, "Error probability analysis of NOMA-based relay networks with SWIPT," *IEEE Commun. Lett.*, vol. 23, no. 7, pp. 1223–1226, 2019.

[17] Y. Yuan, Y. Xu, Z. Yang, P. Xu, and Z. Ding, "Energy efficiency optimization in full-duplex user-aided cooperative SWIPT NOMA systems," *IEEE Trans. Commun.*, vol. 67, no. 8, pp. 5753–5767, 2019.

[18] H. Huang and M. Zhu, "Energy efficiency maximization design for full-duplex cooperative NOMA systems with SWIPT," *IEEE Access*, vol. 7, pp. 20442–20451, 2019.

[19] Y. Ye, Y. Li, D. Wang, and G. Lu, "Power splitting protocol design for the cooperative NOMA with SWIPT," in IEEE Intl. Conf. Commun. (ICC), pp. 1–5, 2017.

[20] Z. Zhang, H. Qu, J. Zhao, and W. Wang, "Energy efficient transmission design of cooperative NOMA with SWIPT network," in IEEE 4th Intl. Conf. Signal Image Proc. (ICSIP), pp. 566–572, 2019.

[21] S. Zhao, C. Mei, and Q. Zhu, "Joint time and power allocation algorithm in NOMA relaying network," *Mobile Inf. Syst.*, vol. 2019, 2019.

[22] O. Maraqa, A. S. Rajasekaran, S. Al-Ahmadi, H. Yanikomeroglu, and S. M. Sait, "A survey of rate-optimal power domain NOMA with enabling technologies of future wireless networks," *IEEE Commun. Surveys & Tuts.*, vol. 22, no. 4, pp. 2192–2235, 2020.

[23] X. Shao, C. Yang, Y. Song, T. Li, and Z. Han, "Game theoretical approaches for cooperative UAV NOMA networks," *IEEE Wireless Commun.*, vol. 28, no. 2, pp. 96–105, 2021.

[24] W. Mei and R. Zhang, "Cooperative downlink interference transmission and cancellation for cellular-connected UAV: A divide-andconquer approach," *IEEE Trans. Commun.*, vol. 68, no. 2, pp. 1297–1311, 2019.

[25] J.-B. Kim and I.-H. Lee, "Non-orthogonal multiple access in coordinated direct and relay transmission," *IEEE Commun. Lett.*, vol. 19, no. 11, pp. 2037–2040, 2015.

[26] M. B. Shahab and S. Y. Shin, "Time shared half/full-duplex cooperative NOMA with clustered cell edge users," *IEEE Commun. Lett.*, vol. 22, no. 9, pp. 1794–1797, 2018.

[27] M. Kim and J. Lee, "Outage probability of UAV communications in the presence of interference," in IEEE Global Commun. Conf. (GLOBECOM), pp. 1–6, 2018.

[28] H. Xie, F. Gao, S. Zhang, and S. Jin, "A unified transmission strategy for TDD/FDD massive MIMO systems with spatial basis expansion model," *IEEE Trans. Veh. Technol.*, vol. 66, no. 4, pp. 3170–3184, 2016.

[29] "CVX: Matlab software for disciplined convex programming." http: //cvxr.com/cvx/. Accessed: 2022-12-30

Innovation and Technological Advances for Sustainability – Al-Naemi et al. (Eds)

Supervised machine learning for intrusion detection: A survey and performance evaluation

Zina Chkirbene & Ridha Hamila
College of Engineering, Qatar University, Qatar

ABSTRACT: Cloud computing is a cost-effective way to host and deliver services over the internet, but the large volume of data transmitted makes the cloud network a target for malicious attacks. To protect against these threats, various intrusion detection systems (IDSs) have been developed to identify different types of attacks on the network. However, some of these systems are not always efficient due to the level of human involvement they require. Machine learning approaches, which have been widely used in IDSs, have the potential to improve efficiency by building well-trained models for anomaly detection using large datasets with a variety of attack types. In this paper, we examine the current state of research on using machine learning techniques for IDSs and evaluate the suitability of different algorithms using the UNSW dataset. We compare the accuracy rate and classification performance of several algorithms to determine their effectiveness.

Keywords: Security, machine learning, intrusion detection, cloud computing

1 INTRODUCTION

Cloud computing offers a reliable and cost-effective way to deliver scalable internet-based services, but network security remains a major issue that can impact its credibility and adoption. Cloud environments are vulnerable to various security risks and vulnerabilities, such as: (1) a lack of control over cloud-hosted resources; (2) the absence of security assurances in service level agreements (SLAs) between cloud providers and consumers; and (3) the possibility of sharing resources with adversaries or malicious users due to multi-tenancy [?].

To address the security concerns of cloud networks, several solutions have been proposed, including network intrusion detection systems (NIDS). However, most NIDS systems are signature-based, which has limitations. For example, it can be challenging to quickly observe and analyze behavioral changes and attribute them to specific parts of the network, such as protocols, operating system versions, or individual users. The proliferation of protocols and the wide range of data in modern networks make it difficult for IDSs to accurately detect intrusions. This is because establishing a baseline for attack detection requires periodic analysis of the condition of an online node to identify any abnormal activity [?].

This inspires researchers to create new anomaly detection techniques capable of overcoming these limitations especially with modern networks[?]. Recently, machine learning algorithms have been proposed as a new technique for network anomaly-based detection [?]. By training on a large dataset with a variety of attack types, these algorithms can build well-trained models that improve the effectiveness of anomaly detection. Machine learning methods can be classified as supervised, unsupervised, or hybrid [1].

DOI: 10.1201/9781003496724-39

In this paper, we evaluate the effectiveness of several important supervised machine learning methods using the UNSW dataset. We compare the accuracy and execution time of decision trees, ensemble trees, and Support Vector Machines (SVMs). The rest of the paper is organized as follows: Section II presents relevant prior work in the field. The UNSW dataset is described in Section III. Section IV presents the simulation results, and Section V concludes the paper.

2 RELATED WORK

The supervised anomaly detection techniques are trained using labeled data to create a trained model. In this section, some supervised algorithms for network anomaly detection systems will explored and investigated.

2.1 *Decision trees*

The decision tree (DT) is a well-known machine learning classification algorithm according to the division method. The DT is composed of features and leaves, where leaves denoting the class label [?]. The DT can be as basic as a single model or as complicated as a combination of models, making it well-suited for use in anomaly detection systems [2].

The authors presented a system for detecting anomalous intrusions based on the J48 decision tree in [?]. The classes were classified as normal, known attack, or unknown attack using the Kyoto 2006+ dataset.This dataset was collected from several honeypots used to monitor real-world traffic during the last three years. According to the authors, the J48 decision tree method has a high rate of accuracy. The proposed model was constructed using the data mining tool WEKA and obtained an overall accuracy of 97.23% percent.

The authors developed a unique fog computing Intrusion Detection System (IDS) [3]. Fog nodes provide a security risk since they generate a vast volume of data and have limited computing capabilities. The research involved their investigation using Classification And Regression Trees (CART), a technique for creating binary trees that prioritizes the features that provide the most information. By using the KDDCUP99 dataset, the proposed IDS system was evaluated against numerous comparisons to alternative algorithms, including decision trees, which has the highest detection accuracy while keeping a reasonable detection time in comparison to the others.

2.2 *Support vector machines*

Support vector machine (SVM) is another supervised learning algorithm for data classification. SVM is a method that begins by transforming the input vectors into a multi-dimensional space. Then, the appropriate data separation is defined to better classify the instances. Due to the long training procedure, SVM is not recommended for large datasets.

The mutual information (MI) algorithm was used by the authors to choose features in [?]. A classifier-based intrusion-detection system was then developed utilizing the Least Squares Support Vector Machine (LSSVM) technique (LS-SVM). Two random variables are utilized to determine the correlation of the feature with the class label using an algorithm. Using the average distances of each data point to its K-nearest neighbors, entropies of the data were calculated. An evaluation of the solution's performance was carried out using three datasets: NSL-KDD, KDD Cup 99, and Kyoto 2006+.The IDS that used the features selection technique performed well in detecting network intrusions.

In paper [4], The authors proposed an approach for detecting anomalous traffic using the SVM algorithm. In particular, a technique for computing the normalized entropy of data characteristics was presented. When these numbers deviate from their typical values due to attacks, the system defined a threshold to detect any malicious network activity. A classifier for detecting network anomalous traffic was created using a support vector machine. The system used the particle swarm optimization (PSO) technique to increase the accuracy of the SVM. To examine the suggested method's performance, the authors used the KDDCup99 and DARPA datasets. The results indicated that the proposed technique was highly accurate in detecting several types of attacks.

2.3 *Ensemble tree*

This machine learning method combines multiple models to create the most accurate prediction model. The ensemble approach is based on the concept of integrating all of the learners who are less accurate into a single, more accurate learner. Bagging, Boosting, and Stacking are all instances of this type of strategy. In [5], the authors investigated the bagging ensemble approach by utilizing Naive Bayes, the partial decision tree algorithm (PART), and adaptive boost to determine the trail. The ensemble approach outperforms PART, Naive Bayes, and Adaptive Boos.

3 UNSW DATASET

The dataset is one of the basic components of the supervise learning algorithm since they allow for the evaluation of the system's efficacy and performance in identifying various attacks. Numerous datasets are available to evaluate IDS's performance [6]. However, A well-trained model requires a current dataset that contains the most recent attacks [7]. The KDD99 dataset is widely used for intrusion detection model.However, several limitations have been reported, including the volume of the training and testing sets, and particularly the problem of class imbalance.

In 2015, the UNSW-NB15 dataset was released by the Australian Center for Cyber Security's cyber-security research group. The dataset includes nine unique attack types that are modeled by current real-world network activity [8]. The UNSW-NB15 dataset was generated using the *PerfectStorm* software, which monitors network traffic anomalies. Three virtual servers constitute the IXIA testing platform: two for regular traffic and one for abnormal behaviors.The UNSW-NB15 dataset includes a variety of features generated using Bro-IDS, Argus, and 12 more *C*-developed algorithms. The Argus process generates binary pcap files through a server, then extracts the features via a client installed on the router. The features of the UNSW-NB15 dataset were obtained using Argus, Bro-IDS, and 12 other *C*-developed techniques. The Argus software is composed of two components: a server that generates binary pcap files and a client that extracts the features. Three log files were created using the Bro-IDS: one for the connection, another for the HTTP request, and another for the FTP request. The result of these tools is saved to a database file, and then used to matching features using flow features [?]. Fundamental, content, and temporal features are the three types of matching qualities. Two sets of characteristics were also extracted: general-purpose and connection-related characteristics. The latter two characteristics were included for the purpose of labeling. There are 45 features in the dataset with five different data types, namely; binary, float, time-stamp, integer, and nominal.

The dataset has 10 traffic packets types, the normal and nine attacks types [9]. The abnormal categories include the following forms of attacks. Attacks include fuzzers, analysis, backdoors, denial-of-service attacks, exploits, generic, reconnaissance, shellcode, and worms as presented in Figure 1. Table 1 offers information about the classes and their associated instances. The dataset has two sets which are the training and testing set. The training set contains 175,241 instances, and 82,332 instance in the testing set. The attacks with their notations in the paper are presented in Table 2.

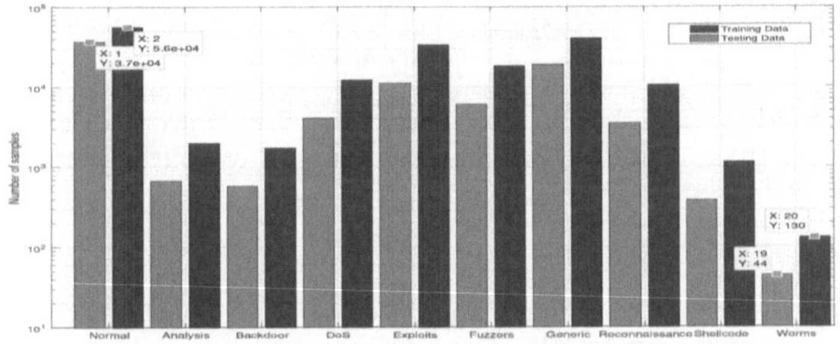

Figure 1. Data statistics.

Table 1. The UNSW-NB15 dataset's classes description.

Class	Number of instances in Training data	Number of instances in testing set
Normal	56000	37000
Generic	40000	18871
Exploits	33393	11132
Fuzzers	18184	6062
DoS	12264	4089
Reconnaissance	10491	3496
Analysis	2000	677
Backdoor	1746	583
Shellcode	1133	378
Exploits	33393	11132
Fuzzers	18184	6062
Worms	130	44

Table 2. Classes notation.

Number	Normal	Analysis	Backdoor	DoS	Exploits	Fuzzers	Generic	Reconnaissance	Shellcode	Worms
Class (c_i)	0	1	2	3	4	5	6	7	8	9

4 SIMULATION PERFORMANCE

4.1 *Definitions*

- **False Negative (FN)** and **False Positive (FP)** are the miss detection and false alarms, respectively.
- **True positive (TP)** indicates the packet that have been successfully detected (as normal) and **True Negative** (TN) shows the number of packets that were appropriately rejected.
- **True positive rate TPR** True positive rate (TPR) called also sensitivity or the probability of detection. It measures the proportion of normal packets that are correctly identified.

$$TPR = \frac{TP}{TP + FN} \tag{1}$$

- **Accuracy** is a critical characteristic for categorizing machine learning approaches. (i.e., how correctly these techniques make decisions in comparison to previously unknown data). The accuracy is indicated in terms of TN and TP.

$$Accuracy = \frac{TN + TP}{number\ of\ packets}. \tag{2}$$

- **Confusion matrix** is a particular table that summarizes supervised learning's performance. The matrix is divided by columns that represent predicted classes and rows that represent actual classes.

4.2 *Simulation results*

Figures 2, 3 and 4 show the confusion matrix of decison tree, SVM and ensemble tree for 10 classes classification, respectively. First, we remark that decision tree has 3 low detectable classes (2, 3, 9), 4 low detectable classes for the SVM (2, 3,4,9), and 5 low detection for classes (2, 3, 5, 9, 10). In particular, these classes have low number of instance that describes each of them. When the training data with a certain class is small in compared to the training data for other classes, the machine learning model will have difficulty identifying any attack against that class. The imbalanced data problem is a classifying difficulty in which one class has a much lower prior probability/instances in the training set than the other. Thus, the decision tree strategy significantly decreased the number of minor classes while increasing the number of classes that were really identified, showing that the system correctly predicted all received instances.

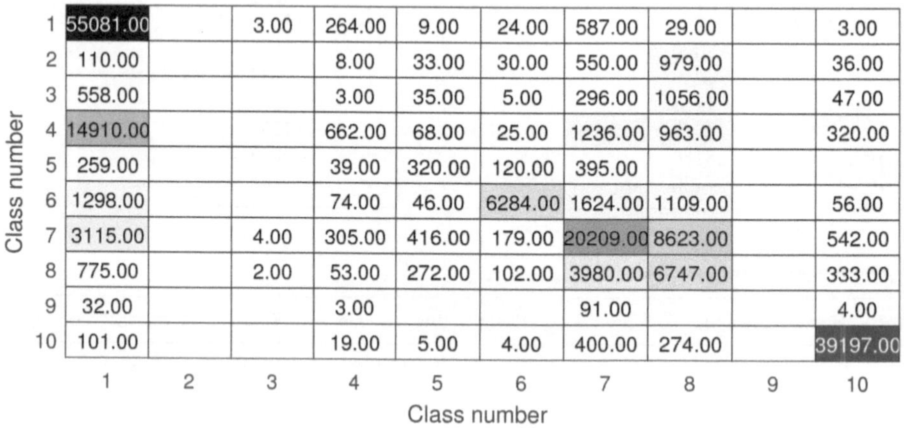

Figure 2. Confusion matrix of decision tree.

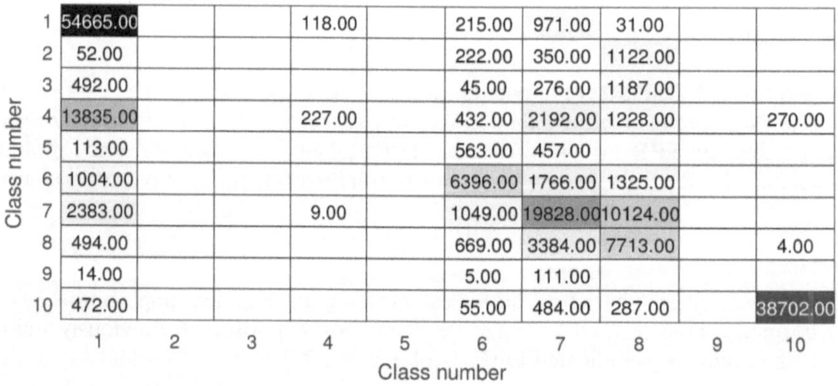

Figure 3. Confusion matrix for SVM algoritm.

Figure 5 shows the accuracy, TPR, and execution time for decision tree, support vector machine, and ensemble tree. The accuracy and TPR values are used to evaluate these approaches' ability in accurately detecting typical packets. We notice that both Dt and SVM algorithms are more accurate than the ensemble tree. The accuracy of the DT method is 74%, the SVM algorithm is 63%, and the ensemble tree algorithm is 53%, but the TPR of these algorithms is less than 50%. This shows that these algorithms reject almost half of the data received, introducing more latency into the network when users are unable to get their packets. Additionally, the ensemble tree takes the longest time to execute.

Figure 6 depicts a comparison of accuracy performance of Classification and Regression Trees (CART) with the J48 decision tree, the fine tree, the medium tree, and the coarse tree, the fast decision tree, the random tree, for the different attack classes. CART algorithm has better performance in terms of accuracy rate compared to the other algorithms.

	1	2	3	4	5	6	7	8	9	10
1	55400.00			86.00		177.00	337.00			
2	82.00			20.00		47.00	559.00	1038.00		
3	487.00			1.00			378.00	1134.00		
4	15751.00			251.00		317.00	812.00	1053.00		
5	362.00			66.00		496.00	209.00			
6	1709.00			108.00		6194.00	1276.00	1204.00		
7	3948.00			230.00		185.00	19719.00	9311.00		
8	623.00			36.00		174.00	4385.00	7046.00		
9	64.00			3.00		4.00	59.00			
10	39268.00			21.00		40.00	441.00	230.00		

Figure 4. Confusion matrix for ensemble tree algoritm.

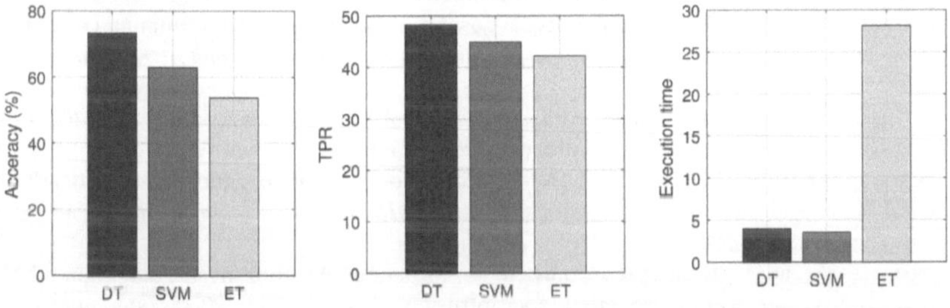

Figure 5. The accuracy, TPR and execution time for decision tree, SVM and ensemble tree.

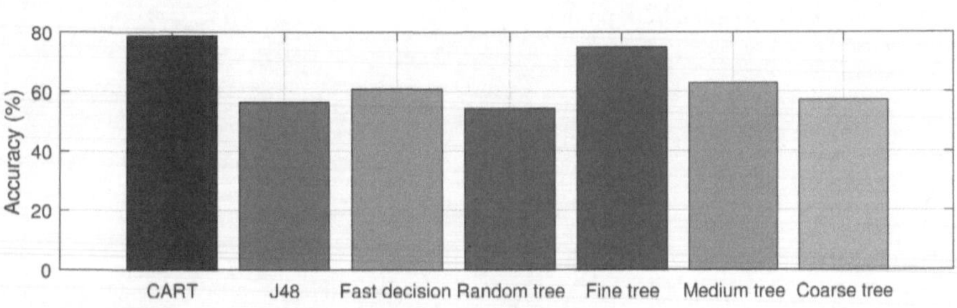

Figure 6. Comparison of various decision tree algorithms.

5 CONCLUSION

The fast development of Internet traffic and the creation of massive amounts of valuable digital data attract attackers seeking to illegally get economic rewards and other benefits. This paper explores the machine learning techniques used for NIDS and also conducts experiments on the UNSW data set. We evaluated the features of various strategies, as well as their benefits and drawbacks. We have presented comparisons of the features of some representative techniques. The decision tree algorithm shows a great performance in intrusion detection and classification for UNSW dataset.

ACKNOWLEDGMENT

This work was supported by Qatar University Internal Grant QUHI-CENG-21/22-3. The publication of this article was funded by the Qatar National Library. The findings achieved herein are solely the responsibility of the authors.

REFERENCES

[1] Chkirbene Z., Erbad A., Hamila R., Mohamed A., Guizani M., and Hamdi M., (2020). Tidcs: A dynamic intrusion detection and classification system based feature selection, *IEEE Access*, vol. 8, pp. 95864–95877.

[2] Agarwal A., Sharma P., Alshehri M., Mohamed A. A., and Alfarraj O., (2021). Classification model for accuracy and intrusion detection using machine learning approach, *PeerJ Computer Science*, vol. 7, p. e437.

[3] Huang C., Zheng L., Wang S., Leung V. C. M., Lin T., and Peng K., (2018). Intrusion detection system based on decision tree over big data in fog environment,' *Wireless Communications and Mobile Computing*, vol. 2018, pp. 1–10.

[4] Lei Y., Network anomaly traffic detection algorithm based on SVM, (2017). *Proceedings – 2017 International Conference on Robots and Intelligent System, ICRIS 2017*, pp. 217–220.

[5] Baik S. and Bala J., (2004). A decision tree algorithm for distributed data mining: Towards network intrusion detection, in *International Conference on Computational Science and Its Applications*. Springer, pp. 206–212.

[6] Chkirbene Z., Erbad A., Hamila R., Gouissem A., Mohamed A., and Hamdi M., (2020). Machine learning based cloud computing anomalies detection, *IEEE Network*, vol. 34, no. 6, pp. 178–183.

[7] Moustafa N. and Slay J., (2015). UNSW-NB15: A comprehensive data set for network intrusion detection systems (UNSW-NB15 network data set), *2015 Military Communications and Information Systems Conference, MilCIS 2015 – Proceedings*, pp. 1–6.

[8] Chkirbene Z., Erbad A., Hamila R., Gouissem A., Mohamed A., Guizani M., and Hamdi M., (2020). Iterative per group feature selection for intrusion detection, in *2020 International Wireless Communications and Mobile Computing (IWCMC)*. IEEE, pp. 708–713.

[9] Chkirbene Z., Erbad A., and Hamila R., (April 2019). A combined decision for secure cloud computing based on machine learning and past information, in *2019 IEEE Wireless Communications and Networking Conference (WCNC)*, pp. 1–6.

Innovation and Technological Advances for Sustainability – Al-Naemi et al. (Eds)

A survey on feature selection algorithms for machine learning

Ziad Almesleh, Zina Chkirbene & Ridha Hamila
College of Engineering, Qatar University, Qatar

ABSTRACT: Wireless communication is one of the most substantial types of communication nowadays, leading to the IoT revolution and increasing demand for IoT networks. On the other hand, this demand also leads to an increase in the number of resource-constrained devices, which creates a new problem related to vulnerabilities to attacks. The handling of attacks on IoT devices must be closely monitored due to resource limitations for these devices, which makes research in this field important. Many intrusion detection systems have been improved by using machine learning models that classify data as normal or abnormal and use evaluation parameters such as accuracy, f1-score, and precision to assess their performance. One of the most important steps in building a good intrusion detection system is feature selection, which can help improve the reasonable classification of data. This paper discusses the latest research on using machine learning techniques for intrusion detection systems. It examines various prominent algorithms and provides further analysis on their appropriateness. CICIDS2017 dataset is used to evaluate the robustness and determinism of intrusion detection system models.

Keywords: Intrusion detection system, IoT, machine learning, feature selection

1 INTRODUCTION

The Internet of Things (IoT) connects a huge number of heterogeneous devices with different services on the same network [1]. A Wireless Sensor Network (WSN) device is a resource-constrained device that communicates through wireless infrastructure and is connected to the internet or computer network to transfer data and establish an IoT environment. IoT networks are high-speed networks with different applications such as intelligent healthcare systems [2], smart cities, and vehicular networks, where IoT devices have different communication procedures and different services. This prompts the use of machine learning and deep learning in IoT security to combat attackers and design intrusion detection systems (IDS). The main techniques for IDSs are Signature Intrusion Detection Systems (SIDS), which detect common and known attackers (stored in a database), and Anomaly Intrusion Detection Systems (AIDS), which are the focus of much research because they aim to achieve the best accuracy, False Acceptance Rate (FAR), and False Rejection Rate (FRR) for IDS models. Finally, Hybrid Intrusion Detection Systems (HIDS) use both of the previous IDS techniques together. The FAR and FRR values are the most important for assessing the performance of IDSs. These values depend on the false positive rate and the false negative rate. A false positive is a false alarm that occurs when normal behavior for the system is detected as an attack by the IDS. False negatives are more dangerous than false positives, because they occur when the IDS fails to recognize an attack on the system. Many parameters must be taken into consideration in the design of IDSs for IoT networks, especially for heterogeneous and constrained devices [3]. One of the main parameters is lightweight IDS. Lightweight techniques are used to enhance IDSs by reducing the dimensionality and feature size by identifying the most significant features, which decreases computational complexity. Higher detection, lower

DOI: 10.1201/9781003496724-40

false positive rate, and false negative rate are also important parameters for IDS design to enhance the accuracy of the system and increase the detection rate. Finally, decreasing the training time by reducing the training set or choosing the most appropriate machine learning or deep learning algorithm is important. Feature selection is a critical step in the machine learning process, as it involves selecting the most relevant and significant features from a dataset to be used in building a model. It can help improve the accuracy and performance of the model, as well as reduce the computational complexity and time required for training. In particular, feature selection plays a vital role in the success of machine learning models and continues to be an active area of research and development.

The rest of the paper is organized as follows: Section II presents related works in the field. The CICIDS2017 dataset is illustrated in Section III. Section IV is the simulation performance, and Section V is the conclusion

2 RELATED WORK

Choosing and categorizing features is essential for designing and implementing an intrusion detection system (IDS) model. This section will delve into the PCA algorithm, as well as several machine learning, to assess their suitability for use in an IDS.

2.1 *Principal component analysis*

Principal Component Analysis (PCA) is a technique used to reduce the number of features in a dataset, which can be useful for implementing lightweight models by decreasing computational complexity. This is particularly important for deep learning-based classifiers, as their computational complexity tends to increase exponentially with the number of features. PCA is a linear transformation method that aims to eliminate correlated data and group uncorrelated data together. The primary metric used in PCA is variance, with higher variance values indicating more information [4]. The main steps involved in calculating PCA include normalization, calculation of the covariance matrix, determination of eigenvalues and eigenvectors, and calculation of contribution rates based on the eigenvalues. In [5], the authors propose a model that combines PCA with a Deep Convolutional Neural Network (DCNN) for anomaly detection. One advantage of this model is its lightweight structure and superior efficiency in feature extraction, making it a promising option for use in Wireless Sensor Networks (WSNs) with limited storage.

2.2 *Random Forest (RF)*

A supervised machine learning algorithm is a type of algorithm that is used for classification and regression tasks in machine learning. It relies on the structure of multiple decision trees to classify a large number of input variables based on their importance [6]. In [7], the authors present a novel approach for anomaly detection that utilizes an ensemble learning (EL) algorithm with different stages. One of these stages involves using a hybrid method called Correlation Feature Selection-Random Forest (CFS-RF) for ensemble learning. Another stage involves using two classifiers (random forest (RF) and support vector machine (SVM)) as adaboosting and bagging EL algorithms for intrusion detection. The three datasets (*NSL_KDD*, *UNSW_NB*2015, and *CIC_IDS*2017) were separately used for the evaluation, and the result of accuracy was 99% for all datasets, and FAR values were 0.004, 0.008, and 0.0012, respectively. Also, the authors in [8] focuse on the investigation of DoS/DDoS attacks in the IoT because the major reports for IoT attacks are “Denial of Service” (DoS) and “Distributed Denial of Service” (DDoS) attacks. A new architecture of DoS/DDoS detection and DoS/DDoS mitigation components is proposed. The Bot-IoT dataset is used to evaluate this model, where the result for the Looking-Back-enabled Random Forest classifier achieves an accuracy of 99.81%.

2.3 *Support Vector Machine (SVM)*

Support Vector Machines (SVMs) are a type of algorithm that seeks to find a hyperplane in a high-dimensional feature space that maximally separates different classes of data. These algorithms can achieve good results even with small training sets due to the decision hyperplane being determined by only a small number of support vectors. However, SVMs can be sensitive to noise near the hyperplane. They are effective at solving linear problems, but for nonlinear data, kernel functions may be used to map the data into a higher-dimensional space where it can be more easily separated. Kernel functions are commonly used in both SVMs and other machine learning algorithms. SVMs are statistical classifiers that are capable of predicting high-dimensional distributions [9]. They use a combination of linear and non-linear equations to do so. In [10], the authors proposed using an SVM to detect intrusions using the KDD99 dataset, which includes various protocols. They first divided the dataset into subdatasets based on protocol type (TCP, UDP, and ICMP) and selected features for each subdataset based on the characteristics of those protocols. They then trained SVM models on the three subdatasets and obtained an average accuracy of 89.02%.

In [11], the authors present an IDS that utilizes an OSVM model. This model utilizes a whale optimization algorithm (WOA) to select optimal kernels for the SVM model. When tested on the KDDCup 99 dataset, the model achieved an accuracy of 94.09% and a detection rate of 95.02%.

2.4 *Convolutional Neural Network (CNN)*

Deep learning is a subfield of machine learning that involves using multi-layered neural networks to learn from unlabeled data. Because of their ability to learn and adapt, deep learning algorithms are considered to be highly effective and have the potential to improve the performance of an intrusion detection system (IDS) [12]. Convolutional Neural Networks (CNNs) are a type of deep learning algorithm that consists of three layers: a convolutional layer, a pooling layer, and a fully connected (FC) layer. CNNs are commonly used for image recognition tasks. In [13], the authors propose a method for intrusion detection that involves converting the data's features into a colored image and using a CNN to classify attacks. This IDS was evaluated using the Aegean Wi-Fi Intrusion Dataset (AWID2) and achieved good results, with an F1 score of 99.73% and a false positive rate of 0.24%. In [9], the author presents an IDS that utilizes a hybrid deep neural network model with a CNN for spatial feature extraction and a Long Short-Term Memory (LSTM) for temporal feature extraction. This model was tested on three datasets (CIC-IDS2017, UNSW-NB, and WSN-DS) and achieved accuracies of 99.64%, 94.53%, and 99.67%, respectively.

3 SIMULATION PERFORMANCE

In this section, we evaluate the performance of the machine learning algorithms namely RF, SVM, and CNN using PCA for features selection. that were used and investigate their performance in terms of accuracy and detection rate, as well as false and positive rates. The performance of the algorithms is investigated in terms of precision, recall, and the F1 function [14].

3.1 *Used dataset*

CIC IDS2017 dataset was issued by Canadian Institute for Cybersecurity (CIC) in 2017. It contains an eight files contain 2,830,743 records for five days traffic data include the latest attack and benign data. Each record contains 78 various features with their labels. For perfect IDS the data files should be combined in a single file, because the files didn't have the same attack. The combination for all files given a file contains 3119345 records with 83 features and 15 classes (1 normal + 14 attack). Table 1 shows the dataset records with each class [15]. In order to evaluate the simulation performance in an optimal way, the simulation performance parameters can be summarized in the confusion matrix. The confusion matrix in Figure 1 shows the actual and predictable value in the matrix with correct and error detection; for correct detection, the actual value is the same as the predicted value, whereas for error detection, the actual value is different from the predicted value. True positive (TP) and true negative (TN) are the correct detection values; on the other side, false positive (FP) and false negative (FN) are the error detection values.

Table 1. The classes of CICIDs2017 dataset records.

Class	Number of records
BENIGN	2359087
DoS Hulk	231072
PortScan	158930
DDoS	41835
DoS GoldenEye	10293
FTP-Patator	7938
SSH-Patator	5897
DoS slowloris	5796
DoS Slowhttptest	5499
Bot	1966
Web Attack – Brute Force	1507
Web Attack – XSS	652
Infiltration	36
Web Attack – Sql Injection	21
Heartbleed	11

	Actual Positive	Actual Negative
Predicted Positive	TP	FP
Predicted Negative	FN	TN

Figure 1. Confusion matrix.

The parameters and equations for the evaluation value can be written as:

$$Accuracy = \frac{T_N + T_P}{T_N + T_P + F_N + F_P}. \quad (1)$$

$$Precision = \frac{TP}{T_P + F_P}. \quad (2)$$

Where the accuracy value determines the correct prediction and the precision value determines the number of times a true positive prediction has occurred.

Feature reduction is an important process to reduce the large size of data [16], as shown in Figure 2, where the dataset consists of 78 features that can be reduced to 10 features. The variance from the first ten PCA features is more than 75% higher than the overall dataset variance, where the first PCA feature variance is about 24% and the second PCA feature variance is about 14.7%. Table 2 shows the summary for the first ten PCA feature variance.

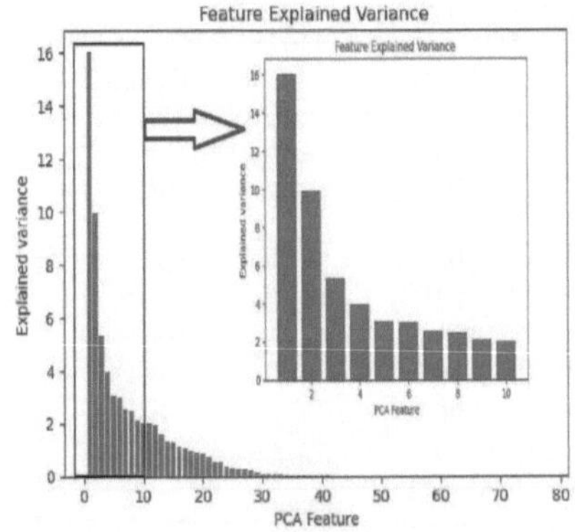

Figure 2. PCA features variance.

Table 2. Variance percent for first ten PCA feature.

	Variance%	Cummulative variance%
PCA1	23.910857	23.910857
PCA2	14.795879	38.706737
PCA3	7.945367	46.652104
PCA4	5.905768	52.557872
PCA5	4.509193	57.067064
PCA6	4.437906	61.504971
PCA7	3.787981	65.292952
PCA8	3.647121	68.940073
PCA9	3.098056	72.038129
PCA10	2.983309	75.021438

Random Forest classification and PCA feature reduction were applied to the CICIDS 2017 dataset. When 10 trees are used in the random forest classification, the average training set accuracy percent is 99.92% and the test set accuracy percent is 99.15%. Figure 3 and Table 3 show the confusion matrix and the classification report for attacks, where the lowest precision percent was for WebAttack; also, WebAttack is the lowest percent for recall and f1-score, whereas the precision, recall, and f1-score percents for BruteForce and Bot reach 100%.

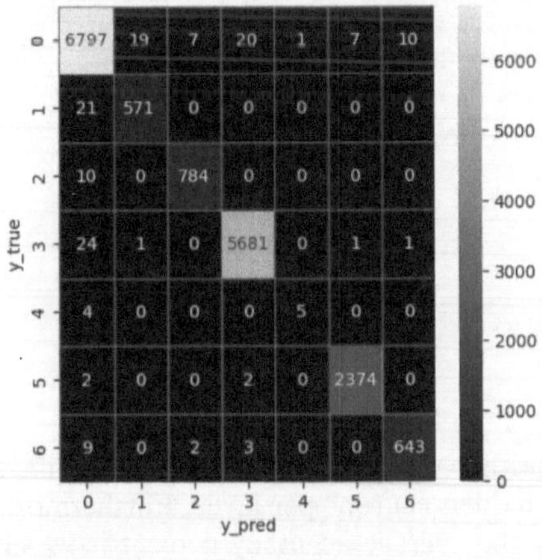

Figure 3. RF confusion matrix.

Table 3. Random forest classification report.

Label	Class	Precision%	Recall%	F1%	Support
0	**BENIGN**	99	99	99	6861
1	**DoS**	97	96	97	592
2	**PortScan**	99	99	99	794
3	**BruteForce**	100	100	100	5708
4	**WebAttack**	83	56	67	9
5	**Bot**	100	100	100	2378
6	**Infiltration**	98	98	98	657
macro average		96	92	94	1699
weighted average		99	99	99	1699

Figure 4 and Table 4 present the confusion matrix for the attack classification SVM model and detailed classification report, respectively, where the accuracy percent for the training set is 89.29% and the percent for the test set is 88.92%. The evaluation parameters for DoS and WebAttack classes differ significantly, with precision, recall, and f1-score for DoS being 92%, 2%, and 4%, respectively, and the same performance parameters for WebAttack being 100%, 11%, and 20%, respectively.

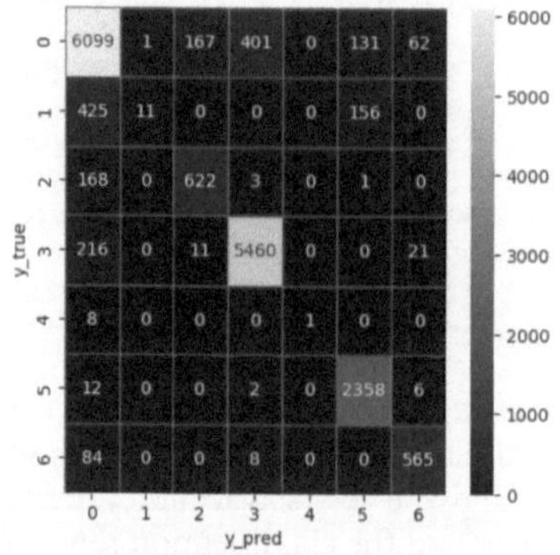

Figure 4. SVM confusion matrix.

Table 4. SVM classification report.

Label	Class	Pecis	Recall	F1	Supp
0	**BENIGN**	0.87	0.89	0.88	6861
1	**DoS**	0.92	0.02	0.04	592
2	**PortScan**	0.78	0.78	0.78	794
3	**BruteForce**	0.93	0.96	0.94	5708
4	**WebAttack**	1.00	0.11	0.20	9
5	**Bot**	0.89	0.99	0.94	2378
6	**Infiltration**	0.86	0.86	0.86	657
macro average		0.89	0.66	0.66	1699
weighted average		0.89	0.89	0.87	1699

CNN deep learning algorithm was applied for CICIDS 2017 dataset classification with two hidden layers and 128 hidden neurons per layer. Furthermore, the Adam optimization algorithm is applied, where the average accuracy percent is 99.43%. Figure 5 and Table 5 present the confusion matrix for the attack classification CNN model and classification report, where the least precision value is 97% for DoS, whereas the least Recall and F1-score values are 0.6% and 75% for WebAttack.

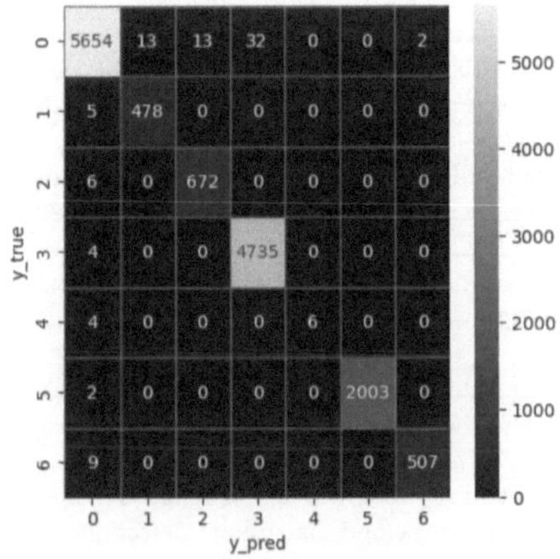

Figure 5. CNN confusion matrix.

Table 5. CNN classification report.

Label	Class	Pecis	Recall	F1	Supp
0	**BENIGN**	0.99	0.99	0.99	5714
1	**DoS**	0.97	0.99	0.98	483
2	**PortScan**	0.98	0.99	0.99	678
3	**BruteForce**	0.99	1.00	1.00	4739
4	**WebAttack**	1.00	0.60	0.75	10
5	**Bot**	1.00	1.00	1.00	2005
6	**Infiltration**	1.00	0.98	0.99	516
macro average		0.99	0.94	0.96	14145
weighted average		0.99	0.99	0.99	14145

4 CONCLUSION

This article shares a review of an IDS using a machine learning algorithm for IoT networks, starting with feature reduction and the effect of feature reduction on the IDS's performance. PCA is one of the most popular feature reductions applied to IDS to reduce computational complexity and feature dimensioning based on variance. Machine learning algorithms such as RF, SVM, and CNN are used to classify data, with the simulation of the CICIDS2017 dataset achieving approximately 99.15%, 89.29% and 99.42% accuracy for RF, SVM, and CCN classification, respectively.

ACKNOWLEDGMENT

This work was supported by Qatar University Internal Grant QUHI-CENG-21/22-3. The publication of this article was funded by the Qatar National Library. The findings achieved herein are solely the responsibility of the authors.

REFERENCES

[1] Chkirbene Z., Gouissem A., Hadjidj R., Foufou S., and Hamila R., (2018). Efficient techniques for energy saving in data center networks, *Computer Communications*, vol. 129, pp. 111–124. [Online]. Available: https://www.sciencedirect.com/science/article/pii/S014036641830121X

[2] Chkirbene Z., Mohamed A., Erbad A., and Guizani M., (2020). Smart edge healthcare data sharing system, in *2020 International Wireless Communications and Mobile Computing (IWCMC)*. IEEE, pp. 577–582.

[3] Chkirbene Z., Erbad A., Hamila R., Gouissem A., Mohamed A., Guizani M., and Hamdi M., (2020). Weighted trustworthiness for ml based attacks classification, in *2020 IEEE Wireless Communications and Networking Conference (WCNC)*, pp. 1–7.

[4] Zhao R., Gui G., Xue Z., Yin J., Ohtsuki T., Adebisi B., and Gacanin H., (2021). A novel intrusion detection method based on lightweight neural network for internet of things, *IEEE Internet of Things Journal*.

[5] Mihoub A., Fredj O. B., Cheikhrouhou O., Derhab A., and Krichen M., (2022). Denial of service attack detection and mitigation for internet of things using looking-back-enabled machine learning techniques, *Computers & Electrical Engineering*, vol. 98, p. 107716.

[6] Chkirbene Z., Hamila R., Erbad A., Kiranyaz S., Al-Emadi N., and Hamdi M., (2021). Cooperative machine learning techniques for cloud intrusion detection, in *2021 International Wireless Communications and Mobile Computing (IWCMC)*, pp. 837–842.

[7] Amaran S. and Mohan R. M., (2021). Intrusion detection system using optimal support vector machine for wireless sensor networks, in *2021 International Conference on Artificial Intelligence and Smart Systems (ICAIS)*. IEEE, pp. 1100–1104.

[8] Halbouni A., Gunawan T. S., Habaebi M. H., Halbouni M., Kartiwi M., and Ahmad R., (2022). Cnn-lstm: hybrid deep neural network for network intrusion detection system, *IEEE Access*, vol. 10, pp. 99 837–99 849.

[9] Li B., Li Y.-R., and Zhang X.-L., (2019). A survey on laplacian eigenmaps based manifold learning methods, *Neurocomputing*, vol. 335, pp. 336–351.

[10] Chkirbene Z., Erbad A., Hamila R., Mohamed A., Guizani M., and Hamdi M., (2020). Tidcs: A dynamic intrusion detection and classification system based feature selection, *IEEE Access*, vol. 8, pp. 95 864–95 877.

[11] Vembu G. and Ramasamy D., Optimized deep learning-based intrusion detection for wireless sensor networks, *International Journal of Communication Systems*, p. e5254.

[12] Chkirbene Z., Eltanbouly S., Bashendy M., AlNaimi N., and Erbad A., (2020). Hybrid machine learning for network anomaly intrusion detection, in *2020 IEEE International Conference on Informatics, IoT, and Enabling Technologies (ICIoT)*. IEEE, pp. 163–170.

[13] Aminanto M. E., Wicaksono R. S. H., Aminanto A. E., Tanuwidjaja H. C., Yola L., and Kim K., (2022). Multi-class intrusion detection using two-channel color mapping in ieee 802.11 wireless network, *IEEE Access*, vol. 10, pp. 36 791–36 801.

[14] Chkirbene Z., Erbad A., Hamila R., Gouissem A., Mohamed A., and Hamdi M., (2020). Machine learning based cloud computing anomalies detection, *IEEE Network*, vol. 34, no. 6, pp. 178–183.

[15] Panigrahi R. and Borah S., (2018). A detailed analysis of cicids2017 dataset for designing intrusion detection systems, *International Journal of Engineering & Technology*, vol. 7, no. 3.24, pp. 479–482.

[16] Chkirbene Z., Erbad A., Hamila R., Gouissem A., Mohamed A., Guizani M., and Hamdi M., Iterative per group feature selection for intrusion detection, in *2020 International Wireless Communications and Mobile Computing (IWCMC)*, 2020, pp. 708–713.

Innovation and Technological Advances for Sustainability – Al-Naemi et al. (Eds)

A comprehensive survey on client selections in federated learning

A. Gouissem
College of Computing and Information Technology, University of Doha for Science and Technology, Qatar

Z. Chkirbene & R. Hamila
Electrical Engineering, Qatar University, Qatar

ABSTRACT: Federated Learning (FL) is a rapidly growing field in machine learning that allows data to be trained across multiple decentralized devices. The selection of clients to participate in the training process is a critical factor for the performance of the overall system. In this survey, we provide a comprehensive overview of the state-of-the-art client selection techniques in FL, including their strengths and limitations, as well as the challenges and open issues that need to be addressed. We cover conventional selection techniques such as random selection where all or partial random of clients is used for the trained. We also cover performance-aware selections and as well as resource-aware selections for resource-constrained networks and heterogeneous networks. We also discuss the usage of client selection in model security enhancement. Lastly, we discuss open issues and challenges related to clients selection in dynamic constrained, and heterogeneous networks.

1 INTRODUCTION

In recent years, there has been a significant increase in the size and diversity of data sources available for analysis, including data generated by social media platforms, internet of things (IoT) devices, and a wide range of other sources. The growth in data size and sources created new challenges in terms of storage, processing, and analysis [1].

Classical analytical methods, which rely on statistical analysis and and human-based decisions, are nowadays limited in their ability to process and analyze data at scale, particularly when the data is complex or unstructured. The integration of machine learning (ML) techniques, on the other hand, can facilitate faster and more precise data analysis, resulting in better decision making and increased efficiency. Additionally, it can also help to uncover previously undetected patterns and trends that may not have been apparent using other methods. ML attract a lot of attention thanks to its ability to be applied in a wide range of areas, including image and speech recognition, relay selection [2], natural language processing, intrusion detection [3–5], anti jamming, [6–8] and even self-driving cars [9].

While ML can facilitate advanced and efficient decision-making, it relies on the collection and centralization of data, which may not be possible due to privacy concerns [10]. To address this challenge, Federated learning (FL) is used to operate on decentralized datasets that are distributed across multiple devices. In particular, FL is a machine learning method that allows data to be trained across multiple decentralized devices, such as smartphones or laptops, without the need to centralize the data [11].

FL is an emerging field in machine learning, with a growing body of literature exploring its potential benefits and challenges. Review papers such as [11–14], provide an overview of the concept and its applications, the state-of-the-art techniques in FL as well as the current

DOI: 10.1201/9781003496724-41

challenges and open issues. However, while these papers have provided a good introduction to the field, there are still many challenges that need to be addressed. One of these challenges is the issue of client selection. In Federated Learning, the selection of clients to participate in the training process is critical to the performance of the overall system. The clients selected should be representative of the entire population and should have the necessary resources and capabilities to contribute to the training process. However, selecting the appropriate clients is a non-trivial task, particularly in large-scale networks with a large number of diverse clients.

Another challenge is the issue of privacy and security. Federated learning has the potential to improve privacy and security for the individuals or organizations contributing data. However, there are still many open questions regarding how to protect the privacy and security of the clients and their data in a federated learning setting. Additionally, there is a need for techniques that can detect and prevent malicious clients from joining the network, which can compromise the overall security of the system. Adaptivity is also a key challenge in Federated Learning (FL) systems. In order to be effective, FL systems must be able to adapt to changing network conditions and client capabilities over time. However, many current FL techniques lack this ability and are unable to adjust accordingly. This limits the effectiveness of the FL system, and it is an area of active research to improve the adaptivity of these systems.

This paper is a review of client selection techniques in FL. The goal of this paper is to provide a comprehensive overview of the state-of-the-art client selection techniques in FL, including their strengths and limitations, as well as the challenges and open issues that need to be addressed in this field. This review aims to provide a comprehensive understanding of the current state of client selection in FL and to provide insights for researchers and practitioners working in this field.

2 CONVENTIONAL FL SELECTION

Several studies [15,16] have made substantial advancements in various aspects of FL, such as minimizing communication costs and addressing issues caused by non independent and identically distributed (i.i.d) data distributions. However, most of these techniques rely solely on the selection of random clients which make the model performance non optimal. In this section, the random selection as well as full and partial client selection will be thoroughly examined. The potential drawbacks of each method will also be discussed.

2.1 *Random selection*

Random selection in federated learning refers to the process of randomly selecting a subset of devices or participants from a larger group to contribute data or computation to a FL model. This method of selection aims to provide a representative sample of data from the entire group and can be used to improve the generalization of the FL model. The authors in [16] propose a method for training deep neural networks in a FL manner using iterative model averaging and conduct a thorough evaluation of its performance using different datasets. The proposed approach in [16] considers the challenges of client availability as well as unbalanced and non-IID data. However, the client selection is performed randomly. In fact, the process begins with the server selecting a group of clients randomly. Each selected client then uses the global information and their own data to perform calculations locally, after which they send the update back to the server. The server takes in these updates and updates the global information, the cycle is then repeated with different clients. This increases the processing parallelism, where a larger number of clients can work independently between each round of communication. It also increases the computations per client in order to reduce the communication cost. The simulation results show that this approach is effective even when dealing with unbalanced and non-IID data and can reduce the number of communication rounds, by 10–100 times, as compared to other methods like synchronized stochastic gradient descent.

Consequently, most of the proposed schemes for FL models rely on just selecting a random subset of participants from a large pool of available clients in each round of FL. This random selection strategy when combined with other algorithms and aggregation schemes can solve specific problems related to security, cost or data distribution. The use of random selection in Federated Learning (FL) has its limitations. It does not take into account the variations in data, resources, and systems among clients, which can impact the efficiency of federated training. By randomly selecting clients, there is a risk of selecting clients with over-represented data or clients that are slow due to limited computation speeds or network bandwidths, which can act as bottlenecks in each round of federated training. This can lead to delays and inefficiencies in the training process [17].

2.2 *Full and partial state knowledge selection*

In FL, full state knowledge means that all devices have complete knowledge of the global model, while partial state knowledge means some devices have limited knowledge and may not have access to the same information. In [18], have demonstrated convergence in federated learning models when clients fully participate in the training process. However, these studies only apply to unbiased client participation. In [19] evaluated the convergence of federated learning when clients have the ability to join or leave the training process or send incomplete updates to the server. Furthermore, most of the client selection techniques available in the literature assume the knowledge of the client-side state information [17,20–26]. However, assuming full knowledge of the client-side state information in federated learning may not be feasible or practical to obtain such information in real-world scenarios. For example, in a large-scale federated learning system with thousands of clients, it may be difficult to collect and transmit detailed state information for each client. Additionally, obtaining such information may be a privacy concern for clients. This could lead to a low adoption rate, and less participation from clients.

3 PERFORMANCE AWARE SELECTION

In Performance aware selection is a technique that chooses a specific group of devices based on their expected performance during the training process, these devices are selected to improve the performance of the model. Thus, the clients can be selected based on their statistical utility and their potential to improve the global model accuracy. Such decisions can be made based on the model updates using various measures [21,22,26,27]. In this section, various methods for selecting performance-aware federated learning techniques will be discussed.

3.1 *Active federated learning*

Active Federated Learning aims to improve the model's performance by selectively training on clients with valuable updates. The authors in [22] highlight that FL training and update process consumes a large amount of the clients' bandwidth which makes minimizing the transmission costs crucial to the success and efficiency of FL. Since the data on each client is highly variable, Active Federated Learning is proposed in [22] to allow clients to be selected in each round based on the current model and client data to maximize efficiency, rather than being chosen uniformly at random. A cost-effective sampling method is also proposed to reduce the number of required training iterations by 20–70% while maintaining the same model accuracy. Active Federated Learning is a method that capitalizes on the stability and availability of most clients during the training process, by using client-side state information to guide the selection of clients. One such approach is ClusterFL [27] that aims to improve the performance of the FL model by identifying and removing clients that are not performing as well as others or are less related to other clients within the same cluster. This approach utilizes the inherent clustering structure among clients based on their local data distributions to identify these clients and exclude them from the training process. However, Active Federated Learning can be computationally expensive to evaluate the model's performance on a large number of clients before deciding which clients to select for the next round of training.

3.2 *UCB-CS*

Upper Confidence Bound for Cooperative Stochastic Bandits (UCB-CS) is a method for selecting clients in Federated Learning that balances exploration and exploitation using the UCB algorithm to select clients that are expected to provide the most valuable updates to the model. In [17] a new approach for selecting clients in FL that is based on the multi-armed bandit algorithm called UCB-CS is proposed.It addresses issues such as communication efficiency, noise, variations in local loss values, and error floors that are present in prior methods. The authors show that UCB-CS can achieve robustness to error floor, faster convergence, and reduced state estimates of local loss values, all without requiring additional communication compared to random selection. Additionally, it ensures fairness among clients by achieving more uniform performance. The work in [17] is performed under the assumption of independent local losses for each client which might be unpractical in some scenarios since clients with similar characteristics may have similar local losses. However, UCB-CS assumes that the rewards (or updates) provided by clients are independent and identically distributed, which may not be the case in some Federated Learning scenarios.

3.3 *Contribution control mechanisms*

Contribution control mechanisms are methods for controlling the participation of clients to improve the training process. The authors in [26] propose an adaptive mechanism for selecting clients based on threshold recomputed at each round. By training the model locally and computing the local contribution of each device, the device itself can decide whether it should participate or not in the training according to a preset contribution threshold.

The work in [21] presents a study on the convergence of federated learning with a focus on biased client selection and its impact on training progress at each client. The authors found that bias towards clients with higher local losses improves convergence rates. They propose a new client selection strategy called Power-Of-Choice which results in faster convergence and better performance when compared to standard methods. The authors in [28] proposed a technique named Pyramid that aims to accelerate the FL training process while maintaining a high model accuracy. PyramidFL performs client selection and prioritization while taking into account the dissimilarities between chosen and unselected clients. It also fully utilizes the data and system heterogeneity within the selected clients to optimize their contribution into the global model. To achieve this, PyramidFL first selects clients based on their utility from the global model and then improves the selection by optimizing client utility locally. The authors also report that PyramidFL was able to improve final model accuracy and speed at the same time. However, the model's performance in PyramidFL may also be limited by the quality of the clients in the lower levels of the pyramid, as these clients are likely to have less computational and communication resources.

4 RESOURCE AWARE SELECTION

Resource-aware selection in Federated Learning is a technique that aims to select clients based on their available computational and communication resources. The selection can also be performed while aiming to optimize the limited clients resources [20,25] and communication constraints [23].

4.1 *Resource constrained networks*

The authors in [25] proposed a federated learning (FL) protocol called FedCS, which is designed to mitigate the training bottleneck in mobile edge computing (MEC) frameworks. In the FedCS protocol, a server coordinates training in a cellular network comprising participating mobile devices with heterogeneous resources. The server gathers information about wireless channel states and computing capabilities from a subset of randomly selected participants, and then selects the maximum possible number of participants that can complete training within a prespecified deadline for the subsequent global aggregation phase. A greedy algorithm is used in [29] to solve the maximization problem by iteratively selecting participants that take the least time for model upload and update. Simulation results showed that FedCS can achieve higher accuracy than an FL protocol that only accounts for training deadline, but FedCS has only been tested on simple deep neural network (DNN) models. There are limitations to the use of FedCS for more complex models, such as the difficulty in estimating the number of participants to be selected and potential bias towards selecting participants with devices that have better computing capabilities, which may not hold data representative of the population distribution.

4.2 *Reinforcement learning based selection*

One of the major challenges in implementing FL in mobile edge networks is that such networks are very dynamic and hard to predict [30]. Therefore, static user selection techniques might fail to track the network and constraints variations. In particular, these participating mobile devices have limitations in terms of energy, processing power, and wireless bandwidth. The FL server must therefore determine the most efficient use of the resources to minimize energy consumption and training time. To solve this issue, the authors in [31] proposed the use of Deep Q-Learning (DQL) to optimize the allocation of resources for training deep neural network models. The system includes mobile devices that collaborate to train the DNN models required by a FL server. These mobile devices are limited in terms of energy, processing power, and wireless bandwidth. The Double Deep Q-Network (DDQN) algorithm can also be used as suggested in [32]. The DDQN algorithm is applied to an optimization problem in which the FL server acts as the agent, the state space includes the energy and CPU state of the mobile devices, and the action space includes the amount of data and energy units that are used from the mobile devices. The rewards of this optimization problem is computed from factors such as data accumulation, energy consumption, and training latency. The results of the simulation studies show that this method can reduce energy consumption by about 31% when compared with other commonly used methods such as the greedy algorithm and can decrease training latency by up to 55% compared to random selection methods.

Other research works, such as [12], have focused on the difficulties of implementing Federated Learning in mobile edge networks and proposed potential solutions and use cases. For instance, [33] proposed a framework for effective resource management for FL in the edge. But, these studies, including [33,34], did not delve into the key challenges of creating incentives and optimizing network performance in edge-based FL.

4.3 *Heterogeneous networks*

The authors in [24] propose a technique named Oort that considers both data and system heterogeneity at the same time. Such approach creates a compromise between data usage and system efficiency, and it has been shown to have superior time-to-accuracy performance compared to random selection. However, this technique considers the differences between chosen and unselected clients, and it doesn't take into account the variations within the chosen clients.

Another approach, called hybrid-FL, has been proposed in [35] as a way to generate an approximately IID dataset using only a limited number of privacy-insensitive participants. This is done by selecting participants whose data can form an approximately IID dataset, then training a model on the collected IID dataset and merging it with the global model trained by all participants. The simulation results showed that this approach can improve classification accuracy even with a small percentage of participants sharing their data. However, it is important to note that this approach has the potential to compromise user privacy and security. To mitigate this, an incentive and reputation mechanism should be implemented to ensure that only trustworthy participants are allowed to upload their data.

4.4 *Game theory and incentive mechanisms*

Incentive mechanisms are generally used to motivate a person or a device to take specific actions. Incentives can be either positive or negative. Positive incentives encourage actions by offering rewards, while negative incentives discourage bad behavior by imposing penalties. The use of game theory to design incentive mechanisms has been extensively studied in other fields such as crowdsensing [36], edge computing [37], device to device communication [38], opportunistic networks [39] and others, where users might be regraded for using or providing specific resources, and penalized for using others. However, in FL, it is challenging to precisely quantify the value of each client's training data and to model the final learning performance of the system. As a result, it is difficult to model the utility function of each participant, which makes it hard to directly apply existing incentive mechanism design works to FL. In contrast, [40] aims to comprehensively review resource optimization and incentive mechanisms for FL in edge networks.

The authors in [40] propose a novel approach to developing an incentive mechanism for Federated Learning (FL) using game theory by using the framework of a Stackelberg game in which FL users are the followers and the base station is the leader. In this game, FL users can strategically set the number of local iterations to maximize their own utility. The base station, as the leader, then uses the best response strategies of the users to optimize FL performance. This approach offers a new perspective on how to incentivize FL participants and improve overall FL performance. The paper also identifies key open research challenges and provides guidelines for FL in edge networks. However, edge network architecture may be complex and distributed, which can make it difficult to optimize resources and implement incentive mechanisms. Moreover, FL based on a Stackelberg game can raise privacy and security concerns, as data is exchanged between devices and organizations.

5 FL CLIENT SELECTION BASED SECURITY

FL Client Selection Based Security ensures that only authorized clients and devices can access protected resources, and that sensitive data remains confidential. In this section, several techniques for selecting clients using security concepts in FL will be presented.

5.1 *Poisoning attacks in FL*

Federated Learning (FL) is known to be vulnerable to poisoning attacks [41–43]. The adversary's goal and capabilities might differ from one model to another [44]. Based on this goal, poisoning attacks can be classified into targeted and untargeted attacks. In untargeted poisoning attacks, the objective is to lower the global model's accuracy on any test input [42]. In targeted poisoning, the objective is to reduce the accuracy of some test inputs while maintaining high accuracy for the remaining inputs. Therfore, untargeted assaults may drastically degrade the performance of the global model for any test input. Poisoning may be carried out either directly on the model (model poisoning) or by changing the training data. In untargeted poisoning attacks, the goal is to minimize the accuracy of the global model on any test input [42]. In targeted poisoning attacks, the goal is to minimize the accuracy on specific test inputs, while maintaining high accuracies on the rest of the test inputs. Untargeted attacks, however, can completely deteriorate the global model performance for any test input. The poisoning can be either performed on the model itself (model poisoning) attacks, [42], or by manipulating the training data.

Several techniques have been proposed to combat such attacks. The main of focus of this survey in this regard is the investigation of the different techniques in the literature that perform the identification and isolation of the malicious nodes by selecting the most trustworthy ones data owners. In particular, several techniques exist in the literature to select the legitimate data owners. The different techniques in Section II and Section III perform the clients selection before the client training itself to avoid useless computations. However, poisoning attacks defense mechanisms perform the selection after the reception of the updates and by trying to identify anomalies in the reports.

5.2 *Krum*

Key-value Retrieval Under Misbehaviors (Krum) aims to improve the robustness of FL in the presence of Byzantine or adversarial clients. Krum works by selecting a small number of clients to update the global model, and then using the median of their updates to update the global model. The authors in [41] investigate a FL model when there are Byzantine workers present. The paper presents a Byzantine resilience client selection technique that allow the isolation of malicious nodes reports. In particular, one way to ensure this resilience is to consider a majority-based approach, but it has prohibitive computational costs. An alternative approach is the use of the KRUM metric that creates a combination of majority-based and squared-distance-based methods by choosing the vector that is closest to its $n - f$ neighbors. Krum has the advantage of having a local time complexity of $O(n^2d)$ and is shown to converge in the presence of Byzantine workers.

5.3 *Multi-Krum*

An extension of the Krum method entitled Multi-Krum is also presented in [41] and combines the resilience of Krum with the convergence speed of the standard averaging and performs well compared to other aggregation rules like the geometric median. The major issue in both Krum and Multi-Krum is that the number of nodes to be isolated from the aggregation is setup in advance and fixed by the administrator whether there is a small or large number of malicious nodes in the network. Such approach might waste legitimate nodes time and resources if the number of malicious agents is smaller than the number of isolated clients. On the other hand, isolating few nodes results in an increased risk level.

5.4 *Bulyan*

Another method used to select benign clients and gradients is proposed in [42] and called Bulyan. The Bulyan algorithm insurance Byzantine-resilience aggregation by using recursive rules based on Krum or Brute, Krum, a Medoid, the geometric median or any other Byzantine–resilient rule. In fact, Among the set of all the users, the client with closest gradient (based on KRUM or any other rule) is first selected. This client is removed from the set and the process is repeated again till obtaining the right number of clients. This approach is proven to be able to significantly reduce the space in which an attacker can cause the model to drift towards suboptimal models. Through experimental testing with different datasets, the authirs demonstrate that Bulyan successfully avoids convergence to ineffective models, instead resulting in models that are similar to those produced by a non-attacked averaging scheme.

5.5 *Trimmed-mean*

Instead of selecting the same set of clients in all dimensions, the Trimmed-mean technique proposed in [15] and [46] allows the selection of a different set of clients at each dimension. In particular, for each coordinate of the gradient, the β smallest and the β highest reports are removed and not considered in the aggregation. According to [45], Trimmed-mean reaches an error rate that is close to optimal when m is approximately $m \leq \beta \leq \frac{n}{2}$ for objective functions that are strongly convex.

5.6 *Other techniques*

Several other techniques exist in the literature to protect the FL training from poisining attacks. For example, the authors in [47] investigated a special type of data poisoning attack where Byzantine nodes may report high data sizes in order for the model aggregation to be unbalanced in their favor. A data balancing approach based on weight-truncation is proposed to protect the FL training. In [48], an enhanced version of the Alternating Direction Method of Multipliers (ADMM) algorithm is presented to ensure the convergence of decentralized learning algorithms to a suboptimal solution even when there is Byzantine data falsification. Additionally, the study includes an analysis of the speed of convergence of the algorithm. The authors in [49] analyze the dependency between iterations and aggregated gradients in order to identify and separate out Byzantine reported updates. A robust version of gradient descent is presented, which uses geometric median for model aggregation rather than traditional averaging. The mathematical analysis and simulation results in the paper support that this proposed method converges uniformly to the true gradient function. However, all of the above techniques and several others in the literature process all the received reports to generate a Byzantine-free report without actually selecting or identifying the malicious ones. Such approaches might lack accuracy in some scenarios since the malicious nodes are nodes are not isolated and can still take part of the model direction decisions.

6 OPEN ISSUES AND CHALLENGES

In this section, we will discuss the open issues and challenges in FL to further improve the performance and effectiveness of Federated Learning systems.

6.1 *Resource constraints challenges*

One of the main challenges in FL is dealing with clients that have limited resources, such as computation power and network bandwidth. Random selection of clients may result in selecting clients that are not capable of contributing effectively to the training process due to their limited resources. In addition, there is a limited number of studies that consider the dynamic nature of wireless networks. Numerous client selection strategies rely on static network circumstances, which may not correctly reflect the dynamic nature of wireless networks. Furthermore, many client selection solutions do not account for the heterogeneous capabilities of the clients, which might result in reduced performances. In addition, numerous solutions rely on a centralized approach, which is sometimes impractical in large-scale networks due to the substantial computational and communication overhead.

In general, most of the client selection approaches described in the literature for constrained networks are limited by their inability to adapt to dynamic network conditions, their disregard for client heterogeneous capabilities, and their centralized structure, which might be impractical for large-scale networks.

6.2 *Security challenges*

Another challenge in FL is ensuring the security and privacy of the clients' data. Random selection of clients may result in selecting clients that are not trustworthy or that have vulnerable systems, which can compromise the security of the FL model. The available client selection techniques in the literature have several limitations. One limitation is that many techniques rely on a centralized approach, which can make the system vulnerable to attacks on the central entity. Additionally, many techniques do not take into account the varying security capabilities of different clients, which can lead to suboptimal performance. For example, the trustworthiness of the clients should be different from one group to another based on their history and contribution to the global model. Furthermore, many techniques fail to consider the privacy concerns of the clients, which can lead to leakage of sensitive information.

Another limitation is the lack of consideration for the dynamic nature of security threats and vulnerabilities, which can change over time. Additionally, many client selection techniques do not have the ability to detect or prevent malicious clients from joining the network in advance, which can compromise the overall security of the network. In particular, most of the selection techniques in that regard let all the nodes train their models and then decide which ones to consider which might waster the client resources.

6.3 *Heterogeneity and dynamicity challenges*

In FL, clients can have different data distributions, system variations, and resource constraints. Random selection of clients may result in selecting clients that have over-represented data or clients that are not representative of the population, which can negatively impact the performance of the FL model. In particular, most of the current client selection techniques in literature have several limitations regarding heterogeneity and adaptivity. One limitation is that many techniques fail to take into account the dynamic nature of wireless networks and the dynamic capabilities of different clients, and rapidly changing network resources which can lead to suboptimal performance and communication issues. In summary, Selecting clients dynamically based on the fast changing and heterogeneous state of the system and clients, the clients' resources and the model's performance is still an open challenge

6.4 *Performance trade-off challenges*

Most of the client selection techniques in the literature focus on a single objective such as heterogeneity, security, or resources. The performance trade-off challenges in Federated Learning aim to balance between these objectives. There was limited efforts in designing a multi-objective model that can handle several selection problem at once at the same time to create Performance trade-off by balancing the communication cost, model's performance and clients participation.

7 CONCLUSION

The selection of clients to participate in the training process is a critical factor for the performance of Federated Learning (FL) models. The clients selected should be representative of the entire population and have the necessary resources to contribute to the training process. However, choosing appropriate clients is a non-trivial task, particularly in large-scale networks with a large number of diverse clients. Poor client selection can lead to a decrease in the performance of the FL system, and increase in communication and computation cost. Furthermore, the clients selected should be trustworthy and reliable, as they may have malicious intent and attempt to poison the model by providing incorrect or biased data. In this survey, we have presented the state-of-the-art client selection techniques in FL, including their strengths and limitations, as well as the challenges and open issues. We covered conventional selection techniques, performance-aware selections, resource-aware selections, and model security techniques. We also discussed open issues and challenges related to resource constraints, heterogeneity and dynamicity, and performance trade-offs. We believe that addressing the open issues and challenges related to client selection will play a crucial role in the success and widespread adoption of FL in various applications.

REFERENCES

[1] Xiaoming Li, Hao Liu, Weixi Wang, Ye Zheng, Haibin Lv, and Zhihan Lv, (2022). Big data analysis of the internet of things in the digital twins of smart city based on deep learning, *Future Generation Computer Systems*, vol. 128, pp. 167–177.

[2] Ala Gouissem, Lutfi Samara, Ridha Hamila, Naofal Al-Dhahir, Lazhar Ben-Brahim, and Adel Gastli, (2019). Machine-learning based relay selection in af cooperative networks, in *2019 IEEE Wireless Communications and Networking Conference (WCNC)*. IEEE, pp. 1–7.

[3] Zina Chkirbene, Aiman Erbad, Ridha Hamila, Ala Gouissem, Amr Mohamed, Mohsen Guizani, and Mounir Hamdi, (2020). A weighted machine learning-based attacks classification to alleviating class imbalance, *IEEE Systems Journal.*

[4] Zina Chkirbene, Aiman Erbad, Ridha Hamila, Ala Gouissem, Amr Mohamed, Mohsen Guizani, and Mounir Hamdi, (2020). Iterative per group feature selection for intrusion detection, in *2020 International Wireless Communications and Mobile Computing (IWCMC)*. IEEE, pp. 708–713.

[5] Zina Chkirbene, Aiman Erbad, Ridha Hamila, Ala Gouissem, Amr Mohamed, Mohsen Guizani, and Mounir Hamdi, (2020). Weighted trustworthiness for ml based attacks classification, in *2020 IEEE Wireless Communications and Networking Conference (WCNC)*. IEEE, pp. 1–7.

[6] Ala Gouissem, Khalid Abualsaud, Elias Yaacoub, Tamer Khattab, and Mohsen Guizani, (2022). Accelerated iot anti-jamming: A game theoretic power allocation strategy, *IEEE Transactions on Wireless Communications*, vol. 21, no. 12, pp. 10607–10620.

[7] Ala Gouissem, Khalid Abualsaud, Elias Yaacoub, Tamer Khattab, and Mohsen Guizani, (2021). Game theory for anti-jamming strategy in multichannel slow fading iot networks, *IEEE Internet of Things Journal*, vol. 8, no. 23, pp. 16880–16893.

[8] Ala Gouissem, Khalid Abualsaud, Elias Yaacoub, Tamer Khattab, and Mohsen Guizani, (2020), Combating jamming attacks in multi-channel iot networks using game theory, in *2020 3rd International Conference on Information and Computer Technologies (ICICT)*. IEEE, pp. 469–474.

[9] Vaishali M Deshmukh, B Rajalakshmi, Gopi BS Krishna, and Gourav Rudrawar, (2022). An overview of deep learning techniques for autonomous driving vehicles, in *2022 4th International Conference on Smart Systems and Inventive Technology (ICSSIT)*. IEEE, pp. 979–983.

[10] Zina Chkirbene, Aiman Erbad, Ridha Hamila, Ala Gouissem, Amr Mohamed, and Mounir Hamdi, (2020). Machine learning based cloud computing anomalies detection, *IEEE Network*, vol. 34, no. 6, pp. 178–183.

[11] Syreen Banabilah, Moayad Aloqaily, Eitaa Alsayed, Nida Malik, and Yaser Jararweh, (2022). Federated learning review: Fundamentals, enabling technologies, and future applications, *Information Processing & Management*, vol. 59, no. 6, pp. 103061.

[12] Wei Yang Bryan Lim, Nguyen Cong Luong, Dinh Thai Hoang, Yutao Jiao, Ying-Chang Liang, Qiang Yang, Dusit Niyato, and Chunyan Miao, (2020). Federated learning in mobile edge networks: A comprehensive survey, *IEEE Communications Surveys & Tutorials*, vol. 22, no. 3, pp. 2031–2063.

[13] Qiang Yang, Yang Liu, Tianjian Chen, and Yongxin Tong, (2019). Federated machine learning: Concept and applications, *ACM Transactions on Intelligent Systems and Technology (TIST)*, vol. 10, no. 2, pp. 1–19.

[14] Yanci Zhang and Han Yu, (2022). Towards verifiable federated learning, *arXiv preprint arXiv:2202.08310.*

[15] Yassine Laguel, Krishna Pillutla, Jérôme Malick, and Zaid Harchaoui, (2020). Device heterogeneity in federated learning: A superquantile approach, *arXiv preprint arXiv:2002.11223*.
[16] Brendan McMahan, Eider Moore, Daniel Ramage, Seth Hampson, and Blaise Aguera y Arcas, (2017). Communication-efficient learning of deep networks from decentralized data, in *Artificial intelligence and statistics*. PMLR, pp. 1273–1282.
[17] Yae Jee Cho, Samarth Gupta, Gauri Joshi, and Osman Yağan, (2020). Bandit-based communication-efficient client selection strategies for federated learning, in *2020 54th Asilomar Conference on Signals, Systems, and Computers*. IEEE, pp. 1066–1069.
[18] Xiang Li, Kaixuan Huang, Wenhao Yang, Shusen Wang, and Zhihua Zhang, (2019). On the convergence of fedavg on non-iid data, *arXiv preprint arXiv:1907.02189*.
[19] Yichen Ruan, Xiaoxi Zhang, Shu-Che Liang, and Carlee Joe-Wong, (2021). Towards flexible device participation in federated learning, in *International Conference on Artificial Intelligence and Statistics*. PMLR, pp. 3403–3411.
[20] Ahmed M Abdelmoniem, Atal Narayan Sahu, Marco Canini, and Suhaib A Fahmy, (2021). Resource-efficient federated learning, *arXiv preprint arXiv:2111.01108*.
[21] Yae Jee Cho, Jianyu Wang, and Gauri Joshi, (2020). Client selection in federated learning: Convergence analysis and power-of-choice selection strategies, *arXiv preprint arXiv:2010.01243*.
[22] Jack Goetz, Kshitiz Malik, Duc Bui, Seungwhan Moon, Honglei Liu, and Anuj Kumar, (2019). Active federated learning, *arXiv preprint arXiv:1909.12641*.
[23] Shaoxiong Ji, Wenqi Jiang, Anwar Walid, and Xue Li, (2020). Dynamic sampling and selective masking for communication-efficient federated learning, *arXiv preprint arXiv:2003.09603*.
[24] Fan Lai, Xiangfeng Zhu, Harsha V Madhyastha, and Mosharaf Chowdhury, (2021). Oort: Efficient federated learning via guided participant selection, in *15th {USENIX} Symposium on Operating Systems Design and Implementation ({OSDI} 21)*, pp. 19–35.
[25] Takayuki Nishio and Ryo Yonetani, (2019). Client selection for federated learning with heterogeneous resources in mobile edge, in *ICC 2019-2019 IEEE international conference on communications (ICC)*. IEEE, pp. 1–7.
[26] Monica Ribero and Haris Vikalo, (2020). Communication-efficient federated learning via optimal client sampling, *arXiv preprint arXiv:2007.15197*.
[27] Xiaomin Ouyang, Zhiyuan Xie, Jiayu Zhou, Jianwei Huang, and Guoliang Xing, (2021). Clusterfl: a similarity-aware federated learning system for human activity recognition, in *Proceedings of the 19th Annual International Conference on Mobile Systems, Applications, and Services*, pp. 54–66.
[28] Chenning Li, Xiao Zeng, Mi Zhang, and Zhichao Cao, (2022). Pyramidfl: A fine-grained client selection framework for efficient federated learning, in *Proceedings of the 28th Annual International Conference on Mobile Computing And Networking*, pp. 158–171.
[29] Maxim Sviridenko, (2004). A note on maximizing a submodular set function subject to a knapsack constraint, *Operations Research Letters*, vol. 32, no. 1, pp. 41–43.
[30] Zina Chkirbene, Amr Mohamed, Aiman Erbad, and Mohsen Guizani, (2021). Energy-efficient networks selection based deep reinforcement learning for heterogeneous health systems, in *2020 IEEE International Conference on E-health Networking, Application & Services (HEALTHCOM)*. IEEE, pp. 1–6.
[31] Tran The Anh, Nguyen Cong Luong, Dusit Niyato, Dong In Kim, and Li-Chun Wang, (2019). Efficient training management for mobile crowd-machine learning: A deep reinforcement learning approach, *IEEE Wireless Communications Letters*, vol. 8, no. 5, pp. 1345–1348.
[32] Hado Van Hasselt, Arthur Guez, and David Silver, (2016). Deep reinforcement learning with double q-learning, in *Proceedings of the AAAI conference on artificial intelligence*, vol. 30.
[33] Jihong Park, Sumudu Samarakoon, Mehdi Bennis, and Mérouane Debbah, (2019). Wireless network intelligence at the edge, *Proceedings of the IEEE*, vol. 107, no. 11, pp. 2204–2239.
[34] Xiaofei Wang, Yiwen Han, Chenyang Wang, Qiyang Zhao, Xu Chen, and Min Chen, (2019). In-edge ai: Intelligentizing mobile edge computing, caching and communication by federated learning, *IEEE Network*, vol. 33, no. 5, pp. 156–165.
[35] Naoya Yoshida, Takayuki Nishio, Masahiro Morikura, Koji Yamamoto, and Ryo Yonetani, (2019). Hybrid-fl: Cooperative learning mechanism using non-iid data in wireless networks, *arXiv preprint arXiv:1905.07210*.
[36] Dejun Yang, Guoliang Xue, Xi Fang, and Jian Tang, (2012). Crowdsourcing to smartphones: Incentive mechanism design for mobile phone sensing, in *Proceedings of the 18th annual international conference on Mobile computing and networking*, pp. 173–184.

[37] Yang Liu, Changqiao Xu, Yufeng Zhan, Zhixin Liu, Jianfeng Guan, and Hongke Zhang, (2017). Incentive mechanism for computation offloading using edge computing: A stackelberg game approach, *Computer Networks*, vol. 129, pp. 399–409.

[38] Peng Li and Song Guo, (2015). Incentive mechanisms for device-to-device communications, *IEEE Network*, vol. 29, no. 4, pp. 75–79.

[39] Yufeng Zhan, Yuanqing Xia, Jinhui Zhang, and Yu Wang, (2017). Incentive mechanism design in mobile opportunistic data collection with time sensitivity, *IEEE Internet of Things Journal*, vol. 5, no. 1, pp. 246–256.

[40] Latif U Khan, Shashi Raj Pandey, Nguyen H Tran, Walid Saad, Zhu Han, Minh NH Nguyen, and Choong Seon Hong, (2020). Federated learning for edge networks: Resource optimization and incentive mechanism, *IEEE Communications Magazine*, vol. 58, no. 10, pp. 88–93.

[41] Peva Blanchard, El Mahdi El Mhamdi, Rachid Guerraoui, and Julien Stainer, (2017). Machine learning with adversaries: Byzantine tolerant gradient descent, *Advances in Neural Information Processing Systems*, vol. 30.

[42] Rachid Guerraoui, Sébastien Rouault, *et al.*, (2018). The hidden vulnerability of distributed learning in byzantium, in *International Conference on Machine Learning*. PMLR, pp. 3521–3530.

[43] A. Gouissem, K. Abualsaud, E. Yaacoub, T. Khattab, and M. Guizani, (2022). Federated learning stability under byzantine attacks, in *2022 IEEE Wireless Communications and Networking Conference (WCNC)*, pp. 572–577.

[44] Zina Chkirbene, Habib Ben Abdallah, Kawther Hassine, Ridha Hamila, and Aiman Erbad, (2021). Data augmentation for intrusion detection and classification in cloud networks, in *2021 International Wireless Communications and Mobile Computing (IWCMC)*. IEEE, pp. 831–836.

[45] Dong Yin, Yudong Chen, Ramchandran Kannan, and Peter Bartlett, (2018). Byzantine-robust distributed learning: Towards optimal statistical rates, in *International Conference on Machine Learning*. PMLR, pp. 5650–5659.

[46] Cong Xie, Oluwasanmi Koyejo, and Indranil Gupta, (2018). Generalized byzantine-tolerant sgd, *arXiv preprint arXiv:1802.10116*.

[47] Amit P. and Danny H., (2020). Towards Realistic Byzantine-Robust Federated Learning, *CoRR*, vol. abs/2004.04986.

[48] Li Q., Kailkhura B., Goldhahn R., Ray P., and Varshney P. K., (2017). Robust federated learning using admm in the presence of data falsifying byzantines, *arXiv preprint arXiv:1710.05241*.

[49] Chen Y., Su L., and Xu J., (2017). Distributed statistical machine learning in adversarial settings: Byzantine gradient descent, *Proceedings of the ACM on Measurement and Analysis of Computing Systems*, vol. 1, no. 2, pp. 1–25.

Innovation and Technological Advances for Sustainability – Al-Naemi et al. (Eds)

Enabling new generation services with multi-level delegation architecture of slicing technology

Amal Kammoun
Universite Sorbonne Paris Nord – L2TI, Villateneuse, France

Abdulhalim Dandoush
University of Doha for Science and Technology, Doha, Qatar
Esme Research Lab, France

Gladys Diaz
Universite Sorbonne Paris Nord – L2TI, Villateneuse, France

Nadjib Achir
Universite Sorbonne Paris Nord – L2TI, Villateneuse, France
INRIA, France

Nabil Tabanne
MEDIATRON Laboratory, Sup'Com, University of Carthage, Tunisia

ABSTRACT: 5G/6G network slicing is identified as key enabler technology for allowing a diversity of sustainable applications while satisfying user's requirements. A major problem of the traditional networking technologies is the use of the "one size fits all" approach that handles all types of services. In this work, we focus on drawing a new architecture born of the need of a multi-level services' orchestration for seeking flexible intelligent management of the new generation use-cases such as massively deployed cloud and edge-cloud based IoT applications for (i) environment monitoring, (ii) Agriculture development, (iii) and new generation Augmented Reality applications in medicine or e-learning. This paper proposes a novel multi-level Delegation Architecture for Network Slicing Orchestration (DANSO). DANSO proposes a multi-level delegation for slices management and optimization framework for resources allocation. Our proposal is based on three pillars: (i) definition of levels to fulfill support of network slicing deployment and management via delegated decisions (ii) split of orchestrator, manager and control roles and (iii) placement-independence of entities in (ii). DANSO has the following advantages. First, it provides an optimization mechanism for the deployment of network slices. second, it takes into consideration a negotiation process for the requested slice. Third, it manages the life-cycle not only for the Virtual Network Function (VNF) level but also for the dynamic deployment and suppression of the whole slices. last, it considers network slicing over several administrative domains.

Keywords: Intelligent management, Network slicing, Network Architecture, Network Function Virtualization, Orchestration, Software Defined Networking

1 INTRODUCTION

Nowadays, services are on the top of digital transformation for sustainable development involving new uses cases such as e-health, autonomous vehicles, smart grid, and smart traffic light. Also, several types of users (e.g., end-user, enterprises, IT systems, etc.) are involved in

DOI: 10.1201/9781003496724-42

this transformation and have application requests with different constraints in terms of resources and quality of services.

The telecommunication and computer network is the nerve that connects all parts of the system together. The faster and more accurate the information transfer, the better the productivity. This vital nerve is based on allocating huge resources such as cables, bandwidth, routers, computing power, storage spaces, energy, and antennas. It is worth mentioning that the more efficient our consumption of resources, the more we reduce waste and move closer to a society that is environmentally and nature friendly.

In fact, the major problem of the traditional networking technologies [1,2] is the use of the *"one size fits all"* approach that handles all types of services and users requirements over a single architecture [3,4]. This is the main cause of the lack of flexibility and scalability of the network solutions. Therefore, service providers and network operators have to face the dilemma about providing efficiently emerging services while satisfying the users requirements without increasing linearly their capital expenditures (CAPEX) and operating expenses (OPEX).

Recent efforts have proposed new system aiming at meeting recent challenges such as [5]. The key concepts for the 5G vision are the Software Defined Networking (SDN) [6], the Network-Function Virtualization (NFV) [7] and network slicing technology [4].

In addition to the flexibility, agility and cost efficiency, network slicing is also the key technology for a multi-domain and multi-tenant environment. It receives the attention of several research groups and standardization bodies such as NGMN [5], ONF [8], ITU-T [?] and 3GPP [9]. In this paper we define a network slice as a customized virtual network running on top of a shared physical infrastructure that may cross several domains. It is composed of a set of connectivity (links), storage, and computing resources dedicated for one application to meet specific needs. According to the literature, a network slice may include one or more network service. Each network service is composed of one or more Physical Network Function (PNF) or Virtual Network Function (VNF).Most of the current network slices deployments are based on centralized orchestration and management solutions which are aligned with the ETSI Management and Orchestration of Network (MANO) Functions Virtualisation (NFV) specification [10]. However, we believe that the coexistence of both centralized and decentralized approaches may enhance, according to the use-case and services requirement, the network performance and provide more flexibility and rapidity for the slice deployment. In this paper, we present DANSO as a multi-level SDN/NFV architecture that defines a multi-level delegation and aims at creating network slices and ensuring their management on run-time. This architecture takes advantage of several proposals in the literature and presents a global vision to deal with the slice creation, control and management. The design of DANSO is based on three pillars:

1) Definition of levels to fulfill support of network slicing deployment and management via delegated decisions to cross-layer control and management functions.
2) Splitting of orchestrator, manager and controller roles into entities defined in these levels.
3) Independent placement for deployment of entities in (2).

DANSO defines four different levels identified as:

(A) *Network Store Layer:* maintains a catalog of pre-defined services ready to be deployed for network slices.
(B) *Control System Layer:* defines the required modules to deploy and configure slice's resources.
(C) *Slice Service Layer:* contains the set of deployed slices and manages the slices services and their life-cycle.
(D) *Virtual Infrastructure Layer:* constitutes the DANSO capacity for deployments (i.e. an abstracted view of resources). DANSO assumes that forwarding elements and servers have been negotiated with infrastructure provider(s) for all the pre-defined use-cases presented in the catalog.

According to DANSO deployment process, the user has access to a business application catalog maintained by the "Network Store Layer". Thus, the user selects the slice that fits its requirements. The "Control System Layer" combines the required network functions according to service and infrastructure requirements, manages the admission control of slice deployment request, and defines the description of network slice to be deployed. Also it stores information to maintain an updated the global view of both: slice deployments and virtual infrastructure capacities (virtual resources). The "Service Slices Layer" takes decisions about slice deployment (creation of new slice or aggregation into an existing slice). Slices deployments are performed into virtualized infrastructure such that described in [?].

The remainder of this paper is organized as follows. Section 2 present the motivation behind our proposal after over-viewing some related work in the context of the network slicing. Our proposed architecture is introduced and described in section 3. In section 4 we describe key architecture operations such as the slice creation and management. Last, we conclude our work in section 5.

2 RELATED WORK

In this section, we present selective key standards and research proposals in the context of control, management and orchestration of network slicing. As part of the 5G-PPP program [11], several project have been interested in the SDN/NFV paradigm such as SONATA [12], 5GEX [13], 5G NORMA [14] and 5G SLICENET [15]. For instance, 5G NORMA has presented a new architecture for mobile networks that proposes multi-service and multi-tenancy support. It is based on the composition and decomposition of network functions which allows the support of different requirements of 5G services. 5G SLICENET is an End-to-End (E2E) architecture for cognitive network-slicing and management of network slices in a virtualized multi-domain multi-tenant network. Most of the existing projects does not provide an optimization mechanism for the deployment of network slices. However, an efficient mapping of network slices functions and services on the underlying infrastructure is an important aspect that may allow an optimal use of resources and a maximization of deployed network slices and operator's profit. In addition, current proposals in the context of 5G-PPP, does not propose any negotiation mechanism between the users and slice providers in case of slice request drop (e.g., due to capacity limitation). A negotiation can lead to an agreement between the two actors which can lead to an increase of the accepted services.

The recursion of network slices is an important aspect for the future network architecture that is also not considered by several projects. Finally, MANO has proposed a centralized solution for the management and orchestration of network resources and functions which poses some limitations mainly with the reaction response time and the scalability while dealing with several administrative domains (i.g., implication of several cloud, mobile or broadband network operators). In other words, due to the centralization, the delay as well as the overhead communication will increase in the system.

3 DANSO ARCHITECTURE

In this section we detail the DANSO architecture depicted in Figure 1. It is a multi-level architecture which is composed of i) a network store that contains a catalog of several network services ready to be used by the users (III-A), ii) several systems modules to ensure optimization and dynamicity for slices deployment and management, and also to update needed information (e.g. data bases about network topologies, resources and functions states, etc.) (III-B), iii) a set of slices that forward users' flows (III-C) and iv) servers and forwarding elements that constitute the virtual infrastructure (III-D). Orchestration and management of all the actions in the system is delegated to several components. Thus, for

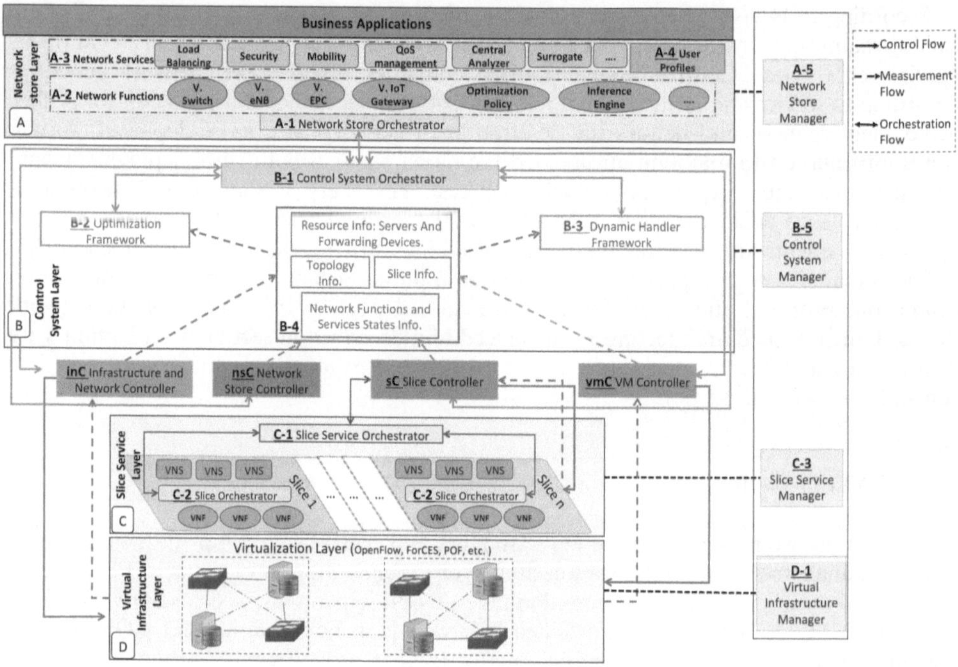

Figure 1. DANSO Architecture overview.

each level a "Manager" and an "Orchestrator" are associated. Also, "Controllers" are defined as being cross-layer entities. The general role of these entities are as follows:

- *Orchestrator:* plays the role of northbound interface for each layer. It is a decisional entity which receives command or information request, it translates the conditions of the order into configuration tasks, then it forwards them to be treated at adequate controller. To deal with an E2E treatment, orchestrator coordinates also the communications between entities at different layers.
- *Manager:* is responsible for the supervision of operations executed at its layer. It ensures correctness of the creation and management processes. It handles the life-cycle of processes and services involved in its layer.
- *Controller:* plays the role of executor and programmer entity. When creating a network slice, it executes the associated process (e.g., optimization of slices placement, mapping of network components into virtual infrastructure, VM allocation, install forwarding rules, etc.) in accordance with the orders coming from the "Control System Orchestrator".

Details of this proposal are described in the following subsections.

3.1 *Network store layer*

The *Network Store Layer* provides a catalog of use cases which are exposed to end users. It has the same utility as the utility of the application store for the software platform (e.g. App Store, Play Store, etc.). Thus, following a user's request, appropriated Virtual Network Functions (VNF) and Virtual Network Services (VNS) are chosen of this catalog to build the chain of NFV (combining VNF(s) and/or VNS(s)). As an example of Network Store we can mentioned the project T-NOVA [16], which provides a marketplace for virtualized network functions. In the following we describe each element of the Network Store layer:

3.1.1 *Business applications*
This represents the higher northbound interface of DANSO. This sub-layer encompasses several use cases ready to be deployed by the proposed architecture upon users request, and by referring for example, to Ericsson [17], NGMN [5] and 3GPP [9]. As examples of those applications, we can list smart wearable, sensor networks, and drones connectivity.

3.1.2 *Network store orchestrator (A-1)*
It coordinates the communication between the *Network Store Layer* and other layers. It receives demands from the *Control Orchestrator (B-1)*, orchestrates them, and then provides, from the *Network Store Layer*, the needed information, functions or services.

3.1.3 *Network functions store (A-2)*
It serves as a database that contains the metadata needed for the function provisioning. This metadata describes the required resources for a network function.

3.1.4 *Network services store (A-3)*
This store provides to end-users value-added services. They are implemented on the fly by the network system on the executed slices in order to extend the executed application.

3.1.5 *User profiles (A-4)*
This component is a database that encodes users policies. The "System Control orchestrator" uses this component to apply policies to each user's request according to their profile.

3.1.6 *Network store manager (A-5)*
It is responsible for the management of the *Network Services* and *Functions Stores*. It is in charge of on-boarding new NF and NS to the *Network Store Layer*, deleting, modifying or updating functions and services of this level.

3.2 *Control system layer*

We define in this level the fundamental modules supporting the optimization of deployments and the management of events introducing changes in the behavior of deployed slices.

3.2.1 *Control system orchestrator (B-1)*
Receives the user's request, translates it through its translator module and then communicates it to the controllers set. In addition, it coordinates the communication between the controllers as well as between controllers and the *Control System Manager*. Once the message is received, "Control System Orchestrator" decides to which controller it should be delivered.

3.2.2 *Control system manager (B-5)*
Defines management functions based on two fundamental modules: the *Optimization Framework (B-2)* and the *Dynamic Handler Framework (B-3)*:

- In a virtualized environment, resources can host and execute several functions and virtual machines simultaneously. Since the capacities of servers and network devices are limited, efficient optimization strategies should be implemented. For this purpose, the ***Optimization Framework-OF (B-2)*** takes into consideration users requirements as well as the network system state in order to optimize the service chain creation, the VM allocation, functions and services assignment to VMs and the deployment of the slices over the network infrastructure.
- In order to deal with events that occurs in the system such as the congestion of a slice, the failure of a VM implementation, the user mobility, etc., we need to monitor in real time if

the deployed slices. The ***Dynamic Handler Framework-DHF (B-3)*** analyzes periodically the information database and detects any events that can trigger any scalability process, deletion or modification of slices as well as VMs and VNFs.

3.2.3 *Information base (B-4)*

This database gathers the information collected by the controllers. This information relates to several network aspects and it is classified into four categories as following. The first database (*Resources information: servers and forwarding device information*) exports information related to the state of the servers and forwarding devices, such as the current processing capacity of the servers, the available link bandwidths, the number of VMs per server and also the list of implemented VNF at the VMs. If any change occurs in this information, controllers have to update the database with new measurement. The second database (*Topology information*), exports the network topology which is reported by the forwarding devices. The third database (*Virtual Network Functions states information*) export the state of the VNFs.

3.3 *Slice service layer*

The slice service layer contains the set of the deployed slices over the underlying infrastructure. Each network slice is isolated from other slices and it is dedicated for a specific type of service. This level defines several components as follows:

3.3.1 *Slice service orchestrator (C-1)*

This orchestrator manages the set of slices orchestrators. It focuses on the management of the network services life-cycle. Basically, it manages the network service view associated with each network slice. During the operational phase, it communicates with the *slice Controller (sC)* to identify the functions and services to be added or removed from slices. Finally, since it communicates directly with each Slice Orchestrator (C-2), it ensures isolation among the network slices.

3.3.2 *Slice orchestrator (C-2)*

It is in charge of the end-to-end life-cycle and management of each slice. Basically, it communicates its slice information and requests to the *Slice Service Orchestrator (C-1)*. It also implements controller commands and decisions (e.g. packet forwarding rules) into slices forwarding devices.

3.3.3 *Slice service manager (C-3)*

It oversees the deployed network slices as well as their life-cycle management. It has also an overall coordination role between *Slice Service Layer* elements and *Control System Layer.*

3.4 *Virtual infrastructure layer*

The virtual infrastructure level encompasses all software and hardware resources that form the virtualized environment. It includes the connectivity between data centers and clouds systems as well as the computing, storage and network capabilities. Within this level, we find the ***Virtual Infrastructure Manager (D-1)*** that controls.

3.5 *Controllers: Cross-layer entities*

3.5.1 *Network store controller (nsC)*

It is in charge of receiving the user's request description. Then, it communicates with the *Network Store Layer* in order to transform this demand into a service chain.

3.5.2 *Slice controller (sC)*

It is responsible to execute several tasks:

- to map a new slice request to an already created slice, if it finds a deployed slice for this type of service, or it will ask for the creation of a new slice,
- to assign a dedicated *Slice Orchestrator (C-2)* for every created slice,
- to interact with the *Network Store Controller (nsC)* in order to request the instantiation of VNFs from the *Network Store Layer (A)*,
- to communicate with the *VM Controller (vmC)* to ask for the setting up of a new virtual machine for a VNF or to ask for the release of all the allocated resources of a slice or of some virtual machines identified by the system.

3.5.3 *Infrastructure and network controller (inC)*

It is responsible for the control and the supervision of the entire underlying infrastructure (concerning computer and network resources). It is in charge of determining which resources will be allocated for a slice while taking into account the description of logical resources as well as the state of the virtual infrastructure.

3.5.4 *VM controller (vmC)*

It is responsible for the programming of virtual machines with different network functions and services.

4 DANSO OPERATIONS

DANSO architecture considers several network aspects like the end-to-end slice creation and management. Therefore, several modules and algorithms are executed in order to provide for the end users the best QoS and QoE.

4.1 *Slice creation*

Several steps are required for the slice deployment. Those steps start by the analysis of the user's request, then the creation of the slice description according to the user's requirements and finally the slice implementation.

4.1.1 *Request analysis*

When the *Control System Orchestrator (B-1)* receives the user's demand, it starts by defining the set of its requirements and also the information about the user profile in order to have a complete description of the user's request. Following this step, the *Control System Orchestrator (B-1)* prepares a slice description ready to be implemented according to that demand. The sequence of those actions is schematized in Figure 2.

4.2.2 *Slice description*

After analyzing the request, the *slice controller (sC)* searches for an existing deployed slice that responds to this demand. Two cases can be envisaged:

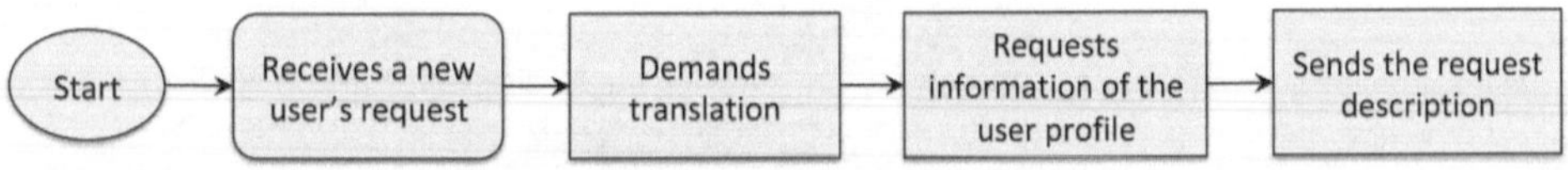

Figure 2. Request analysis algorithm executed by the control system orchestrator (B-1).

(a) First, the *slice controller (sC)* may find some deployed VNFs that can be reutilized for the user's request. In fact, due to the limited capacities of the system resources, the reutilization of some VNF may be considered. However, only some types of VNF, such as routing, can be shared by several users because of the isolation constraint. In this case, the *slice controller (sC)* will study the performances of the VNF which will be reutilized. In our previous work [18], we have proposed an algorithm for the performance evaluation in terms of reliability, availability and latency.
(b) Second, the *slice controller (sC)* will create a new slice description while specifying the needed information for its implementation.

Figure 3 presents the sequence of the described actions for the creation of the slice description.

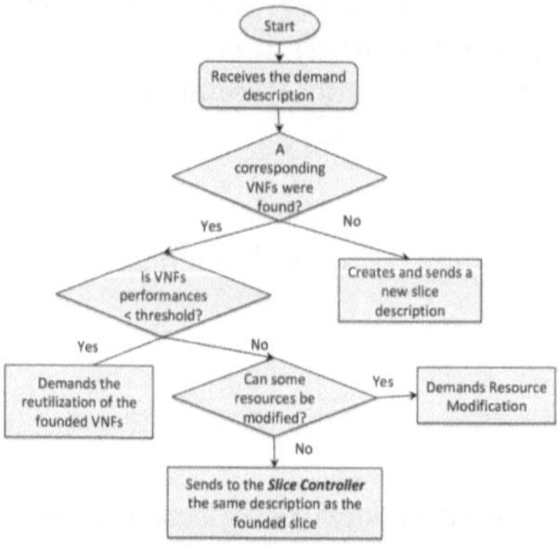

Figure 3. Slice description algorithm executed by the slice controller (sC).

4.1.3 *Slice implementation*
After defining the required operation for the user's request, three actions are possible as shown in the Figure 4.

(a) If the defined action is the reutilization of some already implemented VNFs, then the *slice controller (sC)* will create a slice which is composed of the reutilized VNFs and the new created VNFs. Then it will inform the *Slice Orchestrator (C-2)* and the *Slice Service Orchestrator (C-1)* about this new operation.
(b) If the action is the modification of the slice, then the *slice controller (sC)* will modify the identified VNFs and map the request into the created slice.

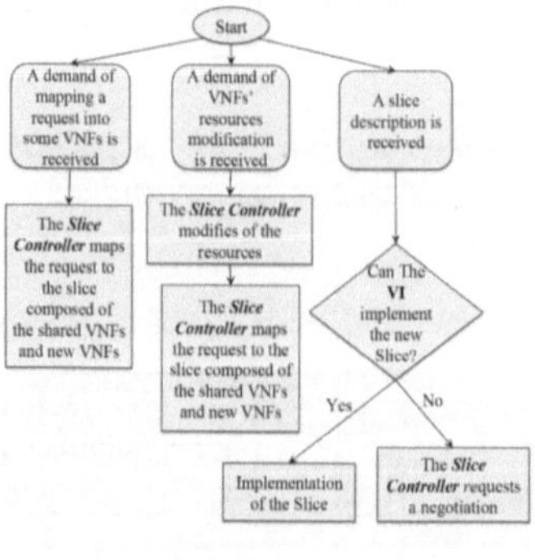

Figure 4. Slice implementation algorithm executed by the slice controller (sC).

(c) Finally, if the action is to create a new slice, then the *slice controller (sC)* studies the *Virtual Infrastructure (VI)* performances to determinate if VI is able to support another slice implementation or not. If the VI can implement a new slice, then the *infrastructure and network controller (inC)* reserves the needed network resources and implements the list of VNF defined in the already prepared slice description. On the other hand, a **request negotiation** will be executed.

4.1.3.1 *Request negotiation* In the case of request negotiation, the *Control System Orchestrator (B-1)* proposes to the user a new description for its demand and a negotiation will take place in order to find a compromise between them. If the slice can be modified then the new user's request will be sent and treated by the *Control System Orchestrator (B-1)*. If the request cannot be modified, then *Virtual Infrastructure Manager (D-1)* may ask other cloud providers for additional resources, as in [19]. However, this case is not considered in this paper and it can be the object of future work. The sequence of the mentioned actions is presented in Figure 5.

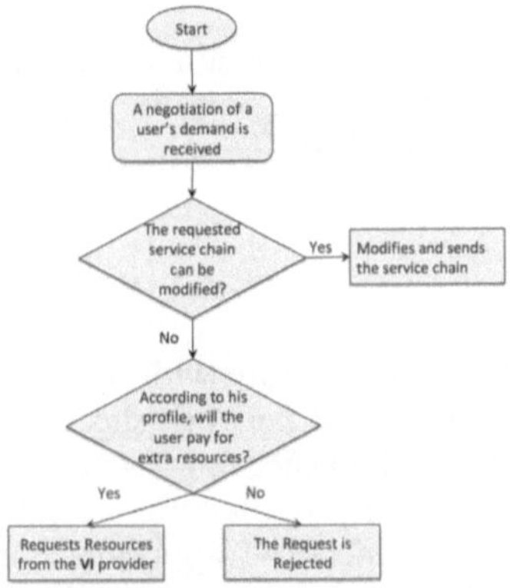

Figure 5. Request negotiation algorithm executed by the slice controller (sC).

4.2 *Slice management*

During slices execution and network operations, many unexpected events can occur. To deal with those sudden events, we propose the ***Dynamic Handler Framework (DHF)*** that surveys the network state, analyzes it and informs the controllers in case of any dysfunction. The DHF may detect several types of events and for each event it triggers the adequate action. Figure 6 presents the sequence of actions to be triggered for two events: slice modification and suppression. In the following we detail some of possible actions.

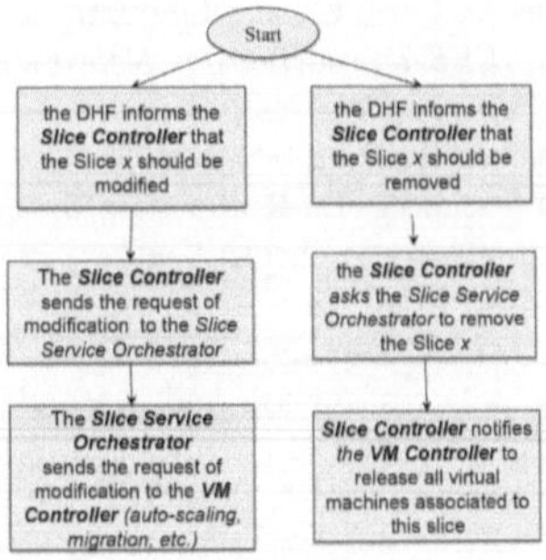

Figure 6. Slice management algorithm.

4.2.1 *Slice suppression*

A slice removal can be required by several reasons: QoS is no longer assured for the slice. In the case of a slice removal, the *DHF* notifies the *slice controller (sC)*. This latter will ask the *Slice Service Orchestrator (C-1)* to remove the identified slice. This action will take place through an exchange between the *Slice Service Orchestrator (C-1)* and the *Slice Orchestrator (C-2)* of the concerned slice.

Finally, the *slice controller (sC)* notifies the *VM controller (vmC)* to release all virtual machines associated to this slice.

4.2.2 *Slice and resources re-provisioning*

Resources re-provisionning can be envisaged to answer with several situations such as "Slice congestion" or "User's mobility". In the case of Slice congestion, the*DHF* asks the *slice controller (sC)* to either increase the performances of this slice (through the migration the auto-scaling of some VNF) or to create another slice with the same description and move some user flow's to this new slice.

4.2.3 *Slice modification*

Concerns all changes involving the current chain of services (e.g. addition, removal or migration of NFVs). User may modify its slice by requesting new functionalities (e.g. enhanced security or additional bandwidth).

5 CONCLUSION

This paper has presented DANSO architecture with a multi-delegation approach for the orchestration, control and management of network slices. Each component of this architecture is described. The slice creation and slice management operations are explained. As a future work, we aim at prototyping our proposed architecture. We also would like to consider a specific 5G use case, such as critical IoT application, in order to highlight quantitatively the efficiency of our proposal compared to other solutions.

REFERENCES

[1] Dandoush A., Alouf S., and Nain P., (2009). Performance Analysis of Centralized Versus Distributed Recovery Schemes in P2P Storage Systems, in *NETWORKING 2009*, L. Fratta, H. Schulzrinne, Y. Takahashi, and O. Spaniol, Eds. Berlin, Heidelberg: Springer Berlin Heidelberg, pp. 676–689.

[2] Lifetime and availability of data stored on a p2p system: Evaluation of redundancy and recovery schemes,' *Computer Networks*, vol. 64, pp. 243–260, (2014). [Online]. Available: https://www.sciencedirect.com/science/article/pii/S1389128614000772

[3] Chahbar M., Diaz G., and Dandoush A., (2019). Towards a unified network slicing model, in *2019 15th International Conference on Network and Service Management (CNSM)*, pp. 1–5.

[4] Chahbar M., Diaz G., Dandoush A., Cerin C., and Ghoumid K., (2021). A comprehensive survey on the e2e 5g network slicing model, *IEEE Transactions on Network and Service Management*, vol. 18, no. 1, pp. 49–62.

[5] NGMN 5G White Paper, NGMN Alliance, (2015), v. 1.0.

[6] Kreutz D., Ramos F. M. V., VerÃ-ssimo P. E., Rothenberg C. E., Azodolmolky S., and Uhlig S., (Jan 2015). Software-defined networking: A comprehensive survey, *Proceedings of the IEEE*, vol. 103, no. 1, pp. 14–76.

[7] Mijumbi R., Serrat J., Gorricho J. L., Bouten N., Turck F. D., and Boutaba R., (Firstquarter 2016). Network function virtualization: State-of-the-art and research challenges, *IEEE Communications Surveys Tutorials*, vol. 18, no. 1, pp. 236–262.

[8] ONF, https://www.opennetworking.org/.

[9] *Feasibility Study on New Services and Markets Technology Enablers, Stage 1, rel-14, V14.2.0*, 3GPP TR 22.891, (2016).
[10] *Network Functions Virtualisation (NFV); Management and Orchestration*, GS NFV-MAN 001, (Dec. 2014), v. 1.1.1.
[11] 5gppp, https://5g-ppp.eu/.
[12] Dräxler S., Karl H., Peuster M., Kouchaksaraei H. R., Bredel M., Lessmann J., Soenen T., Tavernier W., Mendel-Brin S., and Xilouris G., (May 2017). Sonata: Service programming and orchestration for virtualized software networks,' in *2017 IEEE International Conference on Communications Workshops (ICC Workshops)*, pp. 973–978.
[13] Bernardos J., Gerö B. P., Di Girolamo M., Kern A., Martini B., and Vaishnavi I., 5gex: realising a europe-wide multi-domain framework for software-defined infrastructures, *Transactions on Emerging Telecommunications Technologies*, vol. 27, no. 9, pp. 1271–1280. [Online]. Available: https://onlinelibrary.wiley.com/doi/abs/10.1002/ett.3085
[14] Rost P., Mannweiler C., Michalopoulos D. S., Sartori C., Sciancalepore V., Sastry N., Holland O., Tayade S., Han B., Bega D., Aziz D., and Bakker H., (May 2017). Network slicing to enable scalability and flexibility in 5g mobile networks, *IEEE Communications Magazine*, vol. 55, no. 5, pp. 72–79.
[15] Wang Q., Alcaraz-Calero J., Weiss M. B., Gavras A., Neves P. M., Cale R., Bernini G., Carrozzo G., Ciulli N., Celozzi G., Ciriaco A., Levin A., Lorenz D., Barabash K., Nikaein N., Spadaro S., Morris D., Chochliouros J., Agapiou Y., Patachia C., Iordache M., Oproiu E., Lomba C., Aleixo A. C., Ro-Drigues A., Hallissey G., Bozakov Z., Koutsopoulos K., and Walsh P., (June 2018). Slicenet: End-to-end cognitive network slicing and slice management framework in virtualised multi-domain, multi-tenant 5g networks, in *2018 IEEE International Symposium on Broadband Multimedia Systems and Broadcasting (BMSB)*, pp. 1–5.
[16] Xilouris G., Trouva E., Lobillo F., Soares J. M., Carapinha J., McGrath M. J., Gardikis G., Paglierani P., Pallis E., Zuccaro L., Rebahi Y., and Kourtis A., (June 2014). T-nova: A marketplace for virtualized network functions, in *2014 European Conference on Networks and Communications (EuCNC)*, pp. 1–5.
[17] Ericsson, Ericsson white paper, 5g systems, https://www.ericsson.com/assets/local/publications/white-papers/wp-5g-systems.pdf , 2017, accessed: 03-04-2018.
[18] Kammoun A., Tabbane N., Diaz G., Dandoush A., and Achir N., End-to-end efficient heuristic algorithm for 5g network slicing, in *2018 IEEE 32st International Conference on Advanced Information Networking and Applications (AINA)*, May 2018, accepted.
[19] Assis and Bittencourt L., (2016). A survey on cloud federation architectures: Identifying functional and non-functional properties, *Journal of Network and Computer Applications*, vol. 72, pp. 51–71. [Online]. Available: http://www.sciencedirect.com/science/article/pii/S1084804516301436

Innovation and Technological Advances for Sustainability – Al-Naemi et al. (Eds)

Secure and privacy-preserving federated learning-based resource allocation for next generation networks

Abdulhalim Dandoush
Division of Information Technology, College of Computing and Information Technology, University of Doha for Science and Technology, Doha, Qatar
ESME Sdurial L2TI at Sorbonne University, Paris Nord, France

Ala Gouissem
Division of Information Technology, College of Computing and Information Technology, University of Doha for Science and Technology, Doha, Qatar

Bekir Sait Ciftler
Division of Data Science and Artificial Intelligence, College of Computing and Information Technology, University of Doha for Science and Technology, Doha, Qatar

ABSTRACT: This paper surveys recent research on federated learning-based resource allocation for next-generation networks in order to identify research gaps and potential future directions. We start by outlining the main challenges and requirements for secure and privacy-preserving resource allocation in these networks. The existing solutions and techniques proposed in the literature are then reviewed. The strengths and limitations of these solutions are analyzed, and trade-offs in terms of security, privacy, efficiency, and scalability are discussed. The research gaps and promising directions for future research are identified, such as integrating multiple solutions (e.g., game theory or optimization techniques with federated learning) and developing new models and algorithms for AI-based resource allocation that address specific challenges and requirements of next-generation networks. The goal is to inspire further research in this critical and rapidly evolving field.

1 INTRODUCTION

Resource allocation is crucial for next-generation networks, as it determines the efficient and fair distribution of limited resources among users and applications [1–4]. The complexity of these new paradigm networks (e.g., Software Defined Mobile Edge) is making it difficult to use traditional network optimization methods that rely on static models. A more effective approach is to use data-driven methods based on machine and deep learning to optimize resource scheduling, allocation, and orchestration. For example, deep learning can be used to learn the characteristics of network conditions, and deep reinforcement learning can be used to make decisions based on the dynamic environment. However, using these learning methods raises fundamental challenges related to security and privacy, as the decision-making process may be vulnerable to attacks [5], and the data used to train the AI models may contain sensitive information. In fact, these methods require access to user data, which may be sensitive or restricted due to regulations. With the growth of privacy concerns due to recent data privacy legislation (e.g., the General Data Protection Regulation (GDPR) adopted by the European Commission in 2018), the concept of federated learning (FL) has attracted a lot of attention. Federated learning is a technique for training machine learning

DOI: 10.1201/9781003496724-43

models on decentralized data. In federated learning, multiple clients, each with their own data, collaborate to train a shared model without sharing their data. Instead, clients send updates to the model parameters to a central server, which aggregates the updates to produce a new global model.

In this work, we present a survey of the recent research on AI-based resource allocation for next-generation networks, focusing on secure and privacy-preserving approaches.

Our goal is to identify the research gap and potential directions for future work in this field. To this end, we first provide an overview of the main challenges and requirements for secure and privacy-preserving resource allocation in next-generation networks. We then review the existing solutions and techniques proposed in the literature related to the application of FL for network resource allocation, coupled with traditional optimization, game theory, and machine learning (ML) based approaches. We analyze the strengths and limitations of these solutions and discuss their trade-offs in terms of security, privacy, efficiency, and scalability.

The remaining of this paper is organized as follows. In Section 2.1, we overview the use of optimization and FL models for prediction and making decisions in the new generation networks. The integration of game theory techniques with FL is discussed in Section 2.2. We provide in Section 2.3 the advantages of using the FL approached powered by Deep Reinforcement Learning (DRL) techniques over the integration of traditional optimization or game theory techniques. Then, some key studies are surveyed. In Section 3, we present some research gaps and introduce future directions. Last, Section 4 concludes our work.

2 LITERATURE REVIEW

2.1 *Hybrid FL and operational research approaches*

In this section, we present an overview of the use of optimization and FL models together for making decisions on scheduling and resource allocation problems in 5G and 6G network orchestration and management. These decisions are critical to ensuring efficient and effective use of network resources, while also meeting the Quality of Service (QoS) and Quality of Experience (QoE) requirements of the new generation of applications [6–8].

Optimization techniques are methods for adjusting the parameters of a model in order to minimize some measure of error or loss [9]. Examples of optimization techniques include gradient descent and its variants, such as stochastic gradient descent (SGD) and Adam. These optimization techniques are used in both centralized and federated learning settings. Hybrid systems are possible, where optimization techniques are used in conjunction with federated learning to train models on decentralized data. In this case, the optimization method would be used to change the parameters of the model on the clients' devices, while federated learning would be used to collect the updates from the clients and make a global model

User devices train ML models required by an edge or cloud server using their local data (i.e., collected by smartphone sensors). However, the implementation of Federated Learning in a large and complex new generation network poses challenges related to communication costs, scheduling of resources, resource allocation, privacy, and security due to the involvement of various services and also hardware devices with different constraints and QoS requirements.

In 5G and 6G networks, optimization techniques can be used to train models that optimize the scheduling of tasks, such as data transmission, in order to reduce the communication overhead and increase the accuracy of the global model [2]. Recent research in this area has shown that optimization techniques can be used to adapt the learning rate of the optimizer based on the non-iid (independent identically distributed) distribution of data across clients [10], and to leverage statistical information from multiple clients to improve the performance of the global model [10,11]. Recent studies have shown that optimization techniques can be also used to optimize the parameters of the federated learning model by minimizing a loss function that measures the difference between the predicted and actual output [12,13], and to optimize the federated learning model for personalized recommendation systems [14].

In [15], the authors have proposed an optimization method for FL based on 5G with enabled edge computing, considering an architecture of three levels: (i) client, (ii) edge cloud, and (iii) central cloud. The updates of the local model parameters are performed on the client-edge, and a global aggregation is run between the edge and the cloud. The study has first formulated an optimization problem to find a trade-off between the performance and the cost of the FL model. Then, it provided a framework for the Open Radio Access Network (O-RAN) to deploy the FL-trained model. Compared with some works in the literature, results from the optimized federated learning approach they proposed show how the algorithm can effectively balance accuracy and cost.

The study [16] presents a multi-agent approach to optimize federated learning in a distributed industrial Internet of Things (IoT) environment. The main contribution of this study is that it proposes a multi-agent framework for FL that can reduce the communication overhead while improving the efficiency of the learning process. The different agents handle the coordination and communication among the devices participating in the federated learning process. The system uses Q-learning to adapt the agents' behavior and improve their overall performance. The authors formulate a stochastic optimization problem that aims at minimizing FL evaluation loss while satisfying delay and long-term energy consumption requirements. The goal of the optimization problem is to select the optimal devices to participate in the local training and also allocate computing and frequency band resources for training and transmitting the model parameters. As in [15], the system consists of three levels. The local models can be aggregated by a neighboring edge power machine every FL epoch or by a robust cloud server with less frequency for constructing the global model.

Overall, distributed optimization techniques used as input to federated learning in mobile networks can provide the ability to improve privacy, communication efficiency, scalability, handling non-iid data, and adaptability in the model training process. However, it appeared recently that the heterogeneous hardware capacities of the end-user and their dynamic states have an important impact on the performance of the FL process [17]. This challenge is rarely addressed in the literature. The authors in [17] have collected real dataset from 136k users' phones and provided the "Flash" platform allowing to assess FL algorithms in heterogeneity-aware and -unaware configuration. They found that heterogeneity reduces the accuracy and convergence time of the global model. Another challenge is that the optimization algorithms require a high convergence time to find an optimal solution, and this may not be adapted to the context of mobile edges. Using simple heuristics or local ML models can be more efficient.

2.2 *Federated learning meets game theory*

As detailed in the previous part, several algorithms based on operational research have been proposed in the literature to manage the network resources. However, one of the main limitations of such approaches when it comes to FL is the non-consideration of the self-interest of the clients, which may lead to suboptimal resource allocation. To overcome these limitations, incentive mechanisms provide a way to motivate clients to participate in federated learning by aligning their self-interest with the overall goal of the system. These mechanisms can be used to encourage clients to share their data, participate in computation, and ensure that resources are allocated in an efficient and fair manner [18]

Game theory has been widely used in resource allocation [19,20], and designing incentive mechanisms in various fields, such as crowdsourcing [21] and edge computing [22]. However, when it comes to federated learning, the application of existing works on incentive mechanism design is challenging. This is because it is difficult to quantify the value of each client's training data and model the final learning performance of the federated learning system. This makes it hard to model the utility function of each participant, such as the parameter server and clients, in FL.

Due to the heterogeneity nature in distributed FL networks, different clients might have different types and amounts of resources (data, computational power, communication resources, ...). Therefore, the objective of incentive mechanisms is to make sure the clients take part in the training and that these resources are well utilized in the learning process. The FL incentive mechanisms in the literature are designed based on the data contribution, reputation, or available computational and communication resources for each client.

It is important to evaluate the contributions of each data provided by clients to encourage their participation based on the assessment of the value of data. The data contribution of clients can be evaluated in qualitative ways. In particular, Shapley value is a widely used method for assessing data value as in [23]. Incentive-aware reward schemes and metrics, such as the contribution index [24] and an auction mechanism [25], have also been proposed to evaluate the contributions of different clients in federated learning. These methods aim to overcome the challenges of incentive mismatches and encourage clients to provide high-quality data. One proposed solution is the Federated Learning Incentivizer (FLI) [26], which dynamically adjusts clients' contributions to match them with rewards and minimize inequality among clients. Theoretical analysis and extensive evaluations show that FLI can produce near-optimal utility and minimize inequality among clients.

Quantitative measures may be used to assess the value of the data provided. for instance, the authors in [27] make use of crowdsourcing and deep learning to incentivize clients using game theory. This is done by analyzing the client decision problem as the Nash equilibrium of a noncooperative game. Stackelberg game is also combined in [28,29] with DRL where the clients decide on the amount of data to participate with, based on an announced reward towards an optimal payment without the need of collecting clients' private information. The data size, training cost, and communication delay are also considered for the incentivization in [30,31].

Blockchain technology can also be used in FL to secure the training process and incentivize clients. For example, the authors in [32] design a collaborative training framework in which clients jointly participate in deep learning model training, with the sharing of local gradients for each client in the form of transactions in the blockchain. A technique called FLChain is also proposed in [33], as a decentralized, public, auditable, and healthy FL ecosystem with trust and incentives for clients. In both cases, these methods aim to address issues such as misbehavior and single point failure by providing audibility and punishing malicious clients.

In FL, a client's reputation is an essential criterion for selecting the best clients to cooperate with, since these clients are more likely to provide trustworthy training. This idea ensures that customer reputation is updated once each training task is completed and is taken into account when choosing clients for further training. When it comes to federated learning in the home appliance, Zhao et al. [34] presented a blockchain-based reputation system in which consumers' reputations are stored on the blockchain and rise or decrease depending on the quality of their contributions. In a similar manner, the authors in [35] presented a system in which all nodes, including those at the network's edge, fog nodes, and cloud servers, may rate one another. Using a multiweight subjective logic model to build a composite reputation value [36], Kang et al. developed a blockchain-based reputation system that identifies attackers and unreliable clients using RONI and FoolsGold techniques and updates reputations appropriately. These efforts are made in an effort to encourage the evolution of federated learning in a healthy manner by rewarding trustworthy and standards-maintaining practices.

The training process for federated learning is also constrained in many other aspects, particularly, in terms of bandwidth, storage space, and energy consumption. Because of the heterogeneity of these devices, it is necessary to provide an efficient incentive system for client resource allocation. For example, the authors in [37] aim to balance the time delay of each iteration and investigate the impact of heterogeneous clients on FL convergence from a computing resources perspective, assuming the server is cost constrained and incentivizes the clients' use of their CPU power based on a Stackelberg game. To design the incentive mechanism and identify the best trade-off between model training time and parameter server's payout in a dynamic network environment, the authors in [38] suggest a DRL-based technique, that considers the complexity of the federated learning algorithm in the fast-changing nature of the learning environment using DRL.

Allocating communication resources is the subject of several other studies. As an incentive mechanism, the authors in [39] model the relationship between the base station and the clients as an auction game about the communication resources. At the base station side, the client selection problem is described as a social welfare maximization problem that can be solved using a primaldual greedy method. In a similar work, the authors in [40] present a new crowdsourcing method by creating a communication efficient cost model that takes into account the efficiency of communication during the exchange of model parameters. Incentives are added in the form of a value-based pay approach, such as a bonus, that is directly correlated with the degree to which individuals engage in federated learning.

2.3 *FL powered by DRL*

Deep Reinforcement Learning (DRL) has been proven to be a powerful approach for optimizing resource allocation in federated learning systems. DRL is better than traditional optimization-based or heuristic approaches, game theory, and incentive mechanisms in this context for several reasons. Firstly, DRL can learn from experience and adapt to changing environments, making it more suitable for handling the dynamic and complex nature of federated learning systems. Traditional optimization-based or heuristic approaches are often limited by their assumptions and inability to adapt to changing environments. Secondly, DRL can consider the long-term consequences of its actions, whereas traditional approaches, such as game theory and incentive mechanisms, are often limited to short-term solutions. Thirdly, DRL can handle partial observability and non-linear constraints that traditional approaches often struggle with. In this context, DRL can provide near-optimal solutions for the resource allocation problem in federated learning systems, which is a challenging problem due to its complexity and uncertainty.

In [41], the authors present an experience-driven computational resource allocation approach for FL using DRL. The proposed method improves the energy efficiency of FL by lowering the CPU-cycle frequency of mobile devices faster in the training group. They formulate an optimization problem to minimize the total system cost, defined as a weighted sum of training time and energy consumption. The proposed DRL-based algorithm can converge to near-optimal solutions for this problem, despite the non-linear constraints and uncertainty of the network quality.

The authors propose a novel resource allocation method called "concurrent federated reinforcement learning" for IoT edge computing in [42]. The method addresses the limitations of traditional central server and edge host decision-making approaches by incorporating the privacy protection of FL, the problem-solving power of reinforcement learning, and concurrency in the form of joint decision-making. The proposed approach can provide state-of-the-art performance in system-wide utility, speed of task completion, and resource utilization.

A DRL-based approach for resource allocation in mobility-aware FL networks is proposed in [43]. The authors propose using DRL to solve the problem of the model owner in deciding how much energy to recharge workers and which channels to use for global model transmissions to maximize the number of global model transmissions while minimizing energy and channel costs. The simulation results show that the proposed DRL approach outperforms conventional algorithms.

In [44], the authors present a DRL-based approach for intelligent multitimescale resource management for multiaccess edge computing in 5G ultra-dense networks. The authors propose using DRL to improve the ability to emerge resource-intensive IoT applications by addressing the challenges of time-varying network environments and heterogeneous resources of network devices in 5G ultradense networks. The proposed approach presents a comprehensive system model and discusses the challenges and solutions of applying DRL in ultradense edge computing (UDEC) systems.

The use of DRL in these works is motivated by the need to address the complex and dynamic nature of FL systems, improve their performance and efficiency, and overcome the limitations of traditional approaches. DRL's ability to learn from experience and adapt to changing environments makes it well-suited to address the complex challenges FL systems face, such as energy constraints, network resource costs, data privacy, and decision-making under uncertainty. The proposed DRL-based approaches in these papers can improve energy efficiency, resource allocation, network management, and data privacy in FL systems.

3 RESEARCH GAP AND FUTURE WORKS

Research gaps and future directions related to the use of algorithmic tools for FL resource allocation include further exploration of optimization techniques that can adapt to non-iid data distributions in federated learning settings. Other research gaps consist on investigating methods for addressing communication costs and scheduling of resources in large and complex 5G and 6G networks, developing techniques for ensuring privacy and security in federated learning for 5G and 6G. Incorporating edge computing into federated learning for 5G and 6G networks is also an area that requires further research. Other areas of research that are important to explore include examining the trade-off between accuracy and cost in federated learning for 5G and 6G networks and developing methods for personalizing recommendation systems using federated learning.

Designing incentive mechanisms for federated learning is an essential yet unexplored research area. Its objective is to ensure the right number of efficient participants. Moreover, the proposed models should be heterogeneity-aware as the end-user and edge computing, power/battery capacity and communication resources are varying from a large space. In addition, the local models to be trained on the user devices must require a reasonable amount of resources.

Decentralized federated learning is one of the next directions for the design of incentive mechanisms in FL. In the current incentive driven FL models, there is usually only one main server and several clients. In decentralized FL, however, a collection of entities collaborate to optimize their own global model. Previous efforts have overlooked incentive mechanisms for this kind of topology, which can degrade the efficiency of the training.

The second direction is federated learning algorithms driven by incentives. Current research tends to concentrate on the training mode of federated learning, but the incentive mechanisms are not completely integrated into the algorithm itself. The incentive mechanism may also be used to alter the parameters of the federated learning algorithm, resulting in more efficient and effective federated learning.

The third direction concerns incentive mechanisms motivated by security concerns. Specifically, the incentive mechanism design for federated learning has not yet taken into account scenarios in which clients may engage in malicious activities during the learning process. These behaviors include introducing corrupted data, uploading poisoned models, and prematurely halting local training to conserve resources. These behaviors can result in a biased learning process and a model of inferior quality.

Table 1 summarizes the advantages and limitations of the different investigated techniques in this paper.

Table 1. Summary of the advantages and limitations of the three different approaches.

	Advantages	Limitations
FL & OR	• Provide the global optimal solution. • Simple implementation using open source solvers (e.g. Ciplex).	• May require unrealistic assumptions. • Inability to adapt to changing environments. • Time consuming and thus difficult to deploy in real-time applications.
FL & GT	• Increase clients motivations to participate in the training.	• It is hard to model the utility function of each participant.
FL & DRL	• Dynamic adaptation to fastly variable environment.	• Ensuring the right number of participating clients. • Availability of the labeled data at the client side.

4 CONCLUSION

In conclusion, we provided a recent survey on the application of Federated learning (FL) in resource allocation for the new generation networks. In particular, this paper has reviewed the literature on the use of FL in resource allocation, focusing on the coupling of FL with operational research, iterative algorithms, game theory, incentive mechanisms, and reinforcement learning. Figure 1 presents the advantages and limitations of those three approaches. The research gap and future works have been identified as well. The literature review shows that several techniques existent in the literature have the potential to efficiently perform resource allocation among the clients. We also identified the current research gap and suggested promising areas for future investigation, such as combining various FL methods, and creating novel models and algorithms that more effectively tackle the unique demands and obstacles of next-generation networks. The research gap analysis reveals that there is still a need for more research to fully understand and utilize FL clients' capabilities, mainly in the areas of computational efficiency, decentralized incentive mechanisms, and security approaches. Therefore, future work should focus on developing more effective algorithms and incentive mechanisms to improve the performance of FL in resource allocation.

REFERENCES

[1] Alwarafy A., Ciftler B. S., Abdallah M., Hamdi M., and Al-Dhahir N., (2022). DRL-based Joint RAT Association, Power and Bandwidth Optimization for Future HetNets, *IEEE Wireless Communications Letters*, vol. 11, no. 7, pp. 1503–1507.

[2] O. Fares, Dandoush A., and Aitsaadi N., (2022). OPR: SDN-based optimal path routing within transit autonomous system networks, in *IEEE International Conference on Communications, ICC 2022*, Seoul, Korea. IEEE, pp. 1058–1063. [Online]. Available: https://doi.org/10.1109/ICC45855.2022.9839243

[3] Kammoun A., Tabbane N., Diaz G., Dandoush A., and Achir N., (2018). End-to-End efficient heuristic algorithm for 5G network slicing, in *32nd IEEE International Conference on Advanced Information Networking and Applications, AINA 2018, Krakow, Poland, May 16–18, 2018*, L. Barolli, M. Takizawa, T. Enokido, M. R. Ogiela, L. Ogiela, and N. Javaid, Eds. IEEE Computer Society, pp. 386–392. [Online]. Available: https://doi.org/10.1109/AINA.2018.00065

[4] Neglia G., Alouf S., Dandoush A., Simoens S., Dersin P., Tuholukova A., Billion J., and Derouet P., (2016). Performance evaluation of train moving-block control, in *Quantitative Evaluation of Systems – 13th International Conference, QEST 2016, Quebec City, QC, Canada, August 23–25, 2016,*

Proceedings, ser. Lecture Notes in Computer Science, G. Agha and B. V. Houdt, Eds., vol. 9826. Springer, pp. 348–363. [Online]. Available: https://doi.org/10.1007/978-3-319-43425-4_23

[5] Robust decentralized federated learning using collaborative decisions, in *2022 International Wireless Communications and Mobile Computing (IWCMC)*. IEEE, (2022), pp. 254–258.

[6] Kang J., Xiong Z., Niyato D., Xie S., and Zhang J., (2019). Incentive mechanism for reliable federated learning: A joint optimization approach to combining reputation and contract theory, *IEEE Internet of Things Journal*, vol. 6, no. 6, pp. 10700–10714.

[7] Chahbar M., Diaz G., and Dandoush A., (2019). Towards a unified network slicing model, in *15th International Conference on Network and Service Management, CNSM 2019, Halifax, NS, Canada, October 21–25, 2019*, H. Lutfiyya, Y. Diao, A. N. Zincir-Heywood, R. Badonnel, and E. R. M. Madeira, Eds. IEEE, pp. 1–5. [Online]. Available: https://doi.org/10.23919/CNSM46954.2019.9012745

[8] Chahbar M., Diaz G., Dandoush A., Cérin C., and Ghoumid K., (2021). A comprehensive survey on the E2E 5g network slicing model, *IEEE Trans. Netw. Serv. Manag.*, vol. 18, no. 1, pp. 49–62, [Online]. Available: https://doi.org/10.1109/TNSM.2020.3044626

[9] Kang J., Xiong Z., Niyato D., Xie S., and Zhang J., (2019). Incentive mechanism for reliable federated learning: A joint optimization approach to combining reputation and contract theory, *IEEE Internet of Things Journal*, vol. 6, no. 6, pp. 10700–10714.

[10] Chen Y., Ning Y., Slawski M., and Rangwala H., (2020). Asynchronous online federated learning for edge devices with non-IID data, in *2020 IEEE International Conference on Big Data (Big Data)*, pp. 15–24.

[11] Zheng Z., Zhou Y., Sun Y., Wang Z., Liu B., and Li K., (2022). Applications of federated learning in smart cities: recent advances, taxonomy, and open challenges, *Connection Science*, vol. 34, no. 1, pp. 1–28, [Online]. Available: https://doi.org/10.1080/09540091.2021.1936455

[12] Konečný J., McMahan H. B., Ramage D., and Richtárik P., (2016). Federated Optimization: Distributed machine learning for on-device intelligence, *CoRR*, vol. abs/1610.02527, [Online]. Available: http://arxiv.org/abs/1610.02527

[13] Jakkam Reddi S., Kumar S., Zaheer M., Charles Z., Garrett Z., Rush K., Konecny J., and Mcmahan H. B., (Mar. 11 2021). *Federated Learning with Adaptive Optimization*, uS Patent App. 17/100,253.

[14] Mortaheb M., Vahapoglu C., and Ulukus S., (2022). Personalized federated multi-task learning over wireless fading channels, *Algorithms*, vol. 15, no. 11, [Online]. Available: https://www.mdpi.com/1999-4893/15/11/421

[15] Firouzi R. and Rahmani R., (2022). 5G-enabled distributed intelligence based on O-RAN for distributed IoT systems, *Sensors*, vol. 23, no. 1, p. 133.

[16] Zhang W., Yang D., Wu W., Peng H., Zhang N., Zhang H., and Shen X., (2021). Optimizing federated learning in distributed industrial IoT: A multi-agent approach, *IEEE Journal on Selected Areas in Communications*, vol. 39, no. 12, pp. 3688–3703.

[17] Yang C., Xu M., Wang Q., Chen Z., Huang K., Ma Y., Bian K., Huang G., Liu Y., Jin X. et al., (2022). FLASH: Heterogeneity-aware federated learning at scale, *IEEE Transactions on Mobile Computing*.

[18] Zhan Y., Zhang J., Hong Z., Wu L., Li P., and Guo S. (2021), A survey of incentive mechanism design for federated learning, *IEEE Transactions on Emerging Topics in Computing*.

[19] Gouissem A., Abualsaud K., Yaacoub E., Khattab T., and Guizani M., (2021). Game theory for anti-jamming strategy in multichannel slow fading IoT networks, *IEEE Internet of Things Journal*, vol. 8, no. 23, pp. 16880 16893.

[20] Combating jamming attacks in multi-channel IoT networks using game theory, in *2020 3rd International Conference on Information and Computer Technologies (ICICT)*. IEEE, (2020), pp. 469–474.

[21] Yang D., Xue G., Fang X., and Tang J., (2012). Crowdsourcing to smartphones: Incentive mechanism design for mobile phone sensing, in *Proceedings of the 18th annual international conference on Mobile computing and networking*, pp. 173–184.

[22] Li P. and Guo S., (2015). Incentive mechanisms for device-to-device communications, *IEEE Network*, vol. 29, no. 4, pp. 75–79.

[23] Sim R. H. L., Zhang Y., Chan M. C., and Low B. K. H., (2020). Collaborative machine learning with incentive-aware model rewards, in *International Conference on Machine Learning*. PMLR, pp. 8927–8936.

[24] Song T., Tong Y., and Wei S., (2019). Profit allocation for federated learning, in *2019 IEEE International Conference on Big Data (Big Data)*. IEEE, pp. 2577–2586.

[25] Zeng R., Zhang S., Wang J., and Chu X., (2020). Fmore: An incentive scheme of multi-dimensional auction for federated learning in mec, in *2020 IEEE 40th International Conference on Distributed Computing Systems (ICDCS)*. IEEE, pp. 278–288.
[26] Yu H., Liu Z., Liu Y., Chen T., Cong M., Weng X., Niyato D., and Yang Q., (2020). A sustainable incentive scheme for federated learning, *IEEE Intelligent Systems*, vol. 35, no. 4, pp. 58–69.
[27] Zhan Y., Li P., Wang K., Guo S., and Xia Y., (2020). Big data analytics by crowdlearning: Architecture and mechanism design, *IEEE Network*, vol. 34, no. 3, pp. 143–147.
[28] Zhan Y., Zhang J., Li P., and Xia Y., (2019). Crowdtraining: Architecture and incentive mechanism for deep learning training in the internet of things, *IEEE Network*, vol. 33, no. 5, pp. 89–95.
[29] Zhan Y., Li P., Qu Z., Zeng D., and Guo S., (2020). A learning-based incentive mechanism for federated learning, *IEEE Internet of Things Journal*, vol. 7, no. 7, pp. 6360–6368.
[30] Ding N., Fang Z., and Huang J., (2020). Optimal contract design for efficient federated learning with multi-dimensional private information, *IEEE Journal on Selected Areas in Communications*, vol. 39, no. 1, pp. 186–200.
[31] ———, Incentive mechanism design for federated learning with multi-dimensional private information, in *2020 18th International Symposium on Modeling and Optimization in Mobile, Ad Hoc, and Wireless Networks (WiOPT)*. IEEE, 2020, pp. 1–8.
[32] Weng J., Weng J., Zhang J., Li M., Zhang Y., and Luo W., (2019). Deepchain: Auditable and privacy-preserving deep learning with blockchain-based incentive, *IEEE Transactions on Dependable and Secure Computing*, vol. 18, no. 5, pp. 2438–2455.
[33] Bao X., Su C., Xiong Y., Huang W., and Hu Y., (2019). Flchain: A blockchain for auditable federated learning with trust and incentive, in *2019 5th International Conference on Big Data Computing and Communications (BIGCOM)*. IEEE, pp. 151–159.
[34] Zhao Y., Zhao J., Jiang L., Tan R., Niyato D., Li Z., Lyu L., and Liu Y., (2020). Privacy-preserving blockchain-based federated learning for IoT devices, *IEEE Internet of Things Journal*, vol. 8, no. 3, pp. 1817–1829.
[35] ur Rehman M. H., Salah K., Damiani E., and Svetinovic D., (2020). Towards blockchain-based reputation-aware federated learning, in *IEEE INFOCOM 2020-IEEE Conference on Computer Communications Workshops (INFOCOM WKSHPS)*. IEEE, pp. 183–188.
[36] Kang J., Xiong Z., Niyato D., Zou Y., Zhang Y., and Guizani M., (2020). Reliable federated learning for mobile networks, *IEEE Wireless Communications*, vol. 27, no. 2, pp. 72–80.
[37] Sarikaya Y. and Ercetin O., (2019). Motivating workers in federated learning: A stackelberg game perspective, *IEEE Networking Letters*, vol. 2, no. 1, pp. 23–27.
[38] Zhan Y. and Zhang J., (2020). An incentive mechanism design for efficient edge learning by deep reinforcement learning approach, in *IEEE INFOCOM 2020-IEEE Conference on Computer Communications*. IEEE, pp. 2489–2498.
[39] Le T. H. T., Tran N. H., Tun Y. K., Nguyen M. N., Pandey S. R., Han Z., and Hong C. S., (2021). An incentive mechanism for federated learning in wireless cellular networks: An auction approach, *IEEE Transactions on Wireless Communications*, vol. 20, no. 8, pp. 4874–4887.
[40] Pandey S. R., Tran N. H., Bennis M., Tun Y. K., Manzoor A., and Hong C. S., (2020). A crowdsourcing framework for on-device federated learning, *IEEE Transactions on Wireless Communications*, vol. 19, no. 5, pp. 3241–3256.
[41] Zhan Y., Li P., and Guo S., (2020). Experience-driven computational resource allocation of federated learning by deep reinforcement learning, in *2020 IEEE International Parallel and Distributed Processing Symposium (IPDPS)*, pp. 234–243.
[42] Tianqing Z., Zhou W., Ye D., Cheng Z., and Li J., (2022). Resource allocation in IoT edge computing via concurrent federated reinforcement learning, *IEEE Internet of Things Journal*, vol. 9, no. 2, pp. 1414–1426.
[43] Nguyen H. T., Cong Luong N., Zhao J., Yuen C., and Niyato D., (2020). Resource Allocation in Mobility-Aware Federated Learning Networks: A Deep Reinforcement Learning Approach, in *2020 IEEE 6th World Forum on Internet of Things (WF-IoT)*, pp. 1–6.
[44] Yu S., Chen X., Zhou Z., Gong X., and Wu D., (2021). When deep reinforcement learning meets federated learning: Intelligent multitimescale resource management for multiaccess edge computing in 5G ultradense network, *IEEE Internet of Things Journal*, vol. 8, no. 4, pp. 2238–2251.

Innovation and Technological Advances for Sustainability – Al-Naemi et al. (Eds)

Estimation of the fuel consumption in autonomous vehicles using machine learning techniques

Ghazli Abdelkader
Department of Mathematics and Computer Sciences, Tahri Mohamed University, Bechar, Algeria

Senouci M. Benaoumeur
The Faculty of Engineering, SDU Mechatronics (CIM), Sonderborg, Denmark

Abid Dhiya Eddine
Department of Mathematics and Computer Sciences, Tahri Mohamed University, Bechar, Algeria

Bouache Mourad
Intel AI Academy, Silicon Valley, USA

ABSTRACT: Energy consumption is the responsibility of each individual and that we must all adopt the right reflexes to preserve our resources. So many reasons that can impact our fuel consumption such as nervous driving, under-inflated tires, poorly maintained vehicle...etc. Adopting economical and responsible driving can allow us to reduce our fuel budget while reducing our carbon footprint. The main objective of this research is to study the influence of some parameters on fuel consumption through the use of artificial intelligence approach. This paper presents the application of five machine learning techniques to estimate fuel consumption of the light cars, in particular, Linear Regression (LR), Random Forest (RF), Decision Tree (DT), K-Nearest Neighbors (KNN), Artificial Neural Network (ANN) and finally Deep Neural Network. These models have been developed using a dataset contains lot of criteria such us speed, distance and some weather conditions (rain, sun...). Performance of the algorithms is evaluated by reporting the prediction errors using the three metrics tools, which are median absolute error, mean absolute error and mean squared error. The statistical evaluation procedure finds that ANNs have the lowest median absolute error and mean squared error (0.304 and 0.357) compared to LR (0.401 and 0.919), RF(0.330 and 0.366), DT(0.400 and 1.177) and KNN(0.340 and 0.493) and the DNN have lowest mean absolute error (0.408) and mean squared error (0.305).

Keywords: Fuel Consumption, Artificial Intelligence, Machine learning, Neural Network, Random Forest, Decision Tree

1 INTRODUCTION

The estimate of fuel consumption may vary and be influenced by several parameters and conditions which may even be climatic. Knowing the variation in fuel consumption can help to detect driver behavior or even technical problems or also when you have a problem with a fuel gauge sensor in the vehicle's tank. This problem makes it difficult for the driver to know how much fuel is left in the car. this solution seems very effective for building a dashboard, especially for large companies working in the field of transporting people or goods to manage and predict their energy consumption intelligently and remotely.

Today's vehicles are occupied with several sensors, which are very important for sensing the environment around the vehicle and for automated driving and parking functions. Sensors can

DOI: 10.1201/9781003496724-44

be classified according to their field of application: functional sensors mainly used for control or regularization tasks, sensors assigned to security operations to protect the vehicle against theft and finally monitoring sensors that provide information to drivers about the condition of his car.

Figure 1. Somme vehicule sensors.

As shown in Figure 1, several sensors are present in vehicles [1] [2] [3], several of which may be related to fuel consumption such as speed sensor, throttle and pedal position sensors, mass air flow sensor used to calculate the amount of air that enters the engine during the intake stroke [4] [5] [6]. This paper focus on the estimation of fuel consumption, so to solve this problem, we proposed a machine learning approach to estimate the amount of fuel consumed based on a set of criteria such as distance traveled, speed and some weather conditions (rain, sun...). We will use the Python programming language and some of its libraries such as Pandas [7] for importing and manipulating data from csv file, Scikit-Learn [8] for data splitting and importing machine learning algorithms and Matplotlib [9] for data visualization. Our work will be applied on the Google Collab platform in order to benefit from easy use and high performance.

2 RELATED WORKS

The following section reviews some approaches found in the literature for estimating fuel consumption based mainly on some learning methods and models using some values of certain sensors installed in vehicles.

In 2017, Federico Perrotta and al. in their paper titled Application of machine learning for fuel consumption modelling of trucks, presents the application of three machine learning techniques for fuel consumption modelling of articulated trucks in particular, Support Vector Machine (SVM), Random Forest (RF), and Artificial Neural Network (ANN). The used data in this research are collected by standard sensors (SAE J1939) for modern vehicles. As evaluation metrics, they used the root mean squared error, mean absolute error and R2 score. The RF slightly outperforms SVM and ANN giving higher R2, and lower error.[10]

Alexander Schoen and al in 2018, created a machine learning model for average fuel consumption in heavy vehicles, this approach was based on neural network in conjunction with seven predictors derived from vehicle speed and road grade to find a model for average fuel consumption in heavy vehicles. The predictors of the model are aggregated over fixed window sizes of distance traveled. The results show that a 1 km window is able to predict fuel consumption with a 0.91 coefficient of determination and mean absolute peak to peak percent error less than 4% for routes that include both city and highway duty cycle segments. Other evaluation metrics was used which are root mean squared error and mean absolute error [11].

Jarosław Ziółkowski and al in 2021, presents a multi-faceted analysis for predicting the fuel consumption of motor vehicles with a particular emphasis on the passenger cars. This research is based on the use of artificial neural networks with a database containing 1750 records. They used the linear pearson correlation coefficient (r) and coefficient of determination (R2) in order to determine if the predicted values match the real values. Finally, they arrived to a coefficient r was within range 0.93–0.95, while the coefficient of determination R2 assumed a satisfactory value of more than 0.98 with a structure 22-10-3 of neural network [12].

The paper of Victor Bodell and al, titled comparing machine learning estimation of fuel consumption of heavy-duty vehicles, proposes in 2021 a machine learning approach to predict fuel consumption. This study uses around 40,000 vehicles have diesel engines and a mileage of more than 20,000 km and it uses operational environmental conditions information, such as road slopes and driver profiles. This research used the following machine learning algorithms in the construction of this approach, which is linear regression (LR), K-nearest neighbor (KNN) and artificial neural networks (ANN). Performance of the algorithms is evaluated by reporting the prediction error. The statistical evaluation procedure finds that ANNs have the lowest prediction error compared to LR and KNN [13].

Vehicular fuel consumption estimation using real-world measures through cascaded machine learning modeling, proposed in 2020 by Ehsan Moradi and Luis Miranda-Moreno a machine learning modeling method using large on-road data collected from a fleet of 27 vehicles. The collection of engine performance data was through the On-Board Diagnostics (OBD) interface of the vehicles, this tool was wirelessly connected to a tablet simultaneously collects second-by-second GPS, inertial data (accelerometer, gyro, and compass outputs), and internal engine parameters. These parameters include engine speed (RPM), manifold absolute pressure (MAP), mass airflow rate (MAF), barometric pressure (P), fuel air equivalence ratio (ϕ), and intake air temperature (IAT). In order to develop their models, they used the support vector regressor and the artificial neural network. As a result, the accuracy of models reached 83% [4].

Machine learning and statistical analysis in fuel consumption prediction for heavy vehicles it is another work made by Henrik Almér in 2015. He studied how to use machine learning to predict fuel consumption in heavy vehicles. In this study, he examined data from several different sources describing the characteristics of the road, vehicle, driver and weather conditions. In this thesis, he used data sources available for Scania, and he used R as a programming language based on its built-in algorithms including linear regression, random forests, support vector machine, and artificial neural networks. He used the root mean squared error as an evaluation metric and he got the best results with the random forest with an error equal to 1.29 [5].

3 METHODOLOGY PROPOSAL

The main objective of this paper is to present a machine learning based approach for estimating fuel consumption based on some data measured by some sensors not installed in the vehicle including weather conditions with the aim of offer an alternative measure in the event of failure or malfunction of the sensors installed in the vehicle.

Our methodology consists of using different techniques and machine learning algorithms in order to compare the different results of the different models proposed for the estimation of fuel consumption in light vehicles. To accomplish our objectives, our methodology consists in dispatching the problems in three phases as shown in the following figure.

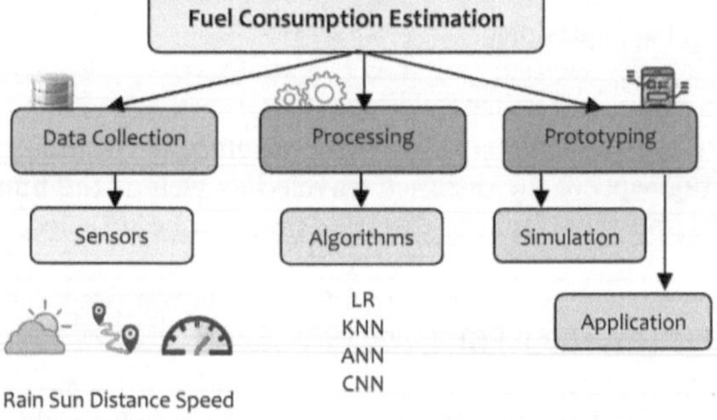

Figure 2. Design of our approach.

3.1 *Data collection phase*

According to several searches in the literature of some approaches aimed at proposing models for the estimation of fuel consumption, several parameters have been introduced by the researchers [14] [15] generally present values captured by sensors whether speed, distance ...etc. the type of fuel also presents an important parameter for the estimation of consumption as well as the meteorological data which requires additional fuel consumption in certain situations. So, this phase is very important since it tries to gather all the heterogenous data influences energy consumption. To avoid overfitting, we will use just some most significant variables.

3.2 *Processing phase*

In order to complete our goal, we will use machine learning methods for regression analysis in order to estimate the fuel consumption for light cars. The Linear Regression, Random Forest, Decision Tree, K-Nearest Neighbors, Artificial Neural Network, and the deep neural network are the algorithms, which we will use. Mean squared error, median absolute error and mean absolute error are our evaluation metrics.

3.3 *Prototyping phase*

This phase is divided into two main parts, the simulation part and the implementation part, the first part use a simulation environment for the preparation of the dataset and the programming of the different approaches and techniques used in order to carry out comparison graphs to measure the robustness of each approach where the python programming language is used for the implementation.

The second part consists in the implementation and the integration of the model in an environment allowing its exploitation in reality. An embedded environment has been chosen which includes both a wide range of sensors installed to retrieve the data as well as an environment for the execution of the code. Our choice fell on an Android smartphone which presents a device that meets our needs and can be used to estimate the consumption of fuel in reality, especially in the event of breakdowns or dysfunctional sensors implanted in our vehicles.

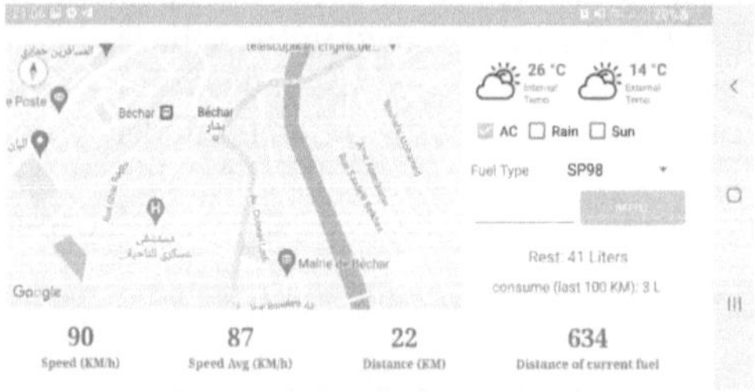

Figure 3. Fuel consumption application.

This application presents a dashboard for fuel consumption, it presents the current speed of the vehicle, its average speed, the distance traveled as well as the number of liters of fuel remaining in the tanks and therefore the distance that can be traveled.

4 IMPLEMENTATION AND RESULTS

The process of our work will be organized in the following steps: dataset collection, data preprocessing, data splitting, training and finally the testing.

4.1 *Dataset*

To implement our approach, we used a supervised learning technique. Indeed, a dataset published by Andreas Wagener on Kaggle website is used. This data illustrated in the following table includes distance traveled (km), consumption (L/100km), average speed (km/h), the temperature inside and outside the vehicle (°C), and anything special that happened, if it was raining, if the AC was on, if it was sunny enough that the car felt hot when he started it and the type of fuel, E10 or SP98, and finally the letters and type of fuel used.

	distance	consume	speed	temp_inside	temp_outside	specials	gas_type	AC	rain	sun	refill liters	refill gas
0	28	5	26	21,5	12	NaN	E10	0	0	0	45	E10
1	12	4,2	30	21,5	13	NaN	E10	0	0	0	NaN	NaN
2	11,2	5,5	38	21,5	15	NaN	E10	0	0	0	NaN	NaN
3	12,9	3,9	36	21,5	14	NaN	E10	0	0	0	NaN	NaN
4	18,5	4,5	46	21,5	15	NaN	E10	0	0	0	NaN	NaN

Figure 4. Dataset overview.

These data were collected in November using fuel type E10 (Gasoline with 10% ethanol), changed to SP98 (or E5, gasoline includes up to 5% ethanol) in winter and again to E10 in spring. The following image represents some of these characteristics and the relationship between them.

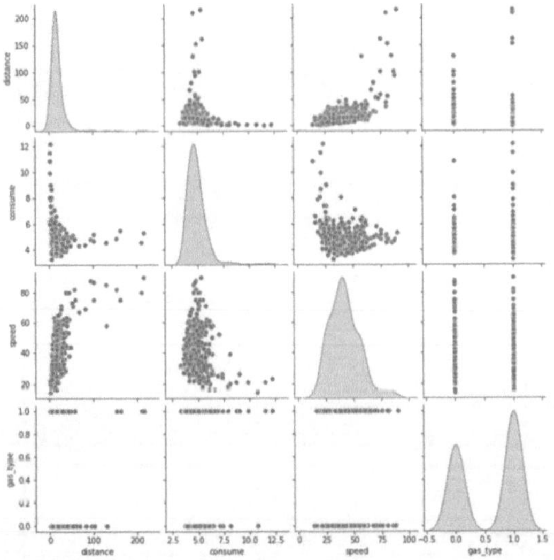

Figure 5. The relation between some characteristics.

4.2 *Data preprocessing and splitting*

For technical reasons, we just used the following characteristics: distance, consumption (consume), speed, inside and outside temperature (temp_inside & temp_outside), fuel type (gas_type), if the air conditioning was on (AC), if it was sunny enough for the car to feel hot when it started it (sun) and finally if it was raining (rain). The fuel type SP98 is represented by the value 1 and the E10 by the value 0.The columns: distance, temp_inside, temp_outside and consume are converted to real values. Regarding data splitting, we split our data into two parts, the first contains 87% for training and the second 13% for testing.

4.3 *Data training and testing*

We have tried several machine learning approaches such as KNN for regression, linear regression, multi-layer perceptron, decision tree and random forest. For the evaluation metrics, we used the mean squared error, the median absolute error and the mean absolute error. After all this and with the removal of the null values, our data will be as follows and it is ready to be used to train our system. Regarding data splitting, we split our data into two parts, the first contains 87% for training and the second 13% for testing.

Table 1. Training results.

	Median AE	Mean AE	Mean SE
KNN Regressor	0.340	0.498	0.493
Linear Regression	0.401	0.628	0.919
Multi-Layer Perceptron	0.304	0.442	0.357
Decision Tree	0.400	0.673	1.177
Random Forest	0.330	0.437	0.366
Deep Neural Network		0.408	0.305

The results of the tests of 13% of the data from the dataset applied to our model are presented in the following figures:

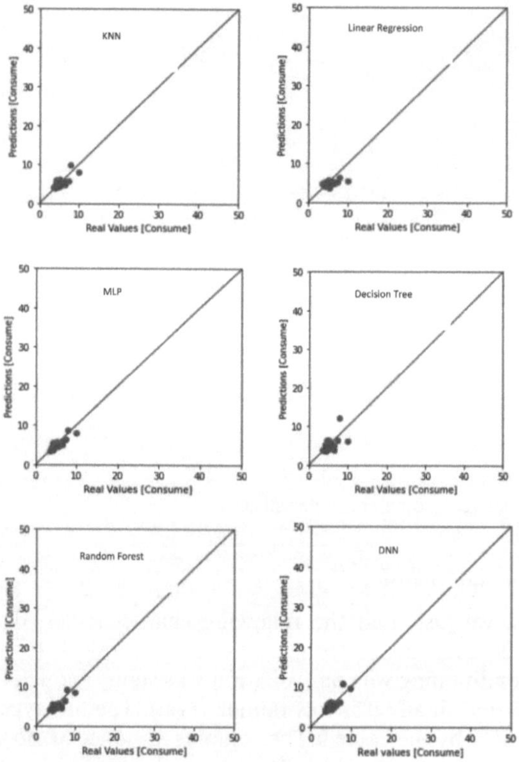

Figure 6. Test results.

From the illustrated results it can be seen that the neural network gave us the lowest median absolute error, while, the deep neural network gave us the lowest mean absolute error and mean squared error. The decision tree gave us the worst mean absolute error and mean squared error; however, the linear regression gave us the worst value in median absolute error. In general, we can say that the best algorithms where the best results were gotten is the DNN and MLP little bit, and we can say also that all the algorithms produce comparable results, the next figure shows the decreased error for our DNN model.

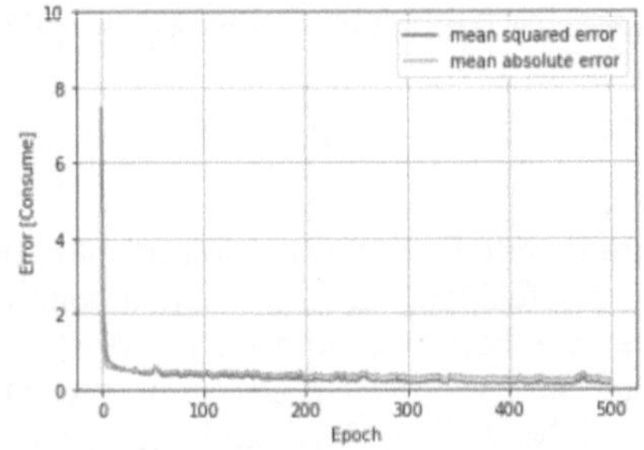

Figure 7. Decreased error in deep neural network.

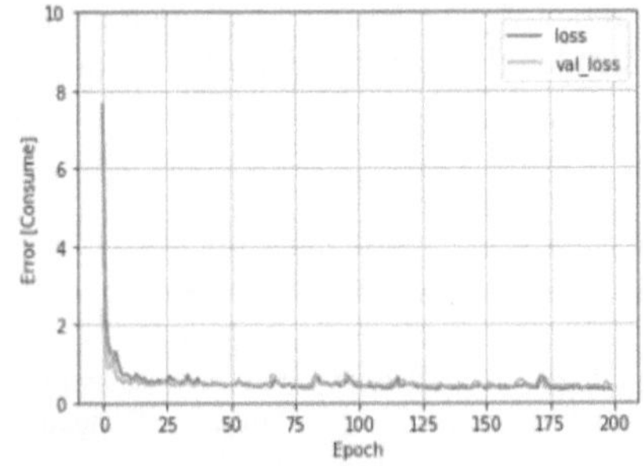

Figure 8. Bad prediction in deep neural network.

5 DISCUSSION

In this section, we present a comparative study of our work with some similar works that we have found in the literature. This comparison was based on certain criteria to test the performances and limitations of our model. The results are grouped in the following table.

Table 2. Comparaison.

REF	MAE RF	MSE ANN	ANN
[10]	3,21	3,46	–
[11]	–	1,66	–
[12]	–	–	0,416
Our	0,437	0,442	0,357

As we can show in the above table, our model based on Artificial Neural Network ANN minimizes at best the mean squared error to 0,357 if we compare it with the model proposed by Jarosław Ziółkowski and all in [12] which have minimized it just to 0,416. In another hand, mean absolute error was well reduced by our approach based on ANN at 0.442 unlike the two approaches proposed in [10] and [11] which present values reaching 3.46 and 1.66 respectively. A random forest implementation of our approach also reduces the mean absolute error to 0.437 compared to the approach proposed by Federico Perrotta and Tony Parry, in their paper titled Application of Machine Learning for Fuel Consumption Modeling of Trucks. Consequently, we can say that our model presents remarkable improvements and minimizes errors by comparing it with some recent methods trying to propose machine-based or deep learning methods for fuel consumption estimating.

6 CONCLUSION

The aim of this study is to inquire into fuel consumption estimation based on distance, speed, temperature (inside and outside), fuel type and any weather conditions (AC, rain, sun). Five machine learning techniques have been used and their performances compared. These are Linear Regression (LR), Random Forest (RF), Decision Tree (DT), K-Nearest Neighbors (KNN), and Artificial Neural Network (ANN) and the last one which is Deep Neural Network. The results of the study gave in terms of predictions, from the comparison of the MSE, Mean AE and Median AE, DNN is the technique that gives the best performance. This allows us to say that the DNN is the best technique used to predict fuel consumption in our case, however, the LR, RF, DT, ANN and KNN demonstrated an acceptable level of accuracy and they can be considered as accurate models in predicting values.

REFERENCES

[1] Fleming W. J., (Nov. 2008). New automotive sensors—A review, *in IEEE Sensors Journal*, vol. 8, no. 11, pp. 1900–1921.

[2] Surendra Kumar Yadav, (2018). *Classification for Sensors in Automotive Cyber-Physical Systems, Master's Thesis*, Faculty of Informatics, Masaryk University, Tchéquie

[3] Fleming W. J., (Dec. 2001). Overview of automotive sensors, in *IEEE Sensors Journal*, vol. 1, no. 4, pp. 296–308

[4] Ehsan Moradi and Luis Miranda-Moreno, (2020). Vehicular fuel consumption estimation using real world measures through cascaded machine learning modeling, *Transportation Research Part D Transport and Environment* Volume 88

[5] Henrik Almér, Machine Learning and Statistical Analysis in Fuel Consumption Prediction for Heavy Vehicles, Master's Thesis at CSC, 2015.

[6] Popa, G.; Ghet,i, M.-A.; Tudor, E.; Vasile, I.; Sburlan, I.-C. (2022). Experimental study regarding long range LiDAR capabilities in sensing safety distance for vehicle application. *Sensors*, 22, 5731. https://doi.org/10.3390/s22155731

[7] https://pandas.pydata.org

[8] https://scikit-learn.org

[9] https://matplotlib.org

[10] Federico Perrotta and Tony Parry, (2017). Application of Machine Learning for Fuel Consumption Modelling of Trucks, *2017 IEEE International Conference on Big Data*

[11] Alexander Schoen and all, A machine learning model for average fuel consumption in heavy vehicles, *IEEE Transactions on Vehicular Technology*, 10.1109/TVT.2019.2916299, 2019

[12] Jarosław Ziółkowski and all, (2021). Use of artificial neural networks to predict fuel consumption on the basis of technical parameters of vehicles, *Energies*

[13] Victor Bodell and all, "Comparing Machine Learning Estimation of Fuel Consumption of Heavy-Duty Vehicles", *International Journal of Transport and Vehicle Engineering* Vol 15, No 2, 2021.

[14] Conrad, Ryan G., "Methods for Estimating Vehicle Fuel Consumption in Urban Road Networks Using Geographical Attributes", *Civil and Environmental Engineering Master's Project Reports.* 43, 2018

[15] Chen, Y.-J. and Chien, C.-H. Fuel Consumption System. *Journal of Computer and Communications*, 3, 153–158, 2015

Innovation and Technological Advances for Sustainability – Al-Naemi et al. (Eds)

Radio resource management and path planning in intelligent transportation systems via reinforcement learning for environmental sustainability

S. Norouzi
Department of Electrical and Computer Engineering, Tarbiat Modares University, Tehran, Iran

N. Azarasa
Department of Civil and Environmental Engineering, Tarbiat Modares University, Tehran, Iran

M.R. Abedi & N. Mokari
Department of Electrical and Computer Engineering, Tarbiat Modares University, Tehran, Iran

S.E. Seyedabrishami
Department of Civil and Environmental Engineering, Tarbiat Modares University, Tehran, Iran

H. Saeedi
College of Engineering, Doha University for Science and Technology, Doha, Qatar
Department of Electrical and Computer Engineering, Tarbiat Modares University, Tehran, Iran

E.A. Jorswieck
Institute of Communications Technology, TU Braunschweig, Lower Saxony, Germany

ABSTRACT: Efficient and dynamic path planning has become an important topic for urban areas with larger density of connected vehicles (CV) which results in reduction of travel time and directly contributes to environmental sustainability through reducing energy consumption. CVs exploit the cellular wireless vehicle-to-everything (C-V2X) communication technology to disseminate the vehicle-to-infrastructure (V2I) messages to the Base-station (BS) to improve situation awareness on urban roads. In this paper, we investigate radio resource management (RRM) in such a framework to minimize the age of information (AoI) so as to enhance path planning results. We use the fact that V2I messages with lower AoI value result in less error in estimating the road capacity and more accurate path planning. Through simulations, we compare road travel times and volume over capacity (V/C) against different levels of AoI and demonstrate the promising performance of the proposed framework.

Keywords: V2X, Path planning, Resource management, MADDPG, AoI

1 INTRODUCTION

Intelligent transportation systems suggest that connected vehicles (CVs) will gradually take over the roads. Meanwhile, path planning remains an important process to increase the driving efficiency. Path planning refers to the selection of a path from a starting point to the terminal point within a particular network under some standard criteria such as path minimization [1]. This is specially the case in traditional path planning algorithms where the traffic volume is assumed static [2]. It means that traffic jams and accidents are not taken into account.

A great deal of progress on path planning has been made in the literature [3], mainly focusing on finding the shortest path/time [4–6]. To improve fuel consumption efficiency, a

DOI: 10.1201/9781003496724-45

path planning scheme using deep reinforcement learning (DRL) is proposed in [7] in which task deadline and fuel consumption of each vehicle in the platoon are considered. A novel three-architecture approach for vehicle path planning (SEARCH) proposed in [8] which exploits Unmanned Aerial Vehicles (UAVs), Vehicular Ad hoc Networks (VANETs), 5G based cellular systems, and Software-Defined Networking (SDN) to address two main issues for navigation management difficulties: lack of local knowledge regarding navigation solutions as well as vehicles that are inflexible in dealing with unforeseeable situations. Based on their findings, the proposed architecture works effectively in terms of reducing the driving time to any destination. Authors in [9], by considering the personalized requirements of vehicle users, propose a digital-twin (DT) enabled path planning scheme to facilitate traffic management using a Q-learning algorithm.

A key subject in intelligent transportation systems is the vehicle communication. According to the existing literature, in [10], we examine Mode 4 of the 3GPP cellular V2X architecture [11]. A distributed algorithm is used to schedule resources and manage interference between CVs. Since path planning information is time-sensitive, it is particularly critical to ensure the freshness of the information received by base station (BS) at intersections. Traditional performance metrics, such as latency or throughput, are not sufficient to measure the timeliness of updates. As such, a new criterion, called the age of information (AoI), is introduced in [12] for this kind of application. In [13] the authors design a radio resource management (RRM) policy to optimize the AoI in vehicular networks that satisfy the demand for real-time and reliable communications. In [14] authors minimize the age of status updates sent by vehicles over a carrier-sense multiple access (CSMA) network using the gradient descent scheme. The authors of [15] investigate the problem of AoI-aware RRM for a platooning system. To solve the corresponding optimization problem, they use the multi-agent reinforcement learning (MARL) algorithm for distributed resource allocation.

As can be seen, no study has been conducted to investigate the impact of RRM on path planning validation. However, this is essential to have accurate and real-time information when planning the optimal path for each CV. Accordingly, RRM provides traffic incident messages that comply with the above requirements. In our work, we model capacity estimation error utilizing AoI levels, followed by formulating an optimization problem to minimize the AoI of vehicles to validate path planning.

The contributions of this work lies in the following key items:

- We formulate an optimization problem for each CV to minimize the AoI.
- A model is developed to describe the spectrum access of multiple CVs as a multi-agent problem and utilize the deep deterministic policy gradient (MADDPG) [16] algorithm.
- We update urban capacities estimation errors according to different AoI values in path planning simulation.
- Numerical experiments demonstrate that the average AoI quantity is maintained within a range of 5–10 milliseconds and fast convergence occurs. Accordingly, path planning performance has significantly improved in terms of travel time (TT) and volume over capacity V/C.

The rest of the paper is organized as follows. In Section 2, the system model is described. We present multi-agent RL-based V2X RRM design in Section 3. Section 4 provides the simulation results and concluding remarks follow in Section 5.

2 SYSTEM MODEL AND PROBLEM FORMULATION

2.1 *Network model*

As depicted in Figure 1, our study is based on the utilization of cellular V2X technology for vehicular communication which consists of one BS and multiple CVs. The BS is located in the center of the environment and is equipped with a single antenna, while CVs are also equipped with a single antenna. The BS is mainly responsible for managing the optimal path by collecting data from all network nodes and building a global traffic information graph of all CVs on the roads. For this purpose, the BS updates the road's capacity with the received V2I information from CVs.

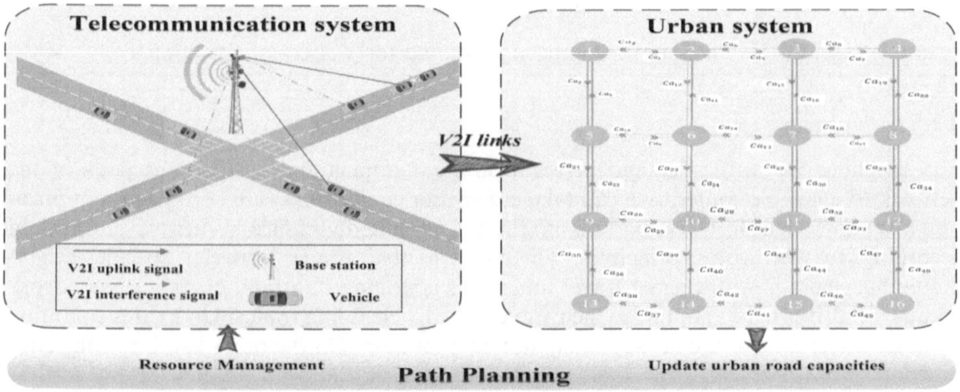

Figure 1. Radio resource management system model.

2.2 *Radio system model formulation*

We assume that $\mathcal{V} = \{v_1, v_2, \ldots, v_V\}$, with $|\mathcal{V}| = V$ indicates the set of CVs, V represents the total number of CVs. $\mathcal{L} = \{l_1, l_2, \ldots, l_L\}$, with $|\mathcal{L}| = L$ indicates the set of roads, where L represents the total number of roads in urban graph model. The time is discretized into equal scheduling slots of length Δt with index t. The system bandwidth is divided into orthogonal sub-channels of size W_n, where indexed by $n \in \mathcal{N}$, and $\mathcal{N}$ is the set of sub-channels. According to the proposed scheme, CV v in road l exchanges highway traffic data with the BS in the center of crossroad via V2I links. Orthogonal frequency division multiplexing (OFDM) is used to handle frequency selective wireless channels. We assume that the channel fading is independent across different sub-channels and remains constant within one sub-channel. We model the V2I channel gain of CV to BS on sub-channel n during one Δt t as,

$$h_v^t[n] = \alpha_v^t g_v^t[n], \tag{1}$$

It is worth noting that α_v^t and $g_v^t[n]$ represent the large-scale fading effect, which is the result of path loss and shadowing, and the small-scale fading, respectively. In addition, we define a binary variable $\rho_v^t[n] \in \{0, 1\}$ that indicates whether sub-channel n is allocated to CV v at time slot t, $\rho_v^t[n] = 1$, otherwise $\rho_v^t[n] = 0$. We can describe the instantaneous rates between CV v and the BS according to the Shannon capacity formula as follows:

$$\begin{aligned} C_v^t[n] &= W_n \log_2\left(1 + \frac{\rho_v^t[n] p_v^t[n] h_v^t[n]}{I_v^t[n] + \sigma_n^2}\right), \\ I_v^t[n] &= \sum_{v' \in V} \rho_{v'}^t[n] p_{v'}^t[n] h_{v'}^t[n], \; v' \neq v, \end{aligned} \tag{2}$$

where $p_v^t[n]$ is the power level used by CV v on sub-channel n for transmitting, $h_{v'}^t[n]$ is the interfering channel to BS from CV $v' \in \mathcal{V}$, and $I_v^t[n]$ corresponds to the total interference power. As previously described, the CV has to ensure timely communication with the BS to exchange any dynamic incidents. As a result, we note A_v^t as the AoI of CV v up to the beginning of the scheduling slot t, it refers to the period of time that has passed since the last successful V2I communication. It is worth noting that this communication inform BS of dynamic traffic capacities. The AoI of CV v evolves according to the following recurrence relation:

$$A_v^{t+1} = \begin{cases} \Delta t, & \text{if} \rho_v^t[n] \cdot C_v^t[n] \geq C_v^{\min} \\ A_v^t + \Delta t, & \text{otherwise} \end{cases}, \tag{3}$$

In this case, $C_v^{\min}$ is the minimum capacity required for V2I communication.

2.3 *Path planning*

For assigning the traffic demand between different origins and destinations pairs of the network to available routes, one can take either user equilibrium (UE) or system optimization (SO) approach. In UE condition, no driver can unitedly reduce his/her travel time and according to Wardrop's principles, all the available routes between specific origin-destination pairs have the same travel time. This condition naturally exists in the network because all drives tend to optimize their travel time by changing routes. Under this condition the total travel time of the network may not be minimum. On the other hand, the objective function of SO problem is the total travel time spent in the network and the traffic flow patterns that solves this problem, minimizes the objective function. In our case study, the number of assigned CVs to each route with both approaches are the same due to consistency of travel times in each system iteration. For this reason, we take the UE assignment approach to simulate the network and use the Dijkstra algorithm for finding the shortest path based on travel time according to the bereau of public roads (BPR) equation as below:

$$T_l(F_l) = T_l(0)\left[1 + \alpha_u \left(\frac{F_l}{Ca_l}\right)^{\beta_u}\right], \tag{4}$$

where l is an index of road, $T_l(0)$, T_l, F_l and Ca_l are free flow time, travel time, traffic flow, and capacity of the road, respectively. α_u and β_u are two parameters of this equation that are mostly regarded as 0.15 and 4 [17]. At last, the Frank-Wolfe (FW) algorithm is applied to solve an iterative first-order optimization. As described earlier, different incidents can affect capacity of a road. In this regard, we denote Ca_l as the real capacity of road,

$$\begin{aligned} Ca_l &= \widehat{C}a_l + \Delta ca, \\ \Delta ca &= \left(-(A_v)/A_{v,(R)}^{\max}\right)\Delta m, \end{aligned} \tag{5}$$

where $\widehat{C}a_l$ and Δca denote the estimated capacity of each road, and deviation from the actual value, respectively. The BS which is responsible for path planning can estimate the roads capacities more accurately with fresh data collected by V2I communications. A_v is a measure of the freshness of information. $A_{v,(R)}^{\max}$ is the maximum AoI between CV v and the BS, and Δm is the maximum amount of error in road's capacities estimation. Thus, the optimization problem for each CV v can be expressed as follows:

$$\min_{\mathbf{r},\mathbf{p}} \left\{\frac{1}{T}\sum_{t=1}^{T} A_v^t\right\} \tag{6}$$

$$\text{s.t. } C1 : C_v^t[n] \geq C_v^{\min}, \quad \forall v \in \mathcal{V}, \forall n \in \mathcal{N}, \tag{7}$$

$$C2 : \rho_v^t[n] \in \{0, 1\}, \quad \forall v \in \mathcal{V}, \forall n \in \mathcal{N}, \tag{8}$$

$$C3: \sum_{n \in \mathcal{N}} \rho_v^t[n] \leq 1, \quad \forall t \in \mathcal{N}, \forall v \in \mathcal{V}, \tag{9}$$

$$C4: p_v^t[n] \leq p_v^{\max}, \quad \forall v \in \mathcal{V}, \forall n \in \mathcal{N}. \tag{10}$$

The objective is to minimize the average AoI for every CV within every T seconds, which leads to the correct estimation of the capacity and its optimal use. Constraint C3 indicates that each CV is only permitted to access one sub-channel per time slot. C4 mandates that CV v's transmit power remain below its maximum value $p_v^{\max}$. In the optimization problem, the sub-channel selection parameter is a binary variable. The objective function is also non-convex. Therefore, the optimization problem is an NP-hard combinatorial optimization problem that is difficult to solve. To address the complexities of the proposed optimization problem, we investigate the state-of-the-art MADDPG method.

3 MADDPG BASED SOLUTION

In a multi-agent environment, agents are intended to maximize their policy functionalities, as outlined below:

$$\max_{\pi} J^v(\pi^v), \ v \in \mathcal{V}, \ \pi^v \in \Pi^v, \tag{11}$$

where $J^v(\pi^v) = \mathbb{E}\left[\sum_{t=0}^{\infty} \gamma^t \tilde{r}_{t+1}^v | s_0^v\right]$ is our conditional expectation, π^v describes the policy of CV v, additionally, the set Π^v includes all feasible policies that are available for CV v. In the proposed model, a CV interacts with the vehicular network environment takes actions in accordance with its policies in order to address the optimization problem, ??. The agents strive to improve their reward 11, which is linked to the objective function. At any time t, the CV observes a state, s^t, and takes appropriate action, a^t. As a result of the CV's selected action, the environment transitions into a new state s^{t+1} and a reward is given to the CV. As a result of our proposed system model, we define the state space, $\mathcal{S}$, the action space, $\mathcal{A}$, and the reward function, r^t, as follows:

3.1 *State space*

Multiple components comprise the state observed by the CV at time slot t: the instant channel information between CV v and the BS, $h_v^t[n]$ for all $n \in \mathcal{N}$, the previous interference from other CVs to CV v, and the AoI of CV v, A_v^t. Accordingly, the state space of CV v is

$$\mathbf{s}_v^t = [h_v^t[n], I_v^{t-1}[n], A_v^t], \ v \in \mathcal{V}. \tag{12}$$

3.2 *Action space*

The action of each CV v is defined as $\mathbf{a}_v^t = \{\rho_v^t, p_v^t\}$. Earlier, we mentioned that $\rho_v^t[n]$ being 0 or 1 indicates if sub-channel n has been selected for a give v and t, and p_v^t represents the power at time t for CV v on sub-carrier n. Using the deep deterministic policy gradient (DDPG) method, the agent is able to select any power within the range of 0 to $p_v^{\max}$. This is the advantage of policy gradient method where we can apply continuous actions spaces and can converge to more accurate results than conventional deep Q learning (DQNs) in which the power has to be discretized.

3.3 *Reward space*

In this study, we use a learning problem that involves agents receiving two reward, a global reward for their cooperation and an individual reward for their performance. Two critics will be trained in the following functions as part of this framework: global centralized critics use observations and actions of all agents as inputs to estimate the global team reward. The local critics, which are specific to each agent, estimate the rewards based on the agent's local observations and actions. The goal is to optimize both global and local rewards simultaneously through the policy. We first discuss MADDPG framework and then design our rewards.

We use π to indicate the set for all CVs policies where CV v policy is denoted by $\pi^v(\pi^v \in \mathbf{p} = \{\pi^1, \ldots, \pi^V\})$ with parameters $\tilde{\theta}^v$, $\tilde{Q}^v$ for CV critic (Q-function) with parameter $\tilde{\phi}^v$ and $\tilde{Q}_G$ for global Q-function with parameter $\tilde{\psi}$. In the first step, we form a neural network, and in the second step, we discuss gradient policies. The number of layers in neural networks for each agent action, critic, and global critic are L_π^v, L_q^v, and L_G, respectively. As a result of the above information, we can develop our neural network in this manner $\tilde{\Theta}_i = \left(W_\pi^{(1)}, \ldots, W_\pi^{(L)}\right)$, $\tilde{\Phi}_q = \left(W_q^{(1)}, \ldots, W_q^{(L)}\right)$, and $\tilde{\Psi}_G = \left(W_G^{(1)}, \ldots, W_G^{(L)}\right)$ as actor, CV critic and global critic. It can be concluded from the above that the gradient policy has the following characteristics:

$$\nabla_{\theta^v} J^v = \mathbb{E}\left[\nabla_{\tilde{\theta}^v} \pi^v(a^v|s^v) \nabla^{a^v} Q_\pi^v(\mathbf{s}, \mathbf{a})\big|_{a^v=\pi^v(s^v)}\right], \tag{13}$$

where $s = (s_1, \ldots, s_N)$ and $a = (a_1, \ldots, a_N)$ are the total state and action vectors. In order to approximate Qfunction for CV v, all actions and states in $Q_\pi^v(\mathbf{s}, \mathbf{a})$ are integrated as inputs. To develop the policy gradient, we utilize two critic networks, which are reformulated as follows:

$$\begin{aligned}\nabla_{\theta^v} J^v = &\underbrace{\mathbb{E}_{\mathbf{s},\mathbf{a}\sim\mathcal{D}}\left[\nabla_{\theta^v} \pi^v(a^v|s^v) \nabla_{a^v} Q_G^\psi(\mathbf{s}, \mathbf{a})\right]}_{\text{GlobalCritic}} \\ &+ \underbrace{\mathbb{E}_{s^v,a^v\sim\mathcal{D}}\left[\nabla_{\theta^v} \pi^v(a^v|s^v) \nabla_{a^v} Q_{\phi^v}^v(s^v, a^v)\right]}_{\text{CVcritic}},\end{aligned} \tag{14}$$

where $a^v = \pi^v(s^v)$ is the action that agent v chooses under policy π^v and $\mathcal{D}$ is the replay buffer. In (14), two terms are used: the first shows the global critic, which receives the actions and states of all CVs and estimates the global Q-function using the global rewards $\tilde{r}_G$, and the second shows the critic of each CV, which receives only its own actions and states. The loss function of the global critic is updated by using

$$\mathcal{L}(\psi) = \mathbb{E}_{\mathbf{s},\mathbf{a},\mathbf{r},\mathbf{s}}\left[\left(Q_G^\psi(\mathbf{s}, \mathbf{a}) - y_G\right)^2\right], \tag{15}$$

where y_G represents the target value of the estimation as follows:

$$y_G = \tilde{r}_G + \gamma Q_G^\psi\left(\mathbf{s}', \mathbf{a}'\right)\Big|_{a'^v=\pi'^v(s'^v)}. \tag{16}$$

The target policy is $\pi' = \{\pi'^1, \ldots, \pi'^V\}$. It is parameterized with $\mathbf{q}' = \{\theta'^1, \ldots, \theta'^V\}$, and the CV loss function and its target update are as follows:

$$\mathcal{L}^v(\phi^v) = \mathbb{E}_{\mathbf{s}^v,\mathbf{a}^v,\mathbf{r}^v,\mathbf{s}'^v}\left[\left(Q_{\phi^v}^v(s^v, a^v) - y^v\right)^2\right], \tag{17}$$

$$y^v = r^v + \gamma Q_{\phi'^v}^v\left(s'^v, a'^v\right)\Big|_{a'^v=\pi'^v(s'^v)}. \tag{18}$$

On the basis of the results presented in [18], we replace the global critic with the twin delayed (TD3) deterministic policy gradient in order to overcome the problem of over-estimation and suboptimal policies resulting from errors in function approximation. In this regard, the gradient of policy is as follows:

$$\nabla_{\theta^v} J^v = \underbrace{\mathbb{E}_{\mathbf{s},\mathbf{a}\sim\mathcal{D}}\left[\nabla_{\theta^v}\pi^v(a^v|s^v)\nabla_{a^v}Q_{G_i}^{\psi_i}(\mathbf{s},\mathbf{a})\right]}_{\text{TD3GlobalCritic}} + \underbrace{\mathbb{E}_{s^v,a^v\sim\mathcal{D}}\left[\nabla_{\theta^v}\pi^v(a^v|s^v)\nabla_{a^v}Q_{\phi^v}^v(s^f,a^f)\right]}_{\text{CVcritic}}, \tag{19}$$

In (19), the twin global critics are updated as follows:

$$\mathcal{L}(\psi_i) = \mathbb{E}_{\mathbf{s},\mathbf{a},\mathbf{r},\mathbf{s}'}\left[\left(Q_{G_i}^{\psi_i}(\mathbf{s},\mathbf{a}) - y_G\right)^2\right], \tag{20}$$

where y_G is updated as follows:

$$y^v = r^v + \gamma(1-\tilde{d})\min_{i=1,2} Q_{\phi_i^{'v_i}}^{v_i}\left(s^{'v},a^{'v}\right)\Big|_{a'^i=\pi'^i(s'^i)}. \tag{21}$$

The agents' critic networks are updated by 17 and 18 and there is a detailed description of the MADDPG algorithm in 3. In TD3, the main idea is to wait for the convergence of value estimates before updating the policy. We are now able to design our rewards, each agent should keep updated connections with the BS and maintain a minimal level of AoI. The global reward should assist the CVs in choosing the proper sub-channel and power to minimize interference between the CVs. To this end, we design individual agent and global reward as

$$r_\ell^v = -\frac{A_v^t}{V} + 0.05G(C_v^t - C_v^{\min}), \tag{22}$$

$$r_g = \frac{1}{V}\sum_{v\in\mathcal{V}} r_\ell^v, \tag{23}$$

where $G(x)$ is the stepwise function used to satisfy $C3$.

4 SIMULATION RESULTS

To evaluate the effectiveness of our formulation, we analyzed a network of 16 nodes, 48 roads, and 240 origin-destination pairs. In this scenario, CV flow of 240 CVs per hour is considered, and the capacity of each road is assumed to be 30 CVs per hour. A summary of the simulation parameters can be found in 1. The simulation results for CVs driving in this scenario are depicted in Figure 2 and 3. We plotted the roads' average travel time and a variable of the roads' quality of service named volume over capacity (V/C) ratio against average AoI for four different values of Δm. V/C is a measure of capacity sufficiency, that is, whether or not the physical geometry and path planning system provides sufficient capacity for the network flow. In the case of low traffic flow in the network, the V/C ratio tends to zero which shows an ideal state of the network. In contrast, as the traffic flow in the network increases, the V/C ratio tends to 1. In the case of an overcrowded state in the network, this ratio passes 1 and the roads' quality of service is very poor.

Algorithm 1: Two time frame modified MADDPG

Generate CVs and establish an environment
Input: For each agent, enter the number of states and actions
Initiate the main global critic networks $Q^{v}_{\phi_1}$ and $Q^{v}_{\phi_2}$
Initiate the target global critic networks of $Q'^{v}_{\phi_1}$ and $Q'^{v}_{\phi_2}$
Set up the policy and critic networks for each agent
for *each episode* **do**
  Provide an update on CV locations and respective channel gains
  Reset the payload of the V2I message and maximum transmission time T to 100 ms
  for *each timestep* t **do**
    for *each agent* v **do**
      observe s^{t}_{v} and Choose an action a^{t}_{v}
    $s_t = [s^1_t, \cdots, s^V_t]$, $a_t = [a^1_t, \cdots, a^V_t]$
    Receive global and local rewards, r^t_g and r^t_l
    Store $(s^t, a^t, r^t_l, r^t_g, s^{t+1})$ in replay buffer $\mathcal{D}$
  Sample minibatch of size S, $(s^j, a^j, r^j_l, r^j_g, s'^j)$, from the replay buffer $\mathcal{D}$
  $y^j_g = r^j_g + \gamma \min_i Q^{g_i}_{\psi'_i}(s'^j, a'^j)$
  Provide an update to global critics by minimizing the loss:
  $\mathcal{L}(\psi_i) = \frac{1}{S}\sum_j \left\{\left(Q^{g_i}_{\psi_i}(s^j, a^j) - y^j_g\right)^2\right\}$
  Update target parameters: $\psi'_i \leftarrow \tau\psi_i + (1-\tau)\psi'_i$
  if *episode mode* d **then**
    Train local critics and actors
    for *each agent* i **do**
      Provide an update to local critics by minimizing the loss:
      $\mathcal{L}(\phi_i) = \frac{1}{S}\sum_j \left\{\left(Q^{i}_{\phi_i}(s^j_i, a^j_i) - y^j_i\right)^2\right\}$
      Provide an update to local critics:
      $\nabla J\theta_i \approx \frac{1}{S}\sum_j \nabla\theta_i\pi_i\left(a_i \mid s^j_i\right)\nabla_{a_i}Q^{g_1}_{\psi_1}(s^j, a^j) + \nabla_{\theta_i}\pi_i(a_i \mid s^j_j)\nabla_{a_i}Q^{i}_{\phi_i}(s^j_i, a^j_i)$
      Provide an update to target networks parameters
      $\theta'_i \leftarrow \tau\theta_i + (1-\tau)\theta'_i$, $\phi'_i \leftarrow \tau\phi_i + (1-\tau)\phi'_i$

Table 1. Parameters for simulation.

Parameter	Value
Frequency of carrier	2GHz
Number of Resource blocks	3
Each Resource block's bandwidth	180kHz
Number of CVs in RRM simulation	20
CVs Speed	10 – 14m/s
BS and CVs antenna heights	25, 1.5m
BS and CVs antenna gains	8, 3dBi
BS and CVs receiver noise figure	5, 9dB
CVs mobility model	Urban case of A.1.2 [11]
CVs maximum power	30dBm
Noise power σ^2	−114dBm
Time constraint of V2I message dissemination, T,	100ms
V2I links path loss model	$128.1 + 37.6\log_d(10)$
Shadowing distribution	Log-normal
Standard deviation of shadowing for V2I links	8dB
Decorrelation distance for V2I links	50m
V2I path loss/shadowing update	Every 100ms[11]
V2I fast fading update	Every 1ms[11]

(*continued*)

Table 1. Continued

Parameter	Value
Fast fading	Rayleigh fading
Replay buffer size for experience	100000
Size of the mini batch	64
Hidden layer number and size of actor networks	2/1024, 512
Hidden layer number and size of critic networks	3/1024, 512, 256
Learning rate of critic/actor networks	0.001/0.0001
Factor of discount	0.99
Target networks soft update parameter	0.0005
Number of CVs in urban simulation	240
Normal roads' capacity	30veh/h
Roads' length	250, 433m
Free flow time	18, 31s
Number of roads	48
α_u	0.15
β_u	4

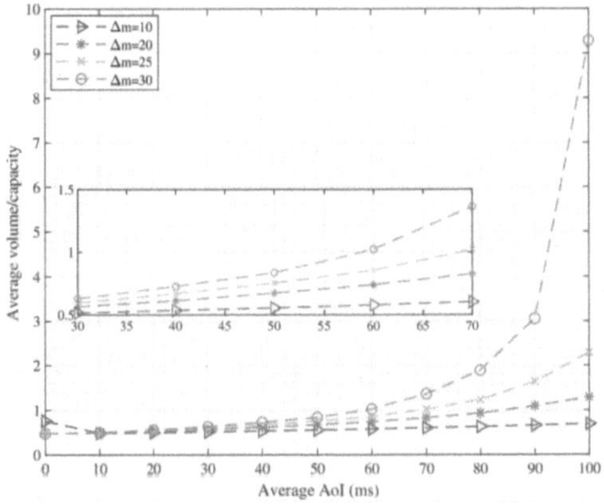

Figure 2. Average V/C with different value of Δm.

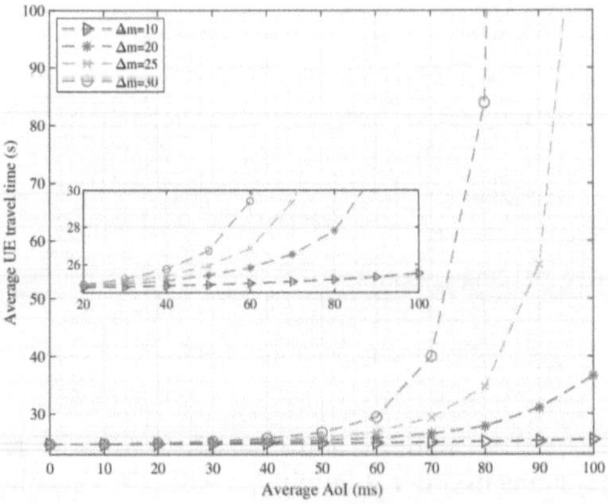

Figure 3. Average roads travel times with different value of Δm.

Figure 3 depicts the average roads travel times versus average AoI values. We can figure out that the average travel time of a CV grows as the average AoI increases. An increase in AoI causes more error in road capacities estimation and, as a result, we face invalid path planning. Moreover, it is shown that in lower values of Δm, the growth of average travel time is more negligible compared to high values of Δm.

Figure 2 plots the values of the V/C ratio against average AoI values. These plots start from a value in the range of zero and 1. As the average AoI increases, the path planning framework fails to update the road capacities accurately and the volume of roads grows, and congestion forms. As a result, the volume of roads gradually grows to the capacity and the V/C ratio passes 1. As mentioned earlier, in lower values of Δm, the increase in the V/C ratio is smaller compared to higher values of Δm.

Figure 4 shows the convergence of proposed MADDPG framework reward. In the proposed framework, individual and global rewards are maximized simultaneously for all agents, resulting in a stable performance. From Figure 5, it is evident that as the number of episodes increases, the AoI quantity decreases. The utilized MADDPG framework maintains the average AoI values from 60 to below 10 ms.

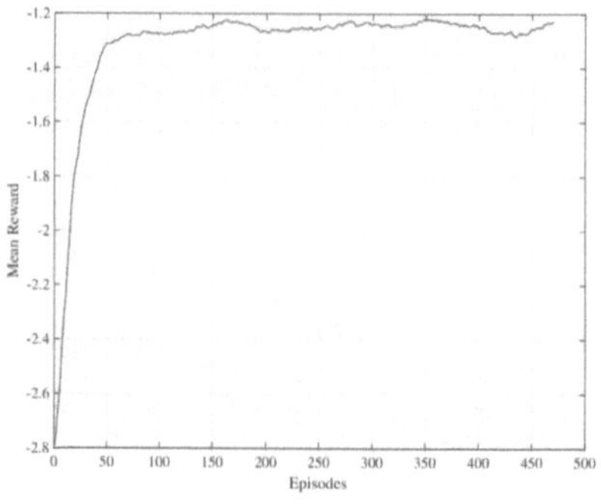

Figure 4. Average reward over learning episodes.

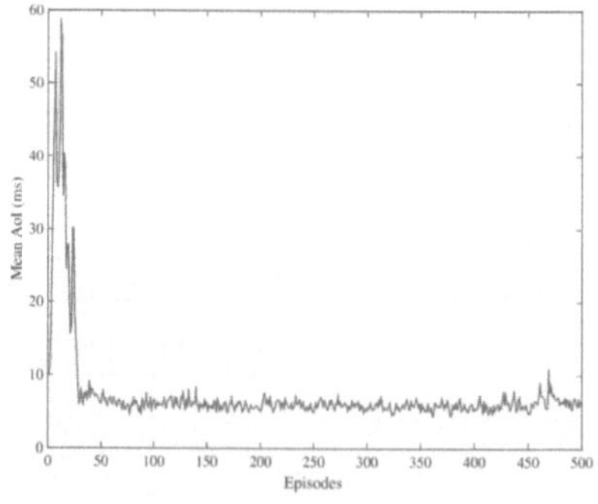

Figure 5. Average AoI over learning episodes.

5 CONCLUSION

As presented in this paper, a joint path planning scheme and radio resource allocation was developed for vehicular systems in order to minimize AoI of CVs while updating urban road capacities for effective path planning. In our work, we were able to confirm that the presented RRM scheme is highly robust and effective in terms of enticing CVs to improve the model of their systems by minimizing their AoI, thereby reducing the error in estimating the capacity. The future work will explore more complicated system models including other technologies such as UAVs to monitor the urban environment for dynamic incidents.

REFERENCES

[1] Wang R., Xu Z., Zhao X., and Hu J., (2019). V2V-based method for the detection of road traffic congestion, *IET Intelligent Transport Systems*, vol. 13, no. 5, pp. 880–885.

[2] Sundar S., Basu A., Kar A., Kumar R., and Kittur H. M., (2018). Shortest path establishment approaches for static and dynamic mobile nodes with quality of service, *Electronic Government, an International Journal*, vol. 14, no. 1, pp. 1–15.

[3] Chu K., Kim J., Jo K., and Sunwoo M., (2015). Real-time path planning of autonomous vehicles for unstructured road navigation, *International Journal of Automotive Technology*, vol. 16, pp. 653–668.

[4] Enayattabar M., Ebrahimnejad A., and Motameni H., (2019). Dijkstra algorithm for shortest path problem under interval-valued pythagorean fuzzy environment, *Complex & Intelligent Systems*, vol. 5, no. 2, pp. 93–100.

[5] Broumi S., Dey A., Talea M., Bakali A., Smarandache F., Nagarajan D., Lathamaheswari M., and Kumar R., (2019). Shortest path problem using Bellman algorithm under neutrosophic environment, *Complex & intelligent systems*, vol. 5, pp. 409–416.

[6] Di Caprio D., Ebrahimnejad A., Alrezaamiri H., and Santos-Arteaga F. J., (2022). A novel ant colony algorithm for solving shortest path problems with fuzzy arc weights, *Alexandria Engineering Journal*, vol. 61, no. 5, pp. 3403–3415.

[7] Chen C., Jiang J., Lv N., and Li S., (2020). An intelligent path planning scheme of autonomous vehicles platoon using deep reinforcement learning on network edge, *IEEE Access*, vol. 8, pp. 99059 99069.

[8] Oubbati O. S., Atiquzzaman M., Lorenz P., Baz A., and Alhakami H., (2020). Search: An SDN-enabled approach for vehicle path-planning, *IEEE Transactions on Vehicular Technology*, vol. 69, no. 12, pp. 14523–14536.

[9] Hui Y., Wang Q., Cheng N., Chen R., Xiao X., and Luan T. H., (2021). Time or Reward: Digital-twin enabled personalized vehicle path planning, in *2021 IEEE Global Communications Conference (GLOBECOM)*, pp. 1–6, IEEE.

[10] Abdel Hakeem S. A., Hady A. A., and Kim H., (2020). 5G-V2X: Standardization, architecture, use cases, network-slicing, and edge-computing, *Wireless Networks*, vol. 26, pp. 6015–6041.

[11] Study on LTE-based V2X Services 3rd Generation Partnership Project (3GPP), *Technical Specification (TS) 36.885 version 14.0.0*, June 2016.

[12] Molina-Masegosa R. and Gozalvez J., (2017). LTE-V for sidelink 5G V2X vehicular communications: A new 5g technology for short-range vehicle-to-everything communications, *IEEE Vehicular Technology Magazine*, vol. 12, no. 4, pp. 30–39.

[13] Wang J., Cao T., Wang X., Zhang M., and Guan J., (2022). Resource scheduling in vehicular networks with age of information and channel awareness, in *2022 IEEE/CIC International Conference on Communications in China (ICCC)*, pp. 344–349, IEEE.

[14] Kaul S., Gruteser M., Rai V., and Kenney J., (2011). Minimizing age of information in vehicular networks, in *2011 8th Annual IEEE communications society conference on sensor, mesh and ad hoc communications and networks*, pp. 350–358, IEEE.

[15] Parvini M., Javan M. R., Mokari N., Arand B. A., and Jorswieck E. A., (2021). AoI aware radio resource management of autonomous platoons via multi agent reinforcement learning, in *2021 17th International Symposium on Wireless Communication Systems (ISWCS)*, pp. 1–6, IEEE.

[16] Sheikh H. U. and Bölöni L., (2020). Multi-agent reinforcement learning for problems with combined individual and team reward, in *2020 International Joint Conference on Neural Networks (IJCNN)*, pp. 1–8, IEEE.

[17] Skabardonis A. and Dowling R., (1997). Improved speed-flow relationships for planning applications, *Transportation research record*, vol. 1572, no. 1, pp. 18–23.

[18] Van Seijen H., Fatemi M., Romoff J., Laroche R., Barnes T., and Tsang J., (2017). Hybrid reward architecture for reinforcement learning, *Advances in Neural Information Processing Systems*, vol. 30.

Innovative solutions for sustainable development

Innovation and Technological Advances for Sustainability – Al-Naemi et al. (Eds)

Evaluating the usefulness of date seed (Phoenix Dactylifera) derived biochar for the removal of caffeine from aqueous media: Isothermal and kinetic studies

Kenesha Wilson, Imane Belyamani, Nora Jamal Saleh & Fatema Jamal Hasan Haji
College of Natural and Health Sciences, Zayed University, Abu Dhabi

ABSTRACT: Caffeine's emergence as a pollutant has been observed in manufacturing practices in the pharmaceutical and personal care product industries. To address the issue, this study aims to explore the potential of using date seed-derived biochar for the adsorption of caffeine from aqueous solutions. A distinct porous structure of the biochar's surface was established by Scanning Electron Microscopy, and a chemical composition rich in aromatic moieties was revealed using Fourier Transform Infrared Spectroscopy (FTIR). The highest efficiency of caffeine removal, 60%, was achieved after 120 minutes at a pH of 2. The rate at which caffeine was adsorbed followed the pseudo-second order model. Both the Langmuir and Freundlich models were suitable for describing the adsorption, and the maximum amount of caffeine adsorbed was 18.94 mg/g, mainly due to interactions between caffeine and the biochar's heterogenous surface. These findings suggest that date seed biochar could be used as an economical and environmentally friendly adsorbent for caffeine from wastewater.

Keywords: biochar, caffeine, date seeds, adsorption

1 INTRODUCTION

Caffeine is commonly found in a variety of beverages, but it is also frequently employed in pharmaceutical and personal care products. Consequently, caffeine is considered one of the growing micropollutants that have been regularly identified in aquifers, soil and potable water [1], due to its high solubility and a half-life of up to 100 days [2]. For the treatment of polluted water, adsorption is an extensively utilized approach where the surface of an adsorbent (solid phase) is brought into contact with the water-containing adsorbate (pollutant molecules), resulting in a drop in the concentration of the pollutant molecules present in the aqueous phase. Biochar is one of the most eco-friendly, economically viable, and efficient adsorbents that have been extensively used for environmental remediation of effluent. Biochar is a material that is high in carbon, permeable, and is produced through a process called pyrolysis, which involves heating biomass in an environment with limited oxygen. The temperature range for the pyrolysis process can vary between 200 to 900°C [3,4]. Biochar made from agricultural waste has been of great interest to the scientific community as a waste management strategy. Date seed waste, also called date pits, is one of the principal types of lignocellulosic refuse that is widespread within the United Arab Emirates (UAE). The UAE has approximately 40M of these trees, each of which gives rise to roughly 15 kg of organic waste annually [5]. Several research studies have indicated that using biochar synthesized from date pits can abate the levels of various types of pollutants including heavy metals, dyes, phenolic compounds, and pesticides [6–9]. The present work demonstrates the capacity of date seeds-based biochar to remediate caffeine-contaminated water samples.

DOI: 10.1201/9781003496724-46

2 MATERIALS AND METHODS

2.1 *Preparation of biochar*

Date seeds were bought at a produce market in downtown Abu Dhabi. The seeds underwent a cleaning process using deionized water to eliminate any impurities and were then left to dry in the sun for a period of 48 hours. Following this, they were crushed and passed through a sieve with a mesh size of 180 micrometers. After crushing, the seeds were dried even further by placing them in an oven set to 60°C for a duration of approximately 24 hours. The seeds were then subjected to pyrolysis with a heating rate of 10°C per minute. The process lasted for 4 hours. The cooled biochar was labeled DSB and was stored in an airtight container until used.

2.2 *Biochar characterisation*

To examine the functional groups, present in the biochar, a Fourier Transform Infrared Spectrometer (FTIR) (Agilent Tech, Cary 630) was utilized, with a scanning range between 400 – 4000 cm^{-1}. The surface morphology of the biochar was assessed using Scanning Electron Microscopy and energy dispersive X-ray spectroscopy (SEM–EDX), with a TESCAN VEGA (LMU INCax-act (Oxford Instruments)) device operating at 20 kv.

2.3 *Effect of pH on adsorption*

Experiments were conducted to investigate how the pH levels influenced the adsorption of caffeine by DSB, within a pH range spanning from 2 to 10. Caffeine solutions were prepared using caffeine powder (Caffeine Reagentplus®, C0750 Sigma-Aldrich, minimum purity of 99%) dissolved in deionized water. HCl and NaOH (0.1 M) were added to the 20 ppm caffeine solutions to adjust the pH levels. Once the target pH level was achieved, 0.1 g of DSB was introduced into the mixture, and the resulting solution was then placed onto a rotary shaker operating at 30 rpm for a period of 24 hours. Caffeine concentration in solutions was determined using a UV-VIS Spectrophotometer (UV-2401 PC double beam) set at λ_{max} = 273 nm wavelength. The adsorption capacity of the DSB and the percentage removal efficiency were calculated by using the equations below:

$$q_e = \frac{(C_o - C_e)V}{m} \tag{1}$$

$$q_t = \frac{(C_o - C_t)V}{m} \tag{2}$$

$$R(\%) = \frac{(C_o - C_e)}{C_o} \times 100 \tag{3}$$

The symbols 'C_o', 'C_e', and 'C_t' represent the initial concentration of caffeine, the concentration at equilibrium, and the concentration at a specific time, 't', following the adsorption process, all measured in mg/L. The character 'm' refers to the mass of biochar utilized in grams (g), while v represents the volume of the caffeine solution measured in liters (L). The adsorption capacity at equilibrium and at time 't' are indicated by q_e (measured in milligrams per gram, mg/g) and q_t (also measured in mg/g), respectively. The efficiency of the adsorption process is denoted as R, represented as a percentage (%).

2.4 *Adsorption kinetics*

To conduct the adsorption kinetic experiments, 0.1 g of DSB was introduced into a 50 mL solution containing 20 ppm of caffeine at the ideal pH and allowed to come into contact for different time periods ranging from 5 to 180 minutes. The solutions were agitated on a rotary shaker set to 30 rpm. Following this, the mixtures were filtered, and the caffeine concentrations were measured. The adsorption kinetics were then analyzed through the application of pseudo-first and pseudo-second order models.

Pseudo-first order (PSO) model is shown in Eq. (4)

$$q_t = q_e\left(1 - e^{-k_{1t}}\right) \tag{4}$$

The pseudo-second order non-linear equation model is:

$$q_t = \frac{q_e^2 k_2 t}{1 + q_e k_e t} \tag{5}$$

2.5 *Adsorption isotherm studies*

To determine the relationship or interactions between the amounts of caffeine adsorbed and the adsorbate, the caffeine concentrations at equilibrium was modelled using Langmuir and Freundlich isotherms.

The Langmuir equation is written as:

$$q_e = \frac{q_m K_L C_E}{1 + K_L C_e} \tag{6}$$

The variable q_e specifies the amount of caffeine adsorbed on the biochar per unit mass, expressed in mg g^{-1}. The symbol q_m represents the maximum adsorption capacity of the biochar for caffeine, also measured in mg g^{-1}. The concentration of caffeine at equilibrium is represented by C_e (mg L^{-1}) The constant K_L is known as the Langmuir constant, which is used to describe the adsorption process and is expressed in L mg^{-1}.

The non-linear form of the Freundlich equation is shown in Eq. (7).

$$q_e = K_F C_e^{1/n} \tag{7}$$

where n (g L^{-1}) is the adsorption heterogeneity factor and K_F (mg g^{-1}) is the Freundlich constant.

3 RESULTS AND DISCUSSION

3.1 *Characterisation of date seeds biochar*

FTIR analyses performed on the DSB before adsorption is shown in Figure 1. The signal observed at approximately 3300 cm^{-1}, can be ascribed to O–H stretching that indicates the

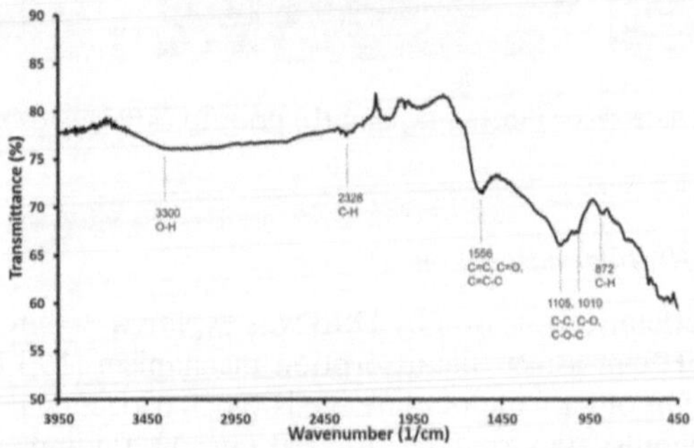

Figure 1. FTIR spectra of date seeds biochar.

presence of phenols, alcohols, carboxyls arising because of the presence of lignin and holocellulose in the DSB [10]. The peak at wavelength 2328 cm^{-1} is associated with C–H bonds. At around 1556 cm^{-1} a broad peak is observed and is likely due to C=O, C=C , and C=C–C normally associated with aromatic rings [11,12]. The bands 1105 and 1010 cm^{-1} could be due to stretches from the bonds, C–C, C–O and C–O–C [13,14]. At around 872 cm^{-1}, the peak seen can be attributed to the C–H out-of-plane aromatic bending [10].

The EDX analysis (Figure 2) showed that the DSB consisted mainly of carbon (88.48%) with small amounts of oxygen (11.24%), and trace amounts of potassium (0.28%).SEM scans of the DSB (Figures 3(a) and (b)) showed the surface to be very porous and suggests efficient adsorption of caffeine molecules would be possible. The higher the porosity and surface area, the more adsorption is likely to occur [15]. The scan of the DSB after caffeine adsorption (Figure 2(c)) showed a less porous structure and clustered large particles which strongly suggests that caffeine was adsorbed on the surface of the biochar thus filling in the porous cavities.

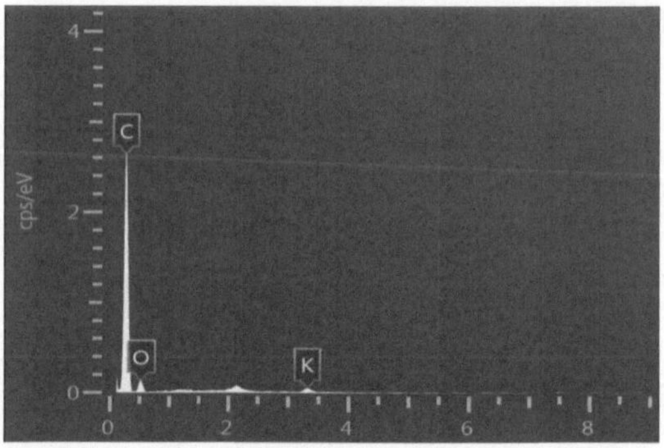

Figure 2. EDX analysis of date seeds biochar.

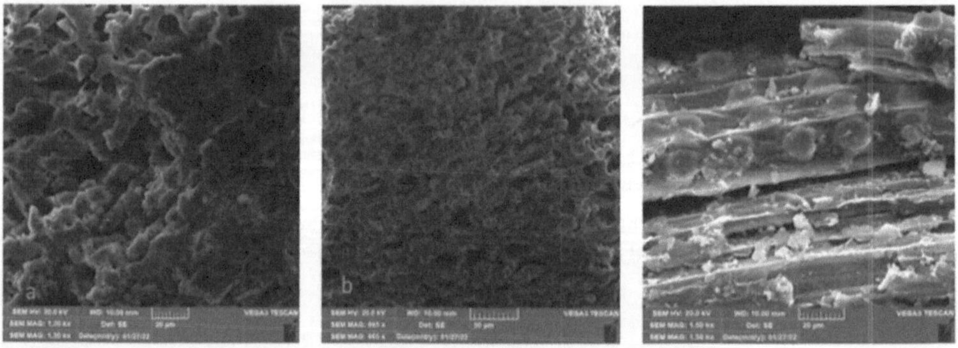

Figure 3. SEM scans of date seeds biochar (a) and (b) prior to caffeine adsorption, (c) after caffeine adsorption.

3.2 *How changes in pH affect adsorption*

The impact of pH on caffeine adsorption by DSB was explored within a pH range of 2-10. It is widely known that pH can affect the adsorption mechanism [16,17]. Figures 3(a) shows that maximum adsorption of caffeine on date seeds biochar occurs at pH of 2.

Consistent with previoulsy reported results, the current findings also suggest that the adsorption of caffeine is more favored at low pH [18]. At low pH, the surface of the

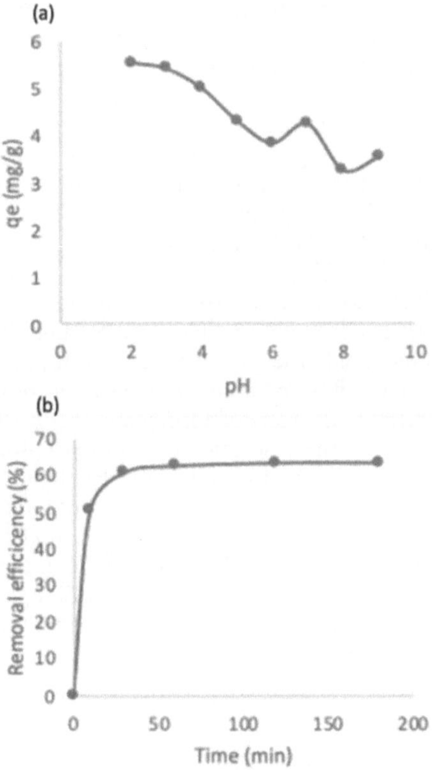

Figure 4. Effect of pH (a) and time (b) on the adsorption of caffeine on date seeds biochar.

adsorbent becomes positively charged as a result protonation in the acidic medium, enhancing its suitability to adsorb negatively charged particles [19]. According to Beltrame [20] caffeine exists in a neutral form at pH values less than 5.5. Therefore, at a pH of 2, it is highly likely that the neutral form of caffeine interacts with the positively charged biochar surface according to interactions such as hydrogen-bonding, π- π interactions [2] and n- π interactions [21], between the heterocyclic rings on the caffeine molecue and the aromatic rings on the exterior of the biochar. Consequently, high adsorption capacity is observed when the pH is low [18,22]. At higher pH, the negatively charged species of caffeine predominates, so adsorption is less favorable as a result of electrostatic repulsions between the similarly charged caffeine molecules and the DSB. Thus, the pH of the caffeine solutions for all subsequent adsorption studies was performed at pH 2. Figure 4(b), shows that as contact time increases, the adsorption capacity also increases until equilibrium is achieved at approximately 120 minutes with a removal capacity of around 64%. A probable reason for this, is that at the beginning of mixing, there is an availabliity of large amounts of active sites on the date seeds biochar. As time progresses and equilibrium is reached, the number of active sites available on the DSB will be reduced [18,23].

3.3 *Adsorption isotherm and kinetic modeling*

The adsorption efficacy was thoroughly assessed by applying the experimental data to both the pseudo-first order and pseudo-second order models. (Figure 4(a) and (b)). Table 1 shows that the experimental data fit both models well, as evidenced by high R^2 values for both fits. As the predicted q_e value obtained from the pseudo-second order model was in better agreement with the experimental results ($q_{e,\ cal2}$ = 6.532 mg/g compared to $q_{e,\ exp}$ = 6.447 mg/g), it can be inferred that the adsorption mechanism is more in line with pseudo-second order kinetics than pseudo-first order. This is consistent with other studies that were done to assess the kinetics of the adsorption of caffeine on various types of biochar [2,18,20,22]. Data that follows pseudo-second kinetics suggests that adsorption is contingent on the number of vacant binding sites [20] and is controlled by a chemisorption process. This indicates that there is electron sharing occurring between the functional groups present on the biochar surface and the caffeine molecules. [19].

Isotherm modeling helps to describe interactions between the adsorbate and the adsorbent. The removal of caffeine by DSB was investigated using the Langmuir and Freundlich adsorption models. If the thickness of the adsorbed layer is monolayer and the adsorbent has an equal affinity for all identical and equivalent localized binding sites on the adsorbate, then the adsorption process would hold true to the Langmuir model. The Freundlich model defines the heterogeneity and multilayered nature of the surface of the adsorbent. According to the Freundlich model, when the intensity of adsorption or surface heterogeneity factor (1/n) is greater than zero $\left(0 < \frac{1}{n} < 1\right)$, the adsorption is favorable [24] as was the case with these results; $\frac{1}{n} = 0.7963$. Table 1 presents the Langmuir and Freundlich equilibrium adsorption data and their corresponding criteria, which indicate that both adsorption isotherm models fit the experimental data well. Nonetheless, the Freundlich model showed a slightly better match to the experimental data compared to the Langmuir equation. Similar studies done on caffeine adsorption by various adsorbents, such as pine needles [22], pineapple plant leaves [20] and water hyacinth [25], have also reported that both Langmuir and Freundlich models were able to accurately fit the experimental data.

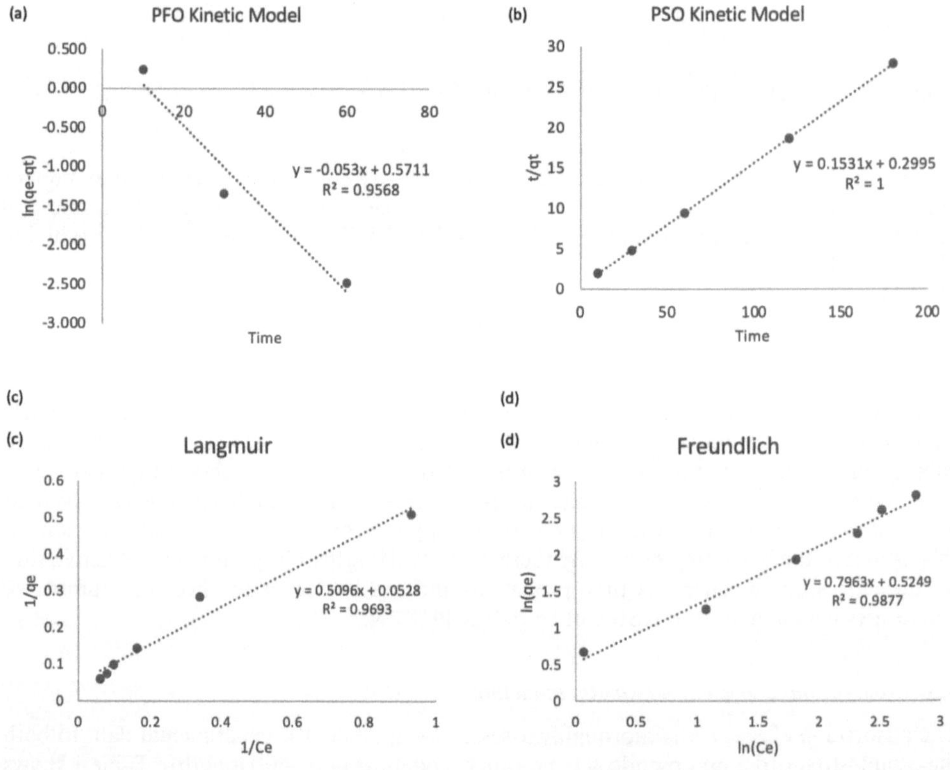

Figure 5. (a) Pseudo-first order plot (b) pseudo-second order plot (c) Langmuir plot (d) Freundlich plot.

Table 1. Kinetics and adsorption isotherm parameters.

Models	Parameters	
Pseudo first order	k_1	0.053
	q_e (cal)	1.770
	q_e (exp)	6.447
	R^2	0.9568
Pseudo second order	k_2	0.0783
	q_e (cal)	6.532
	q_e (exp)	6.447
	R^2	1.000
Langmuir	q_{max} (mg/g)	18.940
	K_L	0.1036
	R^2	0.9693
Freundlich	n	1.256
	K_F	1.690
	R^2	0.9877

3.4 *Adsorption mechanism*

The predicted kinetic and isotherm data presented in this study indicates that date seed biochar is a suitable adsorbent for caffeine molecules. A mechanism that is highly probable may be due to the forging of covalent bonds and electrostatic attraction between caffeine and the adsorbent. In an acidic medium it is plausible for the nitrogen atom in caffeine to engage in nucleophilic addition reactions with the carbonyl groups that exist on the surface of DSB. Due to the complexity of adsorption mechanisms, interactions involve, π–π interactions, hydrogen bonding. n–π interactions are also highly likely due to weak van der Waals forces [26]. On the surface of the biochar, polar groups with hydrophilic character, such as –OH and –COO were identified from FTIR scans. Hence, it can be concluded that caffeine adsorption occurs due to dipole-dipole interactions, where the π electrons and 2-nitrophenol aromatic rings of caffeine interact with π aromatic electrons present on the surface of the adsorbent. [19].

Table 2. Comparative adsorptive removal of caffeine with other published work.

Adsorbent	pH	Kinetic Model	Isotherm Model	Adsorption capacity (mg/g)	References
Water hyacinth biochar	2	Ps2	L, F	2.488	[25]
Oxidized pine needles biochar	4	Ps2	L, F	5.35	[22]
Rice husk charcoal	4	Ps2	L, F	8.0	[27]
Date Palm Seeds	2	Ps2	L, F	18.9	This study

4 CONCLUSION

Biochar made from date seeds was successfully utilized to adsorb caffeine from aqueous solution. The ideal pH for adsorption was found to be 2 and the adsorption capacity reached 18.9 mg/g. The data fitted the pseudo-second order kinetic model, as well as the Langmuir and Freundlich adsorption isotherm models. These findings indicate that utilizing date seeds as an economical option for removing caffeine from industrial wastes, particularly from pharmaceutical and personal care products, could be a feasible solution. Future work will focus on pyrolyzing biochar at different temperatures and modifying the surface of the biochar to enhance the adsorptive capacity. Thermodynamic studies will also be explored to investigate the impact of temperature on the adsorption mechanism.

ACKNOWLEDGMENT

The authors are grateful to the lab technicians at Zayed University in assisting with this project, especially former employee Dr. Cijo Xavier. Thanks also to Dr. Jibran Iqbal for invaluable feedback with designing this work and Dr. Tony Myers for language editing.

REFERENCES

[1] Li S., Wen J., He B., Wang J., Hu X., and Liu J., (2020). Occurrence of caffeine in the freshwater environment: Implications for ecopharmacovigilance, *Environmental Pollution*, vol. 263, p. 114371.

[2] Álvarez-Torrellas S., Rodríguez, Ovejero G., Gómez J. M., and García J., (2016). Removal of caffeine from pharmaceutical wastewater by adsorption: influence of NOM, textural and chemical properties of the adsorbent, *Environmental Technology*, vol. 37, no. 13, pp. 1618–1630.

[3] Ahmad M. et al., (2014). Biochar as a sorbent for contaminant management in soil and water: a review, *Chemosphere*, vol. 99, pp. 19–33.

[4] Abu-Al-Aish A. and Love S., (2013). Factors influencing students' acceptance of m-learning: an investigation in higher education, *The International Review of Research in Open and Distributed Learning*, vol. 14, no. 5.

[5] Giwa A., Yusuf A., Ajumobi O., and Dzidzienyo P., (2019). Pyrolysis of date palm waste to biochar using concentrated solar thermal energy: Economic and sustainability implications, *Waste Management*, vol. 93, pp. 14–22.

[6] Al-Tarawneh A., (2022). Biochar as a Cadmium Scavenger in the Aquatic Environment Remediation: Date Seeds as Raw Material, *Journal of Ecological Engineering*, vol. 23, no. 3.

[7] Tripathy S., Sahu S., Patel R. K., Panda R. B., and Kar P. K., (2021). Efficient removal of Cr (VI) by polyaniline modified biochar from date (Phoenix dactylifera) seed, *Groundwater for Sustainable Development*, vol. 15, p. 100653.

[8] Hassan S. S., Al-Ghouti M. A., Abu-Dieyeh M., and McKay G., (2020). Novel bioadsorbents based on date pits for organophosphorus pesticide remediation from water, *Journal of Environmental Chemical Engineering*, vol. 8, no. 1, p. 103593.

[9] Mahdi Z., Yu Q. J., and El Hanandeh A., (2019). Competitive adsorption of heavy metal ions (Pb2+, Cu2+, and Ni2+) onto date seed biochar: batch and fixed bed experiments, *Separation Science and Technology*, vol. 54, no. 6, pp. 888–901.

[10] Uzun B. B., Apaydin-Varol E., Ateş F., Özbay N., and Pütün A. E., (2010). Synthetic fuel production from tea waste: characterisation of bio-oil and bio-char, *Fuel*, vol. 89, no. 1, pp. 176–184.

[11] Ahmad M. et al., (2012). Effects of pyrolysis temperature on soybean stover-and peanut shell-derived biochar properties and TCE adsorption in water, *Bioresource Technology*, vol. 118, pp. 536–544.

[12] Ali N. S., Jabbar N. M., Alardhi S. M., Majdi H. S., and Albayati T. M., (2022). Adsorption of methyl violet dye onto a prepared bio-adsorbent from date seeds: isotherm, kinetics, and thermodynamic studies, *Heliyon*, vol. 8, no. 8, p. e10276.

[13] Nguyen V. H., Van H. T., Nguyen V. Q., Dam X. V., Hoang L., and Ha L., (2020). Magnetic Fe3O4 nanoparticle biochar derived from pomelo peel for reactive Red 21 adsorption from aqueous solution, *Journal of chemistry*, vol. 2020.

[14] Yao S. et al., (2019). Resource utilization of a typical vegetable waste as biochars in removing phthalate acid esters from water: a sorption case study, *Bioresource technology*, vol. 293, p. 122081.

[15] Hassan M., Naidu R., Du J., Liu Y., and Qi F., (2020). Critical review of magnetic biosorbents: Their preparation, application, and regeneration for wastewater treatment, *Science of the Total Environment*, vol. 702, p. 134893.

[16] Shah N. S. et al., (2019). Hydroxyl and sulfate radical mediated degradation of ciprofloxacin using nano zerovalent manganese catalyzed S2O82−, *Chemical Engineering Journal*, vol. 356, pp. 199–209.

[17] Anastopoulos I. and Pashalidis I., (2019). The application of oxidized carbon derived from Luffa cylindrica for caffeine removal. Equilibrium, thermodynamic, kinetic and mechanistic analysis, *Journal of Molecular Liquids*, vol. 296, p. 112078.

[18] Portinho R., Zanella O., and Féris L. A., (2017). Grape stalk application for caffeine removal through adsorption, *Journal of environmental management*, vol. 202, pp. 178–187.

[19] Fakioğlu M. and Kalpaklı Y., (2022). Mechanism and behavior of caffeine sorption: affecting factors, *RSC Advances*, vol. 12, no. 41, pp. 26504–26513.

[20] Beltrame K. K., Cazetta A. L., de Souza P. S., Spessato L., Silva T. L., and Almeida V. C., (2018). Adsorption of caffeine on mesoporous activated carbon fibers prepared from pineapple plant leaves, *Ecotoxicology and environmental safety*, vol. 147, pp. 64–71.

[21] Tran H. N., Wang Y.-F., You S.-J., and Chao H.-P., (2017). Insights into the mechanism of cationic dye adsorption on activated charcoal: The importance of π–π interactions, *Process Safety and Environmental Protection*, vol. 107, pp. 168–180.

[22] Anastopoulos I., Katsouromalli A., and Pashalidis I., (2020). Oxidized biochar obtained from pine needles as a novel adsorbent to remove caffeine from aqueous solutions, *Journal of Molecular Liquids*, vol. 304, p. 112661.

[23] Bachmann S. A. L., Calvete T., and Féris L. A., (2021). Caffeine removal from aqueous media by adsorption: An overview of adsorbents evolution and the kinetic, equilibrium and thermodynamic studies, *Science of the Total Environment*, vol. 767, p. 144229.

[24] Al-Ghouti M. A. and Da'ana D. A., (2020). Guidelines for the use and interpretation of adsorption isotherm models: A review, *Journal of hazardous materials*, vol. 393, p. 122383.

[25] Ngeno E. C., Orata F., Lilechi D., Shikuku V. O., and Kimosop S., (2016). Adsorption of caffeine and ciprofloxacin onto pyrolitically derived water hyacinth biochar: isothermal, kinetic and thermodynamic studies, *J. Chem. Chem. Eng*, vol. 10, pp. 185–194.

[26] Keerthanan S., Bhatnagar A., Mahatantila K., Jayasinghe C., Ok Y. S., and Vithanage M., (2020). Engineered tea-waste biochar for the removal of caffeine, a model compound in pharmaceuticals and personal care products (PPCPs), from aqueous media, *Environmental Technology & Innovation*, vol. 19, p. 100847.

[27] Bernardo M. M. S. et al., (2017). Study of the removal mechanism of aquatic emergent pollutants by new bio-based chars, *Environmental Science and Pollution Research*, vol. 24, no. 28, pp. 22698–22708.

Innovation and Technological Advances for Sustainability – Al-Naemi et al. (Eds)

Synthesis and characterization of porous composite with tailored porosity fabricated using rice husk by utilizing red mud and coal mine tailing

Vaibhav Pandey
Department of Ceramic Engineering, Indian Institute of Technology, BHU, Varanasi, India

Ashish Singh Pareta, Anupam Tiwari & Praveen Kumar Singh
Department of Mechanical Engineering, Indian Institute of Technology, BHU, Varanasi, India

Ashutosh Gupta
Department of Ceramic Engineering, Indian Institute of Technology, BHU, Varanasi, India

Saroja Kanta Panda
Department of Mechanical Engineering, Indian Institute of Technology, BHU, Varanasi, India

ABSTRACT: Management and use of red mud (RM) and coal mine overburden waste (MW) have drawn attention over the past decade due to their enormous scale and associated environmental problems. The harmful impact of this solid waste and its leachates has damaged the nearby ecosystem and forced industries to look for alternative solutions just rather than dumping. Despite having a lot of research, industries are still looking for a solid solution that can be implemented. Mass conversion with economic feasibility is the major obstacle faced by these wastes. The current work examines the potential use of these wastes for fabricating porous ceramic bricks by incorporating multiple quantities of their mixture with rice husk. The physical qualities of bricks have been studied in relation to variations in the MW: RM ratio, sintering temperature, and Rice husk amount. The developed porous composite possesses apparent porosity, compressive strength, and thermal conductivity in the range of 26–54%, 0.83–6.83 MPa, and 0.37–0.67 W/mK respectively. The developed porous composite with tunable pore microstructure is seen as a potential material for thermal insulation and other similar applications.

Keywords: Red Mud, Coal mine overburden, Rice Husk, Porous Composite, Thermo-mechanical property

1 INTRODUCTION

In the era of growing industrialization, demand for raw materials and energy resources has tremendously increased. This increase in demand has also increased the production of industrial wastes which are generated as a secondary product along with raw material and energy harnessing. The massive production of waste (including municipal, electronic, and industrial waste) has become a major global concern in recent decades [1–3]. In terms of overall solid waste generated, industrial waste makes up a smaller portion. However, the presence of hazardous and harmful substances as well as valuable metals makes their treatment necessary.

Since the Industrial Revolution, numerous nations have relied on coal as one of their primary fossil fuels to drive their economies. Governments have since come to the realization

DOI: 10.1201/9781003496724-47

that using coal is the key to industrialization, and as a result, coal mines have grown significantly. However, the operation of coal mines is criticized for having a negative influence on the environment, particularly because a significant quantity of waste is generated during the mining and processing of coal. These wastes come with a number of risks, which includes instability and erosion of MW heap slope [4], accidents due to spontaneous combustion and subsequent explosion of coal gangue [5], acid mine drainage which subsequently harms nearby soil fertility and ground water quality [6], air pollution [7], and finally discernible effect on the landscape's appearance [8].

Aluminum is one of the most used materials among metals. Global production of aluminum was around 60 million in the year 2015 making it the second largest in the world in terms of metal production [9]. Red mud is a solid waste produced after refining bauxite ore in the Aluminium industry. Around 1.5 tons of RM is produced for each ton of aluminum metal. Due to the heavy demand for aluminium, production of aluminum is increased enormously leaving behind large bauxite residue with global production of RM approximately 100 MT annually [10]. Mostly Bayer process has been used for the extraction of alumina which uses sodium hydroxide to dissolve the aluminum silicate. RM mainly consists of oxides of iron, silicon, and alumina with some traces of heavy metals and other potential pollutants. Due to heavy alkalinity disposal of RM is a major concern that leads leaching of pollutants into soil and water [11].

It is urgent to improve the comprehensive utilization of RM and MW due to the humongous generation and environmental impact of RM and MW on dumping. Considerable work for the potential utilization of RM and MW has been carried out by various researchers over the world. In this regard, various approaches have been developed for the alternate utility of RM and MW. Both RM and MW consist of rare earth and other valuable metals like titanium, iron, etc. Studies have been done on the recovery of these metals [12,13]. But due to large production, just recovery of valuable metals from these wastes is insufficient for proper waste management of these wastes. Researchers are continuously looking for a solution that can effectively treat the harmful impact of this waste as well convert it into a valuable product with a high mass conversion rate. Most pronounced utilization of RM and MW comes out from the ceramic industry having a product range from bricks [14], tiles [15], concrete additives [16], construction materials [17], and cement [18] to porous ceramic [19]. RM based composites was also used in various applications e.g. as adsorbent to treat wastewater by removing inorganic pollutants [20], as air filter to treat pollute gases [21] and, as lightweight aggregates [22].

In view of the demand for the potential utilization of RM and MW due to their adverse effect on the environment, this paper is an effort to utilize RM in ceramic material production for construction and other applications. Many researchers gave their input in the production of RM and MW-based brick by different approaches [10]. Geo-polymerization is the most used approach till now due to relatively less energy intensiveness. However, research showed that because the Fe_2O_3 content of red mud is not exploited, the RM based geopolymer products shows lower strength compared to fired products. The firing route is another approach that is mostly followed by researchers to prepare various ceramic products like fire bricks, ceramic tiles, insulation materials, etc. However, red mud products take a lot of energy and are therefore rather expensive. Instead, the possibility of using red mud in conjunction with other garbage through a fire route has been explored in an effort to cut costs. Also, the use of RM with other waste like Fly ash improves the moldability of green compacts. Ceramic composites made via the firing route using MW require additional binders because of their non-plasticity. Thus, mixing RM with MW helps in improving the binding ability and moldability of the mix. Insulating bricks are the most frequently used refractory for heat insulation due to its high porosity, light weight, and exceptionally low thermal conductivity. Common ingredients in insulating bricks include various forms of alumina, silica, fireclay, kaolin, calcium silicate, expanded vermiculite, perlite diatomite, and aggregates. Efficient thermal insulation requires a sufficient level of porosity in the prepared ceramic composites. Various methods with their advantages and disadvantages are used to achieve a different level of porosities e.g., polymeric impregnation, sol-gel technique, sacrificial fugitive method, foaming through blowing agents, additive manufacturing, etc. The sacrificial method is mostly known for providing sufficient porosity without compromising the strength of the porous composites compared to other methods. Also, most of the fugitive materials which are being used are biowastes like sawdust, rice husk, coke fines, etc. So, using these biowastes not only helps in introducing porosity to the sample but also gives a way to tackle bio-generated waste disposal issues. One of the agricultural biowastes which are widely used as fugitive materials is Rice Husk (RH). Each year, millions of tons of rice are produced worldwide. The literature states that per metric tonne of rice, 200 kg RH is created [23]. Due to its vast surface area and high carbon content, rice poses an environmental pollution hazard, making RH disposal a significant task in the rice industry. There have been attempts to appropriately utilise RH without harming the environment. Some scientists used RH directly to strengthen it with other materials, but the majority of them employed it as ash to take use of its high silica concentration.

Keeping in mind the harmful impact of this industrial and agricultural waste on the environment, the present work is an attempt to utilize these wastes to fabricate porous composites. RM and MW are used as major ingredients of the insulation brick and RH is used as a pore former. The different mixture ratio of RM and MW is prepared and the physical and mechanical properties of each composition are evaluated and compared. RH with different weight percentages is used for each composition to get various levels of porosities and insulation properties are measured accordingly. The present work comes with an effective and economic method to develop a low-cost product that can be widely used, thus giving an efficient solution to the mass conversion of RM and MW waste and reducing their harmful impact on the environment.

2 EXPERIMENTAL

2.1 *Materials*

RM and MW were as used as the primary ingredient for the fabrication of insulation brick and were obtained from Hindalco Industries- Renukoot, India, and NCL, Singraulli, India respectively. The RH was collected from the local agricultural land near Varanasi, India. Both RM and MW was initially ball-milled to get finer particle and was sieved through a 600-micron-sized sieve for uniformity. In the studies, the RH of size below 63 microns has been used to achieve narrow and consistent porosity of the bricks. Sucrose solution (Sucrose: Water: 6:4) was used as a binder.

2.2 *Preparation of porous composites*

Three compositions of were prepared by mixing dried powders of RM and MW homogeneously in varying proportions (RM: MW; 2;1;1;1.1;2). Each mixture composition was then added with a different RH amount (5 to 20 %). After adding the required amount of binder solution, the MW-RM-RH mixture was compacted uniaxially at 20 MPa to fabricate a cylindrical green compact with a diameter of 10 mm and a thickness of 20 mm. The addition of sucrose solution provides the necessary plasticity for molding. The green compacts were then heated slowly to 110°C in an electric oven for two hours to remove surface moisture. Compacts were then sintered at temperatures of 1000, 1100, and 1200°C at a rate of 5°C/min with a holding timing of two hours.

2.3 *Characterization of raw materials and porous composites*

Initially, raw MW was physiochemically characterized in order to quantitatively and qualitatively evaluate it. An X-ray Fluorescence (XRF) device was used to detect the presence of oxides in raw MW (S8 Tiger, Bruker model). With a step size of 0.02° and a 2 range of 5–80°, the XRD (Rigaku Miniflex 600 Desktop X-Ray Diffractometer) data of raw MW and sintered porous MW-RM composite are studied. The bulk density is determined using known test protocols, while the apparent porosity is evaluated using Archimedes' displacement method and water immersion. Following formula was used to calculate bulk density, and apparent porosity [24]:

$$\rho = \frac{m}{V} \tag{1}$$

$$\tau = \left(\frac{w - m}{w - s}\right) \times 100 \tag{2}$$

where, ρ is bulk density of sample in gram per cc, m is dry weight of composite in grams, V is volume of prepared composite, τ is apparent porosity of the sample, w is soaked weight measured after soaking composite in water for 24 hours, s is the suspended weight of the sample. On the Universal Testing Machine (INSTRON), compressive samples thoroughly prepared in accordance with ASTM C1161-18 are tested at a crosshead speed of 0.2 mm/min. For the compressive testing, porous composites of 10 mm diameter and 20 mm thickness have been prepared. The heat conductivity of porous MW-RM composite was evaluated using a TPS-500 Hot Disk equipment (Goteborg, Sweden).

3 RESULTS AND DISCUSSION

The XRF data (Table 1) reveals that both MW and RM exhibits mineralogical similarities with varying composition. Silica and alumina are the major oxides in both the raw samples with minor oxides of titanium, sodium, potassium, calcium, and magnesium. Iron oxide content is considerably high in RM samples, giving them red color. The XRD data (Figure 1) shows different phases of silica, alumina, titania, and other minerals found in MW and RM. Because red mud is produced during the Bayer process, which involves leaching bauxite ore with sodium hydroxide, it comprises several mineralogical forms of aluminium hydroxide (gibbsite, boehmite, diaspore), as well as hematite, goethite, and other minerals.

Table 1. XRF analysis of MW and RM.

Oxides	MW	RM
SiO_2	49.40	10.81
Al_2O_3	24.81	20.59
Fe_2O_3	3.44	25.58
K_2O	2.46	0.15
MgO	1.76	0.18
Na_2O	1.07	7.72
TiO_2	1.10	11.87
CaO	1.00	2.05
BaO	0.17	NA
P_2O_5	0.17	0.47
ZrO_2	0.02	0.11
MnO	0.05	0.03
V_2O_5	0.03	NA
Cr_2O_3	0.01	0.13
CuO	0.01	0.02
ZrO_2	NA	0.11

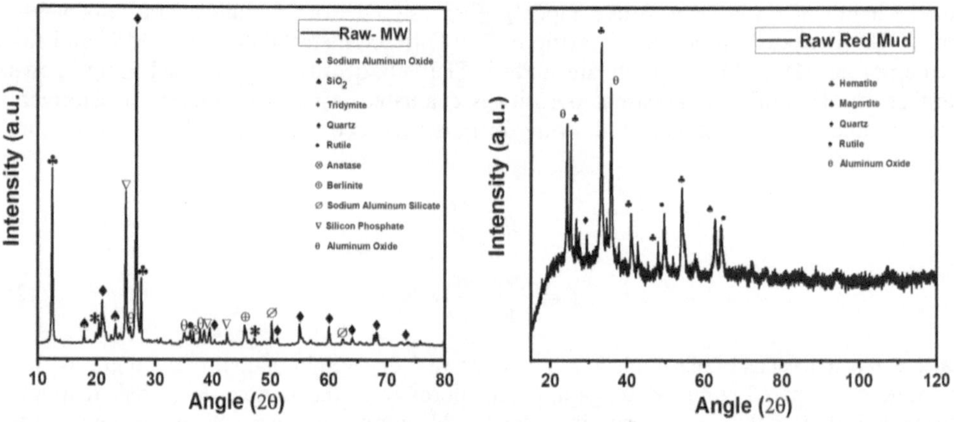

Figure 1. XRD analysis of raw MW and red mud.

The prepared porous MW-RM composite posse's porosity is in the range of 26–54% (Figure 2). RH is used as pore former which directly influences the porosity of the MW-RM composites. An increasing trend is observed for apparent porosity with the increase in weight percentage of RH. Porosity enhancement using RH is limited to the maximum amount of RH in composites beyond which the composite is unable to hold its shape during sintering due to excess burnout gas pressure. In the present case, the maximum RH limit in the MW-RM composite is found to be 21 weight %. A reverse trend is observed for sintering temperature which reduces porosity in the MW-RM composites. An inverting trend is observed for sintering temperature. The apparent porosity of the MW-RM composite decreases with an increase in sintering temperature due to the consolidation of both grains and pores. High fire temperatures cause localized raw material fusion, resulting in poor porosity, whereas adding more RH increases the amount of volatiles produced during firing, leading to the formation of extremely porous structures. Additionally, data demonstrates that porosity decreases as RM content increases because red mud particles form more dense and compact structures than MW particles. The compressive strength of fabricated porous MW-RM composites was found to be in the range of 0.83–6.83 MPa (Figure 3). The results indicate that compressive strength rises with temperature and RM content, and falls with the addition of RH. The presence of iron oxide in RM samples helps in producing iron and aluminum silicates at high sintering temperatures which ultimately results in high compressive strength of composites. However, at increasing red mud concentrations, the MW content decreases noticeably, which lowers free silica availability (which is necessary for the synthesis of the corresponding silicates) and, consequently, compressive strength. The thermal conductivity test demonstrates that the prepared porous MW-RM composites have the capability to be used as an insulator. The thermal conductivities are found to be in the range of 0.37–0.67 W/mK (Figure 4). Thermal conductivity measurement shows thermal insulation properties of composites decrease with an increase in firing temperature and red mud content. High temperature and the addition of red mud cause localized fusion, a decrease in porosity, and a decrease in the composite insulating properties. Further, the addition of RH increases the thermal insulation due to an increase in the porosity of composites. Figure 5 shows the XRD plot of sintered porous composites prepared at different MW: RM ratios with 10% RH addition and sintered at 1100°C. It is identified that the constitution phases quartz (SiO_2), alumina (Al_2O_3), and hematite (Fe_2O_3) are present in a significant amount. These phases can be found in the composites both free and mixed form. X-ray diffraction (XRD) plot comparisons reveal that the SiO_2 peak intensity drops noticeably with an increase in the RM content of porous composites. This occurs because free silica, when heated, creates different iron and aluminium silicates. The data also demonstrates that the iron silicate is developed at lower temperatures, viz. 1100°C, and that its formation may be the fundamental reason for providing strength to the porous composites.

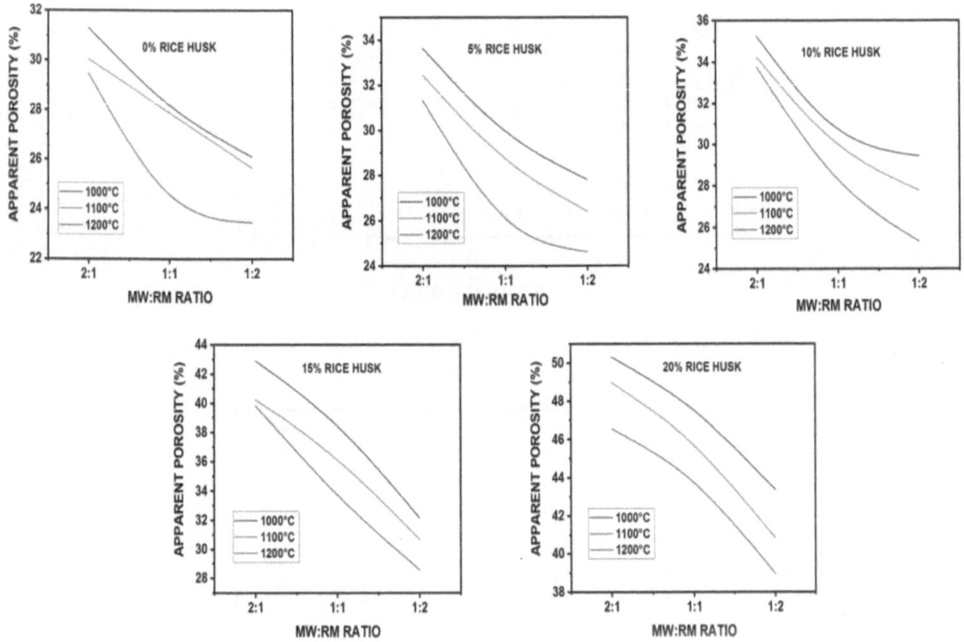

Figure 2. Variation of apparent porosity with MW:RM ratio for different temperatures and RH weight %.

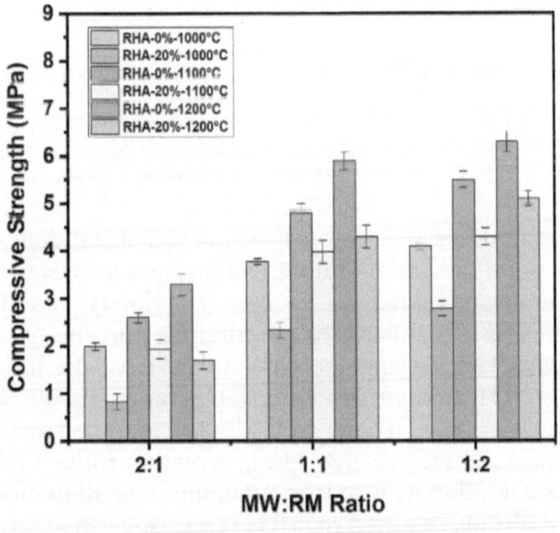

Figure 3. Variation of compressive strength with MW:RM ratio.

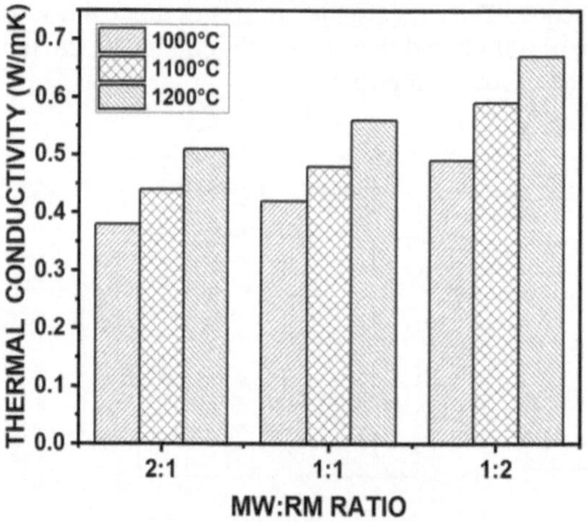

Figure 4. Variation of thermal conductivity with MW:RM ratio.

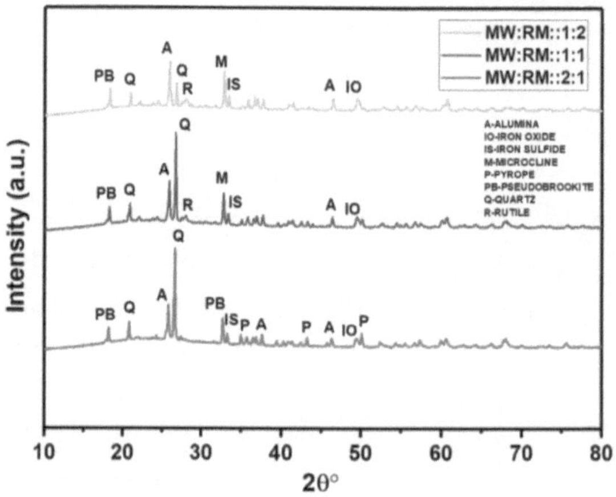

Figure 5. XRD analysis of sintered porous composites.

4 CONCLUSION

In summary, porous composites with good thermo-mechanical characteristics have been successfully fabricated using industrial and agricultural-grade wastes. Red Mud and Coal Mine overburden waste are industrial wastes used as primary ingredients for fabricating porous composites. Rice Husk is utilized for tailoring the porosity of composites. Different compositions are prepared by varying the MW to RM weight ratio and RH amount. Various parameters like RH amount, sintering temperature, MW: RM weight ratio are investigated to evaluate their effect on composite physical and chemical characteristics. After burning, the mineralogical phases like free silica, hematite, rutile, and alumina present in MW and RM lead to the development of iron, titanium, and aluminium silicates as well as other harder phases. Results also showed that RH is a trustworthy pore-forming agent. Due to the localized particle fusion and hard phase production, the high-temperature firing of composites produces materials with improved strength, and bulk density but decreased thermal conductivity. Red mud content increases the mouldability property of MW and strength of the sintered composites but also decreases the porosity. The apparent porosity and compressive strength of the developed porous composites are found to be between 26–54%, and 0.83–6.83 MPa, respectively. The findings show the composites have thermal insulation capabilities with low thermal conductivity up to 0.37 W/(mK) and good thermal resistivity properties with sufficient mechanical properties.

ACKNOWLEDGMENT

The authors would like to thank the Central instrument facility, IIT (BHU), Varanasi, Department of Ceramic Engineering, IIT (BHU), Varanasi and Department of Mechanical Engineering, IIT (BHU), Varanasi for providing the experimental facility.

REFERENCES

[1] Mandal, A. K. and Sinha, O. P., (2017). Effect of bottom ash fineness on properties of red mud geopolymer. *J. Solid Waste Technol. Manag.* 43, 26–35.

[2] Verma, H. R., Singh, K. K. and Mankhand, R., (2017a). Comparative study of printed circuit board recycling by cracking of internal layers using organic solvents dimethyl formamide and dimethylacetamide. *J. Clean. Prod.* 142, 1721–1727.

[3] Verma, H. R., Singh, K. K. and Mankhand, T. R., (2017b). Liberation of metal clads of waste printed circuit boards by removal of halogenated epoxy resin substrate using dimethylacetamide. *Waste Manag.* 60, 652–659.

[4] Song, Y.-S., Cho, Y.-C. and Kim, K.-S., (2015). Monitoring and Stability Analysis of a Coal Mine Waste Heap Slope in Korea. In: Lollino, G., Giordan, D., Crosta, G.B., Corominas, J., Azzam, R., Wasowski, J., Sciarra, N. (Eds.), *Engineering Geology for Society and Territory*, vol. 2. Springer International Publishing, pp. 217–220.

[5] Ciesielczuk, J., (2015). Chapter 16 – Coal Mining and Combustion in the Coal Waste Dumps of Poland. In: Sokol, G.B.S.P.V. (Ed.), *Coal and Peat Fires: A Global Perspective*. Elsevier, Boston, pp. 463–473.

[6] Banerjee, D., (2014). Acid drainage potential from coal mine wastes: environmental assessment through static and kinetic tests. *Int. J. Environ. Sci. Technol.* 11, 1365–1378.

[7] Adibee, N., Osanloo, M. and Rahmanpour, M., (2013). Adverse effects of coal mine waste dumps on the environment and their management. *Environ. Earth Sci.* 70,1581–1592.

[8] Bian, Z., Dong, J., Lei, S., Leng, H., Mu, S. and Wang, H., (2009). The impact of disposal and treatment of coal mining wastes on environment and farmland. *Environ. Geol.* 58, 625–634.

[9] USGS, 2015. *Mineral Commodity Summaries 2015*. U.S. Geological Survey, Washington, DC.

[10] Arup Kumar Mandal, Himanshu Ranjan Verma and O.P. Sinha; Utilization of aluminum plant's waste for production of insulation bricks. *Journal of Cleaner Production* 162 (2017) 949–957

[11] Balomenos, E., Gianopoulou, I., Panias, D., Paspaliaris, I., Perry, K. and Boufounos, D., (2013). *Efficient and Complete Exploitation of the Bauxite Residue (Red Mud) Produced in the Bayer Process.* EMC 2013, 285e292.

[12] Chenna Rao Borra, Bart Blanpain, Yiannis Pontikes, Koen Binnemans and Tom Van Gerven, (2016). Recovery of Rare Earths and Other Valuable Metals From Bauxite Residue (Red Mud): A Review, *J. Sustain. Metall.* 2:365–386.

[13] Koen Binnemans, Peter Tom Jones, Bart Blanpain, Tom Van Gerven and Yiannis Pontikes, (2015). Towards zero-waste valorisation of rare-earth-containing industrial process residues: a critical review, *Journal of Cleaner Production* 99 17–38.

[14] Nivetha, Johnson Daniel and Vemuri lakshminarayana, (2017). *Study on Strength and Behaviour of Red Mud Bricks, jr. of Industrial Pollution Control* 33(s2) pp 1227–1230.

[15] Wei Wang, Weijie Chen, Haitao Liu and Chao Han, (5 June 2018). Recycling of waste red mud for production of ceramic floor tile with high strength and lightweight, *Journal of Alloys and Compounds* Volume 748, Pages 876–881.

[16] James Joyce P., Tobias Hertel, Andrei Goronovski, Alan H. Tkaczyk, Yiannis Pontikes and Anna Björklund, (November 2018). Identifying hotspots of environmental impact in the development of novel

inorganic polymer paving blocks from bauxite residue, *Resources, Conservation and Recycling* Volume 138, Pages 87–98.
[17] Harekrushna Sutar, Progress of Red Mud Utilization: An Overview.
[18] Ri-Xin Liu and Chi-Sun Poon, (20 January 2016). Utilization of red mud derived from bauxite in self-compacting concrete, *Journal of Cleaner Production* Volume 112, Part 1, Pages 384–39.
[19] Wei Wang Weijie Chen and Haitao Liu, (1 June 2019). Recycling of waste red mud for fabrication of SiC/mullite composite porous ceramics, *Ceramics International* Volume 45, Issue 8, Pages 9852–9857.
[20] Ardau C., Lattanzi P., Peretti R. and Zucca A., (2013). Treatment of mine wastes with transformed red muds (TRM) and other iron compounds: leaching column tests, *Proc. Earth Planet. Sci.* 7 467–470.
[21] Yang S., Zhang Y., Yu J., et al., (2014). Multi-functional honeycomb ceramic materials produced from bauxite residues, *Mater. Des.* 59 333–338.
[22] Molineux C. J., Newport D. J., Ayati B., et al., Bauxite residue (red mud) as a pulverised fuel ash substitute in the manufacture of lightweight aggregate, *J. Clean. Prod.* 112 (Part 1) (2016) 401–408.
[23] Ashutosh Gupta, Vaibhav Pandey, Mayank Kumar Yadav, Kalyani Mohanta and Manas Ranjan Majhi, (2022). A comparative study on physio-mechanical properties of silica compacts fabricated using rice husk ash derived amorphous and crystalline silica; *Ceramics International*, 48 (23), B. 35750–35758.
[24] Vaibhav Pandey, Mayank Kumar Yadav, Ashutosh Gupta, Kalyani Mohanta, Panda S. K., Singh V. K., (2022). Synthesis, morphological and thermomechanical characterization of light weight silica foam via reaction generated thermo-foaming process; *Journal of the European Ceramic Society* 42 6671–6683.

Innovation and Technological Advances for Sustainability – Al-Naemi et al. (Eds)
© 2025 The Author(s), ISBN 978-1-032-80372-2
Open Access: www.taylorfrancis.com, CC BY-NC-ND 4.0 license

Free funding and free marketing: An addition to crowdsourcing and business model canvas

Noureddine Lamairia
Management Department, University 8 May 1945, Guelma, Algeria

ABSTRACT: The study aims to define the concept of free funding and free marketing brought by the researcher, presenting a framework based on observation and corresponding to all ways of free funding and free marketing, as well as we reviewed the link and importance of the concept of free funding and free marketing with some concepts that are in line with it such as crowdsourcing and clarifying the difference in first, As well as the relationship of the framework and its importance in the business model Canvas and thus maximizing the effectiveness and efficiency of new entrepreneurial projects. All this will maximize opportunities in the establishment of entrepreneurial projects and its sustainability, particularly in the first stages.

Keywords: Free Funding, Free Marketing, Crowdsourcing, Business Model Canvas, Innovation, Entrepreneurship

1 INTRODUCTION

With digital transformation and the adoption of startups, especially on information technology, communication and interactive marketing, and with the high risk ratio for new entrepreneurial projects, it has become important to achieve innovation, whether in product, management or marketing with the lowest costs to contain both risk and competition, and for this we first consider that the framework based on free funding and free marketing as new concepts applied by startups that rely on crowdsourcing or idea holders who seek to achieve funding with the lowest risks and marketing that achieves effectiveness and efficiency. Second, apply this framework to business model canvas, especially for new entrepreneurs, or in general the strategy of a modern startup that is characterized by competitiveness, cost leadership and excellence. In the end, it achieves effectiveness and efficiency from the beginning of the business, based on financing and marketing innovation.

Many studies have talked about the topic of crowdsourcing and the business model canvas, we will know about them in the presentation, but there has never been talk about free funding in the literature, only there are different articles talking about free marketing but as a concept of free tools available on the Internet that are used in order to be more effective for digital marketing strategies (free online tools) [1] [8] [16]. Starting from this consideration, the paper aims to fill these gaps by realizing a broader concept of free marketing and figuring out how the concept and free funding affect startups bmc and the concept of crowdsourcing as well.

To design our framework, the researcher relied on the descriptive and analytical approach to the variables of the study and what was mentioned in the theoretical framework, as well as attending workshops, competitions and entrepreneurship exhibitions with observations, including Google Startup weekend, Fintech weekend, and e-commerce & digitization exhibitions … etc. and conduct interviews with project holders and why they benefit, using the researcher's background as an e-marketer and holder of small startup projects based on rewards.

The paper is organized as follows. In the first section, we talked about the concept of free funding and free marketing, In the second section we look at crowdsourcing, its types and its relationship to the framework, in the third section we talked about the business model Canvas and its relationship to the framework. Finally, we summarize the discussion and highlighted the importance of using the proposed framework.

2 FREE FUNDING AND FREE MARKETING

2.1 *Free funding*

Free Funding is a concept that explains obtaining funding to start or grow a business from various free sources, most of which are creative ideas competitions, entrepreneurship awards, competitions and honors for modern startups, as well as sources of solidarity with idea holders and entrepreneurs from web platforms, places of worship and charities. It differs from crowdfunding in that it is easy to obtain and without a share in capital (e.g., the participation of guiddini.com.dz project owners in startup competitions since their emergence and the launch of events to bring sponsors to narrate their experiences or present their large companies, made them believe in Guiddini entrepreneurs and finance their activities, and thus the company has raised capital to start through free funding from crowd).

2.2 *Free marketing*

Free Marketing consists of all free practices to market or advertise the product such as attending specialized exhibitions that open the opportunity to display and sell products, holding specialized events with free funding through sponsorship and shopping for your company or product with the invitation of influencers on social media pages and content creators. Also, having a series of content pages that belong to the entrepreneur will also build brands, activate trust, build the first visitors and market your products later.

Networking with a public page with website owners and podcast content creators is later useful in marketing the product by advertising it, mentioning its features and then selling it, especially if there are affiliate links. (e.g., piaca-dz.com start in publishing content about the Algerian entrepreneurial ecosystem in Facebook groups helped spread the company's name, attachment to its brand and increase sales later due to the volume of trust and credibility).

Digital marketing tools are included in free marketing because they give a great benefit, some of the tools can send a lot of emails and are useful in marketing via e-mails such as Mailchimp. There is something useful in knowing the most prominent topics of famous content and hashtags (e.g., viral finder and Hashtagstack), and there is something useful in identifying the marketing strategy of competitors and their customers and the nature of their content (e.g., the tool of SPYFU and SERPSTAT), and therefore the matter has a direct character on the competition and improve competitiveness and create competitive advantages in the product, price and promotion. And many of them [1]:

- Buffer is a social media scheduler that can save time by using forever-free plan. Three different social media platforms may be added to the accounts using this package. (For example, Twitter, LinkedIn, and Facebook). Up to 10 posts can then be scheduled each week from that point.
- Google Analytics is a free tool that assists small businesses in monitoring website interaction and activity is Google Analytics. It can use to assess the efficacy of your SEO campaign and get comprehensive client demographic data. and also evaluating companies existing marketing strategy's effectiveness.
- Google Alert considers that is one of the best free marketing tools available for businesses. It offers receiving alerts anytime for the company when mentioned online and monitor the competitors' mentions.

- Help a Reporter Out (HARO) an excellent tool that connects journalists to reliable sources, can help to generate awareness for the company. When someone from a company is cited as a source in a reputable newspaper, HARO it may uses to gain free publicity.
- Typeform helps by getting consumer input is crucial to expanding any company and enhancing goods and services. By Typeform enabling companies to design stunning forms, surveys, and quizzes for clients and website visitors.

3 CROWDSOURCING

3.1 *Definition of crowdsourcing*

The concept of crowd-sourcing is one significant step further. It is the process of obtaining required services, ideas, or content by soliciting contributions from a large group of unidentified people, and especially from an online community, rather than from conventional employees or suppliers [2]. Crowd sourcing has established itself as a viable option for businesses seeking new ideas and solutions to micro-tasks from the crowd. Scholars have been increasingly interested in crowd sourcing in tandem with the expansion of this technique among innovative firms, as evidenced by the number of relevant articles published since 2006. Crowd sourcing can be viewed as both an open innovation practice and a co-creation support tool from a theoretical stand point [3]. In addition, crowdsourcing can be defined as a category of participative online job in which individual or a group of people offers the task to be done by other groups of individuals with different skills which gives mutual benefit to both crowdsourcers and crowd workers via a flexible open call [4].

3.2 *Types of crowdsourcing*

It is noted from the last two definitions that they focused on limiting crowdsourcing to one of its forms, which is microtasking, while crowdsourcing is more comprehensive, and for this we put all types of crowdsourcing as follows [5]:

- Crowdtesting is inviting and receiving feedback from a subset of your target audience to identify issues or problems with a product (often apps and websites). This gives more visibility into how the product is being used and what challenges the audience may be having when interacting with products.
- Crowdfunding is a type of crowdsourcing that allows individuals to pool their money and resources in order to help fund the production or development of creative projects. Kickstarter, Indiegogo, and GoFundMe are all crowdfunding platforms that allow individuals to fundraise for their projects. depending to leverages the power of social networks and communities.
- Collaborative knowledge is the idea of shared information in an open and accessible way. It is a term often used when referring to large groups of people such as researchers, students, and members of a community. Collaborative knowledge in crowdsourcing is the use of a collective intelligence approach to form and solve problems.
- Open innovation in crowdsourcing is the process of engaging diverse, external people in order to generate new ideas and solutions. Open innovation relies on a team or community approach so that different perspectives and knowledge can be shared.
- Creative and design crowdsourcing involves soliciting feedback from a large group of people to develop ideas, products or services. Creative crowdsourcing can include anything such as websites, apps, or even video games. In addition, the creative crowd is also used to create graphic designs such as logos, posters, etc.
- Microtasking is a type of crowdsourcing where individual tasks are assigned to small groups like writing a blog post or doing research. Microtaskers can either be individuals or an organization that divides the work amongst its members.

3.3 *Importance of crowdsourcing*

Among the points of importance of crowdsourcing are the following [6]:

- Crowdsourcing brings together communities around a common project or cause
- Efficient way of solving time-intensive problems
- Deeper engagement by communities, who resonate and build loyalty to the product or solution

3.4 *Dropping the framework on crowdsourcing*

After getting to know both free funding and free marketing, it can be said that free funding is complementary and in addition to crowdfunding, and free marketing is complementary and very similar to Crowdtesting in using digital marketing tools like Charlie and similar to Creative Crowdsourcing Platforms in using graphic design such us Canva [7] [8]:

- An increasingly common type of marketing collateral is visual content. Visual content, which ranges from personalized infographics and illustrations to completely developed sales presentations, Canva is an excellent approach for new firms to establish credibility and develop a polished image. Canva is one of the most user-friendly and simple. It makes anything for free, including posters, brochures, infographics, and photos for social media.
- The top CRM (Customer Relationship Management) software can helps keep track of present and potential customers. Charlie is kind of a quick and simple way to actually get to know someone, Charlie is a tool that looks at future schedule or runs an email address to give companies a comprehensive run-down on a person—social media accounts, biography, interests, important news, … etc., Charlie will tell Marketers everything when entering the email address of the individual whose information they want ad-hoc.

In our theorical framework, we focused on freeware, freelancing also represents crowdsourcing, especially in small tasks that are required quickly or need a specific profile. The following figure can thus summarize the scope of our framework, and a comprehensive description of the crowdsourcing process.

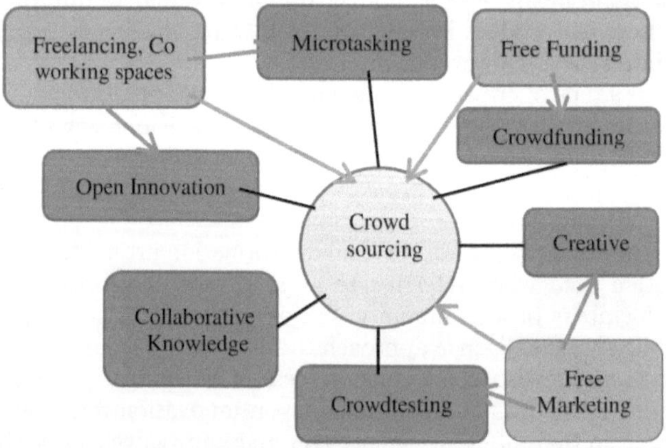

Figure 1. Dropping free funding and free marketing on crowdsourcing.

From Figure 1, It is clear that both free funding, free marketing, freelancing, and co working spaces can be direct actors in crowdsourcing or indirectly influence one of the types of crowdsourcing, which means not dispensing with the types of crowdsourcing, but adding free funding and free marketing would add efficiency and more competitive advantage.

Also, the figure provided for the description of the framework does not represent a system of crowdsourcing (CSSs) because the crowdsourcing system may represent the business model as a basis for companies, that is, the system on which the company builds and builds all its business and operations, For example: to develop a platform that links the organization as a job providers with the crowd workers to perform some tasks in exchange for rewards or prizes [9] .All that our framework seeks is for small businesses, modern startups or entrepreneurs' ideas to benefit from the concept of free funding and free marketing at the beginning of their projects.

4 BUSINESS MODEL CANVAS

4.1 *Business model canvas definition*

The business model canvas defined as a tool to help entrepreneurs understand a business model in a concise and organized way. It presents the structure of a business plan on single page, allowing individuals a quick overview of its key components, thereby facilitating a quick grasp of any business initiative. The canvas can be used to accelerate the development and testing of business ideas before they reach the stage of a formal business plan [10]. Business model Cnvas is becoming one of the trending topics by academics and practitioners today. It deals extensively on how to design new business models and especially how to make the transition from an existing business model to a newly designed business model or try to find new breakthroughs to find new values that can provide new competitive advantage for the business undertaken [11].

4.2 *Business model canvas building blocks*

The Osterwalder and Pigneur's BMC consists of the following building blocks [12]:

- Customer segments it refers to those different groups of people or organization the company aims to reach and serve solution in exchange for their payment
- Channels it describes how the company will reaches its customer segments.
- Customer relationships it defines how the company do retain and how grow the targeted customer segments.
- Value proposition it describes the bundle of products and services that creates value for a specific customer segment. A value proposition could include characteristics such as newness, performance, customization ... etc.
- Revenue stream it represents the cash the company generates from each segment (e.g., subscription).
- Key resources it is the most important asset required to make the business model work and run the startup.
- Key activities it describes what the company must do to make the business model work, such as sales, marketing, platform and networking activities.
- Key partnership it constitutes the network of suppliers and partners that makes the business model work. Partnerships may be strategic alliances between noncompetitors as well as competitors (coopetition).
- Cost structure it describes all costs (fixed, variable ... etc.) incurred to operate the business model.

4.3 *Benefits of BMC*

The benefits of business model canvas include [13]:

- Visualize and explore business model.
- Clear expression and communication of ideas or concept to the team, stakeholders or potential investors.

- Involving the team and stakeholders in ta way raises engagement and commitment to the business.
- Incorporate user feedback into the business model.
- When having multiple product lines or services, get a clear view on the potential conflicts and dependencies.
- A business model canvas can serve as the companies a more detailed business plan.

4.4 *Dropping the framework on businrss model canvas*

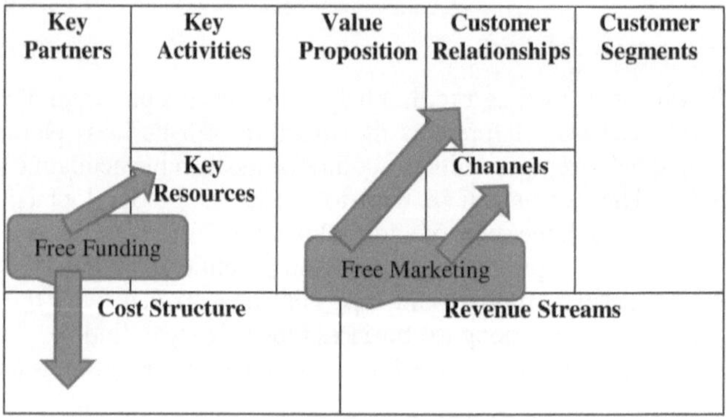

Figure 2. Apply the framework to BMC.

The figure illustrates by Osterwalder and Pigneur's business model canvas [14], the concept of free funding influences the business model canvas by reducing costs and providing financial resources, while free marketing helps in establishing relationships with customers and providing channels. Thus, the importance of free funding is on the left side of the business model canvas, while free marketing is important on the right side of the business model canvas as shown in Figure 2.

The value of the framework in linking the business model canvas, free funding and free marketing differs in that the work here aims to enhance the elements of the business model canvas. As for D'Andrea *et al.* CBM (Crowdsourcing Business Model) concept [15], it aims to integrate crowdsourcing as a whole as a comprehensive strategy for the company and a business model that reflects the company's vision, perceptions and actions in order to compete in the market, such as the company's competitive strategy (e.g., diversification or cost leadership).

The importance of the proposed framework is also reflected in the achievement of creativity and comprehensive innovation due to the effectiveness of the idea of free funding and free marketing, both of which help projects and entrepreneurs to achieve innovation in financing, and marketing innovation, which ultimately enhances the competitiveness of projects.

5 CONCLUSION

Based on the information gathered our framework focused on new concept which is free funding and free marketing and their relation and influence in crowdsourcing, business model canvas and we conducted the following results:

Free funding is a concept that explains obtaining funding to start or grow a business from various sources, most of which are creative ideas competitions, entrepreneurship awards, competitions and honors for modern startups, as well as sources of solidarity with idea holders and entrepreneurs from web platforms, places of worship and charities. It differs from crowdfunding in that it is easy to obtain and without a share in capital.

Free marketing consists of all free practices to market or advertise the product such as attending specialized exhibitions that open the opportunity to display and sell products, holding specialized events with free funding through sponsorship and shopping for your company or product with the invitation of influencers on social media pages and content creators.

Having a series of content pages that belong to the entrepreneur will also build brands, activate confidence, build the first visitors and market products later.

Networking with a public page with website owners and podcast content creators is later useful in marketing the product by advertising it, mentioning its features and then selling it, especially if there are affiliate links.

Both free funding, free marketing can be direct actors in crowdsourcing or indirectly influence one of the types of crowdsourcing, which means not dispensing with the types of crowdsourcing, but adding free funding and free marketing would add efficiency and more competitive advantage. The Figure 1 provided for the description of the framework does not represent a system of crowdsourcing (CSSs) because the crowdsourcing system may represent the business model as a basis for companies, that is, the system on which the company builds and builds all its business strategy and operations.

The importance of free funding is on the left side of the business model, while free marketing is important on the right side of the business model.

The value of the business in linking the business model, free finance and free marketing differs in that the work here aims to enhance the elements of the business model. As for D'Andrea *et al.* CBM concept, it aims to integrate crowdsourcing as a whole as a comprehensive strategy for the company and a business model that reflects the company's vision, perceptions and actions in order to compete in the market

The importance of the proposed framework is also reflected in the achievement of creativity and comprehensive innovation due to the effectiveness of the idea of free funding and free marketing, both of which help projects and entrepreneurs to achieve innovation in financing, and marketing, which ultimately enhances the competitiveness of projects.

All that our framework seeks is for small businesses, modern startups or entrepreneurs' ideas to benefit from the concept of free funding and free marketing at the beginning of their projects and achieving sustainability in the early stages.

We highly recommend that project holders adopt free marketing and free funding, as each of the mentioned companies has done for all new project holders in order to achieve effectiveness, efficiency and sustainability. We also recommend that the government seek to support startup and innovation entrepreneurs with more radio and television programming that helps define their projects, and award challenges.

Future works will test the impact of free funding and free marketing on the startups community.

REFERENCES

[1] Johnson J., *Free Marketing Tools Small Business*, https://www.uschamber.com/co/grow/marketing/free-marketing-tools-small-business[retrieved: Novembre, 2022].

[2] Orlowsk M., (March 2015). Challenges for workflows technology to support crowd-sourcing activities Conf. *Business Intelligence and Technology*, France, p. 4.

[3] Awasthi L. S., (January 2016). Role and challenges of crowdsourcing in management and technology, in *International Journal of Advance Research And Innovative ideas in Education*, vol. 2, pp.4625–4631.

[4] Salleh S. S., Arshad N. H., Janom N., Marzuki N., (2014). E-managemnet concept and process of crowd sourcing platform certification, in *e-Learning, e-Management, and e-Services*, pp. 128–133.

[5] Clickworker, "About Crowdsourcing", https://www.clickworker.com/about-crowdsourcing[retrieved: October, 2022].

[6] Hargrave M., *Crowdsourcing: Definition, How It Works*, https://www.investopedia.com/terms/c/crowdsourcing.asp[retrieved: November, 2022].

[7] Miller D., *Free Marketing Tools*, https://www.entrepreneur.com/growing-a-business/10-free-marketing-tools-every-entrepreneur-should-know-about/279338[retrieved: October, 2022].

[8] Lee K., *Marketing Tools*, https://buffer.com/library/free-marketing-tools/#even-more-free-marketing-tools[retrieved: November, 2022].

[9] Salleh S. S., Arshad N. H., Janom N., Marzuki N., (2014). E-managemnet concept and process of crowd sourcing platform certification, in *e-Learning, e-Management, and e-Services*, pp. 128–133.

[10] Holdford D., Pontinha V., Wagner T. D., (August 2021). Using the business model canvas to guide students in bulding business plans, in *American Journal of Pharmaceutical Education*, pp. 1–13.

[11] Sundah D. I. E., (January 2018). Developing enterpreneurial competencies for successful business model canvas, in *Journal of Physics Conference Series*, [Digests 2nd International Joint Conference on Science and Technology, p. 4, 2017].

[12] Ojasalo J., Ojasalo K., (March 2018). Service logic business model canvas, in *Journal of Research in Marketing and Entrerpreneurship*.

[13] Ziemba A., *An Introduction to the Business Model Canvas*, https://www.boldare.com/blog/business-model-canvas[retrieved: November, 2022].

[14] Osterwalder A., Pigneur Y., *Business Model Generation: A Handbook for Visionaries, Game Changers, and Challengers*. Wiley, NJ: Hoboken, 2010.

[15] D'Andrea A., Ferri F., Grifoni P., (January 2015). CBM: An integrated crowd-sourcing business model, in *Journal of Contemporary Management*, vol.4, pp. 47–58.

[16] Chaffey D., *The Best Free Marketing Tools That Every Hands-on Marketer Should Use*, https://www.smartinsights.com/marketplace-analysis/customer-analysis/8-free-online-marketing-tools-that-every-hands-on-marketer-should-use/[retrieved: January, 2023].

Innovation and Technological Advances for Sustainability – Al-Naemi et al. (Eds)

Root cause analysis to eliminate clogging in sewers caused by roots in Qatar

Aiyad Gannan, Nada Arafeh & Khaled Walid
Department of Maintenance and Mechanical Engeering, University of Doha for Science and Technology, Doha, Qatar

ABSTRACT: The purpose of this paper is to present a prototype designed to solve a common issue that occurs in Qatar sewage systems which explains having unfinished sewers maintenance in most neighborhoods, it is due to clogged house sewers. House sewers are often clogged by many components but roots are the most common, this results in house floods, loss of time, effort, cost and a long shutdown time of the system. Therefore, a new mechanism was needed, studied, designed, and 3D-modeled to physically test its size and functionality and most importantly, solve the main problem of unclogging sewers from roots. Collected data was based on research and interviews with civil and maintenance engineers which led to generating mechanical design concepts using SolidWorks and 3D printers, other concepts included material properties, stress analysis, and environmental considerations. The mechanism was designed particularly to overpower the roots resistance and cut through them, leaving a clean inner pipe and an uninterrupted wastewater flow, and to eliminate chances of flooding. Due to the purpose of this study and the nature of its operation, a stress analysis study was made on the drill head to validate the scenario of fabricating the mechanism using the chosen material.

Keywords: sewers blockage, design, 3D Printing, Simulation, root cause analysis

1 INTRODUCTION

Typically, when clogged drain is noticed at home, the first thing that comes to mind is to grab a plunger. Many people do not know that traditional plunging is nearly ineffective in some situations, such as when the house's main sewer line is clogged. If this occurs, significant flooding and plumbing issues throughout the house may be experienced. Drainage pipes transport waste out from sinks, bathrooms, and other fixtures across the property. These lines all connect to the main sewage line.

This main pipe transports all waste from houses to the street sewage system. Once the main sewer gets clogged, drains throughout the house stop functioning. A crucial reason of clogged sewers are roots from trees and plants around houses [1].

This work was based on valuable and verified sources for the literature review to gather information and initiate ideas. Sources were from the internet: websites, electronic books, companies' journals, and meetings with civil, and mechanical engineers in Qatar.

2 BACKGROUND

First, The richest destination of water and nutrients in soils around houses are sewers due to the dump of waste and wastewater. Roots make their way to sewer lines either from pipe ends or break into the pipe to get the needed nutrients. Once the roots enter the sewer lines, roots

DOI: 10.1201/9781003496724-49

either break the pipe or clog it, either way; it is a constraint of household domestic water usage because it can cause floods, downtime, extra expenses, and more complications [4].

Therefore, the most common reason for clogged sewers are roots, especially around houses where trees and other plants are planted. In Qatar, the 6-inch diameter pipes are used in the sewage system which allows enough room for roots to fit inside the pipe and clog it. Before finding a solution to unclog pipes, a clog must be indicated at first. There are the traditional signs to indicate clogs in pipes meaning that the main sewer could also be causing the problem [3].

This work aims to solve the blockage caused by roots and improve wastewater flow in sewers to avoid floods, water overflow, expensive water bills, and other consequences of clogged sewers. Due to the incompliance that the current solutions offer, a new mechanism was needed to shred roots and open the closed channel.

There are three common methods of cleaning clogged roots which are somewhat effective, but all have a drawback that raises the need for a new mechanism with fewer impacts on the people, environment, and finances [2].

3 PROPLEM DEFINITION

Before Like any mechanical system, sewers require maintenance procedures to ensure clear and uninterrupted waste flow of the system. A critical problem affecting the effectiveness of sewage handling is block of flow by various objects such as build-up grease, debris, foreign objects, or tree roots. Almost all houses in Qatar have planted trees in the yards and usually less maintenance is performed for sewers unless there is a damage.

Roots are a common cause of blocked sewers and are an important part of a plant, normally grow underground and follow or extend to where there is water i.e., a small crack or opening in a sewer line will attract roots growth due to high amount of nutrients. When roots are attracted into the sewer, they multiply and eventually clog pipes and therefore cause block of flow or overflow due to the flow blockage [5].

As a result, household use is affected due to downtime and other reasons. This project was to investigate, Design, and 3D Print a mechanism to unclog 6-inch house sewers. This project also includes a basic maintenance and testing measures to ensure the best quality of the mechanism in case it was implemented and fabricated.

This mechanism suggests the new prototype design which combines the pros of existing solutions and eliminates the cons. This work also states a scenario and basic considerations for future implementation of the idea.

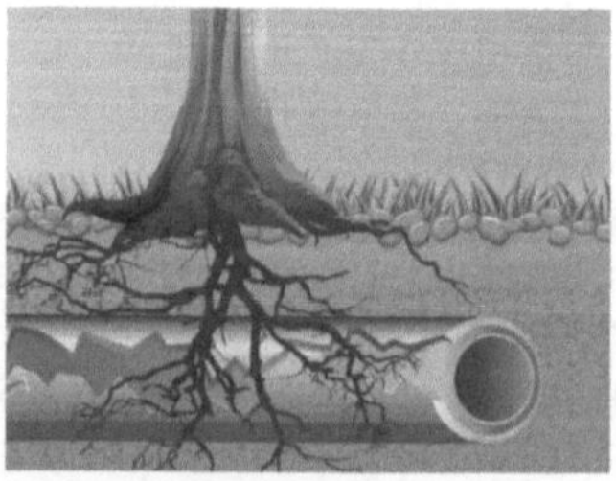

Figure 1. Sewers blockage by extended tree.roots

4 METHODOLOGY

4.1 *Technical approach*

The proposed design was dimensioned to fit and operate easily inside 6-inch sewers. SolidWorks Software was used. After the research and constructing an objective and criteria table, the design target specifications were defined. To identify the target specifications, research was made on existing solutions to unclog roots from house sewers to create a mechanism combining the pros and eliminating the cons [6]. It was decided that the new mechanism consists of seven components.

4.2 *Generating design concepts*

One of the existing solutions of unclogging sewers from roots is the Snake Method, in this method the operator pushes a flexible shaft through the pipe, an extracting head is attached at the end of the flexible shaft, the head will perform a clamping mechanism, the roots then will get caught to the head, the operator will retract the shaft from the pipe to clean the head and repeat the process till the pipe is unclogged.

The Snake Method is effective, yet the procedure can take up to 5 hours to finish. The proposed design is a modification to the Snake Method, attaching an inspection camera and designing a drill head to shred roots will speed up the process of opening the clogged flow.

The generated design was made to the following characteristics: Ease of movement and use, eliminate shaking (provides steadiness) to prevent the pipe from breaking, portable, eco-friendly, fit for pipe safety and dimensions, easy to clean and sanitize, effective in cutting roots and opening the flow, provide external manual grip and control for operator, visibility and monitoring inside the sewer, time saving, moderate effort, noncorrosive material, hollow insides for a lightweight prototype.

All seven components were required for their function and role in this project which is explained below.

This was a crucial part to the project as it was challenging to generate a new design. Designing parts was done using SolidWorks Program. Parts were designed based on the estimated dimensions while considering the pipe diameter of 6 inch.

Table 1. Design concept.

Item	Description
Casing	Enclosing the inner components, casing is the foundation which has other components were built and assembled on Casing is the connecting component between the drill head and flexible shaft
Drill head	Used to shred and cut the roots that are clogged in the sewer with the attached blades Inspection camera will be inserted in the middle of the drill head
Wheels	Wheels allow the mechanism to slide inside the pipe providing protection to the inner pipe and smooth movement. Four wheels were designed and printed to provide the smooth movement on the four sides of the casing
Inspection Camera	Attached to the tip of the drill head where it has a circular hole designed for it Provides visibility and monitoring option for the inside pipe
Flexible Shaft	Used to provide manual control of inserting it into the sewer and how deep it should go
Support block	A block designed to contain and hold the flexible shaft to provide steadiness and support while the shaft rotate

4.3 *Components design*

The design phase started with a couple of sketches until the best design of each part was constructed and finalized. SolidWorks was used due to its ease of use and built-in features that were needed later for a simulation study for this 3D model.

Figure 2. Proposed design of drill head and casing.

The designed components were: Drill head and blades, Casing, Wheels and wheel shafts, Support block for flexible shaft. The blades were designed to be sharp, pointy, and short. The blades are 2 mm in thickness with 45° degrees chamfered ends, which gives the blades the sharpness needed to. The drill head holds 30 evenly distributed blades; divided over 5 rows, each row contains 6 blades rotated 45° from the row before it. As a result, providing 360° coverage around the drill head.

Each designed part was assembled to build up the mechanism. The total dimensions of the assembly are the following:

(1) Maximum Diameter (height and width): 14.5 cm
(2) Maximum Length: 40 cm.

5 RESULT DISCUSSING

5.1 *3D printing approach*

Since it is a prototype to new mechanism, the main purpose was to present the new design of physical prototype to observe and partially test it. The purpose was to showcase the new mechanism and if the dimensions will succeed to fit inside 6-inch diameter sewers. It was decided that the project would be executed by the help of 3D printers. Since it is a new suggested design, an innovative technology was preferred as a safer option.

5.2 *Selecting parameters*

Custom made parameters can be set for different materials, where temperature and speed can be chosen and monitored. In this project, no specific parameters were chosen but there are recommended specifications set by default for PLA filament. The parameters selected for this work are the default settings. Speed and scale of printing can be adjusted to save time for larger parts which take the most time [6].

At the start of the printing phase, smaller scales were chosen as a faster test run to make sure that the printer will not have an issue generating the G-code and to ensure that there is no issue in the STL file.

5.3 *Material selection and 3D printing*

For this mechanism, PLA material was chosen to print all parts for the following reasons [7]:

(1) Polylactic acid, or PLA, is by far the most popular FDM 3D printing material
(2) Available and relatively inexpensive
(3) Comes in hundreds of vibrant colors and blends
(4) One of the eco-friendliest options for 3D printers
(5) Available in soft and hard forms
(6) Familiarity with the usage

Material specifications of PLA filament were chosen from Fill30 Pro brand which is a common company to produce 3D printer filaments. The filaments have a few specifications such as diameter, color, weight, print temperature, bed temperature [6]. The specifications vary depending on the material and filament. Each filament material such as PLA, can have multiple specifications which include:

(1) Material: PLA
(2) Diameter: 2.85 mm
(3) Colors: White, Red, Blue, Orange (chosen based on availability in classroom 5.120)
(4) Weight: 750 g
(5) Print Temperature: 180 – 210°C
(6) Bed Temperature: ± 35 – 60°C

5.3 *Real life mechanism scenario*

5.3.1 *Material selection*

Sewer pipes are humid, poorly ventilated, and make good housing environment for bacteria. Because of brass and copper properties, they are used in less-fictional applications. The recommended metal for the proposed solution is 304 Stainless Steel; this grade of stainless steel is used in industrial, medical and household application [9].

304 Stainless Steel is the most common form of steel due to excellent corrosion resistance and its ease to be sanitized. The following sections will discuss the best suitable material for each part of the design [8].

5.3.2 *Casing*

The casing is what holds all the components together, it needs to hold the weight, and be able to sustain the forces from other components.

The casing consists of four –10mm thick- walls. According to the 304 stainless steel mechanical properties; each 10 mm plate has tensile strength up to 750 MPa (108778 Psi), and proof stress of 230 MPa [8].

Stainless steel in general has high density with an average of 7.8 g/cm^3 by using 304 stainless steel for the casing; no additional protective layer will be needed. Stainless steel is one of the easiest metals –after gold and silver- to be sanitized [8].

5.3.3 *Drill head*

The drill head will be the first part of the mechanism that will impact with the roots rotating at a maximum speed of 2000 rpm. The drill head will be responsible in cutting and shredding through the roots, in very humid and unventilated circumstances. Tree roots in the sewer system have the same physical properties of wet wood.

The roots in sewer system have has a very low hardness (<50 KN) and toughness <1 N.m. The high tensile strength and proof stress of 304 stainless steel, rotating at maximum speed 2000 rpm will shred and drill through the roots with minimal resistance. A stress analysis study was made to ensure that the drill head will withstand the objected forces [5].

5.3.4 *Wheels*

The high-speed rotation and the process of eliminating roots will cause a high vibration within the mechanism. Silicone wheels are required to provide grips with the interior walls of the pipe.

Silicone is type of rubber; because of its high elasticity the rubber silicon wheels provide high friction with the touching surface.

In addition to the wheels support, they will make the mechanism movement even and leveled, and prevent damages that may occur from vibration and rotation [10].

5.3.5 *Motor and power source*

Sewer systems can be accessed through manhole, usually located outdoors and far from any source of power. The mechanism is portable; all components run on rechargeable batteries.

The following table will compare most 3 suitable portable drill motors with respect to their Rpm, power, & torque.

Table 2. Table type styles.

	Technical Description		
Item	RPM	Max Power [Nm/s]	Torque [Nm]
DEWALT 18V XR	550–2000	460	27
BOSCH PSR 18 LI-2	0–1350	370	38
WORX WX372	0–1800	400	50

Figure 3 Labels Shredding the roots require high rotational speed and high power. The sharp edges of the drill head will cut its way through the roots while rotating in high speeds. Dewalt 18V XR provides the highest speed while being cordless and portable at the same time.

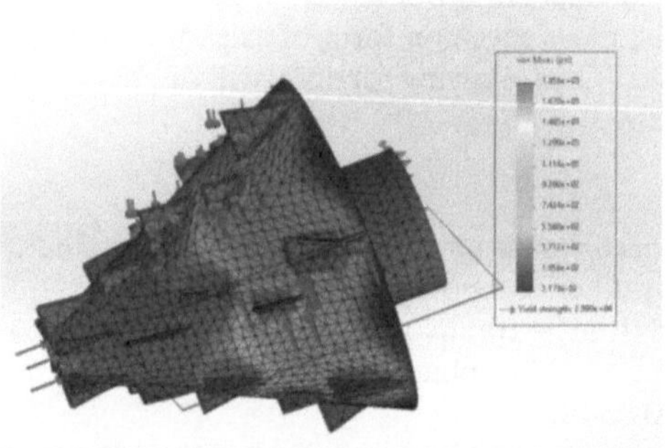

Figure 3. Stress analysis.

5.3.6 *Flexible shaft*

The flexible shaft is what delivers the rotational movement and power from the motor to the drill head. Flexible shafts allow the mechanism to enter the sewer pipes from tight confined spaces such as manholes; it can deliver up Rpm with maximum torque of 550 Nm.

5.3.7 *Pipe inspection camera*

The process of eliminating roots in sewer pipes happens underground. The camera will give a live feed of the operation. The drill head has an opening at its end, this hole is designed to hold 23mm pipe inspection camera, and this will allow the mechanism to provide a live feed of the process:

5.3.8 *Mechanism specifications*

Height (cm)	Width (cm)	Length (cm)	Power (Nm/s)	Torque(Nm)
14.5	14.5	40	460	27
Max Rpm	Process Duration (hr)	Interior Visuals	Portability	
2000	0.5 – 2	Yes	Yes	

The table above shows the mechanism specifications and durability. The system is 14.5 x 14.5 Cm, which will give 5 mm clearness in total for the shreds to escape behind. The drill head will be rotating at a maximum of 2000 Rpm and with a power of 460 Nm/S; the mentioned power with the drill head design will give the ability for the mechanism to shred its way through the roots to improve the flow of the wastewater in sewer pipes.

6 RECOMMENDATIONS

a) *Design modifications*
b) *Casing*
The designed mechanism has sharp corners; sharp corners absorb the forces with very minimal distribution of the forces. Converting the sharp corners to fillet corners will significantly improve the forces distribution along the casing, which will expand the life expectancy of the casing and the mechanism in general.
c) *Drill head and blades*
The drill head and blades are what absorb the first direct impacts while rotating at a speed up to 2000rpm.
The cone shape naturally distributes the forces evenly along the head. Since the blades will shred the roots and get impacted from the root's resistance, the blades should be welded with fillet joint with the head, fillet welding adds strength, and extra material to help keep the blades from breaking.

7 CONCLUSIONS

This work did aim to solve clogging in sewer pipes caused by tree roots. A new mechanism was designed and a prototype was 3D printed, to give a visual on the design. The 3D printed prototype successfully proved that the dimensions of the design are fit t go inside 6" PVC sewer pipes.

The designed mechanism will go inside sewer pipes and the designed drill head will rotate to shred and cut its way through the tree roots. The mechanism design gives a life feed visual from inside of the sewer pipe using 2mm pipe inspection camera that can be implemented on the beginning of the drill head.

The project was successful in the end of the mechanism. The proposed design succeeded the theoretical and practical aspect. Simulation has been done on specific material to study the Impact on the most effected part with specific forces, loads, stress analysis and still passed the impact during the highest rotating speed.

REFERENCES

[1] Edwards. R. A., Rowe R. N. and Trought M. C. T. (1999) Control of tree growth in the urban environment using geotextile fabric bags. *Arboriculture Research and Information* Note no 148/99/EXT.
[2] John F., Ahrens O. and Neal R., (1970) *J. Water Pollution Control Federation*, Vol. 42, No. 9.
[3] Thomson G., (2008) 71st Annual Water Industry Engineers and Operators Conference.
[4] Östberg J., Martinsson M., Stål Ö. and Fransson A., (2012) *Urban Forestry & Urban Greening*, volume 11, issue 1, pages 65–71.
[5] Mattheck C. (2000), J. Biomechanical Study on the Interactions of Roots with Gas and Water Pipelines for the Evaluation of Tree Sites.
[6] Gopinathan J. and Noh I. (2018) Recent Trends in Bioinks for 3d Printing.
[7] Bradshaw S., Bowyer A. and Haufe P. (2010) *J. The Intellectual Property Implications of Low-cost 3d Printing*, vol.7 issue 1.
[8] Armstrong E. and Eash T. (1969) *Advances in Cryogenic Engineering*, *Volume 14*, pp. 64–70.
[9] Blackwood D. and Pereira B., (2004), *Journal of Materials Science: Materials In Medicine* 15.
[10] Kuliczkowska E., Kuliczkowski A. and Parka A. (2022). *J. Engineering Failure Analysis* vol. 135.

Innovation and Technological Advances for Sustainability – Al-Naemi et al. (Eds)

The use of Carob powder as biocoagulant to improve the quality of wastewater

Arris Sihem, Ayat Asma, Boutamine Khadija, Maadadi Ayoub &
Bencheikh Lehocine Naradine
Laboratory on Engineering and Environmental Process (LIPE), University Constantine 3, Constantine, Algeria

ABSTRACT: Coagulation/flocculation is a major phenomenon that occurs in industrial and municipal water treatment to remove suspended solids. The use of conventional metal-based coagulants/flocculants is valued for its high performance, low cost, ease of use, availability and effectiveness. However, it is currently raising new concerns. The alarming issues of prolonged effects on human health and increased pollution of aquatic environments by the large quantities of non-biodegradable sludge generated are becoming hot topics. As a result, alternative coagulants and flocculants, called biocoagulants and bioflocculants because of their biological origin and biodegradability, have recently been developed for water and wastewater treatment. The objective of this study is the treatment of wastewater from the STEP of Ain Baida Harriche – Ferdjioua, Mila wilaya, in the region of Constantine in Algeria, after pretreatment steps (screening, grinting and oil removal) using the carob as a biocoagulant. The characterization of the carob powder was made through the determination of the surface group functions by the Bohem method which showed the acid character containing carboxylics and phenols. FTIR spectroscopy analysis indicated the presence of polysaccharides and proteins causing coagulation. The value of zero-point pH of carob was equal to 5.5 confirming the low acid nature. Results showed that the biocoagulation bioflocultion treatment could be technically and economically feasible since efficiencies of turbidity, chemical oxygen demand, and phenol removal of 78.80, 80.18 and 78.41% respectively could be achieved using an optimal coagulant dose of 20 mg/L of carob and pH = 8.

It can be concluded that using the carob as a low-cost material into the coagulation/flocculation process can offer some advantages such as high treatment efficiency and economic savings on overall treatment plant operation costs.

Keywords: biocoagulation, carob, wastewater, biomaterial, COD, turbidity

1 INTRODUCTION

The development of human society is seriously threatened by water scarcity. The best strategy considered for meeting current and future water needs are wastewater recuperation and reuse. On the other hand, water industrial pollution plays an important role in the world water crisis. In developing countries, it represents an especially and dangerous problem. In principle, wastewater, whether domestic or industrial, is collected by a complex sewerage system and then treated in a wastewater treatment plant before being discharged into the natural environment. In principle, wastewater, whether domestic or industrial, is collected by a complex sewerage system and then treated in a wastewater treatment plant before being discharged into the natural environment. Thus, the treatment or purification of wastewater

DOI: 10.1201/9781003496724-50

aims to reduce the carried pollutant load in order to return quality water to the aquatic environment, while respecting the natural balance and its future use (fishing, recreation, food, agriculture or industry, ... etc. Since the "International Drinking Water Supply and Sanitation Decade", many efforts have been made in the drinking water supply and sanitation sector in Africa [1]. These efforts have been made to treat wastewater discharged from industries and agricultural activities. The dilution of wastewater in rivers, which was common practice, has given way to the use of sophisticated and highly efficient treatment plants that use the latest technologies [2]. Unfortunately, these treatment plants face several physical and biological malfunctions, the smooth running of the operation is hampered, the use of several treatment techniques including: coagulation-flocculation ... etc.

Coagulation and flocculation is often the first stage of solid–liquid separation in a wastewater treatment plant, named the pre-treatment step. Nonetheless, it can also be applied as a main or even as a post-treatment [3]. This process eases the elimination of suspended matter and colloids by assembling them into a floc, where separation is carried out by decantation, flotation and/or filtration systems. Coagulation is a destabilisation process of colloids, it consists in decreasing and neutralising the electric charge which favours the approach of the particles in order to agglomerate them. Flocculation is the agglomeration of destabilised particles into micro floc, then into more dense floc for settling or filtering from the solution.

Among the compounds used in treatment drinking water and wastewater, we classify coagulants positively charged compounds, and flocculants negatively charged polymers (polysaccharides) [4].

The chemical compounds commonly used as coagulants/flocculants include aluminium salts ($Al_2(SO_4)_3$), iron salts ($FeCl_3$ or $Fe_2(SO_4)_3$) [5,6], magnesium carbonate [5,7], hydrated lime [5], and polymers (polyaluminum sulfate chloride, and polyferric sulfate, aluminum chlorohydrate, polyaluminum chloride (PAC)) [8,9]. Ferric chloride and aluminium sulphate are the most widely used metal-based coagulants and are characterised by their excellent performance in reducing suspended solid in water. However, Kurniawan and al, reported that the excessive concentrations of aluminium reduce pH of water, have a low biodegradability and can be accumulated to food chain [10]. Some indications regarding the impacts of the conventional coagulants/flocculants on human health, including central nervous system failure, dementia, Alzheimer's disease, and severe trembling after consumption of water, have been reported. Another limitation of using metal salts as coagulant agents is the resulting large volume of sludge and the relative high coagulant cost [11,12].

The safest and the more appropriate alternative to these harmful chemicals is the exploitation of bio-based coagulants, which can be extracted from plants, animals or microorganisms.

Important characteristics of these sources that allow their use as bio-coagulants are their contents of polysaccharides [13,14], protein polymers and some functional groups, such as hydroxyl and carboxyl groups, which favor the mechanism of adsorption, polymer bridging and charge neutralization. The use of coagulants of plant-based instead of inorganic coagulants and/or organic polymers reduces the required doses of coagulant, and reduces the concentrations of metals in the treated wastewater [15]. Due to their low toxicological risk and high biodegradability, they are safe for human health and aquatic life, and even the sludge generated during treatment can be valorized as biofertilizer in agriculture. Thus the use of biocoagulants is part of a sustainable development.

Natural coagulants such as cactus mucilage, moringa seeds, oleo verra, acorn have been shown to be effective as indicated by a significant amount of research [12,16].

The carob tree (Ceratonia siliqua L.) is a leguminous evergreen shrub of the Leguminosae family. Today it is mainly found in Spain, Portugal, Morocco, Greece, Italy, Turkey, Algeria, Tunisia, Egypt, Lebanon, Syria and Cyprus. It grows naturally on plateaus and at altitudes of up to 1700 m. It adapts to different types of soil but is not suited to humid and acidic soils. It is an evergreen sclerophyllous tree, which resists very well to arid zones but is not very tolerant to cold (−5°C), it is tolerant to salt (40 mmol of NaCL/L) [17]. The fruit is an indehiscent pod, elongated, compressed, straight or curved, thickened at the sutures, 10 – 30 cm long, 1.5 – 3.5 cm wide, and about 1 cm thick. In many Mediterranean countries the fruit is used in popular beverages and confectioneries [17].

The objective of this study is to highlight the effectiveness of carob powder as a biocoagulant in the treatment of wastewater from the Ain Baida Harriche-Ferdjioua plant, to optimize the biofloculation process by considering the effect of pH and dose on the reduction of turbidity, COD and phenols, and to characterize the carob powder (Boehm's method, pH_{PZC}, FTIR) in order to understand the mechanisms involved.

2 MATRERIAL AND METHODS

2.1 *Investigated wastewater*

The wastewater samples used in this study were collected at the Ain Beida Harriche wastewater treatment plant (WWTP) in Ferdjioua, located in the municipality of Mila, eastern of Algeria. This plant has an operating capacity of 9600 m^3/day. It was designed in 2013 to receive municipal wastewater in order, to protect the Beni Haroun dam, to fight various forms of hydric diseases and also to provide adequate water for irrigation of agricultural soils around MILA municipality.

2.2 *Wastewater samples characterisation*

The samples were collected in a 20 L Jerry can containing the pre-treated water (T = 12.1°C, pH = 7.93), just after the de-oiler, in dry weather. These samples were stored at 4°C for analysis and use. Table 1 presents the results of the analysis of different physico-chemical parameters characterizing the pre-treated wastewater.

Despite the pre-treatment applied in the treatment plant, i.e. a screening, de-sanding and oil removal stage, the pre-treated wastewater showed a very high organic load in terms of COD, a particulate load in terms of suspended solids and turbidity, while the nitrogen fraction in the form of ammonia, nitrate and nitrite (the sum) is low. The concentrations of chlorides and phenols are relatively high, which is not the case for ortho-phosphates. This wastewater will be treated by coagulation/flocculation/settling using carob as a biocoagulant.

Table 1. Wastewater characterization.

Parameters	Units	Values
Temperature (T)	°C	18.3
pH	/	7.62
Turbidity (Tur)	NTU	264
Conductivity (Cond)	μs/cm	2670
Salinity	%	1.4
Suspended matter	mg/L	980
Volatil matter	mg/L	810
Chlorure	mg	293.94
Phenol	mg/L	229.6125
Hardness	mg/L	4000
Ortho Phosphate ($\mathbf{PO_4^{-2}}$)	mg/L	0.018
nitrate ($\mathbf{N{-}NO_3^-}$)	mg/L	0
nitrite ($\mathbf{N{-}NO_2^-}$)	mg/L	0.381
Ammonium ($\mathbf{N{-}NH_4^+}$)	mg/L	4.4
COD	mg/LO_2	703.83
BOD_5	mg/LO_2	25

2.3 *Biomaterial preparation*

For the preparation of biocoagulant the carob was rinsed with distilled water to remove impurities, the carob fruits opened and the pits removed, dried in the oven at 50°C for 48 h, until constant weight, fruits ground, sieved through a 0.315 nm sieve, and stored in a desiccator for use. It was used as a raw coagulant to treat wastewater without any chemical pre-treatment.

2.4 *Coagulation/floculation process*

The most commonly used method for evaluating and optimizing the coagulation-flocculation processes is the jar test. This study consisted of a series of simultaneous batch experiments involving three stages, namely, rapid mixing, slow mixing, and sedimentation. Beakers of 1 litre were filled with 500 ml of sampleswater.

The aim was to optimize the dose of coagulant and the pH (by jar-test) for which the lowest turbidity and DCO of the raw water would be obtained.

The experiment started by mixing the contents of the tank completely to confirm uniformed initial turbidity, temperature, and pH values. The wastewater used in these tests had a pH of 7.62 and a temperature of 18.3°C. A standard jar test apparatus with a digital feedback control system (TM6C Model) was used to simulate the biocoagulation/flocculation process.

A volume of 500 ml beakers containing the wastewater samples were stirred at 160 rpm for 10 min after addition of biocoagulant. For the flocculation stage, the samples were stirred at 45 rpm for 20 min, the formed flocs were left to settle for 25 min. After settling, the turbidity, COD, and phenols of the supernatant were determined.

The same steps taken to optimize the pH, varying from 2 to 12, while maintaining the optimal dose already determined. The pH value was adjusted to the desired value with HCl and NaOH before adding coagulant.

The turbidity, the chemical oxygen demand, and phenol removal efficiency (Turbidity %, COD %, and phenol %) were calculated using (1), (2), and (3) equation respectively.

$$\text{Tur}(\%) = \frac{\text{Tur}_0 - \text{Tur}}{\text{Tur}_0} \times 100 \tag{1}$$

Where Tur_0 and Tur are the initial and final supernatant turbidity, respectively, which was measured by a HANNA turbidimeter (HI88713).

$$\text{COD}(\%) = \frac{\text{COD}_0 - \text{COD}}{\text{COD}_0} \times 100 \tag{2}$$

Where COD_0 and COD are the initial and final chemical oxygen demand (mg/l), respectively. COD was measured according to the Afnor standard method.

$$[\text{Phenol}](\%) = \frac{[\text{Phenol}]_0 - [\text{Phenol}]}{[\text{Phenol}]_0} \times 100 \tag{3}$$

Where [phenol] and $[\text{phenol}]_0$ are the concentration of phenol before and after biocoagulation biofloculation process. Phenol concentration was measured using a spectrophotometer (UV-1601) at a wavelength of 270 nm.

3 RESULTS AND DISCUSSION

3.1 *Characterization of carob*

The qualitative characterization of the used biocoagulant is very important, it gives essential information for the understanding of the interactions involved in the study. The acidic or basic character can be identified by Boehm method to identify the surface functions [18,19], the contact pH [20], and FTIR (Fourier transform infrared) spectroscopy analysis using IR spectrophotometer (IR JASCO V-630 model) in the range of 400 – 4,000 cm^{-1}.

The method of bohem showed that the carob had an acid nature containing carboxylic and phenolic groups with rates of 0.0075 and 0.00037 mol/g, respectively.

The pH of the zero charge point of the carob was $pH_{PZC} = 5.56$, confirming its low acid character.

The results of the FTIR analysis of the carob powder, and the coagulated wastewater are presented in Figure 1.

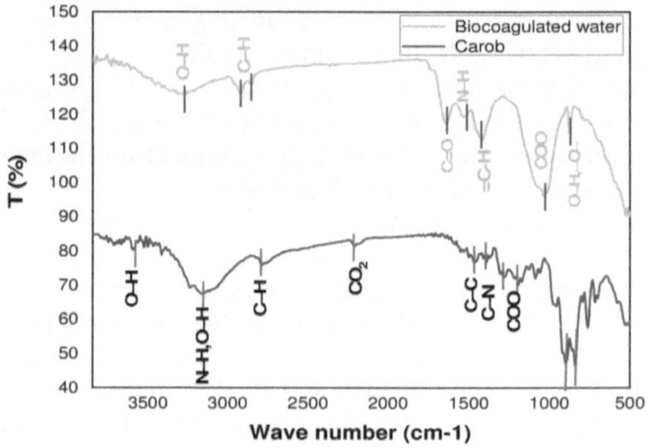

Figure 1. FTIR spectroscopy analysis of Carob powder.

The most relevant functions on the spectra are presented in Tables 2 and 3.

Table 2. Result of FTIR analysis of Carob powder.

Biocoagulant		
λ (cm^{-1})	Groupement	Vibration
837.71	α-D-galactopyranose	O–H,–O–
898.66	β-D-mannopyranose	O–H,–O–
1192.76	Acide carboxylic	COO
1330.42	Amines	C–N
1463.71	Alcanes	C–C
2212.92	Carbone dioxyde	CO2
2791.46	Alcanes	C–H
3160.76 and 3589.84	Alcools, phenols, acides, amines, glycoproteines	N–H,O–H

Table 3. FTIR analysis of biocoagulated wastewater.

biocoagulated wastewater		
λ (cm^{-1})	Groupement	Vibration
869.73	α-D-galactopyranose	O–H,–O–
1017.27	Acide carboxylic	COO
1408.75	Aromatiques	=C-H
1513.24	Amide	N–H
1641.91	Aldehyd	C=O
2850.27, 2922.59	Alcanes	C–H
3292.86	Alcools, phenols, acides, amines, glycoproteines	N–H,O–H

Following the STM sheets which identifies the peaks in the spectrum by the corresponding functional groups, the spectrum could be explained as follows:

The broad absorption band at 3283.21 cm^{-1} for raw carob and at 3370 cm^{-1} for saturated carob corresponds to the hydrogen elongation vibrations of the O-H hydroxyl groups (of carboxyls, phenols, or alcohols) [21,22]. This band also corresponds to the O-H elongation vibration of cellulose, pectin and lignin.

The broad absorption band at 1229.4 cm^{-1} for raw carob corresponds to the acidic group [23] confirming the acidic character of carob. Faint bands at 969.73 cm^{-1} indicate the presence of α-D-galactopyranose and *β*-D-mannopyranose, this confirming that the major compound in carob powder was galactommanan which a colloidal polysaccharide, agent responsible for coagulation flocculation phenomena [21,24,25].

3.2 *Optimization of biocoagulant dose*

In order to optimize the dose of biocoagulant, six samples were prepared by adding increasing amounts of the carob to the pre-treated wastewater (Tables 3 and 4) to have samples of different biocoagulant concentrations (0, 10, 20, 30, 40, and 50 mg/L). The wastewater used in this test had a pH of 7.62 and a temperature of 18.3°C. The results obtained for the three process evaluation variables, COD, turbidity and phenols, are presented in Figures 2 and 3.

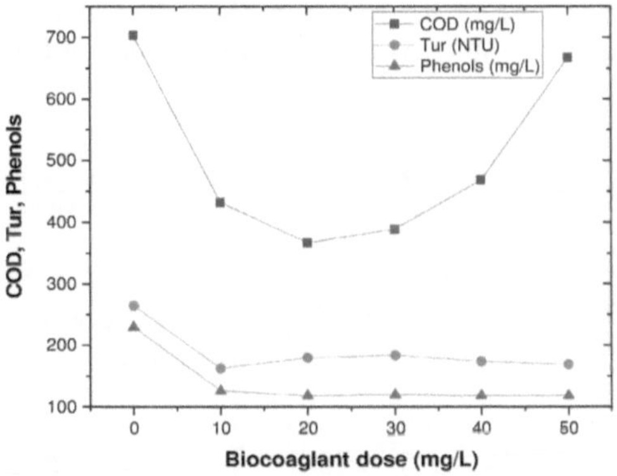

Figure 2. Optimisation of Carob dose as biocoagulant. Conditions : DCO_0 = 703.83 mg/L, Tur_0 = 264 NTU, $[phenol]_0$ = 229.61 mg/L, pH = 7.63, T = 18.3°C.

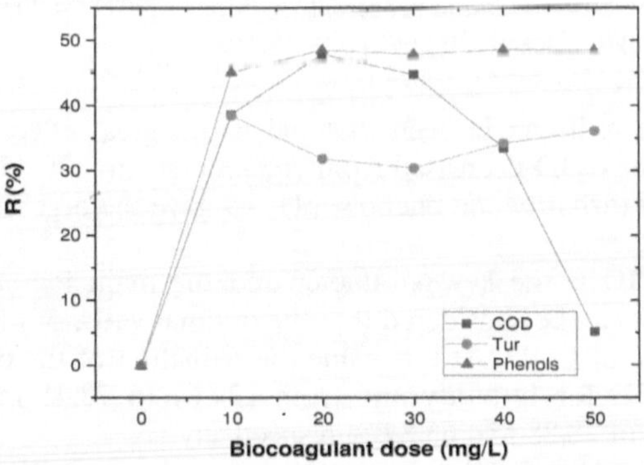

Figure 3. Effect of carob dose on efficiency of biocoagulation. Conditions : DCO_0 = 703.83 mg/L, Tur_0 = 264 NTU, $[phenol]_0$ = 229.61 mg/L, pH = 7.63, T = 18.3°C.

From the Figures 2 and 3, it can be clearly seen that the dose of carob significantly affects the coagulation rate. It shows that the COD decreased with the increase of the dose, passing from 703.83 to 366.93 mg/L, and the optimum was reached at a dose of 20 mg/L, giving a reduction efficiency of 47.86 %. Beyond this point, the increase in the amount of carob acted adversely on coagulation and increased the COD in solution, thus reducing the treatment rate, which reached 5.18 % at a dose of 50 mg/L with a residual COD of 667.36 mg/L.

Turbidity and phenols had a similar behavior, in fact the contents decreased when the dose of carob increased until reaching the optimum at 20 mg/L, with residual concentrations of 180 NTU and 118.25 mg/L for turbidity and phenols respectively. The corresponding yields for this reduction were 31.81 and 48.5 % for turbidity and phenols, respectively.

The increase in COD beyond the optimum value could be explained by the high kinetic energy generated by the excessive presence of carob molecules that prevented the neutralisation of the surface charge, making the surface phenomena unstable. Also, this amount of carob in solution had an additional COD, due to its solubility, which was added to the real COD of the wastewater. It should be noted that the turbidity remained constant, confirming the solubility of carob in the studied wastewater [26].

3.3 *Optimisation of pH solution*

The pH is a key parameter in the surface phenomenon, its optimization is important. For this purpose the wastewater pH was varied from 2 to 12 to cover the total interval and examine its influence. The dose of bio-coagulant was fixed at the previously determined optimal dose of 20 mg/L. The results obtained are represented in Figure 4.

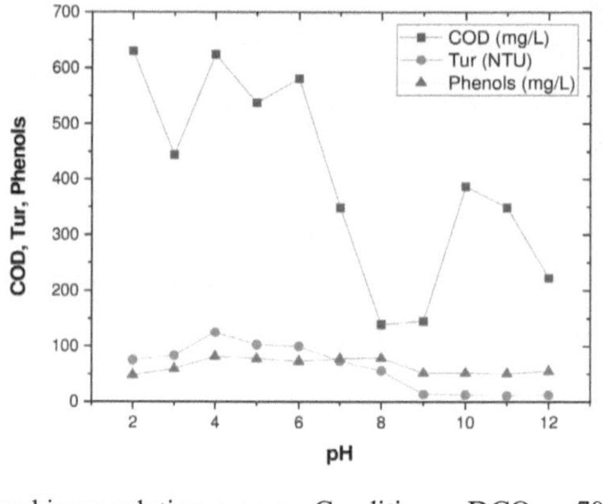

Figure 4. Effect of pH on biocoagulation process. Conditions : DCO_0 = 703.83 mg/L, Tur_0 = 264 NTU, [phenol] = 229.61 mg/L, Dose = 20 mg/L, T = 18.3°C.

According to Figure 4, it can be seen that pH had a great effect for the coagulation-flocculation process. The COD decreased from 703.83 mg/L to 139.45 mg/L with a yield of 80.18 % at pH = 8. For turbidity and phenols, pH = 9 gave the best yields: 94.79, and 77.55 % respectively.

Considering that COD is the key parameter and the main factor to be controlled in wastewater treatment, it can be considered that the optimal value of pH is equal to 9, giving the maximum reduction of COD. At this value the reduction of the other two parameters, from 264 to 55.75 NTU for turbidity and from 229.61 to 78.41 mg/L for phenols was appreciable with yields of 78.88 and 65.84 % respectively.

Generally the pH affects the dissolution of the solute and changes the surface charge of the materials. In conventional coagulation with ferric chloride, the process is well controlled due to the knowledge of the $FeCl_3$ species distribution diagram as a function of pH, which help us in practice to limit the pH study range from 5.5 to 8.5 where Iron is in the Fe^{3+} (neutralization, destabilization, adsorption and precipitation) and $Fe(OH)^{2+}$ (entrainment) forms, states favoring the coagulation operation. But for the carob powder, it is difficult to pronounce on its different forms according to the pH, however the pH_{PZC} was an alternative parameter giving information about its surface charge.

According to the characterization results, carob had a pH_{PZC} = 5.5. The state of the surface charge of the carob is a function of this pH:

If:pH< 5.5 → the surface is negatively charged;
If:pH = 5.5 → the surface is neutral;
If:pH> 5.5 → the surface is positively charged;

When the reaction medium was highly acidic pH < 5.5, the surface oxygenated sites of carob were hydrogenated and have a positive charge ($S - OH^+$). In weakly acidic conditions, dissociation of the hydrogenated sites ($S - OH^+$) from the proposed the biocoagulant occured and the neutral sites ($S - OH$) were formed. When the pH = pH_{PZC} = 5.5 the surface was neutral. When the reaction medium was basic pH > 5.5, the surface hydroxyl functions of the coagulant started to dissociate and negatively charged surface sites appeared. As a result, the surface of the carob powder became negative ($S - O^-$) [17–19].

$$S - OH + H_3O^+ \rightarrow S - OH_2^+ + H_2O \tag{1}$$

$$S - OH_2^+ \rightarrow S - OH + H^+ \tag{2}$$

$$S - OH + OH^- \rightarrow S - O^- + H_2O \tag{3}$$

Several works have also reported proteins as active coagulation components in plant extracts. Proteins have charged groups, consisting of charged amino acids and charged functional groups, which have the potential for electrostatic interaction to neutralize the charge sites on the pollutant molecules to produce the neutral flocs. In this regard, the researchers opinions vary according to which proteins, polysaccharides, organic polyelectrolytes, can be reported as main coagulant agents. It is difficult to identify a particular class of compounds in the plant extract as the agent responsible for coagulation. Cationic polyelectrolytes in the biocoagulants play a predominant role in entanglement through a net-like structure mechanism [20,22,27].

3.4 *Water quality after treatment under optimal conditions*

The wastewater was treated by coagulation-flocculation-settling under the optimal conditions obtained, namely the dose of carob as biocoagulant was 20mg/L and the pH was 8. A detailed characterization of the treated water was crucial in order to evaluate its quality for possible uses and mainly for irrigation. The characterization of the wastewater after treatment as well as the wastewater standards according to the Algerian official journal are shown in Table 4.

Table 4. Physico-chemical characterization of wastewater treated with carob.

	Unity	pretreated wastewater	Treated wastewater	standards of the official journal
T	°C	18.3	20	20–25
Tur	NTU	264	55.75	5
pH	/	7.62	8	**6.5–8.5**
Cond	µS/cm	2670	2230	**2800**
Salinity	%	1.4	1.1	/

(continued)

Table 4. Continued

	Unity	pretreated wastewater	Treated wastewater	standards of the official journal
SM	mg/L	980	400	**35**
MVS	mg/L	810	40	/
TAC	°F	2.8	10	/
Hardness	mg/L	4000	1.25	**500**
BOD_5	mg/L	25	200	**40**
COD	mg /L	703.83	139.45	130
PO_4^{-3}	mg/L	0.018	231.07	0.1
Cl^-	mg/L	293.94	237.85	500
Phenol	mg/L	229.61	78.41	0.1
NH_4^+	mg/L	4.4	4.95	0.05–0.5
NO_2^-	mg/L	0.381	0.084	0–0.2
NO_3^-	mg/L	0	0	0–50

From Table 4 it can be seen that the use of carob as a biocoagulant improved the quality of the treated wastewater in an effective and appreciable degree.

In the light of these results, especially the considerable decrease in conductivity, the decrease and cancellation of the Nerst potential can be assumed, confirming the neutralization of the surface charge which is one of the processes that control the flocculation coagulation. The increase in orthophosphate concentration after treatment was due to the dissolution of the carob in solution. The increase in BOD_5 is also noticeable, indicating the release of organic molecules such as carboxyls, phenols, alcohols, cellulose, pectin and lignin from the carob. These functional groups appeared on the FTIR spectrum, by absorption bands around 3283.21 cm^{-1}, which would beneficially increase the biodegradability of the sludge from the wastewater treatment.

4 CONCLUSION

During this study, interested was in the treatment of an urban wastewater using carob as a biocoagulant. The experimental work was carried out in three distinct stages where the first part of is, was devoted to: The characterization of pretreated wastewater from the Ferdjioua plant which had a significant COD, BOD_5, turbidity, conductivity, chloride and phenol with values of COD = 703.83 mg/L, BOD_5 = 25 mg/L, turbidity = 246 NTU, conductivity = 2670 μS/cm, Chloride = 293.94 mg/L and phenol = 229.61 mg/L.

The characterization of the carob powder was carried out by the Boehm method which indicated the carboxylic acid and phenolic character, also by infrared spectroscopy which indicated the presence of polysaccharides, carboxylic acids, proteins, polyphenols (tannins) which were agents responsible for the coagulation flocculation. The optimization of coagulation flocculation of pretreated wastewater gave COD, phenol and turbidity reduction efficiencies of 80.18, 78.41, and 78.80 % respectively, obtained under optimal conditions with a carob dose of 20 mg/L and pH = 8. In the light of these results we can conclude that carob powder is an interesting alternative to chemical coagulants due to its excellent efficiency, low cost, low environmental impact, biodegradability, and its added value by the valorization of the generated sludge as biofertilizer in agriculture.

ACKNOWLEDGMENT

Thanks to the engineers of the Laboratory of Engineering and Environmental Process (LEEP-Tamayouz), University Constantine 3.

REFERENCES

[1] Zhao C. and Chen W., (2019). A review for tannery wastewater treatment: some thoughts under stricter discharge requirements, *Environ Sci Pollut Res.* 26 26102–26111. https://doi.org/10.1007/s11356-019-05699-6.

[2] Amokrane A., Comel C. and Veron J., Landfill leachates pretreatment by coagulation flocculation, *Water Research.* 31 (1997) 2775–2782. https://doi.org/10.1016/S0043-1354(97)00147-4.

[3] Rajasulochana P. and Preethy V. (2016). Comparison on efficiency of various techniques in treatment of waste and sewage water—A comprehensive review. *Resour. Technol.* doi: 10.1016/j.reffit.2016.09.004.

[4] Lapointe, M. and Barbeau, B. (2020). Understanding the roles and characterizing the intrinsic properties of synthetic vs. natural polymers to improve clarification through interparticle Bridging: A review. *Sep. Purif. Technol. 231.*

[5] Moran, S. (2018), Engineering science of water treatment unit operations. In *An Applied Guide to Water and Effluent Treatment Plant Design*; Elsevier: Amsterdam, The Netherlands, pp. 39–51.

[6] Weydts, D.; De Smet, D. and Vanneste, M. (2015), Processes for reducing the environmental impact of fabric finishing. In *Sustainable Apparel*; Elsevier: Amsterdam, The Netherlands, pp. 35–48.

[7] Kurniawan et al., (2020). According [Barakwan, R. A.; Hardina, T. T.; Trihadiningrum, Y.; Bagastyo, A. Y. Recovery of alum from Surabaya water treatment sludge using electrolysis with carbon-silver electrodes. *J. Ecol. Eng.* **(2019)**, *20*, 126–133.

[8] Ayat A., Arris S. and Meniai A. H., Landfill leachate pretreatment by coagulation/flocculation process: Optimization by response surface methodology, 2021 *12th International Renewable Energy Congress (IREC)*, Hammamet, Tunisia, 2021, pp. 1–5, doi: 10.1109/IREC52758.2021.9624872.

[9] Sander Y., Ariesanti Y. and Roeslan M. O., (2021). Effect of moringa oleifera leaves on human blood coagulation process, *2021 IEEE International Conference on Health, Instrumentation & Measurement, and Natural Sciences (InHeNce)*, Medan, Indonesia, pp. 1–5, doi: 10.1109/InHeNce52833.2021.9537224.

[10] Kurniawan S. B., Abdullah S. R. S., Imron M. F., Said N. S. M., Ismail N'., Hasan H. A., Othman A. R., Purwanti I. F. (2020 Dec 12). Challenges and opportunities of biocoagulant/bioflocculant application for drinking water and wastewater treatment and its potential for sludge recovery. *Int J Environ Res Public Health.*;17(24):9312. doi: 10.3390/ijerph17249312. PMID: 33322826; PMCID: PMC7764310.

[11] Dao, V. H., Cameron, N. R. and Saito, K. (2015). Synthesis, properties and performance of organic polymers employed in flocculation applications. *Polym. Chem. 7*, 11–25

[12] Benalia, A., Derbal, K., Panico, A., and Pirozzi, F. (2018). Use of acorn leaves as a natural coagulant in a drinking water treatment plant. *Water*, *11*(1), 57.

[13] Bouchareb, R., Derbal, K., Özay, Y., Bilici, Z., and Dizge, N. (2020). Combined natural/chemical coagulation and membrane filtration for wood processing wastewater treatment. *Journal of Water Process Engineering*, *37*, 101521.

[14] Barbosa A. D., da Silva L. F., de Paula H. M., Romualdo L. L., Sadoyama G., Andrade L. S., (2018). Combined use of coagulation (M. oleifera)

[15] And electrochemical techniques in treatment of industrial paint wastewater for reuse and/or disposal, *Water Research.* 145 153–161. https://doi.org/10.1016/j.watres.2018.08.022

[16] Water, S.; World Health Organization. Guidelines for Drinking-Water Quality. Incorporating First Addendum 2006; World Health Organization. Geneve, Switzerland; Volume 1. Available online: https://apps.who.int/iris/handle/10665/43428 (accessed on 31 July 2022)

[17] Zakmout A., Sadi F., Portugal C. A. M., Crespo J. G., Velizarov S., (2020). Tannery effluent treatment by nanofiltration, reverse osmosis and chitosan modified membranes, *Membranes.* 10 378. https://doi.org/10.3390/membranes10120378.

[18] Baptista A. T. A., Silva M. O., Gomes R. G., Bergamasco R., Vieira M. F., Vieira A. M. S., (2017). Protein fractionation of seeds of Moringa oleifera lam and its application in superficial water treatment, *Separation and Purification Technology.* 180 114–124. https://doi.org/10.1016/j.seppur.2017.02.040

[19] Chen F., Liu W., Pan Z., Wang Y., Guo X., Sun S., Jia R., (2020). Characteristics and mechanismof chitosan in flocculation for water coagulation in the Yellow River diversion reservoir, *Journal of Water Process Engineering.* 34 101191. https://doi.org/10.1016/j.jwpe.2020.101191.

[20] Bouaouine O., Bourven I., Khalil F., Baudu M., (2018). Identification of functional groups of Opuntia ficus-indica involved in coagulation process after its active part extraction, *Environ Sci Pollut Res.* 25 11111–11119. https://doi.org/10.1007/s11356-018-1394-7.

[21] AYAT, Asma, ARRIS, Sihem, ABBAZ, Amina, et al. (2021). Application of response surface methodology for modeling and optimization of a bio coagulation process (sewage wastewater treatment plant). *Environmental Management*, vol. 67, no 3, p. 489–497.

[22] Ayat, A., Arris, S., Abbaz, A. et al. (2021). Application of response surface methodology for modeling and optimization of a bio coagulation process (sewage wastewater treatment plant). *Environmental Management* 67, 489–497. https://doi.org/10.1007/s00267-020-01407-0

[23] Buelvas-Caro S. D., del Rosario Aguas-Mendoza Y, Olivero-Verbel R. E. (2020), Clarification of superficial waters of the magdalena river using seeds of Tamarindus indica as bio-coagulants. *Respuestas*, vol. 25, no. 2, pp. 170–176.

[24] Muhamad, N. A. N.; Juhari, N. F.; Mohamad, I. N. (2020). Efficiency of natural plant-based coagulants for water treatment. *In IOP Conference Series: Earth and Environmental Science*, IOP Publishing: Bristol, UK.

[25] Sihem Arris, Asma Ayat*, Mossaab Bencheikh-Lehocine, Hassan-Abdeselam Meniai, (2020). Removal of turbidity and chemical oxygen demand using an eco-friendly coagulant/flocculent (optimization and modeling through the response surface methodology), *Desalination and Water Treatment* www.deswater.com, doi: 10.5004/dwt.2020.26558 1–11.

[26] Priyatharishini, M. and Mokhtar, N. (2021). Performance of jackfruit (Artocarpus heterophyllus) peel coagulant in turbidity reduction under different pH of wastewater. *Mater. Today Proc.* 46, 1818–1823

[27] Azamzam A. A., Rafatullah M., Yahya E. B., Ahmad M. I., Lalung J., Alam M., Siddiqui M. R. (2022). Enhancing the efficiency of banana peel bio-coagulant in turbid and river water treatment applications. *Water*. 14(16):2473. https://doi.org/10.3390/w14162473

Innovation and Technological Advances for Sustainability – Al-Naemi et al. (Eds)

Sustainable utilization of red mud stabilized weak soil as a pavement subgrade material

Binumol Shahulhameed
Senior Lecturer in Civil Engineering, Engineering Department, College of Engineering and Technology University of Technology and Applied Sciences, Shinas, Oman

A. Maria Rajesh
Lecturer in Civil Engineering, Engineering Department, College of Engineering and Technology, University of Technology and Applied Sciences, Shinas, Oman

Shabbir Ahamad Kazi
Lecturer in Chemistry, Engineering Department, College of Engineering and Technology, University of Technology and Applied Sciences, Shinas, Oman

Umar Sabhapathy
Lecturer in Civil Engineering, Engineering Department, College of Engineering and Technology, University of Technology and Applied Sciences, Shinas, Oman

Abrar Abdallah AlRashdi, Bayan Al Ashkhari, Walaa Al Yahyai & Wasin Al Saidi
Student in Civil Engineering, Engineering Department, College of Engineering and Technology, University of Technology and Applied Sciences, Shinas, Oman

ABSTRACT: Industrialization is the vital to the development of any country's economy. However, on the other hand, the production of industrial waste is increasing enormously, which adversely effects the environment and natural resources. Red mud is also a widespread industrial waste produced during aluminum extraction from bauxite ore in Bayer's process. To minimize the impact of this solid waste material, the effective method is to utilize it in construction works with suitable stabilization. This study outlines the characterization of red mud and its utilization as a pavement subgrade construction material based on the laboratory findings and finite element analysis results. Basic properties like specific gravity, particle size distribution, Atterberg's limits, optimum moisture content (OMC) and maximum dry density (MDD) are determined. Engineering property like shear strength is determined and result is discussed in geotechnical point of view. The model footing test has been performed in laboratory under different conditions (dry and partially saturated) to check the suitability of red mud as a pavement subgrade construction material. The finite element (FE) analysis is performed using commercially available software and the result is compared with laboratory model test. The result shows that red mud can be used as subgrade material. However, as the red mud is extremely dispersive, it needs protection against erosion.

Keywords: Sustainability, Red mud, pavement subgrade material, finite element analysis

1 INTRODUCTION

Red mud, commonly referred to as bauxite residue, is a type of industrial waste created during the Bayer process used to refine bauxite into alumina. Iron oxides, which give its red color, are among the many oxide compounds that make up this substance. The Bayer process accounts for over 95% of all alumina production worldwide; for every ton of alumina produced, 1 to 1.5 tons of red mud are also produced (Yan *et al.*, 2022). Due to the material's high level of

DOI: 10.1201/9781003496724-51

production and alkalinity, it might present a serious environmental risk as well as storage issues. As a result, considerable effort is put into developing better strategies for handling it. Red mud is composed of solid and metallic oxides (Yadav *et al.*, 2018). Iron oxides, which make up to 60% of its volume, are the major source that give its red color. The pH value of the red mud ranges from 10 to 13, making it extremely basic (Ramees *et al.*, 2021). The major elements also consist of titanium oxide, silica, and residual alumina in addition to iron.

On the other hand, due to the increase in highway infrastructure, high quality aggregates are being exhausted with considerable rate. To achieve the sustainable development in infrastructure development, there is a global interest in using recycled and industrial waste materials in construction (Freitas *et al.*, 2017). The use of industrial waste in pavement construction is based on technical, ecological, and economic criteria. Several studies were conducted to utilize the waste material like fly ash, blast furnace slag, cement kiln dust, marble dust etc. as a pavement material but study on the red mud is very limited. However, some effort has been made to find out the geotechnical characteristics of red mud.

Red mud is a complex material with a chemical and mineralogical composition that varies widely, depending upon the origin of the bauxite and the alumina production process (Deelwal *et al.*, 2014). Mukiza *et al.* (2018) reviewed the probability of exploiting red mud as a road base material, weak subgrade soil stabilizer as well as a subgrade material and his results proved that red mud illustrates good characteristics as a subgrade material in comparison with the natural soil.

Aluminum industries are a main element of economic basis of a country like Oman. Therefore, making effective use of it must be a big obstacle to the improvement of the community. Contrarily, a better strategy for reducing the pollution from these industries plays a crucial role and can be improved by carrying out the relevant study.On the basis of the experimental results and finite element analysis, an attempt has been made in this study to determine if red mud might be used as a substitute for other subgrade construction materials. The stability of the red mud as a subgrade material has also been tested in a laboratory using a model for subgrade pavement. The Mohr-Coulomb model is used to perform FE analysis using PLAXIS, and the findings are compared to those obtained from testing a laboratory model.

2 MATERIALS AND METHODS

2.1 *Materials*

The red mud used in the present study is collected from Sohar Industrial zone, Sohar, Sultanate of Oman. A typical discharge of red mud at discharge point is shown in Figure 1.

Figure 1. Discharge point of red mud.

The method of extraction, the origin of the minerals, their effectiveness, and the method of disposal are the key determinants of the red mud's chemical composition. The zeta potential, flocculation, and dispersion of red mud particles produced from bauxites with different mineralogical compositions are all due to the hydroxyl groups on their surface (Hind *et al.*, 1999). High-grade road base material using red mud from the sintering process is a better option, that may lead to large-scale consumption of red mud (Qi, 2005). Based on this assumptions,red mud was used as the basis material to build a 15 m wide by 4 km long highway in Zibo, Shandong. A relevant department evaluated the road's subgrade stability and strength and found that it satisfies the highway's strength criteria (Yang *et al.*, 2006).

2.2 *Methodology*

Present study consist both experimental analysis of red mud and numerical method for the analysis of subgrade on red mud. The experimental method refers to investigation of morphological, chemical, mineralogical and geotechnical property of red mud and model test on the red mud as subgrade material.The particle morphology and chemical composition of red mud is evaluated using scanning electron microscope (SEM) equipped with energy dispersive X- ray (EDX) micro analyser. According to BS requirements, the geotechnical characterization of red mud was completed. The X-ray diffraction (XRD) technique is used to determine the mineral phase that is present in the red mud. Using the commercial software PLAXIS, a finite element analysis was carried out, and the results were compared to test results from a laboratory model.

3 RESULTS AND DISCUSSIONS

From the results, it was observed that red mud contains irregular, agglomerated and spherical particle with varying particle sizes. As it is well know that red mud is a highly alkaline, in this research studies were conducted to determine the particle morphology after treatment using acid.

Figure 2. Scanning electron photomicrograph.

Figure 2 shows the scanning electron photomicrograph of the red mud after acid treatment. From this it was clear that the alkali compound of the red mud may be responsible for the agglomeration of the red mud.

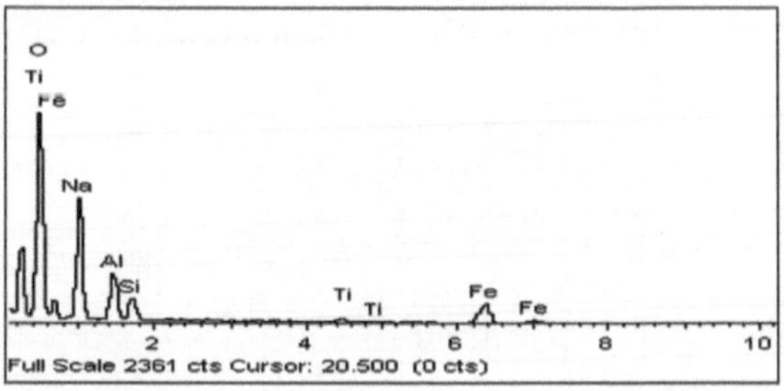

Figure 3. EDX analysis of red mud.

Chemical properties of the particles of the red mud has been evaluated using EDX analysis. Figure 3 demonstrates the EDX analysis of red mud and the percentage of element (by weight) available in red mud is shown in Table 1.

Mineralogical analysis of red mud can be found in Figure 4 and it was observed that the red mud possesses hematite, boehmite, gibbsite, rutile, goethite and sodalite as major minerals.

Table 1. Percentage (by weight) of chemicals present in red mud.

Element	O	Na	Al	Si	Ti	Fe	Total
% by weight	44.7	25.98	7.54	3.85	1.07	16.86	100

The whole chemistry of the material can be demonstrated by it pH value. For that purpose, pH value test of the red mud has been performed and found out to be 11.4 which show that red mud is highly alkaline in nature.

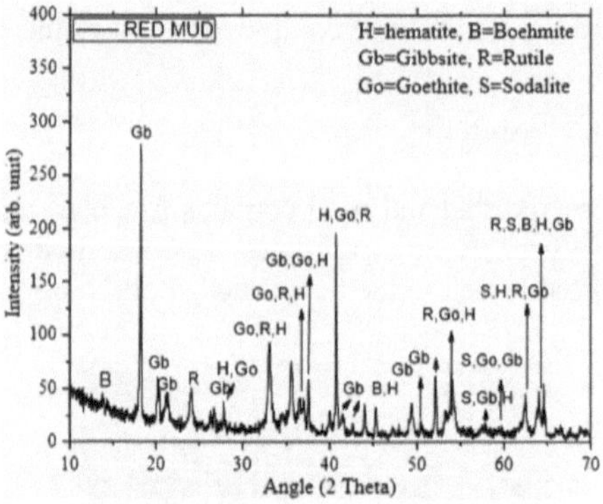

Figure 4. XRD analysis of red mud.

Based on the chemical properties it was found that the red mud can be utilized as a pavement subgrade material along with the conventional subgrade materials in suitable proportion. Toxicity of the components present in the red mud depends on the process involved in its production. Alkalinity has to be balanced and at the same time the iron content must be maintained in the red mud while used for improving the subgrade soil characteristics.

3.1 *Soil properties*

Particle size analysis of red mud is performed, and the curve is shown in Figure 5. Around 90 % soil particles are below 0.075mm, which indicates that red mud is a fine-grained soil. Around 25% of grain size is less than 0.002mm, which indicates the clay size particle.

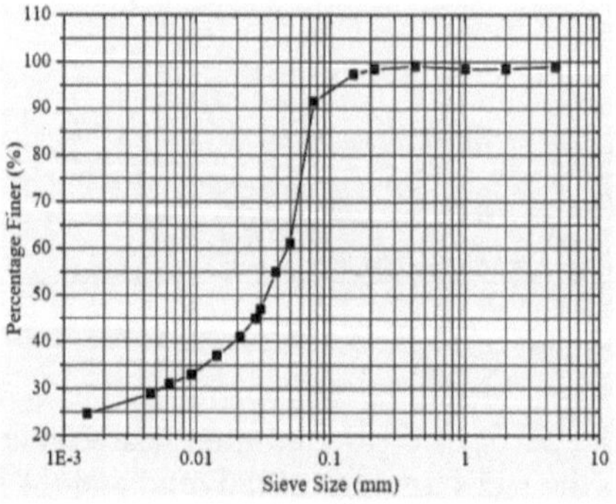

Figure 5. Particle size analysis of red mud.

3.2 *Atterberg's limit*

To classify the red mud, Atterberg's limit test is performed. The liquid limit (LL) and plasticity index (PI) of red mud are found to be 24.75% and 7.25% respectively.The linear shrinkage quantifies the amount of linear expansion to be experience by clayey soil.

3.3 *Specific gravity*

Specific gravity of the red mud has been performed as per BS standads and found out to be 3.33. The high specific gravity of the red mud is due to iron rich minerals.

3.4 *Compaction*

Compaction curve of red mud under light compaction and heavy compaction are shown in Figure 6.

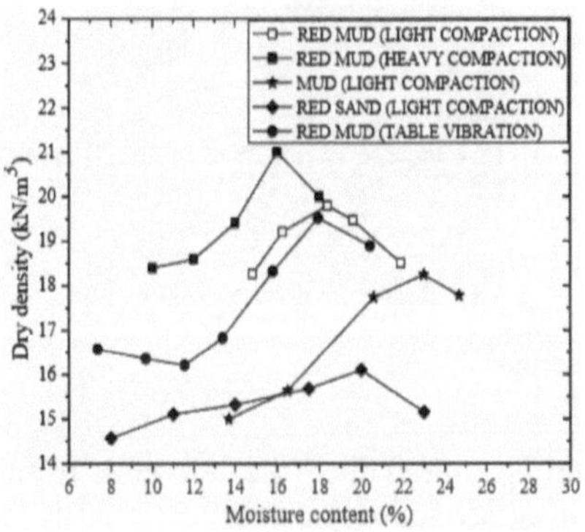

Figure 6. Compaction curve of red mud using different compactor.

It can be observed from the curve that red mud has higher MDD (19.8 kN/m3) and relatively low OMC (18%) at light compaction. The higher MDD of red mud in comparison to other soil is due to the high specific gravity value of red mud.

3.5 *Shear strength*

Stress-strain curve of triaxial shear test for the red mud at different confining pressure is shown in Figure 7. From the triaxial shear test result cohesion and angle of internal friction were found to be $28.8 kN/m^2$ and 34^0 respectively.

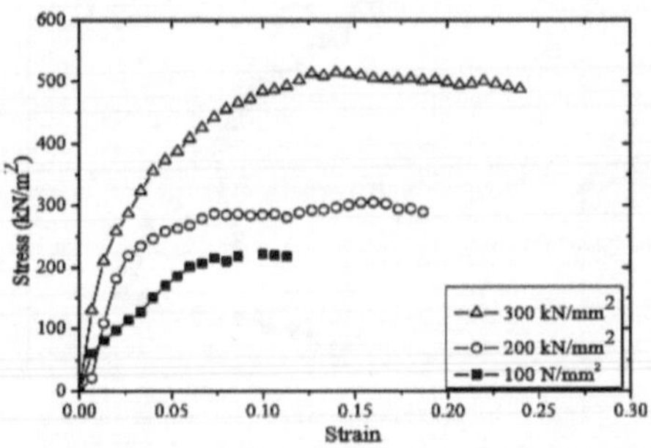

Figure 7. Stress-strain curve of red mud.

6 FINITE ELEMENT ANALYSIS

The model footing test has been conducted in a 750mm x 250mm test tank having depth 500mm as shown in Figure 8. A 20mm thick rectangular plate with length and width equals to 248mm and 75mm, respectively has been used as model footing. The red mud under different conditions such as dry and partially saturated conditions has been used as base material. The load applied on the footing is measured using proving ring and the settlement is measured using two number of dial gauge as shown in Figure 8. Total settlement corresponding to a load is taken as the average of the two dial gauge reading.

Figure 8. Model footing test setup.

In dry condition, red mud is poured into the test tank using rain fall technique while partially saturated condition is developed by adding 18–20% water. The density achieved using rainfall technique was 16.7kN/m3. For comparison, in case of partial saturated condition, soil was compacted to achieve this density.

Figure 9 shows the load-settlement curve obtained from laboratory model test for red mud in dry state and partially saturated state. From the load-settlement curve, red mud in partially saturated condition gives higher bearing strength and longer strain as compared to dry state. As red mud is fine grained, this variation may be due to increase in apparent cohesion during saturated condition.

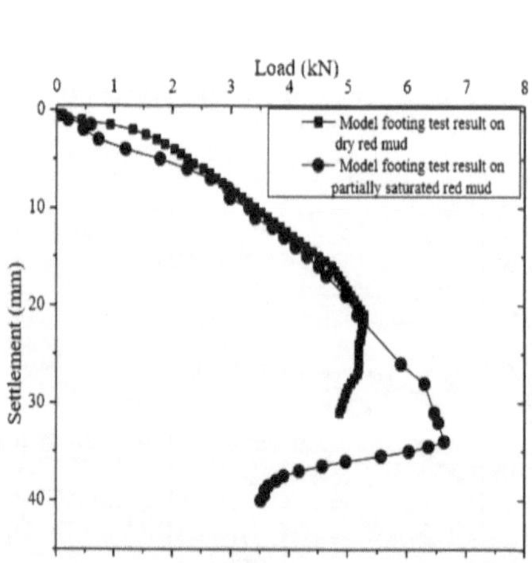

Figure 9. Load–settlement curve of partially saturated and dry red mud.

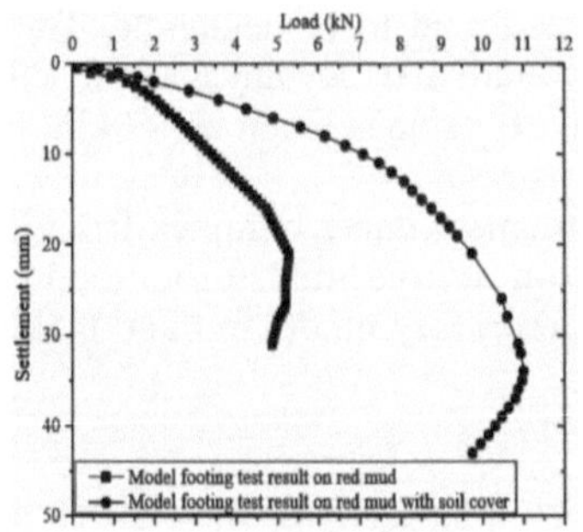

Figure 10. Load–settlement curve of red mud with soil cover at top.

In this study, an attempt has also been made to find out the effect of soil cover used to avoid erosion, on load carrying capacity of red mud. A soil cover of thickness 0.5B (i.e 0.5 times of width of footing) has been used on the top of red mud whose shear strength parameter is given in Table 2.

The load-settlement curve of red mud with soil cover at top along with the curve without soil cover is shown in Figure 10. The load bearing capacity of red mud with soil cover at top is found to be 11kN in comparison to that of 5.28kN without any soil cover.

The FE analysis of the above experimental work was done using software PLAXIS, to compare the results. The soil parameters considered for red mud and topsoil are presented in Table 2.

Table 2. Parameter of mohr-coulomb model.

Material property	Red Mud	Top soil
Unsaturated unit weight(kN/m^2)	16.5	17.3
Cohesion(kN/m^2) (Dry Soil)	1	20
Angle of internal friction (Degree)	34	32.5
Young modulus “E” (kN/m^2)	150	450
Poisson’s ratio	0.14	0.19
Permeability (m/day)	4.6×10^{-6}	6×10^{-5}

It is well known that the FE results depend upon the mesh size and soil parameters. The effect of mesh size and the soil modulus E and ν on the settlement has been studied. It has been found that there is no change in settlement with the increase of element beyond 600, which comes under the category of fine mesh, as shown in Figure 11.

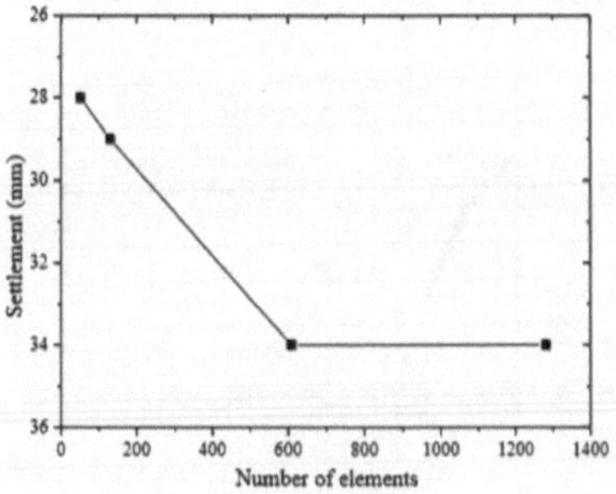

Figure 11. Variation of settlement with number of element (mesh size).

The effect of Poisson's ratio is found to be negligible. The effect of soil modulus E has been studied and it has been found that beyond 600kPa, effect is marginal as shown in Figure 12. For the present study, E value is taken as 150kPa to minimize the error between model test result and FE analysis result.

Figure 13 shows the load-settlement curve obtained from laboratory model test and FE analysis for the footing resting on dry red mud. From load-settlement curve, peak load is found to be 5.28kN in case of laboratory model test and 6.5kN as per FE analysis.

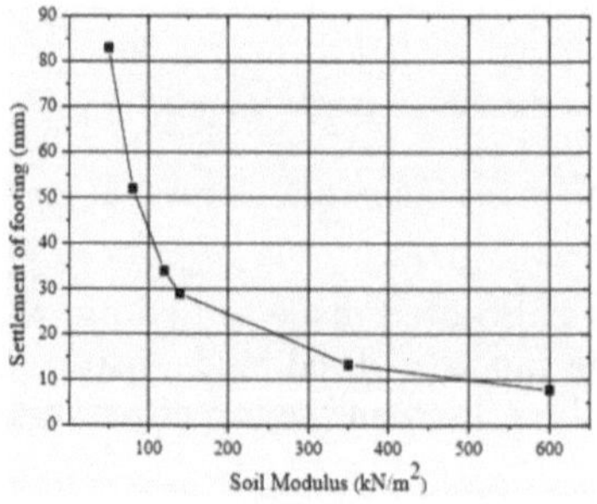

Figure 12. Variation of settlement with respect to change in soil modulus.

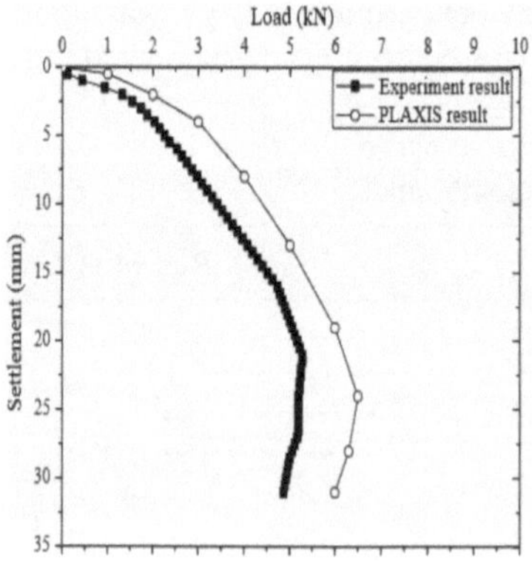

Figure 13. Load–Settlement curve of dry red mud.

Partially saturated condition is developed by adding 18–20% water to the red mud and then it fis compacted in the test box to achieve the density of 16.7kN/m^3. Figure 14 shows the load-settlement curve obtained from the model footing test and FE analysis for the footing resting on partially saturated red mud. From the curve, ultimate load is found to be 6.5kN in case of laboratory model test and 17.05kN in case of FE analysis. The large variation may be due to considered E value.

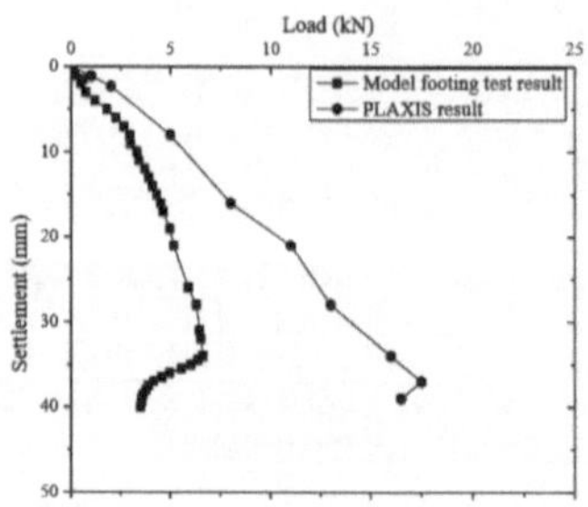

Figure 14. Load–Settlement curve of partially saturated red mud.

7 CONCLUSIONS

Due to presence of iron compound minerals, the specific gravity of red mud is very high as compared to soil.

(1) Red mud has higher MDD in comparison to other materials, due to high specific gravity.
(2) Red mud has low plasticity and low volumetric and linear shrinkage with 90% of particle finer than 0.075mm. But, the angle of internal friction of the red mud is very high as compared to other fine grained soil.
(3) Based on dispersiveness test, red mud is found to be highly dispersive hence need to protect against external weathering using soil cover at top. The pavement with soil cover at top is found to have higher bearing capacity.
(4) Properties of the red mud reflects that we can utilize the red mud as geotechnical material like Backfill material, road sub-grade material embankment material. Red mud is further stabilized to enhance the more strength with lime, gypsum, fly ash etc.
(5) The experimental results of model footing test on corresponding finite element analysis showed that red mud can be used as an alternate pavement subgrade construction material with soil cover at top to protect against erosion.

Research related to the proportion of red mud based on the chemical characteristics are progressing. Moreover, the studies can be continued to include the suitability of adding red mud in other components of the pavement especially the asphalt layers.

ACKNOWLEDGMENT

The authors would like to acknowledge the support of this work from University of Technology and Applied Sciences, Shinas, Sultanate of Oman.

REFERENCES

Bhaskar, M., Akhtar, S. and Batham, G., (2014), Development of the Bricks from Red Mud by Industrial Waste (Red Mud), *International Journal of Emerging Science and Engineering*, Vol. 2 (4), pp.7–12.

Bhuvaneshwari, S., Soundra, B. and Gandhi, S. R. (2007). Stabilization and microstructural modification of dispersive clayey soils, *International Conference on Soil and Rock Engineering*, Columbo, Srilanka.

Collazo, A., Fernandez, D., Izquierdo, M., Novoa, X. R. and Perez, C., Evaluation of red mud as surface treatment for carbon steel prior painting, *Progress in Organic Coatings*, Vol. 52 (4), 2005, pp.351–358.

Emile Mukiza, LingLing Zhang, Xiaoming Liu, Na Zhang, (2019). Utilization of red mud in road base and subgrade materials: A review, Resources, Conservation and Recycling, Volume 141, Pages 187–199

Freitas, L. A. R. U. and Magrini, A. (2017). Waste management in industrial construction: Investigating contributions from industrial ecology, *Sustainability*, *9*, 1251.

Hind A. R., Bhargava S. K. and Grocott S. C. (1999). The surface chemistry of Bayer process solids: a review, *Colloids and Surfaces A: Physicochem. Eng. Aspects*,146, pp. 359–374.

Kang, Suk-Pyo, Properties of alkali-activated slag-red mud soil pavement using recycled aggregates. (2016). *Journal of the Korean Recycled Construction Resources Institute*, Vol.14, No.3, Korean Society of Construction and Recycling Resources, p.276–283.

Mishra, B and Kirkipatrick, D., (2002). Recovery of value added products from red mud, *Minerals & Metallurgical Processing*, Vol.19, pp.87–94.

Mo Zhang, Mengxuan Zhao, Guoping Zhang, Derrick Mann, Kevon Lumsden, Mingjiang Tao, (2016). Durability of red mud-fly ash based geopolymer and leaching behavior of heavy metals in sulfuric acid solutions and deionized water, *Construction and Building Materials*, Volume 124, Pages 373–382, ISSN 0950-0618.

Newson, T., Dyer, T., Adam, C. and Sharp, S., (2006). Effect of structure on geotechnical properties of bauxite residue, *Journal of Geotechnical and Environmental Engineering*, Vol. 132 (2), pp.143–151.

Parlikar I., Saha U. V., P. K. and Khadilkar, S. A., (2011). Technological options for effective utilization of bauxite residue (Red mud) – A review, *International Seminar on Bauxite Residue (Red Mud)*, Goa, India.
Rameez Ahmad Mantoo, Dr. Esar Ahmad and Govind Singh Chauhan, 2021, A study on use of red mud as sub-grade in road construction, *International Journal of Engineering Research & Technology*, Volume 10, Issue 07 (July 2021),ISSN (Online) : 2278–0181
Rout, S. K., Sahoo, T. and Das, S. K., (2012). Utility of red mud as an embankment material, *International Journal of Earth Science and Engineering*, Vol. 05 (6), pp.1645–1651.
Suchita Rai, K. L. Wasewar, J. Mukhopadhyay, Chang Kyoo Yoo, Hasan Uslu Neutralization and utilization of red mud for its better waste management *Arch. Environ. Sci.* (2012), 6, 13–33.
Yadav, Harinder and Garg, Achal, (2018). Study of Red Mud as an Alternative Building Material for Interlocking Block Manufacturing in Construction Industry, 10.17706/ijmse.2015.3.4.295-300.
Yan Ma, Athina Preveniou, Anastasios and Johan Berg Pettersen, (2022). Circular economy and life cycle assessment of alumina production: Simulation-based comparison of Pedersen and Bayer processes, *Journal of Cleaner Production*, Volume 366, 132807, ISSN 0959-6526.
Yang J. K., Chen F. and Xiao B. (2006). Engineering application of basic level materials of red mud high level pavement (In Chinese). *China Munic. Eng.* (5):7–9.
Zhang, Jizhe; Liu, Shengjie; Yao, Zhanyong; Wu, Shaopeng; Jiang, Hongguang; Liang, Ming and Qiao, Yaning, (2018). Environmental aspects and pavement properties of red mud waste as the replacement of mineral filler in asphalt mixture, *Construction and Building Materials*, 180. 605–613.

Innovation and Technological Advances for Sustainability – Al-Naemi et al. (Eds)

A sustainable approach to wastewater treatment using constructed wetland: A greenhouse technology

A. Maria Rajesh & Binumol Shahulhameed
Faculty of Civil Section, Engineering Department, College of Engineering and Technology, University of Technology and Applied Sciences, Shinas, Oman

Raya Said Al Zaabiya, May Darwish Ali Al Balushi, Hamdah Ali Ahmed Al Shehhi, Athari Khalifa Handi Al-Saadi & Nooralsnaa Abdallah Al-Farsi
Bachelor Student Engineering Department College of Engineering and Technology, University of Technology and Applied Sciences, Shinas, Oman

ABSTRACT: The research concentrates to treat wastewater by natural method. Wastewater treatment consumes the prospective to reduce water demand. Constructed Wetlands are a wastewater treatment method for optimizing their performance to meet treated discharge standards. These systems have progressed and effective replacement for conventional wastewater treatment. Natural wetlands act as the "kidneys of the Earth." The demand for water has increased as the urban population, urbanization, and economic development have expanded, particularly in water-scarce regions. The Conventional wastewater treatment methods pollute environmental and many numbers of labour for their operation and maintenances, and huge construction cost. Therefore, this paper summarizes an environmentally friendly technology for treating wastewater. The main aim of the study is to reduce the carbon footprint of the environmental for wastewater treatment process. There are four phases for experimental investigations, the first phase is to optimize the ratio of materials from earlier studies with different parameters such as different filter media, different plantations and different in and outflow. In the second stage, the experimental work is carried out to find the optimal result. The third phase of research is to develop a prototype of wetland system. The fourth phase of research is involved in predicting recycling and reduction technology for solid waste generation and wastewater treatment process. Based on the research, the biological oxygen demand (BOD) was reduced by 68.5 percent, 52 percent for Chemical Oxygen Demand (COD), 79.4 percent for Total Suspended Solids (TSS), 68.5 percent for Total Dissolved Solids (TDS), 32 percent for Total Nitrogen(N), and 38.2 percent for Total Phosphorous in wastewater from the UTAS-Shinas after treatment. The efficiency of wetlands system is compared between before and after treatment of wastewater in UTAS – Shinas. The treated water can be reused in many uses, such as irrigation, showering, washing etc.

Keywords: Greenhouse wastewater treatment Technology, Sustainable Development, Prototype, Waste management and recycling

1 INTRODUCTION

The Climate change has emerged as a serious worry around the world. This phenomena is attributed to an increase in anthropogenic greenhouse gas (GHG) emissions from several human activities, such as carbon dioxide (CO2), methane (CH4), and nitrous oxide (N2O). Each accounts for around 10% of anthropogenic methane emissions sources, including home and industrial, and is regarded to be particularly cost-effective. The carbon footprint (CFP) approach can be used to calculate the amount of global warming caused by wastewater treatment plants and to locate areas that need to be prevented or mitigated (Flores-Alsina *et al.*, 2011, Wiedmann and Minx 2008). Many of the energy and chemicals used in plant operation contribute to global GHG emissions, according to research that looked into the CFP of conventional wastewater treatment plants (such activated sludge systems) (Laura Flores 2017). Because of their low cost, low energy requirement, and simplicity of operation and maintenance, constructed wetland systems are a natural substitute for activated sludge systems for the treatment of urban and industrial wastewater. (2018) Arden and Ma; (2014) Vymazal. They have shown to be a successful choice for wastewater treatment. The best option for treating wastewater without causing pollution is to use constructed wetlands. For lowering global pollution, researching green wastewater treatment technologies is crucial. Utilizing plantations and open space is a key strategy for reducing pollution of the environment. The procedures for Oman's natural wastewater treatment process will be carried out as part of this study. A greenhouse wetland treatment facility with high levels of energy efficiency, minimal maintenance requirements, water conservation, better human health, and environmental friendliness will be taken into consideration. 2016 Biswas and Yek The global community is now seriously concerned about climate change. With the help of the Carbon Footprint (CFP) technique, wastewater treatment facilities' impact on global warming may be calculated, and hotspots can be avoided or controlled. Numerous studies that assessed the CFP of traditional wastewater treatment facilities (such as activated sludge systems) discovered that the energy and chemicals necessary to run the facilities are primarily to blame for their contribution to global GHG emissions. (Arden 2018). For the treatment of urban and industrial wastewater, constructed wetland techniques are green technologies that provide a low-cost, low-energy alternative to activated sludge systems. They are easy to maintain and run. (2003) (Dixon *et al.*) Studies done in the past that compared the effects of artificial wetland systems to traditional wastewater treatment methods found that the former was the more environmentally friendly option due to less chemical and electricity use. (2003) Green F. B. Researchers and engineers working on ponds focused on modifying pond design, understanding more about the mechanics, and enhancing pond removal and energy efficiency. Many methods have been introduced to address odour nuisances, sludge accumulation, and poor effluent quality. Earlier research comparing the environmental implications of constructed wetland systems and traditional methods found that the former was the most environmentally friendly wastewater treatment option, owing to its minimal electricity and chemical use. Nonetheless, the majority of these studies focused on wastewater treatment systems in cities. To the best of the authors' knowledge, only one study has looked into the environmental effects of winery wastewater treatment in artificial wetlands (Flores *et al.* 2019a). However, using emissions rates from the literature, direct GHG emissions from wastewater treatment were estimated in this study. The importance of considering genuine GHG emissions measured on-site in full-scale effluent, on the other hand, cannot be overstated. (Brix and Arias 2005) Constructed wetlands are wastewater treatment systems that are engineered to reproduce natural processes in wetland plants, bed medium, and microbiomes. Municipal, industrial, agricultural, landfill leachate, and stormwater runoff are all treated in CWs. (Antwi *et al.* 2018) Artificial intelligence (AI) is being utilized to forecast and enhance the removal efficiencies of heavy metal ions, dyes, nutrients, persistent organic pollutants (POPs), chemical oxygen demand (COD), and other contaminants. (Vymazal 2009) Constructed wetlands (CWs) have been used to treat a range of wastewaters for over 50 years. Furthermore, traditional wastewater treatment method (e.g. activated sludge systems) have discovered that the majority of their contribution to global GHG emissions comes from the energy and chemicals utilized in operation plant. However, the need to treat wastewaters with new technology, the existence of newly created organic compounds in wastewater, and increasingly severe discharge limitations around the world are all posing new difficulties to CW research. The research focus of this review is on research problems that have been reported in recent years. A built wetland (CW) is a wastewater treatment system made up of shallow ponds or channels seeded with aquatic plants that uses natural microbiological, biological, physical, and chemical processes to clean wastewater.

2 MATERIALS

The proposed project site factors that may restrict the maximum area of a treatment wetland for a given source-water volume. The designer must choose the kind of substrate medium to be used in the artificial wetland. Gravel, sand, and pebble were substrates used in this case; they were locally accessible. Site-specific biological circumstances do not typically pose a significant technical challenge for constructed wetland projects, and this was true for this study as well. The foundation for performance-design is the water budget. The temperature controls how quickly process happen in a wetland. The temperature controls how quickly process happen in a wetland. Rainfall is a key factor in the transport of pollutants and is a primary source of inflow in the hydrologic cycle. Rainfall largely dilutes the wastewater. Pollutant concentration is decreased. Rainfall is a major source of inflow in the hydrologic cycle and has a significant role in the transportation of pollutants. The wastewater is primarily diluted by rainfall. It reduces pollutant concentration. Additionally, precipitation has a tendency to mingle with the wastewater in the influent and effluent chambers, diluting it. This diluted wastewater begins to travel both surface- and subsurface-wise over the slope. By determining influent concentrations by sampling and analysis, wastewater was characterized. In order to size the wetland and anticipate its effectiveness in the face of unknowable future fluctuations, it is essential to know the amounts of the contaminants in the water to be treated. The characteristics of wastewater include its physical, chemical, and biological components.

3 EXPERIMENTAL SETUP

The storage Tank is used for inlet and outlet purpose. The capacity of storage tank is 50 Gallon. The inlet storage tank is used for collecting waste waters. The outlet storage tank is used for storage and distribution of Treated waters to Garden. The Glass tank is used as Treatment bed. The inside filling material's transparent purpose here used by Glass material. It is designed Aspect ratio 1 : 0.75. The Gravel, Sand, soil and wetland plants are acted as a filtering media. The Gravel is filled at bottom layer, Sand at middle layer, and Soil at top layer. Then planting reeds and sugarcane in the top layer. Pipes are used to Supply the flow of wastewater and Treated waters.

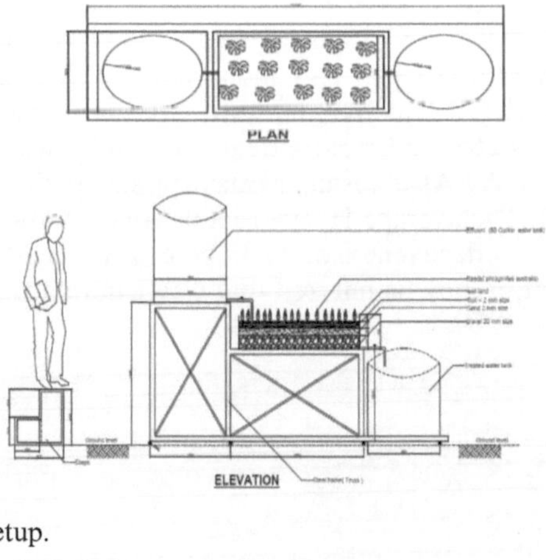

Figure 1. Experimental setup.

In order to support aerobic conditions through photosynthesis and air diffusion, wetland systems are typically no deeper than two feet. To stop the growth of anaerobic zones, the rate of area loading must be controlled. The forms employed in mechanical treatment systems have very low energy intake, which causes very slow flow through them and extremely long hydraulic retention periods. All of the biological treatment mechanisms utilized in any conventional biological treatment technique are applied in wetland treatment; hence, these systems call for the addition of nutrients and trace elements. Winter operations demand considerable planning.

4 EXPERIMENTAL INVESTIGATION

Economic and environmental factors have been examined in the study. The engineered wetland system was discovered to be a low-cost method that might reduce the costs associated with purchasing, operating, and maintaining wastewater treatment equipment. Last but not least, creating wetlands is an environmentally beneficial, low-cost alternative to current wastewater treatment systems. The "kidneys of the Earth" have been referred to as natural wetlands. Biological filters, such as built wetlands, are extremely effective at removing BOD, COD, TSS, and organic nitrogen from water flowing through them on its way to receiving lakes, streams, and seas; nitrates are almost totally removed. It is crucial to keep in mind that wetlands are only one part of a multisystem while assessing their performance. Wastewater from municipal, industrial, and agricultural sources, surface mining runoff, irrigation return flows, urban storm water discharges, leachates, and other contaminants can all contaminate wetland areas.

For the treatment of several types of wastewaters, a pilot scale study was conducted. Wetlands are built on the principles of ecosystems, which process almost all types of wastewater through intricate and regular interactions between supporting materials, macrophytes, and bacteria. The methodology is centered on wastewater collection, the creation of wetland tank models, the investigation of soil and plant properties, and the examination of treated effluent. In order to cleanse the wastewater, a small-scale prototype was constructed in UTAS-Shinas, Oman using resources that were easily accessible, like gravel, sand, and pebbles. A size of Treatment chamber is 150 x 60 x 30 cm constructed as wetland units, and the system was created with a 1% slope between the entrance and output points.

5 RESULTS AND DISCUSSION

The average turbidity of wastewater was 35 NTU, while the average removal efficiency of CW wastewater was 98.39 percent. As a result, CW can be determined to be quite successful at eliminating turbidity. The average TSS in wastewater was 60.8 mg/L, however the finished wastewater from CW had a lower mean TSS and a removal efficiency of 79.4 percent. As a result, it may be stated that CW has the ability to efficiently eradicate TSS. The average BOD of the wastewater was 265 mg/L before treatment, and it was reduced by 68.5 percent following treatment with CW. As a result, it may be stated that CW has the ability to efficiently eliminate BOD. The average Heavy Metals concentration in the wastewater was 5.332 mg/L, with a removal effectiveness of 73.33 percent to 95.70 percent following treatment with CW. As a result, it may be inferred that CW can remove Heavy Metals.

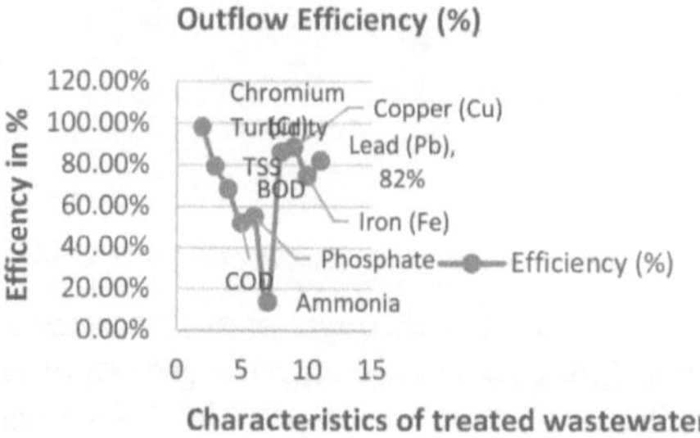

Figure 2. Characteristics of treated wastewater.

Domestic wastewater has been successfully treated in constructed wetlands with vertical sub-surface flow. With an artificial created wetland setting, the concentrations of parameters COD, BOD, N, and PO4 in the influent of wastewater from a sewage treatment facility at UTAS-Shinas were investigated. For an integrated setup, this system might reach removal efficiencies of 68.5 percent, 52 percent, 79.4 percent, and 98.34 percent for BOD, COD, TSS, and Turbidity, respectively. Turbidity removal efficiency were higher in systems that treated household wastewater. In Wetland models are designed ecosystem, which utilizes both complex and ordinary interactions between supporting this research, various pilot-scale vertical flow-built wetlands (VFCWs) are set up in this study to see how design and operational variables affected the performance of tidal-flow VFCWs. As a result, the study divides into four phases.

6 CONCLUSION

Constructed wetlands (CW) are a sustainable alternative to traditional wastewater treatment methods that can be used to treat wastewater efficiently. They are also a potential technique for environmental remediation and secondary wastewater treatment. The effluent had a 7.2 pH on average. The wastewater's PH was 7.0 after CW treatment. Therefore, it may be said that CW has little impact on the PH. While effluent from CW was reduced and had an average removal effectiveness of 98.39%, the average turbidity of wastewater was 35 NTU. Therefore, it may be said that CW removes turbidity quite effectively. The mean TSS in wastewater was further reduced to 60.8 mg/L in the final wastewater from CW, which also had an average removal efficiency of 84.5%. Therefore, it may be said that CW can successfully eradicate TSS. After being treated with CW, the wastewater's average BOD decreased, with a removal effectiveness of 93.6%. Therefore, it may be said that CW can successfully remove BOD. After being treated with CW, the wastewater's average Heavy Metals removal efficiency ranged from 73.33% to 95.70%. Therefore, it can be said that CW is a successful heavy metal removal method. Constructed wetlands have a number of benefits, including a long system lifetime, ease of maintenance, cheap operating costs, environmental friendliness, and a wide range of toxins that can be treated. Additionally, with minimal sludge generation, CWs may efficiently treat numerous pollutants in secondary and tertiary wastewater. Additionally, they offer an environmentally sustainable method with minimal energy usage and operating expenses.

ACKNOWLEDGMENT

The authors are thankful to the management of UTAS-Shinas, Sultanate of Oman and the Head of the Engineering and Civil Engineering Section for their continuous support and encouragement in carrying out this research.

REFERENCES

Adewuyi GO, Olowu, RA (2012) Assessment of oil and grease, total petroleum hydrocarbons and some heavy metals in surface and groundwater within the vicinity of NNPC oil 52 | Treatment of petroleum refinery wastewater with constructed wetlands depot in Aata, Ibadan metropolis, Nigeria. *IJRRAS* 13(1):166–174. Retrieved May 14, 2016, from Miklas Scholz, in *Wetland Systems to Control Urban Runoff*, 2006

Al-Isawi, R., Ray, S., and Scholz, M. (2017). Comparative study of domestic wastewater treatment by mature vertical-flow constructed wetlands and artificial ponds. *Ecological Engineering*, 100, 8–18. https://doi.org/10.1016/j.ecoleng.2016.12.017

Antwi, P., Li, J., Meng, J., Deng, K., Koblah Quashie, F., Li, J., Opoku Boadi, P., (2018). Feedforward neural network model estimating pollutant removal process within mesophilic upflow anaerobic Sludge blanket bioreactor treating industrial starch processing wastewater. *Bioresour Technol* 257, 102–112. https://doi.org/10.1016/j.biortech.2018.02.071.

Asuman Korkusuz. E., Meryem Beklİoglu and Göksel N. Demİrer (2014), Treatment efficiencies of the vertical flow pilot-scale constructed wetlands for domestic wastewater treatment, *Turkish J. Eng. Env. Sci.*, Volume 28, Pages 333–344.

Biswas, W. K., Yek, P., (2016). Improving the carbon footprint of water treatment with renewable energy: a Western Australian case study. *Renewables Wind. Water, Sol.* 3, 14. https://doi.org/10.1186/s40807-016-0036-2.

Braden, S., Ma, X., (2018). Constructed wetlands for greywater recycle and reuse: a review. *Sci. Total Environ.* 630, 587–599. https://doi.org/10.1016/j.scitotenv.2018.02.218.

Brix, H. and Arias, C. A. (2005). The use of vertical flow constructed wetlands for onsite treatment of domestic wastewater: New Danish guidelines. *Ecological Engineering* 25, pp. 491–500.

Chen Y.-h. et al., Constructed landscaping combination constructed wetlands system used for sewage treatment, (2010). *2020 International Conference on Digital Manufacturing & Automation*, Changcha, China, pp. 430–433, doi: 10.1109/ICDMA.2010.409.

Choi, Hyeseon, Nash Jett DG. Reyes, Minsu Jeon, and Lee-Hyung Kim. 2021. Constructed Wetlands in South Korea: Current status and performance assessment. *Sustainability* 13, no. 18: 10410. https://doi.org/10.3390/su131810410

Christy K. Benny, Saswati Chakraborty, Dyeing wastewater treatment in horizontal-vertical constructed wetland using organic waste media, *Journal of Environmental Management*, Volume 331, (2023), 117213, ISSN 0301-4797,

Dixon, A., Simon, M., Burkitt, T., (2003). Assessing the environmental impact of two options for small-scale wastewater treatment: comparing a reedbed and an aerated biological filter using a life cycle approach. *Ecol. Eng.* 20 (4), 297–308.

Eva Gonzalez-Flo, Xavier Romero, Joan García, (2023). Nature based-solutions for water reuse: 20 years of performance evaluation of a full-scale constructed wetland system, *Ecological Engineering*, Volume 188, 106876, ISSN 0925-8574.

Favero L., Mattiuzzo E. and Planland D. F., (April 2077). Practical results of a water budget estimation for a constructed wetland, *Wetlands*, Vol. 2, pp. 230239.

Florentina Zurita and Alejandra Carreón-Álvarez (2015), Performance of three pilot-scale hybrid constructed wetlands for total coliforms and *Escherichia coli*removal from primary ffluent – a 2-year study in a subtropical climate, *Journal of Water Health*, Volume 13, Issue 2, Pages 446–458. https://doi.org/10.2166/wh.2014.135.

Green, F. B., Bernstone, L. S., Lundquist, T. J., Oswald, W. J., (1996). Advanced integrated wastewater pond systems for nitrogen removal. *Water Sci. Technol.* 33 (7), 207–217.

Günter Langergraber, (2006), University of Natural Resources and Applied Life Sciences, Austria, *Simulation of the Treatment Performance of Outdoor Subsurface Flow Constructed Wetlands in Temperate Climates*, Pages 210–219.

Haiyan Li, Yihui Zhang, Liyuan Wu, Yi Jin, Yongwei Gong, Angzhen Li, Jinxi Li, Fei Li, (2021). Recycled aggregates from construction and demolition waste as wetland substrates for pollutant removal, *Journal of Cleaner Production*, Volume 311, 127766, ISSN 0959-6526.

Hussein, A. and Scholz, M. (2018). Treatment of artificial wastewater containing two azo textile dyes by vertical-flow constructed wetlands. *Environ Sci Pollut Res* 25, 6870–6889. https://doi.org/10.1007/s11356-017-0992-0

Jiang Yu, Wen-qing Chen, Yong-hong Shui, Jian-quan Liu, Wai-tim Ho and Shu-yan Zhang, (2011). Research on wastewater treatment through integrated constructed wetlands, *2011 International Conference on Multimedia Technology*, Hangzhou, pp. 1197–1201, doi: 10.1109/ICMT.2011.6002928.

Mthembu, (2013), Durban University of Technology, South Africa. Constructed wetlands: A future alternative wastewater treatment technology, Vol. 12 No. 29 (2013) Florentina Zurita and Alejandra Carreón-Álvarez (2015), Performance of three pilot-scale hybrid constructed wetlands for total coliforms and *Escherichia coli*removal from primary ffluent – a 2-year study in a subtropical climate, *Journal of Water Health*, Volume 13, Issue 2, Pages 446–458. https://doi.org/10.2166/wh.2014.135.

Nikhil Savio, Diksha Pandey, R.K. Srivastava, (2022). Potentialities of plant based hybrid wetland systems for the treatment of household waste water using Canna indica, Agave americana, Pistia stratiotes and Tagetes erecta, *Materials Today: Proceedings*, ISSN 2214-7853.

Oluseyi Ewemoje and Abimbola Sangodoyin (2011), Developing a Pilot Scale Horizontal Sub Surface Flow Constructed Wetlands for Phytoremediation of Primary Lagoon Effluents. *11th edition of the World Wide Workshop for Young Environmental Scientists (WWW-YES-2011)* – Urban Waters: resource or risks?, Jun 2011, Arcueil, France. hal-00607955

Rousseau D. P. L., Lesage E. A., P. A, (May 2018). Constructed wetlands for water reclamation, *Desalienation*, Vol. 218, pp. 181–189.

Song B., Zheng Y. and Zhao X., (2009). Simulation study on constructed wetland treatment on villages and small towns' sewage, *International Conference on Environmental Science and Information Application Technology*, Wuhan, China, 2009, pp. 302–304, doi: 10.1109/ESIAT.2009.416.

Swald, W. J., Golueke, C. G., Tyler, R. W., (1967). Integrated pond systems for subdivisions. *J. Water Pollut. Control Fed.* 39 (8), 1289.

Tanveer Saeed, Md Jihad Miah, Asheesh Kumar Yadav, (2022). Development of electrodes integrated hybrid constructed wetlands using organic, construction, and rejected materials as filter media: Landfill leachate treatment, *Chemosphere*, Volume 303, Part 3, 135273, ISSN 0045-6535.

Vandana Patyal, Dipika Jaspal, Kanchan Khare, (2022). Screening of low-cost waste materials for removal of phosphorus from wastewater in Constructed Wetlands, *Materials Today: Proceedings*, Volume 71, Part 2, Pages 265–269, ISSN 2214–7853.

Vymazal J., (June 2005). Horizontal sub-surface flow and hybrid constructed wetlands systems for wastewater treatment, *Ecological Engineering*, Vol. 25, pp. 478–490

Wang D.-Q. et al., (2009). Research progress on phosphorus removal in substrates of constructed wetlands, *2009 3rd International Conference on Bioinformatics and Biomedical Engineering*, Beijing, China, pp. 1–6, doi: 10.1109/ICBBE.2009.5163343.

Wibowo Y. G., Wijaya C., Halomoan P., Yudhoyono A., and Safri M., (Jul. 2022). Constructed Wetlands for treatment of acid mine drainage: A review, *Jurnal Presipitasi : Media Komunikasi dan Pengembangan Teknik Lingkungan*, vol. 19, no. 2, pp. 436–450, https://doi.org/10.14710/presipitasi.v19i2.436-450, (IEEE)

William J. Mitsch, Karen M. Wise, (1998) The Ohio State University, Columbus,USA, Water Quality, Fate of Metals, and Predictive Model Validation of A Constructed Wetland Treating Acid Mine Drainage, Pages 1888–1900

Xu C. L., Daiand J. K., Wang W. J. (June 2018). Study on water quality of dispersed domestic sewage, *China Environmental Protection Industry*, Vol. 12. pp. 29–31.

Yajiao Li, Jiaping Wang, Xingchen Lin, Hui Wang, Huaien Li, Jiake Li, (2022), Purification effects of recycled aggregates from construction waste as constructed wetland filler, *Journal of Water Process Engineering*, Volume 50, 103335, ISSN 2214–7144

Innovation and Technological Advances for Sustainability – Al-Naemi et al. (Eds)

A carbon credit quality assessment and verification model towards carbon neutrality

Nur Maisarah Shahril Khuzairi
Research and Development Green Genesys Research, Selangor, Malaysia

Zaihisma Che Cob
College of Computing & Informatics, Universiti Tenaga Nasional (UNITEN), Selangor, Malaysia

Aaron Sanjay Dass
Group Sustainability Malayan Banking Berhad, Kuala Lumpur, Malaysia

ABSTRACT: Climate change due to anthropogenic carbon dioxide and other greenhouse gases emissions has become a serious global concern. Transitional solutions to reduce emissions, such as carbon offset, have given rise to carbon emission trading, commonly known as the carbon market. However, the carbon credits used for trading have their fair share of criticism, primarily in terms of their quality, validity, and efficacy. This paper explores the carbon market ecosystem and proposes a carbon credit quality assessment and verification model that encompasses five dimensions: (i) source, (ii) objective, (iii) location, (iv) timeframe, and (v) fidelity and its various factors to aid in addressing these concerns transparently and efficiently. Thus, achieving real-world environmental impact and global carbon neutrality goals.

Keywords: carbon credit model, carbon market, carbon neutrality, quality assessment, carbon emission trading

1 INTRODUCTION

Over the last decade, climate change due to anthropogenic carbon dioxide (CO_2) and other greenhouse gases (GHG) emissions has become a serious global concern. Hence various efforts [31] including negative emissions technologies have been developed to curb and reduce carbon emissions, such as green hydrogen, blue carbon, air capture carbon technology and etc. However, such efforts have yet to yield any significant and sustainable results.

While we wait for such technologies to catch up to current needs, transitional solutions to reduce emissions, such as carbon offset, have given rise to carbon emission trading or commonly known as the carbon market. This is a fundamental transformation strategy for reducing carbon emissions and achieving a climate-resilient economy [1–3]. Notably, research has shown that carbon credit trading has had positive results in reducing carbon emissions or slowing down the growth of carbon emissions [1,4,5].

According to United Nations Data [6], a total of 482 large organizations have pledged some form of carbon neutrality target and net zero commitment by 2050, and together, these organizations account for annual revenues of more than USD 16 trillion, exceeding the gross domestic product (GDP) of China. The majority of companies explicitly intend to utilize offsets to achieve their targets.

This growing demand for carbon credits by organizations to offset their carbon emissions to fulfil their Environmental, Social, and Governance (ESG) requirements has created a situation where stakeholders have strong reservations regarding the integrity and

 DOI: 10.1201/9781003496724-53

transparency of the production and the pricing of such credits. Core questions about a carbon credit's quality, validity, and efficacy in relation to real environmental impact and pricing need to be addressed transparently.

Based on the literature reviewed, there is no efficient model currently proposed to understand what constitutes a high-quality carbon credit. To fill this research gap, this paper discusses the concept of carbon credits and the carbon market ecosystem. A model is proposed that enables stakeholders to classify the quality of carbon credits produced, enabling the development of a more transparent, ethical, and sustainable carbon market for real-world high-impact results.

The rest of the paper is structured as follows. In Section 2 will overview the concept of carbon credits and the carbon market. Section 3 discusses the carbon credit quality assessment and verification model and its relationship to each dimension. Section 4 indicates the role of the model in the ecosystem, and finally Section 5 highlights the conclusion and direction of our future works.

2 OVERVIEW

2.1 *Carbon credits*

Carbon credits are defined in two contexts: (i) permits that are issued part of the so-called cap-and-trade program administered in certain nations, where organizations that pollute are awarded credits to continue emitting carbon dioxide as part of their business process to a specific limit, and they may sell the unused credits to other organizations that need them; (ii) certificates issued by an offset producer to organizations and buyers seeking to offset a certain amount of carbon dioxide emissions into the atmosphere for various reasons, including fulfilling one's ESG responsibility, Nationally Determined Contributions (NDC) attainment and etc. These carbon certificates are used as a commodity to be traded in the carbon market to organizations or other buyers that want to offset or reduce their carbon emissions.

The trading of carbon credits enables organizations and businesses to offset their carbon emissions on the one hand while continuing to profit from their business as usual on the other [7]. For example, an organization produces 10 $GtCO_2$-eq per year, and in order to claim carbon neutrality, they have to purchase the equivalent amount of carbon credits sold in the market to compensate for their emissions. In contrast, those who sell carbon credits have an obligation to reduce and offset the same amount of 10 $GtCO_2$-eq of carbon emissions purchased by the organization via the carbon market.

Conceptually this exchange process of carbon credits is a measurable, verifiable, and sustainable approach toward carbon neutrality as it can reduce global carbon emissions while boosting the economy [8].

2.1 *Carbon market*

The carbon market is where carbon credits are traded and exchanged. The primary purpose of the carbon market is to promote carbon emissions mitigation, which subsequently will reduce the rise of overall carbon dioxide in the atmosphere [9]. Carbon market's principle is similar as the standard buy-and-sell system, where one party (demand side) pays another party (supply side) to obtain permits for carbon emissions. The demand side can use the carbon credits purchased in lieu of mitigating their emissions to achieve climate targets while supply side is obligated to reduce or remove the equivalent carbon emissions as traded, thus this buy-and-sell system can lower the global carbon emissions volume.

Various studies have been conducted to examine the effectiveness of the carbon markets on emission reduction, and findings show that emissions decrease significantly [10] where governmental policy is implemented alongside carbon markets. A study by Bi, Xiao and Sun

[11] used a dynamic Computable General Equilibrium (CGE) model to study the implementation of a carbon market in China. The results revealed that although the short term carbon mitigation effect is less prominent but in the long run overall carbon emissions are reduced due to the interprovincial transaction in the carbon market.

Carbon markets can be criticized as an easy way out of climate change. However, if implemented effectively, it is a more sustainable mitigation approach toward climate change. This is because it reduces and removes carbon emissions and promotes the economy's growth, which later turns into supporting more carbon abatement projects and initiatives. Despite the well-documented benefits of carbon markets as a mitigation approach to reducing carbon emissions. Carbon credits have raised some concerns about their quality and integrity.

2.2 *Previous work on carbon credit trading model*

Existing works in the field of carbon credits mainly focus on models that promote the implementation of carbon trading or its effectiveness in reducing carbon emissions. Woo and colleagues [12] proposed a theoretical framework that uses Blockchain technology for carbon credit acquisition and implementation. The purpose of this framework is to encourage participation in the carbon credit market and create a resilient trading system. Another study also developed a stochastic game theoretical model to study the necessary trade-offs to maximize profits and minimize carbon emissions [13]. Also, a virtual power plant operating model considering carbon trading and green certificate trading is suggested by Zhang and colleagues [14]. The results showed that it increases system revenue, reduces system carbon emissions, and improves the overall efficiency of renewable energy under the carbon emission quota. Kuwae and colleagues [15] also explored carbon credit trading, specifically on blue carbon. They posited a framework of the Fukuoka City blue carbon offset credit system to promote collaboration between stakeholders to carry out environmental, economic, and societal improvements.

Although there is a growing attention around carbon trading and its effectiveness, there are also several challenges that the field is struggling to address, which include the integrity of carbon trading and its offset projects to truly help combat climate change [16]. Also, evaluating the quality of carbon credits used in the trading mechanism is rare [17].

Recent works have challenged the credibility of using carbon credits to offset organizations' carbon emissions [18,19]. This is because global carbon emissions have recently reached their highest recorded level ever, clearly showing that carbon credits trading has not achieved its promised results.

It is posited that the abovementioned issues can be attributed to the lack of guidelines that focus on the concept of carbon credits and should inform qualities related to its different dimensions and factors. Existing work on models and frameworks are a valuable contribution to the body of knowledge and offers opportunities for how research and practice should move forward. However, attention should also be explicitly given to carbon credits on their quality. As the demand for carbon credits is likely to increase in the future, it is essential to distinguish carbon credits based on their quality so organizations can make informed decisions to act responsibly in buying carbon credits to offset their emissions. Also, this is to ensure the carbon market ecosystem upholds its intended goal towards climate change. Hence, this paper aims to address this gap by proposing a conceptual carbon credit quality assessment and verification model to promote high-quality carbon credits.

3 CARBON CREDIT QUALITY ASSESSMENT AND VERIFICATION MODEL

In this section, a quality assessment and verification model for carbon credits based on five dimensions and several factors in relation to each dimension is proposed. The model intends to be a guide as much as a descriptor of quality assurance for carbon credits produced.

Hence it should be used to verify the quality of carbon credits from an inclusive perspective, as this will help the transferability value of such credits between different applications and research toward the primary goal of achieving carbon neutrality.

As shown in Figure 1, the five dimensions of the proposed quality assessment and verification model are: (i) source, (ii) objective, (iii) location, (iv) timeframe, and (v) fidelity. Each dimension will be discussed individually in the following section and exemplify their factors and impact on the quality of a carbon credit.

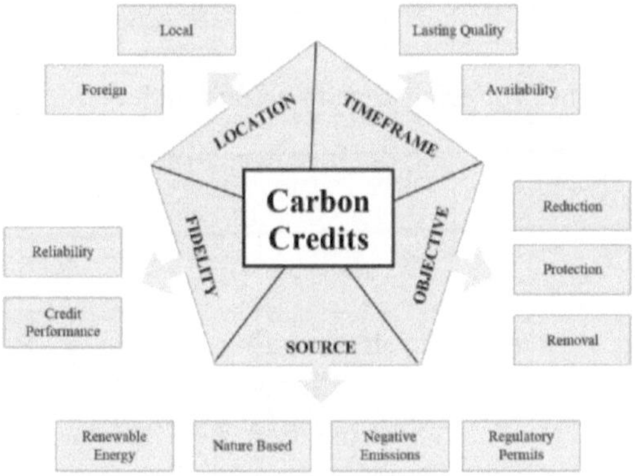

Figure 1. Carbon credit quality assessment and verification model.

3.1 *Source*

From which source the carbon credit created? Understanding the type of production source will enable demand-side insights on making effective and efficient organizational decisions in selecting the type of carbon credits to offset their carbon emissions. Generally, carbon credits fall into four categories: renewable energy, nature-based, negative emissions technologies, and regulatory permits. Each of these categories produces carbon credits differently and have different environmental uplift.

Renewable energy carbon credits are generated through projects that deploy renewable energy, such as solar and hydropower. They aim to reduce carbon emissions by switching to low-carbon or no-carbon energy options. Although usually accompanied by positive environmental uplift. There are, however, concerns about other impacts, such as deforestation, resource depletion, and waste storage.

Nature-based carbon credits are credits sourced through afforestation and maintenance of forests. These nature-based carbon credits through afforestation offer high environmental uplift, such as mitigating climate change by reducing and storing carbon while at the same time preserving natural resources, offering protection from droughts, heat waves, and floods [8,20]. Compared to renewable energy carbon credits, nature-based is more valuable as it focuses on carbon removal from the atmosphere instead of just reducing carbon emissions.

Like nature-based carbon credits, negative emissions technologies also focus on carbon removal. Instead of sourcing it through afforestation, negative emissions look into direct air carbon capture and storage, soil carbon sequestration, bioenergy carbon capture and storage [21]. According to a study by Alcalde and colleagues [22], negative emissions technology such as direct air capture and enhanced weathering is costly, with an estimated cost of up to USD 570/tCO_2 and USD 1600/tCO_2, respectively, as compared to nature-based carbon credits which will only cost an average of USD 108/tCO_2. Although negative emissions carbon credits come with a high price, it has high potential. Developers from a direct air capture plant in Switzerland claim that with a price of USD 500 to USD 600/tCO_2, it can capture 900 tonnes of CO_2 per year, equivalent to carbon emissions from 200 cars [23,24].

Although negative emissions technologies play a promising role in a carbon-neutral future, the technologies are still in the early stages of development for massive commercial deployment. Hence, economically nature-based carbon credits are more practical to trade in the carbon market as they are more cost-effective to produce and maintain.

Regulatory permits are commonly used in "cap-and-trade" situations where the unused carbon credits are traded to other organizations as an offset. This kind of credit would have a much lower value as compared to the above types, as there is no natural environment uplift other than forcing organizations to become more efficient or face high taxes by regulatory bodies.

3.2 *Objectives*

There are three objectives in producing carbon credits: reduction, protection, and removal (as shown in Figure 2).

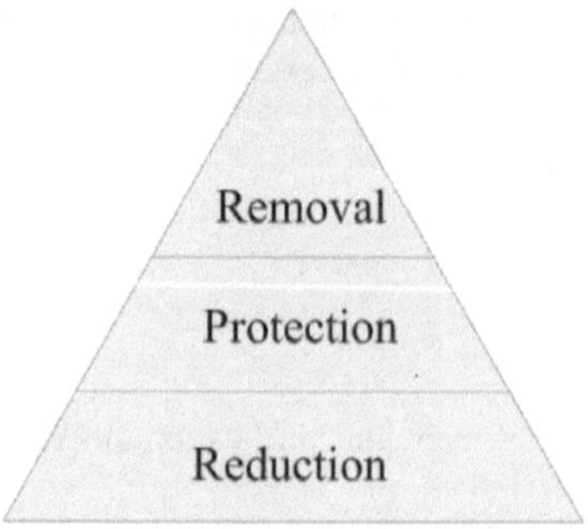

Figure 2. Hierarchy of carbon credit quality based on objectives.

In reduction, the objectives are to reduce carbon emissions on the one hand and sell the reduced quota to organizations that are unable to reduce and are exceeding their emission cap limit. These reduction credits are commonly used and seen in the "cap-and-trade" carbon market. Purchasing such reduction carbon credits gives license to an organization to emit more carbon emissions than their regulatory allowance, relieving the pressure to reduce carbon emissions or adopt low-carbon innovation [25]. In general, the reduction credit does not contribute towards carbon neutrality as when one reduces its emission, it gives another the permit to pollute.

On the other hand, protection credits are considered of better quality as the goal is to protect and preserve existing carbon sinks. Effective protection carbon credits can support the reduction of carbon emissions and prevent any release of carbon emissions. However, there are concerns about reductions being reversed due to natural disasters or project mismanagement. Also, overestimation of actual stored carbon and carbon reduction due to faulty or creative calculations may present obstacles to real environmental progress and carbon neutrality.

Removal credits focus solely on removing carbon emissions from the atmosphere. Various approaches can generate removal credits, including new projects on growing vegetation, carbon capture, and storage technologies. There seems to be a misinterpretation by some that merely contributing and supporting such projects equate to the creation a removal credit.

For example, organizations would claim carbon neutrality simply by supporting removal projects and initiatives by adding projected future removal numbers into the creation of current valid credits for offsets. This kind of creative accounting can seriously undermine the fundamental objectives of carbon neutrality. Thus it is crucial to understand that removal credits will only be available when removal projects are matured with actual carbon removal taking place and accounted for.

3.3 *Location*

The location of projects that are tapped for creating carbon credits is another vital dimension with two essential factors foreign and local. Currently, carbon projects tend to be based in developing countries where large organizations and wealthier nations will purchase these carbon credits at more affordable rates. Such an approach is conceptually a win-win scenario for both demand and host countries [26]. However, this approach has many concerns, such as how the host country will offset its own carbon emissions to reach carbon neutrality when most offsetting projects are done for foreign demand. Another issue is regulation and quality control, as there is no standardized international framework or local government regulation for the oversight of these projects. It becomes complex and challenging to verify the legitimation of such carbon offsetting projects from instances that might jeopardize the value of carbon credits [25]. Such as double claiming, project mismanagement or failure, creative accounting and many more.

Hence, attention and priority should be given to carbon credits created locally. This is because locally produced carbon credits can support the local economy and offset national emissions, enabling a more holistic pathway to achieve carbon neutrality. Furthermore, project developers will have an extra layer of oversight from local buyers, who are ethically conscious of the credits they utilize.

3.4 *Timeframe*

In essence, carbon credits are considered to be time-sensitive, and a significant challenge is the availability and lasting quality of the credits. It is crucial to identify these factors to understand the impact of credits issued. This is because, in some instances, credits sold are immature, as the offsetting will only take place years in the future.

Availability is an essential metric for assessing high-quality carbon credits. In availability, the main concern is to ensure that the carbon credit issued is readily usable. That is, the mitigation activity has already taken place, and those carbon credits are mature to contribute to carbon offsetting. To gauge the availability of a carbon credit, it is therefore vital to understand the context of the mitigation projects and whether the project has been implemented or will be implemented in the future. For example, some organizations are marketing carbon-neutral on their products. However, offsetting for the carbon credits issued will not occur for two years or longer [27], which means that the carbon credits are not currently available to offset the emissions immediately, thus making it an invalid or a low-quality credit.

The lasting quality factor can be expanded to two perspectives, in terms of how long the carbon credit is valid before renewal from the same project and the credit's permanence due to external factors. A single carbon credit has a general pre-defined lasting period and must be renewed based on the type of credit and project. For example, according to Climate Action Reserve (CAR), most non-sequestration projects have a 10-year crediting period and can only be renewed once. In contrast, sequestration projects have a renewal period of up to 100 years. On the other hand, Verra, formerly known as Verified Carbon Standard, suggests a minimum commitment of 20 years to 100 years but varies based on the project's risk of reversal [28]. Thus, a credit's lasting quality regarding its validity for renewal depends heavily on the context of the offsetting project, permanence, and risk level. For example, project types such as reforestation can have high lasting quality but also are exposed to high risk from natural disasters or project mismanagement.

3.5 *Fidelity*

The fidelity dimension considers factors such as the reliability of an issued carbon credit and its actual performance.

Reliability in carbon credits is defined as the trustworthiness of the entire process. Most credits issued are prone to the risk of double issuance and double counting, jeopardizing the reliability of the credit. Such double issuance and double counting risk are not new [29]. Double issuance means one single carbon credit is issued to two or more organizations when the emissions reduction or removal only account for once. This will subsequently lead to double counting. For example, the demand country claims the same mitigation outcomes for their NDC attainment, yet the reductions are still reflected in the host country's inventory. This will undermine the integrity of the credit issued, defeating the purpose of carbon credits trading in the first place. Technologies for transparent tracing, such as blockchain [30] and local sourcing policies, should be explored.

Another factor that must be examined is an issued credit's performance. A carbon credit issued should constitute one full ton of carbon dioxide reduced or removed from the atmosphere [1]. So the credit holder has the right to emit one ton of carbon dioxide. Credits that experience leakages during the mitigation process should be assessed and addressed rigorously.

The above mentioned dimensions and factors in the carbon credit quality assessment and verification model work as a set of performance indicators and metrics to define the quality of an issued carbon credit. This is an especially effective and practical guide for nations and organizations who intend to adopt carbon credit schemes or markets to achieve their NDCs in a transparent and meaningful way.

4 THE ROLE OF CARBON CREDIT QUALITY ASSESSMENT AND VERIFICATION MODEL IN CARBON MARKET

Using carbon credits as a transitional solution towards a carbon neutral society while maintaining global economic growth has credible potential. However, the role of quality carbon credits has yet to be conclusively demonstrated within the current market ecosystem. Hence, as depicted in Figure 3, the carbon credit quality assessment and verification model is positioned between the offset developer or project and the carbon market as a grading and safeguard mechanism to ensure the quality and transparency of issued credits for sale.

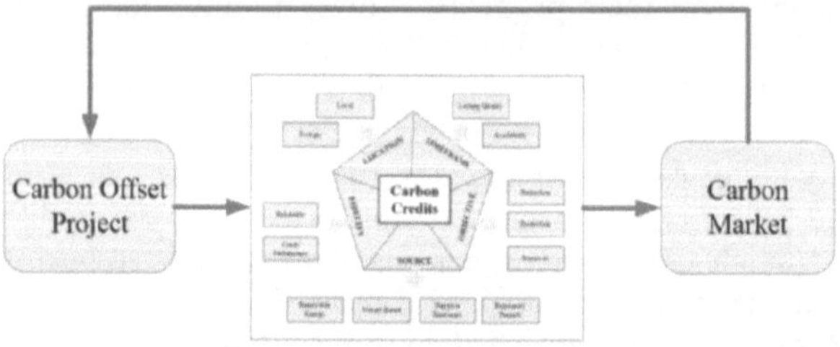

Figure 3. Carbon credit quality assessment and verification model's role in carbon market.

Alternatively, the carbon credit quality assessment and verification model can be positioned as a product verification tool by the buyer to meet in-house ethical and ESG requirements.

5 CONCLUSION AND FUTURE WORK

In recent years, carbon credits have gained much attention as a transitional solution to carbon neutrality. Governments and policymakers are increasingly seeing the potential of carbon credits in supporting economic growth and meeting climate targets. However, there is substantial skepticism about the quality of carbon credits and their impact on reducing carbon emissions.

Therefore, this paper explored the concept of carbon credits, the offset developer, carbon markets, buyers, and the mechanisms of the current ecosystem with its many benefits and concerns. We also proposed a carbon credit quality assessment and verification model to address concerns about quality, transparency, and accountability of issued carbon credit for sustainable ecosystem growth.

The carbon credit quality assessment and verification model examines issued credits based on five dimensions: source (renewable energy, nature-based, negative emissions technologies, regulatory permits), objective (reduction, protection, removal), location (local, foreign), timeframe (availability, lasting quality) and fidelity (reliability, credit performance). When classifying issued carbon credits, all five dimensions and their factors should be examined thoroughly before a determination of quality is made. We also demonstrated the role of the carbon quality assessment and verification model in the carbon credit market ecosystem as a grading and safeguard mechanism for carbon markets or a product verification tool by the buyer to meet in-house ethical and ESG requirements. The model may also be applied in other use case scenarios such as carbon offset project development, non-carbon GHG credit verification, carbon reduction accounting and auditing and carbon registry verification methodology development.

It is firmly believed that the carbon market ecosystem is a high-value transitional solution towards reaching economically sustainable carbon neutrality and net zero goals. This research work and model will aid in addressing above-mentioned concerns transparently and efficiently.

A prototype and algorithm is currently being tested based on the model, and its results will be used to enhance further, refine and validate the carbon credit quality assessment and verification model.

REFERENCES

[1] Dong Z. Y., Yang J., Yu L., Daiyan R., and Amal R., (November 2021). A green hydrogen credit framework for international green hydrogen trading towards a carbon neutral future, *International Journal of Hydrogen Energy*, vol. 47, pp. 728–734.

[2] Hermwille L., (January 2016). Climate change as a transformation challenge. A new climate policy paradigm?, *GAIA – Ecological Perspectives for Science and Society*, vol. 25, pp. 19–22.

[3] Kinley R., (July 2016). Climate change after Paris: from turning point to transformation, *Climate Policy*, vol. 17, pp. 9–15.

[4] Zhang W., Li J., Li G., and Guo S., (February 2020). Emission reduction effect and carbon market efficiency of carbon emissions trading policy in China, *Energy*, vol. 196, 117117.

[5] Pizer W. A., and Chang X., China's new national carbon market, unpublished.

[6] UN Data, GDP by type of expenditure at current prices – US dollars, 2020. [Online]. Available: http://data.un.org/Data.aspx?d=SNAAMA&f=grID%3a101%3bcurrID%3aUSD%3bpcFlag%3a0

[7] Ghosh B., Papathanasiao S., Dar V., and Gravas K., (August 2022). Bubble in carbon credits during COVID-19: financial instability or positive impact ("minsky" or "social")?, *Journal of Risk and Financial Management*, vol. 15.

[8] Chen L., Msigwa G., Yang M., Osman A. I., Fawzy S., Rooney D. W., et al., (August 2022). Strategies to achieve a carbon neutral society: a review, *Environmental Chemistray Letters*, vol. 20, pp. 2277–2310.

[9] Liu Y., and Lu Y., (March 2015). The economic impact of different carbon tax revenue recycling schemes in China: a model-based scenario analysis, *Applied Energy*, vol. 141, pp. 96–105.

[10] Aihua L., Miglietta P. P., and Toma P., (May 2022). Did carbon emission trading system reduce emissions in China? An integrated approach to support policy modeling and implementation, *Energy System*, vol. 13, pp. 437–459.

[11] Bi H., Xiao H., and Sun K., (October 2018). The impact of carbon market and carbon tax on green growth pathway in China: a dynamic CGE model approach, *Emerging Markets Finance and Trade*, vol. 55, pp. 1312–1325.

[12] Woo J., Asutosh A. T., Li J., Ryor W. D., Kibert C. J., and Shojaei A., (2020). Blockchain: a theoretical framework for better application of carbon credit acquisition to the building sector, in *Construction Research Congress 2020: Infrastructure Systems and Sustainability*, M. E. Asmar, P. Tang, and D. Grau, Eds. Arizona: ASCE, pp. 885–894.

[13] Mirzaee H., Samarghandi H., and Willoughby K., (July 2022). A three-player game theory model for carbon cap-and-trade mechanism with stochastic parameters, *Computers & Industrial Engineering*, vol. 169, pp. 108285.

[14] Zhang L., Liu D., Cai G., Lyu L., Koh L. H., and Wang T., (January 2023). An optimal dispatch model for virtual power plant that incorporates carbon trading and green certificate trading, *Electrical Power and Energy Systems*, vol. 144, pp. 108558.

[15] Kuwae T., Watanabe A., Yoshihara S., Suehiro F., and Sugimura T., (April 2022). Implementation of blue carbon offset crediting for seagrass meadows, macroalgal beds, and macroalgae farming in Japan, *Marine Policy*, vol. 138, pp. 104996.

[16] Mendelson R. O., Litan R. E, and Fleming J., A framework to ensure that voluntary carbon markets will truly help combat climate change, *The Brookings Institution*, Report, 2021. [Online] Available: https://www.brookings.edu/research/a-framework-to-ensure-that-voluntary-carbon-markets-will-truly-help-combat-climate-change

[17] Schneider L., Healy S., Fallasch F., Leon F. D., Rambharos M., Schallert B. et al., What makes a high-quality carbon credit?, (2020) [Online] Available: https://files.worldwildlife.org/wwfcmsprod/files/Publication/file/54su0gjupo_What_Makes_a_High_quality_Carbon_Credit.pdf

[18] In S. Y., and Schumacher K., (April 2021). Carbonwashing: a new type of carbon data-related ESG greenwashing, *SSRN Electronic Journal.*

[19] Gurcam S., (March 2022). The neoliberal initiative of the aviation industry to fight the climate crisis: greenwashing, *International Journal of Environment and Geoinformatics*, vol. 9, pp. 178–186.

[20] Moomaw W. R, Masino S. A., and Faison E. K., (June 2019). Intact forests in the United States: proferestation mitigates climate change and serves the greatest good, *Frontiers in Forest and Global Change*, vol. 2.

[21] Fawzy S., Osman A. I., Doran J., and Rooney D. W., (July 2020). Strategies for mitigation of climate change: a review, *Environmental Chemistry Letters*, vol. 18, pp. 2069–2094.

[22] Alcalde J., Smith P., Haszeldine R. S., and Bond C. E., (September 2018). The potential for implementation of negative emission technologies in Scotland, *International Journal of Greenhouse Gas Control*, vol 76, pp. 85–91.

[23] Bourzac K., (October 2017). We have the technology, *Nature*, vol. 550, pp. S66–S69.

[24] Marshall C., In Switzerland, *A Giant New Machine is Sucking Carbon Directly From The Air*, Science.org. https://www.science.org/content/article/switzerland-giant-new-machine-sucking-carbon-directly-air (accessed Oct. 17, 2022).

[25] Kreibich N., and Hermwille L., (July 2021). Caught in between: credibility and feasibility of the voluntary carbon market post-2020, *Climate Policy*, vol. 21, pp. 939–957.

[26] Blum M., (December 2019). The legitimation of contested carbon markets after Paris – empirical insights from market stakeholders, *Journal of Environmental Policy & Planning*, vol. 22, pp. 226–238.

[27] Northern G., (2011). Greenwashing the organic label: abusive green marketing in an increasingly eco-friendly marketplace, *Journal of Food Law & Policy*, vol. 7, pp. 101–134.

[28] Elyse C., Kerr A., Morton S., Seal A., Voehler K., Yan L. et al., (March 2018). Forest carbon credits: a guidebook to selling your credits on the carbon market, *The Executive Office of Energy and Environmental Affairs, Commonwealth of Massachusetts.*

[29] Kreibich N., and Obergassel W., (2016). *Carbon Markets After Paris- How To Account For The Transfer Of Mitigation Results?*. Wuppertal, Germany: Wuppertal Institute for Climate, Environment and Energy.

[30] Tonev I., and Nikolaev N., (August 2020). Review on Blockchain applications for electric power systems, *2020 21st International Sysmposium on Electrical Apparatus & Technologies (SIELA)*, pp. 1–4.

[31] Ali H., Phoumin H., Suryadi B., Farooque A. A., and Yaqub R., (2022). Assessing ASEAN's liberalized electricity markets: the case of Singapore and the Philippines, *Sustainability*, vol. 14, pp. 11307.

Innovation and Technological Advances for Sustainability – Al-Naemi et al. (Eds)

Characterization of the coal mine tailings mortar of Bechar power station

Fezzioui Naima
Department of Civil and Hydraulic Engineering, Faculty of Technology, Tahri Mohammed University Structural Mechanics Laboratory "LMS", Bechar, Algeria

Adel Benidir
National the Centre for Integrated Studies and Research Building (CNERIB), Souidania Algiers, Algeria

Yassine Miloudi
Department of Civil and Hydraulic Engineering, Faculty of Technology, Tahri Mohammed University Structural Mechanics Laboratory "LMS", Bechar, Algeria

Abderrahmane Makkaoui & Rekioui TayEeb
Department of Civil and Hydraulic Engineering, Faculty of Technology, Tahri Mohammed University Laboratory of Reliability of Materials and Structures in south "FIMAS", Bechar, Algeria

ABSTRACT: Mining activity in the town of Bechar in southwest of Algeria was definitively completed in 1975 by generating artificial mountains known as coal slag heaps. These heaps cumulate a volume of 3767330m3, and occupy an area of 167.71 ha. The development of urban fabrics has now resulted in some slag heaps being located very close to built-up areas, or squarely in the middle. The reuse of these slags heaps as materials in the field of civil engineering has been proposed as a solution to take charge of these slag heaps. The objective of this work is to carry out an experimental hygrothermal characterization of a mortar based on slag heaps by comparing it to an ordinary mortar. The results obtained demonstrated the good insulation of this material, as well as its good hygroscopic capacity.

Keywords: Coal Mine Tailings, environment, Mortar, Thermal conductivity, Adsorption

1 INTRODUCTION

The region of Bechar, in southwestern Algeria, experienced very significant economic and industrial development after the discovery of coal and its exploitation in 1918 [1]. This mining production reached its peak between 1942 and 1960. In 1975, the exploitation works were stopped and the mine was closed definitively [2], giving rise to black hills called slag heaps cumulating a volume of 3767330m3, and occupying an area of 167.71 ha. Slag heaps were formed by three different mechanisms: 1- By manual sorting at the level of the mines by separating the Coal from the schist. 2- By flotation at the level of the coal washery, generating heaps of schlamms and schist and so giving rise to washing slag heap. 3- By combustion in the boilers of power stations where the use of coal as fuel has generated refractory residues called Coal ash [3].

Indeed, these mountains of coals pollute the atmosphere and become a real alarming environmental condition. Hills of coal waste, reaching up to thirty meters in height and hundreds of meters in width, have not finished carrying billions of toxic micro particles and sulfur fumes into the air [4]. All of these components can cause not only an increase in air

DOI: 10.1201/9781003496724-54

pollution and groundwater near the slag heaps, but also a deterioration in health (the rate of people who have developed respiratory diseases in these localities would be in perpetual increase) [5].

The development of urban fabrics has meant that some slag heaps are currently located very close to built-up areas, or squarely in the middle as shown in Figures 1 and 2. The reuse of these slags heaps as materials in the field of civil engineering has been proposed as a solution to take charge of these slag heaps [6]. The objective of this work to perform a thermohydric characterization of a slag heap mortar. We chose the mortar; and our choice is practical because this material is the basis of all masonry work. In addition, its presence as a first layer in contact with the interior environment requires that it must acquire certain hygrothermal properties in order to regulate interior humidity and thus contributes to thermal comfort.

Figure 1. Slag heaps in Bechar near the power station (taken from the google earth satellite image).

Figure 2. Slag heaps in the town of Kenadsa: town located at 18 km from the town of Bechar and the city where coal was first discovered.

2 MATERIALS

2.1 *Study site*

Slag heaps of the power station of Bechar have been selected for this study. They were formed from 1949 to 1960 and currently occupy 350,567.11 M^3. A sampling campaign has been carried out in the immediate vicinity of the thermal power station. These heaps are the result of the combustion of coal used for electricity production. They are composed of Coal ash clinker of very variable grain size and ashes.

2.2 *Chemical analysis*

The results of the chemical analysis are given in the Table 1.

Table 1. Chemical analysis.

	slag heaps on the north side	slag heaps on the south side
SiO2	77.23	80.59
CaO	Null	Null
MgO	Null	Null
Al_2O_3	3.82	10.33
Fe_2O_3	1.84	7.34
SO_4	Nul	Null
C	2.29	Null
Carbonates	Null	Null

We note the absence of chemical elements harmful to the concrete (Sulphates), nevertheless we note the dominance of silica. Magnesia and lime are expressed in zero quantity.

2.3 *Mineralogical properties*

Quantitative minerals evaluations of materials was analyzed by the XRD (X-ray diffractometer). The analysis of phase composition was performed by using the HighScore Plus program which detects the peaks and makes a correspondence with database tables already integrated into the software.

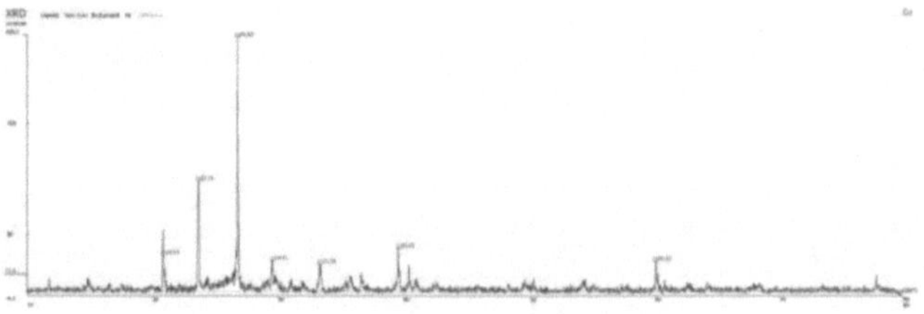

Figure 3. DRX of slag heaps of the Bechar Djedid power station.

The analysis of the heap examined is shown in Table 2. It shows that our material is composed of 87% SI O2 quartz; and 17% calcium carbonate.

Table 2. Table caption.

Component	SiO2	CaCO3
% Mass	87%	17%

The Fourier transform infrared spectra (FTIR) This test aims to characterize the infrared absorption spectrum (wavelength between 2.5 and 25 μm) of a sample. The device is a Fourier Transform Infrared Spectrophotometer (FTIR). This test is used to identify crystalline, non-crystalline, and even organic materials.

Analysis by infrared spectroscopy of the sample of slag heap showed the existence of a C-H bond (C diagonal) of an elongation nature, the absorption bung with a wavenumber of 3300 – 3310 cm-1 with a low and high intensity.

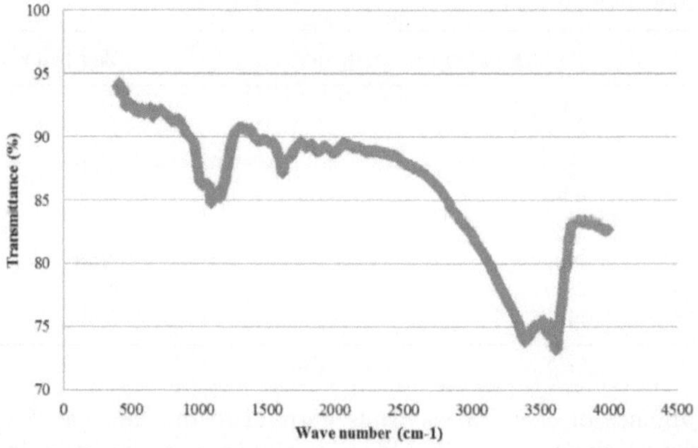

Figure 4. FTIR spectra of slag heaps.

2.4 *Granulometric analysis*

For the experimental study, the first step consisted of creating the granulometry graph of studied slag heaps according to NFP 18-560 [8]. The results are given in the Figure 5.

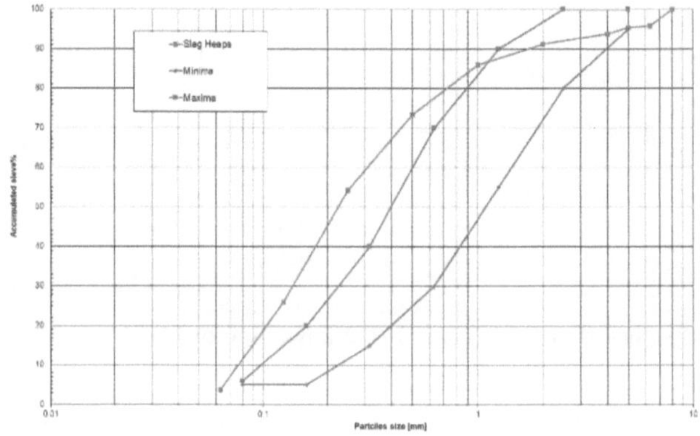

Figure 5. Granulometric curves.

The fineness modulus is equal to 1/100th of the sum of the refusals, expressed in percentages on the different sieves. It was determined and its value is: =

From the granulometric curves, the coefficients of uniformity and curvature were determined:

$Cu = 0{,}35/0{,}09 = 3{,}88 > 2$
$Cc = = (0{,}18)^2/\ 0{,}35x0{,}09 = 1{,}02$

The sand from slag heaps has a spread out grain size, denser with a well-distributed continuity. It is natural and light aggregates.

Table 3. Densities and fineness modulus.

Fineness modulus[-]	1.5
Apparent density [kg/m3]	789
Absolute density[kg/m3]	1730

3 MIX PROPORTION

We will carry out an experimental characterization of the slag heap by comparing it with an ordinary sand mortar.

The formula adopted for making the mortar is the following:

The mortar is composed by weight: in one part of cement, in two parts of slag sand for the case of the slag mortar and clean sand of good grain size for the case of the reference mortar and between 0.5 and 0.6 of water (tap water).

The cement used was an ordinary cement with a specific weight of 3.13 gr/cm3 and a compressive strength at 28 days of 55 MPa.

The mortar samples were prepared, the dimensions of the samples were made according to the standards required for each property to be determined:

For sorption isotherms: cubes of dimension $2.5X2.5X2.5cm^3$ EN ISO 1271 [9].

For capillary absorption EN ISO 15148 [10] and for measurements of thermal conductivity: specimens of $16X8X4cm^3$.

Figure 6. On the left: cubes for adsorption isotherm determination on the right: measurement of the thermophysical properties of the two mortars with the CTmeter device.

3.1 *Bulk density*

The Table 4 illustrates the results of the density of the two mortars studied. The apparent density (at 23°C and 50%HR) of sand mortar is higher than slag mortar which exhibited a visibly porous texture (Table 4). This can be justified by this sand of coal mine tailings which gives a relatively porous mortar compared to that made with sand.

Table 4. Bulk density of the two mortars NF EN 1602 [11].

	Bulk density
Coal mine tailings mortar	1596.52 ± 5 kg/m^3
Sand mortar	2198.85 ±5 kg/m^3

3.2 *Thermal properties*

The thermal properties of the two mortars are given in Table 5.

This Table 5 shows the contribution of slag heaps in increasing the insulation of the mortar. Indeed, the coal mine tailings resulted in a 73% decrease in thermal conductivity of

Table 5. Thermal properties of the two mortars.

	λ (W/m.K)	Cp(J/kg.K)	$a \times$ 10-7 [m^2.s-1]	E J/m2 . K 1 . s 1/2]
Coal mine tailings mortar	0.486	1415.2	2.16	1048.96
Sand mortar	1.854	2009	4.19	2861.82

mortar. The ordinary sand mortar gave a higher specific heat than that of the slag heap, and presents a significant thermal inertia justified by the values of thermal diffusivity and thermal effusivity which are much higher compared to those of the slag mortar.

3.3 *Capillary absorption*

The Figure 7 shows the difference in mass compared to the surface of the sample as a function of the square root of time of the two mortars according to standard EN ISO 15148 [10]. The graph obtained is of type B, therefore the capillary absorption coefficient is calculated as follows:

$$A_w = \frac{m_{ft}}{\sqrt{86400}}$$

m_{ft}: the value of mt 24 hours after the start of the test

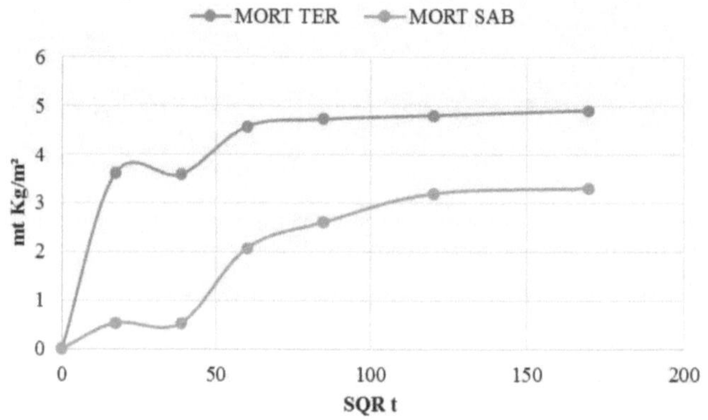

Figure 7. Capillary absorption of the two mortars as a function of the square root of time.

A significant increase in absorption for the slag mortar was noticed for the first 5 minutes of the experiment, which is estimated to be 85% of the increase caused by the sand mortar. The gap was just as wide for the 25 minutes after the start of the experiment. After an hour, the gap starts to get smaller between the two mortars.

3.4 *Adsorption isotherm*

The adsorption isotherms are obtained according to standard NF EN ISO 12571 [9]. Before starting the adsorption test, the material specimens were dried during 24 h in an oven at 105°C until constant weight. 6 six jars are used in which 3 samples are placed. Each jar contains a saturated saline solution prepared to give a constant relative humidity. The humidities tested are: 19, 34, 43, 77, 80 and 90%RH.

The slag mortar reaches equilibrium more quickly than that of sand, it reached a water content of around 6.7% while this value was only around 2.6 for ordinary mortar. The absorption behavior of relative humidity varies from humidity to humidity. Indeed, for high humidity, the mortar absorbed 57% of the total saturation water content.

Twelve models were chosen [12] to fit the adsorption isotherms of the two mortars. It was found that the Oswin model gives a better result for the slag mortar and the Caurie model is the one that has adjusted the experimental results of the sand mortar compared to the other models. However, Smith's model gave a relatively better result than the others by adjusting two points for the slag mortar, while the others were very far from adjusting our results.

According to Oswin's model, the equilibrium adsorption water content for the slag heap mortar is as follows:

$$W_{ads} = 1.257\left(\frac{HR}{1 - HR}\right)^{0.356}$$

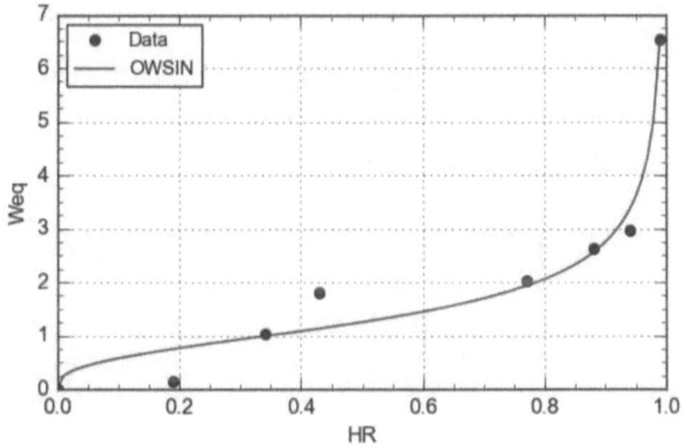

Figure 8. The model fit curve in adsorption of the slag mortar.

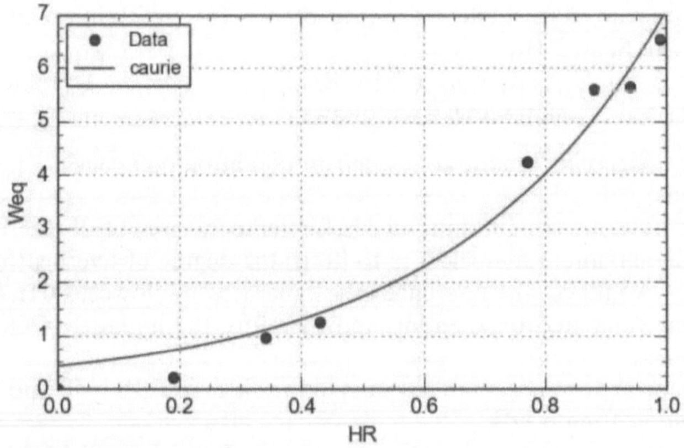

Figure 9. The model fit curve in adsorption of ordinary mortar.

According to the Caurie model, the equilibrium adsorption water content for sand mortar is as follows:

$$W_{ads} = \exp(-0.86 + 2.81.HR)$$

4 SIMULATION

In order to verify the impact of the slag mortar on the energy behavior of buildings in the different Algerian climatic zones, a numerical simulation with TRNSYS-COMIS was carried out, including its behavior with that of certain conventional mortars characterized in document DTR C3-2 [13]. Five cities with different climates were selected for the study:

El Bayadh, Cold semi-arid climates with short, hot, arid summer and very cold, dry, windy and partly cloudy winter. Adrar, subtropical desert climate, with sweltering sunny summer, and cool, dry, and mostly clear winter. Bordj Badji Mokhtar, hot desert climate, with long, extremely hot summers and short but hot winters. Oran, Mediterranean climate, with short, warmand muggy summer and long, cold and windy winter. Constantine witch is influenced by Interior Mediterranean climate, with short, hot, dry, and mostly clear and long, cold, wet winter.

Figure 10 shows the simulated building. It houses 3 people who use two 100Watt PCs from 2pm to 11pm and a 150Watt TV from 9am to 2pm and 5pm to 11pm.

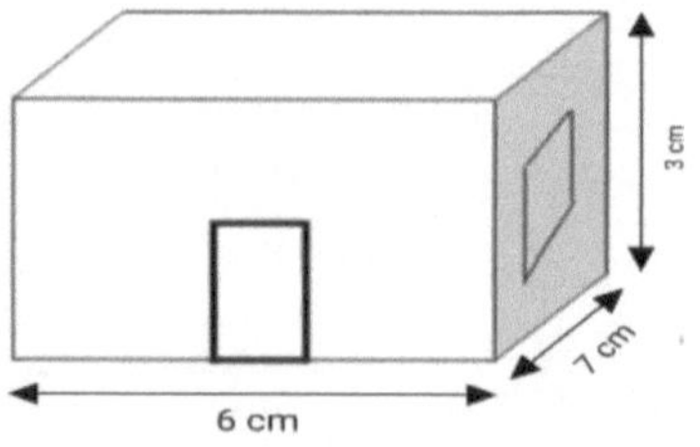

Figure 10. Simulated building.

The building envelope is composed as follows: external wall: Brick 10cm, air space 3cm, Brick 10cm.

Roof: Cement mortar 2cm, hollow flooring blocks 16 cm; Reinforced concrete 4 cm, Waterproofing 1 cm; Sand 2 cm.

Floor: Floor tile3 cm, Cement mortar 2cm, Reinforced concrete 4cm and stones 40cm.

The results obtained are summarized as follows:

- For the different Algerian climatic zones, the air conditioning needs are much higher than the heating needs
- Indeed, we can have air conditioning needs for temperatures of 28°C, that is to say we must refer to the maximum temperature to judge the degree of overheating.
- For all the cases examined, the reference case generated the lowest QHEATs.
- The sand mortar generated overheating and therefore the highest QCOOLs for the different cities studied
- The slag heap mortar gave acceptable results in the cities of El-Bayad and Constantine for both cooling and heating needs,
- For cities with extreme hot and dry climates such as Bourdj Badji Mokhtar, and Adrar, slag mortar significantly contributed to air conditioning needs, where it was ranked first in terms of QCOOL

Table 6. Descriptions of cases examined.

Cases	Mortar used	λ (W/m.K)	Cp (J/kg.K)	ρ (kg/m³)
Case 2	Sand mortar	1.854	2009	2198,85
Case 3	Slag mortar	0.487	1415,2	1596,52
Case 4	Battard mortar [13]	1.15	1080	1900
Case 5	Cement Mortice [13]	1,4	1080	2200
Case 6	Lime Mortar [13]	0,87	1080	1800

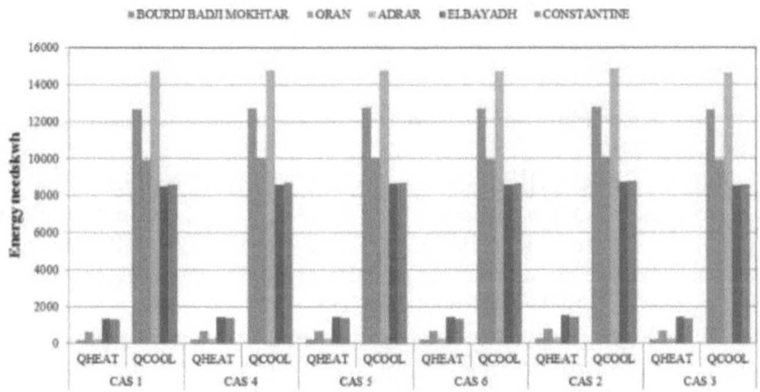

Figure 11. Energy heating and cooling requirements for the different cases examined in Table 6 for different climate.

- For the city of Oran, the slag heap mortar gave acceptable results for QCOOLs
- The lime mortar identified by the DTR also gave important results for the majority of cities.
- The mortar effect varies from city to city, we recorded percentages of the order of 7% to 44% difference with the reference case for QHEATs.
- The impact of mortars was more significant on heating requirements than on cooling requirements. We calculated deviations of the order of 2 to 7% from the reference case.

5 CONCLUSION

The aim of this work is to valorise mining waste to see the possibility of integrating them into construction materials by characterizing their thermo hydric properties. The slag heaps of the Bechar power station were chosen to be integrated as sand in the mortar. The DRX chemical tests showed the presence of quartz in these heaps. Infrared spectroscopic analysis of the heap sample showed the existence of a C-H (diagonal C) bond of elongation nature. The thermophysical characterization of the slag mortars showed good thermal performance, with a thermal conductivity of 0.48W / mK against 1.87 W / mK for the sand mortar. However, for specific heat, the cement mortar exhibited a higher specific heat. The calculation of the thermal effusivity and diffusivity of the two mortars shows the rather large inertia of the cement mortar against that of the slag mortar.

For the hydric behavior, it has been noticed that the slag heap mortar exhibits good hygroscopic behavior, in terms of the speed at which it absorbs moisture. Adjustment of the absorption isotherms showed that Oswin's model fitted the experimental results of the slag heap better while the Caurie model fitted the results of the sand mortar better.

The results of the simulations with the TRNSYS software showed the contribution of the slag heap mortar in the reduction of heating needs. Their impact on air conditioning energy performance was less significant, in fact, percentages of reduction in these needs were calculated in the order of 2 to 7% against 7 to 44% for QHEATs.

ACKNOWLEDGMENT

The authors gratefully acknowledge National the Centre for Integrated Studies and Research Building (CNERIB).

REFERENCES

[1] Projet de traitement des charbons du sud oranais. Société française industrielle et commerciale des pétroles. Paris, Février 1950

[2] OFFICE NATIONAL DE RECHERCHE GEOLOGIQUE ET MINIERE – Etude des conditions de prise en charge des terrils des villes de Béchar et de Kenadsa – Rapport final, 3 volumes, inédite 2003.

[3] Makkaoui A., Rikioui T., Mekki K., Abdelaziz Y., Hamouine A. (2004). Pyramides noires à Kenadsa et Bidon II, sous une ébauche de valorisation. Annales de l'université de Béchar, N° 5, année 2009, PP: 28–36 *pathways of disease, Circulation.* Circulation, **109**: p. 71–77.

[4] Mostadi A. and Biara R. W. (2018). Une Friche Industrielle A L'entrée De La Ville De Kénadsa– Une Perception Négative D'un Territoire Inhospitalier. *Revue d'Economie & de Gestion* Vol 2, N 2, pp. 37–51

[5] Gaulier C., Billon G., Lesven L., Falantin C.a. (2020). Leaching of two northern France slag heaps: Influence on the surrounding aquatic environment. *Environmental Pollution* 257 113601

[6] Bella N., Dahane M., Mebarki T., Benmalek A., Benallal M. (2022). Valorisation of Coal Mine Tailings in Roads Construction. *Algerian Journal of Environmental Science and Technology* March edition. Vol. 8. No 1

[7] Ait Oukaci L., Mekki K. Valorisation Des Terrils De Charbon Dans la Fabrication Des Matériaux En Génie Civil, Projet de fin d'étude, Bechar University, 2006-2007.

[8] NFP 18-560 Granulats. Analyse granulométrique par tamisage. Octobre 1978

[9] NF EN ISO 12571 (2000), Hygrothermal performance of building materials and products – determination of hygroscopic sorption properties. 2000. p. 50–758.

[10] EN ISO 15148, Performance hygrothermique des matériaux et produits pour le bâtiment – – Détermination du coefficient d'absorption d'eau par immersion partielle, Octobre 2003

[11] NF EN 1602 Produits isolants thermiques destinés aux applications du bâtiment – Détermination de la masse volumique apparente. Septembre 2013

[12] Miloudi Y., Fezzioui N., Benidir A., Labbaci B., Ait Oumeziane Y., Bennaceur S. (October 2019). *Modelling of Temperature Effects On Sorption Behaviour Of Compressed And Cement Stabilized Earth Blocks. Magazine of Concrete Research* 73(3):1–36

[13] DTR C3-2, (1998). Rapport Technique, Document Technique Règlementaire, Règlement Thermique des Bâtiments d'Habitation – Règles de Calcul des Déperditions Calorifiques", DTR C32'.

Innovation and Technological Advances for Sustainability – Al-Naemi et al. (Eds)

Dubai World Expo 2020 impact on sustainable development: Youth perspective on the contemporary human milieu in the UAE

Sahera Bleibleh, Boshra Hassan, Aya Mousa & Rauda Alneyadi
Department of Architectural Engineering, United Arab Emirates University, Al Ain, United Arab Emirates

ABSTRACT: The witnessed Dubai World Expo 2020 event in the last six months, represented in nonstop activities, events, forums, and various cultural experiences, requires multiple in-depth studies to grasp its unfolding lessons. This research explores the impact of Dubai World Expo 2020 on youth perception of world spatial sustainable development and challenges. The primary focus of this project is to identify the result of such mega-development on the future of the city based on different dimensions such as career path, network, attraction, identity, and innovation from the younger generation's perspective. This mission is a key step in driving the intertwined relationship between space and people in future development. Thinking of youth as future leaders, it's important to understand their perception and how this mega event contributed towards shaping their perspective. The outcome of this preliminary study contributes to understanding the human milieu of the UAE context as influenced by the recent mega-development.

Keywords: Dubai World Expo 2020, Sustainable development, youth, human milieu, UAE

1 INTRODUCTION

World Expos, also known as International Registered Exhibitions (IRE), are a global gathering of countries dedicated to finding solutions to today's pressing issues by presenting a journey within a shared theme through engaging and immersive activities. Tens of millions of tourists attend World Expos, which allow countries to build amazing pavilions and have an impact on the host city for years to come. In 1851, London hosted the first World Expo, known as the Great Exhibition [1]. The goals of mega-event planners have increasingly included sustainable development, one of which is youth involvement. Such a goal has not only been maintained internationally but has also become a key component of the newly established Sustainable Development Goals (SDGs). Since the preparations for Dubai World Expo 2020 began in 2015, Young Connect (YC) has been formed, and numerous initiatives have been developed to allow the youth to engage in the event's development. These efforts included programs that encouraged adolescents to broaden their goals and perspectives [15].

This research spots light on the impact of mega events that have spread over the last six months; however, they may have forever impact on our communities and the country's future development. This provides significance in engaging youth in sustainable practices within the built environment. Such unusual events need to be analyzed in different aspects to maximize the benefits and learn from any rising challenges.

Building on the findings of the correlation between Dubai World Expo 2020, youth, and sustainable development, this pilot study attempts to learn from the event spatial lessons and explore how youth can contribute to shaping future development to achieve sustainability

DOI: 10.1201/9781003496724-55

goals. This mission is a key step in meeting sustainable development goals through the eyes of youth.

There has been no similar study exploring mega-event effects on the community, specifically for youth. Also, the expected outcomes lie in the identification of lessons that can be applied for future development.

2 PLACE, DESIGN, AND SPACE OF DUBAI WORLD EXPO 2020

Urbanization Expo 2020 Dubai, created by the American company HOK, is arranged in the shape of a flower and has as its theme "Connecting Minds, Creating the Future." The three thematic districts that make up the city are called Opportunity, Mobility, and Sustainability. Circulation arteries with symbolic names like Horizon Avenue, Sky Avenue, Sunrise Avenue, and Hope Avenue, among others, define and crisscross the areas. Al Wasl Plaza Figure 1, the centerpiece, is capped by a transparent dome measuring 130 meters in circumference and 67 meters high. Al Wasl means "the link" in Arabic. A dynamic 360-degree projection screen is formed by its steel trellis covered in canvas, which takes design cues from the Expo 2020 Dubai emblem. The domed plaza's primary function is to host the opening and closing ceremonies in addition to the nightly, massive audio-visual shows. The dome is transformed into a vibrant lighthouse that unites all pavilions and directs tourists thanks to bright projections that are visible from both within and outside the plaza [2].

Figure 1. Al Wasl Plaza (by the author, 2022).

The leadership of Expo 2020 tasked the design firm SWA group with crafting a remarkable environment in order to create a setting that welcomes the globe and can accommodate distinctive pavilions from 192 nations [2]. They pushed and assisted SWA in developing a landscape that:

- Captures the essence of Dubai
- Reacts to the harsh weather
- Reveals the natural landscape's magnificent and delicate beauty and provides visitors with Sensory pleasure and rest
- Provides enticing gathering locations at various dimensions and encourages human comfort
- Promotes accurate wayfinding
- Complements and unifies the many pavilions' architectural forms

For the numerous wonders of Expo 2020, the countryside provides an unforgettable experience setting [3].

2.1 *Specifications of the open space*

Rethinking The SWA-designed open space network at Expo 2020 delivers sensory enjoyment, unique placemaking, experiential connectedness, and mitigation of adverse climate extremes. Visitors may experience Dubai's special cultural and environmental traits at this time when it is welcoming the globe through the Expo's network of open spaces. The terrain itself serves as a source of inspiration for the interpretive features, which include pavement patterns and forms that are meant to represent wadi, sand dunes, and mountains, where water is utilized sparingly to support planted materials most of the time but occasionally lavishly for spectacular effects and to demonstrate the host country's welcome hospitality. Open areas moderate the harsh weather while simultaneously providing visitors to the Dubai World Expo 2020 site with comfort and relief in subtle and obvious, natural and manmade ways. In addition to tree canopies, creative buildings also add shade thematically designed to reflect elements of local identity, as shown in Figure 2.

Figure 2. Shaded pathways along Expo 2020 (by the author, 2022).

Additionally, cooling may be accomplished by strategically incorporating water and plants, which lower the surrounding temperature while simultaneously providing the aesthetic and acoustic delights of reflection, shadow, gurgle, and whisper. More than half of the plants utilized on the property are indigenous and hardy species. Paving materials also go above the guidelines of the Solar Reflectance Index (SRI) to stop the development of heat island effects.

2.2 *The sequence*

The entrance from the metro station is made up of a striking collection of landscaped areas, starting with a broad Welcome Plaza and narrowing to a central, tree-lined promenade that serves as both the location of the leadership pavilion and the starting point for a formal procession to Expo 2020 incredible domed Al Wasl Plaza. Visitors may reach the three zones without having to go back to the center through linking pathways.

Visitors may also reach significant landscaped areas between the pavilion communities from The Loop, including the Ghayath Trail to the north, the Oasis (home to the iconic fountain "Surreal"), two sizable parks, Jubilee Park to the south, and Al Forsan Park. Each of these parks allows for additional programming and has a distinct personality, giving visitors diversity in their relaxation and an understanding of the complexity of the Expo's themes and their roots in Middle Eastern culture. All of these public areas will be preserved as part of Dubai World Expo's legacy and used as the foundation for District 2020, a new human-centered city inside a city.

3 EXPO OPEN SPACE – DESIGN ELEMENTS

3.1 *Jubilee Park*

The Jubilee Park takes its design cues from wadis, which are periodic rivers that develop in the desert during periods of strong storms and support a diverse range of plants that frequently persist even throughout the dry months. Wadis give the desert life by forming lush Oasis along their course that encourage plant growth. The color, patterns, and sinuous curves of the paving patterns at Jubilee Park symbolize the movement of water in the environment and serve as a metaphor for the wadi. The Wadi route here meanders around the area to create several landscaped areas; it first follows the perimeter of the park's main performance area before continuing to the main arrival plaza next to the loop boulevard [3].

Some of the park's key components are located here, including dining and snack options, souk market kiosks, and the observation tower. The main arrival plaza is connected to the outdoor event stage on the other side of the park by a swath of date palms that hugs the wadi path's margins. The landscaping chosen for this park closely resembles the desert flora of the UAE. A range of soft grasses and plants for seasonal planting are combined with native, drought-tolerant species [3].

3.2 *Al Forsan park*

The Al Forsan Park draws inspiration from culture, while the Jubilee Park is influenced by Dubai's peculiar natural beauty. Al Forsan Park is a green gathering place for various cultural and social events. It is nestled between the Opportunity and Sustainability Thematic Districts on both sides and the UAE Pavilion on the south. The mosaic, a type of art that is prevalent across the Islamic world, serves as design inspiration. In order to create a sequence of places that bring together various programs, topics, and age groups, Al Forsan Park draws inspiration from this art form both physically and conceptually [3].

Individual rooms come together in this area to form a unified design that is defined by repeating abstract geometric patterns. The park may be interpreted as a metaphor for the UAE, a country made up of seven emirates, each with its own unique personality and a long history as a meeting place for many peoples and cultures. Each "mosaic space" inside the park is likewise created to accommodate the demands of all age groups, from somatic places like restaurants, snack bars, and restrooms for visitors to unwind to more interactive exhibits that highlight and engage the Expo experience. On the northwest and southeast sides of the park, there are snack-in areas, and a polygonal outdoor performance area will accommodate outdoor performances of plays, concerts, and talks [3].

In order to prepare for a future mosque building that would serve the residents of the future city of Expo 2020, the vegetation on the park's north side has been purposefully kept simple. A spacious open grass offers a location for outdoor recreation and sporting events. The flora species chosen for this park offer a rich pallet of colors and an overall pleasurable sensory experience, with a focus on soft textures and aromas, much like the brilliant tiles of a beautiful Middle Eastern mosaic [2].

3.3 *Ghayath Trail*

Ghayath Trail, a crescent-shaped landscape that protects the UAE pavilion, is located just across from Al Forsan Park on the inside of the ring promenade. Ghayath Trail, which connects many large-scale sculptures and was designed as an entertaining "art walk," meanders through a dense canopy of native (Ghaf) trees while offering a quirky, adventurous view of the surroundings. The circuitous route continues through the ring promenade, linking to the Al Forsan Park arrival plaza [2].

3.4 *The concourse loop*

The concourse loop Figure 3, a pedestrian promenade with shade, serves as the main pedestrian route. It encircles the property, letting visitors go between the three districts and the site's three main open areas without having to go back to the center. Bird-shaped hanging forms make up the spectacular, ten-meter-high shade structure, which cools and creates eye-catching shadows. The Ghaf trees, the UAE's national tree, border the concourse loop [2,3].

Figure 3. The concourse loop (by the author, 2022).

4 ATTRACTING YOUTH TO VOLUNTARY ACTIVITIES OF EVENTS

In general, young people can develop a lifelong commitment to giving back and supporting their communities when you start recruiting volunteers for them at an early age. Imagine the difference they will be able to bring about as they become older and get more authority to follow through on that promise. Furthermore, youthful volunteers frequently contribute a new viewpoint, passion, and vigor to your organization. They are more likely to be technologically proficient and receptive to new ideas, which means they may help spur progress and offer up new opportunities. Are youthful volunteers such excellent recruits just because of their age? Not completely. It could also be influenced by the generation to which they belong [5].

The generation known as Gen Z, which comprises persons born between 1997 and 2012, is the youth of today. The problems that our globe faces are ones that Gen Z is keenly aware of. They also have a strong desire to change things since they understand that their destiny is in their own control. Eight out of ten members of Generation Z say they hope to change the world in some way throughout their lifetime. They actively volunteer, save money, and contribute it to organizations they are enthusiastic about more than any previous generation [5].

Dubai has made an effort to incorporate young Emiratis and UAE citizens when hosting a major event so they may get the most from being in the host city. For instance, the Volunteer Program has requested volunteers from all around the nation to act as the Expo's core staff and as UAE ambassadors. The concept was first introduced in October 2017 during a Three Year Countdown event in Dubai's Jumeirah Beach Residence district. It involved UAE nationals and locals in brainstorming sessions to help identify the primary function of volunteers during the Expo. Through a platform administered in association with the UAE Ministry of Community Development and Emirates Foundation, the Expo 2020 Volunteer Program received over 160,000 applications from UAE nationals and residents. The 30,000 chosen volunteers, who represent more than 180 countries, range in age from 18 to 79. Emirati nationals make up 61% of volunteers aged 18 to 24 (and 46% of all volunteers), while women make up more than half of the volunteer workforce [6].

For the purpose of educating volunteers on coronavirus safety guidelines, communication and management techniques, and traversing the sizable expo site, the Volunteer Program is collaborating with the Dubai Police and the telecoms provider Etisalat. Additionally, volunteers are studying the background of expos from across the globe, how to provide a welcoming and secure event, Emirati culture and cultural awareness, and fundamental sign language [6].

5 IMPORTANCE OF DIGITAL APPLICATIONS IN MEGA EVENTS

When we discuss digital transformation, we mean the digitalization of all business operations. In other words, it refers to the procedure of integrating new technology throughout an organization. This entails utilizing several pillars, such as mobility, Big Data, or the Internet. Digital transformation may be included in events in a variety of ways. Modern procedures have really been made possible by new sorts of technology. For instance, the e-development tickets in the 2000s were astounding. Digital will inevitably be included in events because of this. It can take many forms, such as the digitalization of event tickets, the development of an event website, or the distribution of tailored thank-you emails [9].

The world is quickly changing these days as a result of our society's ongoing progress. It is crucial to go toward digital, not just for the advantages it offers businesses but also for the influence it will have on customer behavior. We can easily draw attendance to an event by including the digital revolution since we are getting closer to the target audience (our industry's attendees). An event's digitization must be considered at all levels if it is to feel intimately linked to the attendees. The idea is to make digital transformation a strength by designing an atmosphere that is both familiar to attendees and distinctive enough to surprise them and elevate your event [9].

The world's demographics are different now than they were a few years ago. The way events are arranged and planned is unavoidably impacted by shifting demographics. We may estimate that the generations who actively utilize tech in society now range in age from 7 to 57. These generations can be divided into Gen X (age range: 45 to 56), Gen Y/Millenials (age range: 26 to 44), and Gen Z (age range: 6 to 25) [9].

6 USAGE OF ROBOTS IN SERVING AND ENTERTAINING

Robotics may be used in the entertainment sector to provide mechanical repetition and dependability. They are also relatively economical and designed to work in complete collaboration with humans. In addition to having a lightweight structure and being simple to operate for a range of activities, robots are designed to be controllable. Robots are quickly gaining popularity in the entertainment sector since they can perform a wide range of jobs with simple programming [7].

Robots built to appeal to us and interact with us on nearly equal footing in our own physical area have a huge psychological edge over abstract chatbots and other more direct and practical forms of AI communication. They are also natural advocates. Even if making robots is tough and expensive, the recent rise in Natural language processing (NLP), computer vision software, machine learning, and other robotics-related technologies guarantees that we'll be seeing more unusual inventions in the future in the halls of tech expos [8].

7 SHADING AND GREENERY IN ENHANCING WALKABILITY

The ability of trees to provide shade and reduce ambient temperatures is one of its most fundamental advantages. The shade and cooling effects that trees provide are, therefore, a crucial component of any pleasant walking environment, especially in hot climates. Tree-lined paths are crucial for promoting walking in context were doing so is not thought to be desired due to the sun and heat. Additionally, it has been demonstrated that trees' shade shields pavements from deterioration and cracking, enhancing walking comfort and extending the life of the footway. Children say that parks, trees, and natural spaces are better urban locations than more formal play facilities, and they are more inclined to stroll there. Therefore, child-friendly cities also need to focus on green infrastructure. Children are more than twice as likely to walk to a park that is within half a mile of their house than they are to depend on their parents to drive them there. The landscape features significantly affect people's attitudes, behaviors, and attitudes and health. The two basic types of landscape components are hard and soft landscapes. When referring to landscape components, the terms "hard landscape elements" and "soft landscape elements" are interchangeable (trees, palms, bamboo, shrubs, climbers, ground cover, and water plants). In order to give users of pedestrian walkways shelter, artificial shade is frequently added [10].

Shades are crucial because they shield walkers from the sun's rays, enhancing their strolling comfort. Walking is more pleasant in the shade, especially on hot, muggy days. For cities and towns to have walkable streets and areas, urban green infrastructure is a crucial component and tool. Urban greenery influences every aspect of how walkable a city is, from appeal, safety, and environmental quality to how comfortable a roadway is to walk on [12].

8 METHODOLOGY

The research aims to connect the short-term findings with the long-term planning strategic polices as follows:

- Short term: Identify the impact of the Dubai World Expo 2020 on youth awareness, the perspective of youth on the world, the UAE, challenges, and sustainable development.
- Long term: Identify the implementation of outcomes and the role of youth towards more sustainable communities.

8.1 *Research objectives*

- Explore the existing literature about previous mega-events.
- Explore the impact of Dubai World Expo 2020 on youth perception and role towards sustainable development.
- Identify the lesson learned for future spatial development.

This research is a qualitative study utilizing different investigation tools to identify youth perception. The tools are divided into four stages, and this study will represent the last stage. First, the researcher observation experience to explore Dubai World Expo 2020 elements and identify possible investigation areas. Second, the pilot study examines the methodology, research question, context, and fieldwork tools. A survey and semi-structured interview questions were developed to conduct the pilot study prior to developing the formal fieldwork. The interviews were conducted with high school and university students (ages 15–30) in order to determine their viewpoints and examine common emerging themes of interest. Each interview lasted around 10–15 minutes and included a total of 26 respondents. The questions focused on the goal of the visits, activities, experience before and after, outdoor experience, lessons, and thoughts, as well as their view of the event's future as a district. Third, the majority of UAEU students attended Dubai Expo 2020 during planned school and university holidays, as well as family visits, according to the findings of an interview conducted in June 2022. The interviews were analyzed using Nvivo software to discover common emerging themes. Fourth, a structured survey was developed through Google Forms based on the identified themes to explore youth perception further. The final fieldwork is an ongoing process and is expected to be completed by January 2023 [1].

9 FINDINGS

The primary results indicate the following point in a nutshell:

- All youth were impressed by the applied technology they saw and experienced in Expo 2020.
- The most enjoyed robots were the Talabat Robots and the Orange-Colored Robots.
- The pavilions that attracted the most students in terms of technology were Alif, Terra, and UAEU.
- The most digital applications used in Expo 2020 were Online tickets and Google Maps.
- The most attractive outdoor element was the Al Wasl Dome.
- 29.1% of the students spent 1–3 hours outdoors, while 65.5% of them spent more than 3 hours.
- Youth mostly used walking as their main way of moving around Expo 2020.
- The elements that mostly facilitated walking outdoors were shading and greenery.
- The most easily reached services were washing rooms and prayer rooms.
- Most youths were satisfied with their walking experience at Dubai World Expo 2020.

Figure 4 shows the word cloud of the collected responses through Nvivo software. Five main themes have been extracted based on young perceptions through interviews. Architecture, technology, urban replication, walkability, and attractiveness [15].

Figure 4. Nvivo analysis [15].

REFERENCES

[1] Gomez-Villarino, M., and Ruiz-Garcia, L. (2021). Adaptive Design Model for the Integration of Urban Agriculture in the Sustainable

[2] Bleibleh, S., Hassan, B., Mousa, A., and Rouda, A. (2022). The Preliminary Impact of Mega Events on Sustainable Development in the Eyes of Youth: The Case of Dubai World Expo 2020. Manuscript in preparation.

[3] Belogolovsky, V. (2021, Nov 4). *Expo 2020 Dubai Will be Remembered for its Architectural richness.* Retrieved from STIRworld: https://www.stirworld.com/think-columns-expo-2020-dubai-will-be-remembered-for-its-architectural-richness

[4] Archello. (2022). *Dubai Expo 2020 Public Realm.* Retrieved from Archello: https://archello.com/project/expo-2020-dubai

[5] Coates, C. (2020, Feb 27). *The Infrastructure of Expo 2020 Dubai: Creating a Complete City.* Retrieved from blooloop: https://blooloop.com/museum/in-depth/infrastructure-of-expo-2020-dubai/

[6] Izmailova, S. (2021, August 20). *Why Your Organizatión Needs Young Volunteers + How to Recruit and Motivate Them.* Retrieved from wildapricot: https://www.wildapricot.com/blog/young-volunteers#what-can-young-volunteers-do

[7] Ammagui, N. (2021, Sep 27). *Expo 2020 Dubai: By and for UAE Youth.* Retrieved from agsiw: https://agsiw.org/expo-2020-dubai-by-and-for-uae-youth/

[8] Anderson, M. (2022, Oct 27). *Robots in the Entertainment Industry: What are They?* Retrieved from iflexion: https://www.iflexion.com/blog/entertainment-robots

[9] Times Square Chronicles. (2022, Aug 17). *How Robots are Used in the Entertainment Industry.* Retrieved from Times Square Chronicles: https://t2conline.com/how-robots-are-used-in-the-entertainment-industry/

[10] Eventdrive. (2018, April 5). *The role of Digital Transformation in Events.* Retrieved from Eventdrive: https://www.eventdrive.com/en/ressources/blog/digital-transformation-events

[11] Natural walking cities. (2019, Sep). *Walkable Cities are Green Cities: How walkability is reinforced by natural urban infrastructure.* Retrieved from natural walking cities: https://naturalwalkingcities.com/walkable-cities-are-green-cities/

[12] Yourdubaiguide. (2021, September 25). *Inside Expo 2020: Features, Facilities, Services.* Retrieved from yourdubaiguide: https://yourdubaiguide.com/inside-expo-2020-features-facilities-services/

[13] Kasim, Z., Shahidan, M., and Yusof, Y. (2018). Use of landscape environmental setting for pedestrian to enhance campus walkability and healthy lifestyle. *Environmental Impact IV.*

[14] Saleh, A. (2021, August 11). Youth play major role in preparations for Dubai World Expo 2020 . Retrieved from WAM: https://www.wam.ae/en/details/1395302959679

[15] Bleibleh, S., Hassan, B., Mousa, A., and Alneyadi, R. (2024). Dubai World Expo 2020: A Youth-Centric Analysis of Mega Events and their Preliminary Influence on Sustainable Development. In *Resilient Planning and Design for Sustainable Cities.* Springer.

Innovation and Technological Advances for Sustainability – Al-Naemi et al. (Eds)

Characterization of oilfield wastewater and treatment feasibility using membrane biofilm reactor technology

Noura Ali Al Balushi, Shabib Al Rashdi & Syed M. Rizwan
National University of Science and Technology, Muscat, Oman

Shima Masaoud Al Saadi
RWTH Aachen University Aachen, Germany

ABSTRACT: The water crisis increased over the last decades. Many researchers work in treating domestic wastewater through several technologies. Biological treatment by fabrication of biological cells shows great results in treating domestic wastewater. Microbial fuel cells or bio-electrochemical systems are newly developedtechnologies with high efficiency, low cost, and environmental sustainability. Current research has novelty in terms of treating the wastewater produced in the oilfield. A simple design had been fabricated to observe the possibility of biological growth on the outer surface of the membrane. The membrane biofilm reactor (MBfR) is composed of a hollow fiber membrane, feed tank, air compressor, and feed pump. Three samples of produced wastewater that were fed into membrane flow cells were examined at varying flow rates and retention times while at room temperature. To measure the thickness of bacterial adhesion to the membrane, scanning electron microscopy (SEM) is performed. By means of spectrophotometry, the Multiparameter Bench Photometer for Laboratories isused to measure the levels of COD, TP, TN, and oil content in the water. It is advised to use the produced oilfield wastewater in microbial fuel cells for treating wastewater and generating energy as a result of the observation of biological growth on the external surfaceof membranes with various thicknesses.

Keywords: hollow fiber membrane, MABR, produced oilfield & water technology

1 INTRODUCTION

Wastewater management in industry poses a significant technical challenge in order to deliver a high-quality, treated effluent. Oilfield produced water is one of the most challenging waste water applications due to its toxic and inorganic pollutants [1]. Oilfield produced water is extremely saline water that exists under high pressure and temperatures, and normally contains high concentrations of petroleum hydrocarbons, oilfield chemicals, salinity, suspended solids, and heavy metals. Most of these sources are located in remote and/or desert regions, making it even more challenging to determine remediation options [1]. In the absence of effective treatment, it can cause significant environmental impacts [2].At present oilfields in Oman produce more than 1 million cubic meters per day (m3/day) [3]. Majority of the water is still being injected to a deep well and the petroleum development Oman (PDO) has planned to slash down deep well injection up to 22 % by 2025. The majority of diverted water is being discharged to wetland because the quality of effluent is not suitable for reuse. However, the amount of produced water is increasing therefore applying other promising technologies is essential [3]. During the past decades, various technologies have been developed for the treatment of oilfield wastewater such as membrane filtration [4], reverse osmosis [5], oxidation [6],

 DOI: 10.1201/9781003496724-56

ion-exchange [7], etc. Compared with physical and chemical processes, biological treatment is a cost-effective and environmentally friendly technique and more compatible with existing plant facilities and operation. Thus, more and more attention has been paid in the treatment of oilfield wastewater using a variety of biological methods such as aerated filter [8], anaerobic baffled reactor [9], anaerobic tank [10], and biological contact oxidation [11].

This study suggests MBfR as a biologic approach technology for treating oil field wastewater. When compared to conventional biological approaches, this method benefits from inexpensive startup and ongoing costs, rapid processing, and minimal processing area. Through the gas-permeable membrane, oxygen is diffused through the membrane, enabling the biofilm to grow and attach itself to its exterior; pollutants in the wastewater are absorbed by the biofilm as a result of concentration differences and microbial adsorption, which then degrades them. The biofilm stratifies the community structure as oxygen and contaminants move between its different sides, creating an aerobic layer and an anaerobic layer. Each layer has different functions due to its different conditions [12].

Wastewater treatment with MBfR technology demonstrated its feasibility and superiority in treating sewageand synthetic wastewater. Furthermore, MBfR has also been shown to be viable for treating industrial wastewater. MBfRis intended to be used in the treatment of oil field wastewaterdue to its excellent results in the removal of nitrogen and superior-strength organic materials. However, as the requested discharge standard of oil-field wastewater is improved, biological treatment alone cannot meet the necessary discharge standard. Finding a new technology thatcan address this problem and the required discharge standardat the same time is thus crucial. Therefore, using a hybrid system to treat such wastewater is a frequent method [13,14]. This research reports on the impact of feed water composition on bacterial adhesion to the membrane surface, with and without oil-in-water emulsions. It is a preliminary study thatwas performed to determine the feasibility of MABR in treating oil-fields wastewater in Oman. This system will be combined with microbial fuel cells (MFCs) or bio-electrochemical systems (BESs) for water purification.

2 MATERIALS AND METHOD

2.1 *Water sample collection*

Two bulk liquids from two different locations of oil production operated by Petroleum Development Oman Company (PDO) were collected. One produced water sample is from the Musallam station in northern Oman. The sample point was the outlet of hydro-cyclones that was used as a physical treatment for the separation of oil, sand, and water. The second sample was from the south of Oman, particularly from Lekhweir station. This was collected from the effluent of the hydration tank. The average values characteristics of both oilfield wastewater are shown in Table 1.

Table 1. Average values for both water sample.

Parameters	Value
pH	5.9–6.5
COD (mg/L)	319–430
BOD (mg/L)	7–10
Total nitrogen (mg/L)	3.2–54.1
Total phosphorus (mg/L)	0.21–0.30
Total dissolved solids (mg/L)	100000–140000
Conductivity (μS/cm)	189000–255000
Oil in water (mg/L)	400–650
Total petroleum hydrocarbons (mg/L)	<0.5

2.2 *Reactor design*

Membrane biofilm rectors (MBfR) were used in a closed flow-cell rectangular system with a single hollow-fiber membrane to treat generated oilfield wastewater. The silicone membrane was attached in the center of the flow cell and has an outer diameter of 460 μm (Oxymem Ltd., County Westmeath, Ireland). A flow cell with dimensions of 5 cm in width, 2 cm in height, and 10 cm in length was used with various flow rates. Figure 1 depicts the system's components, which include an oxygen-based membrane, a reservoir, a flow cell, an air compressor, and a peristaltic pump. Transparent pipes were used to link a recycle reservoir with a 1 L volume to the flow cell's intake and exit. The rate of wastewater entering and leaving the reservoir was measured using a peristaltic pump. With a pressure of 0.2 bars, air was provided through the dead-end-operated hollow fiber membrane. The inoculum was recirculated for 24 hours before starting the continuous flow. A 108 mL per minute influent flow rate was used to generate a hydraulic retention time (HRT) of 60 minutes. The system operated at 25°C, or room temperature. For 50 to 60 days, the effluent was continuously cycled through the flow cell.

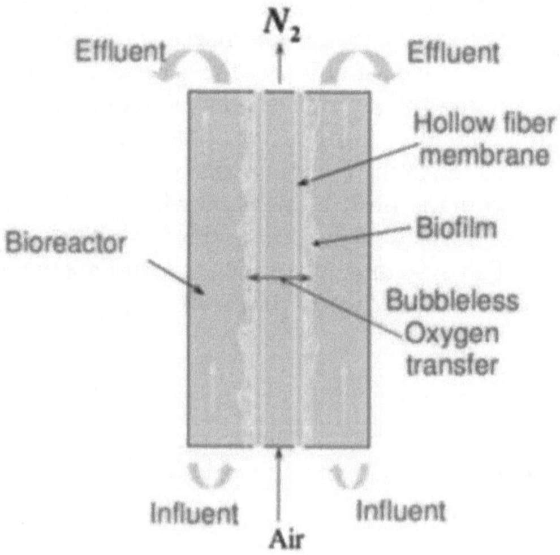

Figure 1. Schematics diagram of membrane-aerated biofilm reactor (MABR) [12].

2.3 *Testing conditions and analytical method*

Two distinct MBfRs were used to analyze the bulk liquids. To examine the quality of the water before and after the trials, total nitrogen as well as COD were measured. In each experiment, the intra-membrane pressure was 0.2 bar. In trials 1 and 2, the influent COD concentrations were 319 mg/L and 259 mg/L, respectively. Each test was performed for 50–60 days using a new membrane material. Since produced water contains high salinity, high chemicals and heavy metals, it required more time for treatment comparedto municipal wastewater. Depending on the qualities of the water, different writers investigated various types ofwastewaters that take varying amounts of time to process [15]. At the end of the experiment, scanning electron microscopy (SEM) was used to scan the membranes to track the development of the biofilm. The USEPA Reactor Digestion Method was used to measure the COD concentrations in the bulk liquid (Hach, Loveland, CO,USA).To measure pH, a glass electrode pH meter was employed. The Hach Methods 10072 and 8048 were used to analyze nitrate (NO3-N), nitrite (NO2-N), and phosphate (PO4), respectively.

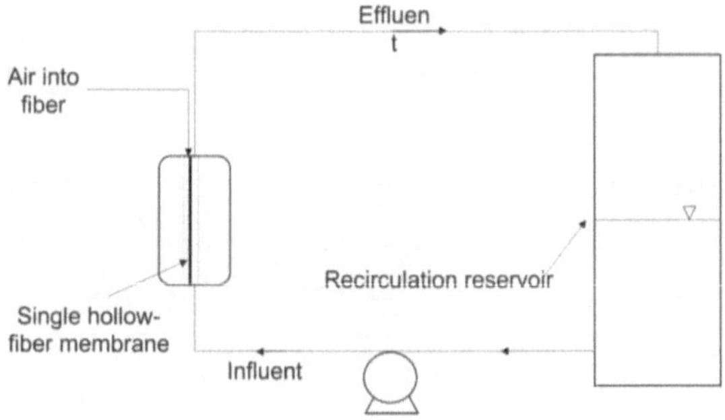

Figure 2. Schematic diagram of MBfR system.

3 RESULTS AND DISCUSSION

3.1 *SEM analysis results*

The growth kinetics of MABR biofilms fed with different compositions of produced wastewater was evaluated using microscopy. In this study, scanning electron microscopy (SEM) images were used to determine the morphology of the membranes. Figures 3, 4 and 5 illustrate SEM images of biofilm growth on membrane surfaces. The pictures show that there is a growth of biofilm on the out surface of the hollow fiber membrane. It is obvious that the growth is very clear when water samples was collected from the hydration tank. In contrast, effluents of hydro-cyclone showed lower growth of biofilm. This is due to the existence of oil-in-water emulsions that inhibit biofilm development and reduce biomass accumulation. Following inoculation, biofilms expanded with different rate to cover the membranes. The thickness of each biofilm grew with time, however it might differ throughout its length depending on the characteristics of the wastewater treated. In overall, biofilm growth means that the water quality is improved using biological treatment since the biofilm is removing the total nitrogen (TN) and chemical oxygen demand (COD) concentrations.

3.2 *Total nitrogen removal*

The effect of pressure on nitrogen removal was studied. It is evident that the concentration of NH4+-N in the first hours increases gradually at 0.2 bar and 0.3 bar because organic nitrogen is transformed into NH4+-N. After that, the concentration of NH4+-N decreases due to the transformation of NH4+-N into nitrate and nitrite. In an aerobic reaction, such as nitrification, the NH4+-N concentration decreases quickly at 0.3 bar due to the high DO concentration. As soon as the aeration pressure is 0.1 bar, the biofilm mainly exists in anaerobic conditions, which

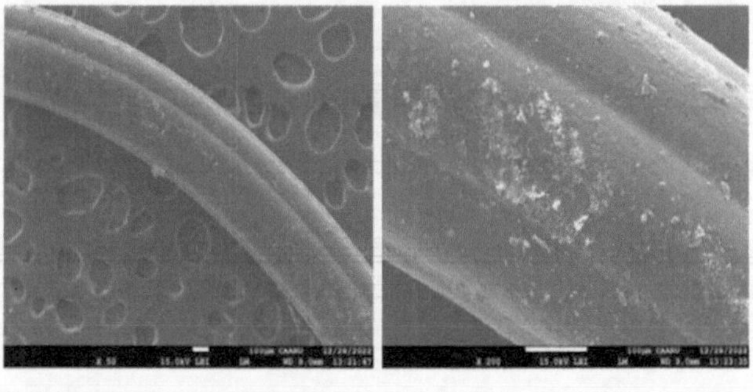

Figure 3. SEM images of hollow fiber membrane.

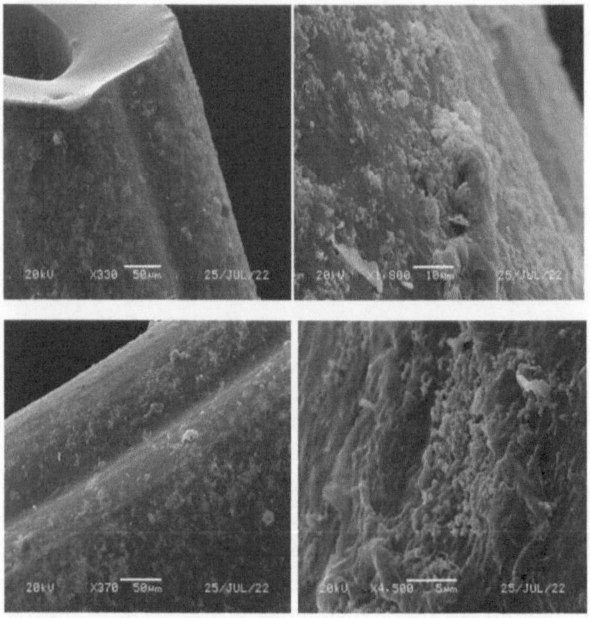

Figure 4. Biofilm growth on membrane treating sample 1 (trial 1).

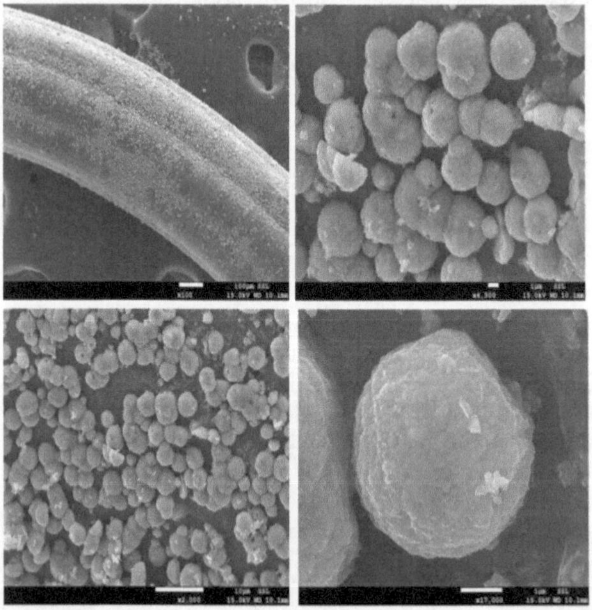

Figure 5. SEM results of biofilm growth on membrane with water sample 2 (trial 2).

means that NH4+-N cannot be transformed into nitrates and nitrites with much ease, thus, NH4 +-N concentration remains at a high level. There is a significant difference between the concentrations of TN at 0.2 bar and those at 0.1 bar and 0.3 bar. The normal elimination of TN is reliant on the cooperative action of nitrification and de- nitrification. It has been found that both nitrifying and denitrifying bacteria are able to live in biofilms when the pressure is 0.2 bar. As a result, TN removal may be accomplished significantly more successfully at 0.2 bar pressure.

3.3 *Chemical oxygen demand and flow rate*

To investigate the impact of feed flow rate on the COD removal process, various feed flow rates were applied in thisexperiment. The findings show that as the feed flow rate is increased (decreasing hydraulic retention time (HRT)), the COD concentration clearly decreases. Different trials were conducted at flow rates from 0.02 m/s to 0.06 m/s. resulting in decreasing the effluent COD from 300 mg/l to 250 mg/l. The biofilm experiences more hydrodynamic shear as the feed flow rate is increased. The mass transfer efficiency of organic matter into and out of the biofilm is enhanced by turbulence brought on by feed flow and the shaking of hollow fiber membranes. In contrast, increasing the flow rate to 0.08 m/s caused detaching of biofilms from hollow fibers due to the intense hydrodynamic shear, hence, high COD concentration in the effluent. The recommended feed flow rate is 0.06 m/s.

4 CONCLUSION

A study of MABR with oilfield wastewater was carried out for 60 days under the parameters of 0.1–0.3 bar aeration pressure and 20 ml/min feed flow rate. The influent quality of the two locations is almost similar. The concentrations of COD, oil, and NO3-N in the effluent decreased slightly and reached 250, 400, and 100 mg/l, respectively. However, the quality of the effluent does not satisfy the discharge request, thus additional treatment is required.

5 RECOMMENDATION

This research is unique in that it looks at how to treat oilfield waste water in a biological cell. To get a better treatment, some recommendations, like a pre-treatment, can be added. For example, ultrafiltration with hollow-fiber membranes that allow operators to meet discharge criteria fixed by local or national environmental regulatory agencies Case studies with different pH, time, temperature, and water samples can be performed to determine the best treatment values.

ACKNOWLEDGMENT

The authors are grateful to the Ministry of Higher Education, Research, and Innovation (Oman) for supporting research and innovation and funding the study. Also, deep thanks to the College of Engineering, National University of Science and Technology, Muscat, Oman.

REFERENCES

[1] Yang, Y., Zhang, X. and Wang, Z. (2002) 'Oilfield produced water treatment with surface-modified fiber ball media filtration.', *Water science and technology: A Journal of the International Association on Water Pollution Research. England*, 46(11–12), pp. 165–170.

[2] Fakhru'l-Razi, A. et al. (2009b) Review of technologies for oil and gas produced water treatment, *Journal of Hazardous Materials. Elsevier*, 170(2–3), pp. 530–551. doi: 10.1016/J.JHAZMAT.2009.05.044.

[3] Prabhu, C. (2018) 'Oilfield produced water output to exceed 1m m3/day from 2019: PDO', Oman Observer, 20 June. Available at: http://www.omanobserver.om/oilfieldproduced-water-output-to-exceed-1m-m3-day-from-2019-pdo/.

[4] Ebrahimi, M. et al. (2010) Investigations on the use of different ceramic membranes for efficient oil-field produced water treatment, *Desalination*. Elsevier, 250(3), pp. 991–996. doi: 10.1016/J.DESAL.2009.09.088.

[5] Melo, M. et al. (2010) Advanced performance evaluation of a reverse osmosis treatment for oilfield produced water aiming reuse, *Desalination*. Elsevier, 250(3), pp. 1016–1018. doi: 10.1016/J.DESAL. 2009.09.095.

[6] Ramalho, A. M. Z., Martínez-Huitle, C. A. and Silva, R. da (2010) Application of electrochemical technology for removing petroleum hydrocarbons from produced water using a DSA-type anode at different flow rates, *Fuel*. Elsevier, 89(2), pp. 531–534. doi: 10.1016/J.FUEL.2009.07.016.

[7] Chang, H. et al. (2019) Potential and implemented membrane- based technologies for the treatment and reuse of flowback and produced water from shale gas and oil plays: A review, *Desalination*. Elsevier, 455, pp. 34–57. doi:10.1016/J.DESAL.2019.01.001

[8] Zhao, X. et al. (2006) 'Oil field wastewater treatment in Biological Aerated Filter by immobilized micro-organisms, *Process Biochemistry*. Elsevier, 41(7), pp. 1475–1483. doi: 10.1016/J.PROCBIO.2006.02.0

[9] Ji, G. D. et al. (2009) Anaerobic baffled reactor (ABR) for treating heavy oil produced water with high concentrations of salt and poor nutrient, *Bioresource Technology*. Elsevier, 100(3), pp. 1108–1114. doi: 10.1016/J.BIORTECH.2008.08.015.

[10] Liu, R. et al. (2010) Microbial diversity in the anaerobic tank of a full- scale produced water treatment plant, *Process Biochemistry*. Elsevier, 45(5), pp. 744–751. doi: 10.1016/J.PROCBIO.2010.01.010.

[11] Guo C., Chen Y., Chen J., Wang X., Zhang G., Wang J., Cui W. and Zhang Z. (2014 Oct). Combined hydrolysis acidification and bio-contact oxidation system with air-lift tubes and activated carbon bioreactor for oilfield wastewater treatment. *Bioresour Technol.* 169:630–636. doi: 10.1016/j.biortech. 2014.07.018. Epub 2014 Jul 21. PMID: 25105268.

[12] Li P. et al., Oil-field wastewater treatment by hybrid membrane- aerated biofilm reactor (MABR) system, *Chemical Engineering Journal*, vol. 264, pp. 595–602, Mar. 2015, doi: 10.1016/J.CEJ.2014.11.131.

[13] Lu M., Zhang Z. Z., Yu W. Y., Zhu W., Biological treatment of oil field- produced water: a field pilot study, *Int. Biodeterior. Biodegrad.* 63 (2009) 316–321.

[14] Li G., Guo S. and Li F., (Dec. 2010). Treatment of oilfield produced water by anaerobic process coupled with micro-electrolysis, *Journal of Environmental Sciences*, vol. 22, no. 12, pp. 1875–1882, doi: 10.1016/S1001-0742(09)60333-8.

[15] Stricker A. E., Lossing H., Gibson J. H., Hong Y., Urbanic J. C. (2011 Jan). Pilot scale testing of a new configuration of the membrane aerated biofilm reactor (MABR) to treat high-strength industrial sewage. *Water Environ Res.*;83(1):3–14.doi: 10.2175/106143009x12487095236991. PMID: 21291022.

[16] Huanqi He, Brett M. Wagner, Avery L. Carlson, Cheng Yang, Glen T. Daigger; (November 2021). Recent progress using membrane aerated biofilm reactors for wastewater treatment. *Water Sci Technol 1*; 84 (9): 2131–2157. doi:https://doi.org/10.2166/wst.2021.443.

[17] Cole A. C., Semmens M. J., LaPara T. M. (2004 Apr). Stratification of activity and bacterial community structure in biofilms grown on membranes transferring oxygen. *Appl Environ Microbiol.*;70 (4):1982–9. doi: 10.1128/AEM.70.4.1982-1989.2004. PMID: 15066788; PMCID: PMC383.

Innovation and Technological Advances for Sustainability – Al-Naemi et al. (Eds)

Circular economy, SDGs, Schwartz's value and sustainability actions in fashion recyclability

Shahid Anjum & Nur Amiratul Izzah Mohd. Noor
School of Business, Universiti Teknologi Brunei (UTB), Gadong, Brunei Darussalamy

Naveeda Qaseem
Westminster Business School (WBS), University of Westminster, London, UK

ABSTRACT: Rise of circular economy (CE) is promising to min_imize resource usage, carbon emissions and waste which has a valuable impact towards the recognition of UN's agenda for Sustainable Development Goals. SDGs involves all stakeholders of society i.e. governments, businesses, NGOs, civil society groups and citizens. The fashion and textile industries are directly and indirectly interconnected with pillars of sustainability i.e. economic, social, peace and inclusivity and environmental as well as all seventeen SDGs, particularly SDG12. Recycling and reuse of textile waste can decrease the negative externalities on the environment by reversing logistics and redistributing textiles at various points of the supply chain as well as by lowering the number of new clothes makers, minimizing landfill trash, and supporting an environmentally friendly lifestyle. This research is combining Schwartz's theory of value in the perspective of fashion industry along with insights from CE and SDGs as guiding base to shape the social actions related to the Theory of Planned Behavior (TPB). It has mapped both of these value and behavior theories to one another. Besides, Brunei is a country with amongst highest social media usage and thus awareness where the people's perceptions have been tested for the knowledge of SDG and CE towards the recycling of the fashion wares and clothing. It has been found that the CE awareness is a huge contributory factor in recognizing SDGs. An empirically tested structural equation model using SmartPLS has found that the knowledge of CE and SGD is crucial in contributing towards the motivation of adoption of recycling and reuse of fashion wares which along with power and security of Schwartz's values and subjective norm of TPB towards usage action with a minimal marginal moderating effects of age, gender, education and income in igniting sustainable actions in the fashion industry.

Keywords: Circular Economy, Sustainable Development Goals (SDG), Fashion Industry, SEM, Technology Acceptance Model

1 INTRODUCTION

Linear economy's take, use, and dispose of system, in which substantial amounts of non-renewable resources are extracted to make products that people use for just a few times before being discarded in landfills, creating waste [1] and [2]. The current linear economic system leaves many economic opportunities untapped, pollutes and degrades the environment, possesses pressures on scarce resources, and strains the economics of the societies

[3] which is particularly true for fashion industry. Clothing and textiles comprise a significant portion of the global economy [4]. Globalization and technology played a significant effect in the fashion industry, revolutionizing it by introducing new trends, influences, and manufacturing processes [5]. Sustainability in the fashion business is, however, a very complicated subject [6]. Garments eventually contribute to climate catastrophes, such as microplastic contamination in our seas [3] which are believed to trigger inflammatory reactions [8]. Approximately 87 percent of all discarded textiles are disposed of through landfill or incineration [7]. Rana Plaza catastrophe compelled to close the gap between economic sustainability and social and environmental performance [9] which resulted in a systematic shift towards circularity is necessary [10]. The circular economy (CE) is an alternative of looking at the connections between markets, customers, and natural resources [11,12]. The circular economy has emerged as a concern in the fashion industry where waste is ow nperceived as a resource [13]. Recycling, reuse and buying used clothing can serve as starting points for avoiding large resource waste, therefore decreasing the negative externalities on the environment. Buying second-hand clothing can also be environmentally friendly and is a vital part of the circular fashion movement. This can benefit the environment by minimizing landfill trash, and supporting an environmentally friendly lifestyle. It is acknowledged as a paradigm that facilitates changes in global source management and contribute to the achievement of sustainability [2,14] as well as call for collaboration at all levels [15].

Sustainable Development Goals (SDGs), focusing on the three pillars of sustainability named as economic, social, and environmental including peace and inclusivity and are vital for fashion industry. The achievement of SDG 12 will also contribute to other SDGs, as all SDGs are interrelated. SDGs have also created an awareness among the public for sustainable environment. Second-hand clothing (SHC) market falls under collaborative consumption involving markets [16] which traditionally, are deemed for consumers with limited financial resources [17] and characterized by poverty. However, it gained popularity among young people in 21st century [18]. Consumers are interested in different products features depending on their behavior, actions, and habits. This is where Schwartz theory of basic values comes into picture [18]. However, when a social norm is upheld within a community, the members are expected to share the same values that reflect consistent behavior [16]. Thinking of public behavior towards sustainability and environmentally responsible consumption, Theory of Planned Behavior (TPB) become relevant. Hence, this paper is to investigate the customer perceived value towards buying SHC and is an attempt to test the Schwartz's theory of value in the perspective of fashion industry along with insights from CE and SDGs as guiding base to shape the social actions and mapping this to TPB. Brunei is a country with amongst highest per capita as well as amongst highest social media usage and thus awareness. People's perceptions in Brunei have been tested for the knowledge of SDG and CE and its impact towards the recycling of the fashion wares and clothing by using a questionnaire survey. Regarding the estimation technique, this study has avoided the use of simple linear regression (SLR) [19] because of it's limitations like regressioneering [21] and regressgineering [22] and have used partial least square (PLS) estimation [20,23] using SmartPLS. It has been found, through the analysis of 213 responses in a country with a small population of only 0.4 million people, that the CE awareness is a contributory factor in recognizing SDGs which both in turn contributes towards the motivation of adoption of recycling and reuse of fashion wares. Moderation effects do not always work, likewise is the case in this study where effects of gender, age, education and willingness to recycle of used clothing has not produced additional impact on willingness to adopt recycled clothing fashions. The paper is divided into four sections. Introduction in section 1, literature review in section 2 and section 3 covering discussion on model development. Section 4 covers the data description and estimation methodolog. Section 5 describes results and section 6 concludes.

2 REVIEW OF LITERATURE

The current linear model of the modern economic system is unsustainable [24] and needs to be changed to a more sustainable socio-technical system. Environmental issues include biodiversity loss, resource depletion, water, air, and soil pollution, as well as excessive land usage, are endangering the planet's ability to support life [25]. Therefore, a Circular economy provides the economic system with an alternative flow model to achieve the UN Sustainable Development Goals. The idea of a circular economy is not completely new and is continuously being developed in order to achieve efficient global resource management and sustainability [2,14]. A circular economy may be seen of as a new viewpoint on the connections between markets, consumers, and natural resources [11,12]. The concept of circular economy is set to provide benefits of EUR 1.8 trillion per year and the creation of over a million employment [26]. It will also play a significant role in reducing green house gases (GHG) emissions [27] and could lead to a USD4.5 trillion increase in global GDP by 2030 and a 1 percent yearly deduction in resource use could save 840 million tons of fossil fuels, minerals, metals and 39.2 trillion liters of water [27,28].

The supply chain for the fashion industry is extremely complicated with a high level of environmental degradation impact. Sustainability in the fashion business, which is predicted to produce 92 million tons of textile waste annually [32], is also a very complicated subject which needs to move towards a reuse-based strategy and reversing logistics [29] in order to avoid large resource waste and decreasing negative externalities on the environment [30]. Buying SHC can also be environmentally friendly and part of the circular fashion movement [31] which can minimiz landfill trash and supporting an environmentally friendly lifestyle. The industry has reached its breaking point because of fast fashion [5]. The fashion industry now targets sustainable principles more frequently than circular economy concepts [10]. Therefore, the circular economy model encourages, reuse and resource sharing techniques in order to maximize a product's life cycle [13]. Shifting from the linear to the circular model requires engagement from all stakeholders [27,29] especially consumers who play a key role in the transition through their attitudes and behaviors [32] and thus using SHC serves as the foundation of sustainable clothing consumption [33]. Over 75 million individuals, mostly women, are employed by the 2.5 trillion USD fashion business globally [35] and SHC market is expected to double in the next five years, reaching 77 billion USD [36] and thus vital for SDGs. Related sustainable development is outlined in 17 SDGs which focus on the three pillars of sustainability: economic, social, and environment including peace and inclusivity. The SDGs apply to all stakeholders in society, including governments, businesses, NGOs, civil society groups, and citizens [34]. The fashion and textile industries are direct and indirectly interconnected with all 17 Sustainable Development Goals, particularly SDG 12. Thrifting or purchasing SHC can benefit the environment by minimizing landfill wastes and encouraging sustainable lifestyles [31]. While clothing reuse is being normalized but discarding is still in high volumes because of fast fashion tears [37] and only 20 percent of the donated clothes are resold. Over the recent years, consumers have voiced their concerns over the environmental impact of the fashion industry, particularly millennials and generation Z. Concerned with their own and the planet's health, they pay attention to eco-products, conscious consumption and value the sustainability, and thus are a vital driving force toward sustainable fashion and lifestyle [38].

3 MODEL AND HYPOTHESIS DEVELOPMENT

Schwartz theory identifies 10 broad values according to the motivation that underlies to them. For this research these 10 broad values are divided into 4 groups: Self-Transcendence, Conservation, Self-Enhancement, and Openness to change. Self-transcendence emphasizes the acceptance of others as equivalent and concern for their wellbeing. These two sub categories of Self-transcendence are named as Benevolence (BEN) and Universalism (UNIV).

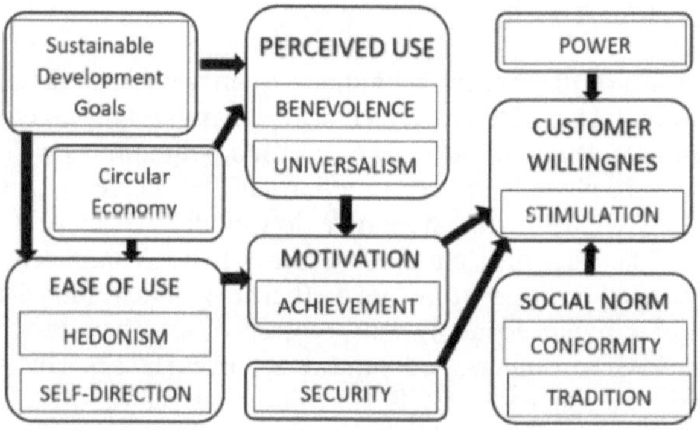

Figure 1. Proposed model.

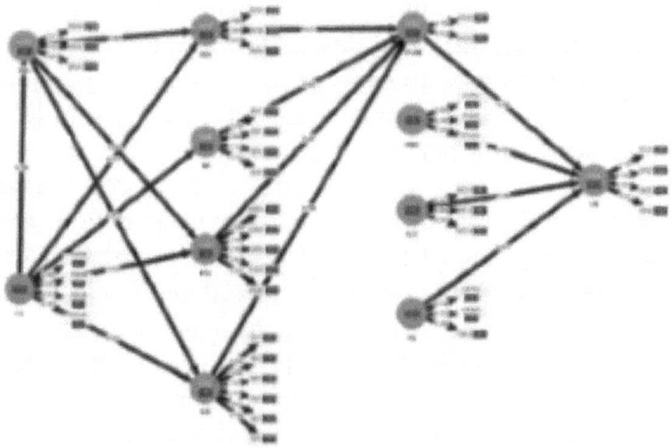

Figure 2. Estimated model 1 (Without link between CE and SDG).

The focus of conservation is on respectful self-repression, the maintenance of old customs, and safeguarding stability in one's relationships with one's environment. This includes features like Conformity (CONF), Security (SEC) and Tradition (TRD). Self-enhancement is pursuit of personal achievement and control over others with either Power (PWR) and/or Achievement (ACVM) characteristics. Lastly, openness to change is about regulating one's own emotions and actions in line with social norms and expectation. Its components include Self-Direction (SD), Stimulation (SIM) and Hedonism (HED). Conventional variables used in the estimation of models of theory of planned behaviour (TPB) [39,40] are perceived usefulness (PU), ease of use (EOU) of technology and subjective norm (SN). Variants of TPB are models of reasoned action [39], decomposed theory of planned behaviour [36] and pure decomposed theory of planned behaviour [41]. A variable named Motivation (MOT), measured through the ACVM, has been proposed and estimated. A mapping between the variables of theory of behaviour and value theory has been made and a combined model is proposed (Figure 1). Various hypothesis proposed are as follow:

- H1 and H2: Awareness of SDG and Awareness of CE have positive (+ve) relationship with EOU and PU respectively and H3 and H4: PU towards buying SHC has +ve relationship with EOU and with MOT and H9: PU towards buying SHC has +ve relationship with EOU as well as H5: EOU has a +ve relationship with MOT
- H6: MOT towards buying SHC has a +ve relationship with CW and H7, H8 and H10: SEC, PWR and SN of buying SHC, all have +ve relationship with CW

Table 1. Model without moderation:* Construct reliability, validity and adjusted R2.

	CA	*rhoA*	*CR*	*AVE*	*AdRSq*
ACVM	−0.196	−0.206	0.619	0.458	0.231
BEN	0.703	0.750	0.832	0.625	0.321
CE	0.559	0.576	0.746	0.427	0
CW	0.613	0.632	0.778	0.473	0.521
HED	0.462	0.499	0.691	0.330	0
PWR	0.325	0.828	0.038	0.386	0.241
SDG	0.897	1.037	0.930	0.815	0
SDR	0.789	0.827	0.851	0.498	0.211
SEC	0.740	0.759	0.849	0.652	0
SN	0.230	0.235	0.619	0.383	0
UNI	0.724	0.736	0.825	0.541	0.078

*Model with Moderating Effects have the same results as well.

Table 2. Model without moderation and linking CE to SDG: Specific indirect effects*.

	Indirect Effects	
Indirect Effects (IE) Path	*IE-1*	*IE-2*
CE → BEN → ACVM	0.017	0.017
CE → SDR → ACVM	0.082	0.080
CE → HED → ACVM	0.134	0.136
CE → UNI → ACVM → CW	0.007	0.008
SDG → BEN → ACVM	0.003	0.003
BEN → ACVM → CW	0.009	0.010
CE → SDR → ACVM → CW	0.024	
HED → ACVM → CW	0.086	
SDR → ACVM → CW	0.058	
CE → BEN → ACVM → CW	0.005	
SDG → SDR → ACVM → CW	0.007	0.007
SDG → HED → ACVM → CW	0.008	0.008
UNI → ACVM → CW	0.025	0.026

4 DATA DESCRIPTION AND METHODOLOGY

The research has used stratified sampling, has targeted Millennials and Generation Z, because of their proximity to Fashion world as well as awareness of SDGs and circular economy issues. The questionnaire has used closed-ended questions with 5-Likert scale and is divided into sections on demogrphy and components of Schwartz value theory. The descriptive statistics shows that female (male) representation is 31.5 percent (pcnt) and respondents age between 18–24 years (y), 25–30y and 30y and up are 64.5, 31 and 5.6 percent respectively. High school education, undergraduates and postgraduates have 11.7, 66.2 and 22.1 percent representation respectively. The questionnaire was designed which has benefitted from studies of Ellen MacArthur Foundation, Boston Consulting, McKinsey, Fashion Revolution, UN reports, among others. Components of Schwartz value theory included in it are awareness of circular economy and SDGs, self-transcendence, conservation, self-enhancement, and openness to change which have 7, 7, 10, 5 and 15 number of items respectively and have values of 0.756, 0.692, 0.583, 0.581 and 0.829 respectively for Chronbach's Alpha. The estimated model values of this, however, may vary from these values depending on the list of value theory based variables to fit into factors of TPB and its variants. Alpha values are satisfactory if they fall between 0.58 – 0.97 [42]. Similarly, values

Table 3. PART 1: Latent variable correlations, square root of AVE for outer loading and HTMT ratio values.

	AVM	*BEN*	*CE*	*CW*	*HED*	*PWR*
ACVM	**0.677**	0.297	0.250	0.465	0.446	0.241
BEN	0.816	**0.791**	0.560	0.535	0.428	0.277
CE	0.753	0.823	**0.655**	0.479	0.489	0.222
CW	1.327	0.805	0.819	**0.688**	0.556	0.370
HED	1.528	0.637	0.877	0.945	**0.574**	0.409
PWR	1.338	0.434	0.505	0.647	0.896	**0.621**
SDG	0.256	0.344	0.544	0.443	0.327	0.184
SDR	1.020	0.744	0.619	0.802	0.806	0.488
SEC	0.652	0.639	0.753	0.772	0.542	0.240
SN	1.077	1.156	0.991	1.353	1.315	0.804
UNI	0.642	0.328	0.443	0.503	0.542	0.180

of alpha greater than 0.5 or less than 0.8 are moderately reliable and thus acceptable [43]. Therefore, all the sections of questionnaire, as are shown in Table 2, are significant, at least moderately if not strongly. Regarding the recycling options, donate, resell, repair/remake and throw intrash comprise 61, 10.8, 8.5 and 19.7 percent respectively. Various measurement approaches are available in the literature for estimation of relationships among various variables named as stepwise linear function [44], logistic regression [45,46], advanced econometric analysis [47,48], efficiency analysis model [49] and data mining approached [50] [51]. PLS is used with SmartPLS because PLS is efficient, suitible to the survey research and easy to use.

5 EMPIRICAL RESULTS AND ANALYSIS

This study is reporting four estimated models named as Model 1, 2, 3 and 4. Model 1 has 2 latent variables (LVs) in the left side (SDG on the upper side and CE on the lower end), followed by 4 LVs in 2nd column from left of diagram. The four variables are, from top to bottom, are HED and SDR (i.e. EOU) as well as BEN and UNI (i.e. PU) which affect the ACVM which in turn affect CW (latent variable on the right hand side) along with three other factors which are PWR, SEC and SN. Model 2, as is shown in Figure 2, has only added an arrow from CE to SDG to model 1 as it has improved the initial model's results marginally. Model 3 and 4 have tried to capture some moderating effects on model 1 and 2 respectively which haven't, however, proven to be successful. Table 2 provides results for construct reliability, validity and adjusted R-Square based on the estimation of model 1 where the composite reliability (CR) scores are within the range of 0.038 to 0.930. Some has exceeded the recommended 0.7 threshold except for PWR, which is 0.325 in Table 2 which also shows the values of average variance extracted (AVE) for Model 2 which are within the range of 0.350 to 0.815, with some are above the recommended values for convergent reliability. AVE values below 0.5 is deemed acceptable providing that the composite reliability (CR) is above 0.6. Table 3 shows indirect effects where column IE-1 equals Indirect Effects for model 1 (i.e. without Moderation effects and where CE and SDG are not linked together). Model 2 results, i.e. considering moderating affects for Model 1, are identical to Model 1 and therefore, are not reported here. Likewise, column IE-2 in Table 4 shows some of the Indirect Effects for model 4 (which is a model with Moderation effects as well as where CE and SDG are also linked together through an arrow going from CE to SDG) and rest of the IE results for Model 4 are shown in Tables 7, 8 and 9. Tables 4 and 5 are two split tables from the same table in order to fit the results in two columns format. Both of these

Table 4. PART 2: Latent variable correlations, square root of AVE for outer loading and HTMT ratio values.

	SDG	*SDR*	*SEC*	*SN*	*UNI*
ACVM	0.091	0.412	0.208	0.239	0.256
BEN	0.301	0.587	0.493	0.449	0.243
CE	0.396	0.447	0.483	0.374	0.287
CW	0.356	0.559	0.543	0.544	0.343
HED	0.264	0.548	0.354	0.518	0.295
PWR	0.176	0.295	0.185	0.323	0.156
SDG	**0.905**	0.269	0.409	0.167	0.141
SDR	0.284	**0.706**	0.414	0.490	0.368
SEC	0.472	0.520	**0.808**	0.405	0.254
SN	0.436	1.106	0.876	**0.619**	0.274
UNI	0.161	0.519	0.332	0.646	**0.736**

Table 5. Model with moderation and added link from CE to SDG.

Indirect Effects (IE) Path	*Indirect Effects*
CE → SDG → HED → ACVM → CW	0.003
CE → SDG → SDR → ACVM → CW	0.003
SDR → ACVM → CW	0.061
CE → SDG → BEN → ACVM → CW	0.000
CE → UNI → ACVM	0.025
SDG → HED → ACVM	0.025
CE → HED → ACVM → CW	0.042
SDG → SDR → ACVM	0.022
CE → SDG → BEN → ACVM	0.001
CE → SDG → SDR	0.043
CE → SDG → HED → ACVM	0.010
CE → SDG → SDR → ACVM	0.009
CE → SDR → ACVM → CW	0.025
HED → ACVM → CW	0.091
CE → BEN → ACVM → CW	0.005
CE → SDG → HED	0.033
CE → SDG → BEN	0.037
SDG → BEN → ACVM → CW	0.001

tables, show results of model 1 where Latent Variable Correlations (above diagnol shown in bold). Square Root of AVE For Outer Loading are the diagnol values which are the bold numbers in both of these tables and HTMT ratio values (below the diagonal line values shown in bold. The same explaination applies to identical results shown for Model 3 and 4 in Tables 7, 8 and 9. This study provides insight into the awareness of CE and SDGs.

Table 6. PART 1: Latent variable correlations, square root of AVE for outer loading and HTMT ratio values.

	ACVM	*BEN*	*CE*	*CW*	*HED*
ACVM	**0.677**	0.816	0.753	1.327	1.522
BEN	0.297	**0.791**	0.823	0.805	0.635
CE	0.25	0.56	**0.655**	0.819	0.876

(*continued*)

Table 6. Continued

	ACVM	*BEN*	*CE*	*CW*	*HED*
CW	0.466	0.537	0.48	**0.687**	0.943
HED	−0.087	−0.160	0.055	−0.158	**0.574**
M-AGE	−0.025	−0.117	−0.062	0.065	−0.076
M-EDU	−0.036	0.087	0.113	0.011	−0.136
M-GEN	0.68	−0.039	0.020	0.043	−0.057
M-W2D	0.027	0.299	0.234	0.372	0.039
PWR	0.091	0.301	0.396	0.353	0.417
SDG	0.413	0.587	0.447	0.562	0.264
SDR	0.207	0.493	0.483	0.538	0.548
SEC	0.24	0.447	0.374	0.544	0.353
SN	0.255	0.243	0.287	0.341	0.519
UNI	0.087	0.160	0.055	0.158	0.295

Table 7. PART 2: Latent variable correlations, square root of AVE for outer loading and HTMT ratio values.

	MAGE	*MEDU*	*MGEN*	*MW2D*	*PWR*
ACVM	0.186	0.12	0.073	0.472	1.568
BEN	0.201	0.14	0.113	0.057	0.443
CE	0.116	0.182	0.168	0.103	0.475
CW	0.198	0.133	0.079	0.175	0.684
HED	0.125	0.233	0.165	0.149	0.583
M-AGE	**1**	0.312	0.039	0.005	0.148
M-EDU	0.312	**1**	0.228	0.035	0.098
M-GEN	0.039	0.228	**1**	0.207	0.139
M-W2D	0.005	−0.035	−0.207	**1**	0.148
PWR	−0.019	0.105	0.028	0.096	**0.674**
SDG	0.066	−0.013	0.052	−0.138	0.180
SDR	−0.298	−0.136	0.01	0.024	0.339
SEC	−0.052	0.002	0.071	−0.081	0.155
SN	−0.195	−0.154	0.022	−0.022	0.311
UNI	−0.134	0.032	0.015	0.006	0.146

Table 8. PART 3: Latent variable correlations, square root of AVE for outer loading and HTMT ratio values.

	SDG	*SDR*	*SEC*	*SN*	*UNI*
ACVM	0.256	0.102	0.652	1.077	0.642
BEN	0.344	0.744	0.639	1.156	0.328
CE	0.544	0.619	0.743	0.991	0.443
CW	0.443	0.802	0.772	1.353	0.503
HED	0.324	0.806	0.54	1.353	0.503
M-AGE	0.074	0.327	0.051	0.384	0.162
M-EDU	0.032	0.153	0.042	0.3	0.042
M-GEN	0.053	0.09	0.08	0.043	0.055
M-W2D	0.161	0.087	0.096	0.419	0.081
PWR	0.201	0.491	0.243	0.707	0,175
SDG	**0.905**	0.284	0.472	0.436	0.161
SDR	0.269	**0.706**	0.52	1.106	0.519
SEC	0.409	0.414	**0.808**	0.876	0.332
SN	0.165	0.490	0.404	**0.618**	0.646
UNI	0.141	0.368	0.254	0.274	**0.736**

6 CONCLUSION

Sustainability has been an issue in the fashion industry and industry seems resistant to change. This study is conducted in Brunei Darussalam by analyzing the customer's value perception using Schwartz's value theory mapped to the TPB. The results have shown that CE aware population is more conscious of SDGs and take them seriously with significant impact in fashion reuse behavior. Both of these factors (i.e. CE and SDG) taken together has shown has huge impact on components of percived usefulness (i.e. BEN and UNI) and components of Ease of Use (i.e. HED and SDR). In short, PU and EOU has significant impact on MOT which along with three other latents named as PWR, SEC and SN determines customer's willingness to embrace the reuse of clothing fashions.

REFERENCES

[1] Ki, C. W., Chong, S. M., and Ha-Brookshire, J. E. (2020). How fashion can achieve sustainable development through a circular economy and stakeholder engagement: A systematic literature review, *Corporate Social Responsibility and Environmental Management*, 27(6), 2401–2424.

[2] Keßler, L., Matlin, S. A., and Kümmerer, K. (2021). The contribution of material circularity to sustainability—Recycling and reuse of textiles, *Current Opinion in Green and Sustainable Chemistry*, 32, 100535.

[3] Chen, X., Memon, H. A., Wang, Y., Marriam, I., and Tebyetekerwa, M. (2021). *Circular Economy and Sustainability of the Clothing and Textile Industry*, 1(12), 3.

[4] Patora-Wysocka, Z., and Sułkowski, Ł. (2019). Sustainable incremental organizational change—a case of the textile and apparel industry, *Sustainability* 2019, Vol. 11, Page 1102, 11(4), 1102.

[5] Ledezma, V., and Ledezma, V. (2017). Globalization and Fashion: Too Fast, Too Furious. *Laurier Undergraduate Journal of the Arts*, 4. http://scholars.wlu.ca/luja

[6] Rosa, A. la. (2019). An Evaluation of Sustainability in Consumption: The Behaviors An Evaluation of Sustainability in Consumption: The Behaviors Behind Purchase, Care, and Disposal of Apparel Behind Purchase, Care, and Disposal of Apparel. https://digitalcommons.unl.edu/textilesdiss

[7] Moazzem, S., Wang, L., Daver, F., and Crossin, E. (2021). Environmental impact of discarded apparel landfilling and recycling. *Resources, Conservation and Recycling*, 166, 105338.

[8] European Environment Agency. (2022). *Microplastics From Textiles: Towards a Circular Economy for Textiles in Europe*. https://circulareconomy.europa.eu/platform/sites/default/files/microplastics-from-textiles-towards-a.pdf

[9] Arrigo, E. (2020). Global Sourcing in Fast Fashion Retailers: Sourcing Locations and Sustainability Considerations, *Sustainability* 2020, Vol. 12, Page 508, 12(2), 508.

[10] Dissanayake, D. G. K., and Weerasinghe, D. (2021). Towards circular economy in fashion: review of strategies, barriers and enablers, *Circular Economy and Sustainability* 2021, 1–21.

[11] BCG. (2018). *The New Big Circle: Achieving Growth and Business Model Innovation through Circular Economy Implementation*. www.wbcsd.org/Archive/Factor-10/Resources/The-new-big-circle

[12] McKinsey and Company. (2016). *The Circular Economy: Moving From Theory to Practice McKinsey Center for Business and Environment Special edition*. www.mckinsey.com/capabilities/sustainability/our-insights/the-circular-economy-moving-from-theory-to-practice

[13] de Aguiar Hugo, A., de Nadae, J., Da, R., Lima, S., Shimomura, Y., and Hosono, S. (2021). Can fashion be circular? A literature review on circular economy barriers, drivers, and practices in the fashion industry's productive chain, *Sustainability* 2021, Vol. 13, Page 12246, 13(21), 12246.

[14] Aguilar-Hernandez, G. A., Dias Rodrigues, J. F., and Tukker, A. (2021). Macroeconomic, social and environmental impacts of a circular economy up to 2050: A meta-analysis of prospective studies, *Journal of Cleaner Production*, 278, 123421.

[15] D'Adamo, I. (2019). Adopting a circular economy: Current practices and future perspectives, *Social Sciences* 2019, Vol. 8, Page 328, 8(12), 328.

[16] Zahid, N. M., Khan, J., and Tao, M. (2022). Exploring mindful consumption, ego involvement, and social norms influencing second-hand clothing purchase, *Current Psychology*, 1, 1–15.

[17] Yan, R. N., Bae, S. Y., and Xu, H. (2015). Second-hand clothing shopping among college students: The role of psychographic characteristics, *Young Consumers*, 16(1), 85–98.

[18] Skimina, E., Cieciuch, J., Schwartz, S. H., Davidov, E., and Algesheimer, R. (2018). Testing the circular structure and importance hierarchy of value states in real-time behavior, *Journal of Research in Personality*, 74, 42–49.

[19] Ismail, Z. and Anjum, S. (2021). Privacy Policy, Training and adaption of EMT to curtail workplace harassment in organizations: An application of TAM, *Advances in Intelligent Systems and Computing Series*. Springer

[20] Anjum, S. and Sharin, N.S.B.H. (2022). COVID-19, SME training support by ASEAN governments and analysis of digital investments in brunei, *AIP Conference Proceedings* 2658, 130001. November

[21] Anjum, S. (2014a). *Statistical Software and Regression Diagnostic Reporting with Fuzzy-AHP Intelligent Zax, Lecture Notes in Software Engineering*, Vol. 2, No. 1, February

[22] Anjum, S. (2014b). Back testing VaR violations, Be-ALAM regression and internal models of portfolio variance forecasting, *Journal of Money, Investment and Banking*, FRDN Inc., 29, Oct

[23] Anjum, S. and Ismail, D. N. Z. (2022). Employee monitoring, productivity, privacy and harassment: Auditing technology training and adoption in public administration, *AIP Conference Proceedings* 2658, 080002;

[24] Korhonen, J., Honkasalo, A., and Seppälä, J. (2018). *Circular Economy: The Concept and its Limitations, Ecological Economics*, 143, 37–46.

[25] Geissdoerfer, M., Savaget, P., Bocken, N. M. P., and Hultink, E. J. (2017). The circular economy – A new sustainability paradigm? *Journal of Cleaner Production*, 143, 757–768.

[26] Ellen MacArthur Foundation. (2013). *Towards the Circular Economy Vol. 1: an Economic and Business Rationale for an Accelerated Transition.* Shared by Business. https://emf.thirdlight.com/link/x8ay372a3r11-k6775n/@/preview/1?o

[27] Accenture. (2018). *Circular x Fashion Tech Introduction to the Global Change Award.*

[28] Rubel, H., Schmidt, M. and Alexander M. F. (2017). *The Urgency—and the Opportunity—of Smart Resource Management. BCG.* https://www .bcg.com/publications/2017/

[29] Sczyka, J. (2020). *Circular Business Models in the Fashion Industry: A Consumer Perspective On Renting Everyday Clothes.* http://urn.kb.se/resolve?urn=urn:nbn:se:uu:diva-413325

[30] Gazzola, P., Pavione, E., Pezzetti, R., and Grechi, D. (2020). Trends in the fashion industry. the perception of sustainability and circular economy: A gender/generation quantitative approach, *Sustainability* 2020, Vol. 12, Page 2809, 12(7), 2809.

[31] Geegamage, T., Ranaweera, A., and Halwatura, U. (2021). *Second-hand Fashion Consumption: A Literature Review.* https://doi.org/10.1111/ijcs.12159

[32] Koszewska, M., Rahman, O., and Dyczewski, B. (2020). Circular fashion – consumers' attitudes in cross-national study: Poland and Canada, *Autex Research Journal*, 20(3), 327–337.

[33] Colucci, M., and Vecchi, A. (2021). Close the loop: Evidence on the implementation of the circular economy from the Italian fashion industry, *Business Strategy and the Environment*, 30(2), 856–873.

[34] UN Alliance. (2021). *Synthesis Report on United Nations System-wide Initiatives related to Fashion 2.* www.unfashionalliance.org

[35] McKinsey and Company, and Business Of Fashion. (2021). *The State of Fashion 2021.*

[36] ThredUP. (2021). *2021 Resale Report.* https://www.thredup.com /resale/static/thredUP-Resale-and-Impact-Report-2021-980436a36adc4 f84a26675c1fcf2c554.pdf

[37] Vehmas, K., Raudaskoski, A., Heikkilä, P., Harlin, A., and Mensonen, A. (2018). Consumer attitudes and communication in circular fashion, *Journal of Fashion Marketing and Management*, 22(3), 286–300.

[38] Culén, A. L., and Srivastava, S. (2017). Gen MZ Toward Sustainable Fashion Practices-Alternative Services and Business Models.

[39] Fishbein, M., and Ajzen, I. (1975). *Belief, Attitude, Intention, and Behavior: An Introduction to Theory and Research.* Reading, MA: Addison-Wesley.

[40] Ajzen, I. (1991). The theory of planned behavior, *Organizational Behavior and Human Decision Processes*, 50(2), 179–211.

[41] Taylor, F. and N. Todd, (1995). Understanding information technology usage: A test of competing models, *Information Systems Research*, 5(2): 144–176.

[42] Taber, K. S. (2018). The use of Cronbach's Alpha when developing and reporting research instruments in science education, *Research in Science Education*, 48(6), 1273–1296.

[43] Ekolu, S. O. and Quainoo, H. (2019). Reliability of assessments in engineering education using Cronbach's alpha, KR and split-half methods, *Global Journal of Engineering Education*, 21(1).

[44] Anjum S. and Qaseem, N. (2021). Basel violations, volatility model variants and value at risk: Optimization of performance deviations in banks, *Economics and Business Letters* 10(3), 240–248.

[45] Anjum S. and Hanafi, E. W. B. A. M. (2020a). Vulnerabilities and Digital Violations in Software Products: Logistic Regression Analysis, *IEEE Xplore, Proceedings of 2020 International Conference on Cyber Warfare and Security (ICCWS). Institute of Electrical and Electronics Engineers.* Denver. pp. 1–6.

[46] Anjum, S. (2020b). DoS attacks, triad and privacy: Software exposures in Microsoft, Apple and Google, IEEE Xplore, *Proceedings of 2020 First International Conference of Smart Systems and Emerging Technologies (SMARTTECH), Riyadh, Edited by Koubaa et al., IEEE Society*. Washington. pp. 53–58, (2020b).

[47] Shamim, F., Anjum, S. and Wakil, A. B. (2015). Banking Risk and Operating Efficiency Measures in the Era of IT, Accounting and Finance Research. *Sciedu Press.* 4 (1).

[48] Shamim, F. and Anjum, S. (2013). Technology diffusion in japanese finance industry: An exploration, *International Research Journal of Applied Finance*, VI (12). December.

[49] Shamim, F., Yamori, N and Anjum, S. (2017). Clicks business of deposit taking institutions: An efficiency analysis, *Journal of Economic Studies.* 44(6). Emerald Publishing. Thomson Reuters.

[50] Anjum, S. (2013). Algorithms for predictive classification in data mining: A comparison of evaluation methodologies, *Journal of Industrial and Intelligent Information (JIII)*, 1(2). June.

[51] Anjum, S. (2014c). Composite indicators for data mining: a new framework for assessment of prediction classifiers, *Journal of Economics, Business and Management*, 2(1). February.

Innovation and Technological Advances for Sustainability – Al-Naemi et al. (Eds)

Impact of training and development on employee performance: Insights for sustainable capacity building of employees in telecommunication organization in Qatar

Dareen Khaled Mekdad
College of Business Management, University of Doha for Science and Technology, Doha, Qatar

Juliet Gladies Jayasuria
College of Business Management, University of Doha for Science and Technology, Doha, Qatar
ORCID ID: 0000-0001-5977-3698

Alissa Alhaj Hossin
College of Business Management, University of Doha for Science and Technology, Doha, Qatar

ABSTRACT: Background: The training and Development teams of every organization aim to drive better employee performance and make a better impact on their businesses. In order to achieve maximum performance, it is critical to consider the capacity building of organizations and institutions. For example, it has been indicated as a means for ensuring the continued existence of organizations by building institutional strength and reducing reliance on key individuals [15]. Effective capacity building allows an organization to continue its work despite staff turnover or changes in the operating environment.

Purpose: This paper focuses on determining the impact of training and development on employee performance by focusing on the insights of sustainable capacity building at one of the organizations in Qatar's telecommunication sector. It also determines the employees' opinions of how training programs could build the capacity of employees.

Research methods: A probability technique was employed in this research by using a simple random sampling method to select participants randomly and a quantitative approach was employed by using a survey-based data collection through a questionnaire. The population of the selected company consisted of 150 employees, the data was received from 122 employees reflecting a response rate of 88.5% of the sample size selected, and it was examined through descriptive statistics, reliability analysis, correlation matrix, and regression analysis.

Results: The findings demonstrated that an alpha value of all variables is above 0.84, which shows that the results for the test are highly reliable and valid. The results proved that there is a high correlation where all independent variables (Training Needs Assessment, Training Content, and Training Evaluation) are correlated with the dependent variables (Task Performance and Work Environment) with coefficients greater than 0.97, respectively. Regression analysis shows the strength of the link between variables, the R Square value in regression statistics table is 0.980, which means that the predictor variables in the model (Training and Development) explain higher and stronger relationship with (Employee performance), the P-value shows a value of 0.0012 which results in accepting the alternative hypothesis which is 'training & development affects employee performance'.

Originality: The originality of the current paper comes in identifying the implementation of training and development with employee performance and how sustainable capacity building helps to upskill and re-skill employees, specifically in one of the Qatari organizations in the telecommunication sector. The researchers have proposed a model to outline a

 DOI: 10.1201/9781003496724-58

sustainable solution of training and development designed to enhance capacity building through employee performance.

Conclusion: The research has outlined the essential aspects of training and development and its impact of employee performance in the telecommunication sector. This study proved that all variables of training and development such as training needs assessment, training content, and training evaluation are correlated with performance. Moreover, this research proves that training and development affects employee performance.

Keywords: Training and development, employee performance, employee perception, capacity building, sustainability, model

1 INTRODUCTION

Contemporary organizations focus on training and development as a paramount interest to build capacity and create evolvement of employee performance. With the increased demands for high-quality services especially in the telecommunication sector due to the market shift to digital technology and artificial intelligence, it is easy to ensure employee capacity building and increase employee performance through effective implementation of training and development. Complex challenges of capacity building are created through the company's poor available resources, including the existing employees and their skills, abilities, and knowledge, which is inadequate to perform efficiently in a high-level context. This research will analyze and illustrate employee perception of implementing effective training and development through capacity building related to employee performance.

Over the last decades, building capacity has been more vital to businesses over the years since it significantly improves performance and is essential for maintaining sustainable development in the long run [7]. However, Qatar's 2030 vision emphasized a future goal that is directly connected to the importance of building workforce capacity through providing high-quality training opportunities that match personnel interests and skills as it highlighted the importance of offering excellent training for Qatar's people as the future economic success will depend on their abilities [11].

Training and development are vital aspects of any organization. It meets all of the requirements for personnel to advance in their jobs by boosting their abilities with recent trends and digitalization technologies utilized to conduct training programs that contribute to the enhancement of organizational capacity building. The organization must carefully design and implement upskilling and re-skilling programs for current and new employees to avoid shortcomings in their personnel performance. However, upskilling and re-skilling play an important role in addressing the employee skills gap and working on improving their ability to perform quality work.

2 RESEARCH PROBLEM

Employee performance is a critical issue in this research study that occurred in one of Qatar's companies in the telecommunication industry. A deficit in training and development in this organization resulted in low employee performance and productivity. In other words, the employees in the selected organization lacked their ability to complete their tasks effectively. The organization did not provide the employees with standard expected responsibilities and capabilities needed by the workforce to execute their jobs. The shortfall in delivering current and future trends in various recent HR practices that focus on the development of employees created a high turnover rate in the company, resulting in a lower retention rate in employee career paths and engagement.

3 RESEARCH OBJECTIVES

This research study aims to better comprehend how various factors of training and development relate to capacity building and employee performance in the telecommunication sector. The study's primary aim is:

- To comprehend and examine the connection between the employees' unique skills and the demographic characteristics of the workforce in the selected telecommunications organization.
- To determine employees' perception of how training and development build the capacity of employees working in the selected organization in the telecommunication sector.
- To examine the opinion of employees on their performance.
- To investigate the association of training and development on employee performance.
- To examine and study how sustainable digital training and development programs affect employees' performance and suggest a sustainable capacity building model.

4 LITERATURE REVIEW

4.1 *Training needs assessment*

Training and development are paramount for any company to remain competitive in the market, especially in an industry that is full of sophisticated use of technology and continuous advancement. Many employees started to view training as an essential protection for job security to build capacity and increase current employee skills to remain with the company, in the long run, to cope with new changes in the environment [2]. Several researchers discussed the importance of the need for training by the personnel in the company and training and development is considered an ongoing process that is necessary for various reasons such as the desire to perform better to achieve organizational goals, the requirement to execute one's job effectively and to understand how to supervise others [6].

4.2 *Training content*

The content that is being delivered through the use of advanced technology has a much greater impact on employee learning than the technology used itself. However, improperly planned and executed training content and material will not achieve the learning outcomes. To meet the trainee's preferred learning or training method and styles to be received during the session, various multimedia technologies could be used to deliver training through the use of visuals such as video or pictures and audio which will facilitate the learning process. Recent research suggests blended learning in which blending technology with traditional instructions or forms in the training program to optimize training effectiveness by combining technology-based and instructor-led based training as it was rated in a survey study of 150 learners as the most effective and sufficient style of training to be offered [1]. E-learning systems are used by organizations to improve employee training and it is a money-saving tool to help employees become more professional at work. In addition, organizations can invest in e-learning to train employee's abilities and for the advancement of employees as well as their development [17].

4.3 *Training evaluation*

Training evaluation refers to determining how successfully employees learn and adapt to a system or program offered. Over the last few decades, various ways for conducting training assessments of systems or software have been created [12]. One of the models to evaluate the training is using the Kirkpatrick model which describes that the training evaluation goes into 4 different processes: reaction, learning, behavior, and impact [5]. Several articles and scholars discussed the purposes behind the evaluation of training and development as most of them see that the purpose is to judge the quality and worth of the training programs towards impacting the enhancement of trainees. In addition to the previous point, researchers suggest that training evaluation should be referred to the trainee's reaction through pre and post-training feedback [3]. The evaluation of capacity building plays a sufficient role, especially its contribution to employee performance. Sufficient funds, workspace, and equipment or materials provided for professional development are transformed into the capacity building through the effective process of implementation of training and other development activities that contributes to building personnel capacity [7]. The objective-driven model is a model of evaluation that focuses on the extent to which training objectives have been met, and the evaluation of the method used to transfer learning [3]. A recurrent element in HRD research is the importance of establishing measurable goals, adhering to a cost-effective strategy to accomplish them, and taking into consideration the evaluation to determine the degree to which they are reached.

4.4 *Employee performance*

Employee performance refers to the outcomes provided by individuals executing their job duties and responsibilities. However, employee performance is evaluated by HR and top management annually to provide insights into current overall organization performance. The importance of manager support in creating a healthy workplace environment and ensuring capacity building will directly affect employee performance and overall company outcome. This is since personnel spends a sizable portion of their life at work performing their jobs [20]. Thus, their cognitive and psychological states, attentiveness, behavior, reactions, and capacities are influenced by their workplace environment. It has a significant impact on both the performance and participation of the personnel. However, the organization benefits greatly from preserving a high level of employee performance and productivity, which in turn increases overall organizational production [13]. In fact, through an effective workplace environment, capacity building will play a sufficient role in enhancing employee capacity by improving their knowledge and abilities through the implementation of training. All organizations need to adopt on-the-job and off-the-job training in their operation standards procedures. Investing in employee capacity building that focuses on transforming the knowledge, abilities, skills, and behavior of employees related to their job is the need of the hour. Focusing on upskilling and reskilling employees to update their knowledge and increase workplace performance is essential to meet current and future trends [20].

5 RESEARCH METHODOLOGY

In this research study, a descriptive research design was employed to investigate and collect the needed information to achieve the aim of this study. In addition, primary research data was collected using a quantitative approach that utilized a survey-based questionnaire to study the relationship between training and development with its impact on employee performance. The population in this study consists of a total of 150 employees in five different subsidiaries of the organization. However, 145 responses were received from employees who are performing their jobs in one of Qatar's companies in the telecommunications sector. A probability sampling technique was used to allow an equivalent chance for every person in the population to be selected to be part of the sample [16]. However, to choose respondents, an arbitrary number of employees with at least one form of training in each subsidiary linked to their role were chosen using simple random sampling which produces statistically valid outcomes for the entire population. 108 employees were a valid number to be picked as a sample size with a 95% confidence level using the sample table created by Krejcie and Morgan [4]. However, 122 valid survey responses from the company and its subsidiaries were received, reflecting a response rate of 88.5% of the sample size selected.

6 METHODS OF DATA COLLECTION

Secondary and primary research was used to collect data and information needed within one-week time frame which will assist the research to illustrate further analysis and results

6.1 *Secondary research*

In this study, a literature review was developed using various peer-reviewed articles from different scholars, published papers with a limited timeframe, and journal articles through the use of several databases to collect data that already existed and examined similar research objectives and topics to support this research study with other valid studies.

6.2 *Primary research*

The primary research used by doing a survey-based approach to employees by adopting an existing questionnaire used by researchers in previous studies. The survey was created and sent out to their work email to all employees. The replies gathered will assist this study in achieving its research aims and goal. The questionnaire's participants had an equal chance of being chosen randomly or by chance within the company, according to the sampling method used.

6.3 *Research instruments*

The developed questionnaire consisted of 26 adapted Likert scale questions and included two constructs: training and development and performance management to achieve the paper's research objectives. The questionnaire was divided into three sections: employee demographic profile, training and development, and employee performance. Using the Likert scale the questions were rated from 1 = strongly disagree to 5 = strongly agree.

Section 1: Demographic profile included standard and generic demographic questions such as respondents' ages, education level, job experience, department, and monthly income.

Section 2: Training and development: consisted of 16 questions in total and it was divided into three sub-factors as follows:

- Training Needs Assessment: This section consists of 6 questions which is primarily used to identify performance requirements and the knowledge, skills, abilities, and other characteristics (KSAOs) of the company's staff to accomplish the set requirements as well as the quality of training and level of satisfaction [10].
- Training Contents: The training content refers to the information and ways in which the information will be presented to the learners (staff/employees) of the company. The main aim of the contents is to provide them with all the knowledge they are seeking after reviewing the training needs assessment and establish training topics and tools to help them move forward with the training process [10,14].
- Training Evaluation: Training evaluation is a systematic approach to gathering data and evidence that will be used to better enhance the training program. Training evaluation provides feedback that helps identify if the desired outcomes have been accomplished. This will also allow specific decisions to be made in the future (long-term) [10,18].

Section 3: Performance Management: Consisted of 10 total questions for employees to answer and it was divided into two categories:

- Task Performance: The efficacy of which job executives implement activities that subsidize the company's methodological/technical core [18].
- Work Environment: The inclusive setting, social features, and physical conditions in which an individual within the organization performs their job. Such elements can influence feelings of welfare, workplace relationships, teamwork, proficiency, and employee health and well-being [14,18].

7 RESULTS AND DISCUSSIONS

In this stage of the study, the techniques and analytical tools are used to interpret and analyze the questionnaire based on each statement's satisfaction level of the employees.

7.1 *Reliability of the questionnaire*

Cronbach's alpha is a tool to measure the internal consistency or reliability of a test. This method is used to test and to determine the reliability of the two constructs: Training and development, and performance management multiple-question Likert scale survey. This test was conducted using the excel tool Anova: Two-Factor without Replication.

The acceptable values of alpha range from 0.84 – 0.97. The closer the range to 1.00, the higher the reliability of this test [21]. Table 1 concludes that this research results for all the dimensions of training and development and employee performance are reliable and valid to use for the research as all test results were greater than 0.8, respectively. This is also confirmed by another study where test of reliability had a safe indicator, and as a result, the findings of that study showed that training evaluation was 0.91, training needs assessment was 0.94, and training contents and delivery methods had an alpha value of 0.88 [19].

Table 1. Reliability of variables.

Variable	Number of items	Reliability (Cronbach Alpha)
Training Needs Assessment	6	0.9645
Training Content	5	0.9774
Training Evaluation	5	0.8443
Task Performance	5	0.9223
Work Environment	5	0.9183

7.3 *Analysis of employees' demographic profile*

Pivot tables were employed to provide the study with precise percentages for each demographic question to examine the descriptive statistics of the personnel in relation to the research objectives.

According to the data from Table 2 frequency distribution, the survey's participants were 89% female and 11% male. In addition, 51% of respondents are between the age group of 26 and 35 and only 3% are between the age group of 18 and 25.

Regarding educational backgrounds, we can observe that 54% of respondents have a bachelor's degree, and 7% have a master's degree.

Regarding the respondents' professional experience, the majority of employees, with a rate of 43%, have between one and five years of experience, and 7% have more than fifteen years of experience.

The departments of the respondents reveal that 65% of respondents work in the educational subsidiary or department, including teaching, childcare, career counselors, academic coaches, etc., and 1% work in administration.

Table 2. Frequency distribution.

Variable	Options	Percentage
Gender	Female	89%
	Male	11%
Age	18–25	3%
	26–35	51%
	36–45	39%
	46–55	7%
Education Level	High School Degree	15%
	Diploma	24%
	Bachelor's degree	54%
	Master's degree	7%
Work Experience	Less than 1 year	7%
	1–5 years	43%
	6–10 years	27%
	11–15 years	16%
	More than 15 years	7%
Department	Education	65%
	Administration	1%
	IT	6%
	HR	2%
	Marketing	2%
	Customer Service	15%
	Operation Manager	5%
	Finance	4%
Monthly Income	Less than 5,000	66%
	5,001–10,000	30%
	15,001–20,000	3%

Regarding the income, we can interpret from the table that 66% of participants generate less than QAR 5,000 per month, while 30% generate between 5,001 and 10,000, 3% make between QAR 15,001 and 20,000.

7.3 *Association of training and development and employee performance*

This analysis underlines the importance of employees' perceptions of the correlation between training and development and employee performance in the organization as well as the effect it has on how managers view employees' ability to accomplish their duties. Table 3 reveals the results of this investigation using the correlation approach.

- The correlation coefficients highlight the areas in which the variables are the most strongly connected to one another. The correlation results indicate that all independent variables (Training Needs Assessment, Training Content, and Training Evaluation) are correlated with the dependent variables (Task Performance and Work Environment) where all coefficients are greater than 0.97, respectively.
- There is a strong positive correlation between training evaluation and task performance (0.9978). Table 3 shows the correlation between the evaluation of training in terms of employees' opinions on how providing adequate training and the level of expected training

Table 3. Association of T&D and employee performance.

Criteria	Task performance	Work environment
Training Needs Assessment	0.9801	0.9954
Training content	0.9865	0.9965
Training evaluation	0.9987	0.9906

provided has an impact on employee performance and productivity. Vasudevan's (2014) research illustrated that the effects of training evaluation on employee performance are high, which is similar to the results achieved in this research in the correlation Table 3 [19]. There is a strong positive correlation between Training Needs Assessment(TNA) and Training content with task performance with a value (0.9801 , 0.9865) and Work environment, which indicates a value (0.9954 , 0.9956). Both illustrate a high relationship. The TNA shows their impact on how the work environment could affect employees to perform better, as well as the effects of TNA on the quality and quantity of implementing their task or job. According to Vasudevan (2014) research, the study indicates a high value of correlation between TNA and employees' job performance. In addition, the training content provided is essential for employees to produce a higher quality of performance and higher productivity [19]. The findings demonstrate how crucial it is to identify employees' training needs in order to boost productivity. Additionally, it suggests that a valid training program with a clearly stated training objective and useful curriculum content may persuade staff members to participate in the training in order to develop their capacities and ability to work effectively. Results also show that better training needs assessments, training content, and training evaluation are associated to higher employee performance and a positive work environment. The interaction between the training as a whole and the way that people conduct their duties has an impact on the workplace.

7.4 *Analysis of employees' perception on training programs and performance*

Employees should understand the importance of training programs. Moreover, this research aims to understand the employee's perception of the training programs that have been implemented in one of Qatar's companies in the telecommunication sector. To determine that, cross-tabulation was used to read the employees' perceptions. Table 4 illustrates the questions that respondents answered about training and development. 65% of the employees perceive that the training programs improve the capacity building of the employees.

Table 4. Cross tabulation of employee perception on training & development.

Statement	Strongly Agree	Agree	Neutral	Disagree	Strongly Disagree
Training Needs Assessment					
1. I feel training is essential for my job title.	30%	38%	19%	1%	13%
2. I understand the purpose behind training for my job.	32%	38%	16%	2%	11%
3. Training reduces the stress of the employees.	25%	37%	23%	3%	11%
4. Enhanced technology used in training programs are useful.	28%	38%	20%	2%	11%
5. Understating the objectives of the training program helps me to perform my task.	30%	38%	18%	2%	11%
6. The company provides right training for the right people.	15%	42%	27%	3%	13%
Training Content					
7. Training programs provided by the organization are adequate for my skills development.	19%	38%	28%	3%	12%
8. Training has significant role in improving work quality.	26%	42%	17%	2%	12%

(continued)

Table 4. Continued

Statement	Strongly Agree	Agree	Neutral	Disagree	Strongly Disagree
9. I am able to understand the tool used by the trainer.	19%	44%	23%	2%	11%
10. Training programs improve employee capacity building	20%	45%	20%	2%	12%
11. The training programs that were offered is organized.	16%	40%	28%	2%	14%
Training Evaluation					
12. Training improves job knowledge in the task.	30%	38%	18%	3%	11%
13. The company's training program meets my expectations.	15%	40%	28%	2%	15%
14. My trainer accepts feedback.	17%	38%	29%	2%	14%
15. I feel lack of training increases the turnover.	8%	25%	35%	16%	16%
16. I had a thought of leaving the organization due to lack of training.	7%	14%	31%	21%	26%

Table 5 illustrates and reads the respondent's results about their performance. The results prove that 59% of participants opinion that the company's improvement plan helps them to work better which illustrates the importance of performance management to enhance employee performance and productivity. Moreover, 61% of participants require a valuable workplace space, this proves that the company needs to enhance its workplace for

Table 5. Cross tabulation of employee perception on performance.

Statement	Strongly Agree	Agree	Neutral	Disagree	Strongly Disagree
Task performance					
1. I don't have the adequate ability to complete the task.	2%	11%	28%	20%	39%
2. The organization is giving adequate attention to complete my task.	10%	41%	31%	7%	11%
3. The company's performance improvement plan helps me to work better.	20%	39%	22%	7%	11%
4. Rewards motivate me to improve my performance.	22%	38%	22%	6%	12%
5. I take extra responsibilities to finish my team task on time.	24%	37%	22%	8%	9%
Work environment					
6. I get adequate communication from my boss.	20%	37%	28%	4%	11%
7. I feel the workplace is valuable for me to perform better.	25%	36%	25%	3%	11%
8. My relationship with my colleagues affects my performance.	16%	30%	26%	12%	15%

(continued)

Table 5. Continued

Statement	Strongly Agree	Agree	Neutral	Disagree	Strongly Disagree
9. Involving employees in decision making improves performance.	25%	42%	20%	4%	10%
10. The organization atmosphere improves my mental health.	24%	33%	27%	4%	12%

participants to improve their performance. In addition, 67% of participants would like management to have their intake in decision-making. This involvement will make employees feel valuable and trustworthy which will help increase their performance and productivity in the organization. These results emphasize how organization atmosphere is significant in improving employees' mental health which impact on their performance as well. Overall, the results prove that organization require a positive workplace environment that involve the employees in decision-making process which increases the employee satisfaction, autonomy resulting in quality performance at work.

7.5 *Regression analysis of training and development impact on employee performance*

To examine and analyze how training and development significantly affects individual performance, regression analysis was conducted. This section of the analysis discusses the hypothesis and demonstrates the substantial effects of training and development on employee performance.

From Table 6, we can note that the R Square value in regression statistics table is 0.980, which means that the predictor variables in the model (Training and development) explain 98% (0.980 x 100) of the variance in the outcome variable (Employee performance) which illustrates a higher and stronger relationship with each other.

As shown in Table 6, the regression for Training and development has a P-value that is less than the conventional two-tailed cut-off (alpha level) of 0.05 ($p < 0.0012$), which means we can treat these regression weights as its association with employee performance are statistically significant. Therefore, the p-value is less than the level of significance which is $p < 0.05$. So, we reject our null hypothesis H_0; we accept our alternative hypothesis, H_a.

Table 6. Regression analysis of training and development on employee performance.

Regression Statistics	
Multiple R	0.990
R Square	0.980
Adjusted R Square	0.973
Standard Error	12.96
Observations	5

Criteria	Coefficients	Standard Error	t Stat	P-value
Intercept	−22.5591	13.9060	−1.6223	0.2032
Employee performance	1.2516	0.1036	12.0806	0.0012

• *H_0 = Training & Development does not affect employee performance*
H_a = Training & Development affects employee performance

The results prove that there is positive relationship between the training and development and performance. The outcome illustrates the need for training at the company which will help them increase their capabilities and encourage higher performance and the quality of tasks provided by the employee as well. Training is necessary for their job to provide higher performance, the quality of content provided in the training could help in building the capacity of employees which motivates employees to engage in training. Moreover, the evaluation of training is important for employees in terms of considering their feedback on training, their opinion of the impact of new knowledge on implementing their tasks, and on turnover rates.

9 FINDINGS

This study findings suggest that training needs assessment, training contents, and training evaluation influence an employee's task performance and work environment when new knowledge and expertise are imparted to them. Employee training and development is a critical component of an employer's ability to remain competitive in the future. The results prove that training and development for employees reduce work stress.

Training content is essential for employees to enhance their capabilities and improve their knowledge to provide a better quality of delivered task or job, representing a percentage of (62%). Respondents see that training needs assessment should be established in the organization as it is the key to understanding employee needs for training to improve their capabilities which represents (65%). The results prove that (61%) of respondents require an enhancement for the workplace space, (67%) prefer to be involved in the decision making which increases employee motivation. (67%) of participants emphasized the need for an organizational atmosphere in improving their mental health which causes an impact on their performance as well.

The correlation results indicate that all independent variables (Training Needs Assessment, Training Content, and Training Evaluation) are correlated with the dependent variables (Task Performance and Work Environment) with a strong positive correlation of (0.9978) which concludes that providing adequate training has an impact on the level of employee performance and capacity building. There is a strong positive correlation between training needs assessment and training content with task performance with values 0.9801, 0.9865 respectively and with that of work environment with values 0.9954, and 0.9956 respectively. This proves the importance of the need for training with the right content to have advanced employee performance and also how these components are related to the workplace environment to motivate employees towards engaging in training programs and encourage higher performance.

The regression analysis results illustrate that training and development positively influence employee performance with a ($p < 0.0012$) which means that the training and development at the company will help them increase their capabilities, and encourage higher performance and the quality of tasks provided by the employee. To support the research findings, the research by Lilian (2019) has also found that employee training has an impact on the employee's performance which leads to better employee morale and reduces turnover as well as absenteeism [8].

9 RECOMMENDATION

The following section provides a brief explanation of the model which is an outcome of this research which provides sustainable solution to capacity building, as shown in Figure 1:

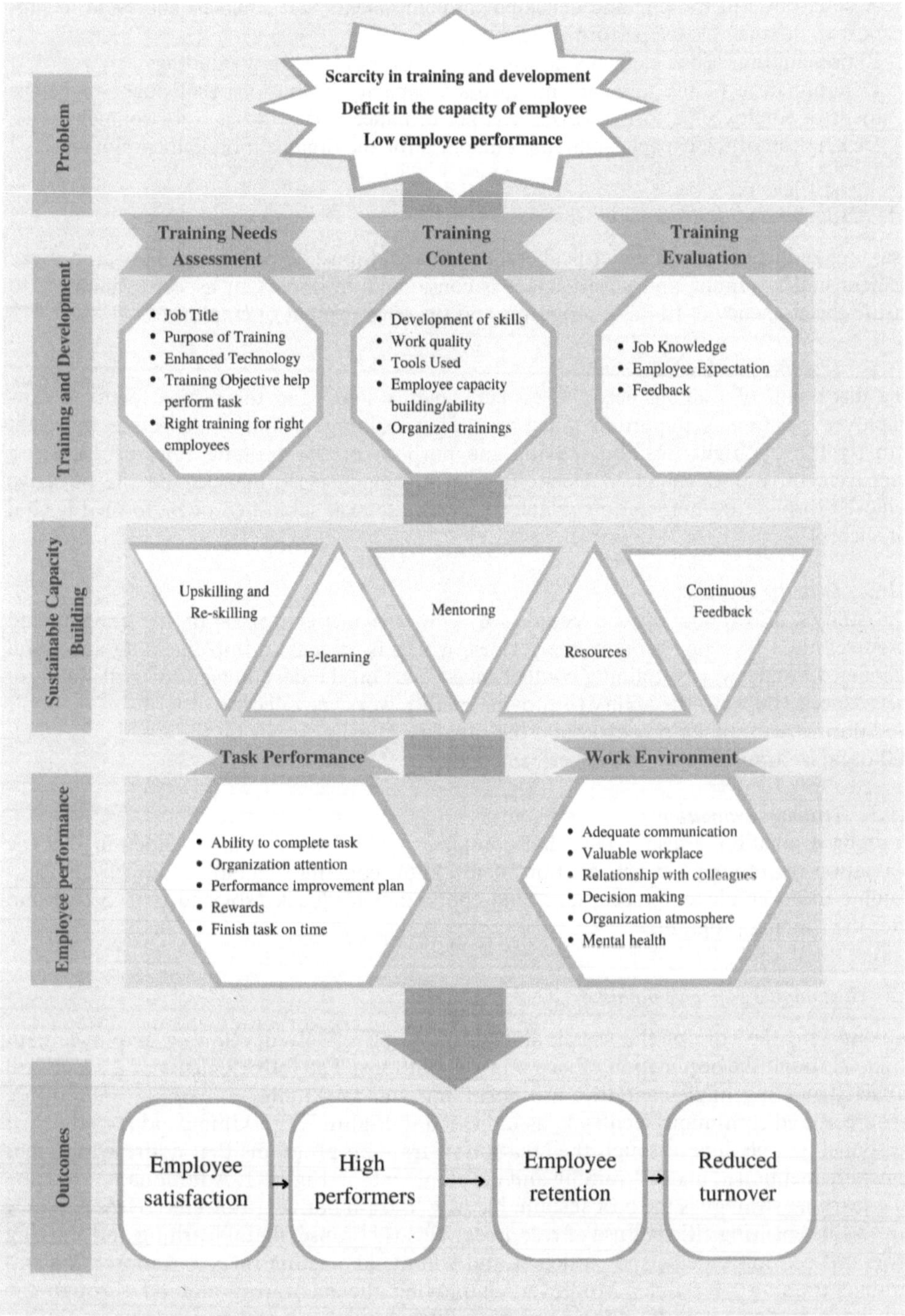

Figure 1. Training and development to drive performance – A sustainable capacity building model.

- A description of training and development components that could be addressed to support to drive employee performance.
- Representation of the elements necessary for successful capacity building
- Determination of the targeted intermediate capacity aspirations that ought to be the focus of employee performance via task performance and workplace environment
- Description of the capacity-building outcomes for the organization and employees

9.1 *Training and development*

Training and development are critical components that must be prioritized. However, every element under training and development is considered an important aspect to improve to ensure the efficiency of training programs and the development of employees.

9.1.1 *Training needs assessment*

The discussion of training needs assessment in this study and the results focus on how employee position is important when providing training programs. Selecting the right training for the right position, having the purpose of the training towards employee improvement, use of enhanced technology in improving performance, and the objective achieved through training and its relatedness with the task submitted or performed is vital for success.

9.1.2 *Training content*

The study's illustration opened people's eyes to the ineffectiveness of the content and resources used in training or by the trainer, which is critical to implementing successful training. In other words, training content must focus on the development of skills in all of their aspects, improve the ability to provide quality work, be effective in terms of the tools used during the training process, build the capacity of employees with the training content, and organize training to improve performance.

9.1.3 *Training evaluation*

Training evaluation is also a part that completes all the other elements, focusing on the evaluation of training's transformation of job knowledge into employee knowledge, how training meets employee expectations, and continuous feedback provided both before and after the training program.

9.2 *Sustainable capacity building*

Moving into the core of the sustainable solution, effectively developing employee capability through the application of sustainable solutions. The four elements of a successful capacity-building implementation are upskilling and re-skilling, E-learning, mentoring, resources, and continuous feedback, as indicated in Figure 1. Upskilling and re-skilling of personnel can be done through the creation of training programs that address all issues that were mentioned under "training and development" in Figure 1. Additionally, utilizing an e-learning strategy is the best sustainable way to train both existing employees and new hires while ensuring efficient use of resources. This is because digital learning and training programs are convenient, and sufficient and will avoid wasting regular resources on traditional training over time. Moreover, employing the mentoring method through the process of implementing upskilling and re-skilling through e-learning will help employees being more comfortable and satisfied enough to continue in the process with full understanding.

In addition, resources and materials should be listed as a link to be used to facilitate the training program for employees, as well as the availability of resources to implement the training itself in terms of funds, equipment, material, trainers, and advanced technology.

Finally, providing feedback for employees continuously will ensure continuous capacity improvement through pre and post-training feedback as well as feedback daily.

9.3 *Employee performance*

Focusing on how to improve employee performance is the main discussion in this research study. The model shown in Figure 1 illustrates the two elements of focus under the employee performance variable. Task performance illustrates the level of employee ability to perform a task, the contribution of organization attention to employee ability to perform a task, and the development of a Performance Improvement Plan (PIP) that will help the employee work on their 3 to 5-year career plan, rewarding employee for achievement as a motivation, and finally their ability to complete tasks on time. In addition, the work environment is the biggest contributor to overall employee performance and capacity building. Focusing on providing adequate communication from top management as well as among co-workers will enhance the overall flow of work and performance, provide a valuable workplace, improve the relationship among coworkers to facilitate daily workflow, involve employees in the decision-making process, and enhance the atmosphere and workplace environment that contribute to the positive mental health of employees.

Outcomes:

Based on the proposed model of a sustainable capacity-building, the desired outcome from this model's implementation is that it focuses on the use of sustainable solutions using e-learning as well as addressing issues that were analyzed in this research under the elements of training and development and employee performance. However, developing this model will contribute to overall employee satisfaction which is the core result behind high-performing employees which leads to retaining employees in the organization in the long term and end up reducing the turnover of employees.

10 IMPLICATION OF FUTURE RESEARCH

This study concluded a noticeable positive relationship between training and development, and employee performance. However, there are still numerous other variables that might play a role along with this construct. It would be beneficial for researchers to conduct research in the areas in the future by including additional sustainable dimensions such as economic growth, quality education, gender equality, and innovation along with other working conditions concerning training and development to create more convincing results that may increase the performance of employees. In addition, future studies could emphasize how training and development could lead to high organizational profit in terms of its effect on employee productivity.

11 LIMITATIONS

This study had three different limitations. The major limitation is that the survey was not distributed equally between the genders of the employees within the organization and only 11% of males who answered the survey cannot represent all the males in the company (the survey statistics showed that 89% of the survey sample were females and only 11% males). Overall this shows that there is inequality in the gender distribution within the organization that can be an indicator of gender bias.

The second limitation of this study is that most of the survey responses came from the organization's education department; one department cannot represent all the other departments in the company. Future research study could be done with more involvement of accounting, technical and human resources department respondents in the organization.

The third limitation of this study is that the respondents of the survey were from one organization due to the time limitation. The results cannot be generalized to all the telecommunication organizations in Qatar. Future research could be done with samples representing various organizations from the telecommunication sector in Qatar.

12 CONCLUSION

In conclusion, this study has outlined the essential aspects of training and development and its impact on employee performance in the telecommunication sector. Sustainable capacity building in the telecommunication sector has been analyzed and clarified that capacity building helps in improving employee performance. This research concludes that the training and development at the company will help employees to increase their capabilities and encourage higher performance and the quality of tasks provided by the employee. This proves the importance of the right training with adequate content is essential to have advanced performance. Also, it was found that these components are related to the workplace environment to motivate employees to engage in training programs and result in higher performance. The study ensures the importance of how implementing sustainable training and development strategies will play a significant role in developing employee's performance. It also enhances the career path of employees through building their capacity to ensure employee's career success within the organization.

The study is based on the information gathered, survey data collected through a questionnaire, data analysis, and findings through quantitative primary research and methodology. Overall, training and development are intertwined with sustainable employee performance through capacity building. The originality of the current paper comes in suggesting a model that supports capacity building with the existing employees in the organization. Moreover, this research was focused on one of the Qatari organizations in the telecommunication sector. Future research could be considered on other organizations to implement sustainable capacity building within their organization to improve the performance of the workforce and bridge the skill gap.

REFERENCES

[1] Bell, B. S., and Kozlowski, S. W. J. (2012). *Advances in Technology-Based Training*. ResearchGate. https://www.researchgate.net/publication/239823550_Advances_in_Technology-Based_Training

[2] Farr, J., and Sullivan, J. (1996). Re-Thinking Training in the 90s. *Journal of Management in Engineering*, *12*(3), 29–33. https://web.stevens.edu/ses/documents/fileadmin/documents/pdf/Training.pdf

[3] Foxon, M. (1989). Evaluation of training and development programs: A review of the literature. *Australasian Journal of Educational Technology*, *5*(2). https://doi.org/10.14742/ajet.2340

[4] Kenpro. (2012, August 25). *Sample Size Determination Using Krejcie and Morgan Table*. Kenya Projects Organization [KENPRO]. https://www.kenpro.org/sample-size-determination-using-krejcie-and-morgan-table/

[5] Kirkpatrick, D. L. (2009). *Evaluating Training Programs*. USA: Berrett-Koehler Publishers.

[6] Kulkarni, P. P. (2013). A Literature Review on Training & Development and Quality of Work Life. www.academia.edu. https://www.academia.edu/36603578/

[7] Lafond, A., and Research, J. (2003). A Guide to Monitoring and Evaluation of Capacity-Building Interventions in the Health Sector in Developing Countries Measure Evaluation Manual Series, No. 7 MEASURE Evaluation Project. https://www.data4impactproject.org/wp-content/uploads/2019/09/ms-03-07-1.pdf

[8] Lilian, N. (2019). Training And Employee Performance A Case Study Of Rider Hotel Mukono Municipality Uganda. https://ir.kiu.ac.ug/bitstream/20.500.12306/6069/1/Namata.pdf

[9] Neelam Tahir, Israr Khan Yousafzai, Dr. Shahid Jan, and Muhammad Hashim. (2014, April 3). *The Impact of Training and Development on Employees' Performance and Productivity A case study of*

United... ResearchGate; unknown. https://www.researchgate.net/publication/314911038_The_Impact_of_Training_and_Development_on_Employees_Performance_and_Productivity_A_case_study_of_United_Bank_Limited_Peshawar_City_KPK_Pakistan

[10] Otuko, A. H., Chege, D. K., and Douglas, D. M. (2013). Effect of training dimensions on employee's work performance: A case of mumias sugar company in kakamega county. *International Journal of Business and Management Invention*, 2(9), 138–149.

[11] Planning and Statistics Authority. (2019). *Qatar National Vision 2030*. Psa.gov.qa. https://www.psa.gov.qa/en/qnv1/Pages/default.aspx

[12] Saminathan, M., and Mohd Suaib, N. (2021). Training Evaluation Models for Skill-Based E-learning System: A Systematic Literature Review. *International Journal of Innovative Computing*, *11*(2), 61–65. https://doi.org/10.11113/ijic.v11n2.323

[13] Satyendra. (2019, May 26). *Impact of Workplace Environment on Employee Performance – IspatGuru*. Ispatguru. https://www.ispatguru.com/impact-of-workplace-environment-on-employee-performance/

[14] Siddiqui, D. A., and Sahar, N. (2019). The Impact of Training & Development and Communication on Employee Engagement – A Study of Banking Sector. *Business Management and Strategy*, *10*(1), 23. https://doi.org/10.5296/bms.v10i1.14592

[15] Straussman, J. D. (2007). An Essay on the Meaning(s) of "Capacity Building"—With an Application to Serbia. *International Journal of Public Administration*, 30(10), 1103–1120. https://doi.org/10.1080/01900690701434661

[16] Taherdoost, H. (2016). Sampling Methods in Research Methodology; How to Choose a Sampling Technique for Research. *International Journal of Academic Research in Management*, 5(2), 18–27.

[17] Tina Cheng, W., and Chen, C. (2015). The Impact of e-Learning on Workplace On-the-job Training. *International Journal of E-Education, E-Business, E-Management and E-Learning*, 212–228. https://doi.org/10.17706/ijeeee.2015.5.4.212-228

[18] Tahir, N., Yousafzai, I., Yousafzai, I., Jan, D., and Hashim, M. (2014). The Impact of Training and Development on Employees' Performance and Productivity A case study of United Bank Limited Peshawar City, KPK, Pakistan. *International Journal of Academic Research in Business and Social Sciences*, 4(4), 86–98.

[19] Vasudevan, H. (2014). Examining the Relationship of Training on Job Satisfaction and Organizational Effectiveness. *Int. J. Manag. Bus. Res*, *4*(3), 185–202. https://ijmbr.srbiau.ac.ir/article_2874_eea2d71f268d93e9f7cdedb9f0e1a56c.pdf

[20] Wassem, M., Baig, S. A., Abrar, M., Hashim, M., Zia-Ur-Rehman, M., Awan, U., Amjad, F., and Nawab, Y. (2019). Impact of Capacity Building and Managerial Support on Employees' Performance: The Moderating Role of Employees' Retention. *SAGE Open*, *9*(3), 215824401985995. https://doi.org/10.1177/2158244019859957

[21] Zaiontz, C. (2020, July 23). *Cronbach's Alpha Basic Concepts | Real Statistics Using Excel*. Real Statistics. https://www.real-statistics.com/reliability/internal-consistency-reliability/cronbachs-alpha/cronbachs-alpha-basic-concepts/

Innovation and Technological Advances for Sustainability – Al-Naemi et al. (Eds)

Ecofriendly conversion of marble waste into value-added tiles for green building

Mo Hamed Kurdi, Khalifa Al Ghafri, Fahad Al Hamad, Khalifa Hassan & Jamal Kurdi
Department of Chemical Engineering Technology, University of Doha for Science and Technology, Doha, Qatar

Michel Michaud
Department of Mechanical Engineering, University of Doha for Science and Technology, Doha, Qatar

ABSTRACT: Sustainable manufacturing is urgently required nowadays, more than ever before, to overcome current challenges in climate changes, depleting natural resources and polluting environment. Based on the Qatar National Vision 2030 (QNV 2030) and the third Qatari National Development Strategy (2023–2027), converting industrial solid waste in Qatar into value-added products is an important target which should be implemented to ensure sustainable consumption and production patterns as set by the United Nations, in the global goal: SDG 12. Away from dumping in landfills, preserving value of waste and byproduct in Qatari industries is an ideal approach to maximize economic gain, keep clean environment, optimize sustainability and reduce carbon footprint. Upgrading existing industrial procedures and increasing recycling rate should be considered for the production of new or improved quality products. Marble mining and processing generate around 200 million tons of waste per year worldwide. Quarrying, cutting, sawing, sizing, carving and polishing marbles produce aqueous slurry, sludge, powder or dust wastes from marble materials. These wastes lead to destructive irreversible impacts on ecological systems, the environment, and public health, particularly when they are discharged arbitrarily to the environment or even dumped into the landfills. Converting waste marble powder into alternative marketable products at a lower cost or better performance compared to that for the current existing products will promote sustainable economic growth and development. Marble waste can be used in different applications such as asphalt, road construction, gypsum plaster, cement, aggregates, concrete, mortar, brick making and other building materials. Also, marble waste can be used for replacing the traditional clay, fillers and other conventional materials to manufacture of high-performance hybrid composites. The socio-economic benefits of producing hybrid polymer composites are not only recycling marble waste but also making a cost-effective sustainable construction material with desired characteristics for green building. The unsaturated polyester is selected as polymeric matrix due to its low cost and good properties. High performance hybrid polymer composites are prepared from unsaturated polyester and then, tested for their mechanical, physical, and thermal properties. In this progress work, we have explored the challenges to produce ecofriendly tiles which are cost-effective, durable and suitable for green building. The disadvantages of marble powder-polyester composite are low modulus and insufficient rigidity, shrinkage particularly during curing, low aging resistance, less water resistance compared to metals or high-performance engineering plastics. This project is still in progress. The plan for the remaining work is to use different additives in small amount, without any significant increase in the cost, in order to overcome the challenges and disadvantages of the produced hybrid composite which can be used for manufacturing tiles. The plan could also include to make some tests, with a fixed percentage of marble powder, but testing different resins, such

 DOI: 10.1201/9781003496724-59

as vinyl-esters and epoxies. For demo testing; two model houses (prototypes) with a size of 60x60x60cm will be constructed. One will be made from the hybrid composite tiles and the second made from traditional tiles. Heat leak test will be conducted in both prototypes by measuring the difference in temperatures which is created between inside the prototype box and the outside surroundings, and the required time to reach thermal equilibrium will be recorded. The longer the time to reach the thermal equilibrium the better the thermal insulation. Reducing the heat leak in green building is one of the most effective method in decreasing the indoor required cooling energy during hot weather. This also can be applicable to the heat leaking out of buildings in a cold weather environment.

Keywords: Marble Waste, Polyester Resin, Hybrid Composites, Tiles, Green Building

1 INTRODUCTION

Based on the Qatar National Vision 2030 (QNV 2030), the state of Qatar has commitments to support sustainable research, economic development and innovation for achieving the 17 global goals and total 169 targets with 231 unique indicators as set by the United Nations [1,2]. To approach the high standard life and well-being of the Qatari society, QNV 2030 has adopted the pillars of sustainable economic growth, social justice and ecofriendly environmental sustainability. Based on aligning the achieved work in the first and second National Development Strategy (2011–2016 and 2018–2022) to the 17 global goals, the third National Development Strategy (2023–2027) will address the future work to cover all uncompleted global goals and targets [1,3]. Using recycled solid-waste to make new products is of vital importance as it helps to keep a healthy environment and sustainability. In 2020, the household solid waste treatment center has treated more than 30,000 tons of car tires in "Rawdat Rashed" and "Umm al-Afaei" landfills, in addition to recycling around 420,000 tons of construction wastes. Also, the center has produced more than 30,000 tons of organic fertilizer and more than 33 million cubic meters of biogas as well as generating around 269 thousand megawatts-hour of electricity. By the end of the Second National Development Strategy 2018–2022, Qatar has recycled 15% of the solid waste as aligned to the SDG 12: Sustainable Production & Consumption [1,3]. For the third National Development Strategy (2023–2027), converting industrial solid waste in Qatar into value-added products is an important target which should be implemented to ensure sustainable consumption and production patterns [1–3] as set in the global goal: SDG 12. Away from dumping in landfills, preserving value of waste and byproduct in Qatari industries is an ideal approach to maximize economic gain, keep clean environment, optimize sustainability and reduce carbon footprint. Upgrade existing industrial procedures and increase recycling rate should be considered for the production of new or improved quality products.

Marble processing industries generate annually around 200 million tons of marble waste worldwide as water slurry, sludge or powder; this includes 15 million tons of marble waste which was created during 2018–2019 in India alone [4]. Marble dust and slurry in water are generated during processing marble by sawing, cutting, sizing, carving and polishing. These wastes lead to destructive irreversible impacts on ecological systems, the environment, and public health, particularly when they discharged arbitrarily to the environment or even dumped into the landfills due to the lack of good regulations and law enforcement [5–7]. Particulates from marble contaminate agricultural land and lead to change in the physicochemical properties of soil with possibility of negative impacts on the micro-organisms and nutrients. The fertility of the soil may become low due to an increase in the alkalinity, water clogging of the soil and the decrease in the concentrations of nitrogen and phosphorus, in addition to a decrease in the cation exchange capacity of soil. This may lead to a reduced crop yield in the contaminated farms [8]. The suspended marble waste in water leads a change in the physico-chemical parameter of water such as water turbidity, increasing the electrical conductivity due to excessive dissolved inorganic salts in ionized form; increasing the alkalinity due to an increase in the concentrations of carbonate and bicarbonate in water, increasing the total dissolved solids (TDS), heavy metal concentrations and increasing water hardness due to an increase in the concentrations of calcium and magnesium. Although these parameters may slightly exceed the desirable limits of WHO standards of drinking water, no significant adversely effects on human health was reported [5,9–15]. After the marble waste slurry is dumped in the landfills, it is naturally dried and forms wind-blown dust. Airborne dried marble powder particles with a diameter of less than 10 microns are generated [16]. Particles with a diameter of less than 1 micron is considered to be the reason for the most dangerous air pollution. Fine and ultrafine crystalline silica existing in marble dust may enter deep into the bronchial tubes and even into the alveolar sacs of human being and animals causing dust-related lung diseases [16–19]. Also, airborne dust deposit over leaves of trees, plants and vegetation create aesthetic problem, and lead to reduce light for photosynthesis, in addition to block stomata and so, prevent leaves from taking up carbon dioxide (CO_2) [16]. Therefore, dumping marble slurry in the landfills requires additional procedures which should prevent generation of fine dried marble dust that causes air pollution. To comply with the strictest environmental regulations, this makes the appropriate disposal of marble dust in landfills technically very difficult and costly.

2 EXPERIMENTAL PROCEDURES

2.1 *Materials*

Unsaturated isophthalic polyester resin, methyl ethyl ketone peroxide, cobalt naphthenate and polyvinyalcohol were bought from AlWakrah Fiberglass company who specialize with fabrication and sales of composite products. Waste marble dust as slurry in water was collected from AKIRE, a marble company in industrial area in Qatar. The catalyst is Methyl Ethyl Ketone Peroxide in Dimethyl Phthalate (BUTANOX M-50). This catalyst is a corrosive and flammable hazard, and students should not breathe vapour or expose the material to heat or ignition sources, and should wear protective gloves eye/face protection and protective clothing because the chemical can cause serious injuries [20]. Students have been trained to use the catalyst in very small amount inside fume-hood with the appropriate personnel protective equipment (PPE).

2.2 *Composite fabrication*

The aim of the first stage of our research, which is presented in this paper, is to prepare the base polymeric hybrid composites at different marble powder contents. The unsaturated polyester, as a cheap polymer matrix, was reinforced by marble waste powder at a range of weight concentration between 16.67% and 50%. The marble slurry was collected from a marble company in the industrial area in Doha. It was dried and grinded to fine powder. Then, the dried marble powder was mixed with the unsaturated polyester resin and catalyst as shown in Figure 1. Using a 3D printer, moulds were fabricated with suitable shapes as required by the testing procedures. Then, the hybrid composite resin was moulded, and was left to be cured for one week as shown in Figure 2. The samples were tested for density, tensile, hardness and thermal conductivity properties.

2.3 *Tiles fabrication*

The aims of the second stage of our research, which will be presented in the next paper, are to optimize the composite formulation as well as to add small amount of additives which will

Figure 1. Slurry waste collected from a marble company in Qatar to prepare hybrid polymer composite which can be used to make tiles for construction of green building.

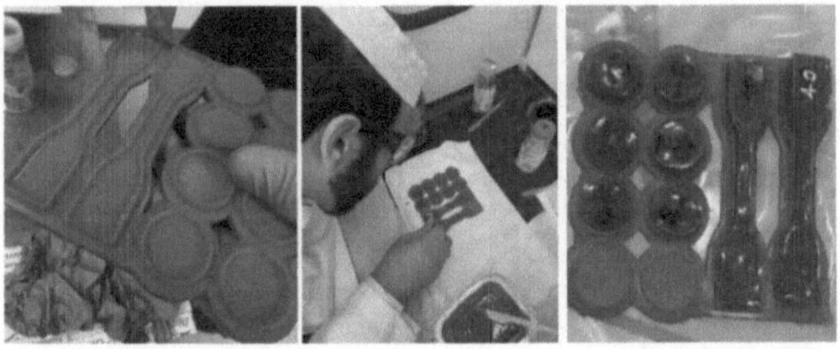

Figure 2. Moulding and curing samples of the hybrid polymer composite to be used for testing the properties and characteristics of the final product at different marble waste concentrations.

improve the performance and quality of the hybrid composite without a significant increase in the cost. The unsaturated polyester was selected as a base matrix material due to its low cost and good properties. The steps to fabricate the hybrid polymeric composite as a tile is illustrated in Figure 3. The marble slurry will be filtered, partially dried at ambient conditions for 24 h, grinded and then further dried in oven at 60 °C for 48 h. The hardener at a ratio of 1:10 in polyester is mixed well with the dried marble powder, and then casted in different mould with a shape and size suitable for testing samples. To prepare tiles, the composite is casted in mould (30 cm x 30 cm x 1 cm) and incorporating with different netted fibers, glass, cotton or jute-recycled material. and then compressed at a pressure of 50 psi for 1 hour. The tile is removed from the mould and cured a room temperature for 10 hours. Compression moulding technique was used in our research because it is simple compared to vacuum infusion moulding technique, and it produces the desired shape of the tile with no issue.

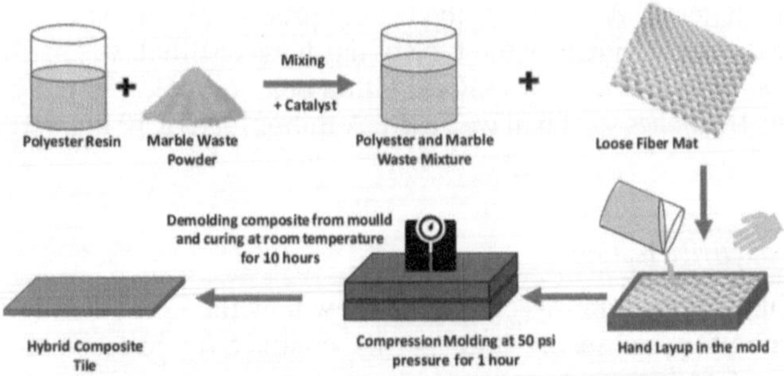

Figure 3. Schematic diagram of hybrid-polymer composite tile preparation.

3 TESTING PROCEDURES

Samples were prepared at different marble powder content in the composite ranging from 16.67% to 50%. The 3D moulds, shown in Figure 2, were printed with a standard size and shape which is appropriate for each testing.

3.1 *Density and porosity*

The density of the composites' samples were determined, using Archimedes' principle, from the weight of each sample and the displaced volume in water medium. The theoretical density (ρct) can be calculated from the following equation:

$$\rho_{ct} = 1/((W_f/\rho_f) + (W_m/\rho_m)) \tag{1}$$

Where ρ and W are referred to density and weight; f and m are referred to filler and matrix, respectively.

The volume fractions of void (V_v) or porosity is determined by using the rule of mixture as follow:

$$V_v = (\rho_{theoretical} - \rho_{measured})/\rho_{theoretical} \tag{2}$$

Where ρ is referred to density of the hybrid composite.

The density of the waste marble dust is 2.8 g/cm^3, and the density of unsaturated polyester resin is 1.15 g/cm^3 as reported elsewhere [21].

3.2 *Tensile testing*

The simplest and most typical kinds of mechanical testing are carried out by a tensile testing equipment. In a tensile test, the sample is pulled with tensile force to record how the sample response to the stress. Tensile tests establish a material's strength and elongation limit in this way. Tensile tests are commonly performed using electromagnetic or universal testing equipment, as they are straightforward to carry out, and are completely standardized. According to ASTM D3039, the composite's samples were prepared to be 25 cm length, 2.5 cm width, and 0.3 cm thickness. (not really ... but your call). The loading rate was set at 5 mm/min using 34TM INSTRON universal testing machine.

3.3 *Rockwell hardness testing*

A hardness test is a technique used to measure the toughness of a material. Hardness is the plastic material resistance to permanent surface deformation before fraction. There are several methods for measuring hardness, each of which tests might reveal different hardness levels for the same material. As a result, the testing procedure may vary, and the results of every test must be labelled to identify the type of hardness test that was performed. Indentec Universal Rockwell (ZHU 187.5, ZwickRoell GmbH & Co. KG, Germany) equipped with indenter consists of 1/6 inches steel ball was used. A minor load of 10 kgf and a major load of 100 kgf was used.

3.4 *Thermal conductivity testing*

The thermal conductivity add on was created to show how the Fourier Rate equation can be used to model straightforward steady-state thermal conductivity in one dimension. Armfield HT11–Linear Heat Conduction apparatus was used to measure the steady-state thermal conductivity of the hybrid composite. A circular sample with a diameter of 2.5 cm and

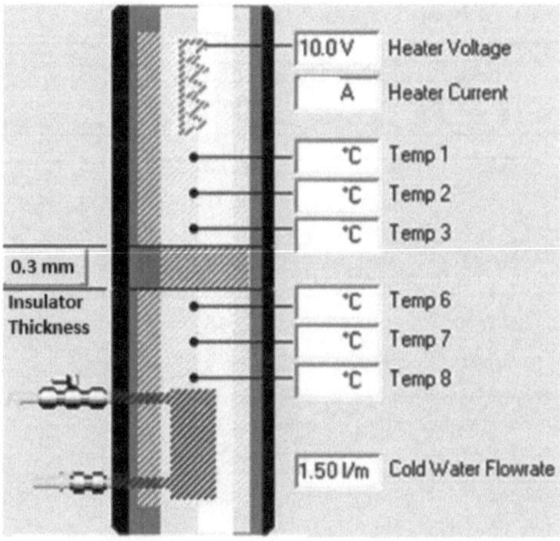

Figure 4. Arrangement of temperature sensors in the Armfield HT11–Linear Heat Conduction apparatus.

thickness between 0.2 and 0.3 cm was used. The arrangement of the apparatus and the six temperature sensors are illustrated in Figure 4. The thermal conductivity of brass is 121 (W/m °C) which was used as a reference for calculation of the thermal conductivity of the sample, using the following equation:

$$K_1 = K_2((A_2 \Delta T_2 L_1)/(A_1 \Delta T_1 L_2)) \tag{3}$$

Where K is the thermal conductivity, L is the axial thickness, A is the cross-section area, 1 and 2 refer to the sample and the brass standard material, respectively [22].

Equation (2) is based on the comparative technique for calculation of the thermal conductivity without the need for accurate determination of the heat flow through sample.

4 RESULTS AND DISCUSSIONS

The physical and mechanical characterizations of the composites are affected by the formulation and additives. In this paper, the content of the marble powder as a filler in the hybrid polymeric composites are varied. Testing results for the samples are presented and displayed to show the influence of marble waste content (wt%) on the composite characteristics. The density, tensile, hardness and thermal conductivity properties of the composite are discussed to find out which additives are needed to improve the quality of the tiles suitable for construction in green buildings. All experiments were repeated at least three times to determine the means and deviation errors.

4.1 *Density and porosity*

The theoretical calculated density, measured density, and calculated volume fraction of the composites' samples are shown in Table 1.

The theoretical density increases with the increase in the marble powder contents in the composite. This is because the waste marble dust has a higher density than the density of unsaturated polyester resin. Also, the measured density increases with increasing marble filler content but at a lower rate. This can be explained by the increase in the porosity of the composite which might be attributed to the compatibility of the marble materials to the

Table 1. Density and porosity of hybrid composites.

Marble Content % in the composite	Theoretical Density (g/cm^3)	measured Density (g/cm^3)	Volume Fractions of Void (V_v)
0	**1.15**	**1.14**	0.87
16.67	1.425	1.316	7.649
23.08	1.531	1.361	11.104
28.57	1.621	1.398	13.757
33.33	1.7	1.435	15.588
37.50	1.769	1.462	17.354
41.18	1.829	1.489	18.589
44.44	1.883	1.51	19.809
47.37	1.932	1.532	20.704
50.00	1.975	1.55	21.519

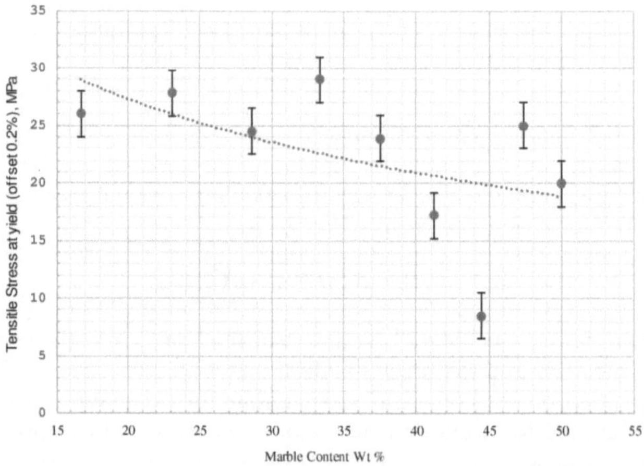

Figure 5. Tensile Stress at yield (offset 0.2%) for different marble content in the composite.

unsaturated polyester resin. We may suggest to add a compatible agent that may reduce the void fraction if a higher density is desired. However, increasing the void fraction may have the benefits of avoiding a high increase in density, as well as reducing the thermal conductivity based on the end applications. According to the standard specifications of tiles (ISO-13006/EN14411), the density has to be greater than 2.0 g/cm^3 [23]. However, for building construction, low density is a good property for alternative tiles as long as other properties are not affected. The water absorption of this composite increase with increasing marble powder content as reported elsewhere [24]. This is attributed to the hydrophilic nature of the marble powder. Adding compatible agent may lead to reduce the increase in the porosity as well as water absorption which will be investigated in our next paper.

4.2 *Tensile strength*

The tensile test is conducted on the hybrid composite samples. The influence of the marble powder content as a filler in the composites on the tensile stress at maximum load is shown in the Figure 5. The tensile stress at yield of the composite sample is between 26 MPa and 24 MPa when marble content is less than 40%, and it decrease to 20 MPa at marble content of 50%. The declining trend in the tensile strength is due to the incompatibility of marble dust with the unsaturated polyester resin which lead to increase in the porosity of the composite.

The influence of marble filler content in the composite on the tensile stress at maximum load (break) is shown in Figure 6. There is a declining trend in the tensile strength at maximum load from 31.5 MPa to 21 MPa. The decrease in the maximum tensile stress is due to the incompatibility of marble material to the resin and the increase in the porosity with increasing marble powder content. According to ISO-10545-4, the minimum modulus of rupture of tiles is 35 MPa [25]. This indicates that the rupture modulus of the composite requires an improvement through incorporating small amount of enhancing additives.

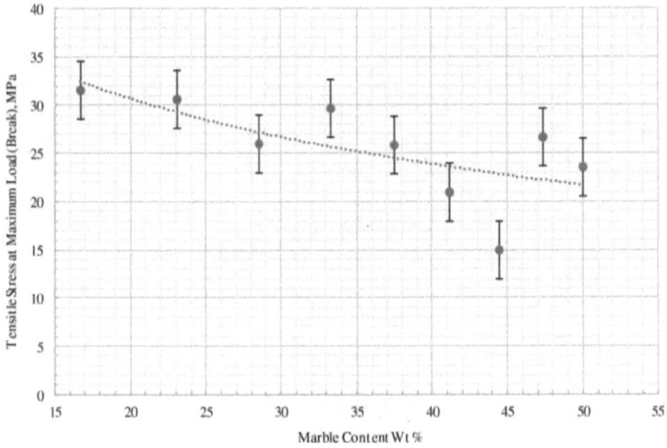

Figure 6. Tensile stress at maximum load (Break) for different marble content in the composite.

The influence of marble filler content in the composite on the tensile strain at break is shown in Figure 7. The tensile strain at break is the ratio of the difference between maximum elongated length and initial length divided by the initial length. There is a declining trend in the tensile strain from 0.02 (mm/mm) to 0.008 (mm/mm). The decrease in the tensile strain is due to the incompatibility of marble material to the resin and the increase in the porosity with increasing marble powder content.

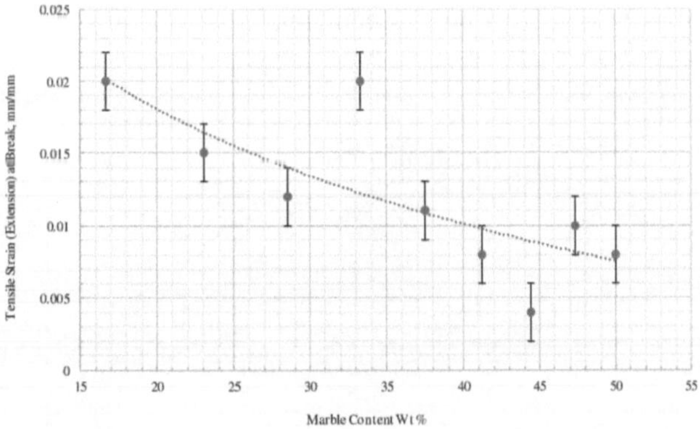

Figure 7. Tensile strain (extension) at break for different marble content in the composite.

4.3 *Hardness*

The Rockwell hardness of the composite increases with the increase in the marble powder content. The increasing trend is from 16 HRB to 19 HRB. The low hardness values (HRB) compared to the hardness of the "Polyester-Glass Fibre-Orange Peel Particulate Hybrid Composite" is attributed to the void existing in our samples [26]. With increasing the marble powder contents, the hardness of composite increases due to the rigidity of marble compared to polyester resin. Mamman and Ramalan (2020) show that the hardness of the hybrid polyester composite is between 28.6 HRB and 40.8 HRB [27]. This is due to the high rigidity of glass fiber existed in their composite compared to the composite used in this work.

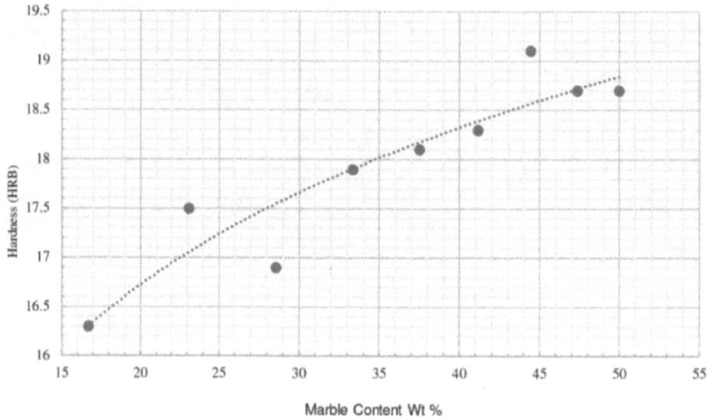

Figure 8. Hardness of composite sample using the Rockwell method with a Rockwell B scale for different marble powder content.

4.4 *Thermal conductivity*

The influence of marble filler content in the composite on the thermal conductivity is shown in Figure 9. With increasing marble content from 16.67% to 50% in the composite, the thermal conductivity of the composite decreases from 0.28 W/m °C to 0.16 W/m °C. This might be attributed to the increase in the porosity of the composite due the incompatibility between the resin and the marble powder. This agrees well with the results reported by Ghazi and Jaddan (2021) for epoxy based composites reinforced with date palm waste particles [28]. A low thermal conductivity improves the thermal insulation of the wall or flooring tiles to make them suitable for construction in sustainable green buildings.

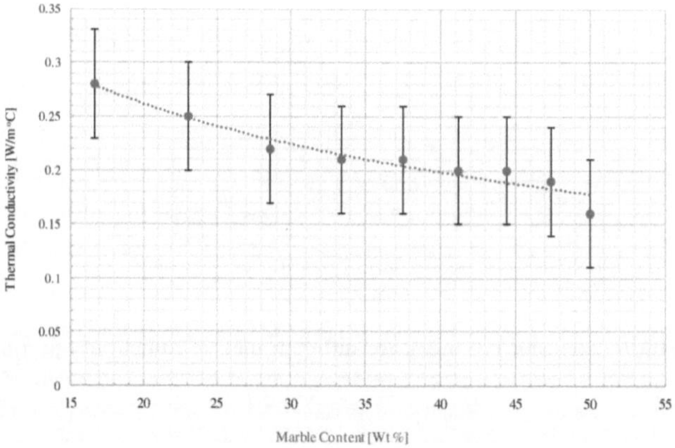

Figure 9. Thermal cnductivity of composite samples for different marble powder content.

5 CONCLUSIONS

Ecofriendly conversion of marble dust waste into value-added tiles for sustainable green buildings has been investigated. In this paper, the base hybrid composites of polyester and marble powder at a weight percentage from 16.67% to 50 % were fabricated, and then, tested for their physical, thermal and mechanical properties. The results indicate producing composites with good properties, but they still need an improvement to obtain cost-effective tiles. Challenges can be overcome by incorporating small amounts of additives such as compatible agent, flame retardant, anti-static and antiaging additives, marble color enhancer with excellent repellent features to oil and water, additive for liquid sealant and vapor transpiration, modifier additive for enhancing mechanical properties. Investigation of such additives will be investigated in the second stage of our research, and will be discussed in our next paper.

6 RECOMMENDATIONS

To meet the Qatar National Vision 2030 (QNV 2030) and the third National Development Strategy (2023–2027), researchers in Qatar have to fulfill the commitments of the state of Qatar to achieve the 17 global goals set by the United Nations. For the global goal: SDG 12, we recommend the investigation of all industrial wastes in Qatar, particularly in the private sections. Converting industrial wastes to value-added products will be a good strategy that should be supported by the government. Upgrading the existing industrial procedures and increasing the recycling rate should be considered for the production of new or improved quality products.

ACKNOWLEDGMENT

We are highly indebted to Dr. Abdulwahab Ibrahim and Philip Fontelo for training students on mechanical testing of composite's samples.

REFERENCES

[1] Qatari Planning and Statistics Authority; Qatar Voluntary National Review; Report on the implementation of the 2030 Agenda for Sustainable Development, 2021. https://sustainabledevelopment.un.org/content/documents/280362021_VNR_Report_Qatar_English.pdf.

[2] The General Assembly; United Nations Statistics Division (UNSD); Global Indicator Framework for the Sustainable Development Goals and targets of the 2030 Agenda for Sustainable Development, after refinement on 53rd session in March 2022; E/CN.3/2022/2. Pages 1 23.; https://unstats.un.org/sdgs/indicators/indicators-list/

[3] The Secretary-General, United Nations Economic and Social Council, Progress towards the Sustainable Development Goals Report, 23 July 2021–22 July 2022, July 7, 2022. https://unstats.un.org/sdgs/files/report/2022/secretary-general-sdg-report-2022–EN.pdf

[4] Anil Kumar Thakur, Asokan Pappu & Vijay Kumar Thakur; (2018). Resource efficiency impact on marble waste recycling towards sustainable green construction materials, *Current Opinion in Green and Sustainable Chemistry*, Volume 13, Pages 91–101, ISSN 2452-2236, https://doi.org/10.1016/j.cogsc.2018.06.005.

[5] Salem, H. S. (2021). Evaluation of the Stone and Marble Industry in Palestine: environmental, geological, health, socioeconomic, cultural, and legal perspectives, in view of sustainable development. *Environ Sci Pollut Res* 28, 28058–28080. https://doi.org/10.1007/s11356-021-12526-4.

[6] Özkaya, K., Ayrilmis, N., and Özdemir, S. (2015). Potential use of waste marble powder as adhesive filler in the manufacture of laminated veneer lumber, *BioRes.* 10(1), 1686–1695. https://doi.org/10.15376/BIORES.10.1.1686-1695.

[7] Kore, Sudarshan, Vyas, Ashok, Kabeer and Syed K. I. (2019). A brief review on sustainable utilisation of marble waste in concrete, *International Journal of Sustainable Engineering*, 13, 1–16, http://dx.doi.org/10.1080/19397038.2019.1703151.

[8] Adewole MB, Adesina MA; (2011). Impact of Marble Mining on Soil Properties in a Part of Guinea Savanna Zone of Southwestern Nigeria; *Ethiopian Journal of Environmental Studies and Management*; Vol. 4 No. 2; Pages 1–8. http://dx.doi.org/10.4314/ejesm.v4i2.1

[9] Deepti Mehta, Deepak Paliwal, Vijayendra Singh Sankhla & Saurabh Tege. (July 2020). Study of Marble Waste and Its Utilization. *International Research Journal of Engineering and Technology (IRJET)*, Volume: Issue: 07, page: 1–5, e-ISSN: 2395-0056, https://www.irjet.net/archives/V7/i7/IRJET-V7I701.pdf.

[10] Ashwini Vaishnav and Nidhi Sharma. (JULY – 2018); The Study of Environmental Impact of Marble Slurry through Remote Sensing and GIS A Case Study of Kishangarh Marble Industrial Area; *International Journal Of Innovative Research & Studies*; Volume 8, Issue VII; Page No: 22–29; DOI:16. 10089.IJIRS.2018.V8I7.18.3154; http://ijirs.in/gallery/5.july%20ijirs%20-%20d714.pdf.

[11] Iqbal, Muhammad & Akbar, Fazal & Ullah, Sana & Anwar, Imran & Khan, Muhammad & Nawab, Asim & Bacha, Muhammad & Wajid and Rashid. (2018). The effects of marble industries effluents on water quality in Swat, Northern Pakistan. *Journal of Biodiversity and Environmental Sciences*. 13. 34–42.

[12] Khan S., Haq F., Saeed K. (2012) Pollution load in industrial effluent and ground water due to marble industries in district Buner, Khyber pakhtunkhwa, Pakistan. *Int J Recent Sci Res* 3(5):366–368.

[13] Mulk S., Azizullah A., Korai A. L., Khattak M. N. (2015 Feb). Impact of marble industry effluents on water and sediment quality of Barandu River in Buner District, Pakistan. *Environ Monit Assess*. 187 (2):8. doi: 10.1007/s10661-014-4221-8. Epub 2015 Jan 24. PMID: 25616784.

[14] Piccini, Leonardo & Di Lorenzo, Tiziana & Costagliola, Pilario & Galassi, Diana. (2019). Marble Slurry's Impact on Groundwater: The Case Study of the Apuan Alps Karst Aquifers. *Water*. 11. 2462. 10.3390/w11122462.

[15] Shah, Wisal & Ahmad, Didar. (2021). Impacts of the marble waste slurry on ground water quality and its reuse potential. *Fresenius Environmental Bulletin*. 30. 2077–2086. https://flore.unifi.it/bitstream/2158/1207605/2/2019-Piccini%20et%20al_Slurry%20marble-Water-11-02462.pdf

[16] Jitin Rahul and Manish Kumar Jain, (2014). An Investigation in to the Impact of Particulate Matter on Vegetation along the National Highway: A Review. *Research Journal of Environmental Sciences*, 8: 356–372. DOI: 10.3923/rjes.2014.356.372; https://scialert.net/abstract/?doi=rjes.2014.356.372.

[17] Butt I. M., Mustafa T., Rauf S. et al. (2021). Pulmonary function parameters among marble industry workers in Lahore, Pakistan; *Research*, 10:938 (https://doi.org/10.12688/f1000research.52749.1).

[18] Ramkissoon, C., Gaskin, S., Thredgold, L. et al. (2022). Characterisation of dust emissions from machined engineered stones to understand the hazard for accelerated silicosis. *Sci Rep* 12, 4351. https://doi.org/10.1038/s41598-022-08378-8.

[19] Sayers, R. R., and Jones, R. R. (1938). Silicosis and Similar Dust Diseases: Medical Aspects and Control. *Public Health Reports (1896–1970)*, 53(33), 1453–1472. https://doi.org/10.2307/4582633.

[20] Fraunfelder, F. T., Coster, D. J. and Drew, Roger. (1990) Ocular injury induced by Methyl Ethyl Ketone Peroxide. *American Journal of Ophthalmology*, vol 110, Issue 6, pp 635–640, December 1990. https://www.sciencedirect.com/science/article/abs/pii/S0002939414770606.

[21] Nayak S. K. and Satapathy A. (2021). Development and characterization of polymer-based composites filled with micro-sized waste marble dust. *Polymers and Polymer Composites*. ;29(5):497–508. doi:10.1177/0967391120926066.

[22] Tritt, T. M. and Weston, D. in *Thermal Conductivity* (ed.Tritt, T. M.) 187–203 (Springer US, 2004).

[23] Technical Specification, Lavish Ceramics Pvt. Ltd., 16 12 2022. [Online]. Available: https://www.lavishceramics.com/technical-specification/. [Accessed 16 12 2022].

[24] Thi Thu Loan Doan, Hanna M. Brodowsky, Uwe Gohs and Edith mäder; re-use of marble stone powders in producing unsaturated polyester composites; *Advanced Engineering Materials*, volume 20, issue 7, 1701061; https://doi.org/10.1002/adem.201701061.

[25] Technical Data, *ALICE TILES*, 16 12 2022. [Online]. Available: http://www.alicetiles.com/technical. [Accessed 16 12 2022].

[26] Mamman R. and Ramalan A., (Oct. 2019). Mechanical and physical properties of polyester reinforced glass fibre/orange peel particulate hybrid composite, *Adv. J. Grad. Res.*, vol. 7, no. 1, pp. 18–26, doi: 10.21467/ajgr.7.1.18-26.

[27] Choudhary, M., Singh, T., Dwivedi, M. and Patnaik, A. (2019), Waste marble dust-filled glass fiber-reinforced polymer composite Part I: Physical, thermomechanical, and erosive wear properties. *Polym. Compos.*, 40: 4113–4124. https://doi.org/10.1002/pc.25272.

[28] Israa F Ghazi and Riyam I Jaddan (2021); Thermal Conductivity Characterization of Epoxy Based Composites Reinforced with Date Palm Waste Particles; *J. Phys.: Conf. Ser.* 1973 012144; DOI 10.1088/1742-6596/1973/1/012144; https://iopscience.iop.org/article/10.1088/1742-6596/1973/1/012144/meta.

Innovation and Technological Advances for Sustainability – Al-Naemi et al. (Eds)

Value stream mapping as a sustainable lean production tool to improve the production process organization

Marwan H. Hilal, Gamal Eldin Farag, Abdulla Alothni & Galal Abdella
Mechanical and Industrial Department, Qatar University, Doha, Qatar

ABSTRACT: Lean thinking is a business idea that was developed in Japan. The primary contribution of lean is the discovery and elimination of waste. Companies use lean to retain their competitiveness by increasing system productivity and process quality. This paper focuses on the benefits of implementing lean-based sustainability on a comprehensive process used in an automation business, starting with quoting a new project, designing the products, manufacturing, programming, client training, and finally handing over. The goal is to use Value Stream Mapping (VSM), one of the important sustainable lean thinking methodologies, to improve the organization's process by minimizing waste and non-value-added processes. First, the process's current value stream map was created using personal interviews, observations, and secondary data from the firm. Following that, a future value stream map based on lean thinking ideas was proposed to improve total lead time for manufacturing and value-added time. Based on the final result, VSM is a helpful and relevant tool that may help managers conceptualize different types of wastes and defects. The results reveal that using various lean thinking techniques reduces the time, material, money, and other wastes caused by the present non-improved process, and this had an impact on the environmental, social, and economic aspects of sustainability.

Keywords: Value stream mapping, current value stream mapping, future value stream mapping, lean thinking, Sustainability, waste

1 INTRODUCTION

1.1 *Lean thinking concept*

Many huge corporations have tried to use lean thinking strategies to remain competitive in an expanding global market. The strategy focuses on waste reduction by eliminating non-value-added procedures (Abdulmalek and Rajgopal 2007). These value-added activities add to the value of the final product or service to the end user (Hines and Rich 1997). The approaches seek to improve production efficiency (productivity) and quality while reducing work-in-process, stock levels, and unnecessary product handling (Hicks 2007). The lean approach takes the industrial system in a new direction of task planning and execution that is both effective and efficient (Grewal 2008).

1.2 *Ease sustainability concept*

Companies with several stakeholders have gradually adopted the sustainability strategy. Its framework contains three aspects which are environmental, social, and economic that must be considered while planning for the future of natural resources and human existence (Feil and Schreiber 2017). Sustainability, according to Rao (2013), is an approach that integrates

DOI: 10.1201/9781003496724-60

process, product design, manufacturing, and planning to detect, monitor, analyze, and control the flow of the process to decrease environmental effects while simultaneously attempting to optimize resource efficiency. The term "triple-bottom-line" (TBL) refers to three separate aspects of sustainability.

- The economic aspect evaluates how the processes can generate the most income with the least amount of loss (corporate profit account).
- The social aspect evaluates how organizational work responsibly adds to the overall success of the operation (people account).
- The environmental aspect measures the process's environmental responsibility.

TBL overview enables firms to concentrate not only on the economic value that they provide but also on environmental and social values while evaluating their operations (Rojek and Nowosielska 2015).

1.3 *Lean and sustainability*

Implement a sustainable approach by adopting appropriate Value Stream Mapping (VSM), a method used for lean thinking to discover waste. The lean approach is a basic strategy to enhance operational efficiency by addressing three major issues: reduce waste, decrease uncertainty, and improve productivity. VSM, according to Rother and Shook (1999), is a diagram of all the activities that take a product through the production flow, from raw materials to final products. Faulkner and Badurdeen (2014) offered sustainable VSM (Sus-VSM) to evaluate sustainability performance that follows the TBL approach. The indicators from TBL are merged into appropriate Sus-VSM metrics using a case study at a company that provides a Building Management System (BMS) for any building upon request of the customer to have complete control and monitor different systems equipment in his building. Suitable criteria were developed using visual symbols as a preliminary analysis of sustainability performance.

1.4 *Sustainability assessment*

Sustainable metrics are used to measure process performance from economic, environmental, and social perspectives based on the contribution of sustainability factors to the processes (Vinodh *et al.* 2015). The goal of identifying sustainability indicators is to evaluate the current status of sustainability conditions. Time and resources are the measures conducted in this research because of the impact on the economy, environment, and society.

1.5 *Structure of the paper*

The rest of this paper is organized as follows: Section 2 presents and describes the proposed methodology for lean-based sustainability improvement. Then, section 3 describes the case study, which includes the use of VSM and the improved VSM with the results and recommendations. Finally, Section 4 concludes all the points mentioned in the research.

2 METHODOLOGY

Figure 1 shows the five key steps of this research: process determination, mapping the existing state, sustainability metrics determination, the study of current waste, and future state model. This paper uses a collection of lean thinking techniques such as VSM, 5S, Brainstorming, Bottleneck, Just-in-time, Gemba's philosophy, and the Pareto chart.

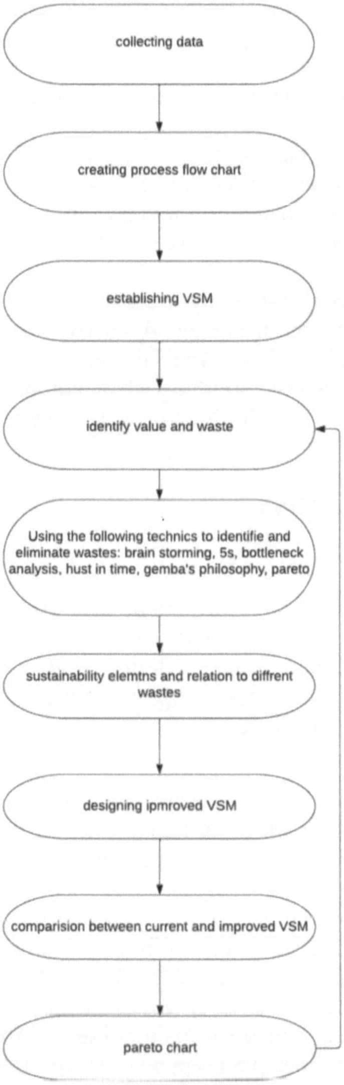

Figure 1. Methodology.

VSM is mapping the parts and information of all components in a logistics system that includes production, suppliers, and customer distribution. VSM is also a mapping tool for a production process or an entire supply-chain network (Seth and Gupta 2005).

5S: is a Japanese approach consisting of five 'S' terms: Seiri (sorting), Seiton (order), Seiso (shine), Seiketsu (standardize), and Shitsuke (maintain) with a strong sense for workplace management. 5S in the workplace aims to organize a workstation efficiently and effectively by recognizing and storing often used goods, maintaining the area and products, and keeping the new order (Wani and Shinde 2021).

Brainstorming: a technique of team problem-solving that depends on the spontaneous contribution of innovative ideas and solutions. This strategy requires an active, unstructured conversation in which each group member is asked to think aloud and provide as many ideas as possible based on their different experience (Bernstein 2017).

Bottleneck: Bottlenecks reduce the capacity and throughput of the production system, resulting in, among other issues, slow productivity, local stock buildup, and, most importantly, reduced overall system productivity. Identifying and removing bottlenecks is the foundation for a company's improvement in production and economic success (Urban 2018).

Just-in-time is an inventory management strategy that ensures the exact amount of inventory products arrive at the exact time required Gemba's Philosophy: In Lean thinking environments, vertical integration is important. They ensure that all levels of the organization remain linked to the front lines, or "the Gemba," where meaningful value is built (Womack 2019).

Pareto Chart: a statistical decision-making strategy that involves selecting a small number of activities with a substantial overall impact. A Pareto chart is often used to display the findings of a Pareto analysis. It is also a graphical tool that assists in breaking down a large issue into its component pieces and determining which components are the most significant. The Pareto analysis is also known as the 80–20 rule (Rahman and Queresi 2010).

3 AN ILLUSTRATIVE EXAMPLE

3.1 *Process description*

BMS (Building Management System) is a system that can control and monitor different systems and machines through one interface, to help in maintenance reporting and reduce the time needed to check or fix any problem. Besides that, BMS is also integrated with the power monitoring system that allows you to optimize energy consumption.

The process starts as shown in Figure 2:

- Quoting for a new project to participate in the tender, if accepted, all the required documents will be sent to the design team, such as engineering drawings and point schedule to start the design. On the other hand, if the offer is rejected because of the high value of the quotation, the requirements should be checked, and the quotation amount should be minimized.
- The PM and the head of the design team will inspect the designs.
- All the drawings and design documents will be submitted to the consultancy
- The consultancy will review and approve the documents if they comply with the project and design specifications. Alternatively, some documents could be rejected, and the consultancy will reply with comments to be modified by the design team.
- The procurement team will order the needed controller and project material.
- Waiting time for the arrival of the materials to the warehouse will be between 6–12 weeks.
- The material will be sent to the site for inspection so we can claim 50% of its price.
- The material will be sent back to the warehouse to start with the assembly and resend to the site.
- The engineers and technicians will start their work (termination, programming, and integration); any defective material will be sent back to the warehouse to request a replacement from the supplier.
- After finishing programming and pre-commissioning, the engineers will raise the inspection for testing and commissioning (T&C).
- If the consultancy approves the inspection, we will start building the graphics and interface for the user. Moreover, if it is rejected, the engineers will modify and resubmit the inspection.
- Final inspection (graphics, user interface) will be raised for the consultant.
- Training for the end user will be given at the last stage of the project when handed over.

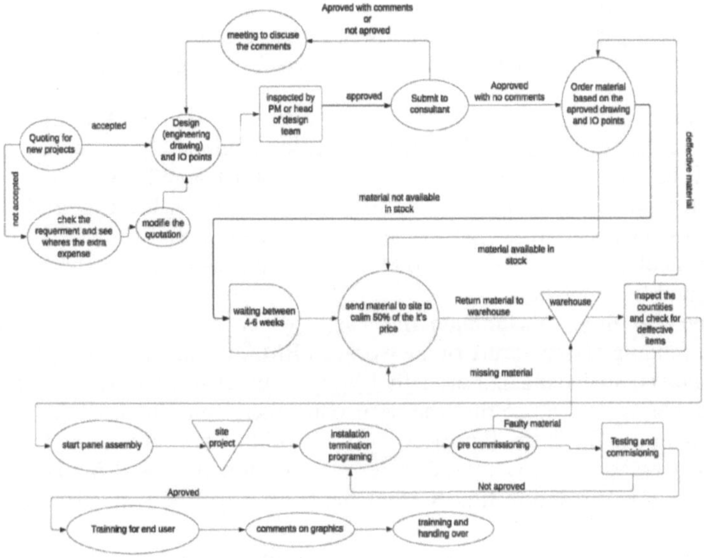

Figure 2. Flow chart of the process.

3.1.1 *Process waste and sustainability indicators*

Potential waste and non-value added will be represented in the following Table 1.

Table 1. Potential waste and non-added value elements.

Types of Wastes					
Defects	Waiting	Unused talent	Transportation/ Motion	Extra process	Work in process
Defective items that are sent from the supplier	Waiting for the material to arrive in 6–10 weeks	The employees who are inspecting the materials that were sent to the site are engineers.	Transportation from the Warehouse to the site and getting them back to the assembly then send it back to the site	Handing over and giving a second training	Material waiting to get assembled in the warehouse
Incomplete requirements specification sent from the consultancy	Waiting for consultancy approval	Using a technician who does not have any training or experience in the BMS field	Inside the warehouse, the assembly line and stored materials are moved from one place to another	Material inspection when sent to the site by engineers	Material waiting to get inspected by the consultancy
	Waiting for the scheduled meeting to discuss the comments and snags			Drawings inspection by the higher management and head of design	
	Waiting for instruction from management to procced				

3.1.2 *Waste identification*

The VSM has been produced. Figure 3 allowed us to visualize the different types of waste that can be eliminated. The following techniques have been used:

- Identify and eliminate most of the waste using Brainstorming, especially the activities done in more than one stage by different people, for example, the material inspection, the revision of the drawings, and the training provided for the end user.
- 5s has been used in the warehouse to rearrange the material location and locate all the frequently used consumables next to the assembly team directly.
- Bottleneck analysis visualized how often the process was stuck between the project's beginning and final delivery.
- Just-in-time was used to eliminate the waste, which has both negative and positive impacts depending on the nature of the work. In BMS, waiting for the material will make you lose clients, so having material that works for all projects (the BMS system has programmable controllers so that the same controllers could be used for many applications) will help you finish small projects faster. In addition, your waiting time will be less, even in bigger projects; any modification or variation proposed in the middle of the project needs a fast reaction time so that the project will be handed over without any delays.
- Gemba's philosophy got our management and project engineer to be more involved in the actual work, and that minimized the wasted time in different stages of the project by answering any concern and fixing any problem directly without the need to go through all the processes needed to seek help from higher management.
- The Pareto analysis helped in future stages to eliminate the wastes that had more impact on the process.

3.3 *Sustainability elements*

3.1.1 *Environmental*

Eliminating or minimizing the defects will reduce the time needed to get a replacement and the energy consumed to ship all the defective material from and to the country of origin; eliminating the unneeded transportation also reduces the energy used (fewer emissions created). Eliminating the extra processes reduces waste, energy, and raw materials. Minimizing the waiting time and WIP will eliminate the energy consumed from heating, cooling, and lighting during the process.

3.1.2 *Human and social*

To ensure a sustainable business, we standardized the updated process and used the unused talent to design, set a schedule, supervise, and set a long-term plan for the fabrication and work. This helped in fair labor practices, community engagement, and equal opportunity.

3.1.3 *Economic*

Using lean and eliminating waste will increase profit, reduce cost, and ensure economic growth.

3.2 *Lean-based sustainability enhancement*

After completing the current VSM and eliminating the waste, the improved VSM has been developed, as shown in Figure 4.

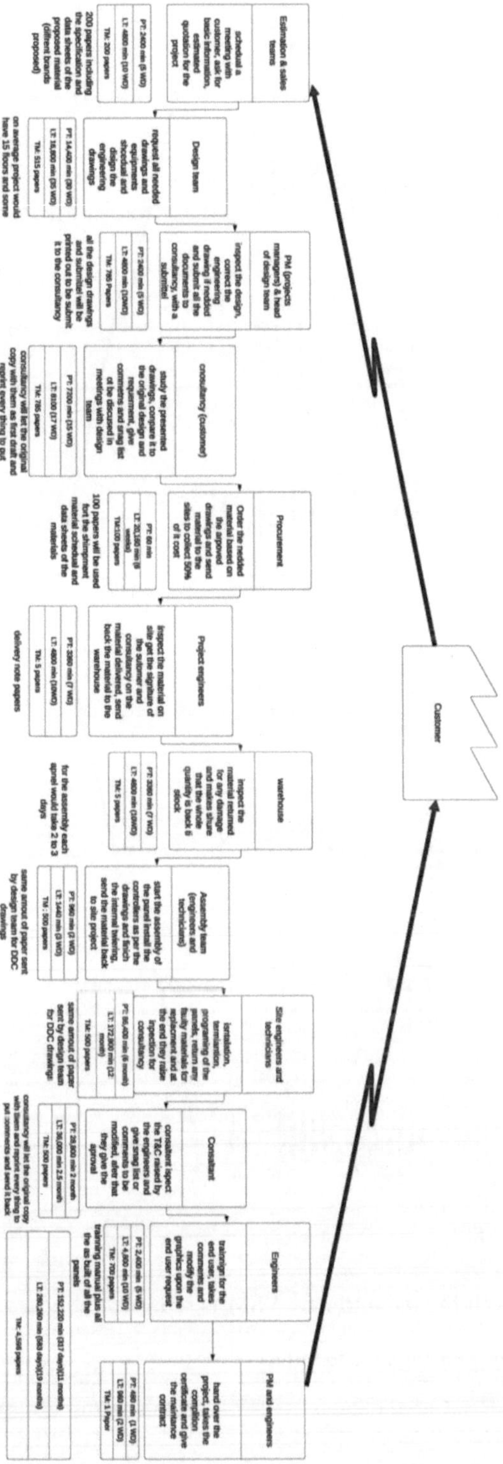

Figure 3. Current VSM.

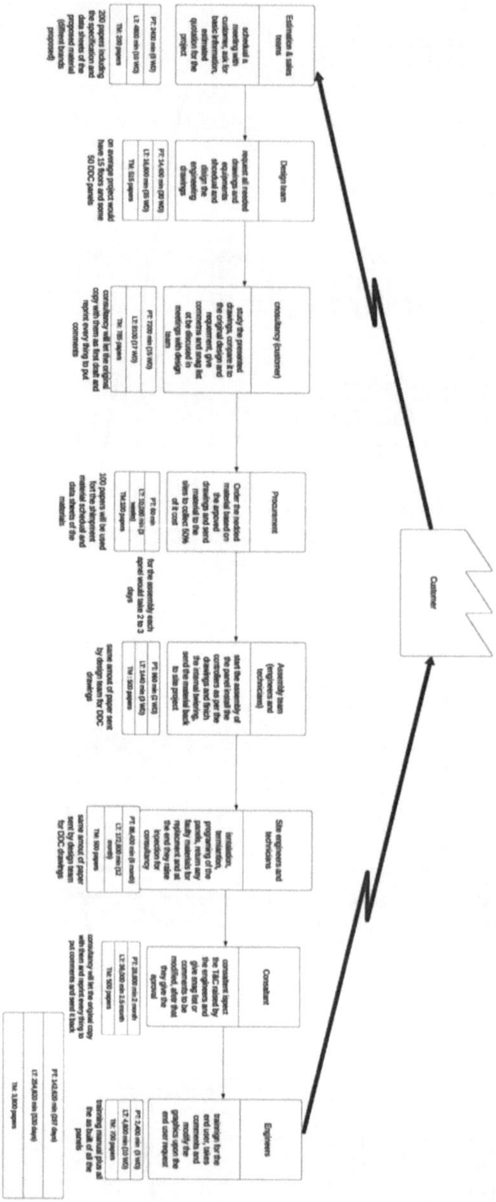

Figure 4. Improved VSM.

A comparison between the old and new VSM is represented below.

Table 2. comparison between current and improved VSM.

Metric	Current Status	Future status	% Improvement
Process time (min)	152,220	142,620	6.3% improvement
Lead time (min)	280,260	264,900	5.4% improvement
Resources (papers)	4,596	3,800	17.3% improvement

Time has been chosen as the sustainability metric for this measurement. The eliminated waste time for the process is 9600 minutes.

To measure the weight of each waste that was removed, the following has been calculated:

- The first processing time is 2400 minutes, calculated from the following wastes: (inspecting the design, correcting the engineering drawings If needed, and submitting all the documents to consultancy, with a submittal).
- The second processing time is 3360 minutes, calculated from the following wastes: (inspect the material on-site to get the customer's signature and consultancy on the material delivered and send back the material to the warehouse), which the Projects engineer does.
- The third processing time is 3360 minutes calculated from the following wastes: (inspect the material returned for any damage and ensure that the whole quantity is back in stock) done in the warehouse.
- The fourth processing time is 480 minutes, calculated from the following wastes: (for handing over the project, taking the completion certificate, and giving the maintenance contract).

The percentage of each waste concerning the total waste reduced is Calculated to show which waste has the largest impact, as shown in Table 3.

Table 3. Percentage of each waste concerning the total waste reduced.

Wastes	Percentage Relative to The Total Waste %
Design Inspection by PM and Head of Design	25%
Material Inspection by Project Engineers	35%
Returned Material Inspection in The Warehouse	35%
Hand Over the Project (Extra Process)	5%

Pareto analysis has been used to prioritize eliminating the elements that will cause approximately 80% of the total waste, as shown in the following Table 4 and Figure 5.

Table 4. Waste weightage.

Causes	min	Cumulative Total	Cumulative %	
Material inspection by Project engineers	3360	3360	35%	80%
Returned material inspection in the warehouse	3360	6720	70%	80%
Design inspection by PM and head of design	2400	9120	95%	80%
Hand over the project (extra process)	480	9600	100%	80%
Total	**9600**			

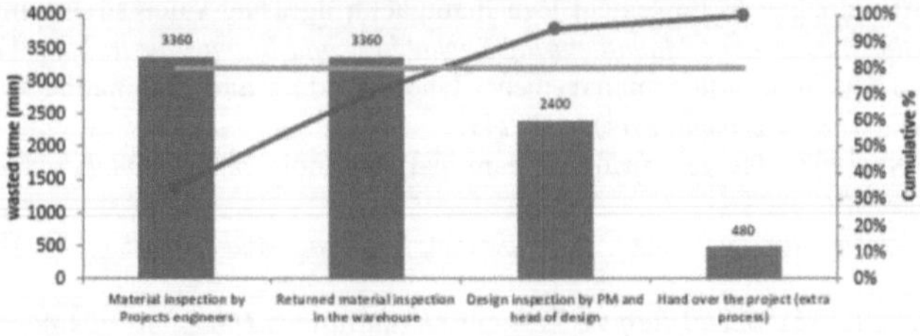

Figure 5. Pareto chart.

The green line in Figure 5 represents 80%, and the red line represents the cumulative percentage. If the material inspection in both the warehouse and on-site were eliminated, 70% of the waste would be eliminated.

4 CONCLUSION AND RECOMMENDATION

This paper demonstrates that a lean-based thinking technique will improve the company's sustainability performance and reduce waste. The major goal is to identify and decrease waste for every element in the process that does not add value to the service. This research also focuses on reducing service lead time and value-added time. According to the current VSM, the process lead time and value-added time are 280,260 minutes and 152,220 minutes, respectively, along with 4596 papers. Furthermore, because of the difference between processing time and lead time, it may be assumed that many non-value-added activities exist, such as waiting for items, transporting materials from one area to another, and unnecessary processes. Finally, after using sustainable lean tools and methodologies and adopting lean thinking, the process lead time and value-added time are reduced to 264,900 minutes and 142,620 minutes, respectively, along with 3800 papers. It suggests that by removing wastes using lean principles and the VSM approach, the production lead time and value-added time, in this case, the study would be reduced by approximately 5% and 6%, respectively, and reducing the percentage of papers used by 17%. More studies may be done by doing in-depth research on VSM in conjunction with computer simulation to validate the suggested VSM approach.

ACKNOWLEDGMENT

This research was partially supported by Qatar University's Associate Professor of Industrial System Engineering Dr. Galal Abdullah. We thank our colleagues from Qatar University who provided insight and expertise that greatly assisted the research, although they may not agree with all of the interpretations of this paper.

REFERENCES

Abdulmalek, F. A., and Rajgopal, J. (2007). Analyzing the benefits of lean manufacturing and value stream mapping via simulation: A process sector case study. *International Journal of Production Economics*, 107 (1), 223–236.

Corinne Bernstein (2017). Brainstorming, https://www.techtarget.com/whatis/definition/brainstorming (accessed 11/11/2022)

Faulkner W. and Badurdeen F. (2014). *Sustainable Value Stream Mapping (Sus-VSM): Methodology to Visualize and Assess Manufacturing Sustainability Performance.*

Feil AA, Schreiber D. (2017). Sustainability and Sustainable Development: Unraveling Overlays and Scope of Their Meanings. *Cadernos EPABE.BR.* 14(3): 667–681.

Grewal, C. (2008). An initiative to implement lean manufacturing using value stream mapping in a small company. *International Journal of Manufacturing Technology and Management*, 15(3–4), 404–417.

Hicks, B. J. (2007). Lean information management: Understanding and eliminating waste. *International Journal of Information Management*, 27(4), 233–249.

Hines, P., and Rich, N. (1997). The seven value stream mapping tools. *International Journal of Operations & Production Management.*

Rao PN. 2013. *Sustainable Manufacturing – Principles, Applications, and Directions.* Ioa(US): University of Northern Ioa.

Rojek M, Nowosielska. (2015). *Social Responsibility of Organizations Directions of Changes.* Wrocław(PL): Publishing House of Wrocław University of Economics.

Rother M. Shook J. (1999). *Learning to See: Value Stream Mapping to Add and Eliminate Muda.* Massachusetts (US): The Lean Enterprise Institute.

Sagar Wani & Dattaji K. Shinde (2021). *Study and Implementation of '5S' Methodology in the Furniture Industry Warehouse for Productivity Improvement*

Seth, D., and Gupta, V. (2005). Application of value stream mapping for lean operations and cycle time reduction: an Indian case study. *Production Planning & Control*, 16(1), 44–59.

Talib, F., Rahman, Z., and Queresi, M. N. (2010). Pareto analysis of total quality management factors critical to success for service industries. *International Journal for Quality Research*, 2, 155–1

Vinodh S., Ruben B. and Asokan P. (2015). Life cycle assessment integrated value stream mapping framework to ensure sustainable manufacturing: a case study. *Clean Techn Environ Policy*. Springer.

Wieslaw Urban (2018). *The Case Study of Bottlenecks Identification for Practical Implementation of the Theory of Constraints*

Womack J., (2019). *Gemba Walks.* Lean Enterprise Institute Inc., Expanded 2nd Edition

Innovation and Technological Advances for Sustainability – Al-Naemi et al. (Eds)

Author index